21st Century Biology and Agriculture

Growth and Development in Plants

- K.V. Krishnamurthy
Former Professor and Head
Department of Plant Science
Bharathidasan University
Tiruchirappalli - 620 024, INDIA

Published by:
Scientific Publishers (India)
5 A, New Pali Road, P.O. Box 91
JODHPUR – 342 001 (India)

Scientific Publishers (India)
4806/24, Ansari Road, Daryaganj
NEW DELHI – 110 002 (India)

E-mail: info@scientificpub.com
Website: www.scientificpub.com

Cover illustration: This textbook series *Biology and agriculture for the twenty-first century* launched by Scientific Publishers, Jodhpur (India), celebrates the native, but critically threatened bird of the western semi-arid grasslands and scrubs of India, the great Indian bustard (GIB) *Ardeotis nigriceps* (Gruiformes: Otididae). GIB has been a part of Indian culture and tradition, known as *gonādh* (*Sanksrit*), because the male GIB's call closely resembles the mooing of cows. Artist: Urvashi Sharma.

ISBN: 978-81-7233-934-0

Typeset by Rajesh Parihar
Printed in India

Series Preface

With great pleasure, the Scientific Publishers (Jodhpur) and I launch the second volume of the postgraduate textbook series under the general title *21st Century Biology and Agriculture* for use in the Indian subcontinent.

This series aims at fulfilling the knowledge needs of postgraduate learners in agriculture and biology, focussing on self-directed learning. Keeping this point in full view, these volumes would dilate on contemporary information in chosen themes in a pertinent, but brief backdrop of historical knowledge, with appropriate textual information laced with relevant illustrations. Most importantly, these books aim to cater to the self-learning needs in passionate and committed learners. By self-learning, the teacher's role turns into mentoring rather than tutoring. Vital details have been 'box'ed so that learners can internalize them easily, swiftly, and forever. Care has been exercised to integrate examples from the Indian subcontinent, so that the learners can relate to concepts and integrate principles quickly. The most critical aspect is the inclusion of specific-case studies and interactive-mode of learning so that learners can learn about the day-today issues and application of theory that surround the nominated theme effectively. Every effort has been meticulously made to see that the books launched under this series are easily readable and user-friendly. Tanay Sharma of Scientific Publishers (Jodhpur) readily agreed to my demand to identify and recruit an efficient copy editor, who has done a neat job.

This book entitled *Growth and Development in Plants,* executed by K.V. Krishnamurthy, a distinguished plant biologist, and was a professor at Bharathisasan University, Tiruchirapalli, Tamil Nadu, of India is the second in this series. Many more similar titles have been enlisted and will be appearing in the near and far future. Scientific Publishers (Jodhpur) and I thank Krishnamurthy for readily and willingly accepting to our request by executing this task, and importantly doing it speedily.

India is a land of varied landscapes. One of the remarkable natural icons of India is the Great Indian Bustard, the populations of which are restricted to Western India. Because this series is being published from Jodhpur, I thought it will be appropriate to remind ourselves of the uniqueness of this splendid bird and make every effort of conserve its diminishing population.

All books to appear in the series *21st Century Biology and Agriculture* will celebrate this Indian icon.

I am confident that learners will benefit from the information and knowledge shared by Krishnamurthy; I am also confident that post-graduate class teachers of biology and agriculture in the Indian subcontinent would find this material appropriate to prescribe this book as a learning resource.

Anantanarayanan Raman
Editor-in-Chief, Textbook Series
21st Century Biology and Agriculture
School of Agricultural & Wine Sciences
Charles Sturt University
Orange, NSW 2800, Australia

Preface

The purpose of the book is to provide a broad explanation on growth and development in plants from seed germination to vegetative growth, maturation and flowering, fruiting and seeding. The book presents the principles and results of previous and ongoing research on plant growth and development throughout the world. Researchers have been interested in plant growth and development for at least the past two centuries. However, their interests have been largely reductionistic and not holistic. They have concentrated their attention all along exclusively on addressing the problem of growth and development from morphological, anatomical, biochemical, physiological, molecular, environmental or genetical angles but not trying to make an integrated study. Also they focused their attention on studying growth and development by exclusively laying emphasis on different hierarchical levels of plant organization i.e. at the cellular, tissue and organ levels, again not attempting to integrate the information obtained from all these hierarchical levels. Some concentrated on in vivo approaches, others on in vitro approaches, some focusing on normal growth and development and yet others on abnormal growth and development. This book tries to integrate all the information drawn from these reductionistic approaches and provides a holistic account on plant growth and development. This is done first through detailed accounts on the various concepts and definitions, on cell division, cell cycle and cell enlargement which form the basis of growth and development, on growth regulators and signaling molecules, and on the attainment of form (morphogenesis) as related to function, at all hierarchical levels of plant organization. A detailed account on vegetative and reproductive growth and development, growth movements and rhythmic growth phenomena is provided. The role of alternative growth strategies and abnormal growth phenomena in the understanding of normal growth and development is also emphasized.

Because of space limitations, the author has concentrated his attention on flowering plants with very little focus on other groups of plants. The author also has not covered the entire gamut of the subject of plant growth and development but has summarized what he has as felt as very important. This book is meant for all those students, researchers and teachers who are curious about how plants grow and develop and what biotic and abiotic factors control growth and development. They will find this book useful in their careers in plant biology,, biotechnology, forestry, horticulture and seed science.

I am very grateful to Dr. A. Raman, Editor-in-Chief, Text book Series 21st Century Biology and Agriculture School of Agricultural and Wine Sciences Charles Sturt University Orange, NSW 2800, Australia and Mr. Tanay Sharma, Scientific Publishers, India for inviting me to write this book. I am very indebted to my wife Brindha, who has gracefully and patiently tolerated my negligence of the family during the writing of this book and who has constantly been the strongest pillar of support to me. I am also indebted to my son Arvind, daughter-in-law Anusha and grandson Sundar for foregoing the time they would have otherwise spent with me. My thanks are also to Sri Darshan Shankar, Dr. Padma Venkat and John Adams for their support and constant encouragement. I am also thankful to those people who have allowed me to use their illustrations in this book.

K.V. Krishnamurthy

17-02-2015
Bangalore

Contents

Basic Concepts and Definitions 1

1.1. INTRODUCTION

According to the Big Bang theory, the Universe in which our solar system exists is believed to have originated about 13–15 billion years ago (BYA). Our solar system and the Earth came into existence around 4.5–5 BYA. Generally, it is believed that life on the Earth first originated in the sea around 3.8–4 BYA and then on the land around 2.5 BYA due to transmigration of life from sea to land (Krishnamurthy 2010). Evolutionary biologists believe that all life started and diversified from a single–celled, **Last Common Ancestor (LCA)**. According to the **Chemoton theory** of Gánti (2003), the LCA was autocatalytic with (i) a translation machinery for synthesis of proteins, (ii) an energy transduction system involving electron transport and high energy phosphate, and (iii) a membrane system enclosing (i) and (ii) (Fig. 1).

During the subsequent diversification of life, multicellular condition (either through a chance aggregation of several spherical cells or by repeated divisions of a single spherical cell) and division of labour resulted. The most important morphogenetic requirements that accompanied the multicellular situations are: (i) fulfillment of geometric and other mathematical and physico–chemical laws, (ii) cohesion between cells, and (iii) communication and transport between different cells of the multicellular structure (Raven 1986, Krishnamurthy 2000). The translation machinery of LCA culminated into a highly advanced eukaryotic condition, while the membrane system culminated into a dynamic Extracellular Matrix (ECM) or cell wall. Consequent to the diverse modes of nutrition of the major groups of living organisms, the metabolic system evolved from the simple energy transduction system to a more complex one. All these changes resulted in the evolution of five kingdoms of living organisms (Whittaker 1969) (Fig. 2) : Monera, Protista, Fungi, Animalia and Plantae. This evolution involved diversifications and convergences (Margulis and Sagan 2003) as well as co–evolution due to several types of interaction between different kinds of living organisms.

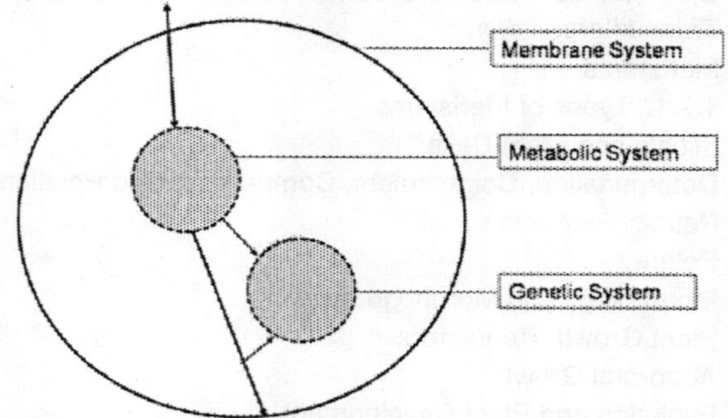

Fig. 1. The Chemoton of Gánti: The LCA of all the life with its three systems and their interactions (Gánti 2003)

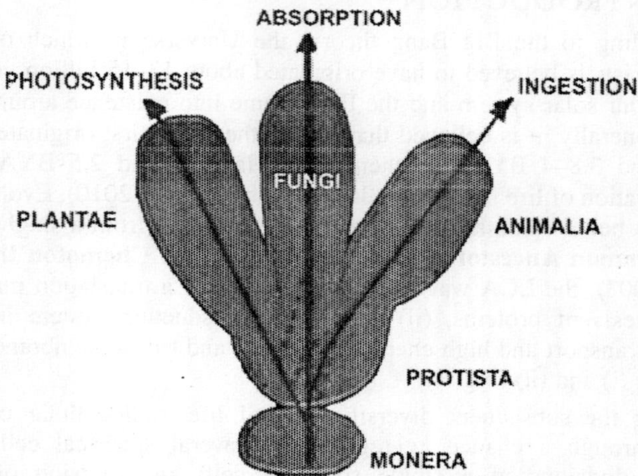

Fig. 2. Diagrammatic representation of the five kingdoms (Whittaker 1969) and the evolution of three major nutritional modes

With reference to the plants, evolution largely depended on and was directed by several growth and developmental strategies. This chapter briefly describes the concepts and definitions relating to growth and development in plants, which are described in detail in the subsequent chapters of this book.

1.2. LEVELS OF STRUCTURAL ORGANIZATION IN HIGHER PLANTS

In the structural organization of a multicellular living organism such as higher plants there are several hierarchical levels, starting from the basic, smallest and least complicated level to the highest, largest and highly complicated level, all of which are interconnected. It can be compared to an automobile antenna made up of several cylinders tucked into one another. The narrowest and the innermost cylinder is comparable to the basic level of structural organization, and the broadest and the outermost cylinder represents the highest level of structural organization separated by cylinders of intermediate levels.

The basic level of structural organization in a plant is constituted of biomolecules, both micro and macro. The next higher level is constituted by the various organelles and other intracellular entities, which together organize the third higher level i.e., the cell. Cells together make up tissues (made of similar cell types) and tissue systems (made of different cell types), which form the fourth higher level of the structural organization. Different tissues and tissue systems come together in a location–specific manner to form different organs of the plant, and several different organs together constitute the highest level of structural organization i.e., the plant.

At each increasing hierarchical level of the structural organization, from the smallest to the largest, there is not only an increasing complexity in form, structure and function, but also in size and volume (Fig. 3). Meristematic tissue cells give rise to various types of cells due to varied differentiation processes, which then aggregate into a homogeneous group of similar cells to form simple tissues or into a heterogeneous group of different cell types to form complex tissues or tissue systems. Parenchyma, collenchyma and sclerenchyma make up the different simple tissues, while xylem, phloem and epidermis are the examples of complex tissues. In this hierarchical organizational set–up, there are two levels of structural integration. The first level of integration is shown by the cell, which integrates several different biomolecules and organelles to form a unit of structure and function. The second level of structural integration is exhibited by the organism, which integrates several different tissues and organs to form an integral whole.

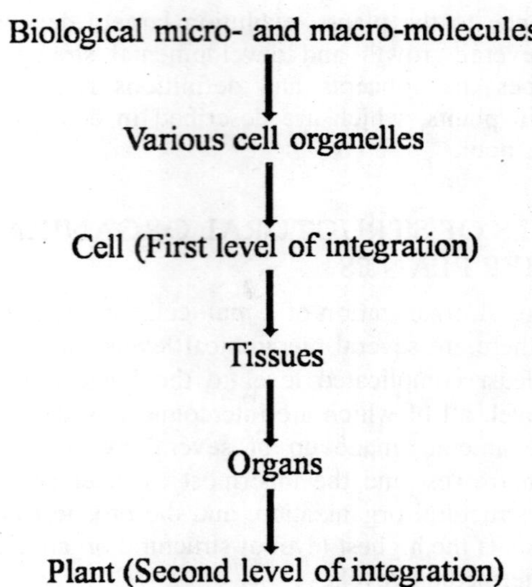

Fig. 3. Levels of biological organization and integration

1.3. DEVELOPMENT

Development is a series of processes by which a multicellular plant or plant organ arises from a single cell. In a sexually–reproducing sporophytic plant, this single cell is often called the **zygote**, a diploid product of gametic fusion. In some categories of plants, where the dominant generation is a gametophyte (for instance, the bryophytes), this single cell is a **spore**. During development, there is a continuous synthesis of large number of molecules that form the raw material for development. In a real sense, the processes that lead to development are invariably not cascaded, but there is a great degree of overlapping throughout the period of development. In other words, development is not an 'ultimate' end–product, but is a 'cumulative' end–product (Twyman 2003). However, it should be emphasized that development is a progressive event from a fairly simple starting point (i.e., a zygote or a spore) to a highly complex plant, with progressive addition of complexity at every developmental step. This gradual developmental strategy is called **epigenesis**. Epigenesis is a term borrowed from animal developmental biology, but it can profitably be used for the plant system as well, without any terminological difficulty. The concept of epigenesis is in complete contrast to the earlier prevalent concept of preformation, which suggests that the embryo is a miniature version of the adult, and development consisted entirely of growth (in contrast to the five processes that together make up development, as detailed below).

Development, according to some, involves the following five major, overlapping processes (Fig. 4):

1. Growth
2. Cell division
3. Differentiation
4. Pattern formation
5. Morphogenesis

Each of these processes is separately described in subsequent pages. However, there are some who consider these five processes as independent and believe that certain types of development may not and need not involve one or more of these processes.

1.3.1. Formal Methods of Representing Development

Two main concepts underlie the formal methods of representing development (Barlow 1994). According to the first concept, development, within limits, is invariant, and its unfolding depends upon the correct set of permissive conditions. Thus, any intervention through incorrect conditions does not permit development to continue. If development only requires YES or NO type of decisions, the final outcome of the developmental sequence would be easily predictable. For example, in a seed, continuation or breaking of dormancy depends respectively on absence or presence of water. If we increase the number of alternate decisions that can be taken at each responsive stage, we can produce a more variable outcome.

The second concept regards the developmental process as essentially self–perpetuating. Here also, decisions are taken as to which new developmental state is to be adopted, but the information relating to the transition is inherent to the system itself, and permissive environmental conditions are regarded as axiomatic. Hence, it is the combination of the current developmental states present in various parts of the developing organism that determines the next set of states. In the example of seed–dormancy mentioned above, the outcome is a seedling because this is the only state prefigured in the previous state. The first concept is essentially physiological and considers the processes as the basic driving force of development. The second concept is essentially structural and considers structural or morphological patterns as very important. However, it should be emphasized that these two concepts of development are not mutually exclusive and they arise from two contrasting views of development.

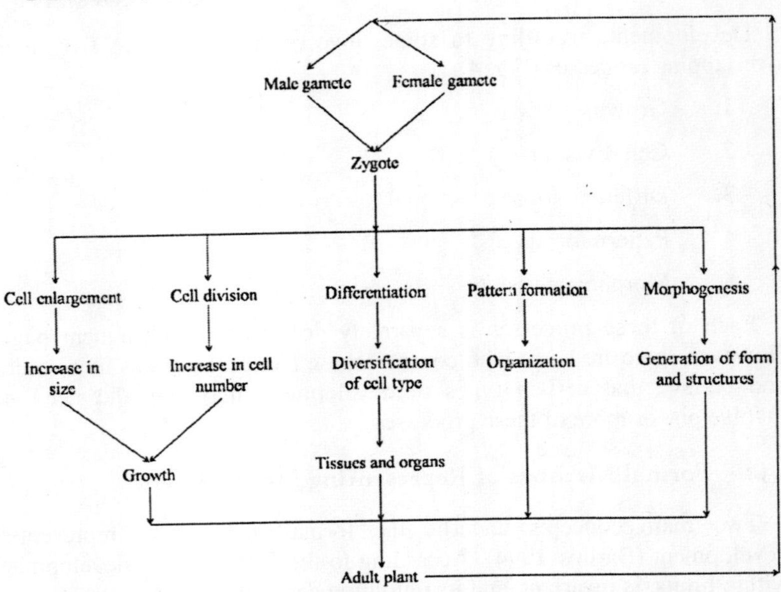

Fig. 4. An overview of development process in plants

1.3.2. Controls in Development

Development is controlled by the genes, the environment and by the interactions of these two. We know that genes govern the synthesis of enzymes, which in turn, control the chemistry of cells, and all these account for development. Now, we have lots of insights into the genetic control of plant development. Barring a very few exceptions, each cell in a developing plant contains the same DNA. Hence, the diversification of the plant body during development depends on the expression of different genes in different cells. **Gene expression** involves two major processes; the **transcription** of DNA into RNA, and the **translation** of RNA into proteins. The information carried in DNA (gene) specifies the amino acids to be incorporated into proteins as well as the sequence of their incorporation during protein synthesis. The gene is also flanked by further information in the form of cis–acting **regulatory elements**, which control where and when the gene is to be transcribed and the rate of its transcription. Consequently, the amount of RNA produced and its spatial and temporal expression pattern are also governed. These flanking regions may also carry **response elements**, which allow gene expression to be regulated by external signals, and thus they link **signal transduction** to the gene expression. Thus, **transcriptional regulation** is often the most important level of control in gene expression. Once the gene is transcribed into RNA, further **post–transcriptional regulation** can take place to effectively control the amount of functional protein formed in a cell. The RNA must be processed and exported from the nucleus, translated and then broken down. All these processes are regulated. Furthermore, all these proteins need processing in various ways before they can function effectively. These processings include folding, chemical modification, cleaving off of some amino acids, conjugation to other molecules and formation of multiproteins. These proteins need to be targeted to different locations in the cell including importation into various cell organelles, or they may also need to be secreted extracellularly. Finally, these proteins also need to be broken down and recycled. Gene expression is regulated at all these levels also.

One should know that all the five cardinal processes mentioned under development are essential for proper development and these are also ultimately mediated by proteins that act either directly or as enzymes to produce other product molecules. In the development of embryo and subsequently in different developing organs of the plants, the pattern of gene expression determines where, when and in what quantity the particular proteins are to be made and thus, governs the properties of each and every cell. Much of the knowledge on the importance of genes in plant development has come to light through the analysis of developmental mutants, which give clues for the functions of specific genes.

There is also an environmental control on the development of plants. Unlike animals, plants cannot move from place to place to find mates, to seek food or to avoid unfavourable environmental conditions. This has introduced a necessity for fine–tuning of their development in order to adapt themselves to the prevailing environmental conditions. For instance, light plays a very important role in plant development throughout its life cycle. In fact, the term **photomorphogenesis** is used to describe the role of light on plant form. Also, light effects are exhibited within constraints placed by an internal timing mechanism known as the **biological clock** that exists in the plants. Plant development is also strongly influenced by temperature regime.

1.4. GROWTH

As stated earlier, development is initiated from a single cell, a zygote or a spore, as the case may be. The whole plant develops from this single cell through the involvement of growth. Growth essentially involves an increase in size and volume, total biomass (weight), amount of protoplasm, and above all, in complexity. Hence, we measure growth by measuring any one of these parameters. For instance, the increase in volume or mass is measured by measuring the increase in length and/or girth.

1.4.1. Types of Growth

Growth involves two events. In the first event, there is an increase in the size of cell only, without an increase in the number of cells (produced through cell division). This type of growth is called **hypertrophy**, although more often this word is used to denote an extreme form of increase in cell size. Hypertrophy may or may not involve or be accompanied by an increase in the number of nuclei. The former situation results in an enlarged cell with more than one nucleus, a condition often referred to as a **coenocyte**. Coenocytic giant cells are found in almost all groups of plants, right from algae to angiosperms. *Caulerpa* and *Vaucheria* are very good algal examples, whereas in angiosperms coencytic condition is found in nuclear endosperm, cells of endosperm haustoria, embryo suspensor, anther tapetum, integumentary tapetum, coenomegaspore, laticifers and developing fibers. Hypertrophied growth of cell, without an increase in nuclear number, but with an increase in DNA content (due to **endopolyploidy** or **polyteny**), characterizes certain special types of cells involved in functions such as secretion, digestion, providing nutrition and hypersensitive reactions (due to infection). Examples are some endosperm haustorial cells, embryo suspensor, anther tapetum, certain types of trichomes, secretary cells and bacterioid–infected root nodule cells. Growth or enlargement without an increase in the number of nuclei or in the DNA content is common in many cells and is often a feature that characterizes a differentiating and maturing cell of any category. Even the cells that divide into two cells first enlarge before actually embarking on cell division. This condition is sometimes referred to as **growth with differentiation**. There are also instances of cells that grow in size without an accompanying differentiation process, often as a result of infection/or an externally induced abiotic stress. Such a condition is referred to as **growth without differentiation**. Details on this phenomenon are discussed later in this book.

In the second event, there is an increase in the number of cells as a result of repeated cell divisions. This type of growth is more well–known than growth without cell division. The growth through the increase in cell numbers is called **hyperplasia** (or **hyperplasy**), although this word is more commonly used to designate an extreme form of cell division. Cell divisions, by adding more cells, contribute to growth. At least apparently, cell number increases in the early developmental phases in a plant (i.e., embryo development) in the absence of visible cell enlargement. However, in later development, there is some amount of growth in the absence of or independent of cell divisions. This is by an increase in cell volume associated with or without change in the shape of the cell. Invariably, there is spatial distribution of growth through cell division and through cell enlargement; the former through specific localization of growth centres (called meristems) and the latter through their derivatives. This spatial division of the location of dividing cells and enlarging cells also results in the determination of shape and pattern in the plants and their parts.

1.4.2. Growth Kinetics

Now, in order to understand the **growth kinetics**, let us consider the growth of a whole organ or plant in relation to time. If one plots a graph of the size or weight of a growing organ or plant against time, a growth curve is obtained. Very often, this growth curve can be fitted with a simple mathematical function in a straight line (linear) or S–shaped curve (curvilinear). These representations are often very useful in the interpretation of growth, although the metabolic and physical processes behind this are very complicated. A very typical S–shaped or sigmoidal growth curve is shown in Fig. 5 A. There are three phases in this curve; a **logarithmic phase**, a **linear phase** and a **senescence phase**. In the first phase, the size (v) increases exponentially with time (t). In other words, the growth rate (dv/dt) is slow to begin with, but the rate increases continuously. Often, the rate of growth is proportionate to the size of the organ or plant; larger the organ or plant, faster is the rate of its growth. During the second phase, the increase in size continues at an approximately constant rate, which is usually the maximum. The constant rate is often indicated by a constant slope in the upper, height curves and by the horizontal part of the lower, rate curves (Fig. 5 B). The last phase is marked by a decreasing growth rate as the organ or plant reaches maturity and begins to senesce. S–shaped growth curves are represented by many species although differently–shaped growth curves are obtained in other species.

Sometimes, some organs from the same plant may show a sigmoidal growth curve and other organs may show a different growth curve. In some species, the linear phase is hardly recognizable making the senescence phase continuous with the logarithmic phase. In some species, the linear phase is greatly extended. In some, such as the fruits of grape, olive and fig, the growth curve is doubly sigmoid. The height growth in trees is either sigmoid or modified sigmoid in many taxa. In some temperate trees which are subjected to distinct seasons, flat regions are often found in the growth curve as a result of the winter dormancies, while the same may be true for tropical trees where there is dormancy during the hot summers. A similar trend is often shown by the growth in girth.

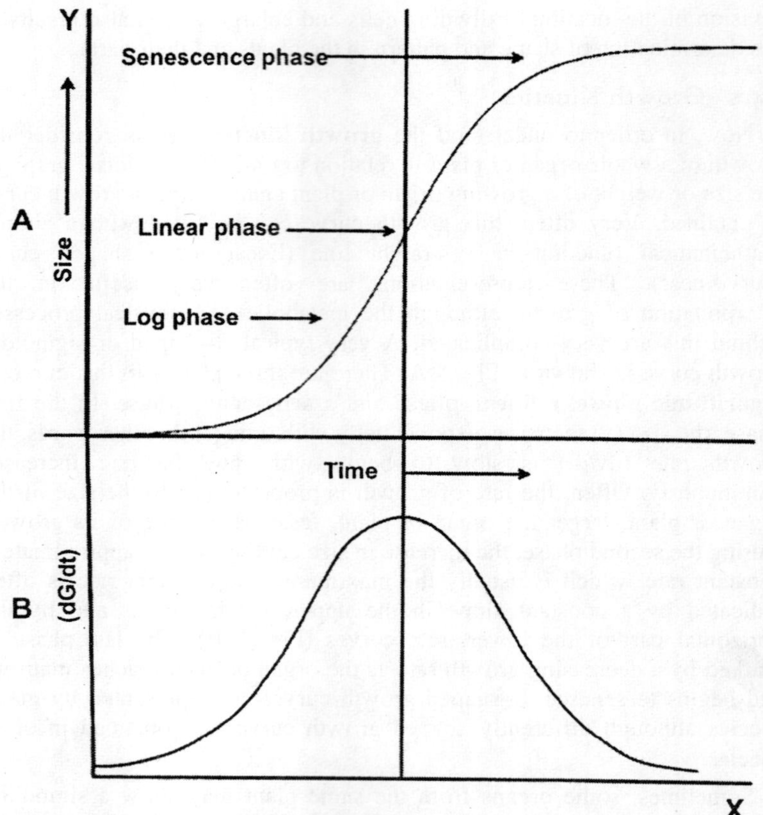

Fig. 5. A - Typical sigmoid growth curve, with its three phases; B - A typical bell-shaped growth rate curve; It is the first derivative (slope) of the total growth curve in A

1.5. ONTOGENY

The development of an individual plant entity from its most initial stage till completion is referred to by the term **ontogeny** or **ontogenesis**. The entity may be a stoma, a trichome, a tracheary element, a tissue or even an organ such as leaf and flower. Usually, the starting point or structure in ontogeny is referred to as an **initial** (in case of cells) or a **rudiment** or **primordium** (in case of an organ). Ontogeny essentially involves a continuum and any step in ontogeny is preceded and followed by specific steps. Ontogeny is characterized by the remarkable changes in form and structure. At every stage of ontogeny, we may perceive or discover the evidence of classical physical relationship of energy and matter in the determination of the eventual form and structure.

1.6. DIFFERENTIATION, DEDIFFERENTIATION, REDIFFERENTIATION AND TRANSDIFFERENTIATION

Biologists have been known to disagree about the meaning or significance of the term **differentiation**. Basically, as used in biology, the word 'differentiate' simply means to make or become different from the original during development. Thus, differentiation is a process through which initial cells undergo structural and functional specialization, and thus, result in distinct cell types (between 10 and 30 different cell types depending on the plant) such as parenchyma, sclerenchyma, tracheary elements and sieve elements. Before differentiation such cells are said to be **undifferentiated**. The structural specializations involve the cytoplasm and/or the cell wall (Refer to Chapter 4). Since structure reflects and determines function, different cell types automatically get functionally specialized after differentiation. In sieve elements, the nucleus is lost, but cytoplasm in a modified form called **mictoplasm** persists. In parenchyma, there is not much differentiation as cytoplasm and cell wall undergo fewer changes (i.e., there is less differentiation). In collenchyma, the cytoplasm undergoes no or little change, but the cell wall experiences various kinds of thickening with primary cell wall materials. In sclerenchyma and tracheary elements, both the cytoplasm and the nucleus are lost due to a process called **Programmed Cell Death (PCD)** (Refer to Chapter 4), while the wall is subjected to secondary thickening with lignin as a major component of cell wall. Such cells are said to have undergone a unique type of differentiation, i.e., they die to do their functions while other cell types die after completing their functions. It should be mentioned here that an analysis of different cell types in plants indicate that some cells such as parenchyma are less differentiated as they undergo differentiation from their progenitors, and little more differentiation in some cells such as collenchymas, and at the end of the spectrum, there are lignified, dead cells of sclerenchyma and tracheary tissue, which may be said to have undergone **terminal differentiation**.

Under normal circumstances, parenchyma and collenchyma cells do not undergo cell division, but under certain conditions such as wounding, infection and in vitro culture, they can resume the capacity to divide (i.e., to become meristematic) (Refer to Chapter 10). Given the proper requirements and conditions, they can even regenerate the whole plant. In other words, these cells possess the property of **totipotency** and the cells themselves are characterized as totipotent. This capacity of cells to resume meristematism is called **dedifferentiation**.

Parenchyma and collenchyma cells also possess the capacity of **redifferentiation**, by virtue of which they get converted into different (but more differentiated) cell types through additional differentiation steps. For instance, a parenchyma cell can become a collenchyma, sclerenchyma or even a tracheary element, while a collenchyma cell can be converted into a sclerenchyma or tracheary element. If a living cell such as a mesophyll cell, without undergoing any type of change in shape or any evidence of dedifferentiation, directly transforms itself into a treachery or phloem sieve element, it is said to have undergone **transdifferentiation**. Terminally–differentiated cells such as sclerenchyma and tracheary elements as well as the enucleated sieve elements do not have the capacity to redifferentiate or dedifferentiate. In other words, they are not totipotent.

These concepts are shown in Fig. 6 and 7.

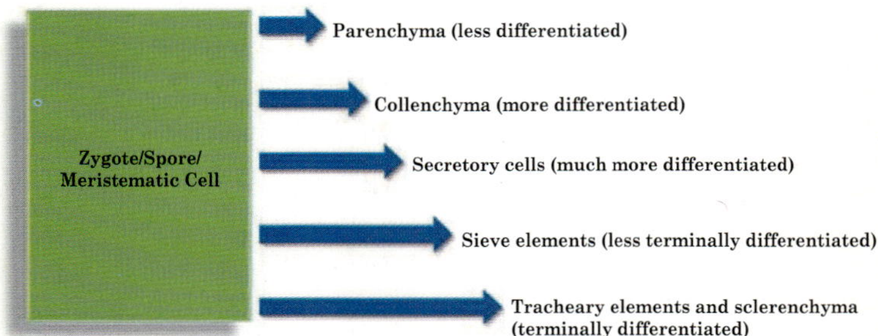

Fig. 6. The degree of differentiation seen in different cell types of plants (indicated by the lengths of arrows), when they get differentiated from progenitor cells

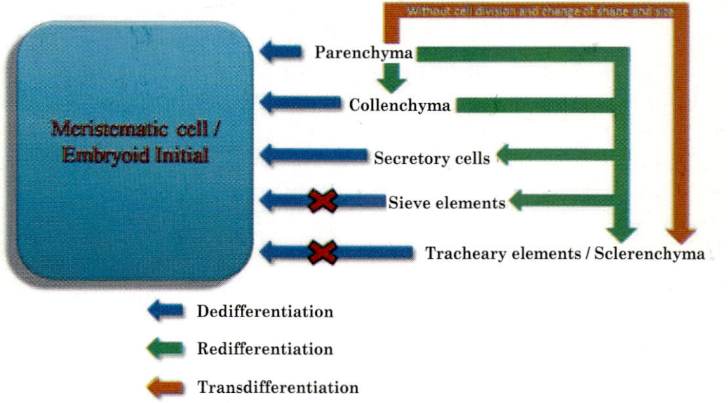

Fig. 7. Concepts of dedifferentiation, redifferentiation and transdifferentiation

In Fig. 7, X indicates the inability of a cell type to get back to meristematism and the lengths of blue arrows indicate the relative ease with which different mature cell types can dedifferentiate; the shorter arrow indicates greater ease.

1.7. MERISTEMS

During early embryonic development, cell divisions occur almost throughout the embryo, but as the embryo matures as well as when it gives rise to the adult plant, the addition of new cells to the plant body is gradually restricted to certain of its regions. Thus, the growth centres formed that remain embryonic, are maintained as it is almost throughout the life of the plant. As a result, the adult plant body becomes a composite structure with both embryonic and matured tissues.

This restriction of generation of new cells to specific regions of the adult body is an evolutionarily–advanced feature, since in many lower plants, almost all the cells are involved in producing new cells. Thus, in more advanced plants, there is a clear–cut division of labour between the different regions of adult plant body. The localization of growth centres in the form of meristems is a strong feature of distinction between higher plants and higher animals as well. The term 'meristem' is derived from Greek 'meristos', which means 'division'. The group of cells (tissues) that organizes the meristem is called meristamtic tissue. There are cells in each meristem that maintain it as a perennial source of new cells and these cells are called **initials**; their derivatives, after a varying number of cell divisions give rise to mature cells of the plant body.

1.7.1. Types of Meristems

Four types of meristems are found in higher plants : **apical**, **intercalary**, **lateral** and **diffuse**. The most common classification of meristem is based on its location in plant body : apical, lateral and intercalary. The apical meristem is located at the apices of the main and lateral shoots and roots. The lateral meristem runs parallel to the long axis of the stem and root, and is of two types, **vascular cambium** and **cork cambium** (**phellogen**). The intercalary meristem consists of a group of meristematic cells intercalated between mature tissue regions and is either derived from apical meristem of shoots (**residual meristem**) or from matured tissues due to dedifferentiation (**resumptive meristems**) (Swamy and Krishnamurthy 1979). The meristems either participate in the longitudinal growth of plant axis (**primary meristem**) or in the latitudinal growth i.e., growth in thickness or girth (**secondary meristems**).

Many developmental botanists distinguish the initials and their most recent derivatives from the partially differentiated, but still dividing the cells under the names of **promeristem** (Evans and Barton 1997; Barton 1998), **protomeristem** or **primary meristem**.

Promeristem is classified into the following categories based on the tissue regions derived from it:

- **Protoderm** - It gives rise to the epidermal tissue.
- **Procambium** or **provascular tissue** - It gives rise to primary vascular tissues.

- **Ground meristem** - It gives rise to ground tissues of cortex and pith.

Promeristems often exhibit different but characteristic patterns of cell division and growth depending on the plant organ. Hence, the different types of promeristems have been given specific names: the **mass meristems**, which grow by cell divisions in all planes; the **rib meristems**, which give rise to parallel, longitudinal files of cells through essentially transverse divisions (i.e., at right angles to the long axis) (as in internodes); the **plate meristems**, in which cell divisions are primarily anticlinal to result in flat organs such as leaf and petal.

After a thorough review of the meristems, particularly with reference to the cell division activity in their constituent cells, Swamy and Krishnamurthy (1978) have shown that all types of meristems whether apical or lateral, are always constituted of a relatively quiescent group of cells and the actively–dividing cells are invariably adjacent to this quiescent zone. Two physiological features characterize this quiescent zone of meristems: (i) the conspicuously drawn out time factor to complete a cell division, and (ii) an equally conspicuous low frequency of dividing cells. The histological and histochemical characteristics of the quiescent zones also present sharp contrast to the meristematic cells. Experimental investigations on the meristems indicate that the quiescent zone is an integral part of meristems without which the normal morphogenesis and organization become greatly disturbed. The quiescent cells not only act as the ultimate source of all the dividing cells of the meristems, but also act as a reservoir of cells to aid in the reconstruction of the meristem, if the dividing cells are damaged for some reason.

1.8. INITIALS AND STEM CELLS

As already stated, the retention of growth centres in the form of meristems at specific locations in the adult plant body and the ability of the meristems to produce new tissues or organs distinguish the higher plants from higher animals. The cells that maintain the meristem as a continuing source of new cells and at the same time maintain some stable form of identity are called **initials,** while their derivatives, after undergoing a few cell divisions give rise to mature cells of the plant body. Invariably, a given initial in any meristem produces two daughter cells after a symmetric or asymmetric division, out of which one continues to act as initial while the other invariably becomes the precursor of an adult cell. The concept of initials and derivatives often goes with a qualification i.e., both are not inherently different from one another, but they do have some differences when we speak of them with specific reference to some categories of meristems such as vascular cambium (Refer to Chapter 5 for more details). Unlike the derivative cell, the initial cell must recreate the complex patterns of transcriptional activity with each mitotic division. But, how this is moulded and maintained is not very clear. It is commonly believed that many cells of the meristem serve as initials because of their specific location in the meristem. In the animals, a fixed number of organs are formed during embryonic development itself. But, in a fully grown higher animal, the adult tissues and organs are maintained throughout its life by the population of cells that reside within these tissues and organs. These cells are called **stem cells** (Weizman 2000). Stem cells of animals are often compared to the initials of plant meristems and in fact, a number of plant biologists have adopted the term 'stem cell' for initials (Veit 2006) or their recent derivatives. Also, there is a confusion about the usage of both these terms (stem cells and initials) while describing the shoot apical meristem. For instance, Fletcher (2004) states, "The stem cells are not permanent initials"; Vernoux et al (2000) say, "It is now generally accepted that the central zone acts as a population of stem cells..... and generating the initials for the other zones, whilst maintaining itself"; Bowman and Eshed (2000) mention, "The central zone acts as a reservoir of stem cells........It should be noted that these cells do not act as permanent initials...", Laufs et al (1998) state, "It is now widely assumed that central cells function as stem cells and serve as initials or source cells for the two other zones of the shoot apical meristem". Finally, while characterizing the shoot apical meristem as a 'group of stem cells', Meyerowitz (1997) designated its central zone as the 'zone of initials'. However, the author of this book advocates the use of the more appropriate and conservative term 'initials'.

A very basic conceptual question that has been raised is about the permanency of the initials. Whether the initials that make up a meristem today can be the initials of the same meristem till the total life of the meristem? Some experiments carried out by Ball and his co–workers (Soma and Ball 1963) and Newman (Newman 1965) on shoot apical meristems have indicated that it is not so. Ball and his co–workers have carried out two types of experiments to come to this conclusion. In one experiment they have put an India ink dot on the surface at the summit of the shoot apical meristem and traced its fate through time–lapse photography. They found that the dot got split into several dots and then got radially displaced from the summit and finally formed a part of the internode or leaf primordia. No dot was found at the summit of the apical meristem after some time. In another experiment, they slightly injured some cells in the summit of the shoot apical meristem through a microneedle; the injured cells included surface as well as subsurface cells. Then, they traced the fate of these cells. Like ink–dotted cells, these injured cells also soon got displaced from the meristem and later on formed a part of the mature tissue derived from the apical meristem. These results indicate that the cells of the apical meristem (i.e., initials) are not permanently located in the meristem and served as initials, but are replaced by newer initials. Newman also concluded that no cell in the shoot apical meristem is a permanent cell of the meristem. The meristem can be compared to an office. At any one time the office is manned by a set of people (the initials in the case of meristems) and these people are replaced by other people once they retire or leave the office (in the meristems, initials are periodically replaced by other initials), but the functioning of the office goes on (i.e., the meristem continues to function). In fact, Newman (1965) has compared the apical meristem to a fountain of water, where the water is coming out of a nozzle pore (comparable to the central zone of the meristem) and splitting to form a radiating pattern (comparable to the derivative cells of the other zones of the meristem from the central mother cell zone). It is implied here that neither the water coming out of the nozzle nor the water that gets split into radiating files are permanent, but get replaced every time by fresh water (by fresh cells or initials in case of meristem).

The same concept of non–permanency of initials can also be extended to other meristems such as vascular cambium, although no experiment has been so far carried out on these.

1.9. DETERMINATION, COMMITMENT, COMPETENCE AND POSITION EFFECT

These terms are related to the process of differentiation. **Determination** (McDaniel 1984) is one of the important aspects of differentiation. Determination means a progressive commitment to a specific course of development that brings about a gradual weakening or loss of capacity to resume division activity or growth. Some cell types are determined earlier and more completely than others, and as stated earlier, some maintain their totipotency even after differentiation. The term **competence** refers to the ability of a cell to develop in response to a specific signal such as light or chemical. It implies that a competent cell is capable of recognizing the right (or even wrong) signal and translate it (or fail to translate it) into a response.

Although a differentiating cell is under the control of gene expression (as already emphasized), its fate i.e., what kind of cell will it become, is determined greatly by its **position** in a developing tissue or organ. Even though distinct cell lineages may be established such as those in a developing embryo or in root apex, it is the position and not lineage that determines the fate of the cells. **Chimeras** have given strong evidence for this. Here, it should also to be pointed out that at the time of division of an initial, it is often impossible to predict which of the two resultant daughter cells will 'inherit' the function of the initial and which will become the derivative. It is also known that a given initial may be replaced by a cell that would be rightly classified as a derivative from an initial through the prior history (Zagórska–Marek and Turzanska 2000). If an undifferentiated cell is displaced from its original position, it will differentiate into the appropriate cell type depending on its newly–acquired position. Laser ablation experiments on *Arabidopsis* root meristems have shown that ablated cells are replaced by cells from other lineages and these replaced cells differentiate according to their new position. This has shown to be largely controlled by **positional information** that is probably conveyed by the signalling events from cell to cell, mediated probably by transmembrane receptor kinases or by plasmodesmata.

1.10. PATTERN FORMATION

The concept of **pattern formation** was first established in the insect *Drosophila,* and it was initially applied to animals, in general. Pattern formation in plant development shows many mechanistic similarities to pattern formation in animals, but yet these are some very fundamental differences between them.

Pattern formation is the developmental process by which cells become organized into specially–patterned tissues (or their progenitors), which in turn, get organized into higher levels of organization such as organs and whole plant. As stated earlier, since development is gradual and progressive, pattern formation also occurs step by step and progressively. The production of a tissue from cells is called **histogenesis**, while the production of an organ from an aggregation and spatial and locus–specific orientation of different tissues is called **organogenesis.** Histogentic pattern (or tissue pattern) may be organized by a single cell type. For example, the parenchyma forming the cortex, pith or pericarp of some fruits (simple tissue). A number of cell types or their progenitors may together organize a heterogeneous histogenic pattern (complex tissue) as in xylem, phloem or epidermis. Tissues of either type together organize the next hierarchical level of pattern formation. In plants, such a gradual pattern formation is first evident in the embryo, thus signalling the basic rudiments of the **body plan**, which subsequently lead to the generation of the structurally–complex patterns of different organs. For the successful establishment and development of a body plan, the concept of pattern formation expects that each cell must behave, in a manner appropriate for its position in the embryo (or meristem), so that right kind of differentiated cell types arise at appropriated locations and cells of the same type form regionally–correct structures. The process by which cells become specialized according to their position is called **regional specification**, which allows cells to adopt the correct spatial organization. Hence, it is an essential requirement for pattern formation. The first event related to patterning that takes place in the embryo is **axis specification** i.e., the establishment of principal body axis with two distinct poles.

1.11. POLARITY

In simple terms, polarity means the orientation of activities of a structure in the space. Since the difference between the two poles of the axis (a proximal pole and a distal pole, or a basal pole and an apical pole) appear in a graded series, the polarity is related to the phenomenon of gradients. For example, in the phloem tissue, there are physiological gradients relating to rates of metabolic processes and concentrations of growth regulators such as auxins and sugars (chemical gradient). There are electrical gradients also. And also, there are gradients in cytological, histological and anatomical aspects depending upon whether the polarized structure is a cell, an organ or the whole plant. These are expressed outward as different morphological features (Prat 1948, 1951; Grebe et al 2001). Thus, polarity is due to an orderly arrangement of polar molecules, all statistically oriented in the same direction, and spirally disposed in the outer cytoplasm (cortex). Polarity is an essential and fundamental component of biological pattern formation (Sachs 1991a, b) and it is also one of the properties that manifests itself very early in the life of plants. Differentiation has its inception in the establishment of polarity in a cell. Sinnott (1960) referred to polarity as one of the basic principles in morphogenesis that has to be accepted 'as given'. This view is probably referable to the fact that the cause of polarity and its usually–irreversible character has proved exceedingly difficult to understand.

Polarity is seen in the egg cell and zygotes of non–vascular and vascular plants. In the angiosperm egg, the nucleus is located at the chalazal end, while there is a large vacuole on its micropylar end, thus exhibiting a clear bipolar organization. The embryo that is derived from it, after fertilization, is also typically bipolar. This bipolar organization continues into the adult plant as well, with a shoot end and a root end. A lot of information on polarity has been obtained from the study of zygotes of the marine algae *Fucus* and *Pelvetia* (Fowler and Quatrano, 1997). The zygote of these algae secretes a uniform cell wall around it and initially, it is spherical and apolar. When subjected to unilateral light (continuously or in pulses) for the first 10 h of the development, the rhizoidal tip emerges after 14 h from the shaded side, which is likely to be the place of sperm entry at the time of fertilization. The directional light signal establishes a **cortical site** to which the secretory apparatus is redirected from its initial symmetrical distribution to an asymmetric distribution (oriented towards the site of polar growth). Polar secretion continues to be oriented towards the rhizoid tip (Fig. 8). The cortical site, responsible for the establishment of polarity, includes dihydropyridine (DHP) receptors, membrane domains that direct ion flow inward, higher levels of free Ca^{2+} and a patch of actin microfilaments (MFs), and these components accumulate several hours before the polar growth. In these algae, the cell wall is required to maintain a stable polarity, and the asymmetry in the wall of the two–celled embryo is essential to establish and maintain rhizoid– and thallus–like characteristics. Rhizoidal cell walls are chemically different from those of the thallus cell.

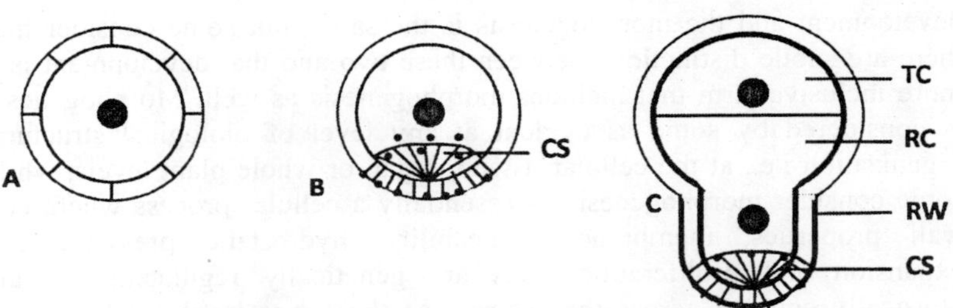

Fig. 8. Establishment of a polar axis and an oriented division in the zygote of brown algae (based on Fowler and Quatrano 1997); A - Non-polarized zygote; B - Establishment of polarity with the formation of cortical site complex (CS) with the concentration of DHP receptors (represented by the lines on the cell wall), F-actin (hatched) and free calcium (black dots); C - Division of the zygote; RC - Rhizoidal cell; RW - Rhizoidal Wall; TC - Thallus cells

Polarity is not only of primary importance in embryogenesis, but also pervades all through development (Bloch 1965). It is to be further noted that bipolar shoot–root relationship is perhaps the most common and the best example of polarity. It is a relatively–specific instance of a general phenomenon. Complex polarity relationships are also seen in dorsiventral organs such as leaves. Hence, according to Bloch (1965), "In a broad sense, polarity designates any internal asymmetrical state within a living system as expressed in differentiation between two ends or two surfaces and also between the ends or surfaces and the central region".

Transplantation (Gulline 1960) and tissue culture studies (Wetmore and Sorokin 1955) indicate that polarity is not only a property of the whole plant, but also of its parts, even if they are severed or isolated from the plant. Polarity in the stem is a very familiar phenomenon. For example, in plants that are propagated by stem cutting, the roots will be formed at the lower end of the stem, and leaves and buds will be formed at the upper end of the stem. Moreover, polarity cannot be changed by reversing a stem cutting by planting it upside down.

In cells exhibiting polarity, it is often related to or rather expressed as, asymmetric cell division (Refer to Chapter 3).

1.12. MORPHOLOGY AND MORPHOGENESIS

Morphology and morphogenesis are two closely–related terms. While morphology refers to the study of form (external or internal), morphogenesis deals with the origin of this form. The term morphogenesis simply means the creation of form (Refer to Box 1.1), and by implication, the differentiation of the associated internal structural features. Some people, especially some plant physiologists, consider that the meaning of development and the morphogenesis is the same, but some consider that there are subtle distinctions between these two and that development is a more inclusive term that includes morphogenesis as well. Morphogenesis is considered by some as evident at any level of biological structural organization i.e., at the cellular, tissue, organ or whole plant levels, while some consider morphogenesis as essentially a cellular process where cell wall properties, membrane permeability, hydrostatic pressure, cell expansion and proliferation rates are genetically regulated, but are physically coupled across the system. At the lower levels of biological organization, morphogenesis reflects the behaviour of many different types of cells that help to shape the tissues and the organs. The morphogenetic changes at the cellular level may involve changes in size and shape, their adherence or dispersal, establishment of new contact with other cells, changes in cell wall and/or organelles, loss of their earlier contents, their quiescence or division activity or even their death due to Programmed Cell Death (PCD). It is to be noted that in the earlier stages of plant development, morphogenesis may be instrumental in driving the developmental programmes, while in the later stages of development, morphogenesis may be regarded as a response to the developmental programmes.

BOX 1.1. AIM AND SCOPE OF PLANT MORPHOLOGY

The study of plant form is called morphology (a term invented by the German naturalist, Goethe), a valid scientific discipline in its own rights, and the study of the way in which it results is called mophogenesis. Morphology may be said to have four main aspects: verbal description of plant form, classification of plant form, genesis of plant form, and relationship between the form and the function. The first three aspects are believed to represent true and meaningful study of morphology, but a few botanists believe that the fourth aspect does not seem to have any purposeful validity. For instance, Troll, an early German botanist, did not feel that one need to deal with morphology causally. The legendary Goethe's view that the root, stem and leaf constituted the fundamental organs of higher plant body and decided the fundamental organizational theme or 'Bau–plan', was the main source of pre–evolutionary ideas of formal morphology comprising the modification of organs, axial and appendicular organs, and the analogies (note that homologies are more emphasized in Systematics). All these helped to integrate the mass of descriptive accounts of higher plant morphology in a more meaningful way. However, the advent of Darwin's theory of organic evolution necessitated the explanation of the ideas of formal morphology in terms of evolution concepts (**evolutionary morphology**). Although some aspects of **formal morphology** could lend themselves to be interpreted by the concepts of evolution, the basic differences in their approach, scope and content are not easily reconcilable, and Goethe's approach was derogatorily labelled as 'idealistic morphology'. However, the superficial similarity between formal and evolutionary morphologies made some post–evolution authors to mix these two more or less indiscriminately, and this resulted in discredit to both.

Initially, evolution was applied only at the species level, but when genetics established the independent inheritance of characters of the species, evolution also began to be analyzed in terms of the plant organs and characters, each being considered as having independent evolutionary trends. However, subsequently this rule of independence for characters (and organs) began to be violated through attempts to determine evolutionary status of a character (or organ) by correlation. This was clearly associated with an increasing trend to ascribe a function–oriented approach to formal morphological features from an evolutionary perspective, and principles were laid down by some morphologists towards formalizing this approach. For instance, homology or analogy was mixed up with evolution. However, it should be emphasized here that the principles of morphology transcend the systematic and evolutionary (phylogentic) boundaries, and that morphology is a unifying discipline.

Although it is possible that the form and the function have some relation with one another, it is questionable, according to some, whether all categories of forms have necessarily an associated and specific functional signification and *vice versa*. Von Goebel, another famous German morphologist, was fond of saying that only those phenomena could be called morphological that could not yet be explained functionally. According to some, the assertive extrapolation that evolution does always have a say in this form–function relationship further complicates the understanding of this relationship, but some others feel that while plant morphology is not obligatorily tied to phylogenetic formulations of form relationships, these linkages in character expression are the most compelling evidence that the diversity of plant form is the result of evolution. More information is needed on the relationship between morphology (the study of form) and morphogenesis (the actual derivation of form) on the one hand and evolutionary origin of form on the other.

The discipline of plant morphology encompasses the whole plant down to the organ level of the organization (often termed by some as external morphology). While morphology occasionally employs the anatomical level of the organization (often termed by some as internal morphology) as morphological markers (to distinguish the different plant organs), it can only do so with histological features that are developmentally correlated with morphogenesis of form generation. Thus, 'the plant's morphology is an emergent property relative to its anatomy' i.e., these two levels of the organization can be relatively independent and the anatomical level need not determine the morphological level. Similarly, the growth habit of a plant is an emergent property relative to the plant's morphology.

(Kaplan 2001; Periasamy and Swamy 1977, 1978)

1.13. PLANT GROWTH REGULATORS

Plant Growth Regulators (PGRs) are also called plant hormones or phytohormones. These are chemical signals that play a major role in regulating growth and development either through promotion or inhibition. The term 'hormone' (from the Greek 'horman', which means 'to set in motion') was borrowed from animal physiology. The basic feature of animal hormones is that they are active at a distance from where they are synthesized. This feature cannot always be equally applied to all plant hormones, and hence some people advocate that the term hormone should not be used with reference to plants. Some of the most important growth regulators are auxins, cytokinins, ethylene, ABA, gibberellins, brassinosteroids, salicylic acid, jasmonates and polyamines nitric oxide. There are natural as well as synthetic PGRs.

1.14. ABNORMAL GROWTH

Any growth deviation from the norm or type is referred to as **abnormal growth**. Abnormal growth in plants is invariably induced due to various kinds of abiotic and biotic stresses. It is sometimes referred to as pathological growth. The abiotic stresses may be physical or chemical. The physical abiotic stresses include temperature, light, radiation, mechanical effects, and the chemical abiotic stresses include chemicals available or introduced into the environment in which plants grow. The biotic stresses are caused by various living organisms, including viruses. The growth or development reactions shown by the plants in response to the biotic and abiotic stresses are often known as **hypersensitive reactions**. Abnormal growth may also be the result of growing plants or their parts (including cells and tissues) under in vitro conditions. The data or the inferences derived from a study of abnormal growth often provide very valuable information on various aspects of normal growth and development, especially on the factors affecting growth and development. Abnormal growth may be due to altered cell divisions, excessive enlargement, modified differentiation patterns, defective polarity, wrong morphogenetic pathways or due to a combination of two or more of these.

1.15. EVOLUTION AND PLANT DEVELOPMENT

Evolutionary developmental botany or the study of the underlying developmental basis for the origin and diversification of the plant structure has developed into a very important discipline in the last thirty years. Evolutionary plant development is also called evolution of ontogeny. This field deals with the modification of development with respect to timing (**heterochrony**), ontogenetic sequence (addition or deletion of specific developmental events), and positional status as well as with an analysis of the rate of evolutionary change (such as gradual, sudden or punctuated) (Friedman et al 2004). In other words, evolutionary plant development should be based on a complete explanation of the data derived from comparative plant morphology, palaeobotany, embryology, molecular biology and genetics, particularly on molecular–genetic information.

The integration of a detailed understanding of phylogenetic relationships among the land plants, extinct and extant, provides a basic framework for the study of the evolution of morphology and ontogeny. Studies on the comparative molecular–genetic bases for the origins and diversification of apical growth patterns, leaves, roots and flowers provide enough opportunities to dissect the evolution of major developmental transitions among the land plants with reference to these plant organs. Perhaps, in future, these studies may lead to the development of new model genetic systems that can span all major land plant groups, and to the integration of functional developmental analysis and functional studies with this comparative genomics (Friedman et al 2004).

Signalling Molecules and Growth Regulators 2

2.1. INTRODUCTION

All the processes that are essential for development such as cell division, cell enlargement, differentiation, pattern formation, polarity and morphogenesis are mediated by proteins (acting either directly or as enzymes to produce other mediating molecules). The proteins are encoded by genes and hence, development is controlled by gene expression to a large extent. Many genes, shown to have important roles in development, encode two particular types of protein: transcription factors and components of **signalling pathways**. Transcriptional factors are important because they control the gene expression, and therefore, act as coordinators or regulators of the developmental processes. Signalling pathways are necessary for cells to perceive external **signals**, either from other cells or from the environment. There are short–range signalling pathways and networks as well as more systemic signals and long–range pathways. An important form of long distance signalling couples the growth of root and shoot. Plants appear to have a much smaller repertoire of signalling pathways than animals as they have a smaller number of distinct cell types and possess plasmodesmal connections that allow them to communicate by the transfer of small molecules and macro–molecules directly from cell to cell. Only a very few secreted proteins are used as signalling molecules in plants, and only a few major signalling pathways are recognized that respond to the hormones or the growth regulators. Recent research shows that sugar not only provides a means of translocating carbon (derived from photosynthesis or starch reserves) to various parts of the plant, but also has the potential to act as a signalling molecule to regulate a variety of growth related processes, including promotions of cell division.

2.2. SIGNALLING

Although it was evident from Darwin's time that informational signals pass from one part of the plant to another, the knowledge about signalling molecules, signal perception and signal transduction and also the word **transduction** in plant science did not exist till about four decades back. Now these words have become very common and are used routinely by many plant biologists (Trewavas and Malhó 1997). Several milestones have been crossed to come to this stage such as the detection of protein kinases in plants, especially of Ca^{2+}–dependent protein kinases (CDPKs) and the isolation of **calmodulin** from plant cells. As a result, we now know that signal perception and transduction pathways are branched (Fig. 9).

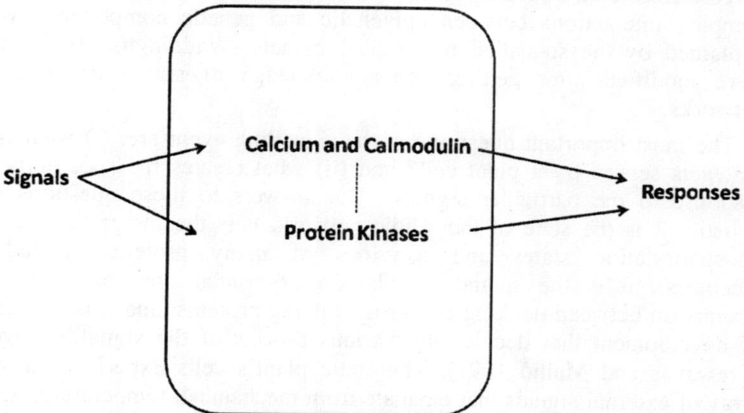

Fig. 9. Two signal transduction pathways common in plants (Based on Trewavas and Malhó 1997)

During this time, we have also come to know that calcium is a **second messenger**, a substance that allows the complex signals arriving at the cell surface to be converted into simple biochemical events inside the cell. Additional second messengers such as cGMP, cAMP and cADP–R are added to the list. More than 70 protein kinases were discovered even by as early as 1997. Now, we also know that the same signal transduction network controls both fast (e.g., stomatal opening and closing) and slow responses (e.g., tropic movements), thus indicating that when a cell perceives a signal, ion flux and gene expression changes become inextricably linked–response phenomena that occur at different rates.

Signal transduction systems are not to be portrayed as simple linear chains of cause and effect, but as complex and intricate systems. It must be realized that every cell contains not only one but two information systems (Trewavas and Malhó 1997): (i) the genetic system involving gene → RNA → protein → phenotype, which is truly applicable only to a very limited number of phenotypic characters specified by a single gene, and

(ii) the epigenetic system involving the interaction of genes and gene products that are 'integrated with networks that transduce a host of internal and external signals'. This integration produces a very complex, but flexible and unstable regulatory network 'interposed between the unitary genetic elements and the phenotype'. Several interlinked proteins are involved and hence, the behaviour of this network can be considered as a good example of **Chaos theory** in action. A map of this network can be constructed including the possible cross–links. Even a very slight change in the environment or modification in a single signal can act as a limiting factor and affect this chaotic network, since the characteristics of this network are the results of connections between the constituents and not derived from their absolute quantitative values (Capra 1996). Initially, the complex interactions between epigenetic and genetic components were explained by the so–called topological models (Waddington 1957), but were modified after getting better knowledge of signal transduction networks.

The most important questions in the signalling event are: (i) what is it that gets sensed by a plant cell? and (ii) what creates the specificity of response to the particular signals? The answers to these questions are holistic; it is the state of the whole network, it is the integration of the phosphorylation states and activities of many proteins (including receptors), it is the spatial distribution of crucial proteins, it is the interaction between docking and cross–linking proteins, and it is the stage of development that decides the various aspects of the signalling event (Trewavas and Malhó 1997). The static plant's cells experience a vast array of external signals that emanate from mechanical, temperature, light, mineral, gaseous, wound, water, physical and electrical influences. The sources of internal signals include a stream of growth regulators, peptides, sugars and other metabolites, steroids, cell wall oligosaccharides, mechanical signals (turgor–related and cell and cell wall–imposed) from other cells and electrical information. Almost all the living cells of a plant can respond to all of these signals and this makes the signal perception and transduction system very complicated (Trewavas and Malhó 1997). However, it is important to mention the concept of **target cells** here.

2.2.1. Concept of Target Cells

Similar to the concepts of signalling molecules and receptor sites, the concept of target cells in plants also had its initiation in insects. Originally, it arose from the knowledge that cells of coleoptiles and etiolated shoots enlarge and extend in response to auxin, and that ethylene would arrest such elongation growth. In these cases, the cortical cells of the shoot were targets for auxin and ethylene. The first substantive evidence for different target types in cortical cells arose from the observations of their differential growth responses to these two growth regulators (Osborne and McManus 2005). For instance, in the young shoot of *Pisum sativum*, the immature cells elongate in the presence of auxin passing basipetally from the meristem, while the addition of ethylene to those shoots arrests the elongation growth and instead expands laterally, with the cell volume remaining essentially unchanged. In *Regnellidium diphyllum*, a water fern, the cortical cells expand and extend with either auxin or ethylene. Thus, three distinguishable cell types (Types 1–3) could be recognized with respect to their responses to auxin and ethylene, although the cells looked similar to the eye. Thus, generally a target cell can be a cell that has a pre–determined competence to respond in a defined way to a specific growth regulator signal. Only a specific growth regulator (or its chemically–related analogue) can evoke a particular response in the cell. Invariably, in plant growth regulators, competence of the target cells can be defined in both qualitative and quantitative terms. A cell must have the ability to perceive, transduce and respond to a signal; otherwise it cannot serve as a target cell for that particular signal. A positive response to the signal is evidence of a cell's specific target state, but a lack of it cannot be taken as an evidence of the non–target condition. There is a threshold level, which is a minimum necessity for a cell to interpret every signal input, and this level must exceed the presence of non–specific inhibitors and homeostatic controls of the target cell response (Osborne and McManus 2005). Here, it should also be emphasized that a cell may perceive, transduce and respond to a particular signal in a different way from its neighbouring cell, which is a clear evidence for their individual and different target states.

2.2.2. Receptors of Signals

For a target cell to recognize a signal, it must have a receptor for that signal. Thus, the internal or external signals are first perceived by appropriate **receptors**. The target cells must possess an array of receptor systems, but so far, comparatively, only a few of these have been precisely characterized (Osborne and McManus 2005). Almost all known receptors are the members of families of related proteins and are located in the plasma membrane, nucleus, cytoplasm or even in the cell wall. Most of the well–characterized receptors are **protein kinases**. In these cases, ligand binding triggers the kinase activity of the receptor, and the subsequent signal transduction occurs through a cascade of protein kinases. Other occupied receptors activate cytoplasm – or plasmalemma–localized members of the GTPase superfamily, which in turn trigger the synthesis or release of a number of secondary messengers. Most of these secondary messengers, directly or indirectly, activate the specific protein kinases, which together control the osmotic motors underpinning growth and movement or phosphorylate transcription factors (Trewavas and Malhó 1997).

Currently, receptors have been identified for ethylene and for a few other growth regulators (see later in this Chapter), and for incident red and blue light (Chapter 8). The receptors i.e., calmodulin and CDPK and

$[Ca^{2+}]_c$ many receptor–like protein kinases have been isolated and sequenced. Receptors for inositol 1, 4, 5–triphosphate (IP_3), cADP–R, fungal elicitor and nod factors (of bacterial nodules) have been detected. But still a lot is to be known on the receptors involved in the regulation of plant development. Plants have evolved elaborate mechanisms capable of detecting the direction of incoming physical, chemical and other signals in order to maximize the capture of basic resources. The vectorial sensing of these signals requires that their receptors be fixed in the cell, and they are commonly assumed to reside on the plasma membrane. However, as mentioned earlier, not all receptors are localized in the plasma membrane.

The signals may be strong or weak, but the plant cells can initiate new developmental programmes in response to even extremely weak signals. In these instances, the physiological response must result from substantial metabolic amplification of the original weak signal, which by itself almost always possesses insufficient energy to actually elicit a response. Amplification may commence with the receptor itself. Phosphorylation of proteins initiating protein kinase cascades would result in further amplification of the weak signal. However, this type of signal amplification is not probably involved in phytochrome receptor signals. Amplification may also explain thresholds in plant and cell developmental responses.

Certain plant cells can impose a developmental programme on their immediate neighbours during particular stages of development (For example, wound healing, lateral root initiation and almost synchronous microsporogenesis in all microspore mother cells of an anther locule). In these cases a single cell can, in response to signals, rapidly induce neighbouring cells to respond for employing receptor protein kinases. There are two ways in which plant cells can employ these receptor protein kinases. Discrete signals can move from one cell to another through the wall symplast, activating receptor protein kinases in the plasma membrane of the neighbouring cell, or possibly the activated receptor protein kinases themselves can move through the plasmodesmata.

The interaction of receptors with GTPases is shown to commence signal interpretation. Once they bind their ligand, the plasmalemma–localized receptors can interact directly with the members of the GTPase superfamily. The G–proteins and the members of the Ras family are among the GTPases with important roles in signal transduction. Interaction of the GTPase with ligand–receptor complex enables it to bind with GTP and become activated. The activated GTPase, with bound GTP, is then released from the receptor and can continue to activate numerous downstream enzymes and/or channels until it is inactivated by the hydrolysis of GTP to GDP (Trewavas and Malhó 1997). The inactive GDP–binding GTPase is competent to interact with another receptor and once again initiate the signalling process. The involvement of GTPases enormously increases the flexibility in signal transduction pathways. Also, individual members of the G–protein family can interact with a number of different receptors. Now, there are evidences available to implicate the involvement of GTPases in blue and red light, growth regulators, fungal elicitors, guard cell control signals involved in stomata closure mediation etc. GTPases are also known to be involved in protein synthesis, secretion and cell cycle and this involvement greatly enhances the fidelity of molecular interactions (Trewavas and Malhó 1997).

2.2.3. Second Messengers, Calcium and Protein Kinases

When plant cells get the signals, their cytoplasm receives the information through changes in the concentration of a number of second messengers. The activated G–proteins signal an array of plasmalemma–bound phospholipases, cyclases and ion channels, which in turn regulate the production of second messengers. Many second messengers operating in one or the other signalling system have been identified and these include phophotidic acid, IP_3, cGMP, cADP–R, Ca^{2+}, H^-, redox reagents such as ascorbic acid or glutathione, H_2O_2 and free radicals. All these second messengers can take part in the transduction of any signal depending on the system and environmental setting. Some of these second messengers such as Ca^{2+} and lysophospholipids, can activate protein kinases, while others mobilize $[Ca^{2+}]_c$ via channel opening. Ca^{2+} is important as it orchestrates direct cooperation among plasmalemma, chloroplast, chromoplast, mitochondria, vacuole and the ER to produce the required and appropriate physiological response. The membranes that separate the cytosol from these other cell components contain channels that are gated, and calcium–mediated signalling takes place when some of these channels open allowing Ca^{2+} to move. Although a variety of Ca^{2+}–binding proteins interpret the $[Ca^{2+}]_c$ signal, calmodulin activates more than 100 proteins. The interactions between calmodulin and these proteins are generally hydrophobic. In addition to calmodulin, the plants possess an array of other Ca^{2+}– binding proteins, two of which are CDPK and a protein phosphatase. Some observations indicate that the plasmalemma may contain a Ca^{2+} sensor whose functions are to sense and to help regulate $[Ca^{2+}]_w$ homeostasis. This sensor is likely to be coupled with a transduction sequence that induces the mobilization of vacuolar Ca^{2+} through the cytoplasm into the wall when required. There must also exist a mechanism for signalling an excess of $[Ca^{2+}]_w$ that can enable the unwanted Ca^{2+} to return to the vacuole. Then, it is possible to visualize that an extracellular Ca^{2+} can act as a messenger.

Post–translational modification of some amino acids such as serine, threonine and tyrosine residues by protein kinases and phosphatases, is a major transduction route for many signals. These modifications provide for extensive interaction and 'cross–talking' among the transduction pathways. Many protein kinases are known, and one of the most important groups of protein kinases is the MAPK cascade, which includes three distinct types of kinases. MAPK cascade is now implicated in the transduction of mechanical signals as well as signals triggered by gibberellins, ABA, ethylene, osmotic stress, auxins, wounding and fungal elicitors. The other important class of protein kinases is the CDPK family. Members of this family are found in microfilaments and plasmalemma. In addition, several protein phosphatases have been detected in plant cells.

Trewavas and Malhó (1997) suggest the possibility of the existence of the '**transducons**' that are the large protein complexes in which signal transduction takes place. Further, they speculate that transducons can consist of an aggregate of receptors, channels, bound calmodulin, protein kinases and phosphatases and Ca^{2+}–ATPases. These may be positioned specifically within a cell and may operate as on/off switches in the metabolic circuitry. They could be coupled together in feedforward or feedback loops by the mobile second messengers, mobile (receptor) protein kinases in membranes and substrate proteins. These interactions are likely to help in improving the rate and accuracy of the information transfer and processing. They can also act as integration and information storage facilities for the cell, characterizing its history and serving as a chemical memory of the past. Thus, genes can be thought of as a kind of read only memory (ROM), and the signalling networks act as random access memory (RAM) of a computer; the keys are punched by environmental and developmental cues (Trewavas and Malhó 1997).

The concept that signal molecules exist in plants, that they are transported from one place to another, and that they elicit growth responses has been established beyond doubt, as is evident from the above account. Today, plant biologists are making all efforts to explore the molecular structure of plant signal molecules, how, where, and when they are delivered, and how they evoke delayed, long–term gene transcription and rapid, short–term metabolic or physical responses. These efforts continue to expose deeper and more complex levels of signal trafficking and cross–talking as well as in the elucidation of more and more signal molecules (including many difficult–to–detect small molecules). Hence, our current understanding of signalling phenomenon is still rudimentary and incomplete (Niklas 2003).

2.3. GROWTH REGULATORS

As noted in the first chapter of this book, growth regulators are also known as hormones or phytohormones. "A plant hormone is an organic compound synthesized in one part of the plant and translocated to another part, where in very low concentrations it causes a physiological (growth) response" (Salisbury and Ross 2004). The response in the target plant organ of hormones is not necessarily promotive, because processes such as growth or differentiation are sometimes inhibited by hormones, especially abscissic acid (ABA). Some botanists do not include organic growth regulators such as 2,4–D that are synthesized only by organic chemists as an auxin, or those synthesized only in organisms other than plants as hormones. Nothing is said about how or how far the hormone needs to move from their place of production before they actually act. Hormones such as ethylene may act in the same cell where they are produced (as in fruit ripening). Hence, many people prefer the word 'growth regulator' to 'hormone'.

Since 1937, gibberellins, ethylene, cytokinin and ABA have joined auxin as phytohormones and together they are regarded as 'classical five' growth regulators (Kende and Zeevaart 1997). Subsequently, several other growth regulators were discovered in plants. It was known in the second half of the 20th century that each hormone affects the responses of many plant parts and these responses depend on the species, the plant part, the developmental stage, the hormone concentration, and interactions among known hormones and various environmental factors (Salisbury and Ross 2004). Hence, it is risky to generalize about the hormones and their effects on growth and development.

2.3.1. Differential Sensitivity to Hormones

In the early 1980s, the concept that differential sensitivity is much more important in determining the effects of a hormone than is the concentration of that hormone within plant cells was greatly emphasized (Trewavas and Cleland 1983). Although a number of researchers were against this concept, Trewavas's research papers forced researchers to consider and measure tissue sensitivity to hormones. Now, both sensitivity and concentration have received importance in hormone action studies (Firn 1986).

If a hormone present in low concentrations is to be active and specific, three major parts of a response system must occur:

1. The hormone must be present in sufficient quantity in the appropriate cells.
2. The hormone must be recognized and bound tightly by each of the groups of target cells that respond to the hormone. Receptor protein molecules, as described earlier, possess the necessary complex structures to recognize and select among much smaller hormonal molecules. Such proteins are called receptor proteins or simply as receptors, as already discussed.
3. The receptor protein must cause some other metabolic change, which then leads to the amplification of the hormonal signal or messenger. Different kinds of amplification processes might occur in sequence before the response to hormone is fully noticed.

2.3.2. Hormones and Gene Activity

Now, it is known that hormones control gene activity. Several genes that are rapidly up–regulated by auxin application have been identified, of which two are important. The so–called SAUR genes from soya bean have been characterized and some of these are regulated by auxin. When soybean seedlings are oriented horizontally, SAUR genes express asymmetrically i.e., the level of SAUR mRNA on the lower side of the hypocotyls was much higher than on the upper side. This probably indicates that asymmetric expression of SAUR genes results from lateral redistribution in gravitropically–stimulated stems, as postulated by Cholodny–Went model. Another family of early auxin–induced genes encodes short–lived nuclear proteins that may act as activators or repressors of genes whose products mediate auxin responses.

How the genes are biochemically controlled by the hormones is not yet very clear. What is known regarding the possible sites of hormonal control of gene activity is shown in Fig. 10.

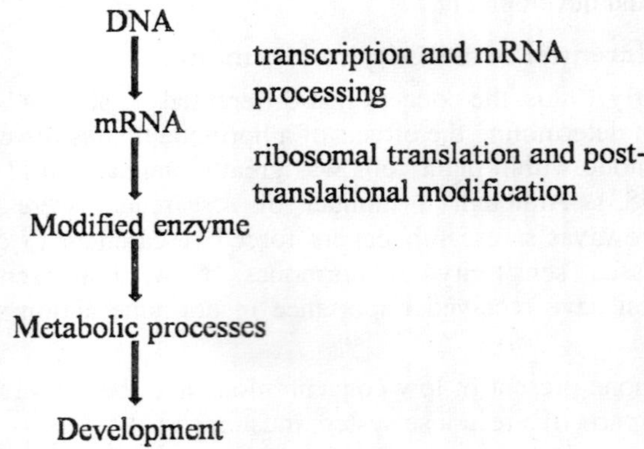

Fig. 10. Possible sites of hormonal control of gene activity

Activation of genes represents a large amplification process because repeated transcription of DNA into mRNA, followed by translation of mRNA into enzymes with high catalytic activity at low concentrations, can lead to the production of multiple copies of an important metabolic product. Subsequently, these products determine what an organism or an organ is composed of and what its phenotypic expression is. There are many control points in the flow of genetic information from DNA to the final product i.e., at the transcription level, at the processing level of mRNA, and at the post–translational level etc., as shown in the above flow chart.

2.3.3. Sites of Hormonal Activity

Hormonal activity is believed to begin with binding of the primary hormone to a receptor protein in the plasma membrane of a target cell. The bound hormone–receptor complex then activates a nearby membrane enzyme named phospholipase C (PLC), which hydrolyses one of the groups of non–abundant membrane phospholipids called phosphoinositides. Phosphoinositides are phospholipids that have inositol or similar lipids. PLC hydrolyzes the inositol to release inositol–1,4,5–triphosphate (IP_3) and diacylglycerol (DAG). IP_3 and DAG can cause a cascade of responses. IP_3 releases the stored Ca^{2+} of vacuole into cytosol. DAG activates an enzyme in the membrane called protein kinase C (PKC), which uses ATP to phosphorylate certain enzymes that regulate various stages of metabolism. Increased levels of CA^{2+} in the cytosol caused by IP_3 also activate certain enzymes, including many protein kinases. Some of such kinases require activation by free Ca^{2+} while others require activation by Ca–calmodulin. Thus, the initial hormonal stimulus finally leads to modified enzyme activity, altered metabolic processes and eventually, a different kind of cell having physiological and morphological differences. Many such changes by various hormones and environmental stimuli interact to help in making a different tissue, organ or plant. The control over activity of certain enzymes after the initial hormone reception seems to be an important key, and secondary or tertiary messengers such as IP_3, DAG, and Ca^{2+} are often involved in hormonal activity.

2.3.4. Different Kinds of Growth Regulators

2.3.4.1. The Auxins

The term 'auxin' (Greek word for "to increase") was first employed by Went (1926), who discovered that probably some unidentified compound caused curvature of oat coleoptile (first leaf) towards light. This compound was found to be available in a relatively–abundant quantity in coleoptile tips (Refer Box 2.1). The critical demonstration was that a substance present in the tips could diffuse from them into a tiny block of agar. The activity of this auxin was by the curvature of the coleoptile caused by enhanced elongation on the side to which the agar block (containing the diffused auxin) was applied.

BOX 2.1. DISCOVERY OF AUXINS

More than 135 years ago, it was Sachs (1878) who first suggested about specific plant organ–forming substances, but the history of discovery of auxins goes back to Charles Darwin. Darwin was fond of birds and was keeping some of them as his pets. He was using the sprouts of the canary grass (*Phalaris* species) to provide the birds with vitamin–rich food. Accidently, Darwin observed that the white coleoptile (first leaf) of this grass was very sensitive and responsive to light (coming from a nearby window). Darwin became curious about this behaviour of the coleoptile and did a few critical experiments on it using light. Further work on this by others like Boysen–Jensen in 1913, Paal in 1919 and Went in 1926 led to the elucidation of the probable mechanisms of phototropism and the discovery of auxins, a name given by Went. Auxins proved to be the foremost of plant hormones. The first hormone to be discovered, indoleacetic acid (IAA), was not identified until 1930s (first purified from urine). The field of plant growth regulators became very prominent with the publication of Went and Thimann's (1937) book on phytohormones. At that time, the term phytohormone was synonymous with auxin. It is impressive that aside from some confusion about the structure of auxin, many of the basic phenomena of auxin physiology were already known at that time.

Types of auxins

Went's auxin is now known to be indoleacetic acid (IAA) (Fig. 11 A), which is still equated with auxin by some physiologists. Plants contain three other compounds that are structurally similar to IAA and cause many of the same responses as IAA, and these should be considered as auxin hormones (Fig. 11 A): **4–chloroindole acetic acid** (**4–chloro IAA**) (found in young seeds of various legumes), **phenylacetic acid** (**PAA**) (widespread among plants and is frequently more abundant than IAA, although it is much less active in causing typical responses of IAA), and **indolebutyric acid (IBA)** (formerly thought as an active synthetic auxin, but now shown in maize leaves and various dicots and so is probably widespread in plants). Not much is known about transport characteristics of these three compounds and whether they, in fact, normally function as plant hormones, although this seems likely (Salisbury and Ross 2004). Three additional compounds found in many taxa show considerable auxin activity. They are readily oxidized to IAA in vivo and are probably active only after conversion. These are yet to be considered as auxins, but now are considered only as auxin precursors. They are **indolacetaldehyde**, **indoleacetonitrile**, and **indole ethanol** (Fig. 11 A). Each one of these has a structure similar to IAA, but lacks the carbonyl group.

Certain synthetic compounds also cause many physiological responses common to IAA and are generally considered as auxins. Of these, **α–naphthalene acetic acid** (**NAA**), **2, 4–dichlorophenoxyacetic acid (2, 4–D)** and **2–methyl–4–chlorophenoxyacetic acid** (**MCPA**) (Fig. 11 B) are the best known. As they are not synthesized by plants, they are not hormones, yet they are growth regulators. Many other kinds of compounds also fit into this category. The term 'auxin' has become much more encompassing because many compounds are structurally similar to IAA and cause similar responses. Each of these known auxin–like chemicals is similar to IAA in having a carbonyl group attached to another carbon–

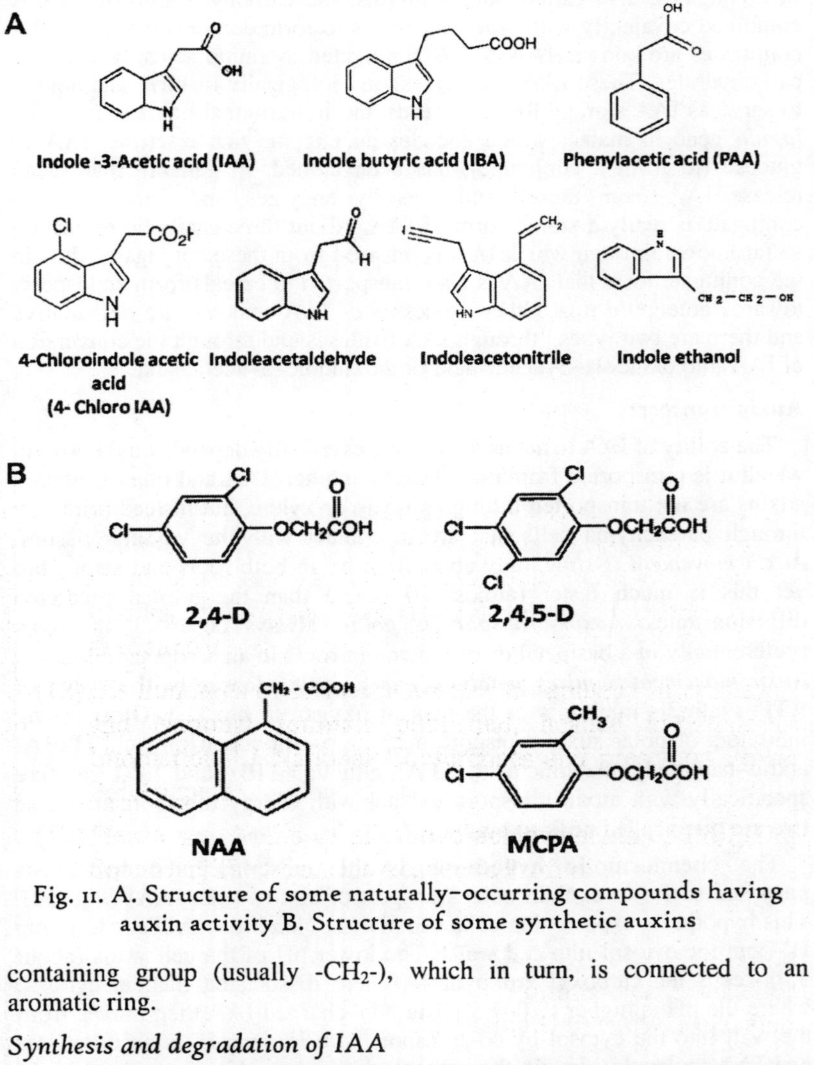

A

Indole -3-Acetic acid (IAA) Indole butyric acid (IBA) Phenylacetic acid (PAA)

4-Chloroindole acetic acid (4- Chloro IAA) Indoleacetaldehyde Indoleacetonitrile Indole ethanol

B

2,4-D 2,4,5-D

NAA MCPA

Fig. 11. A. Structure of some naturally–occurring compounds having auxin activity B. Structure of some synthetic auxins

containing group (usually -CH_2-), which in turn, is connected to an aromatic ring.

Synthesis and degradation of IAA

IAA is most probably synthesized from D–form of the amino acid tryptophan. Three routes for IAA biosynthesis from tryptophan via indole–3–pyurvic acid, tryptamine or indole–3–acetonitrile have been proposed, although the first route is more preferred in most plants. IAA is synthesized in young tissues such as shoot meristems, and growing leaves and fruits. Plants should have a mechanism to control the amounts of IAA.

The regulation of the rate of synthesis is one mechanism and temporary inactivation IAA by formation of **auxin conjugates** is another mechanism. In conjugates, also called **bound auxins**, the carbonyl group of IAA is combined covalently with other molecules to form derivatives. Many IAA conjugates are known. IAA occurs conjugated to amino acids, peptides or carbohydrates. These IAA conjugates are biologically inactive and appear to serve as IAA storage forms in seeds and in hormonal homeostasis. The *lagulu* gene in maize, which encodes an enzyme that esterifies IAA to glucose (to form a conjugate), has been cloned. In general, plants can release IAA from these conjugates by enzymes, indicating that the conjugate is really a stored form of IAA. About three catabolic routes are so far known through which IAA is released from these conjugates. It is in the conjugate form that IAA is also transported in cereals from endosperm towards coleoptile tips. Other processes of IAA removal are degradative and there are two types : through IAA oxidases and through the conversion of IAA into oxindole–3–acetic acid or dioxindole–3–acetic acid.

Auxin transport

The ability of IAA to act as a hormone essentially depends on the way in which it is transported from one place to another. IAA and other synthetic auxins are not transported through phloem or xylem, but instead primarily through parenchyma cells that are in contact with the vascular tissues. *Auxin movement is slow*, only about 1cm h^{-1} in both roots and stems, but yet this is much faster (almost 10 times) than the normal predicted diffusion rates. *Auxin transport is polar*, always occurring in stems preferentially in a basipetal direction and in roots in an acropetal direction. *Auxin movement requires metabolic energy*, as evidenced by the ability of ATP–synthesis inhibitors or the lack of oxygen to block it. Other strong inhibitors of polar auxin transport are 2, 3, 5–triiodobenzoic acid (TIBA) and α–naphthylphthalamic acid (NPA), although TIBA and NPA interfere specifically with auxin transport and not with energy metabolism. These two are often called **antiauxins**.

The **chemiosmotic hypothesis** is the most accepted hypothesis purposed so far to explain polar transport of auxins (like IAA) (Fig. 12). This hypothesis expects that cells use plasma membrane ATPase to pump H^- from the cytosol into cell walls. The lower pH of the cell walls (about 5.0) keeps the carboxyl group of IAA less dissociated than in cytosol, where the pH is higher (7.0–7.5). The non–charged IAA then moves from the wall into the cytosol by co–transport with H^- (one H^- is absorbed per one IAA molecule). Inside the cytosol, the higher pH causes the auxin's carbonyl group to dissociate and attain a negative charge. As the concentration of charged IAA builds up in the cytosol, its wall–ward movement is promoted. However, polar transport through an organ requires that IAA moves out only from the basal end of the cell opposite to that through which it entered. The preferential exit at the basal end assumes that some carrier in that region of the membrane transports charged IAA out towards the cell wall, where the low pH again causes most of them to become non–charged. For this hypothesis to operate, it is necessary to learn how charged IAA is transported out at the basal ends of cells, because such transport requires cells to be polarized so that they absorb at one end and secrete at the other. Direct evidence for the polarized location of IAA transporter at the basal end of the stem cells (of pea plant) was obtained and this transporter is blocked by NPA, which probably explains how that antiauxin prevents basipetal auxin transport. TIBA apparently blocks at the same site. The mechanism of polar transport of IAA and other auxins still requires more study, but polar transport down the stem from young leaves or apical meristem certainly occurs. Certain flavonoids that are abundant in plant cells, especially quercetin, apigenin and kaempferol, are powerful inhibitors of the basal transporter that causes auxin efflux from cells (Jacobs and Rubery 1988). These compounds may act as a part of an auxin–transport control system.

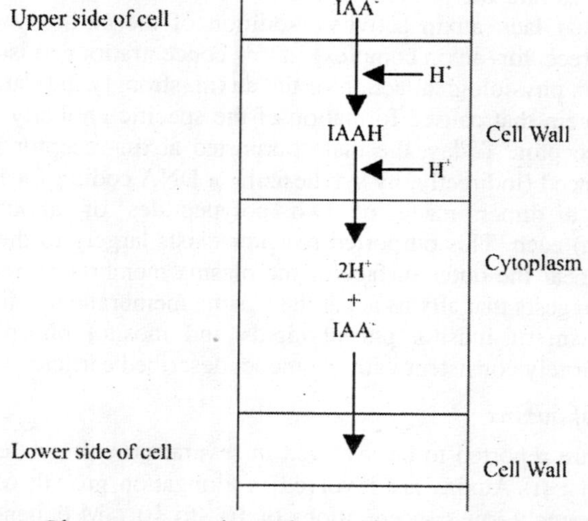

Fig. 12. Chemiosmotic hypothesis of polar auxin transport

Auxin receptors and transduction

Isolation of *Arabiolopsis* hormone response mutants has resulted in the identification of putative hormone receptors and elements of hormonal transduction pathways. An auxin–binding protein (now called ABP1) was the first putative hormone receptor in plants to be identified based on binding experiments. The purification of this protein and the subsequent cloning of the corresponding cDNA showed that the signal is localized in the ER. An ABP1 homologue, an immunologically related receptor protein, is also localized in the plasmalemma. However, it is not known, whether ABP1 plays any role in mediating growth. A number of 'auxin–resistant' mutants of *Arabidopsis* that can grow on high concentrations of auxin have been isolated and these help us understand the auxin transduction pathway. It was shown that the product of one such auxin resistance gene was AXR1 and that it has sequence similarity to the ubiquitin–activating enzyme E1. Another mutant in the *AUX1* gene of *Arabidopsis* shows auxin resistance with respect to root growth and does not exhibit root gravitropism. The product of the *AUX1* has sequence similarities to plant and fungal amino acid permeases, indicating a role for this protein in auxin transport. Auxins not only affect the kinds of proteins formed by gene activity, but also they do it very rapidly. Many evidences indicate that the main control is at the transcription level.

A few binding receptor proteins for auxins have been purified and antibodies have been developed against them. For a true auxin receptor, binding of auxin must occur at low and physiologically reasonable concentrations and the protein should not bind to molecules with similar structures that lack auxin activity. Addition of the antibody (developed against the receptor–auxin complex) in low concentrations to isolated plant part prevents physiological action of the auxin, strongly indicating that the binding protein that caused formation of the specific antibody is indeed a hormone receptor. Today, the main purported auxin–receptor protein has been sequenced (indirectly, by synthesizing a DNA coding for it), and it is apparently a dimer made of two polypeptides of about 20 KDa (KiloDalton) each. This purported receptor exists largely in the ER, but it also exists near the outer surface of the plasma membrane. Accumulating evidence suggests that auxins act at the plasma membrane to cause changes in metabolism of inositol phospholipids and inositol phosphates. This finding is largely consistent with the model described earlier.

Functions of auxins

Auxins are reported to be involved in several growth–related activities in the plants: (i) Auxins are involved in elongation growth of roots, but only at extremely low concentrations of 10^{-7} to 10^{-13} M depending on the species and age; at higher concentrations elongation is inhibited. (ii) Auxins also promote adventitious root development on stems, especially on the basal end or nodal regions of cuttings. (iii) In plants with apical dominance, the lateral buds are believed to be arrested from further growth by auxins present in the shoot apex and young leaves, although this is contested by a few plant physiologists, (iv) Synthetic auxins such as 2, 4–D, 2, 4, 5–T, MCPA, Piclorum and NAA act as herbicides, affecting the growth of weeds, especially the dicots in a cereal–field. The mechanism behind this activity is not yet very clear, although several explanations on the possible modes of action, especially on the vascular cambium, have been suggested.

More information on auxin roles are provided in the subsequent chapters of this book.

2.3.4.2. *The Gibberellins*

Gibberellins were first discovered in Japan in the 1930s from the studies with fungal infected rice plants that grew excessively tall (Box 2.2). T. Yabuta and T. Hayashi isolated an active compound from this fungus and they named it as **gibberellin**. Although the first gibberellin was discovered as early as the discovery of IAA, it did not attract as much attention as IAA did until the early 1950s.

BOX 2.2. HISTORY OF GIBBERLLINS

Japanese farmers first observed the excessively tall rice plants as early as 1880s during the beginning of the rice–growing season. These abnormal seedlings attracted their attention and they noticed that these plants often became unhealthy and sterile. The Japanese gave this disease many names, but the most common was **Bakanae** (foolish seedling). In 1898 the causative organism of bakanae disease was deduced as a fungus belonging to the genus *Fusarium* by Hori. In 1926, Kurosawa discovered that the disease was caused by a substance secreted by the fungus *Gibberella fugikuroi* resulting in the controversy over the identity of the true pathogen. It was Wollenweber (1931) who cleared this controversy by establishing that *Fusarium moniliforme* was the asexual stage of *Gibberella fungikuroi* (Saw.) Wr., an ascomycetous sexual stage. In 1935 Yabuta et al isolated an active compound from this fungus and they named it **Gibberellin A**. Because of World War II, most work on gibberellins was put on hold as a consequence of which the western people did not have an access to these findings.

In the early 1950s Stodola and his colleagues in the U.S. Department of Agriculture and Brian and his colleagues at the Imperial Chemical Industries in England discovered the early Japanese papers. The latter group of investigators discovered an entirely new compound from *Gibberella fugikuroi* and named it gibberellic acid. The former group of workers also isolated a new compound in addition to what they thought was gibberellin A from this fungus, which they called gibberellin X. The Japanese scientists discovered at about the same time that gibberellin A was actually made up of three compounds, which they called GA_1, GA and GA_3. Gibberellin X, GA_3 and gibberellic acid are all the same compound. The latter two are accepted in describing the compound and are synonymous terms today. Radley, West and Phinney independently described compounds similar to gibberllic acid in plants. In 1957, Takahashi et al isolated GA_4 from this fungus. He also showed that GA was identical to gibberellin A. In 1968, MacMillan and Takahashi proposed that gibberellins should be assigned numbers in order to avoid confusion between these chemicals.

Types of gibberellins

As of 1990, around 84 gibberellins had been discovered in fungi and plants. Of these, 73 occur in higher plants, 25 in *Gibberella* fungus and 14 in both. Till 1997, about 112 gibberellins were known and identified (Kende and Zeevaart 1997). Gibberellins exist in angiosperms, gymnosperms, pteridophytes, and probably in mosses, algae, and in at least two fungi. They were also found to exist in two species of bacteria. Some of the known gibberellins are just physiologically inactive precursors of other active ones, and still others are inactive hydroxylated products.

Gibberellins are isoprenoid compounds. Specifically, they are diterpenes synthesized from acetate units of acetyl–CoA by the **mevalonic acid pathway**. All gibberellins are derivatives of the **ent–gibberellane** skeleton. Gibberellin biosynthetic pathway is usually divided into three stages. The first stage involves conversion of **geranyl geranyldiphosphate** (a 20–carbon compound) to **ent–kaurene**, the second stage involves conversion from ent–kaurene to GA_{12}–aldehyde, and the third stage involves conversion from GA_{12}–aldehyde to various gibberellins. The structure of ent–gibberallane with its ring–numbering system, along with structures of six active gibberellins is shown in Fig. 13. All gibberellins are acidic and hence, are named **gibberellic acids (GA)** with a different subscript to distinguish them such as GA_1, GA_3, GA_4, GA_7, GA_9 and GA_{32}. All gibberellins have either 19 or 20 carbon atoms grouped in either four or five ring systems. The fifth ring system is the lactone ring. Although all gibberellins have one carboxyl group attached to carbon 7, some have an additional carboxyl group attached to carbon 4 and can be called gibberellic acids. Only GA_3, the first highly active and long time commercially–available gibberellins, has been historically called GA. The number of hydroxyl groups on ring A, C and D ranges from zero (as in GA_9) to four (as in GA_{32}), with carbon 3 or carbon 13 or both most commonly hydroxylated (Salisbury and Ross 2004). In general, 19–carbon gibberellins are more active than 20–carbon gibberellins. Compounds such as phosphon D, AMO–1618, CCC or cycocel, ancymidol and paclubutrazol inhibit gibberellins synthesis.

The commonly used GA_3 appears to be only slowly degraded, but during active growth most gibberellins including GA_3 are rapidly metabolized by hydroxylation to inactive products. Also, they can readily be converted to conjugates that are largely inactive. These conjugates might be stored or translocated before their release at the proper time and place for the required activity. The known conjugates of GA_s include glucosides.

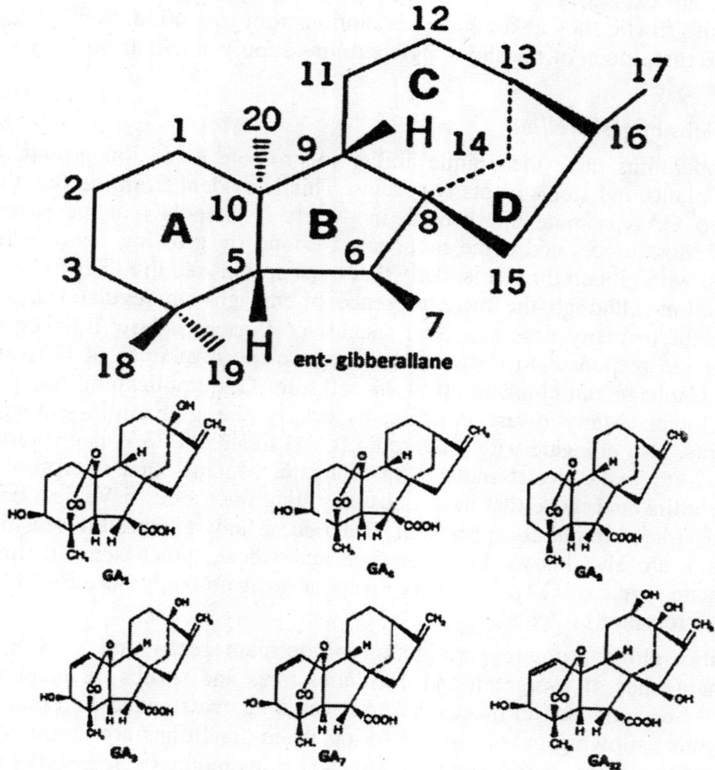

Fig. 13. Structure of ent–gibberallane and six active gibberellins

Synthesis of gibberellins

Which is the place of gibberellin synthesis in plants? It is likely that they are synthesized either in the plant organ where it is found or is translocated. Immature seeds contain relatively high amounts of gibberellins compared with other plant parts, and cell–free extracts from seeds of some species can synthesize gibberellins. Hence, much of the gibberellins of seeds are synthesized by the seed itself and are not transported to seeds from other plant parts. The ability of other plant parts to synthesize gibberellins is not yet clearly established, but it is likely that they synthesize them. The young leaves are thought to be major sites for gibberellin synthesis, as in the case of auxins, and their transport from the leaves occurs not only by diffusion but also through xylem and phloem and is not polar. Roots and stems are also likely to synthesize gibberellins as they can be detected in the xylem exudates of roots and stems, when these organs are excised. However, exogenous gibberellins have little effect on the root growth; they also inhibit adventitious root formation. Some others believe that, much of the shoot's gibberellins supply comes from the roots via the xylem.

Functions of gibberellins

Gibberellins have the unique ability to promote extension growth of intact plants and stems of many species. This is evident from the fact that applied GAs promote this growth in dwarfs or biennials in the rosette stage. Most dicots and some monocots respond by growing faster when treated with gibberellins. It is likely that most species require GA_1 for stem elongation, although the mere presence of this growth regulator is not sufficient in many cases. Several species of Pinaceae show little or no elongation responses to GA_3, but respond well to a mixture of GA_4 and GA_7. Cabbage can elongate up to 2m tall after GA_3 application, and also can flower. Many dwarf mutants, especially the gibberellin–synthesis mutants, also elongate with gibberellin (GA_3) treatment. GA_3 is converted to GA_1 to make dwarf mutants to elongate. Mutants that overproduce gibberellins and those that have unusually long internodes have also been found, for example in species of *Brassica*. Many gibberellin–sensitive mutants are also known in maize, pea and wheat, which seem to have adequate levels of GA_1, but cannot respond to it probably because of the lack of receptor proteins.

Gibberellins promote germination of dormant seeds and growth of dormant buds of evergreen and deciduous trees and shrubs of temperate zones. Seed and bud dormancy is often overcome by extended cold periods in winter, allowing growth in the spring when conditions are favourable. For some species, bud dormancy can also be overcome by increasing the day lengths that occur in the late winter, and for seeds of many species dormancy is broken by brief periods of red light when they are moist. Gibberellins overcome both kinds of seed dormancy and bud dormancy by acting as a substitute for low temperatures, long days or red light. Gibberellins can substitute for long–day requirements for flowering in some species. Gibberellins also overcome some species' need for inductive cold period for flowering. Some gibberellins are much more effective in enhancing flowering than others (Salisbury and Ross 2004).

Gibberellins stimulate mobilization of foods and mineral elements from storage cells of seeds during seed germination. Gibberellins stimulate the conversions of polysaccharides to sucrose and proteins into mobile amino acids or amides, especially in cereal grains. Gibberellins stimulate aleurone layer cells to synthesize hydrolytic enzymes as well as stimulate the secretion of these enzymes into the endosperm, where they digest food reserves of cytoplasm and cell walls. Gibberellins particularly control the production of α–amylase in aleurone cells and probably in scutellum and other parts of the embryo. The scutellum is much more important than aleurone layer in providing enzymes that digest endosperm reserves. Gibberellins have no significant effect on digestion induced by the scutellum even though this organ is believed to produce gibberellins that activate the aleurone layer. Gibberellins have much less dramatic effect on mobilization of food reserves in dicots and gymnosperms than in cereal grains.

Gibberellins (particularly GA_4 and GA_7) also promote parthenocarpic fruit development in some species. This suggests a role for them in normal fruit development (Refer to Chapter 10). Gibberellins produced in young leaves might also renew the activity of dormant vascular cambium in woody taxa. They are also involved in fiber development of some of the cambial derivatives. Other important effects of gibberellins are the delay of aging (senescence) in leaves and fruits, leaf shape determination, especially in heterophyllous taxa and the growth of petals in some taxa.

Mechanism of action and signal transduction

The several effects of gibberellins described above suggest that they have more than one primary site of action. Even a single effect of GAs such as enhanced stem elongation in whole plants is the result of at least three contributing events. First, cell divisions are stimulated in the rib meristem of the shoot apex, which result in the long cortical and pith files (For more details, refer to Chapter 5). Gibberellins promote cell division because they stimulate cells in the **G1** phase to enter into **S** phase and they also shorten the **S** phase duration. Increased cell number leads to more rapid stem growth in length. Second, gibberellins sometimes also promote cell enlargement by increasing the hydrolysis of starch, and some other storage polysaccharides into glucose and fructose. These contribute to the materials for new cell wall formation and also make the cell's water potential momentarily more negative. As a result of decrease in water potential, water enters more rapidly causing cell expansion and diluting the sugars (For more details, refer to Chapter 3). Third, gibberellins often increase cell wall plasticity, up to about 15 times greater than normal. How gibberellins loosen cell walls and also enhance the formation of hydrolytic enzymes leading to stem elongation is not clearly known.

Some progress has also been made in identifying the components of the signal transduction pathway in GA–regulated growth by using *Arabidopsis* response mutants of two types; those that have a dwarf phenotype but do not grow in response to GA, and those that grow tall and slender, mimicking the effects of GA treatment. The *spindly* (*spy*) mutant belongs to the second category and is defective in a tetratricopeptide repeat–containing protein that is likely to be involved in protein–protein interactions.

2.3.4.3. *The Cytokinins*

Discovery, definition and occurrence

In 1913, Haberlandt discovered that an unknown chemical present in vascular tissues of various plants stimulated cell division that caused cork cambium formation. This discovery was apparently the first demonstration that plants contain chemicals, now known as **cytokinins,** which stimulate cell division. Van Overbeek in 1940s found that coconut milk is also rich in compounds that promote cell division. Skoog and his coworkers in the early 1950s tried to identify this cell–division promoting chemical factor and found it to be highly active adenine–like compound. They named it as **kinetin**. Although kinetin itself has not been found in plants (and only artificially synthesized), related cytokinins are found in plants. The most active of these compounds are **zeatin** and **zeatin–riboside**. Fig. 14. shows the structures of the free–base form of the three most–commonly detected and the most physiologically active cytokinins in various plants; zeatin, **dihydrozeatin** and **isopentenyl adenine (IPA)**. Also, shown in the figure are kinetin and another synthetic cytokinin, **benzyladenine,** both of which are highly active. However, there are reports that benzyladenine or its riboside exists in plants. All cytokinins have a side chain, rich in carbon and hydrogen, which is attached to the nitrogen protruding from the top of the purine ring. The cytokinin either exists in the **free–base form** as shown in the Fig. 14 or as a **nucleoside**, in which a ribose group is attached to the nitrogen of position 9. An example is zeatin riboside, a relatively abundant cytokinin in many plants. The nucleosides can be converted to **nucleotides,** a component of nucleic acids.

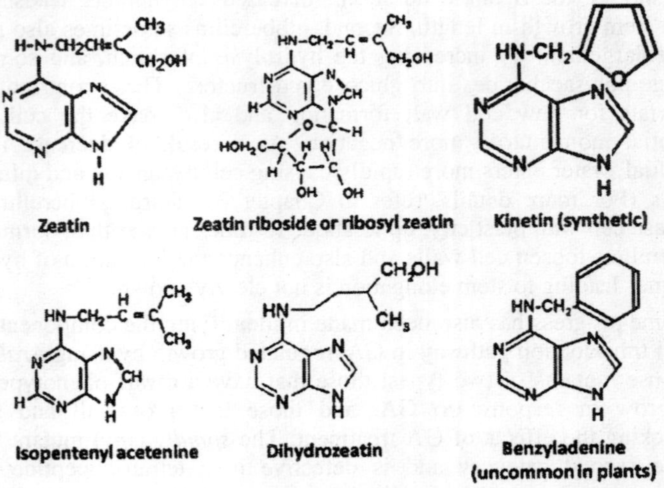

Zeatin | Zeatin riboside or ribosyl zeatin | Kinetin (synthetic)

Isopentenyl acetenine | Dihydrozeatin | Benzyladenine (uncommon in plants)

Fig. 14. Structure of most–commonly detected and the most physiologically active cytokinins and the synthetic benzyladenine.

On the basis of the above, one may be tempted to ask two questions: (i) How can we define a cytokinin? (ii) Can the free bases, nucleosides and nucleotides all be considered as cytokinins? Cytokinins can reasonably be defined as "substituted adenine compounds that promote cell divisions in a tissue system" (Salisbury and Ross 2004), a system that has optimal auxin concentration. Most evidence favours the free–base (of nucleic acids) as an active form; this is based on a study of more than 200 natural and synthetic cytokinins. Cytokinins are known from most groups of plants and are widespread in occurrence; they are also present in pathogenic bacteria and fungi.

Synthesis and transport

Cytokinins are formed from **isopentenyl AMP**, the latter being first converted to **isopentenyl adenosine**, which is then oxidized to zeatin. Zeatin gives rise to dihydrozeatin. The levels of cytokinins in cells are affected by the degree of their degradation by **cytokinin oxidase**. This enzyme is induced by its own substrate and thus, plays a role in regulating cytokinin levels in plants. Cytokinins are inactivated irreversibly by two different reactions: formation of N–conjugates with glucose at the 7– or 9– positions or with alanine at the 9–position and the oxidative cleavage of the N6 side chain of the cytokinin substrate.

In which part of the plant cytokinins are formed? We do not have any direct method to determine this, and only indirect methods are available. One method has been to find out where they are most abundant. In general, their levels are the highest in young plant organs, such as young leaves, seeds and fruits, and root tips. Hence, it was logically assumed that they are synthesized in these organs. If root tips are severed horizontally, cytokinins exude from the xylem. Since roots can store enough cytokinins, their derivation from some other source is unlikely. Hence, there is now also a widespread belief that root tips synthesize cytokinins and transport them through the xylem to all parts of the plants. Shoots are also shown to synthesize some of the cytokinins, which they require. Although sieve elements also show the presence of cytokinins, suggesting their involvement in transport of cytokinins, they are not readily distributed by the phloem. Similarly, it is almost definite that although young leaves, fruits, and seeds serve as transport sinks, they also do not readily transport their cytokinins elsewhere through either xylem or phloem. The tentative conclusion is that except for delivery from roots through xylem, transport of cytokinins within the shoot is rather limited (Salisbury and Ross 2004).

Functions of cytokinins

The promotion of cell division and organ formation is the major function of cytokinins. If a high cytokinin to auxin ratio is maintained, meristematic cells divide and give rise to buds, stems and leaves, but if this ratio is lowered, root formation is favoured, especially in tissue–culture derived calli. These two hormones together are also important in promoting crown gall formation (Refer to Chapter 11). Cytokinins delay senescence and increase nutrient–sink activities. Cytokinins enhance the ability of young tissues to act as sinks for phloem transport. They delay or retard senescence probably by maintaining the integrity of tonoplast membrane and by protecting membranes in general. Cytokinins promote lateral bud development and thus, break apical dominance, but prolonged elongation of buds needs IAA or gibberellins. Cytokinins enhance cell expansion in cotyledons and leaves of dicots through enhanced water uptake by cells and through enhanced cell division (but mainly through cell division). Endogenous cytokinins are needed for normal growth of stems and roots as they do not require exogenous supply of the same and hence, their role in promoting growth of stems and roots is not very clear. Cytokinins promote chloroplast development and chlorophyll synthesis.

Mechanism of action of Cytokinins

The differences in cytokinin effects show that they might have different mechanisms of action in different tissues. However, it is generally agreed upon that a common primary effect is followed by numerous secondary effects that depend on the physiological state of the target tissue or cell. Amplification of the initial effect has to be occurred because cytokinins are present in very low concentrations (0.01–1 μM) in various tissues. Some promotive effect of cytokinins on the synthesis of RNA and enzyme was suspected for a very long time, but no specific effect on DNA synthesis has been observed, although exogenous cytokinins often enhance cell division.

Cytokinins have been known to increase the transition from **G2** to mitosis and they do this by increasing the rate of protein (enzymic or structural) synthesis needed for mitosis. Cytokinins decrease the length of **S** phase. It is suggested that cytokinins target some nuclear proteins, and perhaps, those proteins would enhance cell division directly, by controlling the synthesis of DNA. Therefore, cytokinins might have a so–called target effect in the nucleus only after first enhancing production of one or more nuclear proteins by translation in the cytosol. The popular concept is that whereas auxins and gibberellins mainly influence through an enhanced transcription of some genes and a repression of other genes, cytokinins might act only at transcription, or by influencing mRNA stability, or both. In at least the following three instances, cytokinins affect the amounts of mRNA molecules that code for known proteins. Two proteins and their mRNAs are strongly upregulated (formed faster or degraded slowly): a chlorophyll a or b binding protein, and the small subunit–protein of rubisco. The third protein controlled by cytokinin is the **phytochrome**. Phytochrome and its mRNA are downregulated (formed slower or accumulated in smaller amounts). From the above, it can be summarized that it is not known with certainty whether cytokinins generally act on transcription, on mRNA stability or on translation, since evidence for each is available. Perhaps, cytokinins affect all three processes in different species or plant parts.

Cytokinin receptors and transduction

Whether plants have special receptor proteins that can bind cytokinins, which then will lead to various physiological effects depending on tissue type, is not known clearly. *Arabidopsis* mutants that form calli and shoots in tissue culture without added cytokinin were isolated by activation T–DNA tagging. The gene *CYTOKININ–INDEPENDENT 1* (*CKI 1*), which was tagged in four of these mutants, was found to encode a protein similar to two–component regulators (Kende and Zeevaart 1997). It has a putative histidine kinase and receiver domain, and it resembles ETR1. Thus, in all likelihood, it functions in the transduction of cytokinin responses and may indeed serve as a cytokinin receptor.

2.3.4.4. *Ethylene*

Synthesis and functions of ethylene

Ethylene is a volatile and gaseous hormone, commonly associated with fruit ripening. First it came into recognition in 1910. But, it was proved only in 1934 that ethylene needs to be synthesized by plants for faster fruit ripening. Ethylene is produced probably by most organisms. Ethylene produced by soil fungi has been shown to promote germination of seeds, to control seedling growth and to retard diseases caused by soil–borne organisms. All parts of all seed plants were shown to produce ethylene. In the seedlings, the shoot apex is an important site of ethylene production. Nodes produce much more ethylene than internodes. Stems produce more ethylene when they are laid horizontally. Roots release relatively small amounts of ethylene. The production of ethylene in leaves generally rises slowly until they become senescent. Flowers also produce ethylene until they wither. In many fruits, little ethylene is produced just before climacteric respiration and then it drastically increases in intercellular spaces to promote ripening.

Ethylene (Fig. 15 A) is derived from carbon 3 and 4 of the amino acid methionine, and the immediate close precursor of ethylene is **1–amino–cyclopropane–3 carboxylic acid** (**ACC**). Ethylene synthesis requires ACC synthase and the ethylene–forming enzyme (EFE). Light, oxygen and CO_2 affect ethylene synthesis. Oxygen is required to convert ACC to ethylene. There are two potential inhibitors of ethylene biosynthesis; aminoethoxyvinyl glycine (AVG) and aminoxyacetic acid (AOA).

Ethylene affects plants in waterlogged soils and submerged conditions, as the waterlogged roots are expected to produce less ethylene because of paucity of oxygen. But yet, waterlogged plants exhibit ethylene toxicity. The effects include chlorosis, decreased stem elongation but increased stem thickness, wilting, epinasty, abscission of leaves, decreased root elongation, adventitious root formation, aerenchyma formation, increased susceptibility to diseases etc. All these symptoms of excess ethylene are caused by trapping of ethylene in the roots. ACC also accumulates, but gets transported in the xylem to the shoots, where it is converted into ethylene, as the shoots are well–aerated.

Although ethylene causes epinasty of leaves by promoting cell elongation on the upper side, it usually inhibits elongation of stems and roots, especially in dicots and hence, they become thicker. Ethylene promotes flowering in some species and inhibits it in others. It induces floral senescence and adventitious roots in many taxa and breaks dormancy in some. These results emphasize the different responses of somewhat similar cells to the same hormone. Ethylene also acts as a signal for pathogenesis–related reactions. Many ethylene effects are accompanied by increased synthesis of enzymes such as cellulase and other wall–degrading enzymes, fruit–ripening enzymes, phenylalanine ammonia lyase (PAL), β, 1–3 glucanase and chitinase. Ethylene increases transcription of many nuclear genes, depending on species, tissue, organ and other factors.

Genetics of ethylene activity

It is not clearly known where ethylene acts within cells. It binds with one or more receptor proteins and these proteins are located in the membranes. Considerable evidence suggests that these receptors contain copper at their receptor sites. Studies on ethylene–related mutants have thrown much light on ethylene receptors, transduction pathways and on the probable mode of ethylene action, especially in *Arabidopsis*. Several ethylene mutants have now been found. Some of them have been induced by EMS treatments of seeds. These include ethylene synthesis mutants and response mutants. All of the synthesis mutants, except one, are ethylene over–producers. All these mutants produce triple effects i.e., stunted growth, swelling of axial organs and exaggerated apical hook formation. Three ethylene–insensitive mutants are found lacking these three ethylene responses found in wild plants.

The mutation *ethylene–resistant 1* (*etr 1*) is dominant, and the mutant lacks a number of effects of ethylene (e.g., inhibition of cell elongation, promotion of seed germination, enhancement of peroxidase activity, acceleration of leaf senescence and feedback inhibition of ethylene biosynthesis). The capacity of *etr 1* to bind ethylene in vivo was only one–fifth that of the wild type, indicating that the mutant is deficient or impaired in receptor functions. The *ETR* gene was found to encode a protein having similar sequence to that of bacterial two–component regulators (Kende and Zeevaart 1997). In its N–terminal portion, it contains a sensor domain with a putative input and histidine kinase region (Fig. 15 C) fused to this is a receiver domain, but the equivalent of an output domain is missing. Using protein expressed in yeast, it was shown that the hydrophobic N–terminal region of *ETR 1* binds ethylene and the *etr 1–1* mutation, which is localized in this region and which leads to ethylene insensitivity, abolishes ethylene binding. These results provide strong evidence that ETR 1 is an ethylene receptor and the ethylene–binding site is located in a membrane–spanning region of the N–terminal input domain. The *ETR 2* and *ETHYLENE–INSENSITIVE* (*EIN 4*) genes encode homologues of *ETR 1*, and mutations in these genes confer dominant ethylene insensitivity onto *Arabidopsis* seedlings (Kende and Zeevaart 1997).

The *Arabidopsis* gene, *ETHYLENE RESPONSE SENSOR* (*ERS*) that encodes a second type of putative ethylene receptor has been cloned. The sensor domain of the ERS protein shows great similarity to ETR 1, but it lacks a receiver domain (Fig. 15 D). Thus, *Arabidopsis* contains at least four genes that encode putative ethylene receptors viz., *ETR 1*, *ETR 2*, *EIN 4* and *ERS*. Thus, there is potential redundancy in ethylene perception. Homologues of *ETR 1* and *ERS* have also been isolated from tomato.

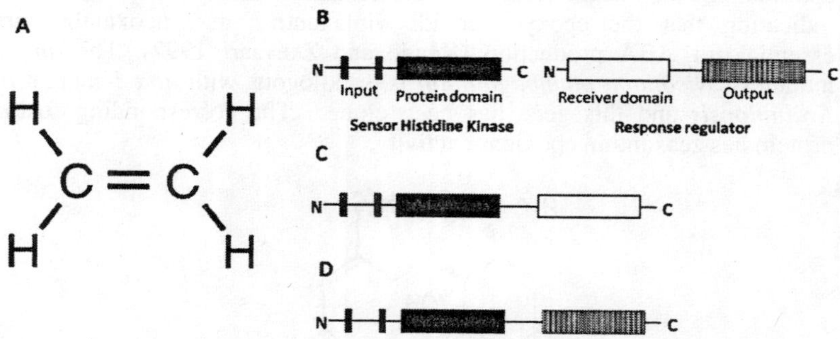

Fig. 15. A. Structure of ethylene; B–D - Two–component signalling systems involved in ethylene activity; B - Bacterial two–component system involving sensor and response regulator proteins; C - The ETR–type two–component system; D - The ERS–type protein; ERS does not have the receiver domain. (B–D: Based on Kende and Zeevaart 1997)

Genes acting downstream of ethylene reception in *Arabidopsis* include *CONSTITUTIVE TRIPLE RESPONSE 1* (*CTR 1*); *ctr1* mutants express the triple–response phenotype constitutively, even in the absence of ethylene or treatment. Genetic analyses have shown that *CTR 1* acts downstream of *ETR 1*, *ETR 2*, *EIN 4* and *ERS* and it is a negative regulator of ethylene responses. *CTR 1* encodes a putative serine or threonine protein kinase that is related to Raf protein kinases. This relationship indicates that the ethylene transduction pathway may be similar to a protein kinase–mediated phosphorylation cascade. A second downstream gene is *HOOKLESS* (*HLS 1*) of *Arabidopsis;* the expression of this is necessary for the appearance of the apical hook, an important response of ethylene effect.

2.3.4.5. *Abscissic Acid*

Abcissic acid (ABA) is a plant growth regulator that was first chemically identified and characterized by Addicott and co–workers in 1963. They named one active compound **abscisin I**, and called a second (a much more active one) compound **abscisin II**. The second compound was proved to be ABA (named in the same year as **dormin** by Wareing, and involved in dormancy). Physiologists agreed in 1967 to call the compound ABA. ABA appears to be universal among vascular plants. They are also found in other groups of plants, but not in bacteria. ABA is a 15–carbon sesquiterpenoid (Fig. 16) synthesized partly in chloroplasts and other plastids by the mevalonic acid pathway. Biosynthesis of ABA in most plants occurs indirectly by degradation of certain, 40–carbon, carotenoids present in plastids with **xanthoxin** as an intermediate. This is supported by the observation that plants, not accumulating carotenoids, also lack ABA. The ABA–deficient *aba 1* mutant of *Arabidopsis* is blocked in the epoxidation of **zeaxanthin** to **antheraxanthin** and **violaxanthin**, indicating that the epoxycartenoids violaxanthin and neoxanthin are essential for ABA production (Kende and Zeevaart 1997). The *aba 2* mutant of *Nicotiana plumbaginifolia* is orthologous with *aba 1* mutant of *Arabiolopsis* and this gene has been cloned. The corresponding fusion protein has zeaxanthin epoxidase activity.

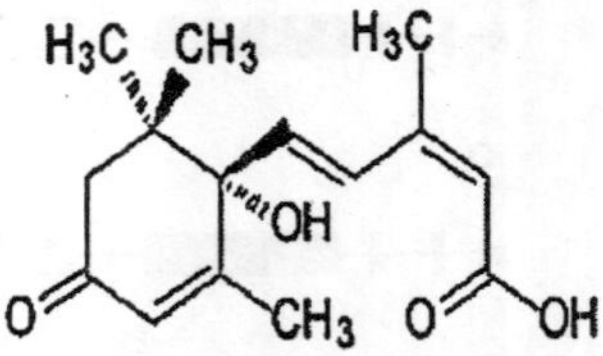

H_3C CH_3 H_3C OH O CH_3 O OH

Fig. 16. Structure of Abscissic Acid (ABA)

Inactivation of ABA can occur in two ways: (i) By forming ABA–glucose ester (ii) By oxidation with O_2 to form **phaseic acid** and **dihydroxyphaseic acid**. Transport of ABA occurs readily in both xylem and phloem, and also in parenchyma cells outside the vascular bundles. There is no polarity of ABA transport in parenchyma cells.

Roles of ABA

ABA gives plant organs a signal that they are undergoing physiological stresses induced by water, salinity, temperature, frost etc. ABA level increases in stressed plants or organs and this increase causes responses that help protect plants against the above stresses. ABA also controls normal embryogenesis and the synthesis of seed–storage proteins through its effect on the activation of transcription (especially in cultured embryos). It also prevents premature germination (vivipary) or growth of many seeds and buds. The most common response of plant tissues to ABA is growth inhibition. Accumulation of ABA was considered by some to increase during bud dormancy and application of ABA to non–dormant bud induces dormancy in it, although additional work strongly speaks against such a hormonal role for ABA. Moreover, radio–labelled ABA has not been shown to move from leaves to the bud when dormancy begins in the bud. Exogenous ABA is a potent inhibitor of seed germination in many species indicating its role in seed dormancy. Also, ABA level decreases in whole seeds when dormancy is broken experimentally. However, the effect of ABA in seed dormancy is not seen in all species and hence, ABA may cause seed dormancy in some species only. This seems reasonable because many other compounds are associated with seed dormancy, especially gibberellins. ABA induces stomatal closure in light or dark for even several days, especially in water–stressed leaves. The role of ABA in causing abscission of leaves, flowers and fruits is controversial. While many have argued for its role in abscission, especially in relation to ethylene, others give more importance to ethylene than ABA. Some even negate the role of ABA in abscission, but argue that ABA causes senescence, which in turn causes abscission.

ABA seems to have three major effects, depending on the involved tissue: (i) Effects on plasma membrane (ii) Inhibition of protein synthesis (iii) Specific activation and deactivation of certain genes (transcription effects). Through the first effect, ABA rapidly inhibits auxin–induced growth. Inhibition of protein and enzyme synthesis could help explain ABA's long–term effects on growth and development. The third effect produces a powerful control over developmental processes.

There are several mutants that are unable to form much ABA (ABA–synthesis mutants). A study of these mutants shows that several genes and enzymes are necessary to synthesize ABA. Specific mention must be made of the *Arabidopsis* gene *ABA–INSENSITIVE 1* (*ABI 1*). A mutation in this gene makes plants insensitive to ABA. *ABI 1* was found to encode a protein with high similarity to a 2C–class serine or threonine protein phosphatase with an N–terminal calcium–binding site. This indicates that the ABI 1 protein functions as a calcium–regulated protein phosphatase and thus, is part of a phosphorylation–dependent transduction pathway that mediates a broad spectrum of ABA response.

2.3.4.6. *Brassinosteroids*

These are also known as **brassins (BRs)**. BRs are 'non–traditional' plant growth regulators. They are a group of naturally occurring polyhydroxy steroids, first isolated from pollen grains of rapeseed plant in 1979. They are now known to be present in several other species including dicots, monocots, gymnosperms, green algae and ferns. By about 1997, more than 60 BRs have been known, of which 31 including two conjugates, have been fully characterized. BRs have been isolated from pollens, seeds, fruits, shoots, leaves, flower buds and galls at levels between 0.5 pg and 30 ng per gram of fresh weight. BR levels are relatively high in pollen (5–190 ng/g fresh weight) (Creelman and Mullet 1997).

The basic structure of a brassinolide is shown in Fig. 17. Of the known BRs, the most abundant are **brassinolide** and its biosynthetically related precursors. Variations in BR structure result from different substitutions in the A/B rings and side–chains shown in this figure, created in oxidation or reduction reactions that occur during biosynthesis. Biosynthesis of brassinolide begins with the reduction of the 5, 6 double bond of **campesterol** to form **campestanol**. The pathway to brassinolide from campestanol may occur via an early oxidation of **campesteronol** to form **cathasterone** at carbon 6 or a late oxidation at carbon 6. Several BRs lacking oxygen at carbon 6 are found in plants.

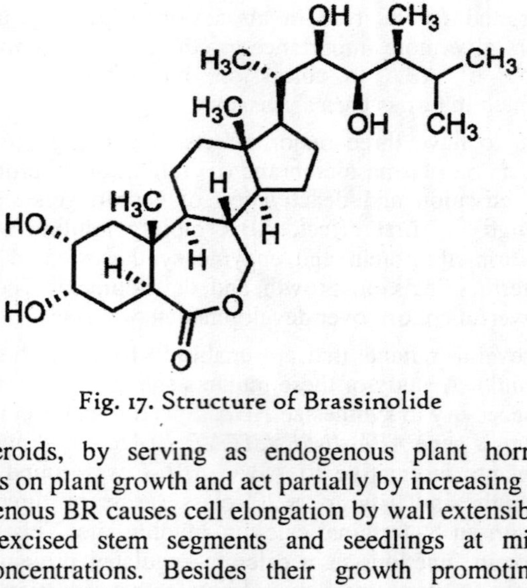

Fig. 17. Structure of Brassinolide

Brassinesteroids, by serving as endogenous plant hormones, have various effects on plant growth and act partially by increasing sensitivity to auxins. Exogenous BR causes cell elongation by wall extensibility and cell division of excised stem segments and seedlings at micromolar to picomolar concentrations. Besides their growth promoting activities, exogenous BRs inhibit root growth, enhance gravitropism, retard leaf abscission, enhance resistance to stress, and promote xylem differentiation. Undisputable evidence for the importance of BRs in some aspects of plant growth and development has been obtained with the isolation of BR–insensitive *Arabidopsis* and pea mutants such as *brassinosteroid insensitive 1* (*bri 1*), *cbb 2* (cabbage; *ika*), and BR–deficient mutants such as *dwf 1–6*/*cbb 1*, *cbb3*/*cpd* and *det 2*. All of the *Arabidopsis* mutants are male–sterile or have reduced male fertility, severe dwarfness and reduced apical dominance. The pea mutants have similar phenotypes showing a reduction in internode and peduncle length, diminished basal branching, and a ridged surface, while BRs play an important role in photomorphogenesis in some plants, but not in peas. We do not have much information on receptors related to brassinosteroids.

2.3.4.7. *Jasmonates*

Jasmonic acid (JA) (Fig. 18) and its methyl ester (**methyl jasmonate, MeJA**) occur in several plant species and the oil of jasmine. Both are derived from linolenic acid, probably as a result of action of the enzyme lipoxygenase, and are cyclopentanone–based compounds. JA is a non–traditional plant hormone. Jasmonates have been found in more than 150 families (including fungi, mosses, and ferns) so far and is ubiquitous in plants. These strongly promote leaf senescence and regulate plant growth. JA levels in plants vary depending on the tissue and cell type, developmental stage, and in response to several different environmental stimuli. For example, levels of JA in soya bean seedlings are higher in the hypocotyl hook, a zone of cell division, and in young plumules than in the zone of cell elongation and more mature regions of the stem, older leaves, and roots. High levels of JA are also found in flowers and in pericarp tissues of growing fruits.

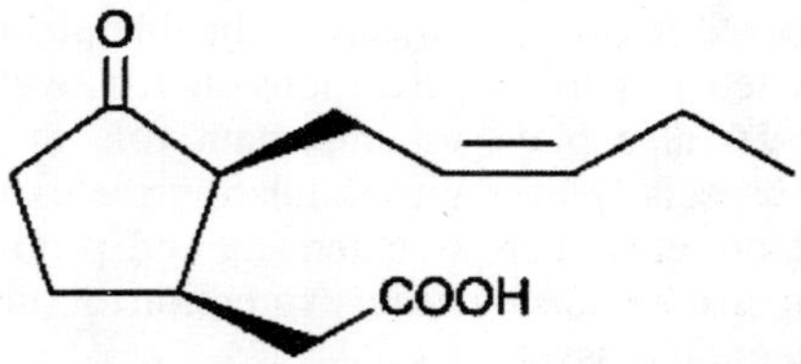

O

COOH

Fig. 18. Structure of Jasmonic acid (JA)

JA levels are rapidly and transiently increased by mechanical perturbations such as those causing tendril coiling, turgor reduction induced by water deficit, and wounding. In addition, JA accumulation can be induced by pathogen–related elicitors such as oligosaccharins. Transcripts of JA and JA–responsive genes accumulate systemically in plants in response to the localized wounding. The systemic signal is apparently released at the wound site and migrates through the phloem to other parts of the plant. **Systemin**, an 18–amino acid peptide, released from the wound site upon hydrolysis of a precursor polypeptide, has been shown to move in the phloem and to induce the synthesis of JA and the expression of JA–responsive genes throughout the apical portion of plants. The tomato mutant *JL 1* may be impaired in its ability to make or release systemin at or near the site of wounding (Creelman and Mullet 1997).

JA and MeJA inhibit the germination of non–dormant, recalcitrant seeds and stimulate the germination of dormant seeds. The increase in JA in seeds is correlated with lipid peroxidation, suggesting that the production of jasmonates may not regulate germination, but rather is a consequence of membrane damage. In those seeds, where JA promotes germination (for example, as in *Arabidopsis*), JA may stimulate seed germination by decreasing sensitivity to ABA. The *jin 4* (*jasmonate insensitive 4*) and *jar 1* (*jasmonate response 1*) mutants of *Arabidopsis* showed increased sensitivity to ABA. JA strongly inhibits root growth and IAA–induced coleoptile elongation. In both, ethylene is not involved. JA plays a role in flowering, fruiting and seeding, as high levels of JA are found in these developing structures. JA may be involved in insect attraction (hence in pollen and/or fruit or seed dispersal), fruit ripening (through ethylene involvement), root carotenoid composition and expression of genes encoding storage proteins, and vegetative storage proteins (as revealed by the expression of genes encoding them). JA has also been implicated in mediating the accumulation of secondary plant products such as **paclitaxel** and related taxanes, alkaloids, anthocyanins and rosmarinic acid. JA or a derivative of it, **tuberonic acid**, has been proposed to play a role in the formation of tubers.

JA affects the expression of certain genes, both of nucleus and chloroplasts. JA application causes chlorosis, and hence, it has been implicated in senescence also (probably wrongly). JA inhibits expression of some genes involved in photosynthesis. Inducible plant defenses to UV light may be mediated in part by alterations in JA levels, as evident by molecular studies. JA also plays an important role in plant insect and disease resistance, especially through modulating the expression of genes encoding cell wall proteins such as extensins and proline–rich proteins, which are PR proteins, as also in the expression of genes involved in phytoalexin and phenolic biosynthesis.

With reference to JA transduction, it has been presumed that it interacts with receptors in the cell to activate a signalling pathway that ultimately triggers changes in the transcription and/or translation of genes participating in the responses mediated by JA. Until about 1997, up to four different classes of JA–insensitive mutants have been identified; *jar 1*, *coi 1*, *jin 1* and *jin 4*. Also promoters of two JA–inducible genes, *Pin* 2 and *VspB* have been analysed. A study of these has indicated that receptor (s) is involved in JA recognition and activity.

2.3.4.8. Oligosaccharins

Oligosaccharins, including free N–glycans, are also non–traditional plant hormones. These are complex carbohydrates (oligosaccharides) that are capable of modulating plant growth and development at low concentrations. Some of these such as oligogalacturonids act as elicitors and evoke pathogen defense responses such as the accumulation of phytoalexins, proteinase inhibitors, lignin, peroxidases, lipoxygenase (LOX) and β, 1–3 glucanases. Also, the effect of elicitors on defense responses is mediated in part via alterations in the levels and/or sensitivities of JA and salicylic acid. Rhizobial lipo–chitin oligosaccharins (i.e., nod factors) have great importance in root nodulation.

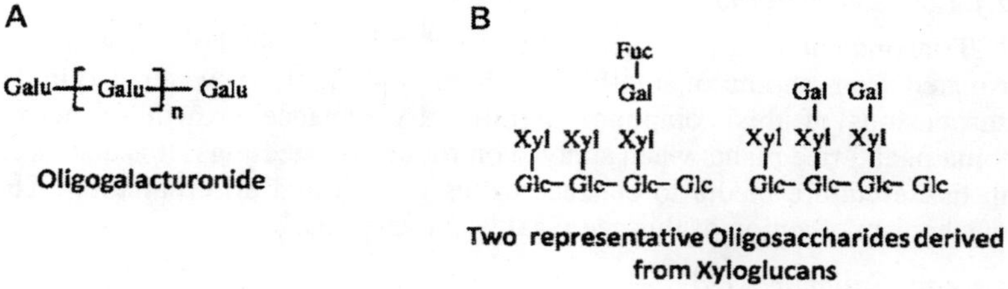

A

Galu—[Galu]$_n$— Galu

Oligogalacturonide

B

Fuc | Gal | Xyl Xyl Xyl / Glc– Glc– Glc– Glc   Gal Gal / Xyl Xyl Xyl / Glc– Glc– Glc– Glc

Two representative Oligosaccharides derived from Xyloglucans

Fig. 19. A - Structure of oligogalacturonide;
B - Structure of two xyloglucan oligosaccharides

The major sources of oligosaccharins are pectins and hemicellulosic xyloglucans. Biologically active pectic oligosaccharins are generated by chemical or enzymic hydrolysis of cell wall pectin and these fragments vary in size with degree of polymerization ranging from 2–20. Exogenous pectin oligosaccharins antagonize auxin action at levels 10–fold to 100–fold lower than those required for the elicitation of plant defense responses (Creelman and Mullet 1997). For example, oligogalacturonides (Fig. 19 A) inhibit auxin–induced growth of isolated pea stem segments, auxin–stimulated rooting in leaf explants and auxin–dependent somatic embryogenesis in carrot cultures. The role of oligogalacturonides in growth and morphogenesis has been studied in tissue culture lines as these form various plant organs depending on available auxin to cytokinin ratio. Oligogalacturonides have been shown to induce flowering or inhibit root formation in tobacco explants and they do this by interfering with auxin–binding sites or by altering auxin metabolism. However, very little is known about the role of these oligosaccharins on whole plant growth and development.

Xyloglucan oligosaccharides, obtained after enzymatic hydrolysis and subsequent purification, are mostly of any one of the following: XXXG, XXFG, and XLFG (Fig. 19 B). Some xyloglucan fragments can modulate plant growth. For example, XXFG inhibits growth induced by auxin, gibberellin, protein and fusicoccin at concentrations of ~1nm. By contrast, at higher concentrations, XXFG and other fragments can promote pea stem elongation in the absence of auxin. These fragments stimulate growth by acting as substrate for XET enzyme. XET may cause temporary cell wall loosening, which in turn helps in cell expansion (Fig.44).

Oligouronides are considered among the early signals for plant defense genes. They are not mobile and therefore, are unlikely systemic signals for defense genes in response to pest attacks. We do not have any information on the receptors of oligosaccharins.

2.3.4.9. *Triacontanol*

Triacontanol is a 30–carbon, saturated primary alcohol. It was first isolated from shoots of alfalfa. It is highly insoluble in water. Colloidal suspensions of this compound significantly enhance growth of maize, tomato, and rice plants when sprayed on foliage of seedlings. It is also used in tissue culture media to enhance callus growth and differentiation. The mechanism of action of this chemical is not very clear.

2.3.4.10. *Salicylic Acid*

Chemically it is a 2–hydroxy benzoic acid having low molecular weight. It is an important plant hormone involved in some known physiological responses, especially in producing insect–attracting chemicals in the inflorescence of some aroids. It is also helpful in promoting resistance to infection by plant pathogens such as viruses, bacteria and fungi, probably causing the production of PR–related proteins.

2.3.4.11. *Turgorins*

These are hormones that control nastic movements. Schildknecht (1983) has first isolated the compounds (such as **Ricca's factor**) at active leaf pulvini of plants such as *Mimosa* and *Acacia karroo*. The later is not sensitive to touch, but exhibits nyctinasty. Two so–called **periodic leaf movement factors** (**PLMFs**) from *Acacia* proved to be the β–glucosides of gallic acid. Several other compounds with closely related structures were also identified from the extracts of other plants (e.g., PLMF–1, PLMF–2, PLMF–3, PLMF–7, S–PLMF2, M–LMF5, PLMF–synth and LMF–synth). A PLMF–1 receptor (predominantly a protein) has been demonstrated in the outer side of the plasma membrane of *Mimosa*.

2.3.4.12. *Polyamines*

These are polyvalent cations that contain two or more amino groups, including the amino acids arginine and lysine. Among the most abundant and physiologically active polyamines are **putrescine**, **cadaverine**, **spermidine** and **spermine**. These compounds exist free or bound to various phenolic compounds such as coumaryl and caffeoyl groups. In contrast to hormones, these are present in larger concentrations (at macromolar concentrations). Among several physiological effects, polyamines promote cell division, stabilize membranes, stabilize isolated protoplasts, promote development of some fruits, minimize water stress of various kinds of cells and delay senescence.

Very little is known about the primary mechanism of action of polyamines, but their positively charged amino groups cause them to combine with negatively charged phosphate groups in DNA and RNA in the nucleus and in ribosomes. As a result of this combination, polyamines often increase transcription of DNA and translation of RNA. Some consider these as not plant hormones (since they are poorly translocated and also since they are too abundant), but others consider them as plant growth regulators needed for certain developmental processes.

2.3.4.13. *Lunularic Acid*

It is mainly present in the gemmae of bryophytes. It prevents the germination of gemmae until they fall off from the mother thallus and the acid is leached out. This acid is not present in algae or in vascular plants.

2.3.4.14. *Batasins*

These are chemicals present in the yam plants and seem to cause dormancy in bulbils. The structure of batatasin I is shown in Fig. 20. These are concentrated in the skin of bulbils and are absent from the core. The breaking down of dormancy causes batatasin to disappear, while they increase during dormancy of bulbils.

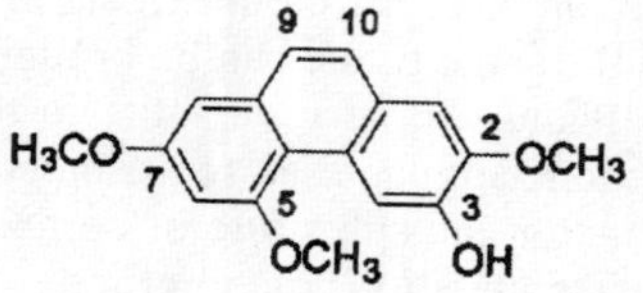

Fig. 20. Structure of Batatasin 1(2) as per Yoon *et al.* 2007

2.3.4.15. *Systemin*

It is a peptide of 18 amino acids (Fig. 21) and has growth regulatory activity. As already stated, it induces the synthesis of jasmonic acid. It is a

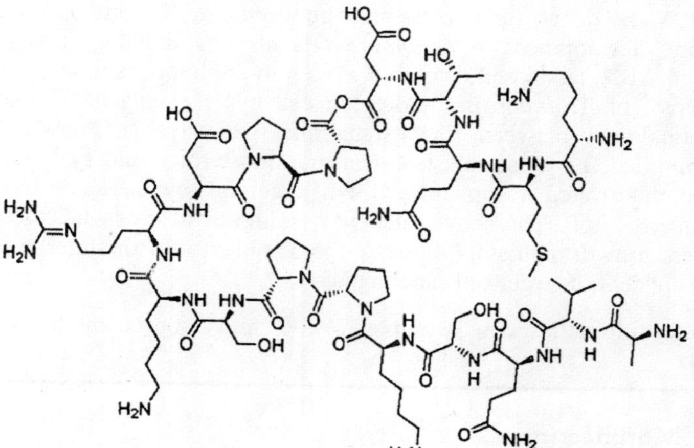

Fig. 21. Structure of systemin

potent inducer of two protease inhibitor genes in tomato and potato. It is readily transported in the phloem and hence, is consistent with the properties of a systemic signal.

2.3.5. Interaction between Growth Regulators

We have already discussed about the individual growth regulators. However, growth regulators do not function in isolation and they often interact with one another, either positively or negatively and sharing the same signal, target cell or receptor to bring about a particular response (Refer to Box 2.3)

BOX 2.3. CROSS–TALK BETWEEN GROWTH REGULATORS

Although for convenience the individual growth regulators can be considered and treated as separate operational entities, it is clear that the same growth regulator can have different effects in different tissues, and also a similar effect in the same tissue can be brought about by more than one growth regulator. Plants are very versatile in the cross–talk of their molecular communication language, as represented by situations where any single growth regulator can substitute or complement in function for another. For instance, elongation of internode by ethylene can be substituted by or complemented with either GA or IAA, where ethylene is the primary signal that gets modulated by addition of GA or IAA. Studies have indicated the level of cross–talk between ethylene and ABA in seed dormancy at an interconnecting junction in their otherwise separate signalling pathways. Molecular evidence for a direct cross–talk between two signals in a common signal transduction pathway (though not identical) has come from works relating to cross–talks between JA and IAA. Evidence for a cross–talk between JA, SA and GA during trichome development in *Arabidopsis* is also available. A cross–talk between ABA, auxin and cytokinin is seen in the hyponastic response of the leaves of *Arabidopsis.* Studies on cell cycle events have revealed commonalities between the signalling pathways of auxin and brassinolide. The cross–talk between brassinosteroids and systemin has been demonstrated in tomato, as their perceiving by the same receptor proteins, SR 160. The above cited few examples of cross–talk between different growth regulators illustrate the complexity of the interpretation of the concept of target cell state in plants.

[Refer full literature in Swarup et al, 2002 and Osborne and McManus, 2005]

2.3.6. Morphactins

Morphactins (derivatives of fluorine–9–carboxylic acid) represent a new class of synthetic bioregulators with high morphogenetic potency

(Schneider 1970; Sankhla et al 1975). These versatile bioregulators exhibit a wide range of influences on plant growth, development and morphogenesis at all stages of plant development. These synthetic growth regulators have high plant tolerance, wide spectrum of activity, depot effect, favourable therapeutic index, subsistence of action, differential activity ranges and rapid metabolism.

Once applied, these growth regulators are transported basipetally from the apical meristem and affect all organs formed from it. They act slowly, transiently retard mitoses and cell divisions and suppress internodal elongation. They remove apical dominance, promote more lateral branches and increase (decrease in some taxa) the number of flower buds produced. They completely or partially eliminate gravitropism and phototropism. They also severely curtail lateral root formation. A shift may occur in flower bud position, in the order of flower bud opening and in floral sex. In some tree taxa, breaking of flower bud may be delayed. Parthenocarpy may be induced and fruit abscission may be enhanced. Morphactins also promote cambial activity and wound callus formation.

The most important morphactins are chlorflurenol–n–hexylester, 2–chloro–ethylphosphonic acid, chlorflurenol–methyl ester, 2–chlor–9–hydroxy–fluorocarboxylate, 2–chloroethane phosphonic acid and IT 3456.

Cell Division and Enlargement 3

3.1. INTRODUCTION

It was emphasized in the first chapter that in the multicellular plants, cells form the ultimate and fundamental units of structure and function. Consequently, cell division becomes one of the most conspicuous features of these plants as it is needed to produce the multicellular condition. The immobility of both plants and their constituent cells as well as the presence of physically rigid to semi–rigid cell walls in plant cells make the cell divisions the main mean to achieve proper organ size and shape in plants. For any given cell division, a new cell wall is deposited and the orientation and position of the new cells are locked in place (Dupuy et al 2010). Actively dividing cells have to pass naturally through a regular sequence of events or **cell cycle** events. Cell division has long been considered to play a very significant role in plant growth, development and organ production. This traditional view is based on the actual observations and analysis of the ontogeny of various plant organs. Contrary to this traditional view, often called as the **cell theory of development** (Evered and Marsh 1989), the alternative **organismal view** suggests that cell division is merely a 'marker' of plant growth and that it does not influence plant growth and / or development (Kaplan and Hagemann 1991). Many people, who believe in organismal view, often consider cell enlargement as more important than cell division in plant growth and development. The processes of cell enlargement, differentiation and formation of tissues are regarded by these people as essentially independent of cell division, which is simply a necessary consequence of their execution. Therefore, there is a paradox, with cell division being viewed as both important and unimportant for plant development (Meyerowitz 1996). The arguments on this topic have tended to be clouded by these two polarized ideas (Fleming 2006), encapsulated in the aphorisms attributed to de Bary (Barlow 1982): "Die pflanze bildet zellen, nicht the Zelle bildet pflanzen (The plant forms cells, not cells the plant)" and to Meijer and Murray (2001), "The plant makes cells and the cells make the plant". The latter authors strongly emphasize that cell division should not be disregarded simply as a secondary consequence of growth. This chapter examines the importance of cell cycle, cell division, and cell enlargement in the growth and development.

3.2. NUCLEAR AND CELL DIVISIONS

It is a prerequisite that the nucleus should divide before a cell divides into two. Nuclear division is called **karyokinesis**, while division of the cytoplasm (i.e., the cell) is called **cytokinesis**. Nuclear divisions are of two kinds:

1. **Mitosis** – During mitosis a nucleus gives rise to two daughter nuclei and each daughter nucleus is morphologically and genetically equivalent to the other as well as to the parent nucleus. In essence, mitosis in eukaryotic organisms, such as the plants, involves the duplication and separation of chromosomes, coupled to the process of dividing one cell into two.
2. **Meiosis** – In meiosis, the parent nucleus undergoes two successive divisions, one of which is a reduction division and the second is a mitosis, also called **heterotypic** and **homotypic** divisions respectively (Swamy and Krishnamurthy 1975a). However, chromosome replication occurs only once. By some precise mechanism, meiosis results in four daughter nuclei, each of which has one half of the number of chromosomes of the

parent nucleus. Mitosis in plants gives rise to **somatic cells** (cells of the vegetative body) and to gamete–producing cells, whereas meiosis gives rise to **gametes** (**sperms** or **eggs**). In both types of divisions (but with some exceptions), the nuclear envelope is lost before the onset of division, but it is again organized in the daughter nuclei.

Usually, karyokinesis is immediately followed by cytokinesis between the two daughter nuclei (Fig. 22 A–E). In this case, the cytokinesis is associated with the **phragmoplast** formation (Fig. 22 D and E). However, in the case of nuclear endosperm and in the micropylar (and in some the chalazal) chamber of helobial endosperm, wall formation takes place once many free nuclei are formed to separate the free nuclei from one another. The mechanism of wall formation had been studied in detail in the helobial endosperm of *Haemanthus*, using electron microscopy. In this process, which usually proceeds centripetally (from periphery toward the large central vacuole), the cell walls formed may enclose single, double or sometimes more nuclei; very rarely 'cells' without any nucleus are also formed. This implies that there is obviously no nuclear control on where and when wall formation has to take place. The first–formed endosperm cell walls grow freely from the embryo sac wall i.e., independent of phragmoplast and cell plate formation, as well as independent of microtubules. The free-growing walls soon branch both towards interior as well as parallel to the embryo sac membrane, as a result of which several endosperm cells are formed. Subsequent cell walls are related to phragmoplasts, in *Haemanthus* after four cell layers are formed. The phragmoplast may be formed between sister or non–sister nuclei.

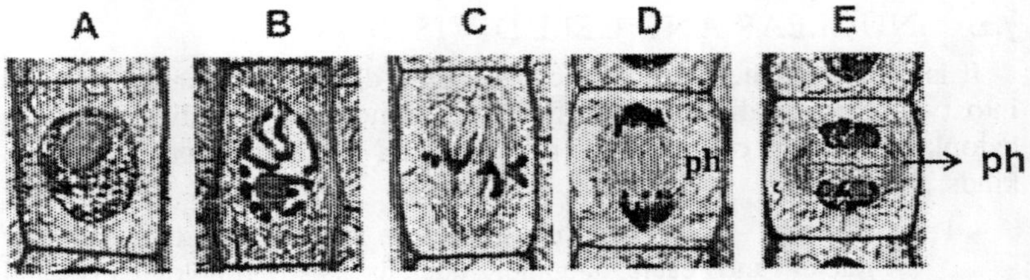

Fig. 22. A–E - Details of mitotic cell division and cell plate formation with cytokinesis associated with phragmoplast (ph) formation

3.3. CELL CYCLE

3.3.1. Cell Cycle Phases

The somatic cells that are actively dividing mitotically pass through a series of regular events, which together are referred to as cell cycle (Fig. 23).

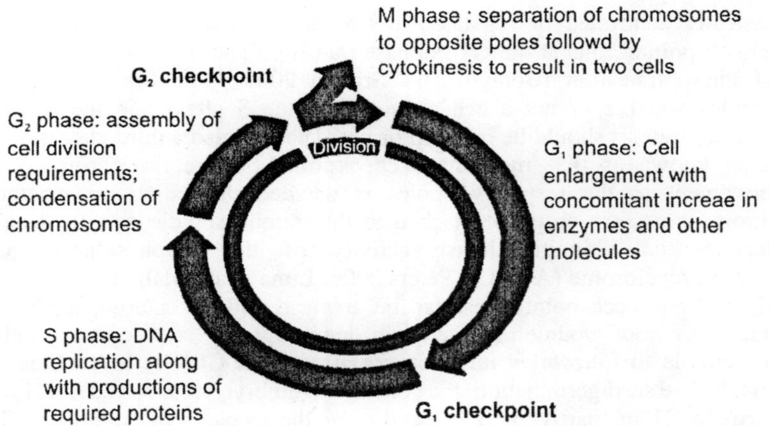

Fig. 23. The cell cycle

The cell cycle is commonly divided into **interphase** and mitosis (Strange 1992). The interphase precedes mitosis and in the majority of cells, mitosis is followed by cytokinesis. It is because of cytokinesis that most of the plant cells are uninucleated. Mitosis and cytokinesis together are referred to as the **M**–phase of the cell cycle. The interphase is divided into three phases respectively designated as **G1**, **S** and **G2** phases, where **G** refers to a gap. The **G1** phase (interval between **M** and **S** phases) occurs after mitosis and during this phase there is an intense biochemical activity leading to an increase in size of the cell and in the number of many cell organelles, endomembranes and other components of the cytoplasm. The **S** phase (abbreviation for synthesis phase) is a period when the DNA replicates. At the time of initiation of DNA replication, the diploid nucleus of the about–to–divide cell has 2C DNA value (Here, C refers to the amount of haploid DNA content), but at the completion of the **S** phase, the DNA content becomes 4C. During the **S** phase the DNA–associated proteins, such as histones, are also synthesized, as also the repair of damaged DNA, if any, is also carried out. Following the **S** phase, the cell enters into the **G2** phase (interval between **S** and **M** phases), which precedes mitosis. During mitosis, the DNA synthesized in the **S** phase is equally appropriated to the two daughter nuclei, thus restoring the original 2C value in them.

3.3.2. Control and Regulation of Cell Cycle

The method of control and regulation of the cell cycle events is basically similar in all eukaryotic cells as the degree of evolutionary conservation of the cell cycle events is very striking among eukaryotes, in species as diverse as yeasts, higher plants and humans. Control and regulation of the cell cycle events in plant cells happen in response to both external and internal changing conditions. In a typical cell cycle, progression is basically controlled or regulated at two crucial transition points called **'check points'**, one at the **G1–S** phase transition and the other at the **G2–M** phase transition (Boniotti and Griffith 2002). The first checkpoint decides whether or not a cell should enter the **S** phase and the second whether mitosis should be initiated or not. There is also a third checkpoint, often known as the **'metaphase checkpoint'**, where the chromosome movement to the respective poles is decided depending on whether chromosomes are properly attached to the mitotic spindle or not. It also depends on ubiquitin ligase activity of the anaphase-promoting complex/cyclosome (APC/C) (Peters 2006; Lima et al 2010). The example of the first check point arrest can be given as follows. During the early stages of root nodule ontogeny in legumes, the root cortical cells susceptible to *Rhizobium* infection are arrested in **G0/G1**. In contrast to this, before seed germination, the cells of the embryo are arrested in **G1**, or partly in **G1** and partly in **G2**, depending on the species. An example of **G2** control is found in lateral root primordia of radish derived from pericycle cells arrested in G2. However, after supplemented with auxin, these cells can go to the M phase and then continue to divide and produce the lateral root primordium (den Boer and Murray 2000).

Studies on crucial plant cell cycle regulators will unlock the mechanisms that regulate the cell division, and provide the much–needed information on the relationship between cell proliferation and plant growth and development. Researches in the last 25 years have shown that progression through the cell cycle depends on the successful formation, activation and subsequent deactivation of **cyclin–dependent protein kinases (CDKs)** at the first two check points mentioned above. Significant progress has been made in the identification and characterization of plant CDKs and **cyclins**. These kinases consist of a catalytic CDK subunit and an activating cyclin subunit (Huntley and Murray 1999, Mirnov et al 1999) and belong to a common class of heterodimeric serine or threonine protein kinases. Intense cloning efforts over the last two and a half decades have identified a large number of CDK–like proteins in diverse plant species. Among these, at least five types are best distinguishable based on their sequences: A–type (PSTAIRE as cyclin–binding motif), B-types (PPTALRE and PPTTLRE as cyclin–binding motifs) and non–classified CDKs (three types respectively with NFTALRE, PITAIRE and SPTAIRE as cyclin–binding motifs). Similarly, cyclins can be classified into nine groups: A1, A2, A3, B1, B2, D1, D2, D3, and D4. But, how many of these CDKs and cyclins are actually involved in the plant cell cycle event is not very clear. Experiments in plants support the relevance for cell division control only for CDC2aA1, CDC2bA1 and CYCB1;1 and, by extrapolation, for their orthologs from other species. There is no evidence for cell cycle functions of PITAIRE or SPTAIRE CDKs. One recently–discovered novel plant cell–cycle regulator is the CDK inhibitor represented by ICK1 from *Arabidopsis*. In contrast to most cell cycle regulators, this diverges significantly from the known CDK inhibitors of other species. It interacts with both a CDK and a cyclin. Its expression is induced by the conditions that are inhibitory to plant cell divisions such as treatment to ABA and low temperature, implying that it may be inhibiting cell division under abiotic stress conditions. ICK2 has also been reported.

Both auxins and cytokinins have also been implicated in the control of cell cycle in plants (den Boer and Murray 2000). Two broad questions regarding hormone–stimulated cell division activity required resolution (Kende and Zeevaart 1997); do the plant hormones influence cell division directly or indirectly, and at what point of the cell cycle do the plant hormones act? Answer to the first question is now beginning to emerge. Auxins alone increase the level of a CDK protein, for example, in the cultured tobacco cells and stem pith explants, but addition of a cytokinin is required for the activation of this kinase. At what point of cell cycle do auxins and cytokinins act is also to be addressed. Both are reported to be associated with progression through the **G1–S** and **G2–M** control points. Auxin stimulates root pericycle cells of radish arrested in the **G2** phase of the cell cycle to enter into mitosis, thereby promoting the formation of lateral root primordia. More definite results have been obtained with cytokinins. Cytokinin–depleted cells accumulate CDK–cyclin complexes that are inactive; cytokinin inhibitor lovastatin blocks the cells in **G2**. Cytokinins have also been shown to regulate the **G1–S** transition. Ascorbic acid has also been shown to be necessary for **G1–S** transition in the cell cycle in root apex (Liso et al 1984, 1988). A completely different aspect of the hormonal control of cell division cycle has been given by Houssa et al (1994). These authors observed that exogenous cytokinin reduced the size of DNA replication units in the shoot meristems of *Sinapsis alba* and *Lolium temulentum* and in the ovules of tomato. On the basis of this data, they proposed that the activation of latent replication origins is a universal effect of cytokinins in the promotion of cell division.

An important result that has emerged in the last 20 years is that the cell cycle length is changed as the major effect of some of the growth regulators and a few other chemicals. G1 phase is the most responsive to signals that change the cell cycle length. If the grass *Dactylis* is exposed to higher levels of CO_2, shoot apical meristem cells increase their rate of cell proliferation resulting in faster growth. This happens partly through an increase in the number of actively–dividing cells, and partly through the shortening of mitotic cycle, particularly the **G1** phase (den Boer and Murray 2000). It is interesting to know that both the shoot and root apical meristems are a mosaic of fast and slow cycling cells and the main difference between these two cell populations is the length of **G2** phase. When the cells are in **G1** phase, in the presence of sufficient stimulus, they have several options. They may proceed further in the cell cycle and progress into the S phase, or they may remain there (i.e., operation of dormancy) and may again resume the cell cycle event once the dormancy is broken. This dormancy state is often called the **G0** phase or **G-zero** phase. In the senescing cells and in the cells with terminal differentiation (such as tracheary elements and sclerenchyma cells) PCD may ensue (Fig. 24) with appropriate changes in the cell cycle events (Fig. 25).

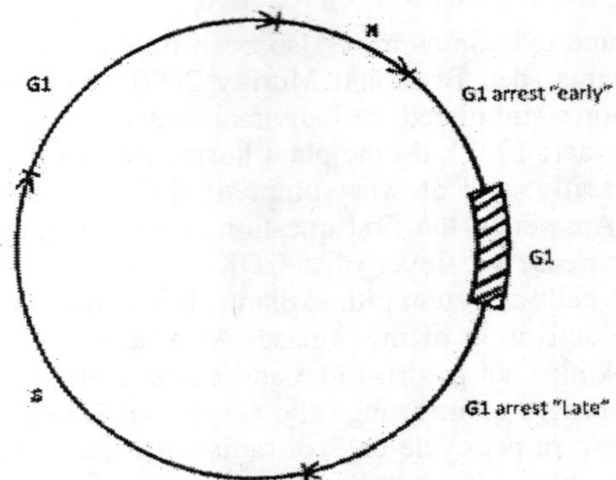

Fig. 24. Cell-cycle model for tracheary elements differentiation as suggested by Dodds (1981a); G1 - Gap I; G2 - Gap II; M - Mitosis; S - Synthetic phase; Hatched lines represent the critical event for initiation of tracheary elements

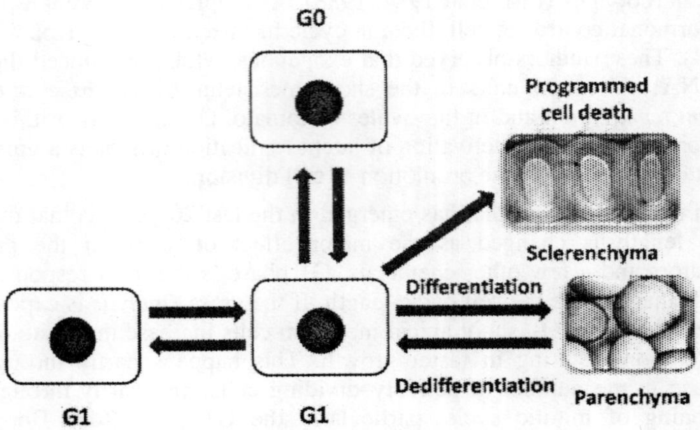

Fig. 25. Options for G1 cells in plants

3.3.3. Modifications of Cell Cycle

Some plant cells show only DNA replication and gap phases but no mitosis. This process is called endoreduplication (D'Amato 1998, Larkins et al 2001) (Fig. 26). Such a DNA replication cycle occurs within the nuclear envelope (i.e., the nuclear membrane is not lost as in the normal mitosis) and without spindle formation. The newly–formed strands of DNA remain in the same nucleus. This type of DNA replication cycle is called an **endocycle** (Nagl 1978, 1981). The single nucleus undergoing the endocycle then becomes polyploidal, a process referred to as **endopolyploidy** or **endoploidy**. In some endocycles, structural changes occur that are similar to those seen in mitosis, and the replicated DNA strands become separate chromosomes. In this case, the cycle is known as an **endomitotic cycle**. During this process, **polytene** chromosomes are formed and these chromosomes contain numerous strands of DNA lying side by side. Some people interpret endocycles as exceptional phenomena and without any functional significance. But, according to others, growth involving endocycles has important advantages as it provides a mechanism to increase the extent of gene expression (Nagl 1981, Larkins et al 2001). Unlike in mitotically–dividing cells, RNA and protein synthesis are not interrupted during an endocycle in these cells so as to enable them to grow in size faster (hypertrophy) and to rapidly assume maturity, and often high physiological activity. Since there is a positive correlation between cell volume and degree of endopolyploidy, it has been suggested that polyploid nuclei may be required for the formation of large cells in plants. It is possible that the reverse may also be true (Kondorosi et al 2000, Edgar and Orr-Weaver 2001). The absolute necessity of endoreduplication of DNA as a prerequisite for cytodifferentiation, at least in some root cells, is an open question (Rost 1994). Apparently, it is a required step in the differentiation of several cell types including tracheary initials, although others consider it as simply a step in a differentiation sequence and not as a prerequisite (Dyer 1976).

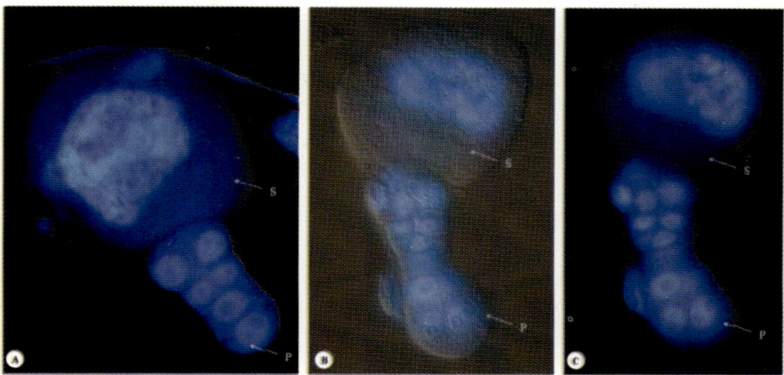

Fig. 26. Endoreduplication in the embryo suspensor cell of *Gagea* (Stained with DAPI and observed under fluorescence microscope) S - Basal suspensor cell; P- Proembryo (Photographs courtesy of Prof. Malgorzata Kozieradzka-Kiszkurno)

3.4. MITOTIC CYCLE AND INDEX

A very interesting aspect of plant development is that almost from the beginning, cell divisions are predominantly restricted to certain regions of the plant body, called meristems. Even within the initial zygotic derivatives, cell divisions are not seen in all cells. For example, the basal cell of the two–celled proembryo becomes the suspensor, without undergoing any division or divides only a few times to become a multicellular suspensor. It is also to be understood that even within the dividing population of cells such as in meristems, there is a variation in the **mitotic index** as well as in the duration of the **mitotic cycle** between the different cells. By mitotic index, we mean the percentage of cells actually undergoing division among the total population of cells at any time of the observation. On the contrary, mitotic cycle refers to the actual time taken for a cell to complete its cell division cycle. Mitotic index and mitotic cycle are very important control events in plant growth and development.

3.5. LAWS OF CELL DIVISION

Ever since **Cell Theory** was proposed by Schleiden and Schwann in the second half of the 19th century, plant anatomists started to emphasize the importance of cell divisions in plant growth and development. Scientists such as Hofmeister, Sachs, Errera and D'Arcy Thompson, have established or discussed a series of empirical rules to describe the broad behaviour of dividing plant cells. Hofmeister (1863) noticed that if a plant tissue grows in different directions, cell divisions in general tend to be perpendicular to the direction that was previously the principal direction of the fastest growth (e.g., the predominantly transverse divisions in the vertical files of rib meristem). Sachs law often called the **rule of rectangular section**, states that in all the tissues, a new cell wall meets the side wall of the parent cell at a right angle (Sachs 1878). In fact, Sachs' law was an extension of Schwendener's hypothesis of 1860 that the final result of cell division in a tissue mass consists in the cell walls being aligned in a system of orthogonal trajectories. Hofmeister and Sachs laws were mainly concerned with the positions of new walls (after division) from a biological standpoint and rather less with their dependence on physical factors. However, in 1886, Berthhold adopted physical aspects such as the principle of minimal areas, comparing the forms of cells and the disposition of partition walls with those which would be assumed by a system of weightless films under the influence of surface tension. Independently, in the same year, Errera (1888) definitely ascribed the properties of a semi–liquid film to the newly–formed cell wall and deduced that it must be subjected to ordinary physical laws and must accordingly assume the form which would be got under the same conditions by a weightless liquid film. In view of the above, Errera's law states that new cell walls follow the shortest path that will divide the parent cell, "as if the nascent wall transiently possessed the surface minimization properties of a fluid" (Dupuy et al 2010). Another basic rule, which we owe to Errera is that "the incipient partition wall of a dividing cell tends to be such that its area is the least possible by which the given space–content can be enclosed". All these laws clearly imply that many of the properties of dividing plant cells might have a physical underpinning, and this idea was emphasized by D'Arcy Thompson (1942) in his book "On Growth and Form". These laws or empirical rules, which govern cell wall placement in plant system (Smith 2001), were deduced from observations of the final arrangement of cells, after growth, in fixed tissues.

D'Arcy Thompson (1942), in considerable detail, dealt with surface tension as a factor determining the division pattern in a tissue mass such as in a developing embryo. According to him, the living cell is a complex field of energy and the energy is distributed internally through the cytoplasm and externally over the surface and any change in the ratio of these two energy components can disturb the equilibrium of the system. Cell division, by keeping constant the ratio of surface to mass, will thus tend to maintain equilibrium. Accordingly, there will be a tendency towards the maximum reduction in the extent of the surfaces created by cell division; cells will tend to divide by walls of minimal area.

Although, for some time, Errera's rules on the cell division were vigorously opposed by a number of biologists, they have steadily gained ground and have started to receive strong support, especially from embryologists working on embryogenesis who believed that segmentations during embryogenesis are purely physical in nature and referable to molecular physics (Krishnamurthy 1994). Errera himself pointed out that his principle could have greater importance in that it enables us to understand something of the inception of form and to relate it to molecular physics. Errera's law has, unfortunately received much less attention from botanists than it actually deserved. Illustrations of Errera's law are encountered everywhere in embryogenesis and meristem behaviours. Critics of Errera's law mainly pointed out at the inoperability of the law in the division of the fusiform initials of vascular cambium, which does not divide by walls of minimal area. But, as Wardlaw (1968) had pointed out, we have to accept Errera's law in its general aspect, recognizing that in different biological circumstances other factors will inevitably become incident and may affect the pattern of wall formation to a greater or less extent; then the value of the law, far from being diminished by seeing exceptions, is enhanced by them.

3.6. PLANES OF CELL DIVISION

The planes of cell division i.e., the orientations of cell division are also important in growth and development. According to Dupuy et al (2010), the plane of cell division is correlated with cell shape. The cells that are longer in the radial dimension undergo anticlinal divisions and the cells that are longer in the tangential direction undergo periclinal divisions. Empirical relationships between cell shape, mechanical force, and direction of growth for the determination of the axis of cell division have been described by early biologists (Théry and Bornens 2006), but the processes that control the positioning of nascent walls remain poorly understood (Lloyd and Buschmann 2007).

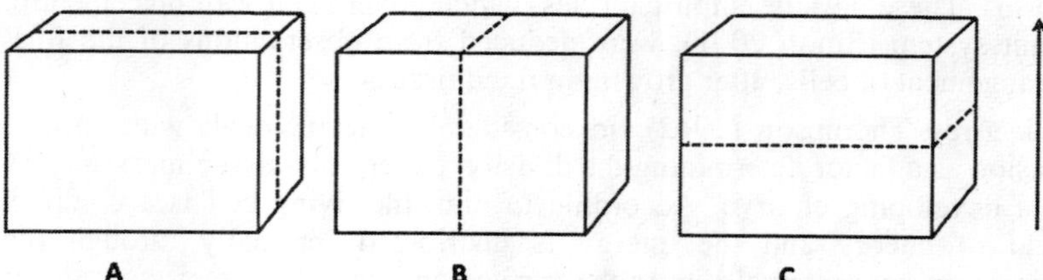

Fig. 27. Diagrams illustrating planes of cell division; A - Periclinal division; B - Radial anticlinal division; C - Transverse division. Division wall is shown by broken lines, while the arrow indicates the direction towards the periphery in A and B, in C, it indicates the longitudinal axis of the plant organ in whose cells these divisions take place.

There are principally four categories of cell division based on planes: **anticlinal, periclinal, transverse** and **diffuse**. Plane of cell division plays a very important role in deciding the further fate of the resultant cells.

3.6.1. Anticlinal Division

If the cell divides by a wall laid at right angles to the surface of the concerned plant part or organ, the division is known as anticlinal (Fig. 27 B). If the anticlinal division occurs parallel to the radius of a cylindrical organ like a stem or root, it is referred to as a **radial, radial anticlinal** or **radial longitudinal division**. Anticlinal divisions are very important in maintaining the integrity and the specific identity of the surface layer of vegetative and floral organs (**tunica** in the shoot apical meristem and epidermis in stem, leaf, floral organs and roots) as well as the specific identity of the fusiform initials of the vascular cambium. Special attention must be paid to the anticlinal (radial) divisions of the fusiform initials. It is known that the vascular cambium forms a continuous circular sheath all around the xylem and cuts off secondary xylem (wood) towards inside and secondary phloem (bast) towards outside. As more and more wood is cut off on the inside, the cambium ring (sheath) has to keep pace with it by expanding more and more in circumference, and in that process has to increase its number of fusiform and ray initials substantially. This is, in essence, effected by the anticlinal divisions of the fusiform initials. In **storeyed cambium**, the fusiform initials divide by radial longitudinal walls (Fig. 28 A). This results in grouping of the initials in horizontal rows, provided that differential elongation of the daughter cells does not occur; if differential elongation happens the initials will crowd past one another and the stratified arrangement will be lost. It is equally clear that if the irregular (i.e., overlapping) arrangement of fusiform initials of **non–storeyed cambia** is due to the differential elongation of the products of radial longitudinal (i.e., anticlinal) divisions, the average length of the initials will increase considerably with each division. This does not occur. Cell multiplication in these non–storeyed cambia is by more or less horizontal (**pseudotransverse**) dividing walls, with a subsequent increase in length of the daughter cells (Fig. 28 B, Fig. 29 A and B). In pseudotransverse type of anticlinal division, the dividing wall is usually laid down near the centre of the cell (less commonly it may be nearer to one end). The length and pitch of the wall are usually very variable, ranging from short and transverse to very oblique and up to half the length of the dividing cell in conifers and considerably more in dicotyledons. In conifers the length of the wall is related to the length of the dividing cell, with long cells tending to have long partitions and vice versa. Again in this group of plants, the direction of slope of the dividing walls tends to be the same in neighbouring fusiform initials, although exceptions occur. After varying lengths of time, reversals in tilt take place. Continued anticlinal divisions in one direction leads to the development of **spiral grain** in the wood and this process is accelerated by a high frequency of anticlinal division. The term **domain** is often used to designate a sector of the vascular cambium where the anticlinal partitions are tilted in one direction. Domains differ in size and the domain pattern changes with time. However, little is known about the direction of slope of the partition in dicotyledons.

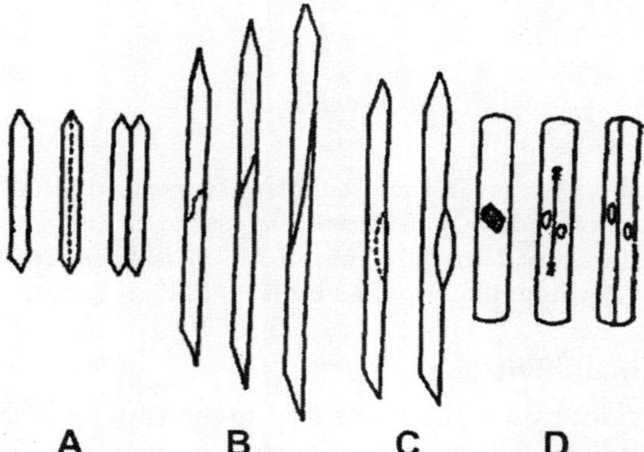

Fig. 28. Types of cell divisions in fusiform initials of vascular cambium; A–C - Types of anticlinal divisions (A - Radial division in stratified cambium; B - Pseudotransverse division in non–stratified cambium; C - Lateral division in non–stratified cambium); D - Periclinal division as seen in radial longitudinal sections

In most cambia, anticlinal divisions occur, which are neither pseudotransverse nor radial longitudinal. The dividing wall is laid down longitudinally and to one side of the fusiform initial. It does not extend the full length of the cell and results in a small cell being cut off to the side of the fusiform initial (Fig.28 C, 29 C). Such divisions, which intercept only one lateral wall of the fusiform initial, have initially been termed **(longitudinal) divisions off the side**, but later they were called **lateral divisions**. Such divisions are a common method of ray initial production in conifers and dicots. The frequency of lateral divisions has to be maintained in each and every species as it is very vital to keep the ratio of fusiform to ray initials constant in an enlarging cambial cylinder.

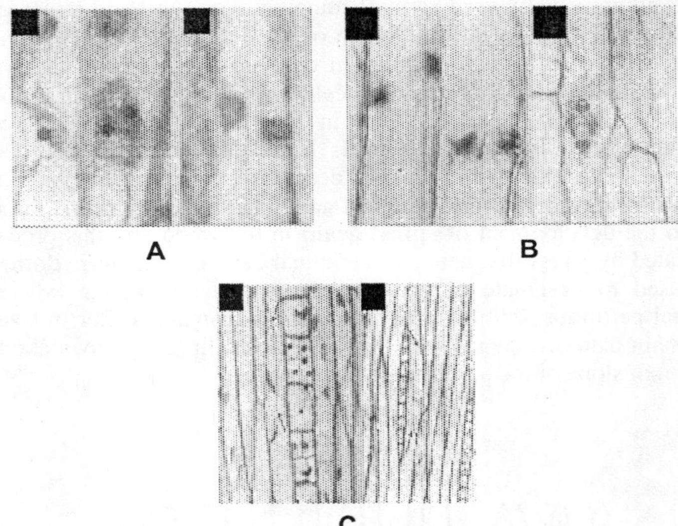

Fig. 29. A and B - Stages in the pseudotransverse anticlinal division of fusiform initial respectively in *Terminalia paniculata* and *Tectona grandis*; C - lateral anticlinal division in *Albizia lebbeck* (arrows) (Photographs supplied by Dr. N. Venugopal)

3.6.2. Periclinal Division

If a cell divides by a wall laid parallel to the surface of the concerned plant part or organ, the division is known as periclinal (Fig. 27A). In cylindrical plant organs, the term **tangential division** or **tangential longitudinal division** is used commonly instead of periclinal division. Periclinal divisions in the flank meristem (other than in tunica) of the shoot apex initiate the formation of leaf primordium, help to differentiate the **mantle** and **core** regions in the floral meristem, distinguish the **parietal**

tissue from the **archesporium** in the young anther and ovule primordia, help in the production of all vertically–aligned cell types of wood and phloem from fusiform initials of the vascular cambium and form the cork cells (**phellem**) and secondary cortical cells (**phelloderm**) from the **cork cambium** (**phellogen**). Special mention must be made of the periclinal division in fusiform initials (Fig. 28 D). Here, the process of cell plate formation is greatly extended in time, since the plate is laid down along the length of a very elongated cell. The nucleus, which is located near the centre of the cell, divides by mitosis and the spindle extends laterally by the addition of peripheral fibers, gradually assuming the form of a disc. The original "connecting fibers" disappear from the cell plate as new **kinoplasmic fibers** (*sensu* Bailey, 1920) are successively added to the growing cell plate. This ring increases in circumference until it intersects the radial walls, when it becomes more or less flattened on two sides. As soon as the cell plate intersects the radial walls of the cell, the fibers in these two sides disappear, leaving two separate aggregations of kinoplasmic fibers, which are connected by the cell plate. These aggregations of fibres, or **kinoplasmasomes** (*sensu* Bailey, 1920), extend across the cell, more or less at right angles to its long axes from one radial wall to the other, and are located in the centre of the protoplast, midway between its tangential surfaces. They move in opposite directions towards the ends of the cell, and as they do so the cell plate is extended until it reaches the ends of the protoplast, thus dividing it into two halves, each containing one daughter cell (Fig. 28 D). Following periclinal division, one of the daughter cells remains as the initial and it expands radially before dividing again. The other daughter cell becomes a xylem or phloem mother cell depending on the side at which it is located. It expands radially and may or may not divide again in the same manner before differentiating into a vascular element.

3.6.3. Transverse Division

When a new cell wall of a dividing cell is laid down at right angles to the longitudinal axis of a cylindrical organ, the division is referred to as **transverse** (Fig. 27C). Transverse divisions are important in megaspore mother cell, zygote, rib meristem of the shoot apex and young internodes etc.

3.6.4. Diffuse Division

In some regions of the meristem, cells may undergo divisions in diverse planes to result in diffuse growth. Such divisions are called **diffuse divisions**. They are vital for the development of the pith and the cortex of stem and root, mesophyll in leaves and the development of pericarp, especially of fleshy fruits. Diffuse growth is important in volume increase.

3.7. ASYMMETRIC CELL DIVISIONS

Another important aspect of cell division is its symmetry. Cell division may be symmetrical and equal so that the two resultant daughter cells are almost equal in size, at least immediately after the division is over, or asymmetrical and unequal so that the two resultant daughter cells are of different sizes from the beginning. **Asymmetric divisions** are very important in plant morphogenesis, as the two resultant daughter cells proceed invariably with totally different pathways of differentiation (Gallagher and Smith 1997, 2000). Asymmetric cell divisions are very important in root hair formation in many taxa, stomatal initiation, companion cell development in phloem, formation of vegetative and generative cells in pollen grains and in the formation of basal and apical cells after zygotic division.

Asymmetric cell divisions are very strong reflections of the existence of strong polarity (see Chapter 1) in the parent cell before division (Stebbins and Jain 1960). This relationship between polarity and asymmetric division is very elegantly demonstrated by observational and experimental data. The stability of polarity has been clearly demonstrated in an experiment with centrifugation of differentiating spores (Bassel and Miller 1982). The normal division of the spore in the fern *Onoclea sensibilis* is preceded by migration of the nucleus from the centre of the spore to one end of it. A highly asymmetric division follows and the longer cell forms the protonema, while the smaller cell forms the rhizoidal cell. Centrifugation of the spore does not change this pattern of division, although the cytoplasm or nucleus becomes displaced or stratified. Only when centrifugation is done just before or during meiosis or cytokinesis, the asymmetric division is blocked.

3.7.1. Zygotic Division

One of the earliest systems in which asymmetric cell division has been extensively investigated is the zygotes of the marine algae, *Fucus* and *Pelvetia* (Fowler and Quatrano 1997). As briefly stated in Chapter 1, the zygote is formed after a sperm enters and fuses with the egg. The zygote initially is spherical and apolar. If these zygotes are subjected to unilateral light (illumination) during the first 10 hr of development, a localized rhizoidal tip emerges from the shaded end and demonstrates the first expression of polarity and preparation for asymmetric division. Light signals establish a **cortical site** to which the secretory apparatus is redirected from its initial symmetrical distribution to an asymmetrical distribution (Fig.8). The cortical site includes dihydropyridine (DHP) receptors, membrane domains that direct ion flow inward, higher levels of free Ca^{2+} and a patch of actin microfilaments (MF). Similar asymmetrical cortical assemblages have been characterized in yeast (bud sites) during cell morphogenesis, but their specific components are different. The division of the zygote, and all subsequent rhizoided cell divisions, although a fewer times only, in *Fucus,* are asymmetric. The asymmetric cell division preserves the asymmetry of the zygote cell wall, resulting in a two–celled embryo in which each daughter cell has a distinct type of cell wall. The rhizoidal cell wall contains the sulphated fucoidan and its binding proteins, while the thallus cell wall does not. Hence, cell wall asymmetries are established in the zygote and are partitioned into the two–celled embryo by a properly–oriented cell division. Asymmetries fixed in the cell wall signal positional information to the cytoplasm that is crucial for subsequent development (Fowler and Quatrano 1997).

The division of the angiosperm zygote is asymmetric and results in a larger, micropylar, **basal cell** and a smaller, chalazal, **apical cell**. These two cells properly organize the suspensor and the embryo respectively and thus, have altogether different fates. This asymmetrical division is foreshadowed by strong polarity exhibited by the zygote. Zygotic asymmetric division is strongly under genetic control. For instance, *gnom* and *emb 30* mutants of *Arabidopsis* have defective zygotic division, which appears near–symmetric. (Mayer et al 1993).

3.7.2. Formation of Root Hair

In some taxa, the asymmetric division gains great significance in the formation of root hair. Root hairs are the projections on the rhizodermal (root epidermis) cells that occur in the zone of differentiation and project out tortuously into the soil. They help in water absorption. In a number of plants, root hair formation follows a well–documented pattern of an early asymmetrical division of the root dermal cell giving size to a small specialized cell or **trichoblast** (Fig. 30). This cell irreversibly develops into a root hair. The nucleus of the trichoblast mother cell moves towards the pole where the trichoblast will be cut off (Sinnott 1960). Studies have indicated that the root hair mother cell specification is initiated only at a stage of the dermal layer before the emergence of the root hair. Following laser ablation of hairless cells of the dermal layer of *Arabidopsis* root, root hairs are not formed by these cells emphasizing the importance of the asymmetrically derived cells in developing into root hairs. The expression pattern of the *GL2* promoter-β-glucuronidase reporter gene fusion construct in the root meristematic region of the seedlings of *Arabidopsis* has been studied. The *GL2* promoter directs the expression preferentially in future ordinary epidermal cells (atrichoblasts) located over the cortical cells (Schiefelbein et al 1997), while not in the prospective root hair initials (trichoblasts) are located over the junction walls of cortical walls. (Fig. 31)

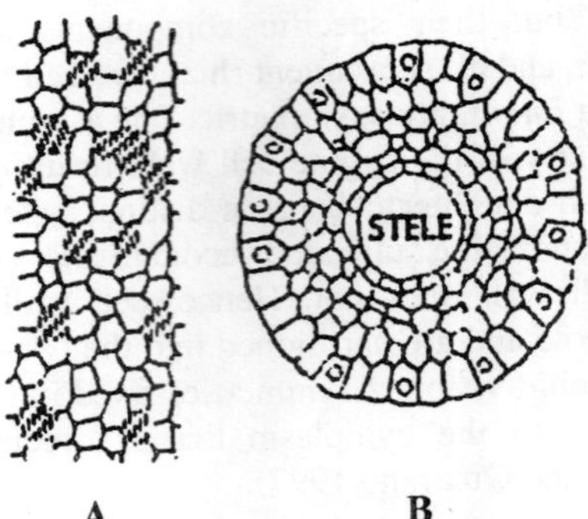

Fig. 30. Paradermal (A) and transverse sections respectively of very young roots of *Hydrocharis morsus-rannae* showing trichoblasts (hatched in A and nucleated in B). (Modified from Cutter, 1969)

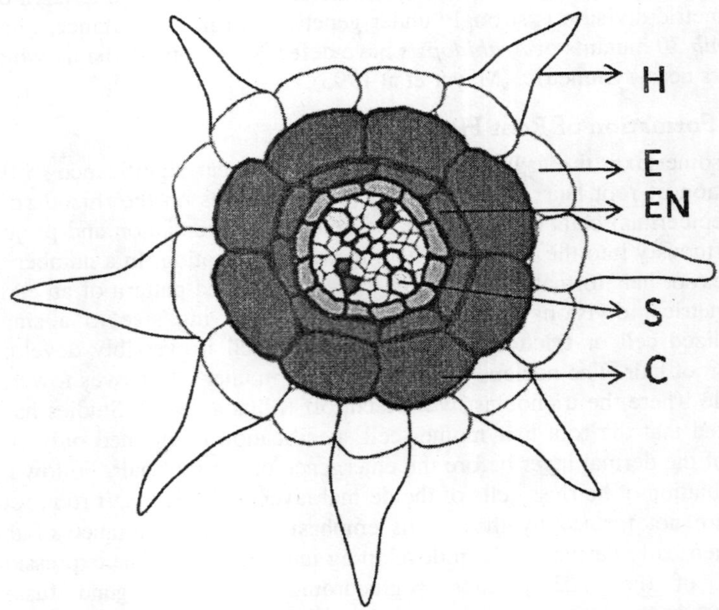

Fig. 31. Diagrammatic transverse section of a root of *Arabidopsis* showing the distribution of different tissue types. The cells of epidermis (E) in between two cortical cells (C) elongate as root hairs (H), while those (atrichoblast) located over cortical cells show the preferential expression of the *GL2* Promoter-β-glucuronidase reporter gene fusion construct (shown with a black dot). The stele (S) is bound externally by pericycle and endodermis (EN).

Cytological and histochemical studies have shown that the developing trichoblasts of *Phleum pratense and Hydrocharis morsus–ranae* have greater activities of one or more enzymes, such as acid phosphatase, peroxidase, succinic dehydrogenase and cytochrome oxidase, than its sister derivative as well as other **atrichoblasts** (Cutter and Feldman 1970 a, Avers and Grimm 1959). Trichoblasts contain much more protein, RNA and organelle–rich cytoplasm. Outgrowths of root hairs in *Hydrocharis morsus-ranae* and *Elodea canadensis* are also associated with a certain degree of endoreduplication of their nuclei (Cutter and Feldman 1970 b).

3.7.3. Formation of Stomata

Stomatal development is a well–studied model for asymmetric divisions. A stoma is initiated by the asymmetric division of an epidermal cell or **meristemoid mother cell (MMC)**, the smaller daughter cell of which is the stomatal initial, while the larger cell continues to function as a regular epidermal cell. The smaller cell is often referred to as a **meristemoid** (Fig. 32); it may directly generate two guard cells by a symmetric division or it may undergo additional divisions before giving rise to guard cells depending on the plant (Sylvester et al 1996). In some taxa, there are **subsidiary cells** around the guard cells, which are derived from **subsidiary cell mother cells** (SMCs) often through asymmetric divisions. It is suggested that a conserved mechanism involving an asymmetric division, perhaps to segregate cell fate determinants, is operating in the initial steps of stomatal ontogeny. It is not clear as to how the process that leads to the placement of the division site altered in these asymmetric divisions, when compared to symmetric. Some indications towards providing answers to this question are shown by studies in *Tradescantia* SMCs (Croxdale et al 1992). These studies have found several differences between asymmetric and symmetric division that may be important (Cleary 1995, Kennard and Cleary 1997). The first sign of asymmetry in the vacuolated SMC is cytoplasmic aggregation at a cortical site nearer to the **guard cell mother cell (GMC)** and this is followed by a migration of the nucleus towards this site. Then microfilaments (MFs) also accumulate at this cortical site between the nucleus of the SMC and the GMC, and simultaneously in the adjacent cortex of the GMC. The nucleus is connected to the newly–accumulated cortical actin by actin filaments. MFs but not MTs are necessary for normal nuclear migration. The polar accumulation of actin persists throughout the mitosis. The SMC mitosis takes place with asymmetrically–shifted MT and MF preprophase bands appearing at an SMC cortex site that predicts the future division plane.

A similar series of events occur in the asymmetric divisions that produce the GMC, suggesting that these events may be general features of plant asymmetric divisions. The signals that initiate these asymmetries, though clearly associated with the GMC, largely remain elusive. It is suggested that as the pressure application to SMC walls leads to cytoplasmic and nuclear migration to the locus of pressure, stretch–activated ion channels and ion fluxes may serve as intermediates in this signalling event. It is further suggested that an extracellular cue originating from the GMC is interpreted by the SMC to orient both nuclear migration and the local accumulation of cortical actin. Either of these asymmetric events can serve to reinforce the GMC–originating spatial cue, thus establishing a more permanent polarity, which might in turn influence the actin cytoskeleton of the adjacent GMC extracellularly, and the SMC cytokinetic apparatus intracellularly (Fowler and Quatrano 1997).

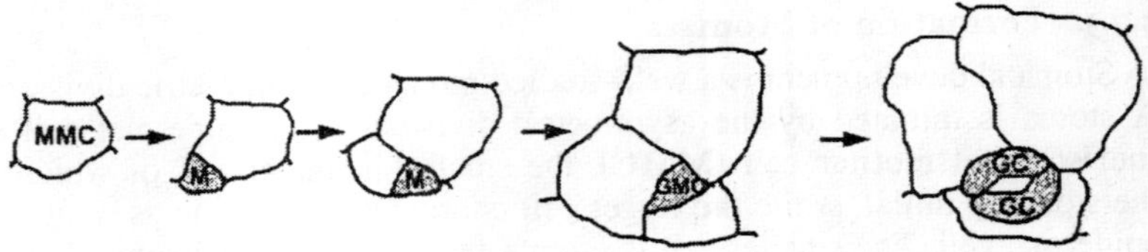

Fig. 32. Stomatal development from an epidermal meristemoid mother cell (MMC) from Meristemoid (M); GMC - Guard mother cell; GC - Guard cell (Source:https://www.york.ac.uk/biology/staff/smb/%28c%29%20Stomatal%20patterning.htm)

3.7.4. Pollen Formation

The cell division in the haploid microspore to form pollen is another very interesting case of asymmetric division. The first mitotic division of microspore, known as pollen mitosis I (PMI), is asymmetric due to the polarity of mitotic spindle. It produces cells unequal both in size and subsequent fate (Fig. 33). Immunoflurescence studies in the orchid *Phalaenopsis equentris* indicate that while the bulk of the MTs appear in radial arrays surrounding the centrally–positioned microspore nucleus, their concentration at the distal surface of the microspore between the plasma membrane and the nuclear envelope is the first indication of nuclear polarity. This distal MT system probably makes the path of nuclear migration to this region just before mitosis. During pollen mitosis I, this MT system disappears and its locus marks the pole of the mitotic spindles. Though actin distribution pattern is unchanged during premitotic stage, its filaments get themselves co–aligned with spindle microtubules during pollen mitosis. These actin MFs are believed to maintain the acentric position of the mitotic spindle at anaphase (Brown and Lemmon 1992). The larger daughter cell produced by this asymmetric division is the **vegetative cell (VC)** and the smaller is the **generative cell (GC).** The vegetative and generative cells have distinct morphological, cytological and gene expression patterns. In tobacco, the *lat 52* promoter is sufficient to direct VC–specific transcription of a nuclear–targeted β–glucuronidase (GUS) fusion, providing a useful marker for VC identity (Twell 1992).

Thus, the asymmetric PMI is necessary to establish the GC fate and a MT structure is involved in anchoring asymmetric factors influencing the division and GC fate. Colchicine disrupts asymmetric division and MT, and hence disrupts the asymmetric division. In two orchid species, specialized MT structures have been observed in the generative pole cytoplasm extending toward the cortical region that the GC will occupy (Brown and Lemmon 1991, 1992).

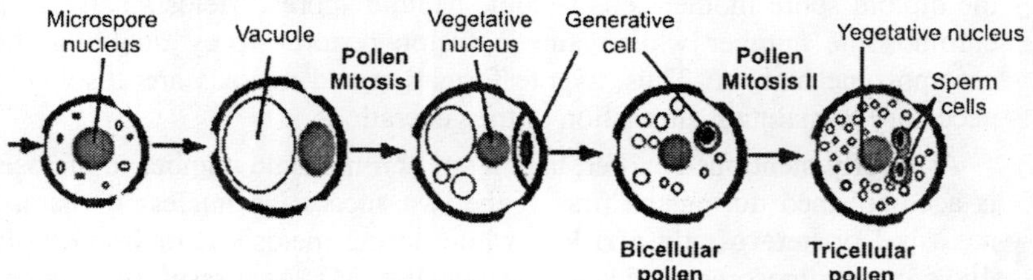

Fig. 33. Male gametophyte development in angiosperms involving asymmetric division of the microspore (Source: http://www2.le.ac.uk/departments/biology/people/twell/lab/pollenis/development)

3.7.5. Endodermis Formation

In the *Arabidopsis* root meristem, asymmetric divisions occur in the endodermis initial cells to give rise to an inner ring of smaller daughter cells that form the endodermis and a ring of larger cells forming the inner cortical layer (Fig. 31) (Scheres et al 1994). This is supported by histological and clonal evidence (Dolan et al 1993). Mutant for the *SCARE CROW* gene has only one layer instead of the normal two layers (cortical and endodermal) present in the wild type. This single layer exhibits cell–type markers of both cell layers (Di Laurenzio et al 1996). Thus, the gene controls specifically asymmetric division responsible for endodermis formation (Scheres et al 1995). The *Arabidopsis short–root (shr)* mutation also does not show the asymmetric division that causes endodermal layer formation, but this mutation produces a single layer expressing only the cortex cell fate (Benfey et al 1993).

3.8. MEIOTIC DIVISION

The life cycle of most plants, particularly of all embryophytes (which include bryophytes, pteridophytes, gymnosperms and angiosperms), is characterized by an alternation of a haploid gamete–producing **gametophytic generation** and a diploid spore–producing **sporophytic generation**. In bryophytes, the gametophytic generation is the free–living, dominant generation, while the sporophytic generation is a much reduced one, largely depending on the gametophyte. In pteridophytes, the sporophyte is a dominant generation with a well–organized plant body, while the gametophyte, although also is independent, is a reduced structure. In the case of gymnosperms and angiosperms, the life cycle is characterized by a dominant sporophytic generation and a highly reduced, fully dependent (on sporophyte), gametophytic generation. Whatever may be the nature of the sporophytic and gametophytic generations, the life cycle is characterized by two cardinal events: (i) fusion of haploid gametes to result in a diploid zygote, and (ii) meiosis during spore formation when the diploid spore mother cells become haploid spores. Meiosis halves the chromosome number while gametic fusion restores it by doubling the chromosome number. Thus, zygote formation and meiosis are absolutely necessary to maintain alternation of the generations.

As briefly mentioned earlier, halving of chromosome number in meiosis is accomplished during the first of the two successive nuclear divisions; meiosis I or **heterotypic division**, while during meiosis II or **homotypic division**, chromosome replication occurs (Fig. 34). As a result of meiosis, four haploid daughter cells are formed. In prophase of meiosis I, five distinct stages, viz. **leptotene, zygotene, pachytene, diplotene** and **diakinesis**, are customarily recognized.

In leptotene, the chromosomes which appear as long, thin threads without any definite structure, duplicate resulting in each chromosome consisting of two chromatids. The next stage, zygotene, is the hallmark of meiosis and distinguishes meiosis from mitosis.

During zygotene, there is pairing of **homologous chromosomes** along their long axes. This pairing is known as **synapsis**, which leads to the formation of pairs of homologous chromosomes called **bivalents**. The formation of **synaptonemal complex** (ribbon–like structure) between homologous chromosomes reflects the pairing process and probably accounts for the physical basis of association between meiotic homologues (Raghavan 2000). For example, the synaptonemal complex of the microspore mother cell of *Lilum longiflorum* is made up of more than 20 proteins, of which at least five are specific for this complex structure.

During pachytene, a shortening and thickening of bivalents takes place, along with a **crossing-over** between homologues of a pair. During crossing–over, bits of the homologous **chromatids** break off and exchange segments in each bivalent, followed by repair of broken free ends to again form complete chromatids, but with recombination of genetic material.

In diplotene, paired chromosome partners begin to separate from one another by splitting along the synaptonemal complex.

During diakinesis, the chromosomes attain their maximum contraction. Diakinesis is followed by mitosis (i.e., homotypic division) in the two haploid daughter nuclei.

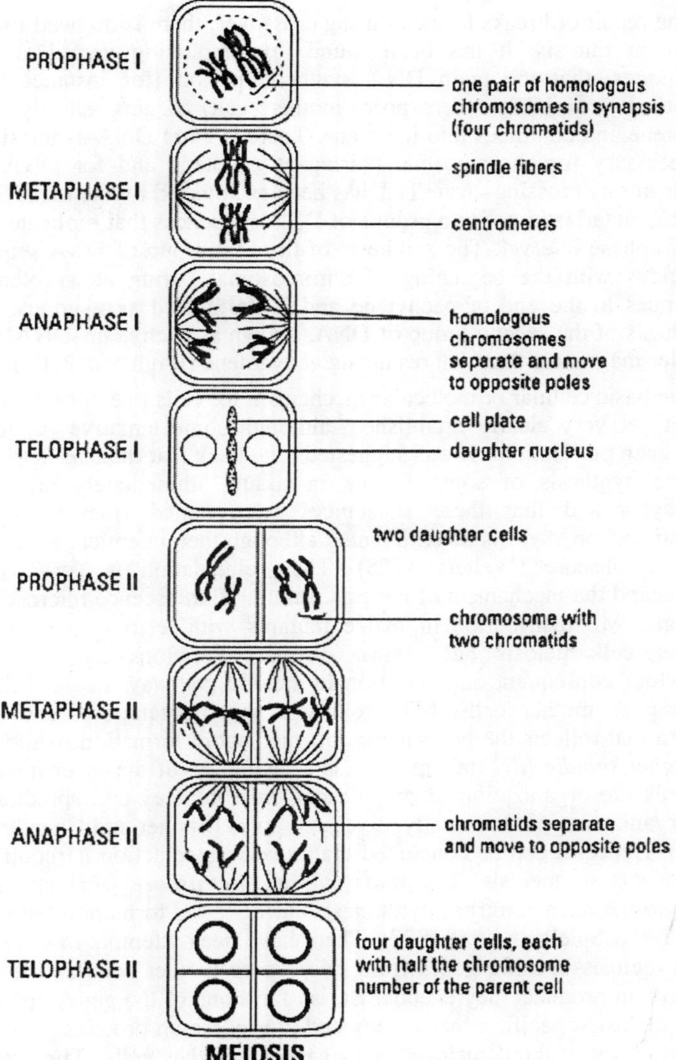

Fig. 34. Stages in meiotic cell division
(Source: www.cliffsnotes.com/study_guide/Sexual-Reproduction-Meiosis.topicArticleId-23791, articleId-23723.html)

There are problems in determining the accurate timing and pattern of DNA synthesis during meiosis and this determination is more complex than in mitosis. This is because the freedom of homologous chromosomes for recombination depends upon an unscheduled round of DNA synthesis for the repair of breaks formed during crossover; there is no need for DNA repair in mitosis. It has been found, by employing varied analytical techniques, that the main DNA synthetic period (for instance, in the meiotically dividing microspore mother cell) begins shortly before leptotene and continues into leptotene. There is extra DNA synthesis, as it is necessary for chromosomal pairing at zygotene and for DNA break repair during crossing–over. This has been established by the identification and characterization of two groups of DNA sequences that replicate during the prophase interval. The synthesis of the first group of DNA sequences coincides with the beginning of chromosomal paring at zygotene and continues to the end of pachytene and is designated as zygotene–DNA. Synthesis of the second group of DNA, known as pachytene–DNA, is of a smaller magnitude than that occurring at zygotene (Raghavan 2000).

The basic cellular or molecular mechanism for cells to enter into meiosis is not yet very clearly established, although some tentative suggestions have been put forward. It was suggested long back that meiosis is triggered by the synthesis of some factors in tissues immediately around the meiocytes and that these substances were called 'meiosis–inducing substances' or 'meiosis determinants', although their chemical identity still remains obscure (Walters 1985). Their elucidation is very vital to understand the mechanism of meiosis. Immunofluorescence microscopy of MT and MF distribution in maize mutants with lesions in microspore mother cell meiosis has shown distinct alterations in cytoskeleton behaviour consequent on aberration in meiotic pathway. In the wild type microspore mother cells, MTs follow a specific temporal and spatial pattern that reflects the polar nature of microspore formation, while in the *divergent spindle (dv)* mutant there is a disruption of the mechanism that controls the organization of cytoplasmic microtubules and spindles with the organization of abnormally–divergent poles (Staiger and Chande 1990, 1991). Hence, it can be concluded that normal gene action is required for key events of meiosis. The *stud (std)* and *tetraspore (tes)* mutants of *Arabidopsis* have perturbed cytokinesis during tetrad formation (Hülskamp et al 1997, Spielman et al 1997). There have been attempts in the past to clone meiosis–specific genes from microspore mother cells and to know the protein products they encode. But so far, none of the genes are shown to be meiosis–specific because they code for common proteins. Very little is known even about meiosis in megaspore mother cells. The effect of meiotic genes in maize has been indirectly inferred from the occurrence of female sterility shown by plants homozygous for meiotic mutations. Some of the meiosis–related mutations in other plants include *ameiotic1 (am1),* which prevents the entry of megaspore mother cell into meiosis; *ameiotic1-prophase1arrest (am1-pra1)*, in which meiosis is blocked in the leptotene to early zygotene stages; *desynoptic*9101 (dsy*9101)*, in which abnormalities in the pairing of homologous chromosomes predominate; *divergent spindle1 (dv1)*, in which spindle orientation in meiosis II is affected; and, *polymitotic (po)*, which disrupts the postmeoitic haploid cell cycle leading to abnormal chromosome divisions, probably unaccompanied by DNA replication. From the picture that has emerged so far, one can say that the molecular basis of meiosis in the plant cells in terms of the action of specific genes that control the onset and exit of meiosis from specific cells is not yet very clear.

In this connection it is pertinent to present the hypothesis proposed by Krishnamurthy as early as in 1976 (Krishnamurthy 1976) that meiotic induction in plants is related to callase investment around the prospective meiocyte in all the groups of plants from algae to angiosperms. The isolation and insulation (as callose is impervious to most macromolecules) of the meiocyte, thus made, is a necessary prerequisite for meiotic induction and continuance. In apomictic and male–sterile taxa, callose is not deposited around meiocytes. Again, isolated meiocytes, where callose wall is removed after callase treatment, fail to undergo meiosis, but undergo only mitosis. Also, many young anthers (in premeiotic stage), whose meiocytes are not yet invested with callose when cultivated *in vitro*, do not embark on meiosis and do not produce microspores.

3.9. CELL ENLARGEMENT

As briefly mentioned in Chapter 1, growth in the plant and in its parts involves not only cell division, but also cell enlargement. Cell enlargement refers to the increase in the size of the cell after its production by cell division. In many instances, before embarking on next division, a cell may enlarge to various extents. Dupuy et al (2010) have stated that a minimum cell volume is required to trigger mitosis and this is consistent with the requirement for a minimum cell volume during progression through the cell cycle. But, as a general rule, during the growth of an organ, cell enlargement becomes dominant over cell division sooner or later, and over time, replaces it entirely.

3.9.1. Types of Cell Enlargement

Korn (1984) recognized three basic types of growing cells in plant tissues:

1. Cells growing in all three dimensions and that too almost equally. In other words, the increase in cell size may be relatively uniform all along its radii so that the cell finally becomes **isodiametric.**

2. Cells whose growth may be more in one plane only, resulting in either length or width increase. Such a growth may be through unipolar or bipolar growth in one plane. Growth in one plane is seen in fibers, trichosclereids, laticifers, tracheary elements, phloem elements, pollen tubes and root hairs.

3. Cells that enlarge only in two dimensions, as in epidermal cells.

In addition to these, we may add one more type, in which growth may be greater in more than two or three directions of a cell to result in armed or stellate parenchyma or irregularly outlined cells (as in the mesophyll cells of *Canna*, pith of *Juncus*, diaphragm cells of many water plants that separate long air chambers or the astrosclereids of plants such as *Nymphaea* or *Gnetum*). In the case of stellate parenchyma or astrosclereids, lateral tension appears to be one of the major factors determining the final shape. Here, the arms are believed to elongate through their entire length (Fig. 56).

3.9.2. Water Uptake and Cell Growth

The mechanic of cell expansion is central to the regulation of cell shape and division (Dupuy et al 2010). But, there has been little focus on the physics of cell growth and expansion within tissues. How do cells grow in volume? Growth in the volume of a cell is mainly caused by an uptake of water. This process necessarily stretches the cell walls as well as the plasma membrane inside the cell wall, and consequently, new cell wall and membrane materials need to be synthesized. The new materials are needed not only to avoid the thinning of cell wall and plasma membrane due to all-round stretching, but also to avoid their rupture, as both these structures cannot stretch beyond a reasonable limit. Thus, growth of the cells results in the increase in surface area of the plasma membrane and cell wall. In most of the cells, this increase in surface area takes place all around the cell, although in cases of pollen tubes, germ tubes, root hairs etc., the wall increases only in the area at the growing tip. The biophysics of the cell growth has been an active area of research for over the last five decades, as much of the current thinking on this aspect of growth goes back to 1965 when Lockhart published his classical paper.

What causes a cell to absorb water and enlarge? It was suggested long back that the cell wall as well as the plasma membrane inside it get extended in small increments gradually through the metabolic activities of the cell, and water enters into the cell wall to fill the void created by this process. However, this earlier suggestion is no longer followed for want of reasonable evidences. The more recent hypothesis suggests that **water pressure** (i.e., **turgor**) drives growth of cells by pushing the plasma membrane and cell wall outward to result in the expansion of the cell. Accordingly, the rate at which water enters into a cell is essentially governed by two factors: (i) the **water potential**[1] gradient, and (ii) the

---

[1] Water potential (Ψ) of any system or part of a system that contains water or could contain water, as being equivalent to the **chemical potential** (μi) of the water in that system or part of that system compared with the chemical potential of pure water at atmospheric pressure and the same temperature (Slatyer and Taylor 1960). It is to be further stated that the water potential of the reference pure water should be considered

permeability of the plasma membrane (as also of the cell wall) to water. Consequently, the rate of cell enlargement is proportional to these two factors. The basic osmotic relations are indicated by the following equation (Salisbury and Ross 2004):

$$\Psi = \Psi s + \Psi p = \Psi s + P \qquad \text{... (1)}$$

where, Ψ = Water potential, Ψs = Osmotic or solute potential, Ψp = Pressure potential, which is a real pressure, and P = turgor pressure. The difference in water potential ($\Delta \Psi$) inside and outside a cell is:

$$\Delta \Psi = (\Psi se + Pe) - (\Psi si + Pi) \qquad \text{...(2)}$$

where, e = External pressure, i = Internal pressure.

If we assume that the pressure on the outside of the cell is negligible and if you include a factor for the relative hydraulic conductivity of the membrane to water (L), then the equation for the rate of cell expansion can be written as :

$$1/V.dV/dt = L(\Delta \Psi) = L(\Psi se - \Psi si - Pi) \qquad \text{...(3)}$$

Where, V is the volume of the cell, dV is an incremental change in the cell volume, and dt is an infinitesimal increment of time. Hence, cell volume, dV/dt, is the rate of increase in cell volume i.e., the rate of cell growth. Growth is a geometric or logarithmic function of time and to express this function, many workers divide dV/dt by the cell volume (i.e., by multiplying by $1/V$ in the above equation) to give a relative growth rate of the cell or r.

Equations (2) and (3) reveal that there are two ways for water potential inside a cell to be more negative than water potential outside the cell, thus making water uptake and, through that, growth of the cell possible. The solutes inside the cell might increase causing osmotic potential inside more negative or might decrease the pressure inside the cell. But because solute concentration inside many growing cells is often kept constant within limitations of measurement, the strategy adopted would be to decrease the turgor pressure. Since we know that pressure in the cell is caused by mechanical resistance of the cell wall (and plasma membrane) for being stretched, any lowering of this resistance would, in that order, relax the cell wall (and plasma membrane), lower the pressure, reduce cell water potential, lead to longer water potential gradient and finally to the movement of water into the cell (Salisbury and Ross 2004).

If a cell enlarges in size due to water uptake, water dilutes the solutes of the cell and thus, lowers the internal solute potential, provided solutes are not absorbed by this cell from the surrounding cells or are newly synthesized in the tissue of which the cell is a part. Such solute

---

as zero. Here, chemical potential refers to the free energy change upon addition of a gram molecular weight of the chemical substance.

accumulation often does follow cell growth, as for example clearly shown in the epidermal cells of sunflower hypocotyl (Salisbury and Ross 2004). The hypocotyl cells of sunflower seedling expand in length 15–fold, but yet the solute concentrations remain almost constant; the increase in solutes closely matched the increase in cell size. Invariably these solutes are sugars (glucose or fructose) and potassium ions. If solute supply or synthesis is absent, cell growth usually stops. Similarly, a mere continuance of water uptake will not promote cell growth beyond a limit. But, at the same time, it should be emphasized that a cell requires water as the driving force for its growth, most probably because it promotes solute absorption. Both water and solute thus absorbed are converted into the protoplast of the cell, which increases its wet/dry mass.

The importance of water to cell growth is also strongly indicated by the noticeable effect of water stress on cell growth, both physical and physiological. The great sensitivity of growth to water stress occurs because the yield threshold (Y) is often very close to the cell turgor.

3.9.3. Intercellular Adjustments during Cell Growth

In a population of cells, the interplay of forces between cells can constrain as well as promote cell expansion. Thus, a very important aspect of differentiating tissues is the unequal increase in size in adjacently–located cells, since such an unequal increase is very much required. In other words, some cells may continue to divide with no or little enlargement; some enlarge slightly, while some enlarge considerably when compared to others. The direction of this enlargement and the type of enlargement may also vary between adjacent cells. For instance, developing fibers increase in length considerably when compared to other xylem and phloem elements, while the parenchyma cells undergo the least elongation. The same holds good for the increase in width of these cells. In such cases, there are considerable **intercellular adjustments**. The formation of intercellular spaces is one such result of intercellular adjustments. Intercellular adjustments in the differentiating tissues involve **coordinated** or **symplastic growth** in some instances and **intrusive** or **interpositional growth** in other instances (Fig. 35). These two types of growth closely involve adjustments in the growth of cell walls of the involved cells. The contiguous growing wall layers, belonging to the two adjacent cells, do not separate, but expand jointly as the two cells expand or they do separate and the growing cell intrudes into the just resulted space (in this case of the two adjacent cells, one grows in size while the other does not). The first is seen in coordinated growth and in this type it is possible that part of the common wall between the two cells may expand and another may not. The second type of growth is common in elongating cambial initials, fibers (Fig. 36), tracheids, laticifers and some sclereids.

Often, the elongating cells grow only at their tips (**apical intrusive growth**), only at one tip or at both tips. The intercellular pectic material through which the elongating cells grow is probably hydrolyzed in advance, facilitating the cells to intrude. Intrusive growth concept has replaced the earlier concept of **gliding** (or **sliding**) growth (Fig. 35 A) according to which a growing cell was supposed to separate from and glide over the wall of adjacent cell. Some people still believe in the existence of gliding growth.

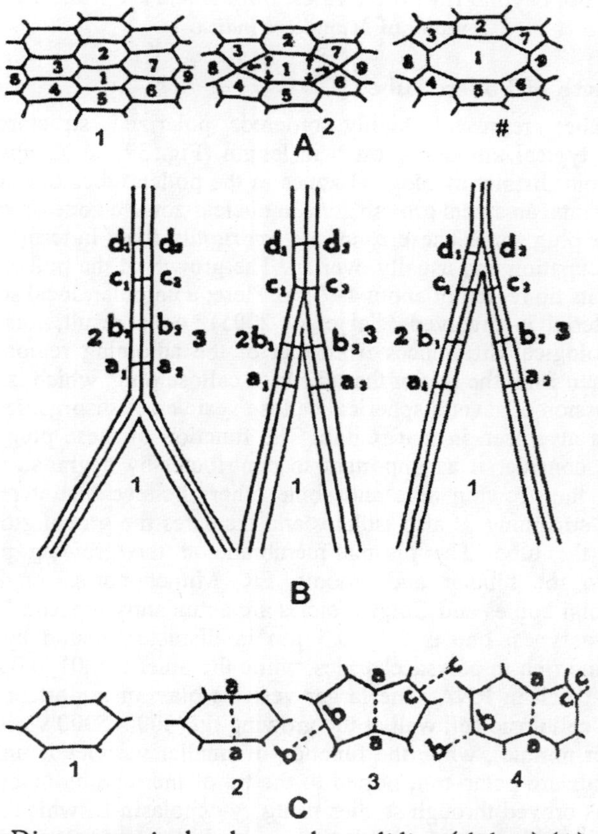

Fig. 35. Diagrammatic sketches to show gliding (sliding) (A), apical intrusive (B) and symplastic growth (C). In A, cell 1, which previously was not in contact with cells 8 and 9, comes to have contacts with them during gliding growth. In B, cell 1 with apical intrusive growth grows through cells 2 and 3 that face each other across a common middle lamella. In C, there is an adjustment of cell shape and cell position after each division (dotted lines) (all modified from Meeuse, 1941)

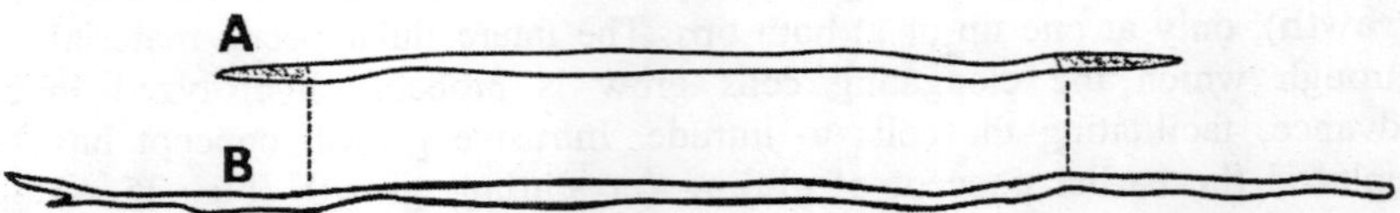

Fig. 36. A - A young fiber before apical elongation; The tip regions that show intrusive growth are stippled; B - A fully grown fiber; The regions of the fiber produced through intrusive growth are indicated as those beyond the dotted lines; Both fibers are from the wood of *Mangifera indica*.

3.9.4. Growth of Pollen Tube

Pollen tubes represent highly ordered, polarized structures and demonstrate typical unipolar growth in length (Fig. 37). A recent model recognizes four distinct cytological zones in the pollen tubes of many, but not all the plants; an apical growth zone, a nuclear zone, a zone of vacuoles and a callose plug zone. These zones are not rigidly fixed in terms of their extent of occupation and usually overlap. The growth of the pollen tube is restricted to its tip region of about 4–7 µm. Here, a unipolar, local secretion of wall material is involved (Cai et al 2005). As a result, this region exhibits cytological differences from that of the adjoining region and is often separated from the rest of the tube by a callose plug, which is formed by the adpression of several spherical callose vesicles. A disorganization of cytoplasm is also seen in that region. The function of these plugs is not clear. Some consider it as important to contribute, by degradation, wall materials to the growing tube and some others believe that it regulates cytoplasmic streaming. It also isolates and insulates the growing tip from the rest of the tube. The plasma membrane at the growing point is connected to the tubular and smooth ER. Mitochondria, amylopasts, secretory Golgi bodies and Golgi vesicles are abundantly present. Vesicles occur in two types; one is 0.1– 0.3 µm in diameter, bound by a unit membrane and rich in polysaccharides, while the other is 0.01–0.05 µm in diameter and rich in RNA. The larger vesicles play an important role in building the cellulosic cell wall at the growing tip (3000–5000 vesicles are produced per minute), while the function of smaller vesicles is unknown. Wall materials are polar–transported to the tip of the tube by cytoplasmic streaming, as proved through studies using cytocholasin B, which inhibits wall deposition, tube growth and the streaming, but not vesicle production. The apical vesicle zone is exempted from the block streaming as otherwise they would be retransferred to the basal part of the tube and would not be available for wall growth at the tube tip. The pollen tube growth is also associated with polar electric currents and by polar distribution of calcium ions, as revealed by proton microprobe analysis, by fluorescence using chlorotetracyclin, which selectively fluoresces calcium ions, and by ^{45}Ca autoradiography. Enzymes such as phosphatases, amylases, dehydrogenases, invertase, oxidases, transferase, pectinase, synthetases, lyases, ligases and lipases are detected in the cytoplasm at the tube tip. Earlier reports of the absence of MTs and MFs in the pollen tube tip cytoplasm are probably erroneous. They are present. MFs are implicated in cytoplasmic streaming and directional movement of cell wall materials and spums. Once the growth of the pollen tube stops, the enzymes disappear implying that they are concerned in some way with pollen tube tip growth. The rate of growth of pollen tube tip is very high and may be up to 600 μm x h^{-1}. The growing tube tip region is translucent as seen under light microscope and is called **cap block**. Such translucence is characteristic of growing tube tips.

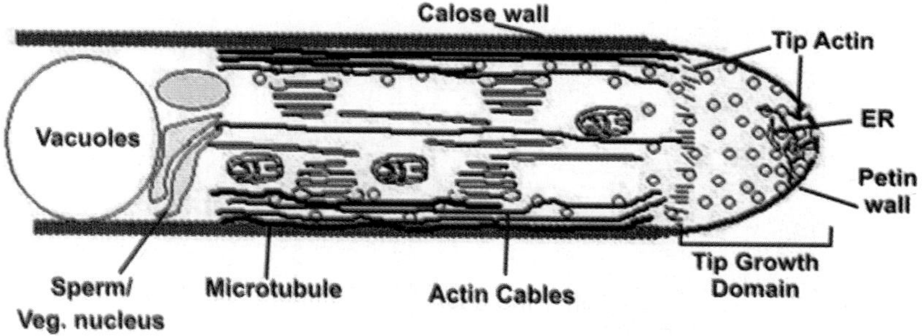

Fig. 37. The growing region of a pollen tube (not to scale)
(Illustration courtesy of Dr. Zhenbiao Yang)

3.9.5. Growth of Root Hair

The root hairs vary from 80–1500 μm in length and 5–20 μm in diameter and are unicellular and unbranched. Similar to the pollen tube, the growth of a root hair is also restricted to its tip. Its tip also displays a polarized organization of cell contents; especially it is enriched with secretory vesicles derived from Golgi bodies. These vesicles carry cell wall precursors. Ca^{2+} influx at its tip probably regulates the secretory process perhaps through its effect on the actin component of the cytoskeleton (Miller et al, 1999; Baluska et al, 2000). Some researchers speak of an **actin cap** (a three–dimensional network of the filaments) at its tip (Fig. 38), while others believe that actin filaments are disorganized and few in number or even absent at the tip.

The cytoplasmic streaming in root hairs (just as pollen tube tip) is described as **reverse fountain streaming**, in which streaming is acropetal along the sides of the hair and basipetal in the central strand. In the subapical part of the root hair, there is an accumulation of a large number of mitochondria, while in the basal region most of the other organelles are found. The nucleus is positioned at some distance from the tip and is controlled by the actin. On completion of growth, the cytoplasmic polarity seen all along in the hair is lost and the position of the nucleus becomes either random or basal. Unlike the pollen tube, no callose is formed to separate the tip part from the basal region of the hair, as it can hinder the absorption and movement of water and sap from the soil to the plant. Schiefelbein et al (1993) have described a mutation in *Arabidopsis* that displays defects in both root hair and pollen tube growth. Compared to wild type pollen tube and root hair, growth of mutant pollen tubes and root hairs is tardy and stymied.

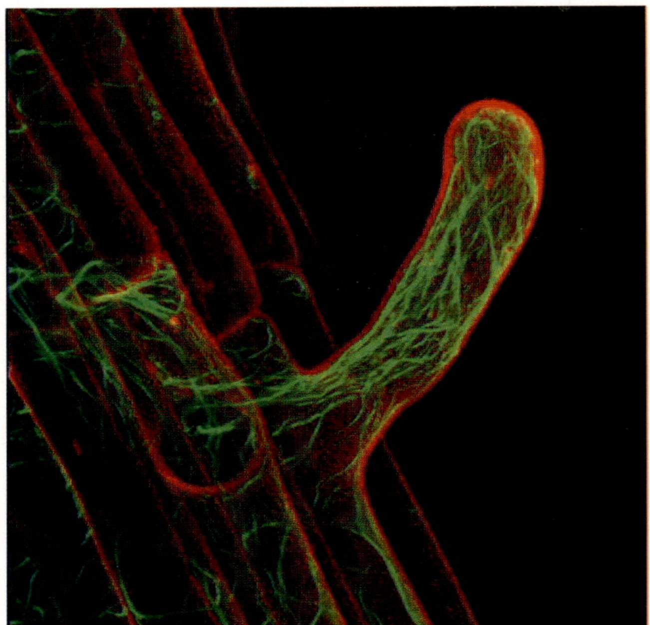

Fig. 38. Root hair of an *Arabidopsis thaliana* seedling expressing GFP fused to the actin binding domain of talin (GFP-mTalin). The root was counterstained with propidium iodide to reveal the outline of the cell walls (red). Image was taken with a Bio-Rad 1024 ES confocal microscope. (Courtesy: The Samuel Roberts Noble Foundation, Inc)

3.9.6. Growth of Seed Hairs

Unlike the pollen tube and root hairs, the elongation growth of cotton seed hairs (Fig. 39) happens via diffuse growth i.e., it occurs throughout the length of the fiber, although it may be more rapid at the tip (Taiz and Zeiger 2002).

A large central vacuole is usually seen in the basal part of the fiber and the organelles appear to be dispersed more or less evenly throughout the cytosol. The primary wall is two layered with an outer layer of pectins and extensins, and an inner layer of xyloglucans and cellulose. Secondary wall is laid down as the tube approaches its final length and this is pure cellulose laid in the helical pattern. Fiber growth stops when the wall rigidity increases and the high turgor of the cell is lost.

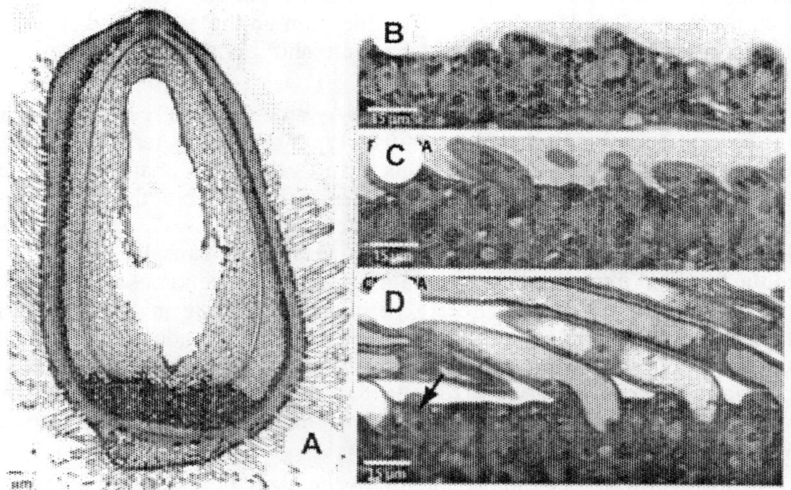

Fig. 39. Cotton hair development; A - L. S. of developing seed showing hairs; B–D - Stages in fibre development (Butterworth et al 2009) (Reproduced with permission from University of Chicago)

3.10. ROLE OF GENOME ON CELL SIZE AND CELL DIVISION RATE

As briefly indicated in Chapter 1, growth in plants involves not only cell division, but also cell enlargement. In this connection, it is to be stated that biochemical, genetic and molecular approaches have dominated the discussions on the role of cell division and cell enlargement in plant growth and development. Biochemical findings have been correlated with the obtained genetic and molecular data to arrive at meaningful conclusions. Still, there is considerable debate over the relative contribution of physical and genetic processes to the coordination of cell division and cell growth (cell size) during plant development (Schopfer 2009). At one extreme, cell division, enlargement and differentiation are all considered by a molecular geneticist to be controlled by DNA and such a control is sufficient to regulate plant growth and development, while at the other extreme, a biophysicist would argue for a purely physical basis for cell division and enlargement to control the growth and development (Green 1980, Lintilhac and Vesecky 1984, Lynch and Lintilhac 1987).

A number of studies have indicated the presence of a positive correlation between the DNA amount (genome size) on the one hand and chromosome volume, centromere volume, length of the synoptonemal complex (SC), chiasmata frequency, nuclear dry mass, nucleolus dry mass, carbon content per cell, epidermal cell size, pollen volume, log seed weight, mass of seeds and mean leaf dimension on the other hand (Bennett 1987, Bennett et al 1981, 1983, Bennett and Leitsch 1997, Rees and Narayan 1988).

This relationship between bulk DNA amount and cell size (and metabolism) is the basis of the **nucleotype theory**, a term coined by Bennett (1971, 1972). He coined this term to describe "that condition of the nucleus (most notably DNA content) that affects the phenotype independently of the informational content of the DNA". On a certain level, he argued, DNA content and cell size must be causally related, due to the physical impossibility of containing very large genomes within small cells. However, so far no direct experimental evidence in favour of this theory is available, but there are several observational evidences. However, the nucleotype may be considered as setting the minimum conditions, or perhaps as exerting a very coarse control of the parameters at the cell level, whereas the genotype is responsible for fine control of these features within these limits.

It has also long been appreciated that the sizes of the nucleus and the cell, and cell division rate are closely linked. For diploid plants, a relationship does exist between the minimum mitotic cell cycle time, the interphase nuclear volume, the DNA content per cell and the size of the cell (Van't Hof and Sparrow 1963). These authors further asserted that this relationship is such that if any one of these variables is known, an estimate can be made for the remaining two. Additional experiments showed a relationship between DNA amount and the duration of the S–phase (Van't Hof 1965). Bennett (1977) clearly showed a strong correlation between genome size and the duration of meiosis in diploid angiosperms. Also, it should be mentioned here that for biophysical reasons, it is impossible to greatly increase the C–value without increasing the minimum time needed for cell division. This is because more DNA not only takes longer time to replicate (i.e., prolonged S–phase), but it also impacts all stages of cell cycle. This delay in cell division is ultimately responsible for generating the positive correlation between DNA content and cell size. At the very least, positive correlation between DNA content beyond a certain point without increasing nucleus and cell size is not possible.

3.11. RELATIVE IMPORTANCE OF CELL DIVISION AND ENLARGEMENT IN GROWTH AND DEVEOPMENT

While discussing the importance of cell division and enlargement in the growth and development, three points need critical consideration. The first point to be considered is what causes a dividing cell to differentiate into a mature cell type. The answer is, if a critical cell cycle factor is missing at the check points, the cell in question will differentiate further without dividing or progressing through S–phase again. This means that at a critical time, the cell differentiation (preceded or closely accompanied by cell enlargement) depends on the absence of a certain cell cycle factor that will allow the cell to recycle. If that factor is missing, the cell will switch to a prescribed maturation path (For more details, refer to Section 3.3 of this chapter). The second point to be discussed is whether cell division is a prerequisite for cell enlargement and maturation events of specific cell types (Aloni 1987) and *vice versa*. In the opinion of Rost (1994), cell enlargement and maturation of the specific cell types do not particularly require cell division in a programming sense. Earlier, Dodds (1981a, b) had expressed a similar opinion. In the stressed cells of intercalary meristems of rice and of wheat leaves, modulations of the cell cycle do not precede any detectable changes in cell growth. These findings provide additional evidence that cell growth in plants is not the only driving force for cell division and *vice versa*. The third and the most important point for discussion is the role of cell division and cell enlargement, independently and together, in deciding the shape, size and architecture of a plant organ and the whole plant. Is cell division informed by cell growth or is cell division the driving force for cell growth? Can cell division be reduced to the surveillance of cell growth? As plant growth is determined by the number of cell divisions and the size of cells (Meyerowitz 1996), some people believe that cell division and cell growth are more likely to be coordinated during plant organ growth. However, the relationship between cell division and cell growth in the context of growth of plant organs is still disputed and far from clear.

It was generally believed for a very long time that the shape of a plant organ is decided by the intensity and pattern of cell division seen in that plant organ during its development (Meijer and Murray 2001). The relationship between cell division and organ growth can be demonstrated by altering the number of cell divisions. Prior to identification of cell–cycle genes and cell–division affecting genes, external means were sought to arrest the cell cycle such as chemical treatment and irradiation. Mutations affecting cell division as well as the expression of cell division (in meristems) affecting genes in transgenic plants have been exploited in recent years to understand the role of cell division in growth and development. Mutated *CLAVATA1* (*CLV1*) and *CLAVATA3* (*CLV3*) genes cause increased cell division and mutated *CURLY LEAF (CLF)* genes convert a determinate floral meristem (with limited cell division) into an indeterminate meristem, with proliferative cell division activity. The dominant *CLV* and *CLF* genes suppress cell division. Over–expression of *AINTEGUMTA (ANT)* increases cell division and hence organ size. From this, it is clear that changes in cell number result in growth–modifications.

The other cell–division affecting genes are *PASTICCINO (PAS)* (which increases cell divisions and often increases the growth of certain tissues) and *ANT* (which reduces the number of cell divisions). Consequently, these genes affect the cell number. Mutations in genes such as *FASS, TONNEAU (TON)* and *TANGLED (TAN)* affect the orientation of cell divisions. Hence, Meijer and Murray (2001) strongly emphasized that cell division should not be regarded simply as a secondary consequence of growth.

Very early studies made on leaf ontogeny demonstrated that changes in the rates and planes, or the patterns of cell division affect leaf development. However, subsequent work provided evidence for the fact that new leaf primordia of irradiated wheat seedlings can be initiated in the absence of cell division (Haber 1962; Foard 1971), although this primordium was limited in size as well as abnormal–looking. Tobacco plants, expressing a dominant negative mutant of *Arabidopsis cdc2a,* had reduced cell numbers (i.e., with reduced cell divisions), but showed no change in plant development (Hemerley et al 1995). In addition, it has been demonstrated that already–existing leaf primordia with down–regulated cell cycle activity (Hemerley et al 1995) and those with a mutation that interferes with correct cell-plate orientation (Smith et al 1996) can develop almost normal leaf shapes. Plant development was also normal in *Arabidopsis* plants over–expressing a mitotic cyclin, despite increased root growth (Doerner *et al* 1996). These observations were used to support the view that during development of plant organs (such as a leaf), shape is acquired independently of the patterns or the intensity of cell division (Kaplan and Hagemann 1991). The expression of *ICK1* results in reduced CDK activity, reduced cell number and inhibition of growth. There was no inhibition of cell expansion, but a strong inhibition of cell division; the former compensating for the latter i.e., inhibition of cell division did not affect plant growth. Apparently, according to these researchers, it is the regulation of cell expansion rather than the pattern of cell division that is responsible for primordium initiation and the final shape and size of the plant and its organs (Reinhardt et al 1998). It was, thus, concluded that cell division, either number or position, is likely to be unimportant for plant organ development.

The importance of cell number in differentiation of the embryo has been discussed (Krishnamurthy 1988, 1994). It was postulated that a progressive increase in the number of cells may be an important factor determining histogenic differentiation in the developing embryo because the various histogenic regions do not appear simultaneously, but differentiate during progressive growth of the embryo (Raghavan 1986b). However, the analysis of literature denies any such relation between cell number and differentiation in embryo. In developing seeds of *Argemone mexicona* subjected to colchicine treatment, cell divisions continue in the globular embryos, resulting in larger globular embryos with many cells at a stage when, in normal seeds, the globular embryos transform themselves into heart–shaped ones even with a smaller number of constituent cells. In other words, the embryonic shoot apex (epiphysis) and cotyledons do not get differentiated in the colchicine–treated ones, in spite of more than sufficient number of cells being produced. In certain citrus varieties, the triploid embryos obtained after cross, do not differentiate a plumule–radicle axis despite the continued mitotic activity in the embryo. In the 'reduced' mutant of tomato, histogenic differentiation in the embryo is either poor or absent, although the embryo often develops without cotyledons and without organized radicle or shoot meristems through continued mitotic divisions (Fig. 40). These examples suggest that the process of differentiation in the developing embryo has no relation to the number of constituent cells at any given stage of embryo development.

The work of Wang et al (2000) showed, for the first time, that expression of a plant cell cycle regulator results in modification of all three processes – cell division, cell enlargement and plant or organ growth. This work further indicated a certain degree of uncoupling between cell size and cell division. According to these authors, if we consider that there exists a normal balance (i.e., coordination) between cell size and cell division in wild–type plants, this balance must have been altered and a new balance is reached here due to ICK1 over–expression. On the other hand, over–expression of cyclins had no significant effect on the cell size. It is, therefore, conceivable, according to these authors, that there is a limit on the capacity of a plant to use cell size increase to compensate for the growth inhibition caused by reduced cell number, and beyond the limit, a reduced cell number inevitably leads to growth inhibition. Thus, cell division and cell enlargement are both vital for growth of plant organs.

Meijer and Murray (2001) have accepted Green's (Green 1976) contention that cell division and expansion are different phases of a continuous process and that both contribute to normal growth and development as well as elaboration of plant structures (Dupuy et al 2010). In both cellular and organismal theories of plant growth and development some degree of coupling between the genetic system and physics of growth is inescapable (Dupuy et al 2010), and this makes it difficult to discriminate between the two theories. The key to distinguish these two theories lies in understanding the precise nature of the coupling between genetic and biophysical processes in living cells.

Observational data have strongly indicated that cell enlargement becomes dominant over cell division sooner or later, and in time, replaces it entirely. Although cell division as such does not significantly contribute to the volume of the growing plant structure, addition of cells is a primary necessity for the growth of multicellular plants. Thus, cell division and cell expansion are different stages of the growth process in a plant, in which cell enlargement determines the final size of the plant and its organs. The enlargement of a cell serves as the transition event between the stages of cell division and maturation of plant parts. Thus, growth and development may approximately be divided into two overlapping stages: growth by cell division, and growth without cell division and with varying extent of cell enlargement.

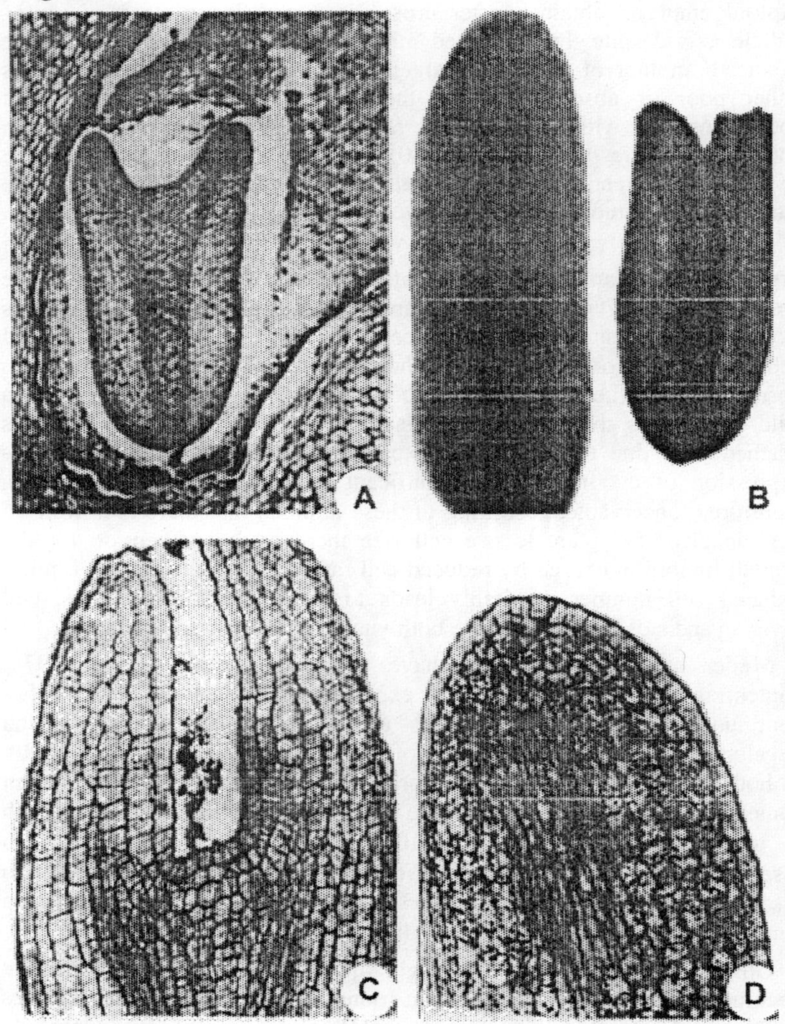

Fig. 40. Normal and mutant embryos of tomato; A - L. S. of normal torpedo stage embryo; B - Whole mounts of mutant embryos; C - L. S. of normal embryo showing normal shoot apical organization; D - L. S. of mutant embryo without cotyledons and organized shoot apical meristem (arrow) (Caruso 1968).

Morphogenesis 4

4.1. INTRODUCTION

It was indicated in the first chapter of this book that morphogenesis refers to the attainment of organic form during and as a result of growth and development. In fact, it was emphasized that morphogenesis forms an important aspects of development. Everyone can recognize the extraordinary variety of forms that are achieved by plants during their development just looking around in the environment. It can also be realized that the plants and the structures of which they are made are dynamic three–dimensional geometric entities having specific and distinctive configurations or forms, a physical reality resulting from their genetic constitution and all chemical, physical and other processes that are involved in growth and development in a specific environment (Wardlaw 1968). The physical configuration or form, i.e., the size, shape, disposition and structure of cell, tissues and organs, is affected in diverse ways and at various stages of ontogeny, and based on the nature and effectiveness of functional activities, indicating strongly the plasticity apparent in the development of the plants. Indeed, genetically identical plants growing under different environmental conditions can assume a wide range of different forms (Steeves and Sussex 1989). As Wardlaw (1968) has emphasized, if our interest is truly in organisms, and not simply in some of their constituent processes, studies of morphogenesis are vital.

In morphogenesis, we should first recognize the importance of Galileo's **Principle of Similitude**, which states that changes in the size of a living organism or a part thereof have unavoidable mathematical, physical and physiological aspects, for example, the relation of surface to volume. Even by the 19th century, it was believed that the form has to be interpreted in terms of physics, chemistry and mathematics. Although we know that morphogenesis cannot be explained solely by these three disciplines, we must make the fullest use of them. It is also important that the acceptable interpretations should be essentially biological ones. Wardlaw (1968) emphasized that "the use of the term morphogenesis can be appropriate only when it demonstrates how a particular physiological process, or a system of biochemical reactions can be closely and directly related to the ultimate visible reality i.e., the manifestation of form" in living organisms (including plants) and their parts.

The morphogenesis, whether in lower or higher plants, begins with an organized basic entity such as a spore, a zygote, a gemma or any other diaspore. Morphogenesis involves all the activities that are involved throughout the ontogeny from this organized entity to the specific and distinctive form and structure that are functionally effective and adaptive, and are characteristic of the adult state. Hence, the living materials for the study of morphogenesis include spores, zygotes, cells, tissues, axial organs and apical and lateral meristems, growing leaves, inflorescence and flowers, all in vivo and in vitro, as well as teratological and abnormal growths caused by abiotic and biotic stresses. Regarding the methods for study of morphogenesis, all relevant genetical, physico–chemical, mathematical, environmental and organismal methods should be used in a correlative and coordinated manner. However, it should be emphasized that the above ideal method will be very difficult to realize, since any morphogenetic development is undoubtedly highly complex.

The attainment of form, as it reflects or is related to function, is very important for the proper functioning of different structures that are recognized at different levels of biological organization. Thus, we can speak of morphogenesis at the cell, tissue, organ and even at the organism level. This chapter attempts to explain morphogenesis at all these levels of the plant organization.

4.2. DIFFUSION REACTION THEORY AND POSITIONAL THEORY OF MORPHOGENESIS

Diffusion reaction theory is the most accepted and discussed theory to explain morphogenesis, whether it be at the cell, organ or organism level. This theory was proposed by Turing (1952) and hence, is also known as **Turing's theory of morphogenesis**. According to this theory, in any biological system, at any level of organization, the chemical substances called **morphogens** are initially distributed in a homogeneous manner, but a regular, patternized diffusion and distribution of these specific metabolites (leading to a biochemical pattern) may eventually be triggered off by random disturbances, affording the basis for the inception of a morphological or histological pattern. In other words, according to this theory, the system (cell, tissue, organ or organism) is held to be a complex, specific, diffuse reaction system, which functions in conformity with the laws of physical chemistry, physics and mathematics to help in the patternized distribution of the morphogens. Turing's model further shows that when small, random deviations in an initially homogeneous field are reinforced by feedback, deviations from the initial concentration of the morphogens can form a stable pattern of peaks and troughs. Such a *de novo* formation of discrete pattern from a uniform field is analogous to many purely physical systems such as the formation of a branched river from a uniform drainage field. The identity of the morphogens is debated, but now, many agree that morphogens are signalling molecules (Tello 2007). Morphogens are synthesized at localized signalling sites and then get spread into the organ/plant part creating gradients in their concentrations.

In considering how complex and unique the reaction systems are, it is important to remember that the constitution of the protoplasm of living cells is the result of a long evolutionary process. The diffusion of morphogens follows the ordinary laws of diffusion (such as Fick's law) i.e., the morphogens move from a region of greater concentration to regions of less concentration, and also proportional to the 'difusibility' of the morphogen. Thus, a gradient in morphogens is created by the **diffusion**. According to some biologists, the formation of gradients of the morphogens can be explained by the alternative **positional theory** (Kerzberg and Wolpert 1998). As per this theory, the propagation of the morphogens is dependent on the closeness of cell–to–cell positions, and the morphogens are propagated along the cell membranes and transferred between the cells that are in contact with one another. Once the morphogen arrives at the cell surface, it binds to receptors and other kinds of molecules (Tello 2007). The diffusion reaction theory considers slow degradation of products of the morphogen and reversible binding (Lander et al 2002), while, in contrast, the positional theory does not consider degradation of products (Kerzberg and Wolpert 1998). Several mathematical models have been proposed so far to explain the distribution of morphogens, taking into account the three biological processes; diffusion, degradation and reversible binding.

4.3. MORPHOGENESIS AT CELL LEVEL

If we assume, as a convenient oversimplification, that all cells immediately derived from a meristem or any diaspore are essentially alike or equivalent, then the successive morphogenetic processes leading to the fully differentiated cells include, according to Wardlaw (1968), the following: (i) The establishment of polarity in the derivative cell; (ii) The sequential formation of general and specific proteins, including enzymic proteins. Some of these proteins are associated with specific cell organelles, which may also undergo changes; (iii) Specific development of particular organelles such as chloroplasts in chlorenchyma; (iv) Vacuolation of cytoplasm; (v) Physical and chemical changes in the cell wall; (vi) Differential growth of different derivative cells; (vii) Changes in the nucleus, in ploidy, DNA level and even its total loss in some cell types such as tracheary elements, sieve elements and sclerenchyma; and (viii) Development of special features in the cell wall such as sieve plate or perforation plate formation, development of primary pit fields and pits, plasmodesmata, wall thickenings etc. Thus, the cell wall, nucleus and cytoplasm with all its organelles, independently as well as in an interactive manner, participate in or control the morphogenetic events in cell differentiation from meristem or diaspore derivatives. Cell morphogenesis is an ordered event involving the sequential evocation of specific genes, which then become the determining and controlling agents. It cannot be separated from the biochemistry and biophysics of the cell and taken place as per the laws of physical science and mathematics.

4.3.1. Role of Cell Wall in Cell Morphogenesis

Morphogenesis at the cell level is the basis to morphogenesis at the other higher levels of biological organization. Cell wall changes are the most important in cell morphogenesis. Among all the characters of the cell, the cell wall (**Extracellular Matrix, ECM**) is the most important character that distinguishes a plant from an animal. It also decides the nutritional mode of plants (or is it the result of the nutritional mode?) (Krishnamurthy 2010). The cell wall largely decides the size and shape of the cell, tissue texture, and also the final form of the plant organ and even the organism in a number of cases. The cell wall is rigid and yet, elastic and plastic. The first feature limits to some extent the size of the protoplast, while the latter two prevent the rupture of the cell when the protoplast increases in volume due to water intake and the consequent turgor change. The various cell types are distinguished from one another essentially on the basis of the nature of the cell walls such as thickness, chemistry, layering, physical characteristics, and ornamentations, if any.

Although originally dubbed as a dead and inactive structure, cells walls, when in contact with living protoplasts, are highly dynamic structures showing spatial and temporal domains that differ in their chemistry, organization and function. Even some consider the cell wall as a 'vital and indispensable organelle' (Krishnamurthy 1999a), as a 'special subcellular compartment outside the plasma membrane', or even as a 'vital extension of the cytoplasm'. The presence of a variety of enzymes in the cell wall adds support to the above statements. Also, chemicals emanating from cell walls are involved in cell–to–cell signalling as well as in cell differentiation. Cell wall also play a vital role in plant defence reactions against abiotic and biotic stresses of all kinds, essentially through the production of stress proteins (both structural and enzymatic), phytoalexins, lignins, callose, oligosaccharins etc. The presence of a cell wall is also an essential prerequisite for plant cell to divide.

4.3.1.1. Primary and Secondary Walls

It was already mentioned in Chapter 3 of this book that cell division in plants involves two steps, karyokinensis or division of nucleus and cytokinesis or division of cytoplasm. Cytokinesis invariably takes place through the formation of a **cell plate**, followed by the deposition of wall materials on either side of it, thus increasing the thickness of the new partition. The, first–formed, thin wall in a cell is called the **primary wall** and the intercellular middle layer (with cell plate) is called **middle lamella**. Many cells may deposit additional cell wall materials (once they attain their full size), inner to the primary wall, and the thus formed wall is called **secondary wall.** The primary wall is often deposited before and during the growth of the cell. There are currently two important models to explain the structure of the primary cell wall of growing cells. According to one model (Fig. 41), there is a network of cellulose microfibrils intertwined with xyloglucan–like hemicelluloses. The xyloglucans are covalently bonded to the cellulose microfibrillar surface and are entrapped within the microfibril. This complex is embedded in a gel of pectins (usually polygalacturonic acid–rhamnogalacturonan1, PGA–RG1). As per the second model, there is a lack of polymers that directly cross–link the microfibrils. Instead, the tightly bound xyloglucan hemicelluloses are sheathed in a layer of less tightly bound polysaccharides, which are embedded in a pectin matrix that fills the spaces between the microfibrils (Fig. 41 B) (Cosgrove 1999).

The secondary wall, as mentioned earlier, is deposited almost always inside the primary wall and invariably after the primary wall has stopped increasing in surface area. Secondary walls are very important in cells that have specialized functions such as mechanical and water–conduction. In such cells, the protoplasts die after secondary wall is deposited; cellulose is often more abundant in this wall than in primary wall, while pectins and structural and functional proteins are absent or very meager. The secondary wall materials that are newly laid down include lignins (in sclerenchyma and tracheary elements; also in cork cells of some taxa), suberins (in endodermis, exodermis and cork cells) and cutins (in epidermal cells). The secondary wall differs from the primary wall in having a fairly ordered arrangement of microfibrils (random in primary wall). In some types of cells, a tertiary wall is also present near the cell lumen.

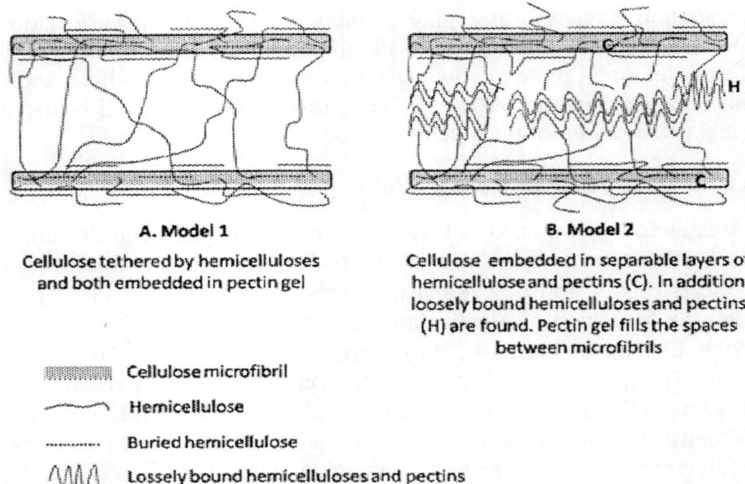

Fig. 41. A and B. Two models of the face view of growing primary wall showing the relative positions of the different cell wall constituents (Diagrams based on Cosgrove, 1999).

4.3.1.2. *Cell Wall Extension Growth*

There are a few explanations on the mechanism of cellulose microfibril deposition in the cell wall when it is being formed. The first one is the **alignment hypothesis** (named by Baskin 2001). According to this widely accepted hypothesis, the orientation of the cellulose microfibrils in elongating primary cell walls and in secondary walls of differentiating tracheary elements normally parallels that of the underlying cortical MTs (Wymer and Lloyd, 1996) and the orientation of cellulose microfibrils is determined by the orientation of MTs. However, this hypothesis is not adequate to explain the deposition of wall materials in non–elongating plant cell walls. Moreover, in some experimentally treated plant cells and in some mutants, there is no correlation between the direction of MTs and the direction of cellulose orientation. The second hypothesis is the **liquid crystalline self–assembly hypothesis** of Bouligand (1976). This was proposed after observing helicoidal cell walls, whose cellulose microfibrils do not match the cortical MTs but have similarity to cholestric liquid crystals. The helicoidal wall structure is likely to arise from a liquid crystalline self–organizing principle. The third hypothesis, called **template incorporation mechanism**, was proposed by Baskin (2001). As per this hypothesis, cortical MTs are not required for cellulose synthesis or for the formation of its microfibrils. The nascent cellulose microfibrils can be oriented by MTs or become part of the cell wall by binding to a scaffold built and oriented around either already incorporated microfibrils or membrane proteins or both. The fourth hypothesis, called the **geometric**

model, is suggested for cellulose microfibril deposition based essentially on helical secondary wall structure of *Equisetum* root hairs (Emons 1994). As per this model, the deposition angle of the cellulose microfibrils with respect to the long axis of the cell is related to the density of active cellulose synthases in the plasma membrane, to the distance between individual microfibrils within a lamella and to the geometry of the cell (Evert 2006).

The orientation of cellulose microfibrils within the primary wall influences the direction of cell expansion. In cells that enlarge more or less uniformly in all directions (such as in pith or cortical parenchyma cells), the microfibrils are randomly laid down. On the contrary, in several elongating cells the microfibrils of the lateral wall (or side wall) are deposited at right angles (i.e., transversely) to the long axis (i.e., elongating axis), but as the cell elongates the outer microfibrils become more or less longitudinal (i.e., parallel to the long axis of the cell) followed by successively inside–located microfibrils. This is as per the **multinet growth hypothesis** (Roelofsen 1959) (Fig. 42). The longitudinal alignment of microfibrils promotes expansion mainly in a lateral direction. However, multinet growth hypothesis cannot explain longitudinal growth satisfactorily in cells whose cell walls have a helical structure. During cell wall expansion or elongation, the helicoidal organization may become completely dispersed, from helical to random.

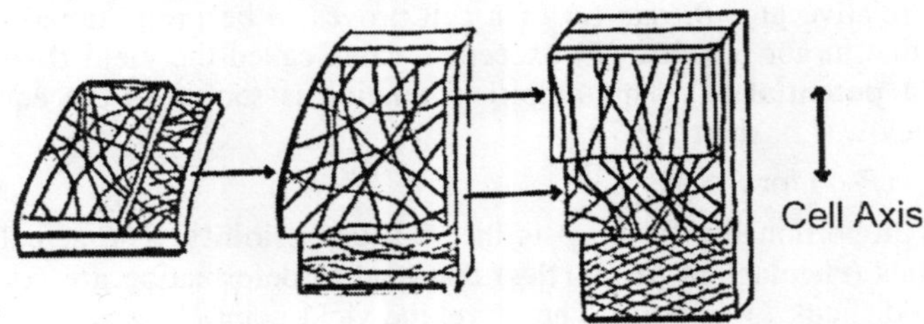

Fig. 42. Multinet growth of primary wall in a growing cotton hair; from near the tip of the hair, from where the tip merges into the tubular part of the hair, and from the tubular part respectively from left to right (Houwink and Roelofsen 1954)

As already mentioned in Chapter 3, plant cells, before they mature, enlarge manyfold in volume by a process that entails irreversible cell wall expansion. Since the size of the cell is governed by the mechanics of the cell wall, it has to be strong enough to withstand the high physical stresses generated by the cells' turgor pressure (P). Given that P of growing cells is typically in the range of 0.3 to ± 9 MPa and wall stress is more or less P x cell radius / wall thickness, it can be calculated that wall stresses are in the order of 10 to 100 MPa (Cosgrove, 1997). As an inevitable consequence of this physical stress, the critical physical event required for cells to enlarge is cell wall relaxation. Cell wall relaxation may be technically defined as the reduction in wall stress at constant cell–wall dimensions. This is the means by which growing cells simultaneously loosen their walls and reduce their turgor and wall potential, thereby enabling them to imbibe water and to expand physically. Thus, wall expansion or extension is a very complex physical process and requires loosening of wall structure resulting in relaxation of wall stress, reduction of turgor pressure, synthesis of polysaccharides and proteins and active respiratory process in the cell (McQueen–Mason 1995).

The research on the physics of cell wall extension growth may be said to have started with the work of Heyn (1931). He oriented a stem segment horizontally, fixed it at one end and then tied a weight at its other end. The weight made the stem segment to bend down. When he removed the weight he could notice that the stem returned part way to its original horizontal position. Heyn came to the conclusion that the weight might have caused the cell walls to stretch **elastically** (as in a rubber band) and **plastically** (as in the stretching of a bubble gum). The former process accounts for the bending back of the stem segment, while the latter for the irreversible bending. Thus, it was Heyn who first introduced the concept of plastic and elastic cell wall extensibility. Plastic wall stretching is achieved as the cell wall is loosened (also called **wall yielding**), so the cellulose microfibrils can slide past each other more easily. This is known as **shear**, which actually involves a breaking of bonds between adjacent microfibrils.

The relative growth rate (*r*) of a cell proves to be proportional to the extent that turgor pressure (*P*) exceeds a value called the **yield threshold** or **yield potential** (*Y*). This empirical relation is shown in the equation given below (Lockhart 1965):

$r = \phi\,(P-Y)$ for $P > Y$

The proportionality factor ϕ is the wall **extensibility**. The actual wall properties (**rheological properties**) that lead to deformation are complex and are difficult to measure. Therefore, the yield point (*Y*) is expressed as the minimum pressure necessary to cause cell growth. Turgor pressure is the cause of existing wall stress and hence is an indirect but equivalent measure of wall stress. The relative growth rate of the cell, thus, depends on five interrelated factors: conductivity of the plasmalemma and cell wall to water that enters the cell, the difference between the cell's inside and outside osmotic potential (solute potential), turgor pressure of the cell, and the two wall–yielding properties i.e., extensibility and yield threshold (Salisbury and Ross 2004).

Earlier, extensibility (and wall relaxation) was measured on dead tissues (Cleland 1981). The most–widely used technique at that time for measuring the mechanical properties of dead cell walls was called the **Instron technique** (Cleland 1984). In this procedure, samples are first boiled in alcohol to remove cell wall proteins and to disrupt the protoplasts. Subsequently, the sample is subjected to two successive extensions at a constant rate of **strain** (deformation), the first extension involving both elastic and a plastic component and the second extension entirely plastic. The **stress** (force) along the walls is then measured as a function of extension. From the slopes of the two curves, plastic and elastic extensibility can be obtained. In more recent years, wall relaxation is being measured in living tissues in order to obtain values of yield threshold and wall extensibility. Most of these efforts are centred on two approaches: (i) A growing tissue is isolated in such a way that it does not get water supply and therefore, no water uptake; and (ii) A living tissue is kept in a pressure chamber and it is prevented from water uptake by increasing the external pressure on the tissue just enough to reduce growth to zero. In both the approaches, wall stress is monitored through measuring turgor pressure with the help of a pressure probe. As wall relaxation occurs, turgor pressure drops exponentially until the yield point (Y) reached at some value above zero. The rate of turgor–pressure decrease can be used to evaluate the wall extensibility (ϕ). In the in vivo (i.e., on living tissues) methodology, natural stresses, instead of externally applied stresses, are used on the cell wall (cell turgor). Turgor applies hydraulic stress in all directions from within the cell while the externally applied pulling stress operates only in one direction. These and other techniques, not discussed here, have demonstrated that cell wall growth is drawn by wall loosening. Although it is possible to measure wall relaxation in isolated wall specimens, as detailed above, the distinction between growing and non–growing walls is most readily seen in extensometer assays of pH–dependent wall creep. **Creep** refers to a time–dependent, irreversible extension, typically due to slippage of the wall polymers relative to one another in growing cells, which exhibit a steady long–term extension. During growth, the primary wall must yield enough to allow an appropriate extent of expansion while at the same time should remain strong enough to constrain the protoplast.

Experimental studies have indicated that wall extensibility and yield threshold may be modulated by hormones such as auxins and gibberellins (abscissic acid and ethylene decrease the extension). The mechanism by which hormones alter the extensibility of cell walls is not very clearly proved. One mechanism proposed is the **acid growth hypothesis** (Brett and Waldron 1990), according to which auxin activates the expression of specific genes that influence the synthesis and delivery of new cell wall polymers for extensibility. This view accounts for the coupling of wall synthesis to wall expansion, but it fails to explain how the secretion of polysaccharides can lead to the wall stress relaxation that is essential for water uptake and expansion of the turgid cell. The acid–induced creep or acid growth is the characteristic of walls from growing cells. The change in pH appears to convert the growing wall from a viscoelastic solid to a viscoelastic liquid, implying the weakening of one or more key bonds linking wall polymers together. There is no experimental evidence for this explanation. The more commonly accepted mechanism is that auxin activates a proton–pumping ATPase in the plasma membrane. Protons are pumped from the cytoplasm into the cell wall resulting in the altering of pH, which in turn is thought to cause a loosening of cell wall structure, triggering turgor–driven extension.

The second hypothesis emphasizes that wall yielding results from a biochemical loosening of the wall to permit turgor–driven extension of the wall polymer network (Cosgrove 1997) (Fig. 43). It is evident from this figure that there are plastic and elastic elements in the cell wall, which can be considered as being in series with each other. As the plastic elements relax, these elements stretch, allowing the elastic elements to shorten and this can happen only if wall stress and turgor pressure are reduced. If water enters almost instantaneously as pressure decreases in response to relaxation of the plastic elements, the elastic elements may shorten only infinitesimally as pressure decreases only infinitesimally. In this steady–state growth process, called creep, wall stress and turgor remain constant. Although this concept draws support from several studies, it neglects the long–term need for integration of the new wall polymers into the expanding wall.

How the processes of loosening, wall relaxation and expansion happen? A large body of literature suggests the involvement of wall enzymes, such as (1→4) β–glucanases or xyloglucan endotransglycosylase (XET), in cell wall loosening processes and matrix glucans show enhanced turnover upon growth stimulation by auxin. The involved polymers that are hydrolyzed are xyloglucan and (1→3) (1→4) β–glucan in dicot and grass cell walls respectively. However, in vitro experiments failed to detect loosening or extension of the cell wall by XET. The hemicelluloses that anchor cellulose microfibrils to the matrix may be cut by the XET enzyme that transfers the newly–formed reducing end to water (i.e., hydrolysis) or to the non–reducing end of another hemicellulose (i.e., a transglycosylation). Alternatively, wall loosening agents may weaken the non–covalent associations of matrix polymers to microfibrils or to other polymers in the matrix (Cosgrove 1997).

Biochemical analysis of acid growth mechanism led to the discovery of a novel family of proteins, named **expansins** that mediate the rheological behaviour of growing cell walls. Two families of expansins are known so far. The acid growth was lost when the wall was treated with proteases and protein denaturing agents, thus supporting the role of these proteins in wall extension. However, not all the cell walls in the plants are susceptible to the action of expansins. The walls from non–growing part of the plant and mature cells do not creep when treated with expansins. The molecular basis for expansin action on wall rheology is still uncertain.

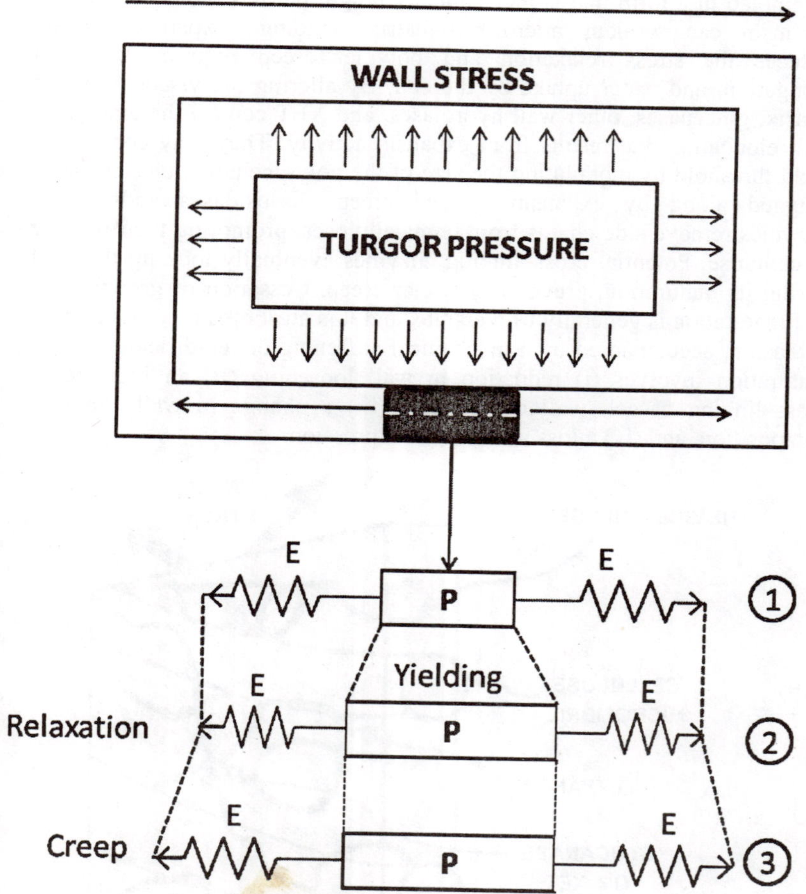

Fig. 43. A mechanistic model for the growing primary cell wall as per Cosgrove (1987). Turgor pressure (small arrows) exerts a force against the primary cell wall and causes a tensile stress in the plane of the wall. (1) Shows both elastic (E) and plastic (P) elements, bear the wall stress. First, the plastic elements yield and relax, but the elastic elements shrink. (2) Hence, there is no net increase in cell wall length. (3) Shows the maintenance of wall stress through the expansion of E, but not of P (creep), resulting in increase of wall length (Diagrams based on Cosgrove 1997).

Based on the discovery and action of expansins mentioned above, Cosgrove (1997) has proposed the following tentative model (Fig. 44) for the control of wall expansion. The primary wall is initially secreted and assembled in a form that is mechanically tough, but has 'hot spots' where expansin can weaken microfibril–matrix bonding. Expansin activity induces the stress relaxation and polymer creep required for wall elongation, and water uptake by the cell. By altering the viscosity of the matrix, glucanases, other wall hydrolases, and XET control the amount of wall elongation that results from expansin activity. They may control the yield threshold by modulating the size of the polysaccharide cluster that is dragged along by expansin–induced creep. Xylosidases and related enzymes remove side chains from hemicelluloses, promoting their binding to cellulose. Potential cross–linking enzymes eventually lock up the wall during its maturation, preventing further creep. Cessation of growth after cell maturation is generally irreversible, and it is attributed to a loss of wall extension, accompanied by a mechanical stiffening or regidification. Wall maturation involves (i) reduction in wall loosening (ii) an increase in cross–linking of cell wall components (iii) a change in wall chemical composition, and (iv) a loss in expansin expression.

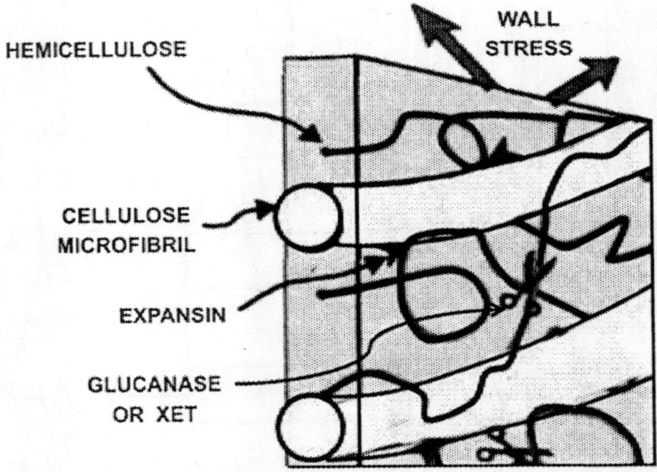

Fig. 44. Potential mechanisms for stress relaxation and creep of cell walls as suggested by Cosgrove (1997); (Diagram based on Cosgrove 1997)

4.3.1.3. *Cell Wall Thickness Growth*

When we consider the mechanism of cell wall growth, it is necessary to understand the differences between growth in surface (or wall expansion) and growth in thickness. Thickness growth of wall is particularly evident in secondary walls, but also seen in primary walls of collenchyma, endosperm etc. According to the classical concept, increase in the thickness of wall happens by two methods of deposition of wall material, **apposition** and

intussusception (Fig. 45). In the former, the building units are placed one on top of the other, while in the latter, the building units are inserted into the existing structure. Lignin and cutin and probably also xyloglucans are incorporated into the already existing wall material probably as per intussusception. If this happens in cellulose microfibrils, the result will be an interweaving of the fibrils.

The importance of wall thickening in the morphogenesis of certain specific cell types is discussed here.

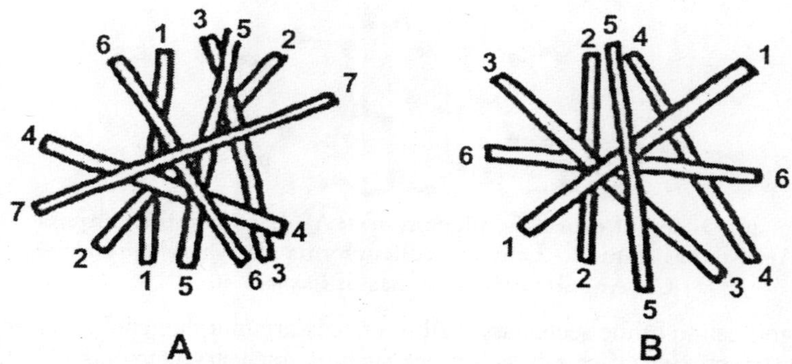

Fig. 45. Diagrammatic representations of growth in cell wall thickness by apposition (A) and intussusception (B); only microfibrils are shown and numbered.

The morphogenesis of collenchyma cells involves changes in primary cell wall thickening, which glistens in fresh sections of plant tissues. The thickening is often unevenly distributed. The collenchyma cell wall is made up of cellulose, pectins and hemicelluloses and in some cases with alternate layers rich in cellulose and poor in pectins, and rich in pectins and poor in cellulose. Based on wall thickening patterns, different types of collenchyma are recognized in plants (Fig. 46). Collenchyma cells can increase in surface area and thickness of their cell walls simultaneously, and they are also capable of developing and maintaining thick primary walls, while the plant organ containing them continues elongation. Since their wall thickenings are plastic and extensible, they do not hinder the elongation of the organ that possesses them. The cells can increase 30–fold. The tensile strength of collenchyma is favourably compared with that of the fibers. In contrast to the elasticity of fibers, the plasticity of collenchyma cells is a major advantage in a growing plant part that requires mechanical support.

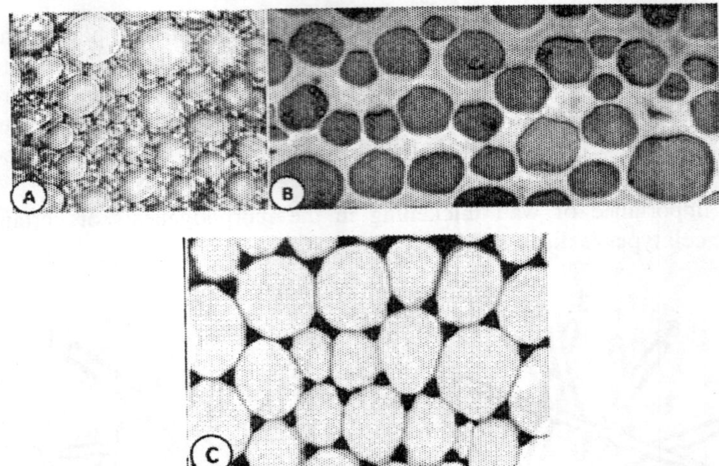

Fig. 46. Different types of collenchymas; A - Lacunar collenchyma of *Amaranthus* stem; B - Lamellar collenchyma of *Cynanchum* species; C - Angular collenchymas of *Ipomoea* stem

Lignification of the secondary wall is a necessary morphogenetic step in the differentiation of certain cell types such as tracheary elements (TEs), sclereids and sclerenchyma fibres (Hepler et al 1970; Wardrop 1981; Savidge 1996, Kaliamoorthy and Krishnamurthy 1998). In the case of TEs, once their expansion is over, cellulosic and other polysaccharide secondary wall materials are laid in characteristic patterns over which lignin is then deposited (Fig. 47).

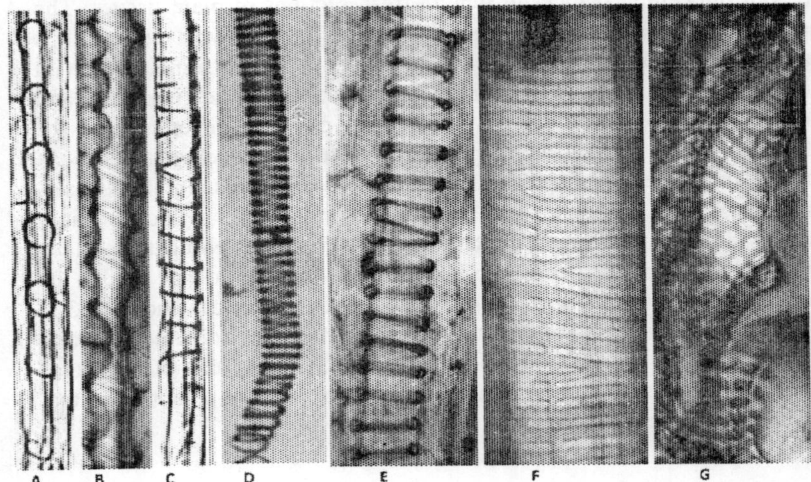

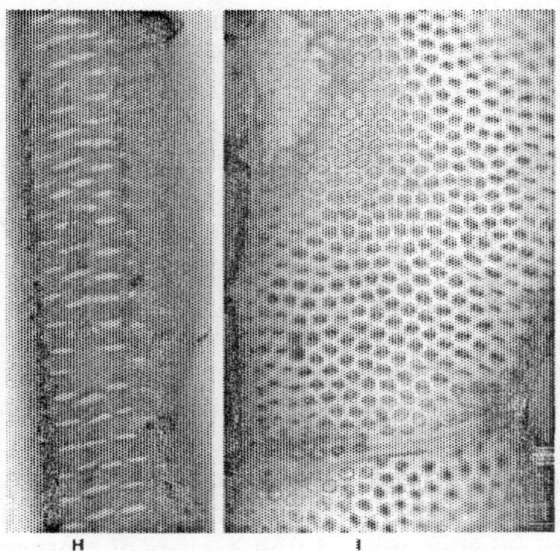

Fig. 47. Tracheary elements showing different patterns of secondary wall: A – Annular (*Commelina* internode); B - Annular–helical transition (*Cynanchum* species); C - Helical (*Commelina* internode); D - Reticulate closely helical (*Bacopa* leaf); E and F - Helical–reticulate transition (*Cissus* stem and *Costus* rhizome respectively); G - Reticulate (*Allium* stem); H - Reticulate pitted transition (*Bacopa* leaf); I - Pitted thickening (*Aquilaria* stem)

Both vesicular secretion containing matrix and adcrusting materials, and lamellar apposition of cellulose microfibrils are involved in secondary wall deposition in the differentiating TEs (Wardrop 1981). The contents of these vesicles are released directly to the developing secondary wall through plasma membrane by **micropinocytosis**. Lignification first appears on the tangential walls and then on the radial walls. Lignin precursor substances are produced either in the procambium or in the vascular cambium, and later diffuse into the zone of the differentiating TEs for conversion into lignin or are synthesized within the differentiating TEs. It is more likely that the latter event takes place. Here, it should be made clear that the secondary wall of TEs contains not only lignin, but also polysaccharides such as cellulose and hemicelluloses and some structural proteins. Secondary wall polysaccharide deposition and lignification are considered to be very closely interrelated during TE differentiation. It has been stated that a cellulose matrix formation in TEs and sclerenchyma requires lignification (Siegel 1956). On the other hand, some workers have claimed that cellulosic secondary wall deposition is not dependent on lignification (Ingold et al 1988, Smart and Amrhein 1985). In *Zinnia* cultures only TEs with a cellulosic secondary wall matrix gets lignified (Northcote 1995). In vitro xylogenic calli subjected to the treatment with metabolic inhibitors such as 2–aminoindan–2–phosphonic acid (AIP), a potential inhibitor of L–phenylammonia lyase (PAL) enzyme, 2,3,5–triiodobenzoic acid (TIBA), an inhibitor of auxin transport, and nifedipine, a calcium channel blocker, were used to resolve this issue (Kaliamoorthy and Krishnamurthy 1998). AIP completely inhibited lignification, although cellulosic secondary wall deposition (along with its characteristic pattern) was unaltered. A similar result was obtained using TIBA and nifedipine. This study proved that secondary cellulosic wall deposition in developing TEs is not dependent on lignifications (Fig. 48). The mechanism of action of Nifedipine, TIBA and other calcium channel blockers such as lanthanum chloride in blocking lignifications was also studied in the xylogenic calli of *Zinnia* (Christopher and Krishnamurthy 2004). Deposition of lignin and other materials for the S1 layer of the secondary wall is reported to begin near the centre of the cell and proceed towards the apices (Wardrop 1964), but Kaliamoorthy and Krishnamurthy (1998) reported lignification to proceed from corners of prospective TEs towards the other regions of the wall (Fig. 49). Hepler and Newcomb (1964) have explained the role of microtubules and fibrillar elements with the cisternae of ER in secondary wall differentiation. There is a close correspondence of alignment between cellulose microfibrils of secondary walls and MTs' presence and orientation. There is convincing ultrastructural evidence for the deposition of the secondary wall material from dictyosome–derived vesicles.

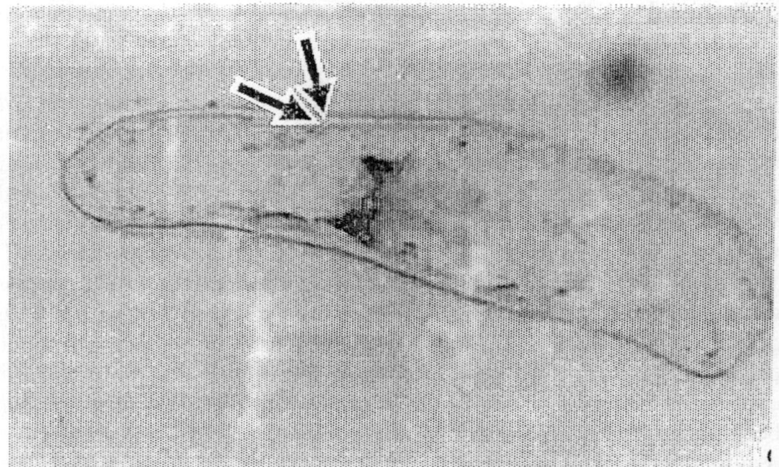

Fig. 48. Prospective tracheary cell of *Cucumis sativus* treated with nifedipine showing the element without (Double arrows) lignification. Cell stained with Phloroglucinol to show the absence of lignin in the secondary wall. (Photograph courtesy of Dr. S. Kaliamoorthy)

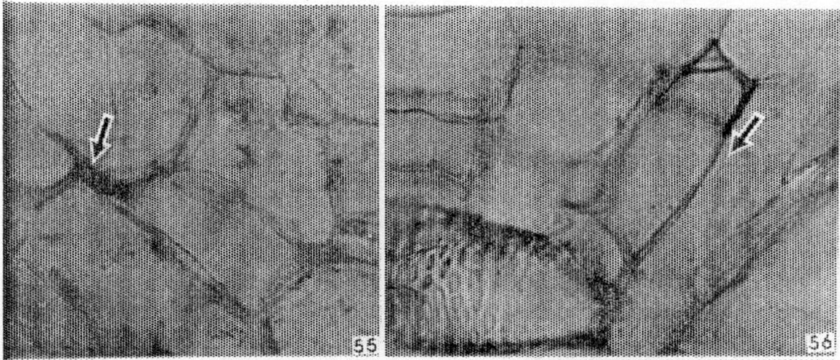

Fig. 49. Xylogenic callus sections of *Cucumis sativus* stained for lignin with Toluidine Blue O showing the commencement of lignifications (arrow) first at cell corner in the prospective tracheary elements and their subsequent spreading to the other parts of the secondary cell wall (Photographs Courtesy of Dr. S. Kaliamoorthy)

Secondary wall formation and lignification events are likely to be similar in sclerenchyma fibres and sclereids, although no critical study has been made. However, the characteristic patterns of secondary wall depositon seen in TEs are absent in sclerenchyma. Barring simple pit areas in the wall, lignification preceded by cellulosic secondary wall deposition is more or less uniform in sclerenchyma.

Cutinization and **cuticularization** are extremely important wall changes seen in the surface cell layer of the aerial parts of vascular plants, as they control water, solute and gas exchange and protect the aerial parts from abiotic and biotic stresses of various kinds. The **cuticle**, also called **cuticular membrane**, consists predominantly of two lipid components; insoluble **cutin** (forming the matrix of the cuticle), and **waxes**, some of which are deposited on the surface of the cuticle as **epicuticular wax**, and others are embedded in the matrix as **cuticular** or **intracuticular wax** (Figs. 50 and 51).

The cuticle often is made of two or more distinct regions; the cuticle proper and one or more cuticular layers. Epicuticular wax is in amorphous form or as crystalline structures of various shapes (tubules, solid rodlets, filaments, plates, ribbons or granules) (Figs. 50 and 51). The cuticular layers are found beneath the cuticle proper and are considered to be the outer portions of the cell wall encrusted to varying degrees with cutin. Wax, pectins and hemicelluloses may also occur in the cuticular layers; pectin bonds the cuticle to the outer walls. This layer is continuous with the middle lamella between the anticlinal walls. Cutin and waxes (or their precursors) are synthesized in the epidermal cells and must be migrated to

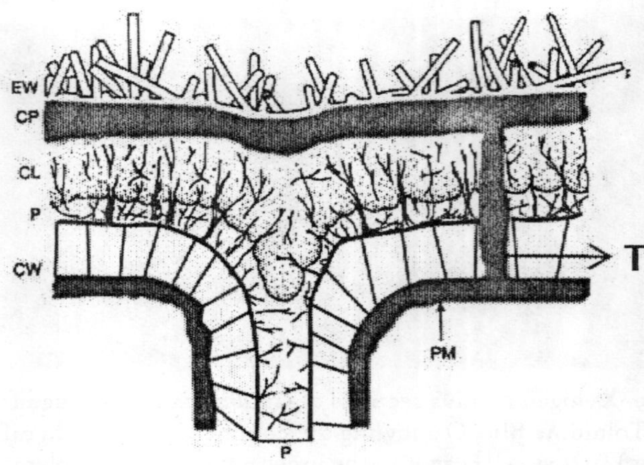

Fig. 50. Typical structure of a plant cuticle CL - Cuticular layer traversed by cellulose microfibrils; CP - Cuticle proper with lamellate structure; CW - Cell wall; EW - Epicuticular wax; P - Pectin layer (Middle lamella); PM - Plasma memebrane; T - Teichode (Modified from Jeffree 1986)

the surface through the cell walls. **Suberization** is another important morphogenetic process that is related to the cell walls of certain categories of cells such as those of cork and exodermis. Since suberin, a mixture of several chemicals including lipids and phenolics, is impervious to water, its presence is very important in arresting water loss. In many roots (especially of monocots), suberin is deposited on the root hypodermal layer cells, except those lying opposite to the root hairs (**passage cells**) providing inward flow of water, but not outward flow from root to soil (Fig. 52).

4.3.1.4. *Cell Wall Ingrowths*

Cell wall ingrowths form another cell wall modification that is of great morphogenetic importance at the cell level. Parenchyma cells with such ingrowths in the wall are called **transfer cells** (Fig. 53). Cell wall ingrowths often develop relatively late in cell maturation and are produced on the original primary cell wall. Hence, such cells may be considered to have specialized form of secondary wall (Pate and Gunning 1972). The plasma membrane of the concerned cell also is extended along with these wall ingrowths, even if the latter are highly tortuous and branched, forming a distinct cell wall–membrane apparatus. This apparatus is bordered by a cytoplasm with a dense population of mitochondria and conspicuous ER. Transfer cells, thus, have a greatly increased (as much as 47%) surface area of the wall–membrane apparatus. Two types of wall ingrowths can be seen in transfer cells; **reticulate** and **flange**. In the former, the ingrowths branch

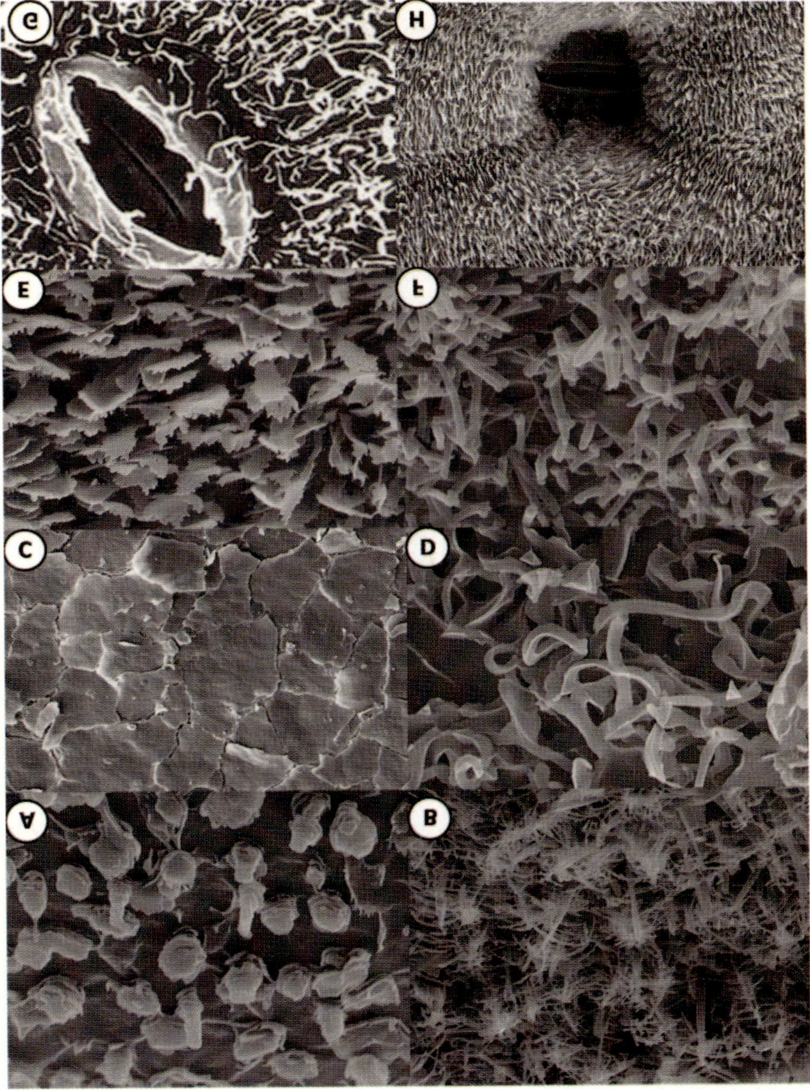

Fig. 51. SEM images of epicuticular wax deposition over epidermal cell walls; A - *Aristolochia gigantea*; B - *Brassica oleracea*; C - *Crassula ovata*; D - *Hebe ochracea*; E - *Iris germanica*; F - *Tulipa* species; H - *Sisyrinchium striatum* (Photographs courtesy by Dr. Hans-Jürgen Ensikat); G - Leaf surface of Rose plant (Courtesy: http://www.biologie.unihamburg.de/bonline/library/webb/BOT311/PlantCellWallsoo/Cutin.html)

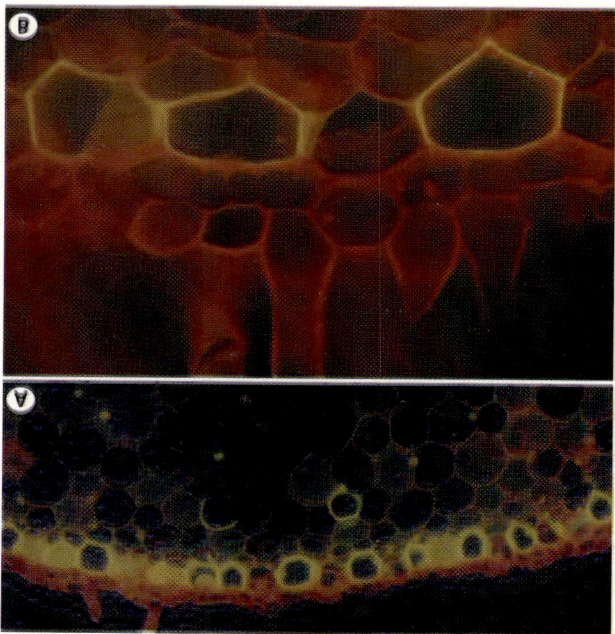

Fig. 52. T.S. of the outer part of the root of *Spathoglottis* showing autofluorescent suberized cells, non–suberized passage cells and root hairs. B is an enlarged portion of A (Photographs supplied by Dr. S. Senthil Kumar)

and fuse latterally to form a complex network of different morphologies, while in the latter the ingrowths are curvilinear and rib–shaped projections (Talbot et al 2002). Although these two types are seen more commonly in transfer cells of different taxa, they may also be seen in the same species and sometimes in the same transfer cell. The transfer cells are believed to play an important role in the short distance transport or transfer of solutes, especially sucrose (Gunning 1977), and hence, the presence of these cells is often correlated with the presence of intense inward (uptake) or outward (secretion) solute flux directions across the plasma membrane. Transfer cells occur in a wide range of locations in the plant body, from bryophytes to angiosperms; in the minor veins of leaves and cotyledons, leaf traces in nodal regions (in association with xylem and phloem), placental cells of ovaries, embryo sac, aleurone cells, endosperm cells (especially haustoria), root nodules, insect galls (nutritive tissue around larval cavities), root–knot nematode-induced giant cells, various kinds of secretory structures such as nectaries, salt glands, glands of carnivorous plants and cyanobacteria–containing cells of some *Hepaticae*, *Azolla* and corolloid roots of *Cycas* (Gunning and Pate 1969).

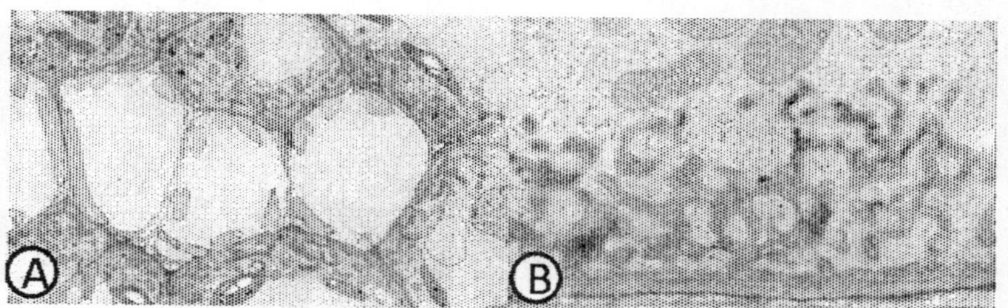

Fig. 53. Transfer cells A. *Anagallis* node B. *Funaria* species (Photographs supplied by Dr. Gunning B.E.S. From "Plant Cell Biology on DVD: Information for students and a resource for teachers". Springer, 2009)

4.3.2. Role of Cell Shape in Morphogenesis

The three–dimensional shape of the cells is also an important factor in cell morphogenesis. Shape of the cells is essentially controlled by the cell wall. Isolated protoplasts (of any cell) without cell wall, as well as isolated cells such as spores, pollen grains and many unicellular algae are often spherical, since the sphere has the least surface area for a given volume. Also, a spherical form is always obtained whenever any body occurs in isolation from others (like a soap bubble, a drop of rain or any celestial body) (D'Arcy Thompson 1942). On the other hand, if cells are grouped together closely into tissues, it is a geometric fact that they often lose their spherical shape and assume a polyhedral shape i.e., the shape having many sides or facets (Fig. 54 A–C), such as soap bubbles in a foam or lead shots in a packed container subjected to 'tension at the surface'. Korn and his associates (Korn 1984) proposed an alternate theory for cell shapes acquired by a group of spherical cells when they come together, to overcome the two problems in D'Arcy Thompson's hypothesis. First, the soap bubble appearance of cells is true only for certain types of parenchyma cells. Second, soap bubbles and compressed lead shots undergo slippage in order to assume states with minimum free surface energy, while plant cells are more or less fixed together with little freedom to slide. Accordingly, it was proposed that cell shape is essentially determined by developmental algorithms and a cell is an ensemble of rods (edges), sheets (facets) and vertices assembled into distinct shapes. Due to **positional specificity**, each edge, sheet and vertex " has both structure in terms of occupancy of space as well as one of several different states for growth or separation" (Korn 1984).

Whatever be the explanation, when several spherical cells are packed together and when all the cells are under some uniform stress (from within or without), the six points of contact between the cells extend into lines, and thus, the spherical cell gets converted into a polygonal one. In the beginning, plant histologists assumed that the resultant cell comes to possess a rhombic dodecahedron or garnet form with 12 lozenge–shaped

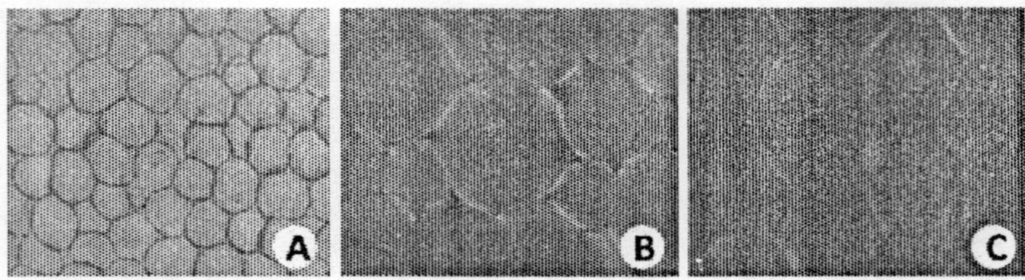

Fig. 54. A–C: Gradual loss of spherical shape of parenchyma cells of onion into polyhedral ones when there is mutual pressure between the different cells

equilateral facets. However, it was subsequently realized that in a relatively homogeneous tissue, the individual parenchyma cells tend to have around 14 facets (Fig. 55) all bounded by vertices with 120° angles with neither the quadrilaterals nor the hexagons being necessarily plane. This structure is called orthic tetrakaidecahedron. This is the ideal packing figure not only because it can be repeated without leaving any small space, but also because it requires a minimum of material for partitioning of space. Moreover, it has an economy of surface in relation to volume. However, plant cells rarely approach this ideal form (Matzke 1940) and may show a variable number of facets as well as a variation in the size of the different facets. Small cells in a tissue invariably have less than 14 facets and large cells more than 14 facets. The fusiform initials of vascular cambium may have 8–32 facets. The presence of intercellular spaces between the cells in a tissue often reduces the number of contacts between the cells and hence, reduces the number of facets. In a tissue, especially in a complex tissue such as xylem, phloem or epidermis, the different facets of a single cell may often be in contact with different cell types, both spatially and temporally. This often results in domain–specific changes in the chemical constituents of the cell wall facets depending upon the type of cell with which the facet is in contact with. For example, an axial parenchyma cell in the wood tissue may be in contact with a ray cell, a fibre, a vessel element, a tracheid, another axial parenchyma cell and/or an intercellular space and the different facets of this cell having contact with other cell types show distinct changes in their chemistry.

The stellate shape of aerenchyma cells (Fig. 56 A) in aquatic or marshy plants, especially those constituting the diaphragm that separates long air canals or chambers, is functionally very important. But for their shape, a diaphragm cannot be organized and cannot function in preventing the collapse of the canals and at the same time maintain flow of the air. However, the functional importance of the shape of some cell types such as astrosclereids (especially in a leaf) (Fig. 56 B and C) is not clear.

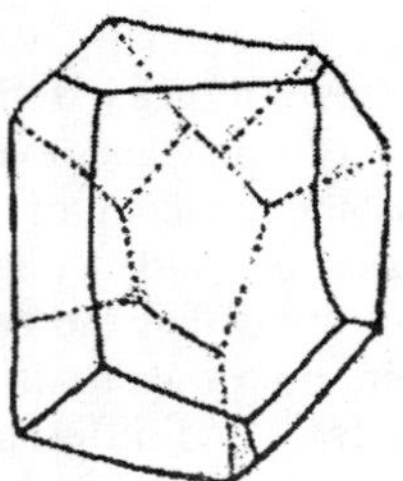

Fig. 55. Three-dimensional shape of a typical parenchyma cell with 14 facets, isolated from onion scale leaf

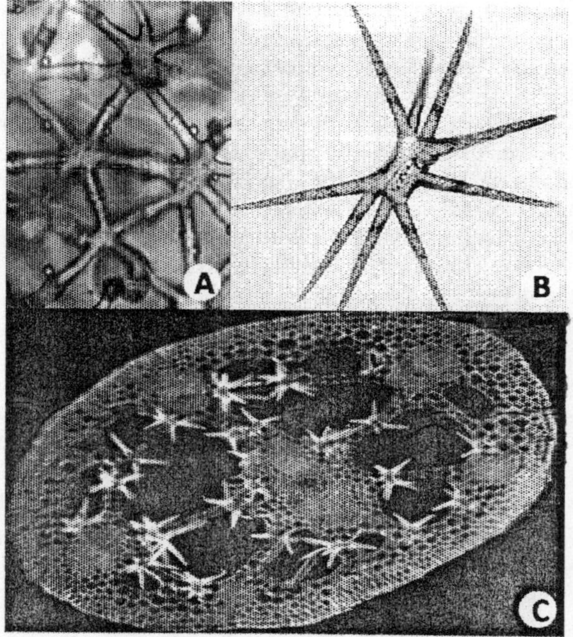

Fig. 56. A. T.S. of portion of the diaphragm of *Juncus* stalk showing star-shaped aerenchyma cells; B. An isolated astrosclereid of *Gnetum* leaf; C. T.S. of petiole of Lotus leaf showing astrosclereids

4.4. MORPHOGNESIS AT TISSUE LEVEL

Tissues and tissue systems are formed as a result of tissue differentiation. They are characteristics of each and every plant part and of great beauty because of the precise regularity, variety and symmetry of their geometric patterns. These tissue patterns, as one may see in a section of a stem, root or leaf, are functionally very necessary and have been regarded as the result of an evolutionary **division of labour** in the interest of the functional economy of each and every species (Wardlaw 1968). Some people suggest that the concept of **functional harmony** during development should replace the older concept of division of labour, although it is to be kept in mind that the two concepts are not mutually exclusive. In a tissue system, morphogenesis is a dynamic process where a large number of cells proliferate and differentiate in a precise sequence of events. Innumerable molecular and physical interactions between the cells are required for the control of cell behaviour during tissue morphogenesis. In fact, tissue morphogenesis is normally a process in which order prevails i.e., differentiating cells typically give rise to characteristic holistic tissue patterns. So, we may say that a cell differentiates as it does because it is part of a particular tissue pattern. A very important aspect of tissue differentiation, where like or unlike cells come together to organize a tissue, is the development of **symplastic** and **apoplastic domains**. Cohesion between cells, one of the important requirements of multicellular condition (Raven 1986) and of a tissue, is attained by the development of specialized intercellular apoplastic matrix material i.e., the middle lamella, which binds the different cells together. Communication and transport between the different cells of a tissue is attained by the development of symplastic connections such as plasmodesmata as well as by the apoplastic extracellular matrix i.e., the cell wall.

4.4.1. Intercellular Spaces

It has shown earlier that close packing of the cells in a tissue is mathematically very advantageous as it requires only a minimum of the material for partitioning of the available space and it has an economy of surface in relation to the volume. However, in a real situation, even the cell types resembling a group of soap bubbles, such as the parenchyma cells, are partially separated from one another by the specific separation of vertices, edges and some facets to form **intercellular spaces (I spaces)** (Fig. 54 A and B), but at the same time exhibit the greatest economy of surface in relation to volume. In other words, during I–space formation, the polygonal cells have their internal corners rounded off, and where the angles are rounded off, the cell walls tend to split apart from one another. Thus, each cell seems tending to withdraw, as far as it can, into a sphere, the 'evolutionarily original' shape. This morphological differentiation of I–spaces is almost always accompanied by several structural as well as chemical specializations in the I–space itself or around it during later stages of tissue specialization. Thus, the formation of I–spaces is an event of great significance and tissue specialization (Knox 1993; Krishnamurthy 2000).

I–spaces occur universally in plant tissues. They are known in ground parenchyma, collenchyma, vascular cambium, xylem, phloem, and leaf mesophyll. Based on the origin, I–spaces may be **schizogenous** (as mentioned in the previous paragraph), **lysigenous** (due to disintegration of cells), or **rhexigenous** (due to cell rupture during expansion). Based on the function, I–spaces may be secretory (gum/resin/other materials) or non–secretory. Based on the degree of cell separation, I–spaces may be of **filled** type (small, filled with solid, liquid or gaseous extracellular material, in addition to air), or of **open** type (enlarged, filled with air) (Roberts 1990) (Fig. 57). Ontogenetically, very early–differentiating I–spaces are created where a cell plate fuses with the parent cell wall during division (Fig. 58) (Jeffree et al 1986), as briefly indicated earlier. According to Roberts (1990), the most interesting aspect about the I–space formation is not its mode of ontogeny, but the fact that it knows when to stop developing further.

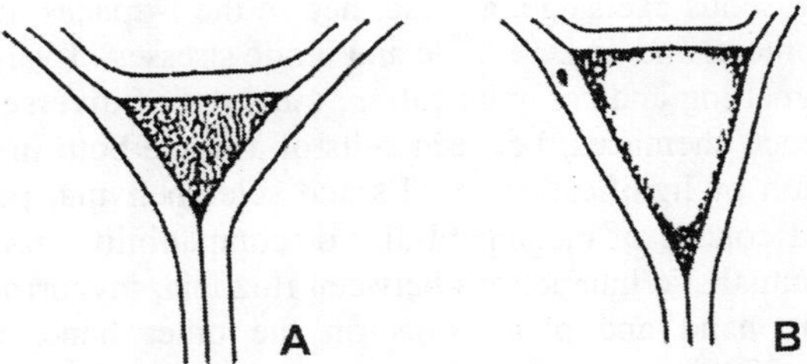

Fig. 57. Diagrammatic illustrations of 'filled' (A) and 'open' (B) types of intercellular spaces; Note that the electron dense materials occupy the corners of 'Open' type of intercellular space (both figures are adapted from Roberts 1990)

Thus, the degree of I–space formation is likely to be controlled and is characteristic of each tissue type. In pea plant, before separation begins, small electron–dense pockets of materials are deposited in the three walls surrounding a future I–space and this deposition is precisely at the point at which cell separation along the middle lamella can be stopped. This material eventually forms the electron–dense material that occupies the corners of the open type I–space (Roberts 1990) (Fig. 57).

The materials (derived either from cell wall or from cytoplasm) reported from I–spaces include cellulose microfibrils, β,1–3 glucans, esterified and non–esterified pectins, xyloglucans and other hemicelluloses, calcium ions, jasmonates, arabinogalactan–rich proteins, threonine–rich proteins, hydroxyproline–rich proteins (extensins), proline–rich proteins, ribonuclease, esterases, acid phosphatases, peroxidases, β,1–3 glucanases, chtinases, receptor kinases etc. depending on the tissue and sometimes in the same tissue at different times (Krishnamurthy 1999a). Hence, the

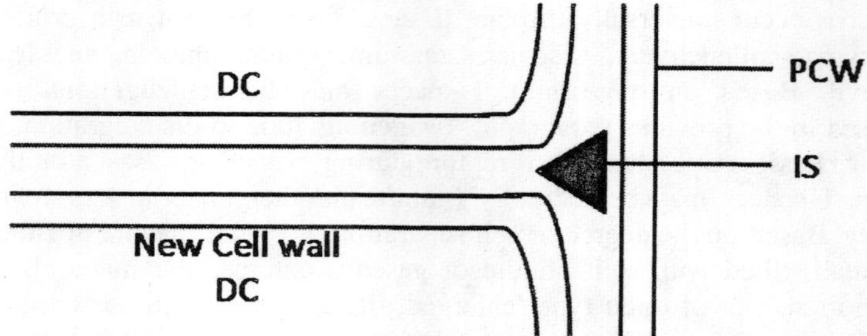

Fig. 58. Diagrammatic structure of mature cell plate and the new cell wall formed between two daughter cells (DC). Note the intercellular space (IS) formed in the place where the cell plate meets the parent cell wall (PCW).

functions of I–spaces must be very diverse. In fact, the following functions, in addition to gaseous exchange, are ascribed to the I–spaces, both during normal development, and under abiotic and biotic stresses of various kinds: intercellular signalling and communication, transport of diverse chemicals (including defence chemicals) between cells of a tissue both in health and disease, initiation of lignification in TEs and sclerenchyma, pollen–pistil interactions and control of incompatibility or compatibility, promotion of symbiotic or mutualistic interactions between rhizobia, mycorrhizae, fungi etc on the one hand and plant roots on the other hand and so on (Krishnamurthy 2000).

4·4·2. Symplastic Domains

A casual mention about symplastic domains has been made in the earlier section of this chapter. In this section, more details are given on symplastic domains and their importance in tissue homeostasis as well as their functions. There would be no development and function of multicellular plants if they do not rely on cell–to–cell communication. Plasmodesmata (PD) are structurally complex channels that span the cell wall and connect the cytoplasm of one plant cell with that of its adjacent cells, and thus, facilitating the effective communication between them. Different structural models have been proposed to explain their organization (Figs. 59 and 60). In the embryos of higher plants, all the cells are initially interconnected through PD and integrated into a single symplast, the domain of common cytoplasm that is bounded by the plasma membranes of connected cells (Gail McLean et al 1997).

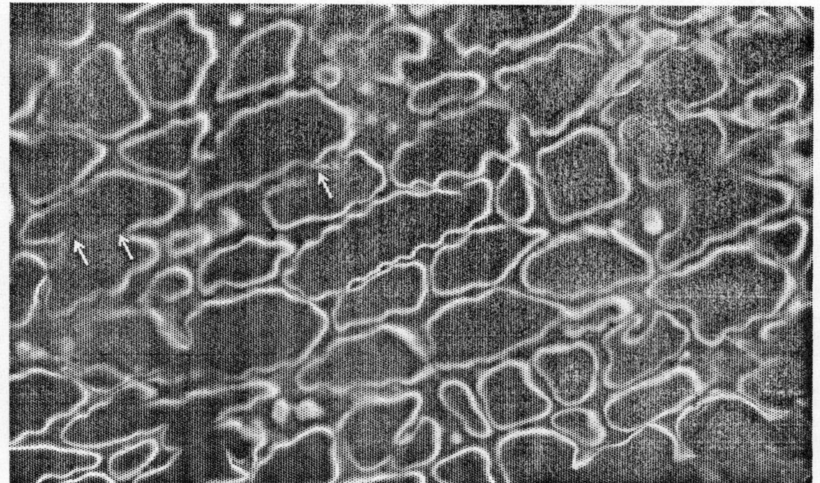

Fig. 59. Plasmodesmata. Phase contrast image of pit fields (thin regions of cell walls) in the thick beaded walls of endosperm cells of coffee seeds showing plasmodesmata (arrows) that connect cells.

As the plant differentiates from the embryo, individual cells or groups of cells become isolated probably by the loss of functional PD. For instance, in a seedling, the root meristem and elongating zones are symplastically isolated, the rhizodermis is symplastically isolated from underlying cells, root hair cells are isolated from other epidermal cells and so on. In *Arabidopsis*, mutants, such as *Knolle* that are unable to develop symplastic domains, have severe developmental defects. Another important instance of symplastic isolation relates to guard cells, which when fully differentiated lose their PD with other cells through degradation of PD, possibly via ubiquitin–mediated proteolysis of plasmodesmal components. This symplastic isolation allows subsets of cells or cell groups to function as distinct compartments. Indeed, as Gail McLean et al (1997) stated, the mature plants can best be described as mosaics of symplastic domains.

The frequency, distribution and functional ability of PD decide the extent or degree of communication and transport within and between these symplastic domains. PD function is decided by its **Size Exclusion Limit (SEL)**, as the size of the plasmodesmata limits the passage of certain larger molecules. This limit is now accurately calculated through the use of fluorescent tracer molecules and confocal laser scanning microscopy. In response to physiological, developmental and/or environmental changes, the permeability of PD between cells, even if they have a broader limit, can be either down–regulated to decrease or prevent symplastic trafficking between cells or up–regulated to allow movement of large molecules and hence regulation of plasmodesmal transport is an important determining

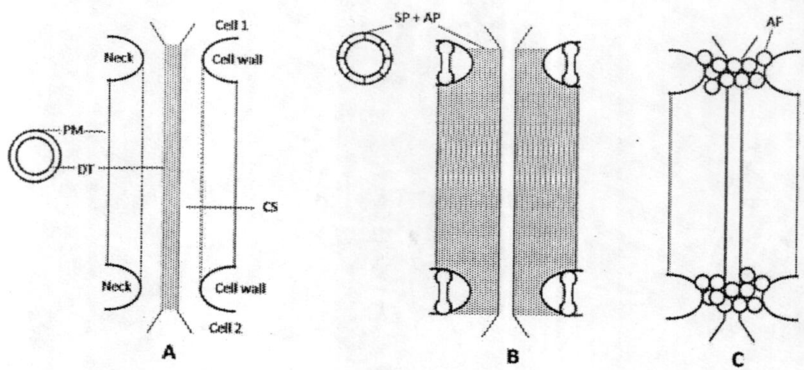

Fig. 60. Models to explain the structure and function of plasodesmata (all shown in diagrammatic L. S.); A - Skeleton model where the membranes delimiting the plasmodesmata are important in their functioning; CS - Cytoplamic sleeve; DT - Tube of appressed ER; PM - Plasma membrane; B - Transmission Electron microscopic model of plasmodesmata; The spokes (SP) and associated proteins (AP) in the neck region are involved in opening the plasmodesmata for movement of cytoplasmic material. An extracellular sphincter (S), associated with the neck region, is also important in the functioning of plasmodesmata. C - An actin-based model for plasmodesmata function. Actins are shown as small circles. These regulate the movement of molecules. (Modified from Gail McLean et al 1997).

factor in symplastic communication and establishment of symplastic domains (Gail McLean et al 1997). PD can also transport endogenous proteins between cells. More than 100 proteins, ranging in molecular weight from 10–100 KDa, apparently move from companion cells into enucleate sieve cells of the phloem. However, such a transport is selective as not all the proteins of companion cells are found in sieve elements.

Symplastic communication is strongly implicated in the regulation of meristematic activity (Refer to BOX 4.1) and development in plants as cell fate appears to depend on its position. For instance, PD disruption between fern prothallus cells induces each cell to differentiate into a separate prothallus suggesting that a signal is transmitted through PD to inhibit cell division and thus, direct all the cells of a gametophyte to develop as a coordinated unit. Thus, positional information and its importance in development and pattern formation are regulated by symplastic communication of specific signals. These signals are possibly growth regulators or their conjugates (proteins) as they can move through PD. The establishment of new symplastic connections and domains is also very vital for plant development. For instance, in some taxa, carpels establish new PD after post–genital fusion, and these connections correlate with the maturation of vascular tissues. Similarly, emerging lateral roots are initially symplastically isolated but later on get connected. Also, in a developing anther, there are PD connections between different sporogenous cells, but these are disrupted when callose walls are formed around each one of them just before meiosis.

BOX 4.1. EVOLUTION OF SHOOT APICAL MERISTEM WITH RESPECT TO PLASMODESMAL NETWORK

Vascular plants have a Shoot Apical Meristem (SAM) not only to produce the shoot with all its component organs such as stem, branches, twings, leaves and flowers or reproductive organs, but also to help in the longitudinal growth of the shoot (For more details, refer to Chapter 5). The structure of SAM differs among plant groups, thus suggesting its probable independent evolution in Lycophytes, ferns, gymnosperms and angiosperms. Several parameters such as structural organization, gene expression patterns and leaf primordial initiation patterns have been used as the basis of such an evolutionary analysis. One such parameter employed is the distribution and pattern of plasmodesmal (PD) network in the shoot apical meristem (Imaichi and Hiratsuka 2007). This parameter was analysed in the SAMs of 17 families and 24 species of angiosperm, gymnosperm and pteridophyte taxa. The SAMs of angiosperms and gymnosperms have low PD density per unit area, with no difference between SAMs showing tunica–corpus organization and those showing cytohistological zonation patterns. On the other hand, SAMs of ferns (including *Psilotum* and *Equisetum*) have average densities of PD (more than 3 times higher). Interestingly, the Lycopods had both fern (in *Selaginella*) and seed–plant (Lycopodiaceae and Isoetaceae) type PD densities. In other words, SAMs with single and plural initial cells have respectively the fern and seed–plant type PD distribution, indicating their probable independent evolution. Further, the fern and seed–plant type PD patterns are equivalent respectively to the lineage–specific network (LPD) (PDs formed in expanding cell plates during cytokinesis) and interface–specific PD network (IPD) (PDs having primary and secondary origins with the later inserted into pre–existing walls) proposed as early as 1996 by Cook et al. Since the coordinated growth of the plant body requires some kind of cell–to–cell signalling and trafficking of regulating proteins and/or mRNAs, PD distribution densities and patterns in apical meristems with single and plural initial cells assume great significance.

For more details, refer Cook et al (1996) and Imaichi and Hiratsuka (2007).

External stimuli may affect symplastic transport by influencing plasmodesmal function, often through their effect on SEL, which is made to either increase or decrease. Anaerobic and osmotic stresses, light quality, temperature etc. affect the symplastic communication and transport. During wounding, PD is plugged by callose so as to isolate the wounded region. Plant viruses can dramatically alter PD function. A maize pathogenesis–related (PR) protein has been localized to PD in order to alter PD function or structure. Thus, it is likely that a systemic acquired resistance may also depend on intercellular transport of signalling molecules, some of which at least may be transported symplastically via PD.

The fact that PD plays an important role in plant development is evident from the movement of developmental regulators between cells. For instance, a plant transcription factor, the maize homeodomain protein *KNOTTED* 1 (*KN1*), moves from cell to cell. Two *Antirrhinum MADS* box family members *DEFICIENS* and *GLOBOSA* also are likely to be transported from cell to cell; so also a third *Antirrhinum* floral regulator, *FLORICOLA*. The fact that important regulators of plant development can move between cells has significant implications for the mechanisms that plants use to program morphogenesis and differentiation. However, as Gail McLean et al (1997) have indicated, the following questions still remain enigmatic: "How does the cell regulate nuclear versus plasmodesmal localization of transported transcriptional factors? How common is macromolecular transport via PD during development?" These and similar questions demonstrate the complexity involved in PD transport of the developmental factors. More studies are needed to resolve this complexity.

4.4.3. Morphogenesis of Epidermal Tissue

The epidermis of the stem, leaf and floral organs is basically derived from the outermost layer of the apical meristem (L1 tunica). Once derived, in the undifferentiated or promeristem condition, it is called protoderm. In roots, it is derived from the outermost layer whose source may be the dermatogen or the outermost layer of the periblem and hence, it is often labelled by many people as **rhizodermis** or **epiblem,** instead of epidermis. But, based on topography the term epidermis is used for the surface layer of both aerial organs and roots of the plant. Normally, the protoderm or epidermis expands only through anticlinal divisions, but periclinal divisions are seen in certain taxa to result in **multiseriate** (or **multiple**) **epidermis** (Fig. 61). Multiseriate epidermis is common in the leaves of *Ficus* and some members of Pittosporaceae, Piperaceae, Begoniaceae, Malvaceae, Arecaceae etc. The **Velamen** of the roots of some orchids and some epiphytic monocots also belongs to the category of multiseriate (root) epidermis (Fig. 62). Multiseriate epidermis not only helps in water absorption, but also in water storage. However, among the layers of the multiple epidermis of leaves, only the outermost layer shares the features of the normal epidermal layer (such as possession of cuticle and stomata).

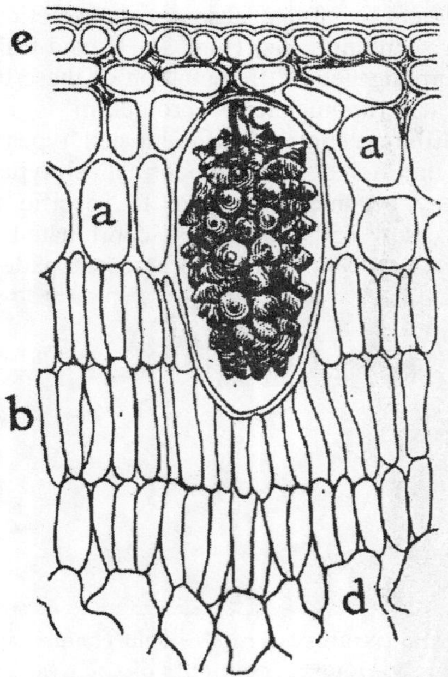

Fig. 61. Cross section of a portion of leaf of *Ficus elastica* showing the multiple epidermis (from e to a) with a cystolith (c), palisade parenchyma (b), spongy parenchyma (d) (Stevens 1911)

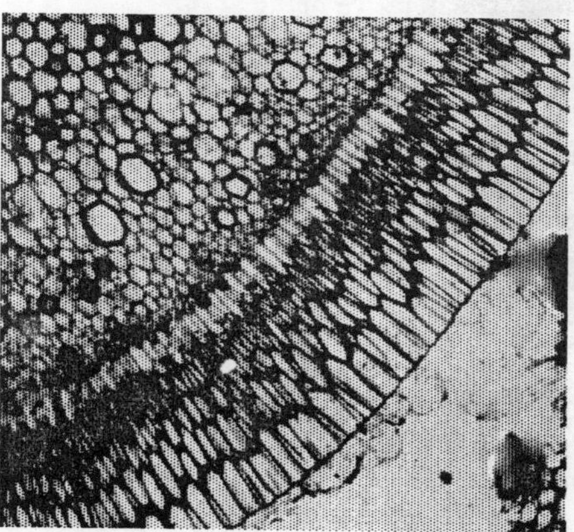

Fig. 62. T. S. of portion of the velamen root of *Vanda* species showing multiseriate root epidermis

The epidermis is a complex tissue system (Fig. 63 A) with diverse cell types such as ordinary epidermal cells, stomata (Fig. 63 B), glandular (Fig. 64) and non–glandular (Fig. 65) trichomes, crystal cells (with silica, calcium oxalate or calcium carbonate), cells with **ectodesmata** or **teichodes** (which help in foliar absorption, excretion etc) (Fig. 50), **hydathodes** (water secreting cells) (Fig. 64 H), and **bulliform cells** (Fig. 66). Hence, the morphogenetic differentiation of these different cell types from a more or less homogeneous protoderm is a matter of great importance. The differentiation of different cell types of the epidermal tissue requires that undifferentiated protodermal cell types are to be first be selected before they become committed to specific fates. Hence, the selection of precursor cells often is coordinated so that mature differentiated cells are distributed in a characteristic pattern.

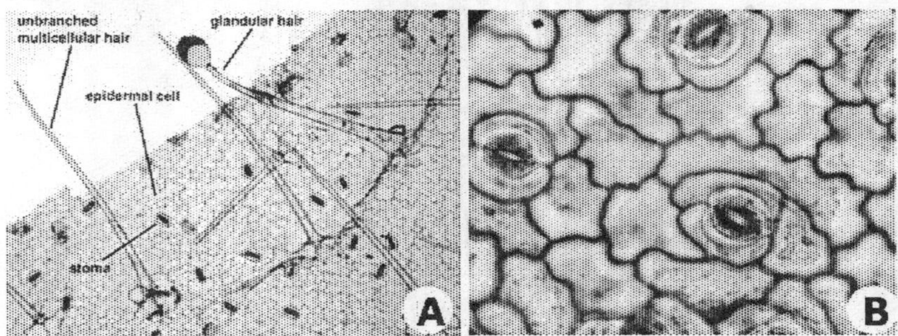

Fig. 63. Epidermal tissue system; A - *Pelargonium* (geranium) leaf epidermis showing various components of the tissue (photograph by John Tiftickjian); B - Epidermal peel of *Basella alba* showing stomata and ordinary epidermal cells

4.4.3.1. Stomatal and Trichome Spacing

One of the simplest tissue patterns is one in which a minimum distance is maintained between differentiated cells in a two–dimensional sheet of cells (Wolpert 1971; Larkin et al 1997). Such a pattern can be produced by different mechanisms. But regardless of how precursor cells are placed, the production of new cells from precursor cells can also contribute to the final spacing pattern (Sachs 1978). The spacing of stomata and trichomes (as well as other distinct epidermal cell types) on the epidermis is a very important morphogenetic event in epidermal tissue differentiation. In a number of taxa, especially in dicot leaves, stomatal spacing appears to be random (whether it is really random needs to be proved beyond doubt), but it appears to be more ordered in other dicots and many monocots. In many taxa, the location of stomatal initials appears to be random to start with, but during their development, non–randomness in stomatal distribution gradually comes about. In other words, a pattern of spatial distribution, with a fixed minimum space existing between them results. Three mechanisms have been suggested in this connection:

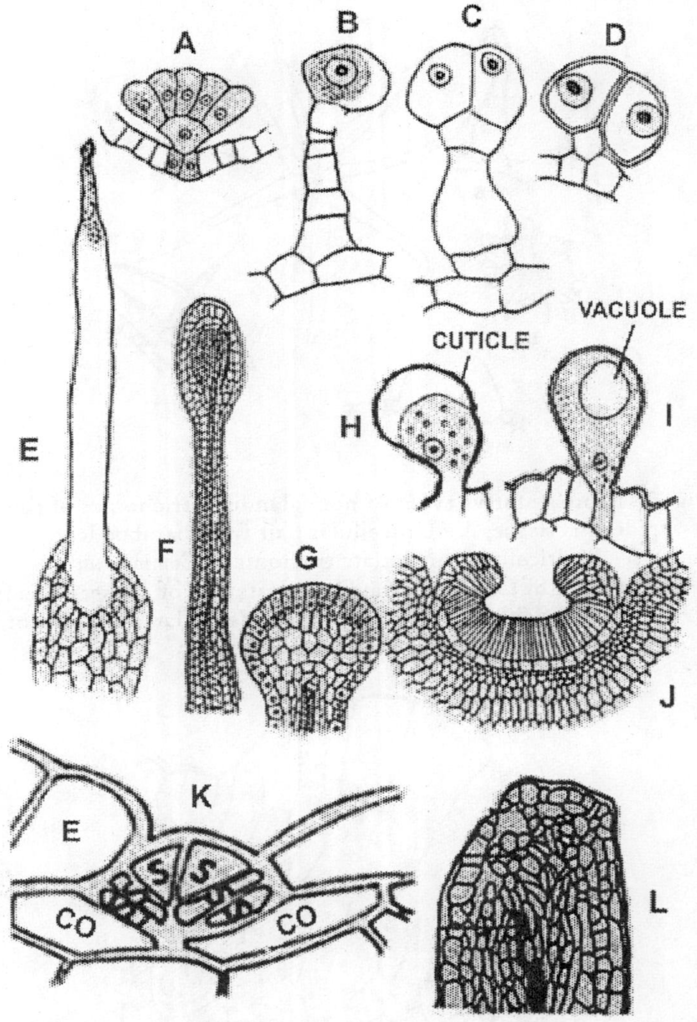

Fig. 64. Some representative types of glandular trichomes of the epidermal tissue; A - L. S. of peltate glandular trichome of *Tecoma capensis*; B–D - Types of glandular trichomes of the leaf of *Hyptis suaveolens*; E - Stinging hair of *Laportea* species; F - Glandular trichome with multicellular stalk in *Drosera indica*; G - Colleter from the petiole of *Jatropha*; H and I - Honey producing glands from the corolla epidermis of *Lonicera japonica*; J - Extra–floral nectary L. S. from *Tecoma capensis*; K - L. S. of salt gland of *Tamarix aphylla* (E - Epidermis; Co - Collecting cells; S - Secreting cells); L - L. S. of hydathode of *Ribes*

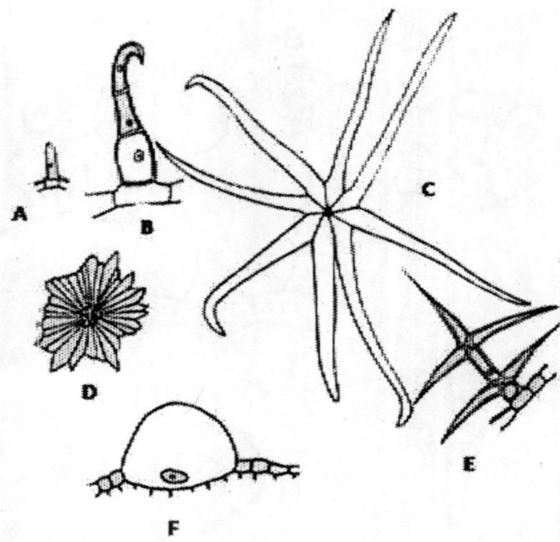

Fig. 65. Representative types of non–glandular trichomes of the epidermis tissue; A - Unicellular hair from bamboo leaf; B - Multicellular uniseriate trichome of *Commelina*; C - Peltate hair of Chrozophora; D - Peltate sale of *Thespesia* leaf; E - Dendroid trichome of *Verbascum;* F - Vesicular Trichome of *Sesuvium* species

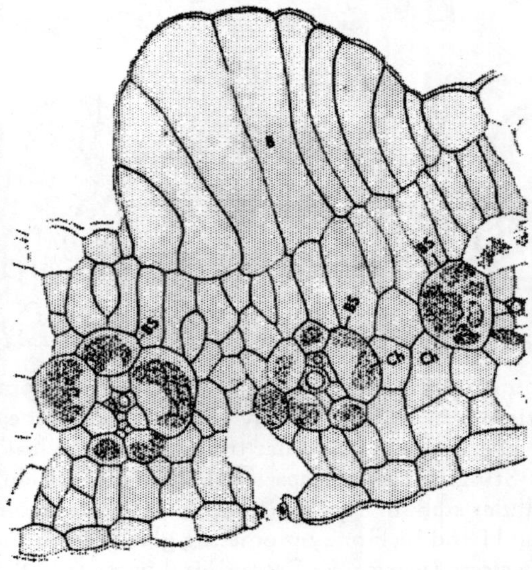

Fig. 66. Transverse section of grass leaf showing bulliform cells (B), bundle sheath (BS) and chloroplasts (Ch)

1. **Cell lineage mechanism**: The spatial pattern of stomata relies on a highly ordered series of cell divisions, usually asymmetric. This is the major driving force behind stomatal patterning in the leaves of many dicots and some monocots. In *Tradescantia*, the pattern of stomatal placement along the entire length of the leaf at the level of the individual initials and the minimum spacing between stomata in a cell file appear to be consistent with the cell lineage theory. A likely involvement of strict cell lineage during the first cell division of the stomatal initial is also underscored by the observation that laser ablation in the developing stomata does not induce stomatal differentiation in neighbouring cells. However, according to others who oppose the cell lineage theory, the facts, that occasional separation of adjacent stomata through variable number of cells and the lack of periodicity in the cell files lacking stomata across the width of the leaf, do not support the cell lineage theory (see Croxdale *et al.* 1992). Studies on stomatal mutants of *Arabidopsis* also lend some support to cell lineage theory. The stomata of this plant are **anisocytic** as there is a pair of guard cells surrounded by three clonally or ontogenetically related subsidiary cells, one of which is distinctly smaller. Hence, each pair of guard cells is separated from another pair by at least one epidermal cell. Cell lineage mechanism is supported by studies on some mutants. In two *Arabidopsis* mutants, *too many mouths (tmm)* and *four lips (flp)* normal stomatal development is disrupted to result in clustering and pairing of stomata respectively on the epidermises of cotyledons and leaves. However, the phenotypes of the two mutants provide the impression that there is indecision in choosing between the effects of the mutations on stomatal patterning and on stomatal production. In a third mutant, *stomatal density and distribution 1–1 (sdd 1–1)*, a two to four–fold density in stomata is promoted. *SDD1* gene perhaps regulates the number of epidermal cells entering into stomatal pathway, thus controlling the spatial pattern. However, the relationship between the abnormal stomatal complexes and the primary gene action effected by these mutations is still unclear.

2. **Lateral inhibition mechanism**: As per this mechanism, the spatial distribution of stomata relies on the interactions or signalling between the different developing stomatal initials. Some kind of inhibitory mechanism from existing stomata or stomatal initials prevents the formation of new stomata within the effective range of the inhibitor. Lateral inhibition, instead of cell lineage mechanism, is said to operate in *Tradescantia*, according to some. Many botanists believe in this mechanism.

3. **Cell cycle dependent mechanism**: As per this mechanism, stomatal patterning is coupled to cell cycle events of a group of initial cells as they are displaced through a specific leaf area. It is commonly observed during stomatal ontogeny in the leaf of many plants that they are arrested at various stages of development. An analysis of such arrested stomata indicated that these inactivated stomatal initials are not randomly located, but are often closely associated with another stoma in an adjacent file (Boetsch et al 1995). On a whole leaf basis, stomatal patterning, based on the cell cycle stage of clusters of initial cells, has found support in the presence of stomatal complexes and stomatal precursor cells in linear discrete groups throughout the leaf length, and differentiation of groups of cells in near synchrony (Chin et al 1995).

Clearly, our current understanding of spacing patterns of stomata is still vague, although it appears that different mechanisms may account for the relation between neighbouring stomata on the leaf epidermis.

The same explanations have been given to the spacing of trichomes in the epidermal tissue (Larkin et al 1996, 1997). The epidermal tissue system of *Arabidopsis* has been analysed from this perspective. In this taxon, the trichome is unicellular and three–branched (Fig. 67). Its single nucleus undergoes endoreduplication (DNA content increases to 16–fold i.e., 32 C level) before settling down at the branching point of the trichome. In this plant, more than 70 trichome mutants have been known and at least 21 genes are involved in the execution of the development of trichome. Of these, two genes that are necessary for the specification of the trichome cell are *GLABRA 1 (GL1)* and *TRANSPARENT TESTA GLABRA 1 (TTG1).* The genes involved in the progressive development of trichome are *GLABRA 3 (GL3)* and *KACTUS (KAK)* that control the enlargement of the committed cell. *GLABRA2 (GL2)* controls the enlargement of the committed cell. *DISTORTED (DIS), GNARLED (GRL)* and *KLINKER (KLK)* control the growth in length of the trichome. *ANGUSTIFOLIA (AN), STICHEK (STI)* and *ZWICHEL* control the branching of the trichome. Two genes are known as negative regulators of trichome patterning in *Arabidopsis* leaf and they are *TRIPTYCHON (TRY)* and *CAPRICE (CPC).* Both the genes are expressed in trichomes and act together. These genes produce small clusters of about four trichomes (Hülskamp et al 1994, Hülskamp and Mathur 2001). The phenotypes of *gl1* and *ttg* mutants have no trichomes on the leaf. *GL1* gene specifies cell fate, maintains cell differentiation and triggers trichome development in cells that normally do not form trichomes (Larkin et al 1993); the last effect is seen when the gene is over–expressed in a *try* mutant (clusters of trichomes are formed here). Thus, trichome formation shows the signs of being the result of an interaction between positive (*GL1*) and negative (*TRY*) gene regulators.

One of the functions of the *TTG* gene in the wild–type plants is to prevent adjacent epidermal cells from becoming trichomes (Larkin et al 1994). It is likely to interact with *GL1* gene.

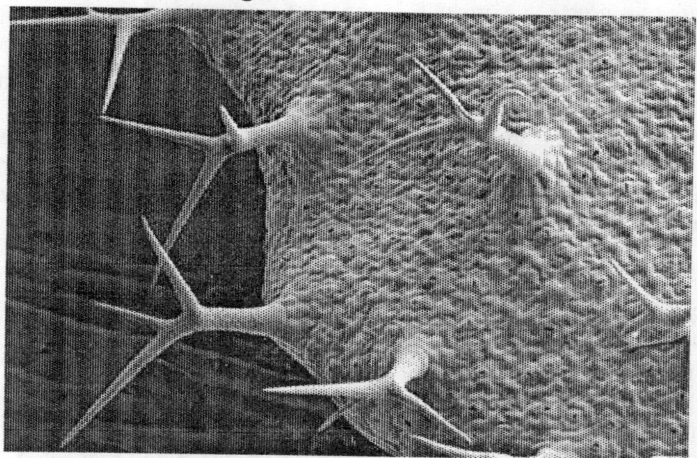

Fig. 67. *Arabidopsis thaliana.* Unicellular, three–branched trichomes (Photograph courtesy of Kim Findlay, John Innes Centre)

4.4.3.2. *Spatial Patterning of Root Hairs*

Regarding spatial patterning of root hairs on roots (Evert 2004), three basic types can be recognized. In type 1, which is seen in most angiosperms (most dicots and some monocotyledons), any epidermal cell of the root can potentially become a root hair. Hence, in this case, the root hairs are randomly produced and hence, are randomly arranged. In type 2, which is seen in Nymphaeaceae and some monocots, root hairs arise from the smaller of an asymmetrically divided epidermal cell, which is called a **trichoblast**. The trichoblast is located at the proximal end (away from the root apex) as in some Alismataceae, Araceae, Commelinaceae, Haemodoraceae, Hydrocharitaceae, Pontederiaceae, Typhaceae, Zingiberaceae etc. or at the distal end of the initial cell, as in Cyperaceae, Juncaceae, Poaceae, Restionaceae etc. In type 3, the rhizodermal cells are arranged in vertical files, composed entirely of shorter hair cells or longer non–hair cells, as in Acanthaceae, Aizoaceae, Basellaceae, Brassicaceae, Salicaceae etc.

In *Arabidopsis,* root hair and hairless cell types are specified in a distinct position–dependent pattern i.e., root hair cells are always positioned over the junction of the radial (anticlinal) walls between two cortical cells (Fig. 31). Several genes are implicated in the root epidermal patterning and these include *TTG1, GL2, WERE WOLF (WER)* and *CAPRICE (CPC).* In *ttg1, gl2* and *wer* mutants, all epidermal cells produce root hairs indicating that *TTG1, GL2* and *WER* are negative regulators of root hair development. The *cpc* mutant does not form root hairs, but transgenic plants over–expressing *CPC* convert all of the root epidermal cells into root hairs. So *CPC* is a positive regulator of root hair development. The expression of *CPC* is controlled by *TTG1* and *WER*, and *CPC* promotes differentiation of hair–forming cells by controlling *GL2*.

Root hairs develop acropetally i.e., younger ones towards the apex of the root. Their initiation sites are intimately linked to acidification of the cell wall. These sites also show an accumulation of expansin and an increase in XET. It is also to be mentioned that root hairs are always positioned opposite non–suberized hypodermal cells, wherever the roots have an exodermis (Fig. 52).

4.4.4. Sclerenchyma Tissue

The factors controlling the development of various types of fibres (Fig. 68) and sclereids (Fig. 69) have been understood to some extent through experimental studies.

Sachs(1972) and Aloni (1976, 1978) have shown that fibre strand development in stems is dependent on stimuli originating in young leaf primordia, since removal of leaf primordia in pea plant prevented fiber differentiation. Similarly, experimental change in position of leaves changed the locus of differentiation of fibers in the subjacent region. In *Coleus*, the induction of primary phloem fibres is polarized in a downward direction from the leaves to the root. The effect of leaves can be replaced by exogenous supply of IAA in combination with GA3 (Aloni 1979). The genetic basis of fibre development has been shown by a study of mutants of *Arabidopsis*. The mutants *interfascicular fibreless1(ifl1)* and *revolute (rev)* do not develop fibres in the interfascicular region (Zhong and Ye 1999). Perhaps, these have their action through their effect on polar transport along the interfascicular region. Wounding or surgical experiments involving removal of portions of organs provide evidence of

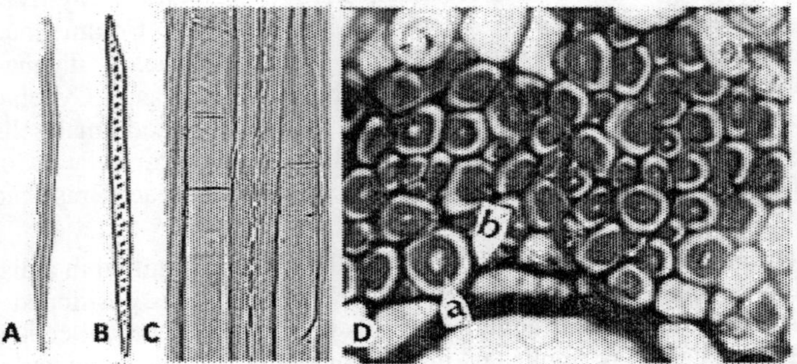

Fig. 68. Different types of fibers; A - Libriform fiber; B - Fiber tracheid; C - Septate fiber; D - A group of gelatinous fibers of *Bauhinia purpurea* in T.S. (a - Gelatinous layer of the cell wall; b - Lignified layer of cell wall)

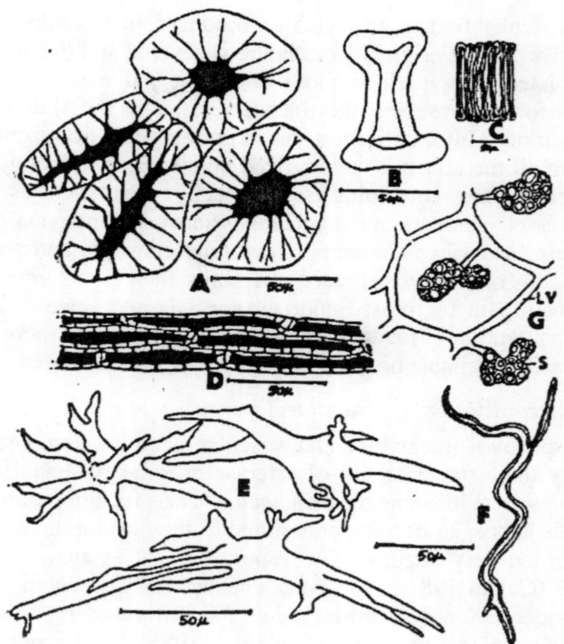

Fig. 69. Representative types of sclereids; A - Brachysclereids from the flesh of pear fruits; B - macrosclereids from bean seed coat; C - A row of sclereids from bean seed coat; D - Columnar sclereids from the dried scale leaves of onion; E - Astrosclereids from the leaves of *Gnetum ula*; F - Branched trichosclereids from the leaves of *Memecylon umbellatum*; G - Terminal sclereids from the leaves of *Salvadora persica* (LV - Leaf vein; S - Sclereid)

the positional control of sclereid development. Auxin levels in leaves influence sclereid development; high concentrations suppress sclereid development, while low concentrations affect lignification and hence, moderate auxin level is required.

4.4.5. Vascular Tissues

The vascular tissues contain xylem that conducts water and sap, and phloem that transports organic substances. In the intact plant, xylem and phloem are present in close proximity to one another, with variable relative arrangement between them such as collateral, bicollateral, concentric or radial (Fig. 70). In the primary parts of the plant, the vascular tissues are derived from the procambium, while in the secondary parts, they are derived from the vascular cambium and hence, are respectively called **primary xylem** and **phloem**, and **secondary xylem** (or **wood**) and **phloem** (or **bast**). Vascular tissues may also be produced in wounded regions of intact plants, when the original vasculature is severed and in calli produced in vitro. Isolated cells, such as those of *Zinnia* leaf mesophyll, can also directly transform themselves into tracheary elements (TEs) under suitable culture conditions. Information on the morphogenesis of vascular tissues derived from all the above is summarized below. Both xylem and phloem are complex tissues containing several cell types such as tracheary elements (vessel elements and tracheids), fibres, parenchyma (axial and ray) in xylem, and sieve elements (sieve tube elements and sieve cells), companion cells or strasburger cells, and fibres and parenchyma (axial and ray) in phloem. But the most important are TEs and sieve elements and hence, more attention is focused on the morphogenesis of these elements, although details, available on other cell types, are also provided.

4.4.5.1. *Differentiation of Vascular Tissues*

The prospective procambial and vascular cambial tissue regions are indicated by the strong activity of esterase enzyme (Gahan 1981). It is, perhaps, an earliest histochemical indicator. The procambial and vascular cambial cells appear to become committed to the development of vascular tissues from the very beginning in response to native auxins, cytokinins, and sucrose (Gahan 1981). However, whether this histochemical reaction provides evidence for **determination** or for **competence** of the vascular meristems remains unclear (Savidge 1985). Determination (or commitment) refers to the genome becoming restricted such that only a single differentiation pathway i.e., vascular differentiation pathway can be followed, whereas competence refers to the capacity of the (prevascular) tissue to respond to an inductive stimulus by following a particular developmental pathway, different from one that it would follow in the absence of induction (Bird et al 1982). On the basis of surgical experimental studies carried out in castor, where a part of the cambial cylinder and derivative tissues were surgically removed and readjusted in their places with inverted polarity (keeping the xylem side outward and phloem towards the pith) and where new xylem was formed on the bark side and new phloem on the pith side, Siebers (1971) concluded that the sites of initiation of cambium and the polarities of tissue differentiation from the cambium were already determined. On the contrary, studies with cultured callus (Details are given in Chapter 10) showed that phloem was formed positionally in association with xylem, and thus, it was proved that the newly produced cells respond to the factors that regulate differentiation process in contiguous tissues. In other words, the cells are competent but not determined for any specific vascular differentiation pathway. That the cambial derivatives are competent, but not determined for any specific differentiation pattern is also evident from the fact that they normally differentiate into a variety of elements of the two different vascular tissues (Savidge 1985). Savidge (1985) has indicated that the differentiation of different cell types from vascular cambium (also from procambium) is based on the relative concentration of auxins, gibberellins and ethylene. He further suggested that a Tracheid–Differentiation Factor (TDF), a Bordered Pit Factor (BPF), a Lignin Factor (LF), and an Autolysis Factor (AF) operate to regulate different aspects of TE differenatiation. High TDF concentration, in the absence of BPF, leads to the formation of libriform fibres. Xylem parenchyma is formed when TDF, BPF, LF and AF are either absent or annulled in the precursor cells (Savidge 1985).

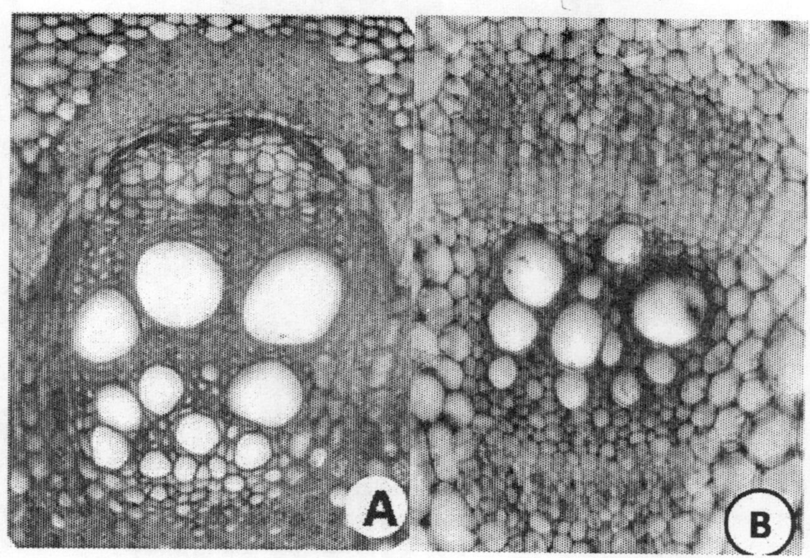

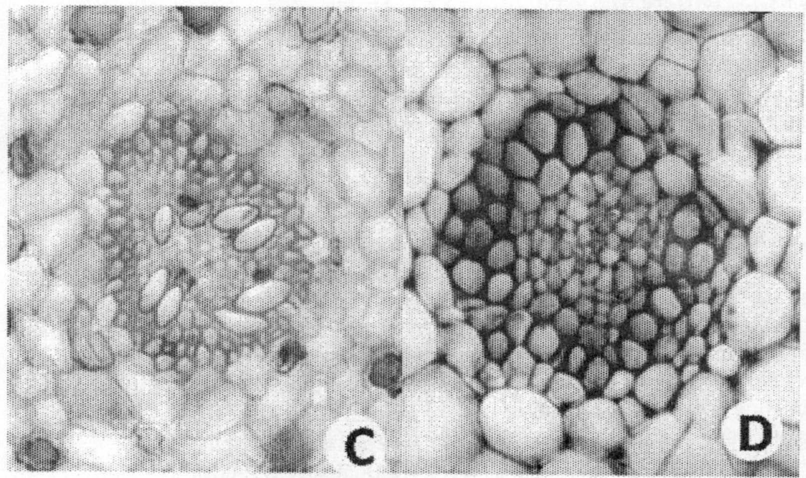

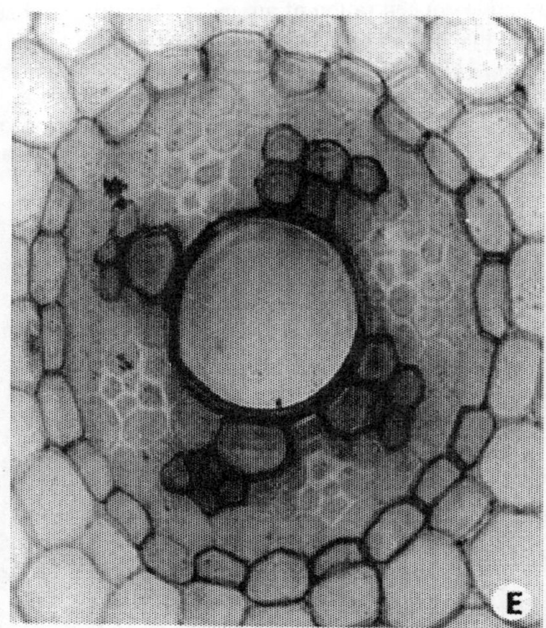

Fig. 70. Different types of vascular bundles; A - Collateral (*Coscinium fenestratum*); B - Bicollateral (*Cucurbita* species); C - Amphicribral (*Cyperus rotundus*) D - Amphivasal (*Polygonatum verticillatum*); E - Radial (*Allium stracheyi*)

4.4.5.1. *Differentiation of Vascular Tissues*

The prospective procambial and vascular cambial tissue regions are indicated by the strong activity of esterase enzyme (Gahan 1981). It is, perhaps, an earliest histochemical indicator. The procambial and vascular cambial cells appear to become committed to the development of vascular tissues from the very beginning in response to native auxins, cytokinins, and sucrose (Gahan 1981). However, whether this histochemical reaction provides evidence for **determination** or for **competence** of the vascular meristems remains unclear (Savidge 1985). Determination (or commitment) refers to the genome becoming restricted such that only a single differentiation pathway i.e., vascular differentiation pathway can be followed, whereas competence refers to the capacity of the (prevascular) tissue to respond to an inductive stimulus by following a particular developmental pathway, different from one that it would follow in the absence of induction (Bird et al 1982). On the basis of surgical experimental studies carried out in castor, where a part of the cambial cylinder and derivative tissues were surgically removed and readjusted in their places with inverted polarity (keeping the xylem side outward and phloem towards the pith) and where new xylem was formed on the bark side and new phloem on the pith side, Siebers (1971) concluded that the sites of initiation of cambium and the polarities of tissue differentiation from the cambium were already determined. On the contrary, studies with cultured callus (Details are given in Chapter 10) showed that phloem was formed positionally in association with xylem, and thus, it was proved that the newly produced cells respond to the factors that regulate differentiation process in contiguous tissues. In other words, the cells are competent but not determined for any specific vascular differentiation pathway. That the cambial derivatives are competent, but not determined for any specific differentiation pattern is also evident from the fact that they normally differentiate into a variety of elements of the two different vascular tissues (Savidge 1985). Savidge (1985) has indicated that the differentiation of different cell types from vascular cambium (also from procambium) is based on the relative concentration of auxins, gibberellins and ethylene. He further suggested that a Tracheid–Differentiation Factor (TDF), a Bordered Pit Factor (BPF), a Lignin Factor (LF), and an Autolysis Factor (AF) operate to regulate different aspects of TE differenatiation. High TDF concentration, in the absence of BPF, leads to the formation of libriform fibres. Xylem parenchyma is formed when TDF, BPF, LF and AF are either absent or annulled in the precursor cells (Savidge 1985).

Experimental manipulations that have been widely exploited to understand the vascular differentiation and the factors controlling it include wounding by a V–shaped cut that severs the vascular bundle or the removal of leaves lying above the stem that is subjected to experimental study. Severing of vascular continuity of the root by root knot nematodes (species of *Meloidogyne*) also promotes the same type of tissue response as in V–shaped cuts (Krishnamurthy, unpublished data). V–shaped cuts confer competence on the neighbouring parenchyma cells to differentiate into new vascular tissues that bypass the wound and link up with the severed tissues below and above the wound (Fig. 71). Similar response was seen in nematode infected roots (Fig. 72). The classical wounding experiment was done by Jacobs (952) on internodes of *Coleus blumei* and this revealed that IAA is the limiting factor in xylem regeneration around the wound. Removal of the leaves above the wound prevented the normal regeneration of new xylem elements, but replacement of the excised leaf on the cut location again promoted xylem regeneration. Surgical removal of the root tips of nematode–infected roots also prevented the differentiation of new vascular tissues (Krishnamurthy, unpublished data). Differentiation of pith or cortical parenchyma cells into xylem elements is predominantly basipetal in progression, paralleling the direction of auxin transport. Based on such experiments in pea seedlings, the **canalization hypothesis** was framed, according to which the flow of auxin through the cells, prior to their differentiation as xylem elements, provides the signal that determines the orderly pattern of xylem differentiation. According to this hypothesis, after some of the parenchyma cells get differentiated into xylem elements, they become preferential conduits for further auxin transport and begin to act as sinks to attract auxin molecules. This has the effect of directing new streams of auxin from young developing leaves (or root apical meristem) towards the vascular tissues, thus causing new vascular elements to be formed towards a pre-existing strand (Sachs 1981).

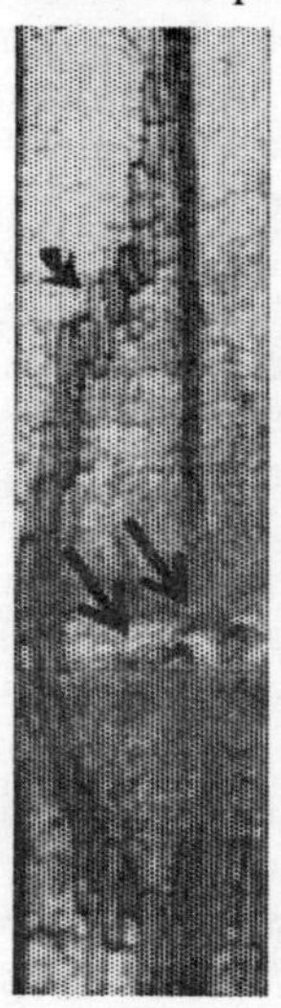

Fig. 71. Regeneration of tracheary elements (single arrows) around wounded vascular strands (double arrow) (After Thompson and Jacobs, 1966)

The above–mentioned process has been analysed from a molecular standpoint by Baima et al (1995). They studied the expression pattern of an *Arabidopsis* gene encoding homeodomain proteins during revascularization in wounded transgenic tobacco internodes. The promotor segment of this gene was linked to the GUS gene, and the reporter gene expression pattern showed the presence of the transgene in single cell or clusters of the cells soon after the wounding. Later, the gene was expressed in the cells along the path of vascular elements formed. The expression of the homebox gene was found to be activated by auxin in leaf tissues, as determined by Northern–blot analysis, thus, supporting a role for auxin in the vascular tissue formation. In the experiments with excised wounded internodes of *Coleus blumei* receiving IAA at their apical ends and a cytokinin at their base, a role for the latter in increasing the regeneration of xylem elements was clearly evident (Baum et al 1991). They showed that in the presence of auxin, there is only a low level of xylem regeneration restricted to the area immediately around the wound, but cytokinin causes a substantial increase in the xylem regeneration above and below the wound.

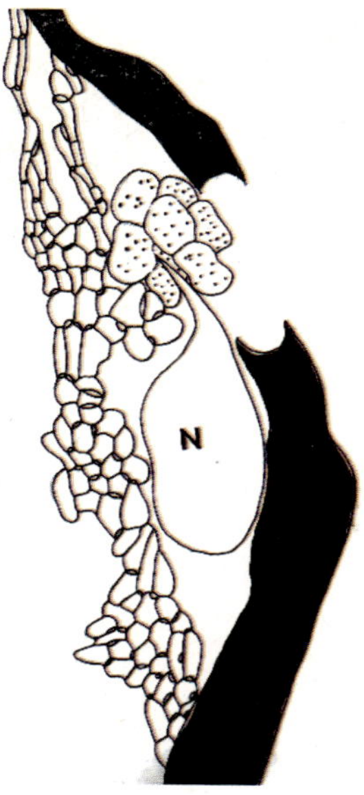

Fig. 72. Partly diagrammatic representation of the L.S. of the root of *Basella alba* infected with the nematode (N), *Meloidogyne javanica*. The original xylem part of the root vasculature (darkened) is injured resulting in discontinuity, by the nematode feeding on specially formed giant cells (dotted cells). The discontinuous vasculature is connected through differentiation of tracheary elements from trans-differentiating parenchyma cells shown on the left side of the figure.

Phloem regeneration was also studied in V–shaped cuts mentioned earlier. La Motte and Jacobs (1963) concentrated their attention on such regenerating phloem in the V–cut *Coleus blumei*. As in xylem regeneration, new phloem elements got differentiated from pith or cortical parenchyma and bridged the severed phloem continuity; also the differentiation is under auxin and cytokinin control. Leaves are shown to mimic the effect of these growth regulators.

In vitro systems also help greatly in understanding the cellular basis of TE differentiation, although some question whether findings of an in vitro system can validate what is happening in natural systems. Organogenic calli (Fig. 73) or specially developed xylogenic calli (Kalimaoorthy and Kirhsnamurthy 1998; Christopher and Krishnamurthy 2004) offer good tissue systems for studying xylem (and phloem) tissue differentiation (Fig. 74), while *Zinnia elegans* mesophyll cell explants form the best study materials for understanding TE differentiation ever since it was introduced (Fukuda 1992). These mesophyll cells undergo **transdifferentiation** into tracheary elements on suitable xylogenic media (with auxin and cytokinin) and in the absence of cell division (Fig. 75)

Fig. 73. Homogeneous parenchymatous organogenic callus of *Cucumis sativus* (photograph supplied by Dr. S. Kaliamoorthy)

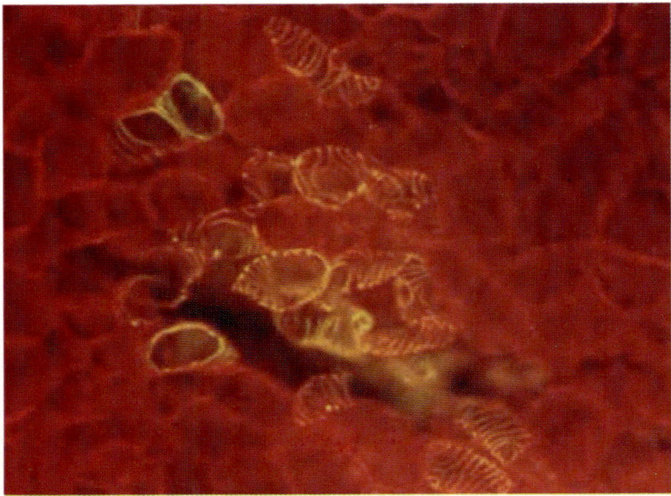

Fig. 74. - Central part of a xylogenic callus of *Cucumis sativus* showing transdifferentiated tracheary elements from parenchyma cells

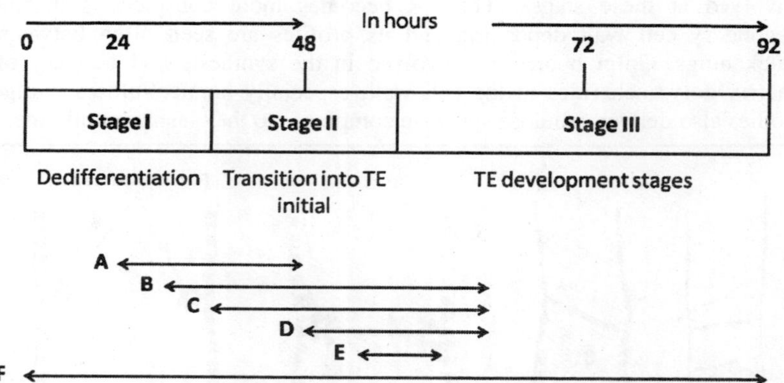

Fig. 75. Stages in the transdifferentiation of leaf mesophyll cells into tracheary elements (TE) and the localization patterns of the transcripts of involved genes; A - Genes operating during dedifferentiation to the initiation of transition to TE; B - Genes operating during dedifferentiation up to the middle of TE development; C - Genes operating right from the transition of mesophyll cells into TE up to the middle of TE development; D - Genes operating from the transition of mesophyll cells into TE initial up to the middle of TE development; E - Genes unique to the early stages of the TE development from the transformated leaf mesophyll cells; F - Genes operating throughout the stages I–III of the TE development from the leaf mesophyll cells (Based on Fukuda 1997)

These cells become highly vacuolated, and each cell has a nucleus and a full complement of organelles. During differentiation the nucleus undergoes dramatic changes in both size and ploidy level very early. Sometimes, it undergoes one or a few mitotic divisions. Endoreduplication probably provides the differentiating TE with additional gene copies to meet the heavy demand for the synthesis of cell wall and cytoplasmic components (Gahan 1988), particularly MTs and enzymes. The expression of tubulin genes begins and this expression continues increasing the amount of MT. After cell enlargement is completed, secondary wall layers are deposited in a pattern that is characteristic of the given type of TE. There is change in the distribution of cortical MTs from random to even-spreading, and also in their orientation from longitudinal to transverse to facilitate expansion of radial wall. During secondary wall formation, these helically arranged MTs shift in orientation several times ending in a flat, S–helix, with concomitant changes in the orientation of cellulose microfibrils. The cortical MTs get concentrated in the bands at the sites of secondary wall thickenings. Secondary wall formation is just preceded by a rapid increase in the brassinosteroids, which are necessary for later stages of TE differentiation. Calcium or calcium–calmodulin system is also involved at these stages. The ER becomes more conspicuous during secondary cell wall deposition and its profiles are seen often between thickenings. Golgi bodies are involved in the synthesis and delivery of matrix polysaccharides to the cell wall, especially hemicelluloses. Golgi bodies also deliver cellulose synthase complexes to the plasma membrane.

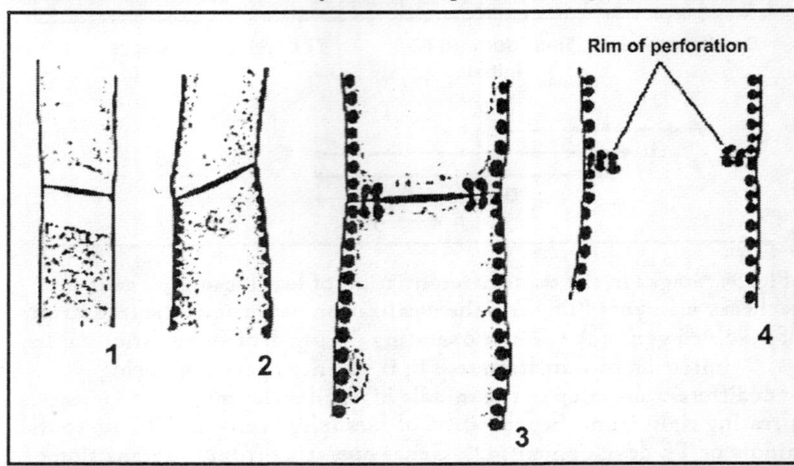

Fig. 76. Stages in the formation of end wall perforation in the vessel element (Esau and Hewitt 1940)

Changes in the enzymes such as xylan synthase, PAL, 4–coumarate co–enzyme A(CoA), ligases, caspases, peroxidases and nucleases, and in other biochemical activities have been closely correlated with TE differentiation. Nearly 400 polypeptides have been identified (Fukuda and Komamine 1983, Fukuda 1997). Transcipts of several genes including those of *TED2*, *TED3* and *TED4* genes were expressed in the differentiating TEs. Recent data suggests the involvement of 380–500 specific genes in the differentiation of TEs, which is divided into three stages (Fig. 75). These genes were also expressed in the differentiating xylem elements of the seedlings of this plant as localized by in situ hybridization (Demura and Fukuda 1994), although the expression of *TED3* transcripts is restricted to the two small regions within the procambium and to immature xylem cells and phloem cells.

In vivo studies have also provided a lot of information on xylem cell differentiation. The cells developing into vessel elements may elongate slightly and broaden. The portions of the primary wall destined to form a perforation plates are not covered by secondary wall material, but get thickened due to swelling of the intercellular substance (Fig. 76) (Esau and Hewitt 1940). In some plants, this swelling is due to the removal of some of the matrix wall materials by wall–degrading enzymes and the subsequent imbibitions of water by the remaining carbohydrates.

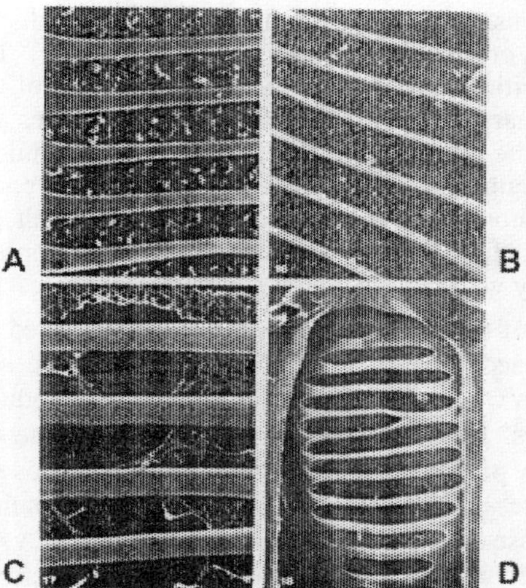

Fig. 77. A - An early stage in the breakdown of the perforation partitions in a scalariform plate of *Ascarina lucida*; B - Microfibrillar web traversing all the openings in a scalariform perforation plate of *Liquidamber styraciflua*; C - Remnants of perforation plate of *Ascarina lucida*; D - Fully formed scalriform perforation plate (Meylan and Butterfield 1981)

Perforations are formed after the laying of lateral secondary walls. The perforation partition develops a granular surface and then numerous small holes before its total disappearance. Most non–cellulosic polysaccharides and some cellulosic components are first removed leaving the partition wall with a mat or web of cellulosic microfibrils (Fig. 77 A–D) (Meylan and Butterfield 1981) during perforation plate formation (Benayoun et al1981), especially in the vessel elements with scalariform perforation plates.

The protoplasm of the cell dies typically after the perforation plate formation in *Zinnia* cell cultures and this is different from the in vivo mechanism described above. The rings of MT are associated with the development of perforation plates, with or without association of the actin filaments. The MTs disappear at the sites of pit formation.

Special attention should be given to the morphogenesis of unique tracheary elements present invariably in the haustorium, the anatomical and physiological organ bridging the parasite and the host plant, of parasitic angiosperms (Fineran 1994). These are called **graniferous tracheary elements** and may belong either to the vessel element or the tracheid category, but tracheids are often more abundant. They are found both in primary and secondary xylem. They are so–named because their lumen contains structural inclusions, which are generally granular, but other configurations, such as amorphous masses of dispersed material having the same origin as granules, may also occur. These elements probably play a critical role in the functional operation of the haustorium. Graniferous tracheary elements are of two distinct types in terms of the composition and the origin of granules; those with granules composed of proteins, often secondarily impregnated with phenolic compounds (Figs. 78 and 79), and those with granules composed of starch grains (Fig.80). Typically, the protein granules appear within the cytoplasm before the bands of secondary wall thickenings begin to differentiate.

On maturation of the cell, the cytoplasm becomes lysed and the protein granules are released from their limiting membrane sac and come to lie freely in the lumen where they persist throughout the conducting life of the tracheary elements. In this respect, they differ from the normal TEs of other plants. The protein of the granule is either loosely arranged or compacted. In the haustorium of Olacaceae, the protein granules are different from those of other parasitic families, and they are structurally more diverse. Their matrix is mostly tubular and at maturity of TEs, the tubules become highly dispersed throughout the cell lumen. Also, there are protein bodies, which are crystalloid and are vacuolar in origin in Olacaceae. Crystalloid proteins perhaps also occur in some taxa of Santalaceae such as *Thesidium*. Starch granules occur in Scrophulariaceae family and in *Ximenia* of Olacaceae. They are of amyloplast origin and are released into the lumen of TEs from their amyloplast confines (Fig. 80). Graniferous TEs have been suggested to have a control over the flow of xylem sap passing through the haustorium. Graniferous TEs are also called by some **phloeotracheids**, as they share some of the features of phloem such as protein granules.

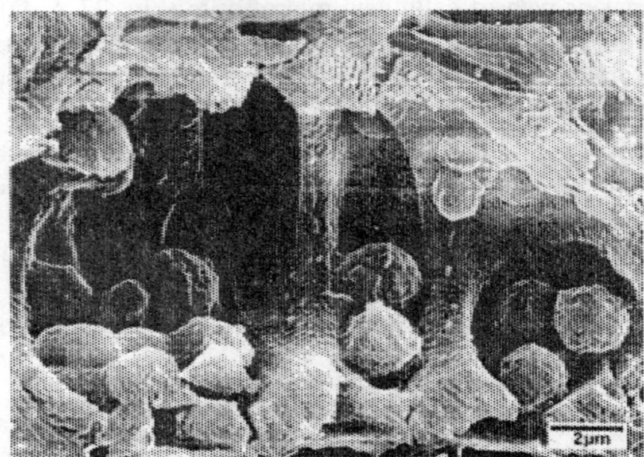

Fig. 78. Scanning electron micrograph of a graniferous tracheid of *Thesium divaricatum* showing protein granules (Photograph courtesy of Dr. Fineran)

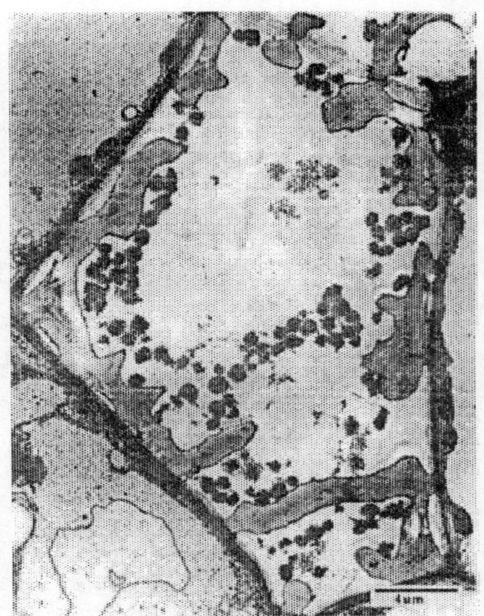

Fig. 79. T.S. of a graniferous tracheid of *Colpoon compressum* as seen under transmission electron microscope, showing several protein granules (Photograph courtesy of Dr. Fineran)

The morphogenesis of sieve element differentiation is very interesting. It is well–known that the mature sieve elements are without nuclei (Fig. 81). Nuclear degeneration is by **chromatolysis** (loss of stainable contents i.e., loss of chromatin) and the eventual rupture of the nuclear membrane or by **pycnotic degeneration** in which the chromatin forms a very dense mass prior to the rupture of the nuclear membrane. Simultaneous with the nuclear degradation, the cisternae of ER begin to form stacks and to migrate towards the wall. The ribosomes disappear from the surfaces of the ER that face one another in a stack followed by ribosomes on the outer surface to become smooth ER. The latter, then, undergo further modification into convoluted, lattice–like and tubular form. Only two kinds of organelles are retained: plastids and mitochondria. **P–proteins** (formerly called **slime bodies**) are also found in mature sieve elements of many plants. The P–proteins are amorphous, granular, fibrillar, filamentous, tubular or crystalline depending on the species and the stage of differentiation of the sieve elements (Fig. 82 A, B). Two major types of plastids, S–type (containing only starch) and P–type (storing proteins exclusively or in addition to starch) are present depending on the species. The mitochondria in mature sieve elements of some taxa lose their internal membrane as stroma becoming thinner. The plasma membrane remains intact, but tonoplasts breakdown to result in what is called a **mictoplasm**.

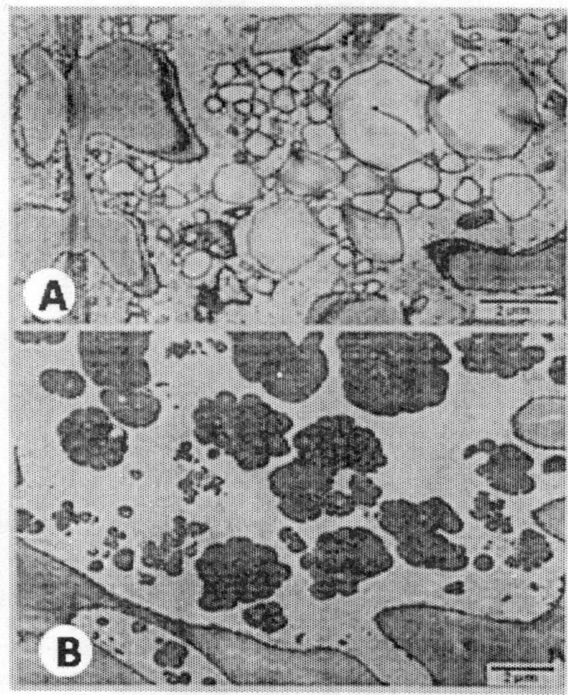

Fig. 80. Ultrastructure of graniferous tracheary elements respectively of *Lathraea clandestine* (A) and *Krameria grayii* (B) showing starch grains (Photograph courtesy of Dr. Fineran)

The future sieve plates in differentiating young sieve elements appear like any other primary wall except that it has fewer plasmodesmata. Each prospective pore site contains a simple plasmodesma in the centre and first becomes distinquishable from the rest of the wall by the appearance of a pair of callose platelets, one on either side of this pore. Cisternae of ER are closely appressed to the plasma membrane bordering the callose platelets at all stages of development and seems to be connected to the plasmodesmal core, the structure of which was already shown earlier in this chapter. Rapid enlargement of the callose platelets delimits the pore sites before wall thickenings occur elsewhere on the young sieve plate. The wall thickenings between pore sites soon overtake the platelets and the pore sites then look like depressions in the plate. The perforation gets initiated with the removal of cellulosic wall material that surrounds the plasmodesma and is sandwiched between the callose platelets. The opposite platelets fuse and then are broken through, giving way to the pore uniformly lined with plasma membrane and callose (Esau and Thorsch 1985). The callose platelets prevent or inhibit cellulose deposition at the sites of future pores, and ER at the pore site may be involved with the laying down of the platelets and later with the release of enzyme activity that leads to the removal of callose and other wall materials.

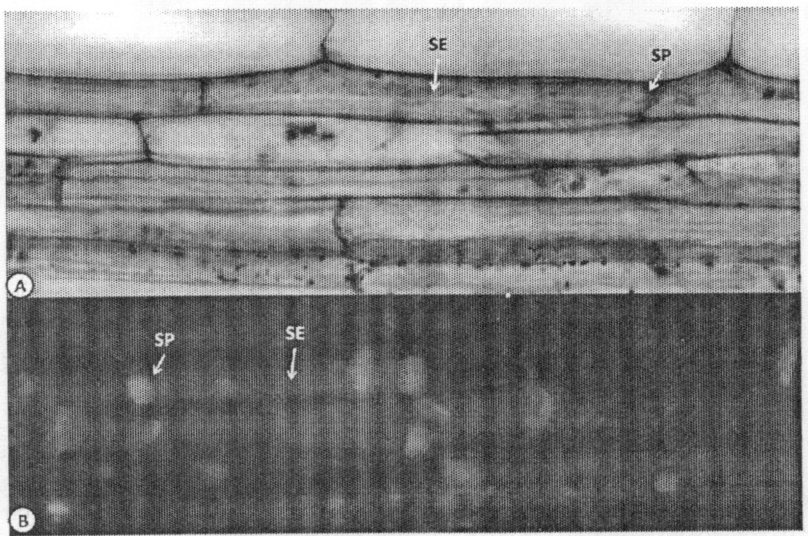

Fig. 81. L.S. of phloem region of *Curcurbita* stem respectively in bright field (A) and fluorescence (B) optics, the later stained with aniline blue. Callose fluoresces in the sieve plate (SP) areas of sieve tube elements (SE); the sieve elements do not have nucleus at maturity. Callose bodies are seen dispersed in the sieve tube element (seen as black deposits in figure A and small fluorescing bodies in figure B)

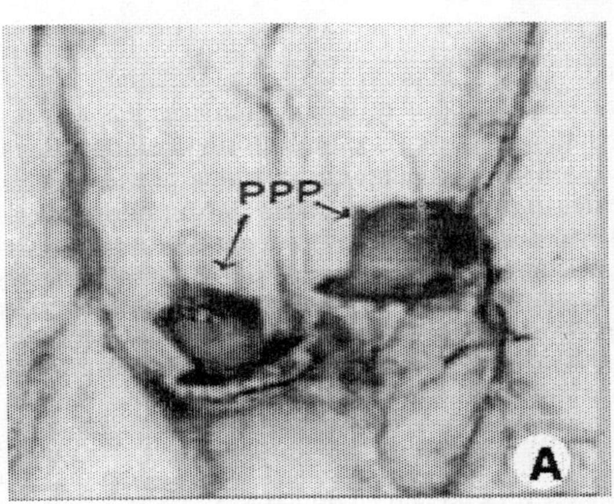

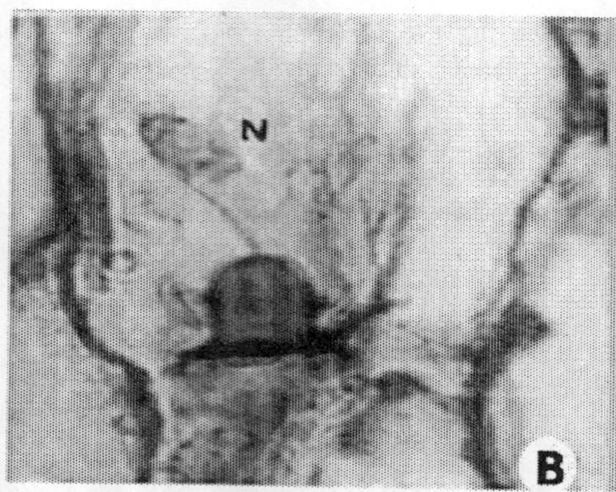

Fig. 82. P - Proteins (PPP) of the sieve elements of *Dalbergia sissoo*; N - Nucleus (Photographs courtesy of Dr. N. Venugopal)

4.4.5.2. *Differentiation of Leaf Vasculature*

The pattern and ontogeny of leaf vasculature i.e., venation appear to guide or limit leaf differentiation and function. All the other leaf tissues differentiate in positions showing a spatial relationship to the vascular system and its pattern. It is not clear how and where leaf vascular pattern is created or revealed, as the data available so far is largely descriptive. A universal theory of leaf vascular pattern formation will have to account for (i) the continuous and acropetal formation (in some cases basipetal formation) of primary and secondary veins in many dicot leaves (Fig. 83) (ii) the formation of generally parallel vascular strands and the strands interconnecting them of monocot leaves (Fig. 84), and (iii) the simultaneous formation of minor veins in both dicots and monocots (Figs. 83 and 84).

Till now no single explanation or theory fully accounts for these disparate spatial and temporal venation patterns, but three general hypotheses, each based on a different approach, have been suggested to explain leaf venation pattern: (i) **Canalization** or **signal flow hypothesis** is based primarily on experimental observations of the inductive effects of auxin on vascular tissue formation (Sachs 1981, 1989, 1991 a, b). (See also pages 138-140) (ii) **Diffusion–reaction pattern hypothesis** is based on computer modelling of interactions among hypothetical diffusible substances (Meinhardt 1996, Koch and Meinhardt 1994). Both hypotheses make use of Turing's diffusion–reaction systems, about which information has been provided earlier in this chapter. (iii) A third hypothesis speaks of the creation of venation patterns based on mathematical modelling for self–organization of two–dimensional space by using topological rules and fractal science in combination with estimates of physiology and transport requirements in a leaf (Kull and Herbig 1995).

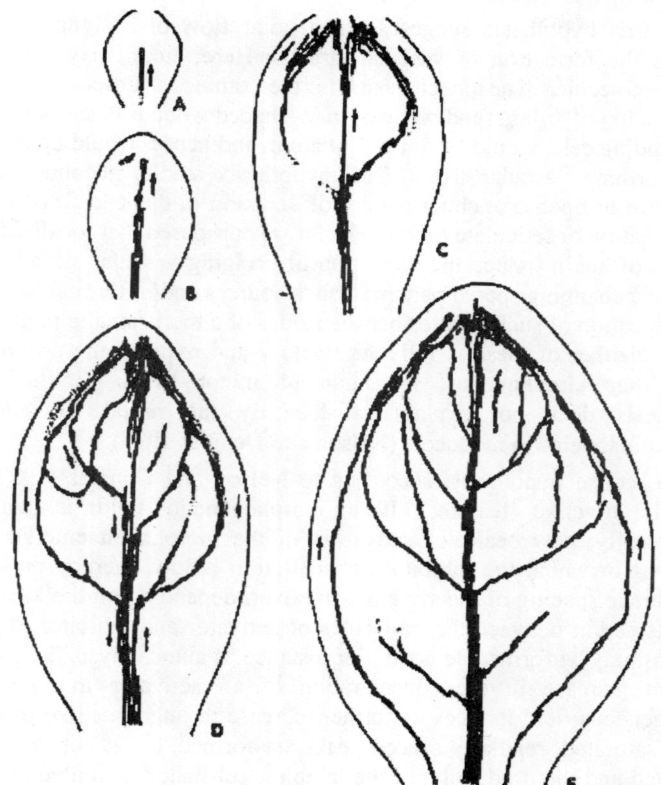

Fig. 83. Successive stages in the differentiation of xylem in the veins of lamina of *Anagallis arvensis*. The first xylem of the mid-vein differentiates acropetally (A–C), while that of upper lateral veins differentiate basipetally (C and D) and that of lower lateral veins differentiates acropetally (E) (De Sloover 1958).

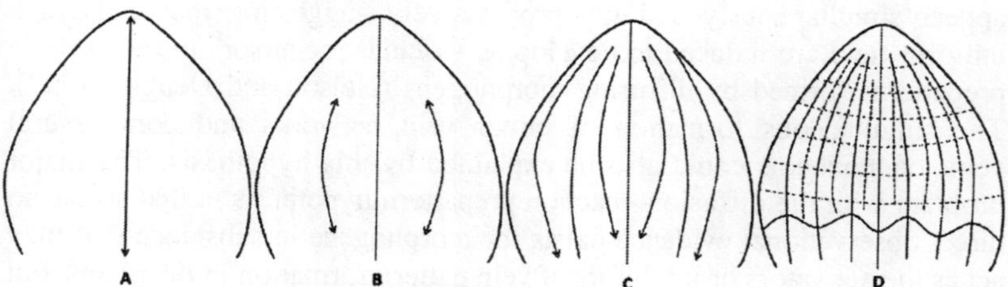

Fig. 84. Vascular pattern ontogeny in the lamina of a grass represented diagrammatically (A–D). Arrows indicate directions of differentiation.

The first hypothesis suggests that a polar flow of a signal molecule induces the formation of vascular strands. Here, auxin may act as this signal molecule. The mechanism is the same as described for the regeneration of xylem (and phloem) in V–shaped wounds described earlier. Surrounding cells would be drained of auxin and hence, would be inhibited from forming vascular tissue. This hypothesis readily accounts for the formation of open branching pattern of venation in dicot leaf. To explain the formation of reticulate minor veins, it was proposed that localized point sources of auxin induce the formation of bridging vascular strands either through a changing spatial pattern of such sources and/or through alteration of the location of such sources between sides of a meristematic region (Fig. 85 A). Neither of these is fully satisfactory and requires the progressive, rather than simultaneous, formation of minor veins. Similarly, this hypothesis does not explain satisfactorily the initial formation of longitudinal veins in monocots (Nelson and Dengler 1997).

The second hypothesis, according to Nelson and Dengler (1997), has several attractive features. If a morphogenetic field is enlarging isotropically, new peaks of activation of the morphogen emerge at the positions in which the inhibitor concentration is low, thereby preserving the average spacing of the system. Moreover, depending on the kinetics of the interaction between the molecules of activator and inhibitor, different patterns may be formed de novo. For instance, if autocatalysis (i.e., a small random increase in the concentration of an activator in an initially homogeneous field induces its further increase through positive feedback) is not saturated, regularly spaced peaks are formed, but if autocatalysis is saturated and the production of the inhibitor substance is limited, a stripe–like pattern results. Combining models of system forming patches with system forming stripes forms a closed network i.e., patches specify where no stripes are allowed (Fig. 85) (Nelson and Dengler 1997). The longitudinal veins of monots get differentiated in a homogeneous two–dimensional field and appear as a series of parallel stripes with new stripes intercalated between the pre–existing ones as they grow apart. New veins appear simultaneously and not progressively suggesting that cells in a uniform field are induced to develop as vascular precursors in response to prepatterns formed by diffusing morphogens (Nelson and Dengler 1997). The simultaneous formation of minor vein networks and commissural veins of monocots could also be explained by this hypothesis. The major problem with the diffusion–reaction prepattern hypothesis is that so far no direct observational evidence exists for morphogenetic substance that may act as the activators or inhibitors of vein pattern formation in the plants, but it is supposed that auxin or other plant regulators (auxin and/or cytokinins) may be involved in this prepattern formation.

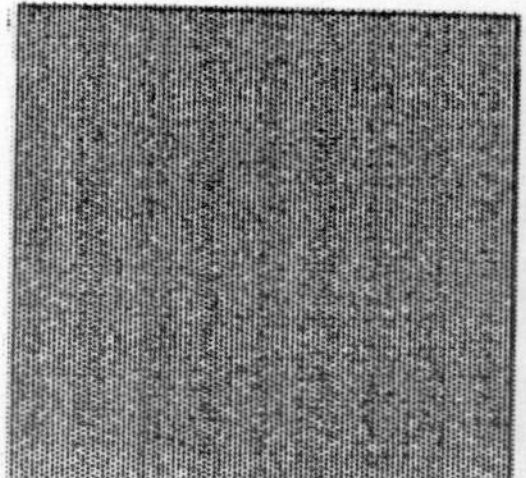
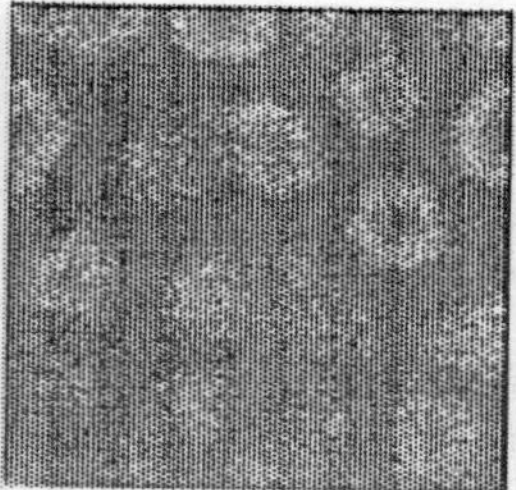
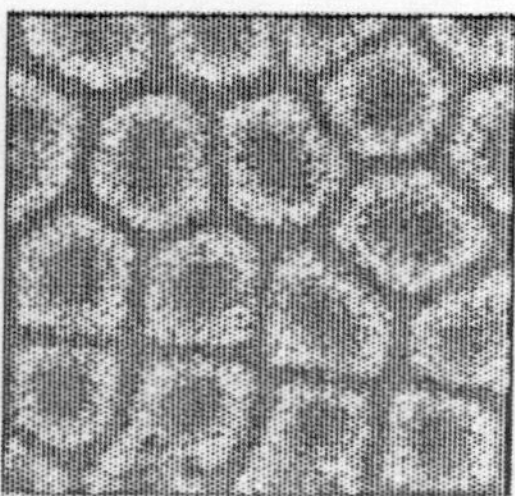

Fig. 85. Diffusion–reaction prepattern hypothesis to explain vascular pattern formation in the leaf lamina. Computer simulation of the system forming patches combined with a system forming stripes. The patches specify where no stripes are allowed, and thus stripes appear at the largest possible distance from other stripes (Figure courtesy of Dr. H. Meinhardt, Max Planck Institute, Tubingen, Germany)

4.4.6. Secretory Tissues

There are several kinds of secretory tissues in plants. The **laticifers**, the cell type or tissue that contains latex, constitute an important and organized secretory system (Hagel et al 2008). They do not discharge their contents outside or into a sink like other secretory tissues/cells. The laticifers occur in nearly 12,500 species belonging to 900 genera and some 22 families, mostly of Dicotyledonae. A few monocot families, *Regnellidium* of Pteridophyta and *Gnetum* of Gymnsopermae also contain laticifers, thus indicating their polyphyletic origin. There are two major categrories of laticifers (Fig. 86):

1. **Non–articulated** - The laticifers in this category originate from single cell, develop along with the plant from the embryo onwards and become coenocytic or rarely remain uninucleate. These laticifers grow by intrusive as well as symplastic growth. The elongating apices may bifurcate or branch repeatedly, thus forming a branched system, or may remain unbranched. They are often associated with phloem tissue.
2. **Articulated** - The laticifers in this category are multicellular with end walls of component cells remaining intact. They become porous or disappear completely. In some, these may anastomose laterally. In some laticifers, the nuclei become greatly enlarged due to the nuclear fusions. The latex may be colourless or variously coloured and varies in consistency from watery to milky. Starch grains of various shapes, salts, organic acids, phenolics, sugars, alkaloids rubber granules, waxes, resins, proteins, essential oils, mucilages, enzymes etc. are present in the latex depending on the plant. The latex may be

colourless, milky white, yellow or orangish. Wounding and infection induce differentiation of leticifers, gum ducts and resin canals, as also chemicals such as ethaphon. We do not have much information as to why some plants develop laticifers, resin canals and gum ducts.

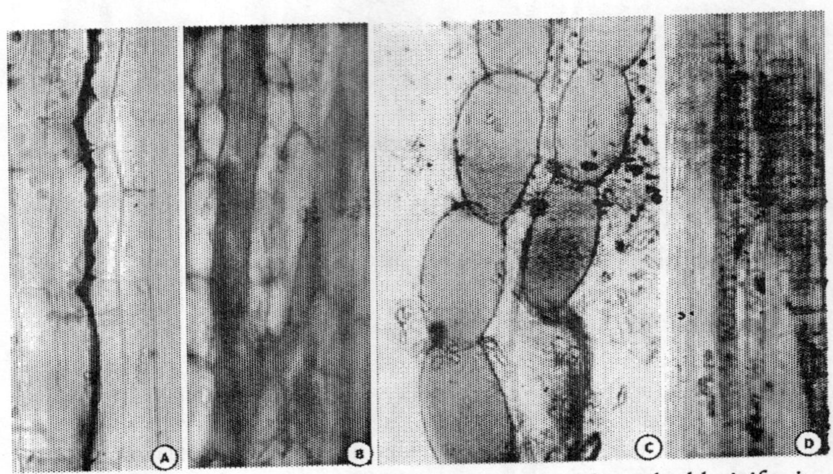

Fig. 86. Types of Laticifers: A - Non-articulated unbranched laticifer in *Vinca rosea* stem; B - Non-articulated branched laticifer in *Jatropha gossypifolia*; C - Articulated unbranched laticifer in banana fruit; D - Articulated branched laticifer in the stem of *Argemone mexicana* (s. septum)

4.5. MORPHOGENESIS AT ORGAN LEVEL

The higher plants have two basic categories of organs: The **axial organs** and the **appendicular organs**. The former include the stem and the root, while the latter the leaf and flower, although some, especially the proponents of the axial theory of flower, consider the flower also as an axial organ. Out of all these organs, it is the leaf and the leafy floral organs that exhibit the greatest variety of morphological form. Hence, this section will concentrate on the morphogenetic aspects of leaf. However, the readers are advised to see the next chapter also in order to understand the structural and developmental aspects of the stem, root and leaf and the chapter after next on the development of the floral organs.

4.5.1. Development of Leaf Form

The form of leaves vary greatly in vascular plants. Leaves may be **simple** or **compound** depending on whether each leaf has a single lamina or more than one lamina (**leaflet**); simple leaves may have a complete

lamina (or **blade**) or may be lobed /dissected to various degrees either **pinnately** or **palmately**. Compound leaves show much more diversity than simple leaves; they may be **pinnately compound** or **palmately compound.** Pinnately compound leaves may be simple pinnate, bipinnate, tripinnate or multipinnate. The main axis or stalk of the simple leaf is called **petiole**, while that of the compound leaf is called the **rachis**. Both simple and compound leaves have an identical origin and early development, but differences between them are set in only during later stages of (lamina) development. Sometimes even within the same plant, leaf form may vary depending upon the age of the plant or environmental factors.

KN1 gene has been implicated in the development of the lobed or dissected leaf of tomato (Hareven et al 1996). Generally, the leaf has about nine lobes borne on a central midrib. Expression of the *KN1* gene is associated with the development of a leaf phenotype with about 2000 lobes in the transgenic plant. Such an excessively dissected lamina is also found in the spontaneously occurring mutants *curl* (*cu*) and *mouse ears (me)*. This observation also suggests that *KN1* may have a role in dissection of lamina.

Pea plant is often used as a model system for understanding the control of form of a compound leaf. Pea leaf has a rachis, which in its basal part bears up to three pairs of opposite leaflets and in its distal part up to four pairs of opposite tendrils (all modified leaflets) and ends in a tendril (a modified terminal leaflet). The leaf also has a pair of large basal stipules. The surgical experiments conducted on this compound leaf at various stages of development indicate a progressive determination of leaf form. Parts of very young primordia (of about 30 μm long) left on the bud are capable of partial regeneration, while the tip of longer primordia (of about 70 μm long) if extirpated, the remaining parts grow into stipules; removal of the two sides of this longer primordium causes the formation of rachis and the terminal tendril from the leftover part. Through other surgical manipulations, it was possible to change a future leaflet into a tendril or *vice versa* (Sachs 1969). From all these experiments, it is evident that the main elements of leaf architecture are laid down early in ontogeny, but the morphology of leaflets is determined later. This conclusion is supported by the leaf morphological variations exhibited by a number of leaf mutants of pea plant. For example, *afila* (*af*) has no leaflets, but a bunch of branched tendrils; the *tendrilless (tl)* has only leaflets, but no tendrils; the *pleofila (af/tl)* has a highly branched system of very small leaflets, but no tendrils; and the *unifoliate (uni)* has one to three leafelets on short petioles, but no tendrils. Genetic analysis of the first three mutants has led to the view that *pleofila* represents the basic phenotype of the pea plant, without interference from either *AF* or *TL* genes. From this basic phenotype, specific interactions between *AF* and *TL* genes are believed to determine pinna identity in the regions of the leaf corresponding to stipules, leaflets and tendrils. *UN1* gene forms a distinct class as it promotes compound leaf development through maintaining a transient state of indeterminacy in the leaf primordium (Lu et al 1996, Hofer et al 1997).

4.5.2. Heteroblastic Development

Heteroblasty refers to the morphological differences between successively produced vegetative parts or organs (particularly leaves) of a plant. Although some degree of heteroblasty is present in all plants when we consider vegetative regions, it is particularly evident through its effect on leaf ontogeny. There is difference in leaf morphology between juvenile and adult plants. Frequently, juvenile leaves show a gradual increase in size and complexity until the adult leaf form and adult size result.

Heteroblasty has been investigated from a genetic perspective. Several dominant mutations in maize, such as the *teopod (tp)*, have been linked to the abnormal expression of juvenile characters such as presence of epicuticular wax (which is restricted only to juvenile leaves in normal plants) and delays in the appearance of epidermal trichomes (Poethig 1988a and b; Bongard–Pierce et al 1996). Poethig (1988 b) has shown that x–ray induced genetic mosaics expressing the wild type allele of the *TP1* gene in the mutant plants possess a mutant instead of a wild–type phenotype. In this respect, the *TP1* gene shares the property of non–cell–autonomous expression with the *kn1* mutation. It has been suggested that the mutant gene is likely to control the production or distribution of a diffusible chemical involved in the maintenance of juvenile status, but this idea has not yet been experimentally proved. In order to know whether *tp* mutation alters the development of the adult leaf through its effect on shoot apical meristem or on specific cells of potential juvenile leaves, comparative clonal analysis of cell lineages in wild–type and mutant *tp2* plants has been done. The results obtained showed that the *tp* mutation does not affect cell lineages of the juvenile nodes of the shoot that arise from the cells in the embryonic shoot apical meristem, but extends the growth of cells near the apex of the meristem. The apparent number of embryonic meristem cells that give rise to each metamer (phytomer) is also unaltered in the mutant. This suggests that the mutation does not prolong the expression of juvenile characteristics in the leaf by altering the patterns of growth and cell division within the shoot apical meristem, but changes the fate of cells normally destined for adult growth to promote juvenility (Dudley and Poethig 1991). Contrary to the *tp* mutation, which prolongs the juvenile phase, the *glossy 15 (gl15)* mutation shortens the juvenile phase and promotes early adulthood. *GL15* transforms juvenile leaf into adult form and it shows homology to homeotic genes that regulate floral organ identity in *Arabidopsis* (Evans et al 1994).

Heteroblasty has also been analysed in terms of changes in the levels of endogenous growth regulators. Studies made in species of *Hedera* have shown that GA converts adult leaves into palmately lobed juvenile leaves. The apical buds of juvenile plants of these species contain higher levels of GA–like substances in contrast to apical buds of adult plants. However, the levels of growth regulators in the leaves of both forms are the same. A similar role for GA as a regulator of juvenile leaf development has been shown in many other taxa, both herbaceous and arborescent. In *Centaurea solstitialis*, the juvenile leaves are simple and entire, while the adult leaves are lobed increasingly. Leaf promordia of different ages of this plant were excised and grown in appropriate culture media containing GA and it was found that the fifth and older primordia develop into juvenile leaves, but younger primordia do not survive. These results show that leaf–type divergence regulated by changes in GA concentration does not involve concomitant changes in the shoot apex, and probably, GA directly acts on the developing leaf to induce juvenile features. GA action is seen in preventing the formation of lobes in the younger primordia and the further growth of lobes already initiated in the older primordia.

4.5.3. Heterophylly

The production of differently shaped leaves in the same adult plant is called **heterophylly**. This is noticed in some water plants, in which depending upon the position of the shoot apex in relation to the water surface, leaves with different morphologies are formed. Submerged shoots have linear and narrow or highly dissected leaves, called **water forms**, while aerial leaves are broad or entire, called **land forms**. Since the shoot apical meristems and the leaf primordia are structurally similar in both forms during inception and early developmental stages, differences must be controlled mostly in later stages of leaf development. ABA has been implicated in the determination of leaf identity in many heterophyllous water plants. Culture of several water plants, in appropriate concentrations of ABA, promotes land form leaf morphology even in submerged shoot parts, thus, demonstrating that endogenous ABA regulates aerial leaf development. ABA is demonstrated in aerial shoots, but not in submerged shoots of *Hippuris vulgaris*.

4.5.3.1. Light and Leaf Form

Leaf form may also be controlled by other environmental factors such as light. Three aspects of light have been shown to affect leaf form; irradiance, spectral composition and photoperiod (Dengler 1994). All the three aspects affect leaf production, size, shape and anatomy. Studies indicate that total daily photon flux is more important than instantaneous photon flux in the modification of leaf growth suggesting that irradiance levels are perceived through their effect on photosynthetic rates in expanded leaves and carbohydrate levels within the shoot have significant effect on the development of newly–formed leaves. This may explain the acceleration of heterobastic leaf shape changes by high irradiance. Reduced irradiance usually increases leaf area and decreases leaf thickness as reported in some dicotyledons. The alterations in leaf area and thickness occur during leaf development through changes in the rates and duration of leaf expansion. The close correlation between leaf area and cell number indicates that control of leaf area is mediated through regulation of cell division.

There is not much evidence for the effect of altered spectral composition of shade light on leaf expansion. Low Red: Far Red induces larger leaf areas in some shade–intolerant terrestrial species however, and regulates leaf shape in atleast two aquatic species (Dengler 1994). Photoperiod probably has the greatest influence on the leaf development through its effect on total daily photon flux. Leaf shape changes have been shown to be associated with the altered phytochrome equilibria induced by day–length changes. These are probably the heteroblastic changes in leaf form that precede flowering. Transfer experiments between controlled light conditions indicate that the degree of plasticity of leaf development becomes limited with time and nutritional conditions experienced by early primordium stages may constrain later ability to fully respond to changing light conditions.

4.6. WHOLE PLANT MORPHOGENSIS

Modern science is essentially reductionistic in its approach to understand any process or phenomenon i.e., the procedures followed are to separate the parts of the whole, study them in detail and then put them together to get a picture of the whole. This is true of studies involving plant morphogenesis as well. Some botanists are especially interested in the morphogenesis of roots, others in leaves or buds and yet others in shoots or flowers, but not in whole plants. Some attack the problems from morphology, some from developmental anatomy and yet others from physiology, genetics or molecular biology and so on, but not from a combination of all these. There is no denial of the fact that such approaches are essentially needed and well–defined and can be satisfying to individual botanists. However, such type of conclusive and thorough studies may claim that they may not be enough as they deal with only a part of the whole, greater integrated system. Hence, in so far as it is possible, if specific investigations are to have their full value and impact, they must be considered in the context of **whole plant morphogenesis**. This is not a new idea, but is reflected in many great works of the past. As an instance, we may cite Agnes Arber's (1950) work, **'Natural Philosophy of Plant Form'**. Thus, whole plant morphogenesis requires 'signal–trafficking and cross–talking across all levels of organization to coordinate the operation of metabolic and genomic networked systems'.(Niklas 2003) of the whole plant.

The whole plant is a complex chimera with contiguous cells or tissues of varied structure, symplastic domains, function and ploidy. In such a situation, it appears as if some kind of complementary physiological mechanism is at work whereby the mutual relationships of the cells or tissues during differentiation, morphogenesis and development are regulated by an overriding holism and by a pervasive tendency towards homeostasis or steady state (Wardlaw 1968). Since the late 19th century botanists have pondered over the problem of whole plant morphogenesis. Contemporary physiology, biochemistry, and developmental biology have admittedly touched on the topic of whole plant metabolism and its relation to growth and development involving the special functions and structural complexities of different organs, the different organization of their biochemical systems and elaborated products and the exchange of metabolic substances between them. However, only fragments of information have been obtained so far to explain meaningfully the operational mechanism of whole plant morphogenesis.

According to Wardlaw (1968), some ideas on whole plant morphogenesis can be obtained by considering the embryo as a miniature whole plant. The embryos from all the major taxonomic groups have many features in common (Wardlaw 1955). These homologies of the organization may be explained, at least in parts, by envisaging the embryo (throughout its development) as a complex and a specific organic reaction system, which obeys the laws of physical chemistry. As development proceeds from the zygote into an embryo, and under the impact of genetical and environmental factors, and of organismal relationships already established such as polarity, correlations and nutrient–supplying capacity, the reaction system becomes modified in characteristic ways (Wardlaw 1968). Most important among these modifications is the establishment of definite gradients in the organization of the embryo. For instance, there is an acropetal gradient of decreasing cell size in the embryo. This cellular gradient is the evidence of an underlying differential but patternized distribution of metabolites in the growing and mature embryo. This gradient has its effect on the further organismal development of the embryo into a seedling and the seedling into an adult plant. Thus, although the reaction system concept continues to be operational in the adult plant, it becomes more and more complicated since the complexity of the adult plant body increases as the embryo becomes the adult, just as the complexity increases from the zygote to the mature embryo. The adult plant becomes a complex of several reaction systems operating at different levels and at different topographic locations in order to maintain the integrity of the whole plant, both structurally and functionally.

One of the greatest differences between animals and plants is that, in the development of the former, the growing embryo sooner or later reaches a stage at which the definitive morphological pattern, characteristic of the adult animal, is established. On the contrary, the situation is different in plants. As Bower (1947) had stated, "The life of the higher plants may be described as an indefinitely continued embryology, the increase in the number of parts being in a geometric ratio". The adult plant and its shoots and roots are capable of potentially indefinite growth and extension, with the formation of an unlimited number of new lateral parts. In other words, the morphogenetic potential of the adult plant continues to remain unaffected throughout its development from the embryo onwards. But, although the plant body is theoretically unlimited, limitations of various kinds are, in fact, imposed by both physical and biological factors (Wardlaw 1968) indicating the finiteness of development of at least some plant parts such as leaves and flowers. This is referred to as the **'notion of holistic finiteness'**.

At this point, it is pertinent to discuss briefly the whole plant determination in order to throw more light on the notion of holistic finiteness. Morphogenetic studies of individual species, growing under relatively constant environmental conditions, have shown that there is a great measure of whole plant determination. The size of the whole plant, the number of branches and leaves that are produced, the branching pattern, shoot architecture and geometry, the location of the inflorescence or flower etc. are so precisely decided as to indicate the operation of whole plant determination in the morphogenesis of the plant. Thus, in most, if not all, plants, whole plant morphogenesis is susceptible to considerably more precise specification than has been previously thought of. Although development in vascular plants shows features of continued embryology as Bower (1947) has emphasized, one may recognize that the morphogenesis of their parts is highly regulated and appears to be obligatorily sequential (Wardlaw 1968).

What is the basis of unity and holistic finiteness, mentioned earlier, in the development of plants? The unified, harmonious and holistic development of the rooted leafy shoot, the morphological developments during the phase of transition to flowering, and the precision and finiteness of the floral development can be ascribed to correlations and to the existence of a unitary reaction system in the shoot apical meristem. This reaction system is envisaged as an **'open–system'**, as understood by physical chemists. The rate of the reactions within this system, according to Wardlaw (1968), undergoes a steady acceleration during the vegetative phase from the embryo onwards. This is a direct result of the steady increase in the nutrient supply, which diffuses and is being drawn into shoot and root apical regions of the plant. As the plant grows in size, these become the characteristic of the concerned species. Under certain environmental conditions critical changes take place in the metabolic materials reaching the shoot apex system, resulting in the modification of the pattern of differential growth in the apex to result in floral transition through the action of sequentially evocated specific genes.

It is pertinent to end this section on whole plant morphogenesis with a question that Kaplan (1984) raised with reference to the developmental basis of great diversity in plant form; "How much of this morphological diversity is merely the result of quantitative variations in growth distribution on a common developmental theme, and how much is the result of qualitatively unique developmental programs, which are distinctive for individual plant groups?" Both quantitative and qualitative variations are important in whole plant morphogenesis, perhaps a right combination of these variations.

4·7· PROGRAMMED CELL DEATH AND MORPHOGENESIS

Programmed Cell Death (PCD) is a very important event in morphogenesis. PCD may involve isolated cells in a number of instances, few cells or almost all cells of a tissue or organ and selectively eliminate them in order to maintain the structural and functional integrity of the concerned tissue, organ or plant. There had been a number of scattered reports on the topic of PCD in plants (and animals) for more than a century. However, during the past two decades more than 40,000 publications have appeared on this topic (including a few books) emphasizing the 'contemporary fascination' that it enjoys among researchers (Krishnamurthy et al 2000). The term 'programmed cell death' was introduced by Lockshin (Lockhshin and Williams 1965). PCD is a physiological process (exhibiting morphologically visible changes) that leads to the selective elimination of unwanted cells in multicellular organisms. It is characterized by the involvement of the dying cells' own machinery as an 'executioner', a sort of suicide. It is reported to play a major role in controlling cell number and turnover, tissue homeostasis and specialization, tissue sculpting and pattern formation, disease incidents and other hypersensitive reactions, defence reactions, and above all, in maintaining the overall integrity of the plant (and the animal). PCD is quite distinct from necrosis. While the former is energy dependent, physiologically active and genetically programmed with the involvement of regulatory genes, stimulatory events, signalling pathways and morphologically expressed distinctive features, the latter is a non–physiological process dissociated from the developmental and morphogenetic events and cell swellings, lysis and leakage of the cell contents (Krishnamurthy et al 2000).

The categories of cells that undergo PCD in plants are as follows:

1. The cells that have served their functions, such as root cap cells, senescing cells of leaf and other plant organs that abciss off, vegetative cell of pollen or pollen tube, synergids, antipodals, anther tapetum and wall cells, cells of stigma and style, endosperm cells including haustoria, and suspensor cells of the embryo.

2. The cells that are unwanted from the beginning such as the stamen and the carpel primordial cells respectively in female and male flowers, non–functional megaspores in monosporic and bisporic embryo sacs, and non–functional microcpores of the members of Cyperaceae and Epacridaceae.
3. The cells that die during redifferentiation into specialized cell types and cells that undergo terminal differentiation such as xylem tracheary elements, cells of certain trichomes, thorns and spines, sclerenchyma cells (sclereids and fibres), cork cells and other suberized cells. The most salient characteristic of these cell types is that they all start their functions only after their death. The other characteristic feature of these cell types and significant changes (morphological and chemical) take place in their cell walls. The cell death results in the terminal differentiation in this category of cells. A sub–category includes cells that undergo 'partial' death during differentiation. The phloem sieve elements form this sub–category. In these cells the nucleus and some other cytoplasmic organelles are lost, but not the cytoplasm and its functional integrity. The functional life of these cells, however, is fixed (from a few days to one or a few years).
4. The cells that are subjected to **hypersensitive reactions (HR)** due to pathogenesis and other environmental and abiotic stresses. Here, cell death is used as a defence mechanism against the stress factor or as a mean of emanating signals for the other nearby cells to build proper defence reactions, so that the spread of infection or stress factor to other cells is often prevented. It is possible that not all the cells that die here may involve PCD and it is likely that necrosis may also be seen, especially due to physical damage.
5. The cells that are present in the wrong places such as the cells that die to form the lobes in leaves and the leaflets in compound leaves, or to result in the holes as in the lamina of *Monstera* and croton varieties.

The mechanism of PCD in the different categories of cells types, mentioned above, appears to be different, at least in some important respects, suggesting that more than one pathway of PCD is likely to operate in plants. Although PCD of plants and animal cells seems to be programmed, there are several operational differences (Krishnamurthy et al 2000). Apoptotic bodies[1] have not been demonstrated so far in plant PCDs;

---

[1] Apoptotic bodies are small structures of animal cells that undergo PCD and consist of small bits of DNA obtained from degraded nucleus, and enclosed by cytoplasmic materials. These are extruded out of the cell by exocytosis before the cell itself is eliminated.

neither the specific involvement of effector caspases. The involvement of regulators and adaptors known in animal PCD is likely to be either absent or the plant PCDs involve a different set of regulators and adaptors. Available evidences from the work done so far indicate that it is premature to assume that animal and plant PCDs are similar. The presence of a cell wall in plant cells is a crucial factor in the exhibition of some of these differences between plant and animal PCDs.

The events during PCD of plant cells that die after performing their functions include cytoplasmic condensation and shrinkage, cell shrinkage, presence of some proteases and signalling events such as the production of reactive oxygen species (ROS), increase in cytosolic calcium, phosphorylation or dephosphorylation changes, chromatin condensation, and presence and activation of endonucleases.

Vacuolar autophagy appears to be the main mechanism, involving lysozymes, by which the cytoplasm is degraded, consumed and cleared. The differentiation of tracheary elements, fibres, sclereids and suberized cells (such as cork cells) appear to follow a very unique mechanism of PCD. Since these cells die to function, the cytological events accompanying PCD are oriented towards making them persist and function almost till the end of the plant part or the whole plant. The cells do not shrink as noticed in other cells undergoing PCD, but instead, often enlarge and elongate to varying degrees during differentiation. Their nuclei are not only metabolically very active almost till the final stages of differenatiation, but often also undergo endoreduplication or polyteny or mitotic division. They are transcriptionally very active and are responsible for the production of several chemicals that form part of the persisting, specialized secondary wall. The presence of some types of proteases in these cells is perhaps related to the cleaning or clearing up of the cytoplasmic debris and to recycle them for use in organizing the secondary cell wall. Cysteine proteases and serine proteases have been reported and these are likely to be involved in autolysis of the cell contents. A 40–KDa serine protease is secreted during secondary wall synthesis and this is believed to coordinate between secondary wall synthesis and PCD in the developing TEs of *Zinnia elegans*. But, a study of an *Arabidopsis* mutant indicates that the two processes are independently regulated. There are reports of endonucleases and the TUNEL–postive reactions in TEs. Fukuda (1997) considers these endomucleases as of the general nucleases category. It is significant to mention here that in many cases, even fully mature and functioning TEs and fibres possess intact nuclei (Venugopal and Krishnamurthy 1984)

PCD is intimately involved in senescence. The natural termination of the life of a plant, its specific parts such as leaves or tissues and cells, as a result of senescence, may be regarded as a normal part of plant development, a sequel to the events of growth, development, morphogenesis and maturation (Greenberg 1996; Krishnamurthy et al

2000). The term 'senescence' specifically refers to the series of changes in a living plant or its organs or tissues or cells that ultimately leads to its death. Since it occurs in an orderly sequence in the life of a plant and as an active degenerative process, it is considered to be genetically controlled or programmed – a process of PCD (Kuriyama and Fukuda 2002). Senescence is controlled by the various chemicals including growth substances as well as by the environmental factors. Sometimes the word **aging** is used, but it is not synonymous with senescence (although very difficult to distinguish it from senescence). Aging is defined as an accumulation of changes that lower the vitality of a living entity without being lethal by themselves. Aging often leads to senescence. Senescing cells are often characterized by a loss of chlorophyll, accumulation of coloured pigments, decrease in the nucleic acid content, increase in the leakiness of cells due to membrane degradation and an overall decline in the metabolism. At the molecular level, many components of the transcriptional and translational systems lose their functions during senescence. There is also a down–regulation of a vast majority of genes, especially genes involved in photosynthesis. On the contrary, thus far, over 100 senescence–associated genes have been identified to be up–regulated in diverse plants (Jing et al 2003) and perhaps these may be involved in the disassembly of macromolecules already present as well as in the transport of mobilized nutrients from the senescing tissues. Also, there are genes that induce the onset of senescence. For example, protease genes have been identified by substractive hybridization methods, and also genes for several enzymes involved in catabolic reactions. Examples for the down–regulation of genes involved in the photosynthesis are provided by *Arabidopsis* and soya bean. There is a reduction in the levels of *Rbcl* and *Rbcs* gene transcripts during senescence of soya bean leaves along with a concomitant decrease in RUBISCO enzyme and chloroplast DNA.

There are evidences to indicate that cytokinins play a vital role in controlling senescence. There is not only a decline in the cytokinin level in the senescing leaves, but also a delay in senescence if cytokinins are externally applied. The cytokinin level can be elevated in transgenic plants by a bacterial gene encoding the cytokinin–synthesizing enzyme called IPT and the leaves of these transgenic plants remain green for a longer period. Senescing plant organs, such as the leaf, fall–off is by a process called **abscission**.

Vegetative Growth and Development 5

5.1. INTRODUCTION

The vegetative body of a vascular plant, in general, consists of a **shoot system** that is invariably aerial in the location and a **root system** that is invariably underground in the location. The main axis of the shoot system is the **stem,** which supports the **leaves** and bears **flowers** or **fruits**. The stem may be branched or unbranched depending on the species and the pattern, and the degree of branching is quite varied. In tropical arborescent taxa, there are 23 basic architectural patterns of the shoot system based on the branching types and crown geometry (Fig. 87) (Hallé et al 1978), and it is likely that an equal or more number of patterns may be seen in herbaceous taxa (Refer to Box 5.1). The main stem and branches are terminated by **shoot apex**, which takes care of **longitudinal growth** (also called **apical growth**) of the plant. Additional shoot apical meristems may be formed adventitiously in various parts of the plant body in some species. The root system generally consists of a **tap root,** which gives rise to **lateral roots** that may further branch to varying degrees to give rise to

secondary and **tertiary roots**. Monocots and pteridophytes generally do not have a tap root, but instead have only **adventitious roots**. The root tips of tap, lateral, secondary and tertiary roots terminate in **root apical meristems**. The root and shoot apical meristems function as growth centres and continue to give rise to vegetative tissues even after the plant grows to maturity through longitudinal growth.

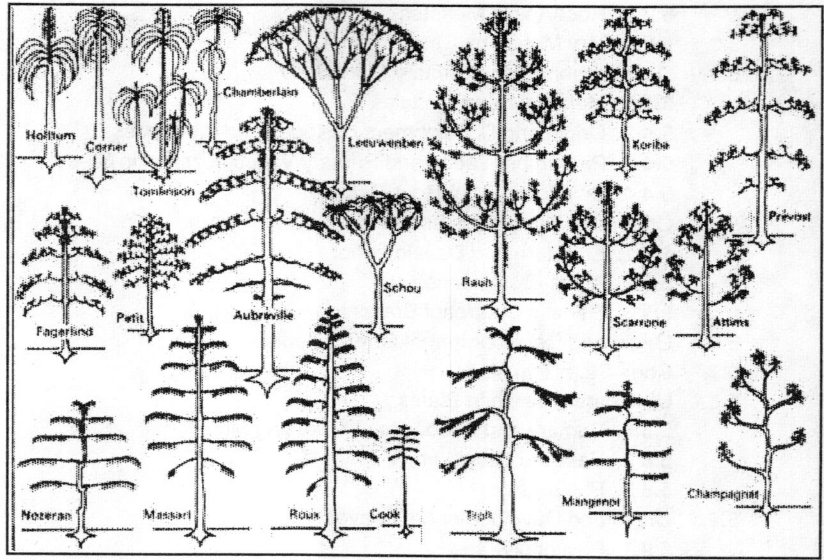

Fig. 87. The basic architectural models of tropical trees named after leading scientists. Root system is stylized while the shoot system is shown in one plane (Hallé and Oldeman 1970 and Oldeman 1974)

BOX 5.1. PLANT ARCHITECTURE

Plant architecture can be defined as the three–dimensional organization of the plant body. Although it is easily noticed and striking in the shoot system, a similar architectural pattern may occur in the root system but in a less–striking form. For the shoot system, this includes the branching pattern, if any, as well as the size, shape and position of leaves and flowers. For a long time, plant architecture has been the only criterion for the systematic and taxonomic classification of plants and as architecture is often species–specific, it continues to be an important means of plant identification even today.

Plant architecture has also got a major agronomic and horticultural importance. It very strongly influences the suitability of a plant for cultivation in the field, garden or plantation as the case may be. It also influences the yield and the efficiency with which a plant can be harvested. In fact, it is one of the main reasons for the success of the 'Green Revolution', which resulted in increased food yield per area, and for which plant architecture was modified. For example, the selection of wheat varieties with shorter and sturdier stems for Green Revolution resulted in greater yield, while still resistant to wind and rain damage. In plantations, better spacing based on the architecture of the crop, resulted in better yield.

Plant architecture is under strict genetic control, although also influenced by environmental parameters such as light, temperature, humidity and nutrient status. The endogenous regulatory principles that control the plant architecture are very striking. Most information on plant architecture and its controlling factors have come from a study of tree architecture, especially of the tropical trees. Such information has been obtained from three different fields, using different approaches: descriptive, theoretical and experimental. In the descriptive realm, shape of the tree crown is established by a combination of deterministic and opportunistic processes. As a result of theoretical studies, simulation of crown shape through computer programming has been obtained. Experimental research, for example, on water transport, has indicated the presence of structural constraints vis–à–vis a mechanism that allows trees to grow tall. Recent experimental works involving wounding, chemical treatment etc. on the various plant parts (that together contribute to the plants' overall architecture) throw much light on the various endogenous factors that are responsible for the plant architecture.

The most important insight that the study of tree architecture has produced is related to the repair of the crown due to various types of damages. This results in a 'reiterative complex' within the crown, which is usually a repetition of the original architecture of the tree from one of the major, undamaged, branches of the crown.

Much more research needs to be done on the tree architecture so as to understand its selection value in the survival of the plants. This again emphasizes the fact that in spite of its long history as a discipline, plant morphology is currently a much neglected area of the study and a revitalization of morphological research will augur well for understanding the form.

Refer to Tomlinson 1983, Raven 1986, Reinhardt and Kuhlemeir 2002 for more information.

5.2. APICAL MERISTEMS

The term apical meristem refers to a group of meristematic cells at the apices of shoot and root. These cells through cell divisions lay the foundation of the primary plant body. As already discussed in the first chapter of this book, meristems are made of initials, which perpetuate the meristems and their initials. Also, the derivatives divide and produce one or more generations of cells before enlarging and differentiating into specific cell types of the mature organ. The change from apical meristem to primary tissue of the mature plant body is gradual and involves integrating the phenomena of cell division, enlargement and differentiation.

The existence of profusely inconsistent terminology in the voluminous literature on apical meristems reflects the extreme complexity of the subject matter. The term 'apical meristem', most commonly, is employed in a broader sense than as composed of initials and their immediate derivatives. It also implies variable lengths of shoot and root proximal to the apex. The terms **shoot apex** and **root apex** are invariably used respectively as synonyms of shoot and root apical meristems, although many botanists consider that the shoot apical meristem denotes only the shoot apex lying distal to the youngest leaf primordium, while shoot apex includes apical meristem along with the subapical region bearing young leaf primoridia and young **nodes** and **internodes** (Fig. 88).

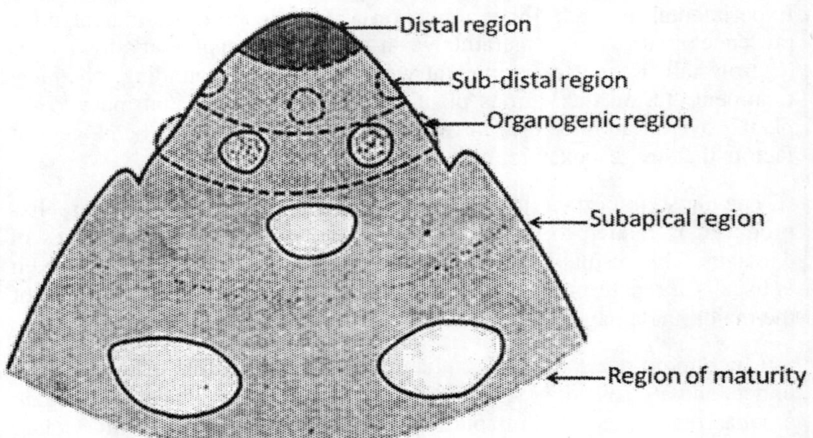

Fig. 88. Diagrammatic representation of a shoot apex as a system of contiguous interrelated zones (Based on Wardlaw 1957)

5.2.1. Shoot Apical Meristem

As already stated, the shoot apical meristem is the most distal region of the shoot apex. It occupies the apex of the main shoot as well of the branches. Externally, the shoot apices differ greatly in their shape and size (width), as well as in the arrangement and sequence of the leaf primordia that they form (Clowes 1961). The width ranges from about 40 µm as in *Luzula* and *Syringa* to about 3300 µm as in *Cycas revoluta*. Most angiosperm species have shoot apical meristems with a width range of 120–300 μm. The width varies in the same species depending on the age of the plant, with seedling apices having less width while mature plants having greater width (Fig. 89 A, B). The height of the shoot apical meristem also varies with the plant, with some water plants having the greatest height (e.g., *Elodea, Hydrilla*). In the vast majority of taxa, the height of the shoot apical meristem varies with the **plastochrone changes**, with the least height during early development and maximum height sometime after leaf initiation (Stein and Stein 1960).

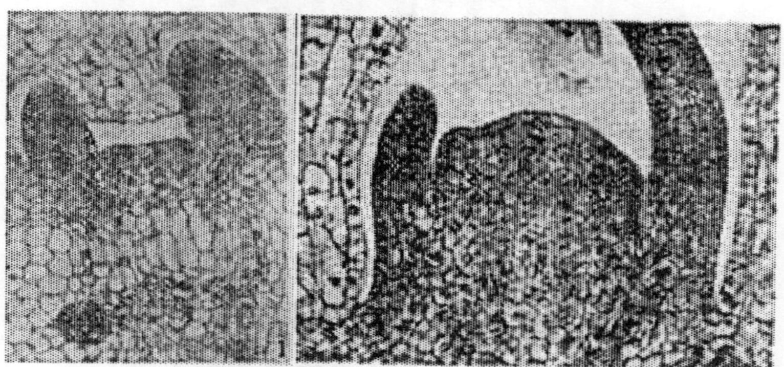

Fig. 89. L. S. of shoot apices of seedling and adult plants of *Commelina benghalensis* respectively

5.2.1.1. *Shoot Apical Meristems with Apical Cells*

The discovery of a single, morphologically and cytologically distinct apical initial cell in many seedless plants (Fig. 90) prompted the proposal of the **apical cell theory** to explain the organization and functioning of the shoot apical meristem. According to this theory, the apical cell is interpreted as a consistent structural and functional unit of the apical meristems governing the process of growth and development. In many pteridophytes, the apical cell is superficially located, large and more commonly inverted pyramidal (Fig. 90). The base of the pyramid is turned towards the free surface and its other three facets are in the body of the shoot apex. Its derivative cells form an orderly pattern, which in turn, is initiated by the orderly division of the apical cell. The successive divisions follow one another in acropetal sequence along a helix. The term **merophyte** is used to refer to the derivatives of the apical cell. Thus, pyramidal apical cells are tetrahedral with the cells being derived from their three inner facets (Fig. 91). In some species such as water ferns, apical cells are three–sided and new cells are cut off from two of its three facets. Such apical cells are seen in the plants whose shoots are bilaterally symmetrical.

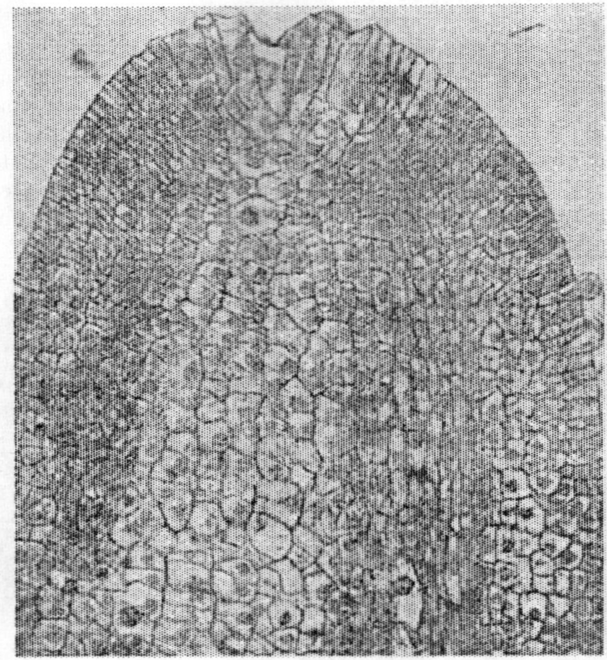

Fig. 90. L. S. of shoot apical region of *Microgramma* showing a single inverted pyramidal apical cell (Photograph Courtesy of Dr. Ann Hirsch)

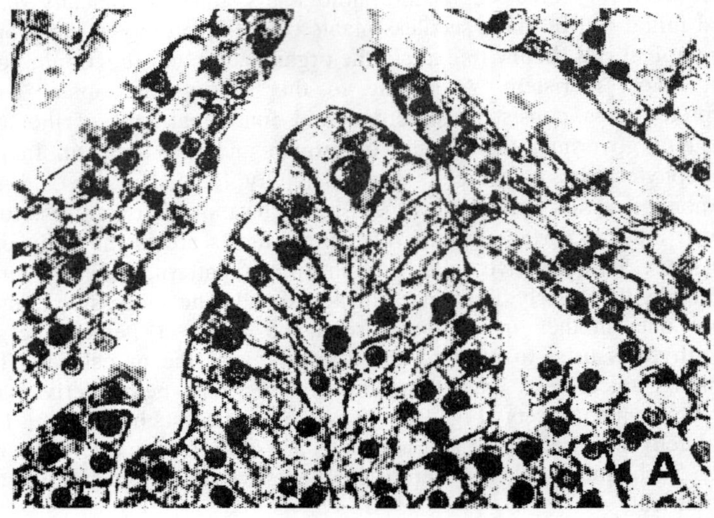

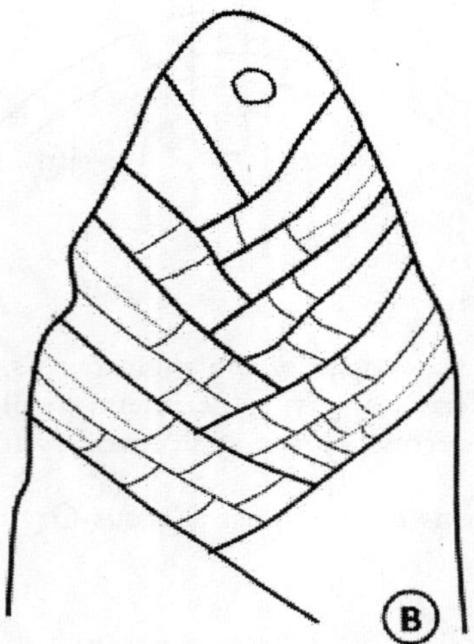

Fig. 91. A. L. S. of shoot apex of *Equisetum* (photograph courtesy of Dr. A.W. Roberts); B. Apical cell and planes of its division and formation of derivatives redrawn from figure A.

Although the apical cell was considered by early plant morphologists to be the ultimate source of all cells of the shoot, such a role for the apical cell began to be questioned later. Some botanists concluded that the apical cell is mitotically active only in very young plants. However, radiolabelled DNA precursor supply to apices with single apical cell indicated that the apical cell remains active mitotically, although not to the intensity seen in its immediate derivative cells.

5.2.1.2. *Shoot Apical Meristems with Histogens*

After studying angiosperm shoot apices of both embryos and adult plants extensively, Hanstein (1868, 1870) proposed that the main body of the shoot axis is produced by a massive meristem of great depth consisting of three tissue generators or **histogens**. This theory of Hanstein is called the **Histogen theory**. These three histogens can be distinguished from one another by their origin, location, developmental course, and the tissue regions that they generate. The outermost histogen is the **dermatogen**, which forms the precursor of the epidermal tissue. The second histogen is the **periblem**, which forms the cortex. The third histogen is the **plerome,** which gives rise to the stele and pith, if present (Fig. 92). The major problem with the concept of histogens is that these three histogens cannot be recognized clearly in the shoot apical meristem, particularly the periblem and the plerome. Hence, the vast majority of morphologists do not accept the presence of histogens, at least for the shoot apical meristem.

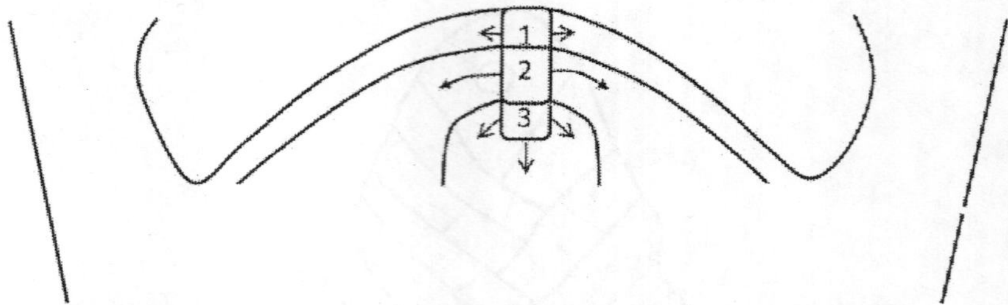

Fig. 92. Vegetative shoot apex of the angiosperms, interpreted as per histogen theory. 1 - dermatogens; 2 - periblem; 3 – plerome, respectively giving rise to epidermis, cortex and central cylinder of the stem

5.2.1.3. *Apical Meristems with Tunica–Corpus Organization*

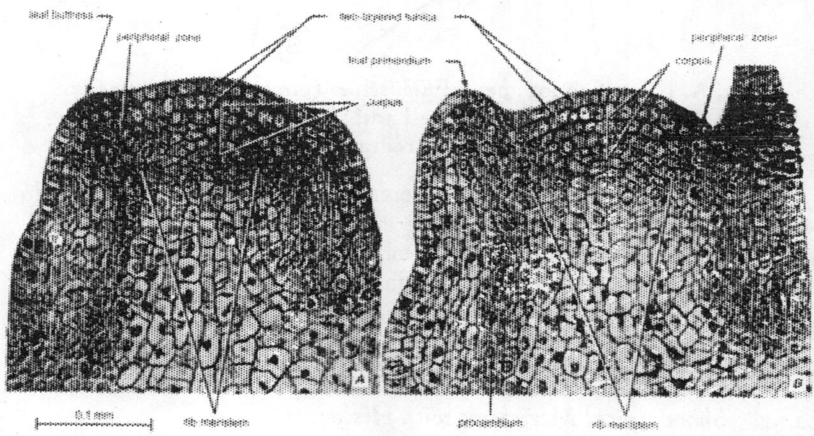

Fig. 93. L. S. of shoot apices of *Solanum tuberosum* exhibiting tunica–corpus organization (After Sussex 1955)

While studying the shoot apical meristem organization of plants, Schmidt (1924) proposed that the shoot apical meristems have a superficial mantle of cells constituting the **tunica,** which in turn, encloses the **corpus** (Fig. 93). This concept is called the **Tunica–corpus theory**. The number of tunica layers vary invariably from one to three respectively, designated as L1, L2, and L3, depending on the species, with two or three layers being more common. Sometimes the number of layers varies in the same plant depending on the developmental stage. Generally, all the cells of the tunica divide only anticlinally, and because of this feature only, its layered structure is maintained. In the corpus, cells divide in all planes, thereby adding bulk to the apical meristem. It has been postulated that each layer of tunica arises from a small group of separate initials, while the corpus has its own initials beneath those of the tunica. The clonal analysis made on experimentally–induced **chimeras** (Fig. 94), as well as on polyploidy induced by colchicine are cited as providing good evidence for 1–3 initials in each tunica layer. Contrary to the histogen theory, tunica–corpus concept explains the organization of the shoot apical meristem, without implying any relation to histogenesis below the apex, although the epidermis of the mature shoot invariably arises from the only or the outermost tunica layer.

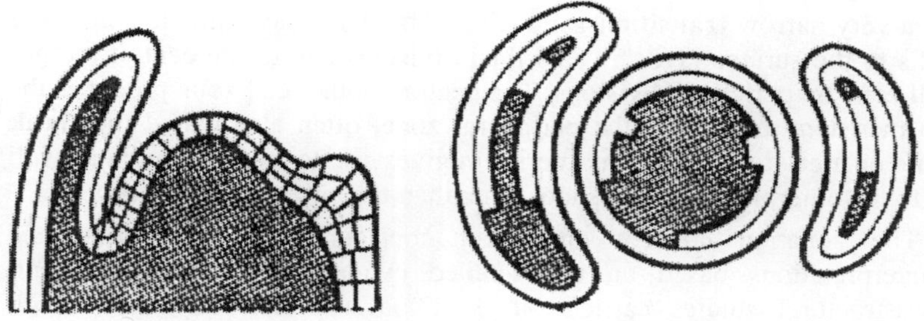

Fig. 94. L. S. and T. S. respectively of the shoot apex of a chimera *Tradescantia filamentosa* var. *albostriata;* shaded regions are the regions of coloured tissue, while unshaded regions are the regions of colourless tissue (Thielke 1957)

Some plant morphologists treat tunica in a very strict sense and apply the term only to the layer(s) that shows only anticlinal cell divisions, while others use the term a little flexibly to denote the layer(s) with predominantly anticlinally dividing cells. Popham (1951) proposed a slightly–modified theory called the **Mantle and core theory**. According to this theory, the mantle includes all cell layers in the apical meristem in which anticlinal divisions are predominant. These divisions are not exclusive, but are sufficiently required to perpetuate the definite cell layers. The core forms the meristem cells enclosed by the mantle. Differences are also found in the cytochemical properties of the cells of the meristem, but these do not necessarily coincide with the arrangement of cells into tunica and corpus. For example, the cells that are weakly stained for DNA and RNA occur both in tunica and in corpus.

5.2.1.4. *Cytohistological Zones in Shoot Apical Meristems*

The tunica–corpus organization could not be found in the shoot apical meristems of gymnosperms, except in the meristems of *Cycas*, *Gnetum* and *Ephedra,* and hence, the concept was not suitable to explain the organization of shoot apical meristem of this group of plants. Hence, alternative explanations were sought. Foster (1938) proposed a **cytohistological zonation concept** for this group of plants (Fig. 95 and

96). This concept was based not only on the planes of divisions of cells, but also on the extent of their cytological and cytochemical differentiation, as well as on the degree of meristematism of the cells forming part of the different zones recognized. Later on, similar zones have also been observed in angiosperm shoot apical meristems. Such zones can easily be superimposed on the tunica–corpus organization in angiosperms. The zones recognized are **apical initials**, **central mother cell zone**, **peripheral zone** and **rib or pith meristem** (Fig. 96a). The last zone lies immediately below the central mother cell zone or is separated from it, in some cases, by a very narrow **transition zone** (Fig. 96b). The apical initials contribute cells to the surface layer by anticlinal divisions and to the central mother cell zone by periclinal divisions. The central mother cell zone produces the rib meristem as well as the peripheral zone, often also called the **flank zone** or **meristem**. The flank meristem gives rise to leaf primordia as well as to the other tissues of the shoot other than the pith.

The zonation concept underwent considerable modifications and reinterpretations based on the detailed cytological, cytochemical and ultrastructural studies carried out later on, especially by the French botanists. These people recognized the following zones: (i) The **waiting meristem** (or *méristéme d' attente*); some call it **apical zone** (*zone apicale*) (ii) The **initiating ring** (or *anneau initial*) and, (iii) the **medullary meristem** (or *méristeme medullaire*). The first zone is equivalent to the apical initial and the central mother cell zone as described by Foster(1938). Some refer to these two zones together as **axial zone**. The cells in this zone are in a relatively quiescent state during the vegetative phase and become

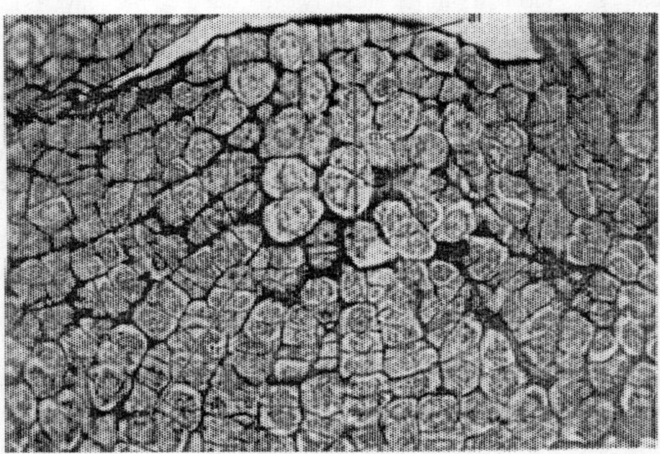

Fig. 95. L. S. of the shoot apical meristem of *Ginkgo biloba*; ai - apical initials; mc - central mother cell zone; tr - Transitional zone (Foster 1938)

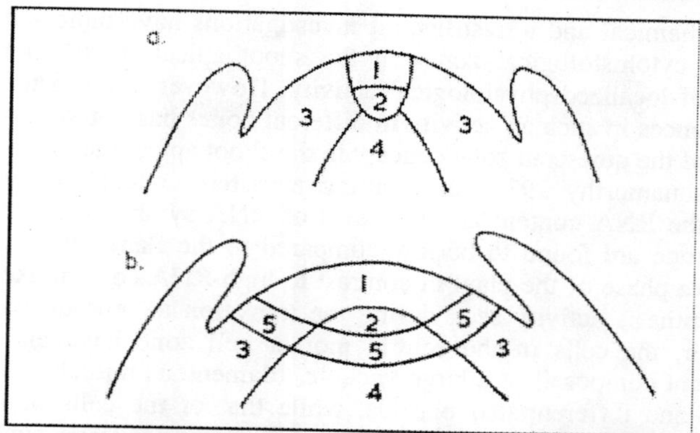

Fig. 96. Diagrammatic representation of zonation pattern in the shoot apical meristems; 1 - Apical initials; 2 - Central mother cell zone; 3 - Peripheral zone, 4 - Rib meristem, 5 - Cambium-like transition zone

very active during the reproductive phase. The second zone corresponds to the peripheral zone or flank meristem and the third to the rib or pith meristem. Counts of mitoses in different zones after feeding the apex with radiolabelled compounds to detect the location and the synthesis of nucleic acids and proteins, surgical experiments, in vitro culture etc. have been undertaken in a very systematic way. Based on these studies, we now know that there is a relative infrequency of mitotic activity in the central mother cell zone (Fig. 97). Recognition of the relative infrequency of mitotic activity has not led to an abandonment of the concept that the most distal cells are the true initials and the ultimate source of all body cells of the shoot. As already stated, a cell functions as an initial not because of any inherent property, but because of its position.

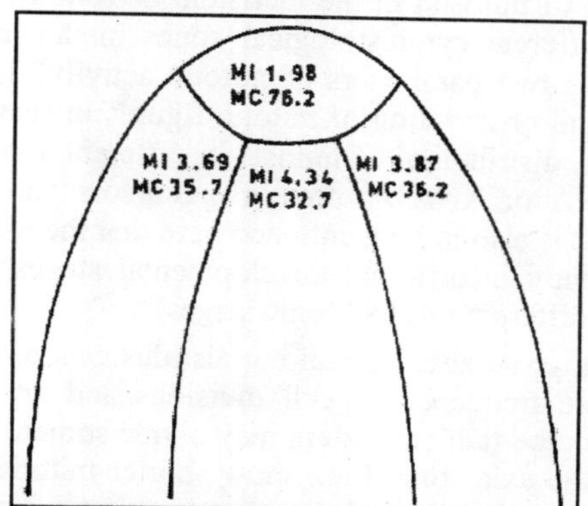

Fig. 97. Representation of mitotic indices (MI) and mitotic cycle (MC) rates in the different zones of the shoot apex (Swamy and Krishnamurthy 1978)

Histochemical and ultrastructural investigations have indicated that the different cytohistological zones in the shoot apical meristem represent regions of localized physiological activity. However, a consistent picture of differences in such an activity in different zones has not yet evolved. In support of the quiescent zone concept in the shoot apical meristem (Swamy and Krishnamurthy 1978) and waiting meristem concept of the French school, the RNA content and the level of RNA synthesis activity in the central zone are found to be low compared to the flank meristem in the vegetative phase of the plant in contrast to high RNA content and level of RNA synthesis activity seen during the transition to reproductive phase. Moreover, the cells of the central mother cell zone have an organelle component composed of a large vacuole, filamentous mitochondria, small nucleoli and differentiated plastids, while that of the cells of the flank meristem has small vacuoles, granular mitochondria, large nucleoli and undifferentiated plastids. These differences are considered to show differences in the propensity of cells of these two regions to divide. Details of the DNA synthesis, and the relative concentrations of RNA and proteins in the cells of central zone and flank meristem suggest difference in the metabolic activity. At the ultrastructural level, attention has been paid to the structure of plastids and mitochondria, density of ribosomes and the proportion of organelle volumes to distinguish the cells of the central region and flank meristem, but differences are not striking enough to attribute any significance to them.

5.2.1.5. *Cell Division Activity in Shoot Apical Meristems*

Cell division activity is not uniform throughout the shoot apical meristem irrespective of whichever concept of apical organization mentioned above one follows. This refers not only to the duration of mitotic activity (**mitotic cycle**) noticed in the cells, but also to the **mitotic index** (frequency of mitosis) in the meristem or zone thereof. In fact, the recognition of different cytohistological zones in the meristem is based primarily on these two parameters of mitotic activity. This conclusion is based on the actual observation of mitotic figures in sectioned meristems, counts of mitosis, distribution of mitosis in different zones, application of radiolabelled DNA or RNA or protein precursors and their subsequent localization etc. It is also to be mentioned here that the rate and duration of mitotic activity vary in different developmental stages of meristem (for example, during different plastochronic stages).

The cells in the axial zone (apical initials plus central mother cell zone) show a negligible frequency of cell divisions and an extended mitotic cycle. The cells in the flank meristem may divide sometimes up to twice as fast as those in the axial zone (i.e., show shorter mitotic cycles) and also divide more frequently, (i.e., show greater mitotic index). One of the consequences of these is that the apical dome increases throughout its development. For example, in *Chrysanthemum segetum,* the actual doubling time of the axial zone cell population is approximately six days, while it is only two–three days in the flank meristem i.e., the axial zone supplies new cells to the surrounding flank meristem at about half the rate at which the later's cells complete their mitotic divisions. An almost similar imbalance in mitotic cycle is seen in almost all other plants investigated thus far (Table 1). A study on the shoot apex of *Arabidopsis* has shown that the mitotic index of the flank meristem is approximately 50% higher than that of the central zone.

Table 1. Cell doubling time in central zone and flank meristem of some shoot apical meristems (Data from various sources)

Sr. No.	Plant	Cell–doubling time (in h)	
		Central zone	Flank meristem
1.	*Chrysanthemum* species (photoflux = 200μm mol/m²)	144	50
2.	*Chrysanthemum* species (photoflux = 70μm mol/m²)	102	32
3.	*Chrysanthemum* species	139	48
4.	*Chrysanthemum segetum*		
	Radiolabelled	139	54
	double labelled	126	52
	colchicine treated	189	48
5.	*Coleus blumei*	250	130
6.	*Datura stramonium*	76	36
7.	*Helianthus annuus*	83	37
8.	*Oryza* species	86	11
9.	*Pisum sativum*	69	28
	radiolabelled	49	31
10.	*Rudbeckia bicolor*	>40	30
11.	*Sinapsis alba*	288	157
12.	*Solanum tuberosum*	117	74
13.	*Trifolium repens*	108	69

From the past literature, it is apparent that a non–uniform mitotic frequency and mitotic cycle duration are the hallmarks of shoot apical meristems.

5.2.1.6. Control of Apical Organization and Function

Invaluable information on the origin, organization and function of shoot meristem has come from genetic and molecular studies of *Arabidopsis thaliana* and a few other plants such as maize. The primary shoot apical meristem in *Arabidopsis* is reported to become apparent relatively late in embryogenesis, after the cotyledons are initiated (Barton and Poethig 1993). However, based on Swamy and Krishnamurthy's analysis of embryogenesis in plants investigated by them as well as in those studied by others, it is clear that the shoot apical meristem is initiated in the midglobular embryo itself in the form of **epiphysis** (Swamy and Krishnamurthy 1977, Krishnamurthy 1994; Raghavan 2000). According to Swamy and Krishnamurthy (1977) and Raghavan (2000) the earliest indication of the shoot apex is the appearance of a pocket of meristematic cells in the upper half of the globular embryo. These cells are the direct descendants of eight–celled proembryo (Fig. 98). Clear structural, cytological and cytochemical differences exist between these progenitors of the apical meristem and the surrounding cells. These differences occur before the shoot apex is enclosed by the cotyledons suggesting that the later are the first formative organs produced by the embryonic apex. The visibility of the shoot apex is invariably associated with a change in the shape of the embryo from globular to cordate and this change is attributed to the continued meristematic activity of the surface layer and the underlying cells in the (shoot) apical part of the embryo. Differentiation of the shoot apical meristem stops when the embryonic apex becomes dome–shaped and often enters into a dormant state or quiescence in the seed. However, in many plants including many legumes, the shoot apical meristem of the embryo is very active and produces a number of leaf primordia, even before germination. The differentiation of the shoot apical meristem in the embryo makes obvious the pre –existing cryptic polarity of the latter.

In *Arabidopsis*, the establishment of the shoot apical meristem requires the activity of *Shoot Meristemless* (*STM*) gene, which is reported to be first expressed in one or two cells of the late globular stage embryo. In fact, it is the group of four cells called epiphysis, which has been established without any doubt as the progenitor of the shoot apex (Fig. 98). The *STM* mutations result in the loss of function. These mutants produce seedlings having normal roots, hypocotyl and cotyledons, but lacking shoot apical meristem. *STM* mRNA is shown to be present in the central and peripheral zones of all adult vegetative shoot apices, but is absent in developing foliar primordia. Another mutant, similar to *STM,* is *zwille* (*zll*). This also does not express the differentiated characteristics of a shoot apical meristem in that the initiation of the meristem is blocked at the torpedo–shaped stage of the mutant embryo. A well–defined shoot apical meristem is also absent in the 'buds' regenerated on cultured explants of the *STM* mutants, indicating that the gene (possibly also *zll*) controls both the embryonic and adventitious shoot apical meristems. A closely similar mutation, referred to as *pinhead* (*pnh*), affects the shoot apical meristem during embryogenesis, and the axillary bud meristem, but not the meristem formed from culture explants.

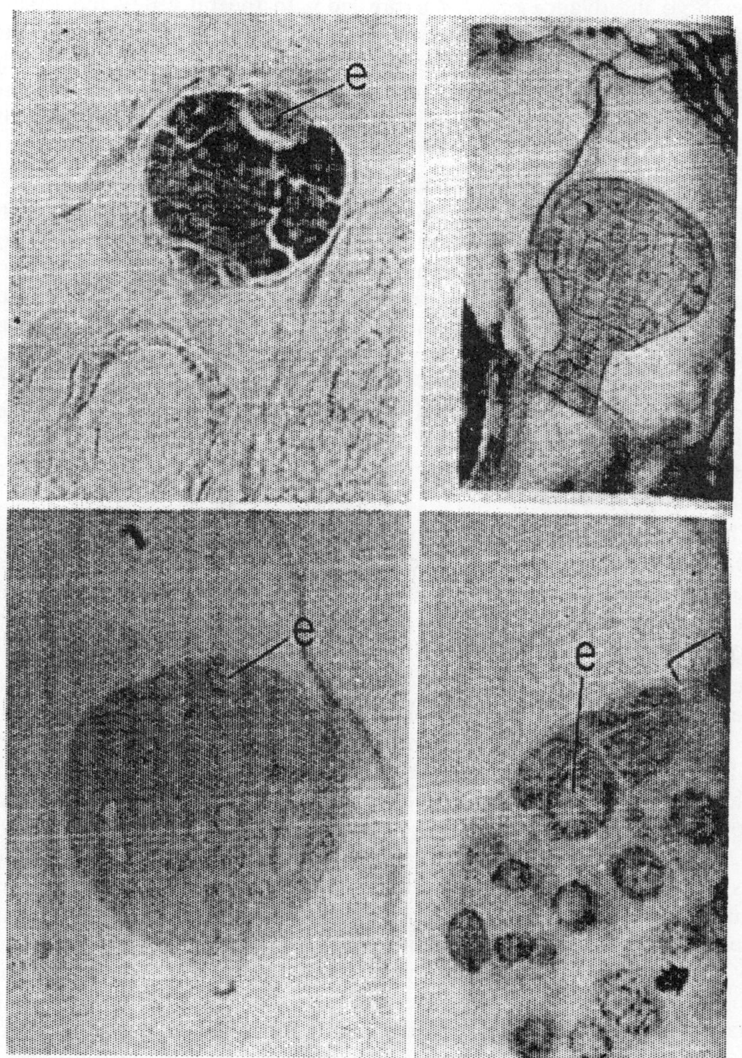

Fig. 98. L. S. of the globular embryos of *Stellaria media* showing distinctive epiphysial cells (e) (Pritchard 1964)

STM gene, as mentioned above, is required for the establishment of the shoot apical meristem and formation of leaf primordia, but the *WUSCHEL* (*WUS*) gene is additionally required for the maintenance of initial function. In *WUS* mutants, the initials undergo differentiation and no longer remain as initials. *WUS* expression is reported to begin at the 16–celled stage of the embryo in advance of *STM* expression. In the fully developed shoot apical meristem, *WUS* expression is restricted to a small group of central cells between the L3 layer (the initial layer of the corpus) and this expression persists throughout the shoot development. Thus, it is stated that *WUS* is not expressed within the initials and that it indicates that signalling must occur between the two groups of cells.

Mention must also be made about *MGOUN (MGO)* gene of *Arabidopsis*, which, like *WUS,* contributes to a precise definition of the shoot apical meristem. Similar to *stm* mutant, cell divisions in the shoot apical meristem of *mgo* embryos are distorted to some extent leading to the formation of a larger than normal meristem. During post–embryonic development, the corpus (central mother cell zone) region of the meristem undergoes proliferation to form a fasciated structure. These effects are likely to imply that *WUS* and *MGO* genes are specifically required for the maintenance of the structural and functional integrity of the shoot apical meristem, particularly its central mother cell zone. However, it is not clearly known how the functions of the central and peripheral zones are differentially regulated.

The embryos of *Petunia* species, which carry a *no apical meristem* (*nam*) mutation, also do not develop an apical meristem. The novel protein apparently encoded by *NAM* gene is believed to function in intercellular signalling and probably inhibits cell divisions around the meristem in the heart to torpedo–shaped embryos.

In addition to *STM* and *WUS* genes, which promote meristem differentiation, there are other genes in *Arabidopsis* that regulate meristem size by repressing the activity of initials. These are *CLAVATA* (*CLV*) genes (*CLV1, CLV2, CLV3*) and *EXTRA COTYLEDON* (*XTC*) genes. Mutations in these genes cause increased production of undifferentiated cells in the central zone, bringing about an increase in size of the meristem (Fletcher 2002). The accumulation of cells apparently is due to a failure to promote cells of the flank meristem towards differentiation. *CLV3* expression is mainly restricted to L1 and L2 layers and to a few L3 cells in the central zone, and probably marks the initials in these layers. The cells expressing *CLV1* underline the L1 and L2 layers. As already stated, *WUS* expression is in the deep–lying central mother cells or 'organizing cells' that confer initial cell identity to the overlying neighbouring cells; the signals from *CLV1* or *CLV3* regions act negatively to dampen such activity. More specifically, it has been proposed that CLV*3* protein secreted by the initials in the apex moves through the apoplast and binds to a *CLV1* or *CLV2* receptor complex and causes the down–regulation of *WUS*, maintaining the appropriate amount of initial activity throughout development. Thus, through this feedback loop, a balance is said to be maintained between the proliferation of promeristem cells and the loss of promeristem cells through differentiation and the initiation of lateral organs in the flank meristem (Fletcher 2002, 2004).

The *STM* gene has been cloned and its predicted amino acid sequence shows a strong homology to the homeodomain (coded by the homobox gene) of the *KNOTTED 1 (KN1)* gene of maize, the initial expression of the transcripts of which is seen in the early differentiating shoot meristem of maize embryos and continues to be seen during post–embryonic stages as well in the shoot apical meristem. In rice, a gene similar to the *KN1* called *OSH1* (for *Oryza sativa* homebox) is found. The rice *OSH1* gene shows the formation of clumps by shoot apices, similar to a teratoma. *OSH1* is probably important in maintaining the cells of the shoot apical meristem in the undifferentiated state.

Many other gene products are essential for the proper organization of the shoot apical meristem in the embryo and in the developing shoot. One such is the product of an *Arabidopsis* gene called *EMBRYONIC FLOWER (EMF)*, which interferes with the development of the typical tunica–corpus organization right from the heart–shaped stage of the embryo. This effect is noticed even in meristems of shoots differentiating from callus. The peculiarity is that the meristem directly forms a reproductive apex. The other genes that control the shoot apical meristem activity are *ULTRAPETALA 1 and 2* (*ULT 1 and 2*), *HANABA TARANAU* (*HAU*) *HAIRY MERISTEM* (*HAM*), *PHAVOLUTA* (*PHAV*), *PHABULOSA* (*PHAB*), *REVOLUTA* (*REV*), *CORONA* (*CAN*), *POUNDFOOLISH, STIP* etc.

In addition to the genetical control for shoot apical meristem organization and activity, there is a hormonal regulation for organization and activity (Refer to Box. 5.2.)

BOX 5.2. HORMONAL CONTROL FOR SHOOT APICAL ORGANIZATION AND ACTIVITY

The shoot apical meristem (SAM) shows a characteristic placement along the proximal–distal axis. This characteristic placement depends on the complex patterns of hormonal synthesis, transport and accumulation, particularly of auxin. Such patterns are evident from the embryo stage onwards. Studies have revealed that relatively low levels of auxin are a prerequisite for the normal organization and activity of SAM, probably through its effect on the expression of *CUC* genes and the activation of *STM* gene. Localized increases in auxin level (particularly in flank meristem zone) trigger leaf initiation, probably accompanied by a down–regulation of *KNOX* genes in leaf founder cells. How reduced levels of auxin promote *KNOX* expression in SAM is not yet clear. Low levels of auxin in SAM may be related to the relatively–unpolarized cell growth in the corpus region. There is also a suppressed GA activity in SAM and this suppression is effected by *KNOX* genes.

Cytokinins are equally important in SAM activity. Attenuation of cytokinin signalling leads to the suppression of SAM activity, while elevated cytokinin levels promote the SAM activity. The importance of cytokinin signalling to homeostatic regulation of the *SAM* is evident from an analysis of *ABPHYL* gene in maize, which encodes an A type–cytokinin response regulator. Loss of this gene function leads to a highly enlarged SAM through a cytokinin–mediated cell proliferation. A link between cytokinin and SAM activity is also revealed by an analysis of *KNOX* phenotypes. This gene expresses to produce a rapid increase in cytokinin levels through an isopentenyl transferase enzyme that mediates cytokinin synthesis.

Refer to Veit (2006) for more information.

2.1.7. *Autonomy of Shoot Apical Meristem*

An important question, relevant to the organization and function of the shoot apical meristem, is whether it can function independently of the more mature regions of the stem. This question has been answered to some extent through microsurgical experiments on the shoot apex and also through the culture of isolated meristems in the appropriate culture media (Fig. 99). It is evident from these studies that the meristem is autonomous to a large extent and it does not need signals from other parts of the plant to direct its functional activity. From an historical perspective, it must be stated that microsurgical experiments on shoot apices were done first in pteridophytes, but then were extended to gymnosperms and angiosperms subsequently. The surgical work on pteridophytes was mainly carried out by Professor Wardlaw and his group, while that on angiosperms was done by Professor Ball, especially on *Lupinus*. The autonomous development of the surgically–isolated meristem is first evident through the production of leaf primordia with the almost simultaneous differentiation of procambial strands, which through basipetal extension get connected to the original vascular system of the stem. In the longitudinally–excised meristem, which separates the central portion from the peripheral portion, epidermal tissue, cortex and even leaves are formed subsequently, leading to maturity of the shoot. A critical analysis of the results of surgical experiments involving removal by puncturing or excision of the distal segment of the meristem have indicated that the central mother cell zone is the most important region of the meristem. Any shoot apical meristem segment that remains after surgical experiments and that has at least a few cells of the central mother cell zone will re–establish the meristem, which will then function normally.

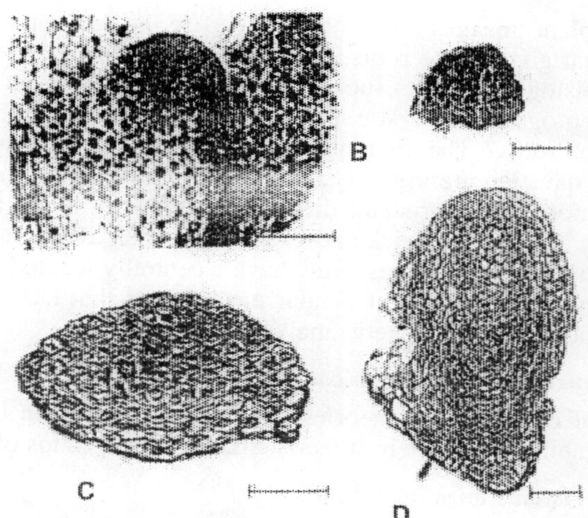

Fig. 99. A. L. S. of shoot apex of tobacco plant; B–D. Development of excised shoot apical meristem of the tobacco plant in culture medium (Smith and Murashige 1970)

Confirmation of the autonomy of the shoot apical meristem is also provided by in vitro culture of meristems. The apical meristem of a few angiosperms have been attempted, the most important of which is tobacco. Such meristems, instead of forming leaf primordia to start with, establish a bipolar axis through the initiation of root(s) at the opposite end.

5.2.2. Root Apical Meristem

In contrast to the shoot apical meristem, the root apical meristem is subterminal as it produces cells not only towards the root axis, but also away from it, for it has a root cap in the distal part of the root apical meristem. It also differs from the shoot apical meristem in that it does not produce lateral appendages such as leaves and flowers. Moreover, branching of the root is initiated beyond the region of most active growth (i.e., the root apex) and also endogenously (in contrast to exogeneous origin of shoot branches and in close proximity to the apical meristem). Since leaves are not produced, the root apical meristem does not show periodical changes in shape and structure (i.e., plastochromic changes are absent). The root meristem also does not produce nodes and internodes and hence grows more uniformly in length than the shoot, where the internodes grow and elongate while nodes do not. Generally, the root may or may not have a specific protoderm, the progenitor of epidermis and hence the outermost layer in the second case is called rhizodermis or piliferous layer.

According to Friedman et al (2004), three basic questions surrounding the origin and evolution of roots remain unresolved: (i) Are roots of different land plant lineages homologous? (ii) Are the developmental gene programmes that give rise to roots and shoots similar? (iii) Do the shared features of histogenesis in shoots and roots arise from common developmental programmes first expressed in land plant sporophytes prior to the origin of roots? The first question is still unresolved and hence, studies on this question are urgently needed. Notwithstanding the fact that the root and shoot apical meristems differ on a number of levels as detailed above, cladistic analysis (Kenrick and Crane 1997, Veit 2006) supports the homology of these two structures. Both have a centrally located 'quiescent centre zone' and the existence of similar mechanisms that has been borne out by recent molecular and genetic analysis.

5.2.2.1. *Organization of Root Apical Meristem*

A number of concepts and theories have been proposed in the past to explain the organization of the root meristem in diverse groups of plants.

Root apical meristems with apical cells

According to the **Apical cell theory,** the root apical meristem is composed of a single apical cell, from which all root cells are ultimately derived. Barring a few pteridophytes, single apical cells are not present in vascular plants. Among pteridophytes, a number of ferns, primitive and

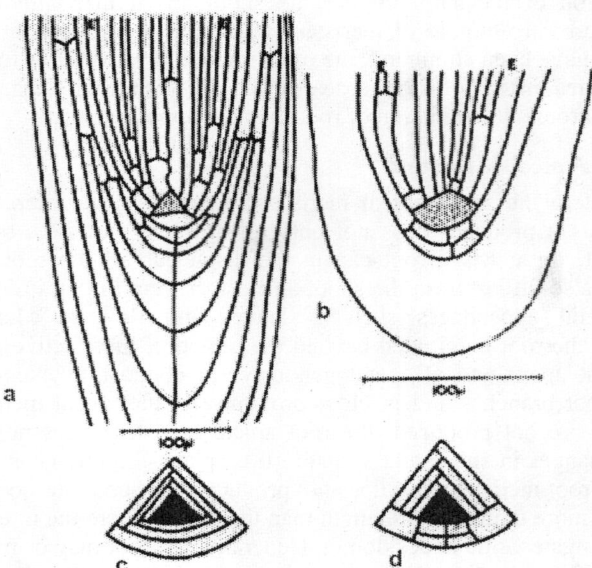

Fig. 100. L. S. of root apices of *Marsilea* and *Azolla* respectively showing body-cap organization and T- division patterns (a and b); apical cell (dark coloured) of these two taxa and its segmentation patterns (c and d); E-epidermis (Clowes 1961)

advanced, and *Equisetum* have single apical cells, which produce all root tissues (Fig. 100). In most ferns, the apical cell is tetrahedral and it cuts off segments on three lateral (proximal) faces and thus, produces the main body of the root. The root cap has its origin either from the fourth facet of the apical cell or from a separate meristem formed early in the root development. The tetrahedral apical cell of *Equisetum* contributes both to the main body of the root and to the root cap, but early root development in *Equisetum* is slightly different from that of the ferns.

Meristems with histogens

According to the histogen theory of Hanstein (1868, 1870), the body of the root arises from a massive meristem comprising of three or four precursor tissue regions or histogens. Each histogen begins with one to several initials at the apex arranged in superposed tiers (Fig. 101). These histogens are the dermatogen (precursor of root epidermis), the periblem (precursor of cortex), plerome (precursor of stele and pith, if present) and **calyptrogen** (precursor of root cap). In many root apices the dermatogen is not found and the surface layer is invariably formed from the periblem. Although not accepted in its original version, the histogen theory is accepted by some in combination with the **promeristem** concept.

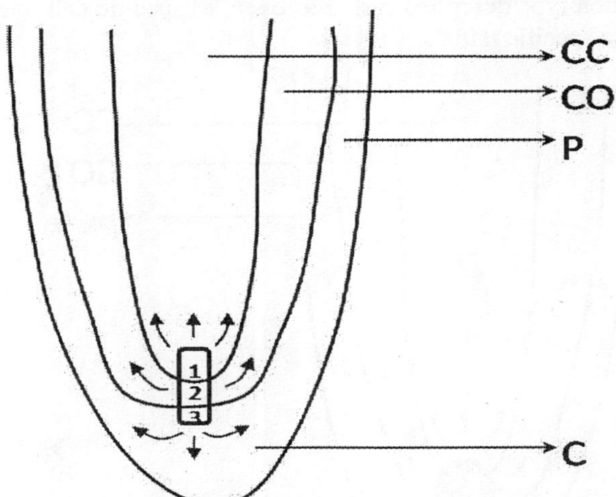

Fig. 101. Diagrammatic L. S. of the root apex interpreted as per histogen theory; 1 - Plerome; 2 - Periblem; 3 - Dermatogens; C - Root cap; CC - Central cylinder; P - Piliferous layer

Open and closed apices

Guttenberg (1960) proposed the concept of open and closed apices with reference to the organization of the root apical meristem. In some gymnosperms and angiosperms, all tissue regions of the root or all except the central cylinder appear to arise from a common meristematic group of cells (Fig. 102) in others, one or more of these regions can be traced to separate initials (Fig. 103). These two kinds of organization are described as the **open** and **closed types** respectively. The distinction between open and closed meristems is not always clear even in the same plant at different developmental stages, and in some cases such as onion, a transitional type between open and closed is seen (Fig. 104).

Popham (1966) categorized the roots of vascular plants into six types and two subtypes. All these incorporate one of two common developmental schemes. In the first, all cell file lineages can be directly followed to a specific tier of initial cells. These initials may contribute to different tissues. For example, in the most common monocot type, there are three initial layers respectively for stele, cortex or epidermis, and root cap (e.g., *Zea mays*), whereas the most common type in dicots also has three layers, but in this case, they are respectively for stele, cortex, and epidermis or root cap. This kind of organization is of the closed type. In the second type, cell lineages do not connect to distinct initial layers, but instead can be followed to a zone of cells lacking any special organization. Both monocots (*Allium*) and dicots (*Pisum*) are examples of this. This is the organization type described earlier as open, as specific cell lineages do not connect to specific initials.

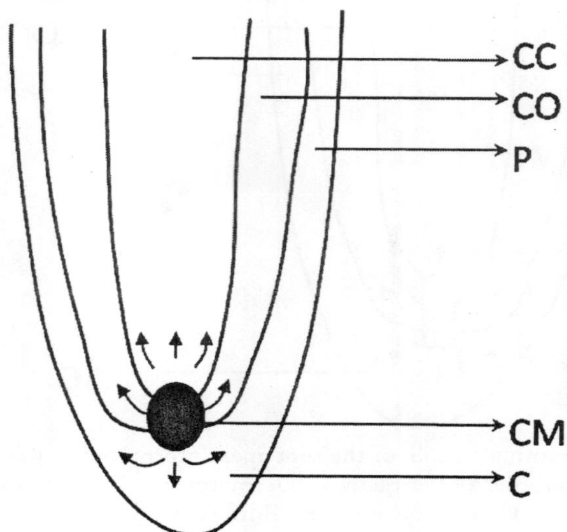

Fig. 102. Diagrammatic L. S. of the root apex showing open type of meristem organization. C - Root cap; CC - Central cylinder; CM - Common meristematic group of cells CO-Cortex; P - Piliferous layer

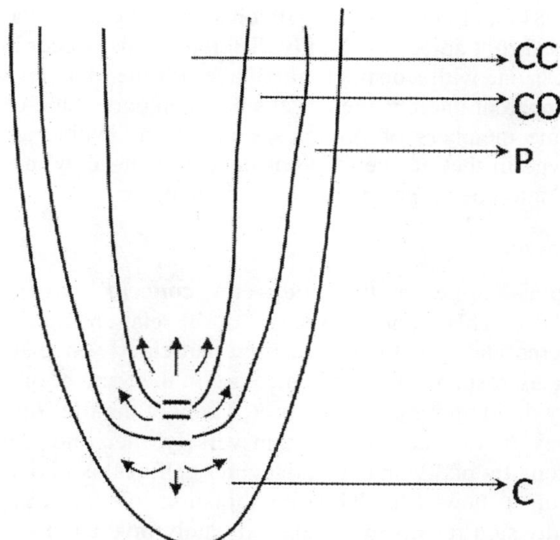

Fig. 103. Diagrammatic L. S. of the root apex showing closed type of meristem organization, where the central cylinder, cortex and root cap plus piliferous layer appear to arise from three separate layers of initials, represented in the figure by the dark horizontal lines; C - Root cap; CC - Central cylinder; CM - Common meristematic group of cells; Co- Cortex; P - Piliferous layer

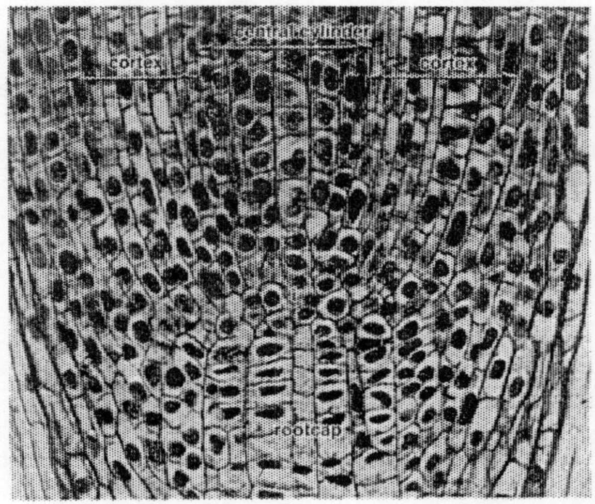

Fig. 104. L. S. of the root apical meristem of *Allium sativum* showing a transitional type of meristem organization (Mann 1952)

Clowes (1981) and Deysson (1980) considered the differences between open and closed root apices in slightly different terms. According to them, a closed apex is one with a distinct initial layer for the root cap and an open apex shows apparent interchange of cells between cortex and cap. The root apices of some members of Asteraceae and Convolvulaceae are of the convertible type in that at one stage of development, they are closed and are converted into open type later in development.

Body–cap concept

The **Körper–Kappe** or the **body–cap concept** was proposed by Schüepp (1917). This concept speaks of a relationship between cell division patterns and growth in root tips, particularly in planes of those divisions that are responsible for the increase in the number of vertical files in the actively dividing region of the root. Many of the files divide into two and where they do so, a cell divides transversely; then one of the two new cells divide longitudinally and each daughter cell of this division becomes the source of a new file. This combination of the transverse and longitudinal division results in an approximately inverted T– or Y–shaped wall pattern, and hence, such divisions are called **T divisions** (Fig. 100 and 120). These divisions are seen in roots with single apical cells (Fig. 100) as well as with a multicellular meristem. The direction of the top strokes (horizontal bar) of the T varies in different root parts. Often, in the cap, it is directed towards the base of the roots, and in the body, it is directed towards the apex. In some roots, there is a clear boundary between the body and the cap, while in others, such a boundary is not clear.

Quiescent centre concept

This is one of the most important and most discussed concepts relating to the root apical meristem organization. Clowes (Clowes 1961) first discovered a **quiescent centre** in the root apex and this discovery has brought about a fundamental change in ideas regarding the behaviour of root apical meristem. According to Clowes, there is a **minimal constructional centre** in the root apex, which largely ceases to be mitotically active and hence, this centre is labelled as the quiescent centre (QC). The positive existence of such a centre on the root apex was proved by extensive researches on normally developing roots by conventional and cytochemical staining methods, surgically–treated roots, on irradiated roots and on roots fed with radiolabelled compounds involved in DNA synthesis (Fig. 105). Mitotic activity was largely supplemented by cells on the margin of the relatively inactive QC. Although the QC is reported (Evert 2006) to arise twice in primary roots, first during embryogeny and then again during the early stages of germination of the seed, Swamy and Krishnamurthy (1975b) and Krishnamurthy (1994) have categorically

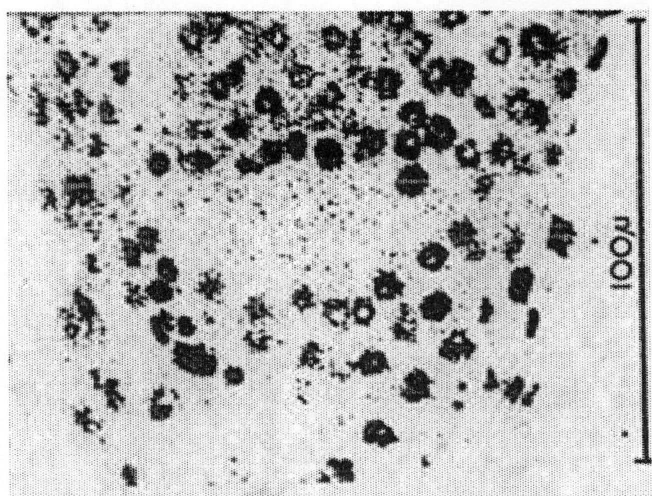

Fig. 105. Autoradiograph of median L. S. of *Sinapsis alba* root apex fed with thymidine-^{3}H showing quiescent centre (Clowes 1961)

proved its continuance, without any break, ever since its inception as **hypophysis** in the midglobular stage at the micropylar pole of the embryo.

The QC of a mature root is hemispherical or discoid in shape and contains as few as four cells as in *Arabidopsis* to more than a thousand. In *Zea mays,* it is a hemispherical mass with a flattened acroscopic surface and is composed of about 1500 cells. In *Helianthus,* it has about 80 cells with a convex acroscopic surface. The report of absence of QC in very thin roots (as in rice) is apparently erroneous as QC is an integral part of the root meristem, however small it may be.

The relatively inactive state of the QC cells does not mean that they have become permanently non–functional. QC cells do divide occasionally and serve to renew the more actively dividing regions around them, the cells of which are unstable and displaced from time to time. By labelling nuclei with tritiated thymidine and by blocking cell cycle at metaphase with inhibitors, one can obtain quantitative data on the duration of the mitotic cycle in different regions of the root meristem. This data indicates that the cells of the QC divide approximately 10 times slower than the adjacent dividing cells (Table 2). Pulse–labelling with thymidine has shown that the differences in the duration of mitotic cycles are largely caused by differences in the duration of **G1** phase. In other words, the mitotic cycle takes a very long time to get completed in the cells of the QC. The mitotic frequency is also very low in the QC i.e., the number of cells undergoing mitosis in this centre is extremely few (Fig. 106).

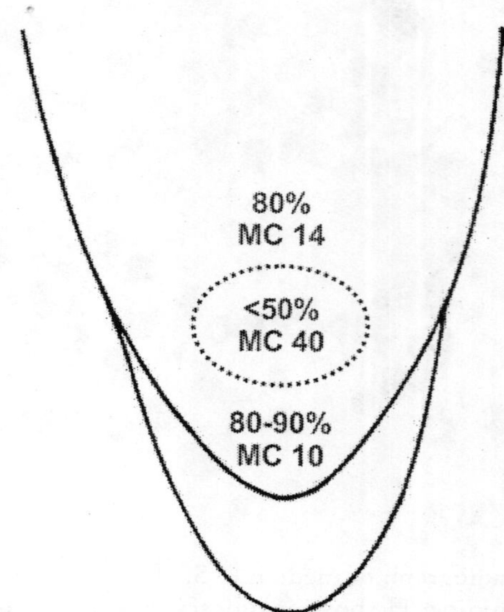

Fig. 106. Representation of low mitotic frequency (figure with percentage marks represent mitotic index) and longer mitotic cycle in the quiescent centre of root apex when compared to subjacent regions; MC - Mitotic cycle (in h) (Swamy and Krishnamurthy 1978)

The paucity of mitotic activity in the QC led Clowes (1961) to suggest that the initials of the root apex are located just outside the QC along its margin, and he designated this group of cells as the **promeristem** of the root apex. However, Swamy and Krishnamurthy (1975b), Barlow (1978) and Steeves and Sussex (1989) have argued that it would be more appropriate to consider the slowly and sparsely dividing cells of the QC itself as the ultimate source of cells for the whole root and hence, as the initials, and to consider the actively dividing cells immediately outside the QC as their derivatives. If we adopt this view, the QC of root is strikingly akin to the central mother cell zone or promeristem of the shoot apical meristem and hence, QC is the promeristem of the root. Some botanists (for example, Kuras 1978) consider the root promeristem as comprising the QC and its immediate, actively dividing derivatives. Even today, there is no uniformity in the use of terms and in the meaning in which these terms are used.

Since QC region has a very low mitotic index and since its few dividing cells also have a prolonged mitotic cycle, the QC often serves as a reservoir of cells to renew the root apex, if the meristem becomes dormant due to unfavourable conditions or becomes damaged. In fact, QC has been shown to resume division during recovery from a period of dormancy induced by cold. The root apical meristem is occasionally damaged as the root forces its way through the soil. When this occurs, cells of the QC divide and once again organize the root apical meristem and root cap. Thus, the QC also serves as a reservoir of genetically–healthy cells. Clowes (Clowes 1961) had shown through studies that destruction of root apical meristematic cells by X–rays prompts cells of QC to divide and regenerate a new meristem. The factor, most likely responsible for the resistance of QC cells to X–ray treatment or cold exposure or most other stress conditions, is the state in which mitotic cycles remain in QC cells i.e., **G1** phase. Dividing cells in **S** or **G2** phases are generally affected by these stress factors, as chromosomes of these are damaged. We have already discussed that the activation of cyclins by protein kinases is the earliest known event in the cell cycle. Some of the cyclin–dependent kinases share a conserved PSTAIR motif (stand for the amino acid sequence of proline, serine, threonine, alanine, isoleucine and arginine) and a high level of this protein has been identified by immunofluorescence studies in the QC of maize roots. This finding has raised the possibility that the PSTAIR proteins may enable QC cells to readily enter cell cycle after an extended interphase.

Table 2. Average duration of mitotic cycle (in h) calculated from metaphase accumulations in dividing nuclei in root apical meristem treated with mitotic inhibitors

Sr. No.	Species	QC	Root cap Initials	Central cylinder	
				Just above QC	200–250 μm above QC
1.	*Allium sativum*	173	33	35	26
2.	*Sinapsis alba*	520	35	32	25
3.	*Vicia faba*	292	44	37	26
4.	*Zea mays*	174	12	28	29

Source: Esau (1977) and Clowes (1969)

The possible causes for the origin of a QC in a growing root have been greatly debated and many views have been expressed. Based on an analysis of growth patterns in root apices, Barlow (1973) and Clowes (1984) have proposed that quiescence in the particular location of the root apical meristem results from antagonistic directions of the cell growth in various parts of the meristem, the root cap or its initials being the most important in suppressing growth. It was argued by Clowes (1978 a, b) that the appearance of QC during embryogenesis coincides with the appearance of the root cap meristem. If the cap is damaged or surgically removed, the QC gets activated to give rise to a new root cap meristem, which in turn, produces a new cap, following which quiescence again resumes in QC. This behaviour prompted Barlow and Adam (1989) to suggest that the activation of QC, after damage or removal of cap, results from an interruption or modification of signalling between the root cap or its meristem and the QC. This signalling candidate is more likely to be a hormone, an auxin. It is suggested that the origin and the maintenance of QC in maize root apices are due to polar auxin supply and the root cap or its initial plays an important role in regulating the polar auxin transport towards the root tip (Kerk and Feldman 1994). High levels of auxin bring about elevated levels of ascorbic acid oxidase (AAO) resulting in the depletion of ascorbic acid within QC. Since ascorbic acid is required for the **G1** to **S** phase transition in the cell cycle in root apices (Liso et al 1984, 1988), Kerk and Feldman (1995) proposed that depletion of ascorbic acid in root apices may be responsible for the formation and maintenance of the QC. Subsequently, Kerk et al (2000) reported that AAO also oxidatively decarboxylates auxin in maize root apices, which forms another mechanism for the regulation of auxin levels within the QC and other root apical tissues. An intact root cap is a must for this metabolic process to occur. Although it is an attractive explanation, the facts that QC does not comprise entirely of mitotically–inactive cells and that some cells of QC divide with a prolonged mitotic cycle, are not explained satisfactorily by this concept. How do these cells selectively escape when adjacent dividing cells are affected by the depletion of ascorbic acid is another major unanswered question. The presence of a QC in root apices without root caps is also not explained by this concept; many radicles also have a QC but no root cap. Moreover, as already explained, the QC has its origin from the hypophysis in the globular embryo stage itself, when a root cap is non–existent.

Meristematic activity around QC in *Arabidopsis* is triggered by glutathione, a diffusible redox agent, which functions together with ascorbic acid in maintaining cellular redox equilibrium. There is an increased level of glutathione in dividing cells and a very low level in QC cells. Thus, there is a link between pool sizes of ascorbic acid and glutathione and regulation of mitotic activity in the root apical meristem. Some botanists believe that the QC may be the seat of hormone biosynthesis in the root and this role itself might prevent QC from active mitotic divisions in its cells. This is evident from the observation that exposing maize roots to TIBA, an inhibitor of polar auxin transport, causes a burst in mitotic activity in QC.

The physical origin of QC must be explained. In case of tap root or primary root, it is the extension of the radicle or embryonic root. As early as 1975, Swamy and Krishnamurthy (1975b) have analyzed in detail the origin of QC in tap (primary), lateral and adventitious roots. In the primary roots, it was shown that the origin of QC is from the hypophysis, a cell first demonstrated and named by Hanstein as early as 1870. It gets differentiated at the micropylar pole of the globular embryo itself, almost simultaneously with the appearance of epiphysis at the opposite pole of the embryo. Embryological works conducted after Hanstein are uniform in recognizing the hypophysis as a morphological entity (Refer to review in Souèges 1934), although there are slight variations in the origin of hypophysis during early embryogeny in terms of cell lineage in different plants, and also there are interpretational differences in tracing its origin even in the same plant (Figs. 107 and 108).

There were also differences of opinion in early literature whether the hypophysis is constituted of a single cell, a single tier of cells or of more than one tier of cells. It is a discoidal single cell to start with (Fig. 108), but divides twice by vertical walls to result in a plate of four cells. It was also demonstrated that the hypophysial cell and its derivative plate of four cells remain histologically conspicuous by their relatively poor stainability, and

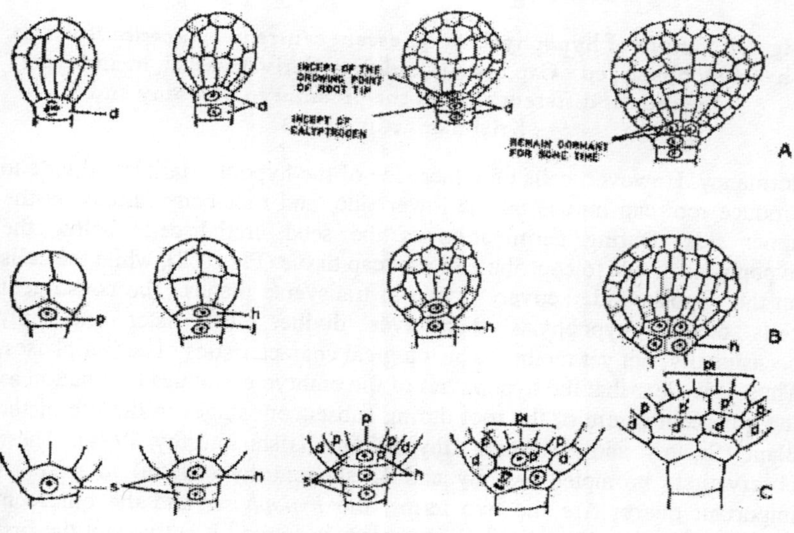

Fig. 107. Early embryogenesis in *Capsella bursa–pastoris* showing the origin hypophysis as per Schaffner (1906) (A). (d - Cells between suspensor proper and embryonal mass), Souèges (1919) (B). (p - Initial cells of hypophysis, h; j - Last cell of suspensor) and Coulter and Chamberlain (1903) (C). d - Dermatogen; d' - Dermatogen intital; h - Hypophysis; p - Periblem; p' - Periblem initial; pl - Plerome) (Swamy and Krishnamurthy, 1975b)

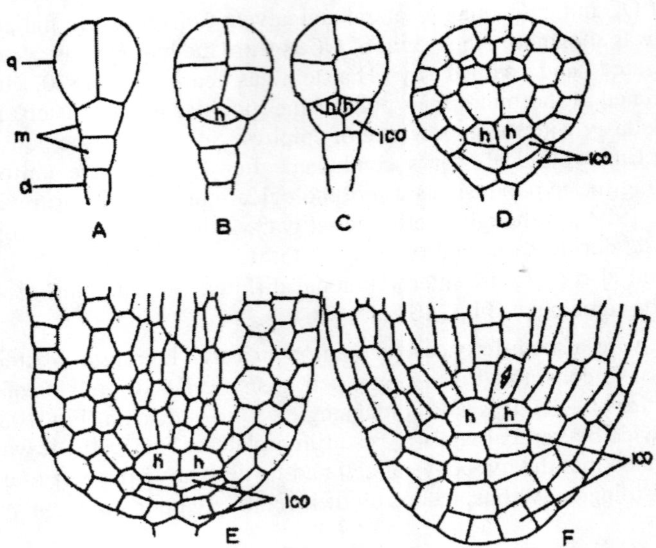

Fig. 108. Origin of hypophysis or quiescent centre in *Sphenoclea zeylanica*; h - Hypophysis; co - Cap initials and their derivatives; ci, m and q are the cells at different tiers of the proembryo. (Swamy and Krishnamurthy 1975b)

dormancy. However, cells on either side of the hypophysial plate divide to produce root cap initials on the lower side, and root body initials on the upper side. During germination of the seed, initial cells below the hypophysis divide to contribute to root cap tissue (Fig. 108), while the cells on the opposite side actively divide in transverse planes. The constituent cells of the hypophysis themselves divide much later and only occasionally, but yet retain the histological characteristics of earlier phases. Thus, it is clear that the hypophysis of the embryo continues to function as the quiescent centre of the root during subsequent stages in the life of the plant (Swamy and Krishnamurthy 1975b; Krishnamurthy 1994). These observations prompted Swamy and Krishnamurthy (1975b) to raise an important query: Are the two terms, the *hypophysis* and the quiescent centre (quiescent centre 'initial') necessary in future? It is true that the first term was coined by Hanstein (1870) at a time when its functional significance even during embryogeny was not clearly known. Hanstein and many embryologists used it merely as a descriptive term because of its unmistakable topographic location. In view of the recent investigation on the quiescent centre of adult roots and from a conceptual perspective, Swamy and Krishnamurthy (1975b) were inclined to suggest the abandonment of *hypophysis* in favour of quiescent centre or quiescent centre 'initial'. They have also emphasized the fact that unlike other 'initials' (for example, the root cap initial, the cotyledon initial and periblem initial) where the original locus becomes progressively diffused histologically due to the establishment of physiological and structural gradients, the derivatives of the quiescent centre retain a histological and physiological delimitation with greater emphasis. In the developing embryo of *Arabidopsis*, the hypophysis (or hypophysial cell), at the early heart–shaped stage, divides to form a lens–shaped cell, which is the progenitor of the four central cells, and a lower hypophysial cell whose derivatives give rise to the central part of the root cap i.e., columella.

In contrast to the clear demonstration of the origin of QC from hypophysis by Swamy and Krishnamurthy (1975b), Raghavan (2000) had stated that tracing the ancestry of the QC back to a specific cell in the early embryos has proved difficult mainly due to the facts that apart from its reduced capacity to incorporate radiolabelled precursors of DNA synthesis, there are no well–defined structural, physiological or metabolic markers that easily identify a cell of the potential QC. However, experiments undertaken by his school to localize poly(A)–RNA in the cells of developing embryos of *Capsella bursa–pastoris* by in situ hybridization using [^{3}H] poly (U) as a probe have helped to trace the origin of the QC from hypophysis. In the globular embryo, all cells other than hypophysis in the future root pole showed autoradiographic silver grains. As would be expected of the QC, the same group of cells of the seedling root of *C. bursa–pastoris* that fail to bind [^{3}H] poly (U) also do not incorporate [^{3}H] thymidine. A cogent evidence for the origin of QC from hypophysis is provided by the *hobbit (hbt)* mutant of *Arabidopsis.* This mutant shows an aberrant hypophysis specification during embryo development and its seedling lacks a recognizable QC and other cells originating from it (Willemsen et al 1998). A precise delineation of cell determinants associated with the non–dividing cells of the QC, the surrounding cells of the columella and the initial cells of the growing root must be taking place during the divisions of the hypophysis.

The importance of hypophysis as progenitor of QC is also brought to light by ablation experiments carried out in *Arabidopsis.* These experiments have clearly demonstrated the importance of hypophysis in determining the behaviour of cells adjacent to it (Scheres and Wolkenfelt 1998). When all the four hypophysial cells are ablated through laser, they were replaced by the initials that otherwise normally give rise to the root vascular cyclinder. Ablation of a single hypophysial cell resulted in cessation of cell division and the progression of differentiation in the columella and cortical cell initials with which it is in contact. These results were believed by Scheres and Wolkenfelt (1998) to indicate that the major role of the hypophysis is to inhibit the differentiation of the contacting initial cells into mature cells through signals that act at the single–cell range (Van den Berg et al 1997, 1998; Scheres and Wolkenfelt 1998) (Compare the results of experiments of Barlow and his group on QC of adult roots described earlier).

The hypophysis is involved in the origin and organization of the QC of the adult tap root. How does the QC of lateral or secondary and adventitious roots originate? Clowes (1961) has studied the ontogeny of the QC in the lateral roots of some aquatic monocots essentially from the point of view of DNA cycle. He concluded that the origin of the QC corresponded with the stoppage of DNA synthesis in the concerned cells, after the lateral root initiation proceeds for a time. Swamy and Krishnamurthy (1975b) reached to the same conclusion through the customary developmental studies made on *Commelina benghalensis*, a monocot. Depending on the location from where the lateral root or adventitious root arises, the tissue of origin of the root may be the pericycle, the endodermis or the parenchyma cells immediately adjoining the peripheral vascular strands of the internode. Even during the early stages of development, two patterns of cell divisions may be generally recognized in the locus of origin. The peripheral cells divide largely by periclinal walls, typical of the root cap tissue, while the inner cells divide in diverse planes to increase in mass (Fig. 109). The duration of this phase, prior to the histological manifestation of the QC 'initial', varies with the locus of origin of the lateral roots. In lateral roots, arising at the junction of root and hypocotyl, the differentiation generally occurs earlier (Fig. 109 C), while in those that arise from internode, the differentiation of QC 'initial' is belated in spite of the larger cell number that is involved (Fig. 109 D). The 'initial' cell is unmistakably located by its topographical position, and also by its basichromaticity. From this stage onwards the division activity becomes localized on either polar sides of the 'initial' (Fig. 109 E), while the 'initial' itself divides only occasionally. With progressive growth of the lateral root, a larger number of the derivatives of the 'initial' becomes involved in the organization of the QC, and the root cap and the root axis regions become defined with greater emphasis (Fig. 109 F) at the time of passage of the root through the cortex.

5.2.2.2. *Control of Root Meristem Activity*

Some recent molecular and genetic analysis made on the root apical meristems reveal, what may be an analogous *WUS*–like activity (noticed in shoot apical meristems) in the QC of root meristems (Kamiya et al 2003, Haecker et al 2004). Artificial activation of this *WUS*–like gene promotes the root meristem, whereas over–expression of CLV3–like peptides leads to the depletion of the root apical meristem (Hobe et al 2003; Fiers et al 2005). In addition to genetical control, root apical meristem activity is also controlled by plant growth regulators (Refer to Box 5.3)

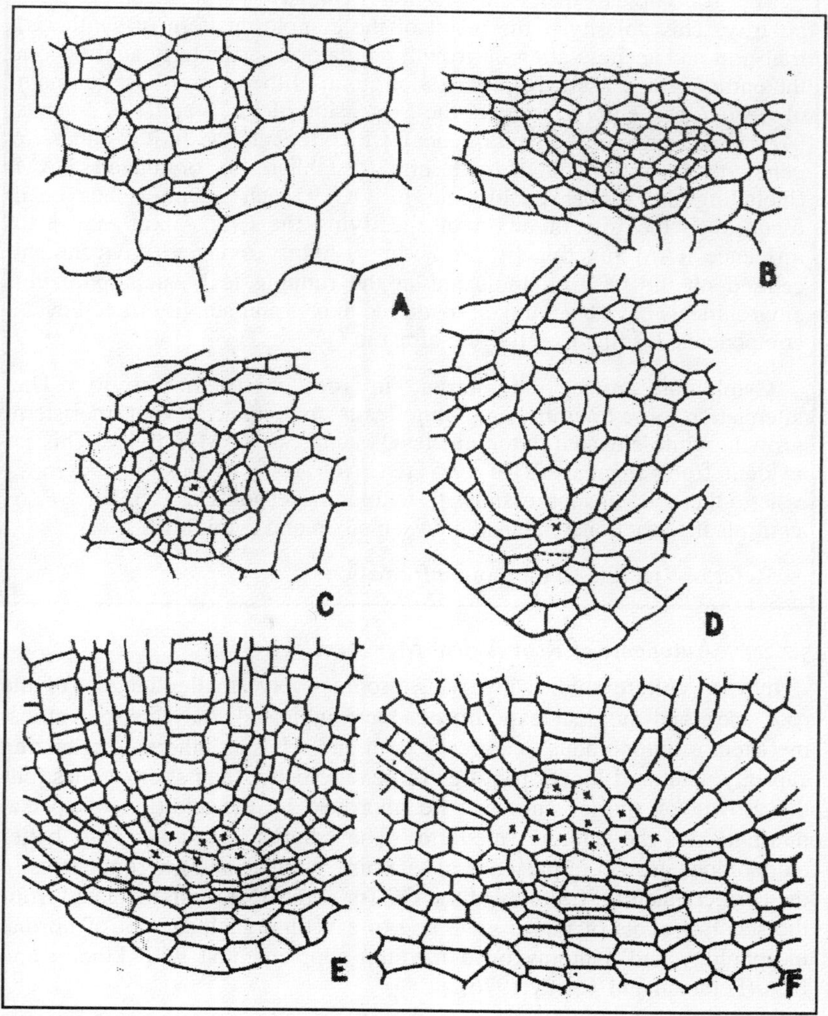

Fig. 109. Stages in the de novo origin and development of quiescent centre in the lateral root of *Commelina benghalensis*; X - Quiescent centre initial and its derivatives (Swamy and Krishnamurthy, 1975)

BOX 5.3. GROWTH REGULATORS IN ROOT APICAL ORGANIZATION AND ACTIVITY

The root apical meristem (RAM) exhibits a typical apico–basal polarity. This polarity is the result of the complex pattern of synthesis, transport and accumulation of growth regulators, particularly auxin, from the embryo stage itself. High levels of auxin in the central basal domain of *Arabidopsis* embryo triggers the expression of *PLT1* and *PLT2* genes (*AP2* class transcription factor genes), which together with *SCR* and *SHR* genes that control radial patterning, establish an organized RAM (including the proper positioning of QC). Much attention has been focused on the role of auxin in specifying the QC. Auxin may also influence RAM function by promoting a highly oxidized environment centred on the QC. As indicated in the running text, such oxidizing environments have been linked to altered expression/activity of cell cycle components, leading to cell cycle arrest in QC.

Cytokinins are also important in root meristem activity. The attenuation of cytokinin signalling leads to enhanced root meristem growth, while elevated cytokinin levels suppress RAM activity. This is evident from an analysis of *PASTICCINO (PAS) 1 , 2, 3* phenotypes, where there is a suppression of cytokinin responses. Probably *PAS* controls the functional balance between auxin and cytokinins.

Refer to Viet (2006) for more information.

5.2.2.3. *Autonomy of Root Apical Meristem*

In vitro culture studies involving progressively smaller lengths of the root apex and surgical experiments have indicated that the root apical meristem is autonomous to a great extent and it has an inherent organizing capacity. Cultured longer root tips of pea are better than shorter ones, but the growth of shorter ones can be improved by the addition of special nutrients to the culture medium. The shorter explants had better autonomous growth in presence of vitamins, sucrose and micronutrients in the correct quantity. Cultures of the QC (of about 1500 cells) isolated from the seedling roots of maize gave rise to a well–organized root of normal morphology and anatomy on a medium supplemented with kinetin and IAA (Feldman and Torrey 1976).

Surgical manipulations on root apices have yielded valuable information on defining precisely the limits of the apical meristem; also, such manipulations have shown that cells are not cut off in the meristem with a predictable destiny. Even small parts of a surgically split (longitudinally) apical meristem can regenerate complete roots, provided that the segment has at least a few cells of the QC (Swamy and Krishnamurthy 1975b).

5.3. INTERCALARY MERISTEM

In a number of plants, especially belonging to the monocots (however, note that intercalary meristem is not a general feature of monocots), intercalary meristems (IM) have been reported to play a prominent role in intermodal ontogeny (Swamy and Krishnamurthy 1979). IM has been variously defined, but the definition given by Esau (1965a) is followed by several authors. She defines it as "an actively growing primary tissue region somewhat removed from the apical meristem". This meristem, as the name indicates, is intercalated between mature tissues.

A survey of available literature on IM indicated that there are two distinct types of IM. The first type is a part of the apical meristem, separated by the matured or maturing tissues, while the second type is a product of dedifferentiation of parenchymatous tissue. The first type is known as **'detached IM'** and the second as **'resumptive IM'** (Jacobs 1947). As an example of the former type may be cited the supranodal meristem of the grasses (Kaufman 1959) and of the latter the IM of the gynophores of peanut plant. Fisher (1970) supports the use of the term 'resumptive', but finds the term 'detached' inappropriate from the developmental standpoint as the IM is a continuum from the apical meristem. However, it should be emphasized that both types of IM are localized centres of intense cell divisions and represent an intensified and endemic expression of primary growth. The longitudinal wave of maturation of tissue derived from the IM is determined by the location of the meristem in the primary plant body.

There appears to be some confusion in the literature on the definition and scope of both **subapical meristem** and IM. Some use the two terms distinctively, some interchangeably and some others use alternative terminologies. The implication that subapical growth is a type of intercalary growth is evident from the following statement of Sachs et al (1959): "The subapical zone is for all purposes an intercalary subapical meristem in its own rights". Similarly, Kaufman's (1965) statement that in grasses intercalary growth first takes place throughout the internode, but later gradually becomes localized to certain specific regions (known as "joints") also implies that subapical growth is a type or a developmental stage of intercalary growth. However, it should be borne in mind that barring some extremely acaulescent plants, subapical growth is seen in all internodes of all vascular plants to varying degrees, whereas IM is seen only in some taxa.

It is pertinent to discuss at this juncture the concept of the **uninterrupted meristem** (UM) of Fisher and French (1978) so as to get a clearer picture on the concept of intercalary meristem. According to them, although the term IM has been broadly used in the previous literature to denote any primary meristematic region removed from the apical meristem, the term should be restricted to the meristem near the base of the internode, but should not be used for similar, actively–dividing meristems located on the upper part of the internode of some monocots, of some Asteraceae, the gynophores of groundnut or the fertile spike of *Ophioglossum*, which should, really be referred to as UM. Again, UM is defined by them as a region of cell division that is progressively confined to the upper part of the internode. They have defended the concept of UM in relation to the entire growth region of the shoot, including the apex, while the previous authors were primarily relating it to a single internode. Their concept of UM is based on the fact that the meristematic zone that progressively gets localized to the distal part of the internode is continuous with the subapical meristematic region of the shoot apex (as in Asteraceae) or the floral or reproductive apex (as in *Ophioglossum*) and that it is not isolated from the apex by mature tissues. Swamy and Krishnamurthy (1979), however, argued that the above concept of UM of Fisher and French (1978) is untenable since the UM of an internode is interrupted from the subapical meristem towards the upper end also by mature nodal tissue (especially in monocots with a nodal plate), as well as maturing/ mature tissue of the basal part of the next younger internode located above it. Hence, Swamy and Krishnamurthy (1979) advocated the involvement of only two types of meristematic activity in internodal ontogeny: (i) The subapical meristematic activity in which division activity is largely restricted to the rib meristem. This activity is seen in all vascular plants (barring acaulescent plants). The wave of cessation of division activity of the subapical meristem is usually from the base upwards in any internode that has only subapical growth. As a result, the wave of differentiation of internodal tissues is invariably acropetal. Even if this division activity does not stop immediately in the distal region of the internode during its acropetal cessation (the so–called UM), the derivation of internodal tissues from such UM is very meager. (ii) The intercalary meristematic activity in which the IM is derived from the subapical meristem when the later gradually becomes concentrated to a strip in any region of the internode for a fairly longer time, or the IM is derived by the dedifferentiation of parenchyma cells. IM is reported at the extreme base of the internode just above the nodal plate, as for example in many grasses and *Equisetum*. In *Avena sativa* (Kaufman 1965) and in *Triticum* (Ordina 1962) it is not located at the extreme base of the internode, but at a slightly 'elevated' location. It is distally located in the stalk of the scapes of many angiosperms. This strip should not be confused with the UM described above, as the former has a prolonged meristematic activity while the UM is a fading subapical activity.

5.4. DEVELOPMENT OF THE PRIMARY PLANT BODY

The primary body of the plant is produced essentially by the apical meristems through longitudinal growth (or apical growth). The primary body of the shoot is produced by the shoot apical meristem and that of the root by the root apical meristem. In some taxa, the primary body is produced also by IM and to a very small extent by UM.

5.4.1. Metamers and Modules

In multicellular plants, there are three levels or units of morphological organization: cell, **metamer** and **module** (Fig. 110) (Barlow 1994). In unicellular plants, however, only the first level of morphological organization is present i.e., the cell. The first level or the cell is the ultimate source of all other cells and of other structural units of multicellular plants. The autoreproduction of cells results in the replenishment of the meristems, which constructs the second structural unit called the **metamer**. Metamers together compose the third and more complicated structural unit called **module**. All these three structural units are related to one another in a hierarchical fashion i.e., the elements that comprise any one level together construct the next and more complex level. The module, which forms the highest level, is ultimately constructed from cells of common ancestry through the intermediate metamer level (Fig. 110). Modules, in combination, build an even higher level of morphological organization, the root and shoot systems of vascular plants. The spatial disposition of the modules decides the particular morphology of the shoot and root systems (Barlow 1994). For example, the various spatial and geometric dispositions of modules in the shoot systems result in the production of 23 basic architectural patterns of arborescent taxa, which has been mentioned already (Hallè et al 1978).

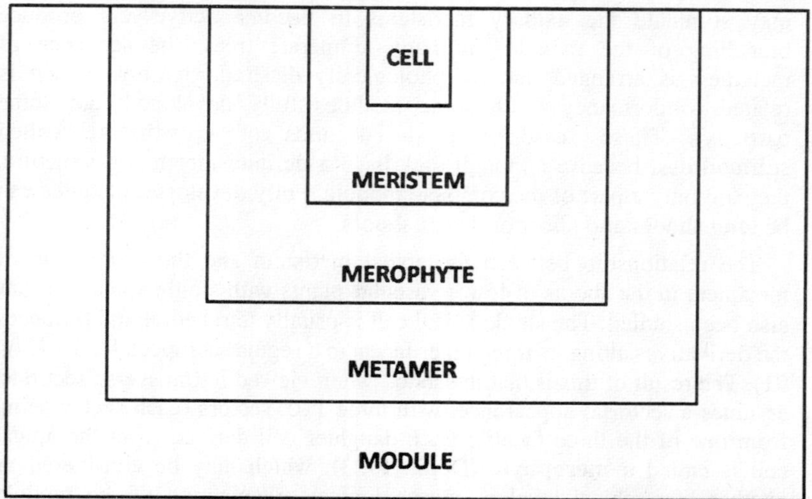

Fig. 110. The hierarchical composition of a module illustrating its interralations to the lower levels of cell, merophyte and metamer (Modified from Barlow 1994)

The metameric organization of the shoot is as follows: The shoot tip of vascular plants produces not only leaves, but also the nodes, the internodes and the leaf–associated buds. This quadripartite set of leaf, node, internode and bud can be called a metamer (shoot). This metamer represents the fundamental unit of shoot construction. Such a metamer of plants was earlier called **phytomer** (even today many botanists prefer this term over metamer). According to Barlow (1994), although phytomer is an attractive term in many ways and is still used in contemporary botanical literature, metamer has the additional features of being a term that implicates a broader biological perspective as it can also be applied to the repetitive structures that are seen during animal development. The anatomical constitution of the metamer varies depending on the phyllotactic pattern of the plant, since each of the different patterns involves different volume fractions of the stem being associated with the metamer. But, as Barlow (1994) had rightly emphasized, conceptually it is the repetitiveness of parts of metamers per se, rather than the details of their construction that is crucial for the definition of the metamer.

A module may be defined as an axis formed of metamers. The growth of axis in the shoot of flowering plants may be terminated by the production of an inflorescence, tendril or thorn or by parenchymatization of the apical meristem (Barlow 1994). The entire shoot system is an aggregation of metamerically constructed modules with terminal and axillary apical meristems of varying age and activity. The growth of the module is seen as long as the apical meristem is active, but once it becomes determinate, it may stimulate the axillary meristems to become active and produce branching of the module. In many temperate trees, the sequence of metamers is arranged into morphologically dissimilar groups, which is related to dormancy (with some modules fully developed and some partially). These seasonally–produced units of growth are called **submodules**, because although they have a definite metameric structure, they are only a part of the complete module. Fully developed modules can be **long shoots and short** or **dwarf shoots**.

The relationship between the apical meristem and the production of metamers in the shoots of lower vascular plants with single apical cell has also been studied. The single apical cell is usually tetrahedral and produces the derivatives along its three inner facets in a regular acropectal helix (Fig. 91). The result of this is that the shoot, when viewed in transverse sections, assumes a sectorial appearance, with three 120° sectors (each sector being from one of the three facets). Each daughter cell derived from the apical cell is called a **merophyte** (Douin 1923), which may be considered as another hierarchical level of morphological organization of the module (Fig. 91). Each merophyte is then subdivided by both longitudinal and transverse divisions. But, despite these divisions, the outline of its original boundary wall could still be recognized in sections. Each merophyte is, therefore, an assemblage of very ordered cells and as they are reiterated structures, they can be considered as equivalent to metamers.

The root system also has a modular character, as evident from its branching pattern. But, when compared to the shoot system, distinct modules are neither so common nor so well–known in roots. In root apices with single apical cells, merophytes are found as described above.

5.4.2. Origin and Development of Nodes and Internodes

5.4.2.1. *Differentiation of Nodes and Internodes*

Simultaneously with the initiation of a leaf primordium, the group of cells that ultimately become part of the node and internode are also specified (Fig. 111). Hence, it should be emphasized that the morphological definition of nodes and internodes is, at least partly, due to their topographical relationship to the adjacent leaf primordium. Only when the meristem region below the apical dome has produced a sufficient number of cells, node and internode can begin to be differentiated. Thus, the plastrochronic changes have a greater significance than merely defining the rate of leaf production (Swamy and Krishnamurthy 1974).

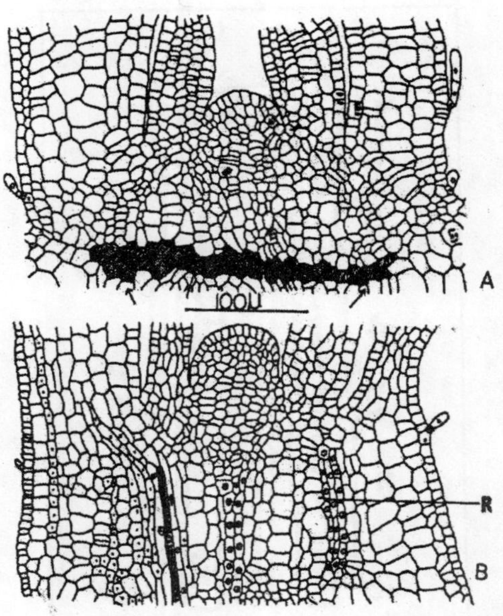

Fig. 111. L. S. of shoot apex of *Commelina benghalensis* showing the initiation of the nodal (hatched in figure A) and internodal regions of the stem through the activity of rib meristem (R in figure B)

The histological definition, in contrast to the positional definition, of nodes and internodes is unnoticed (Barlow 1994), even though the cells within the nodal region eventually become distinguishable from those in the internode on account of the different amounts of growth in the cells of both regions. A remarkable set of observations on young shoots of *Sambucus racemosa* (with opposite decussate phyllotaxy) by Zobel (1989 a, b) has demonstrated that cells in the node and internode differ not only biochemically, but also in their relative amounts of growth. In the shoot apex of this species, the outer pith contains tannin cells. The first visible tannin mother cells occur nine cell layers from the shoot surface. Once formed, a tannin mother cell ceases to divide and becomes polyploidal through the continued replication and division of its nucleus. The tannin cells run the entire length of each internode (Fig. 112). The nodal region is always free of tannin cells. The presence of tannin–containing mother cells and tannin–free cells close to the apex of *Sambucus* suggests that node and internode mother cells may be differentiated at the same time. If so, node and internode are not simply passively defined by their position relative to the accompanying pair of leaf primordial, but are defined before the primordia become visible i.e., the node and internode are prefigured at a very early stage of shoot development (Barlow 1994). Swamy and Krishnamurthy (1974) and Lyndon (1987) arrived at a similar conclusion.

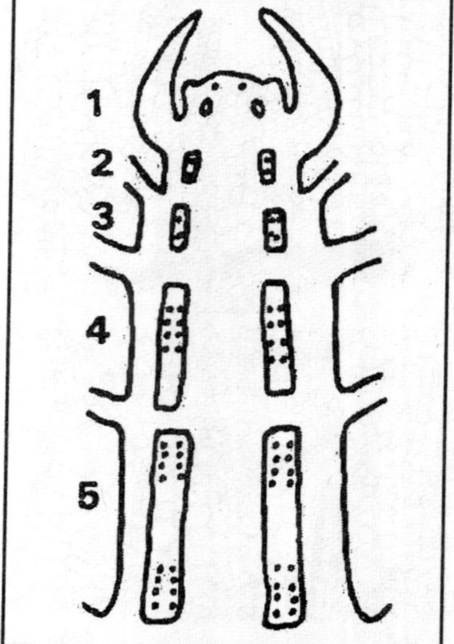

Fig. 112. Distribution of tannin coenocytes (stippled) as an exclusive feature of internodes in *Sabucus nigra*. Location of these helps in demarcating nodes and internodes from the earlier stages of development. [A drawing redrawn by Barlow (1994) from a photograph of Zobel (1985)]

5.4.2.2. *Development of Internode*

Two distinct types of meristematic activity are involved in the progressive development of the internodal primordium once it is delineated. The first is the **subapical meristematic activity** of the subapical zone (Fig. 88), which is one of the five morphogenetic zones recognized by Wardlaw (1957; 1968) in the shoot apices of vascular plants. The second is the intercalary meristematic activity. According to Wardlaw (1957), the subapical meristematic zone is very important because the growth in length of the internode is very conspicuous in this region; the final form and behaviour of the internode is decided in this region only. Subsequent to Wardlaw, several other authors have confirmed the importance of subapical meristem in the organization of the internode (Refer to Romberger 1963). The upper limit of this meristem is the organogenetic region of the shoot apex, which is well–defined (Fig. 88), while the lower limit is rather vague as this region gradually merges with the region of maturation. The extent of this meristematic activity is variable and depends on the plant in terms of the length from the apical meristem towards the base and in terms of the internode number (Refer to full literature in Swamy and Krishnamurthy 1979). In *Gloriosa, Karthalsia, Luzuriage* and *Dioscorea,* this activity extends from 1.9 cm up to 18.1 cm below the apical dome depending on the plant. It extends to at least the seventh internode and perhaps a little beyond it depending on the species, although at decreasing frequencies of meristematic activity. The characteristic features of this meristem are as follows: (i) The initials for this meristem are either fully derived from the apical meristem or partly from the apical meristem and partly from the primary thickening meristem as in many monocots; (ii) it is predominantly composed of the rib meristem. The plane of cell division is almost exclusively transverse. The packets of recently derived cells within a mother cell wall can be seen very clearly. Rib meristem activity is absent or extremely limited in acaulescent plants and short shoots. In some species, several internodes mature at the same time, while in others maturation is stepwise with one internode maturing after another. For example, in many palms, several leaf primordia are produced very rapidly even in a seedling shoot apex, and production of nodes and internodes are formed subsequently. Some authors have shown that the growth of the internode occurs throughout their length at first (in both monocots and dicots), but becomes progressively localized in the upper part (many dicots), in the lower part (many monocots) or rarely at any region in between the upper and lower parts of the internode. This pattern has close relationship with the location of IM and UM. The length of the final internode depends on the duration of division activity of the rib meristem.

Apart from the rib meristem, other aggregations of cells derived from the various zones of the shoot apical meristem and constituting the 'primary meristematic tissue' of the internode are involved in the formation of the internodes. These include the protoderm and procambium. The protoderm is the direct continuation of the outermost layer of the shoot apical meristem, the first tunica layer.

4.2.3. Long Shoots and Short Shoots

A few plants show a differentiation of the shoot system as **long shoots** and **short shoots** (or **dwarf shoots**) to produce a useful division of labour. The long shoots produce height growth and their proliferation adds to the overall framework of the tree. Short shoots usually have a specialized function such as serving commonly as sites for reproductive structures, for example cones in conifers and Ginkgoales and flower spurs in angiosperms (hence, also called **spur shoots**). In some plants, they serve as thorn–bearing structures. The short shoots represent an economizing in axis 'expenditure', which is most successful in the exposed situation or with the deciduous condition (Hallé et al 1978). Short shoots are of two types:

1. **Terminal short shoots,** as in *Terminalia*, in which each of these shoots, produced by progressive eviction of meristems, forms a part of a mosaic of long–lived leafy rosettes as an essential architectural feature of the tree (Fig. 113)
2. **Lateral short shoots** are closely circumscribed on the branch system. These occur characteristically in *Ginkgo* (Fig. 114) and *Pinus*.

Fig. 113. A portion of the aerial region of *Terminalia catappa* showing short terminal shoots (single arrow) and longer lateral shoots (double arrows) again ending in shorter shoots

Fig. 114. Long and short shoots of *Ginkgo biloba;* here, short shoots are lateral (Robert W. Ridge 1987) (Reproduced with permission from International Christian University)

There may also be transition shoots between long and short shoots. These occur in some species such as the species of *Acer*, *Fagus* and *Ulmus*. In these taxa, the long shoots develop from distal nodes, short shoots develop from basal nodes and intermediate shoots develop from intermediate nodes. Usually, short shoots have poorer rib meristem activity and they elongate less than 2 cm per year and do not bear lateral branches. Thus, not only shorter congested internodes are characteristic of short shoots, but also they fail to branch vegetatively. Also, short shoots are usually short–lived and determinate. Lateral short shoots are not found frequently in the tropical environment. Short shoots are extremely rare in monocots and have been reported in one or two species of *Pandanus*.

5.4.2.4. *Structure of Internode and Node*

Internodal structure

The internode of the aerial stem shows diverse variations in its internal organization. Basically, the internode consists of three tissue systems: **dermal, fundamental** and **vascular**. The diversity in the structure is seen mainly as differences in the relative distribution and quantum of the fundamental and vascular tissue systems. In most gymnosperms and dicotyledons, the vascular system appears as a hollow cylinder segregating an outer **cortical** and an inner **pith** or **medullary** zones of fundamental tissues. The vascular system is either in the form of a continuous cylinder or as discrete vascular bundles each separated from one another by a variable quantum of fundamental tissue known as **medullary ray** or **interfascicular region**. In the monocotyledons, a pith and cortex cannot be distinguished easily and the vascular bundles are embedded in the fundamental tissue without any obvious regularity or pattern.

The dermal system i.e., the epidermis is an uniseriate layer of cells. Except in most monocotyledons and in certain very slender herbaceous dicotyledonous taxa, the epidermis is replaced by a cork or similar tissue. The internodal epidermis resembles that of the leaf in most respects, if not all, in terms of the various tissue types; however, the stomatal index in the internodal epidermis is often very low.

The cortex is the zone that follows the epidermis towards the inside. It differs widely not only in its radial dimension, but also in the type, quantum and distribution of its cell types. It may be composed entirely of thin–walled parenchymatous cells, or, in addition, may have chlorenchyma, collenchyma and/or sclerenchyma fibers. Scattered or idioblastic sclereids, secretory structures, laticifers etc. may also be present. The cortex is usually delimited with great difficulty from the centrally–located vascular column (including the pith) called the **stele**. A typical **endodermis** is invariably absent in the internodes of angiosperms and gymnosperms, but is often present in pteridophytes with aerial stems. However, an endodermoidal layer or a **starch sheath** is present in a number of taxa.

The inner core is the stele (Refer to Box 5.4), which invariably consists of the vascular system and the innermost pith region. In dicots and gymnosperms, the vascular system invariably consists of discrete vascular bundles or it is in the form of a continuous cylinder with phloem tissue on the outside and xylem on the inside. In monocots, the vascular tissue is in the form of scattered vascular bundles. As already mentioned, the vascular bundles of internodes are collateral (with one group of xylem and phloem located along the same radius), bicollateral (with xylem between two phloem groups one on either radial side) or concentric (with xylem on the inside and phloem around it, **amphicribral**, or the reverse, **amphivasal**).

BOX 5.4. THE CONCEPT OF STELE AND THE STELAR THEORY

The concept of stele in plant axial organs was first proposed by Van Tieghem and Douliot in 1886 while discussing the organization of tissues in the shoots of *Primula* and *Gunnera*. The main aim of this theory was to provide a unified morphological interpretation of the plant axis i.e., the plant stem and root. In both the organs, the cortex encloses a central core or a column of tissue and hence, both are fundamentally similar in gross anatomy. As stated earlier, this central core of tissue is called the stele, which in many plants is a single unit (**monostelic**), while in some plants it is divided into more than one unit (**polystelic**). Van Tieghem in 1891 defined the external limit of the stele as pericycle and the internal boundary of the cortex as endodermis (it is ontogenetically derived from cortex). However, some including Eames and McDaniels (1947) consider the endodermis as the outer boundary of the stele. The central core often encloses a ground tissue region i.e., the pith.In all roots, both the pericycle and the endodermis are histologically well–defined, but this is not seen in stems. However, the tissue that may go into the constitution of the pericycle is either parenchyma or sclerenchyma. Wherever an endodermis cannot be distinguished by the presence of **casparian thickenings** (special subserized thickenings) on the walls of its cells, its position, according to the stelar theory is said to be indicated by starch sheath.

After the stelar theory was formulated, it gained more favour on the whole as a series of comparative studies was made in different groups of plants. However, with its increasing acceptance, new terms and ideas were introduced to include tissue organizations, which varied somewhat from the forms originally assigned by Van Tieghem. One of the ideas introduced into the concept was that 'the tissue continuity is not a basis of morphological identity and histological structure is not necessarily a key to morphological regions in an organ'.

The original definition of a stele rests upon the supposition that a distinct limit i.e., the endodermis exists between the stele and the cortex. Here, it should be noted that the distribution of the endodermis (or a layer considered equivalent to it) may vary in different regions of the same plant. Thus, there may be various combinations of (i) an endodermis, which surrounds the whole cylinder of bundles (cylinder endodermis); (ii) an internal endodermis, in addition to a cylinder endodermis; (iii) an endodermis surrounding each bundle (bundle endodermis, which in fact surrounds each stele in a polystelic axis); or, (iv) altogether lacking an endodermis. According to Van Fleet (1961), in spite of the absence a morphologically–definable endodermis, a layer of physiologically–distinct character occurs in many plants between the stele and cortex. The implication that the occurrence of an endodermis is related with certain physiological phenomena, casts a new light on its significance as a morphological entity.

With reference to the external boundary of the stele, the pericycle is recognized often as essentially a residual tissue between vascular tissue and endodermis, with its primary function as a generative layer for the phellogen, new vascular tissues, lateral root primordia, vascular cambium etc. However, the pericycle is not always distinguishable in many plants and most 'pericyclic' tissues have been shown to be derived from the primary phloem, particularly in the stem. The pericycle sclerenchyma fibers are demonstrated to be really primary phloic fibers. Therefore, the concept of pericycle, which is an essential component of the stelar theory, needs a thorough re–examination, and this speaks of the weakness of the stelar theory. Hence, many modern investigators have indicated the urgent need for the complete re–examination of the concept of the stele, especially in seed plants.

However, the stelar theory, in spite of its weaknesses, has proved extremely helpful in the description and interpretation of the structure of the axial organs, particularly in the lower vascular plants.

The pith is of variable size. It may be fully parenchymatous, collenchymatous or sclerenchymatous. In some plants, it may be partly parenchymatous and partly collenchymatous or sclerenchymatous. Its cells may contain inclusions such as starch grains, phenolics or other materials.

A number of stelar types are recognized in vascular plants (Fig. 115). They are as follows:

1. **Protostele:** It consists of a core of xylem ensheathed by phloem, either in discrete strands or in a continuous cylinder. The protoxylem is always **exarch** (facing the exterior). Pith may or may not be present, and which may respectively be known as **medullated** or **solid** protostele. This type of stele occurs in the roots of all plants and in the aerial axes of Lycopsida and Psilopsida of Pteridophyta. A number of subtypes are also recognized such as **actinostele**, **plectostele** and **haplostele**. A characteristic feature of the protostele is that no **leaf gaps** are formed at the place of insertion of leaves. Such leaves are often called **microphylls** and the condition is known as **microphyllous**.
2. **Siphonostele:** This type is characterized by the presence of a pith and a non–exarch protoxylem. As seen in transactional view, the vascular tissues of this stelar type appear as a continuous ring of xylem covered by phloem, or as discrete vascular strands. The stranded type has a tubular network of interconnected strands separated by vertical strips of parenchyma as seen in longitudinal sections. If these strips of parenchyma are very closely overlapping one another to form a dissected vascular cylinder, the term **dictyostele** is employed. This type of stele is a characteristic feature of many ferns. It is not advisable to apply this term to the stem steles of dicots and gymnosperms; instead the term **eustele** is applied. In the case of monocots, for the condition of scattered vascular bundles in the ground tissue, the term **atactostele** is applied.

Structure of the node

The anatomy of the node presents a great deal of contrast from that of the internode. The continuity of the vascular cylinder is interrupted at each node by the divergence of the vascular strands to the leaf, branch (or its bud), stipules etc. The departing vascular trace to the leaf is called the **leaf trace** and this departure is always associated with a parenchymatous area

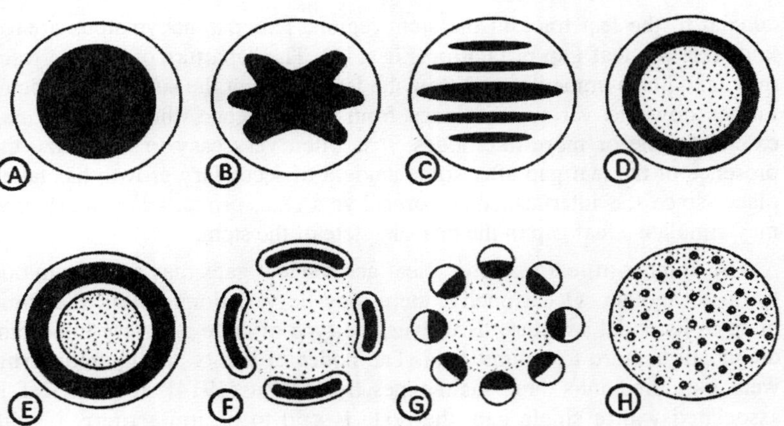

Fig. 115. Diagrammatic transections of types of stele; A. Haplostele; B. Actinostele; C. Plectostele (all protostele types); D. Ectophloic siphonostele; E. Amphiphloic siphonostele or solenostele; F. Dictyostele; G. Eustele; H. Atactostele; xylem is in black and phloem is in white; pith and ground tissues stippled.

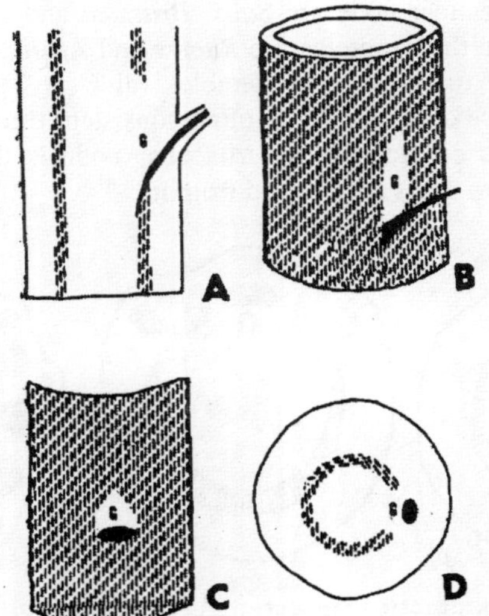

Fig. 116. Diagrams to illustrate leaf trace and leaf gap. Striped area represents the stele and solid black lines in A and B and black areas in C and D indicate leaf traces; G - Leaf gaps; A - Longitudinal sectional view; B - Entire trace; C - Face view of leaf trace and gap D - Transverse sectional view of the same.

adaxial to the leaf trace in the stem region. This parenchymatous area is known as the **leaf gap** or **lacuna** (Fig. 116). The departure of the leaf trace may take place immediately below the leaf to which the supply is made or farther down to various distances from it and hence, the leaf gap may extend to one or more internodes. It is often very easy to recognize the presence of the leaf gap after some amount of secondary growth has taken place, since the interfascicular parenchyma (i.e., primary medullary ray) may simulate a leaf gap in the primary state of the stem.

The number of leaf traces to a leaf and the leaf gaps may vary from one taxon to another. Occasionally, there may be variations within the same plant in its nodal structure. However, the type of node is a very important diagnostic feature to delimit taxa. The nodes of dicots and gymnosperms were classified into three main types by Sinnott (1914). If each leaf is associated with a single gap, the node is said to be **unilacunar**, if with three gaps, **tricacunar** and if with many gaps, **multilacunar** (Fig. 117). In most plants, only one vascular trace is associated with each gap. Hence, generally, unilacular nodes have one trace for each leaf, trilacunar have three traces and in multilacunar, the number of traces is equal to that of gaps. However, unilacunar with two or more traces are also present. Examples for unilacunar node are *Clerodendrum* and *Calophylum*, the examples for trilacunar node are *Salix*, *Brassica* and *Polyalthea,* and for multilacunar node the examples are *Rumex* and *Aralia*. The nodal type is also considered to have considerable value in the phylogeny of dicotyledons, since a given type is often consistent throughout a genus or even family. It is considered that trilacunar node is the most primitive, while the other two types are derived from it.

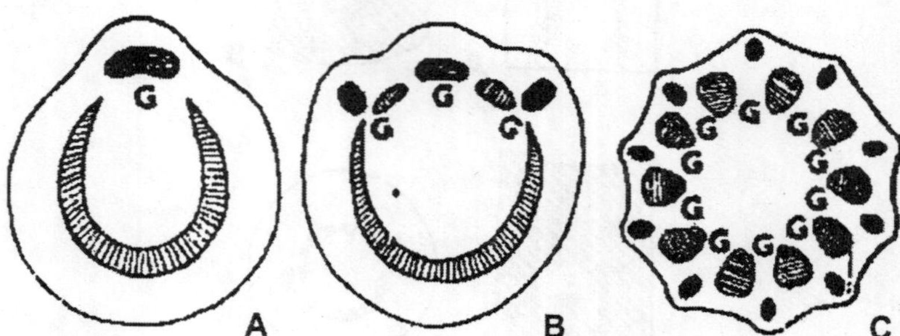

Fig. 117. Diagrammatic transectional representations of unilacunar (A) trilacunar (B) and multilacunar (C) nodes; solid black indicates leaf traces; G - Leaf gaps; Striped - Stele of stem

The nodal region in most monocots is characterized by a **nodal plexus** (Fig. 118) or **plate** where there is a complicated anastomosis of vascular tissue. The vascular bundles of the internodes below and above the node fuse in the nodal plate. Invariably the vascular elements of the nodal plate are formed from ordinary parenchyma cells without change of shape and size transdifferentiation.

Fig. 118. T. S. of mature node of *Commelina benghalensis* showing nodal plexus formed by the anastomosis of intermodal vascular bundles

5.4.3. Procambialization and Primary Vascularization in Stem

Cells derived from the shoot apical meristem begin to differentiate in the nodal and internodal regions of the stem. The most prominent feature of this differentiation is the blocking out of the procambium from which the primary vascular tissue is derived. The height at which this happens varies, but in many plants it occurs at the level of the youngest primordium. The actual sequence of events leading up to the appearance of undisputed procambium is the subject of much controversy. This is partly due to real differences between plants, but much of it must be ascribed to the problems of interpretation of the situation and to the preconceived ideas by some authors. The subject has also been hampered by a confusion of terms used to describe these two processes. This confusion prevailed to a large extent until 1943 when Esau published her review. Cells which give rise to pith and cortex tend to vascuolate rapidly and do not stain to the same intensity that the procambial precursor cells, which are blocked out by the vacuolated cells, stain at the shoot apex.

Procambium is usually identified by the small cross–sectional area of its cells and more elongated nature than the neighbouring cells. However, there is no single dependable way of identifying the procambium and it is obvious that different authors use the term in rather different ways. It is disappointing and frustrating that so little is there to say about the structure of procambial cells. Hence, it is no wonder that the following statement of Shininger made as early as 1979 is valid even now: "We do not have a meaningful concept of procambium at a biochemical or ultrastructural level, and the development of concepts at these levels is important".

The procambium can form a complete cylinder in the stem of some taxa, but more commonly it exists as discrete strands. The ring is called by some **meristem ring, prodesmogen** or **residual meristem**. The last term is more appropriate and satisfactory. The residual meristem may be considered as a continuum of the apical meristem. Some confusion also arises over the distinction between residual meristem and procambium because some plants have procambial cells very high up in the apex leaving no space to have an intermediate kind of cells between it and the apical meristem. Some authors have also given the impression that the residual meristem is a precursor of both procambium and interfascicular parenchyma. The residual meristem and procambium differentiate in relation to the leaf primoridia. The procambium becomes distinguishable from other parts of the residual meristem and its position is clearly related to the leaves. The first procambium to differentiate at any one level is the procambium that is claimed to be the nearest to the leaf above. Later on, further procambium develops and this also is claimed to belong to younger leaves arising higher up on the apex. Thus, it was considered by many that the entire primary vascular system of angiosperms is made up of leaf traces.

Two controversial points arise from the above description of the development of procambium. These are the **foliar nature** of the vascular tissues of the stem and the **acropetal wave of differentiation**. At one time it was said that some parts of the vascular system of angiosperms are **cauline** and other parts **foliar**. Some people, however, considered that it was wholly cauline or wholly foliar. The arguments between exponents of these views are largely arguments over the meaning of words. If 'foliar' means developed in relation to leaves, then the vascular system is foliar. But it has been shown that a vascular system is still produced even when the leaves are systematically removed by surgery; plants without leaves also possess vascular systems. Thus, a vascular system can exist even in the absence of leaves and if this is the meaning of 'cauline', then angiosperms have cauline vasculature. Esau (1965b) has objected that experimentalists have not shown that a procambium does not exist before they puncture leaf primordia or incise the apex. This is a vital criticism, but it may be an unimportant one. Another point which the supporters of cauline vasculature put forward is the existence of procambium above the level of the youngest leaf primordium. It has been shown by indirect means that in *Alstromeria* the median trace to a leaf arises eight plastochrones before the leaf itself arises. The trace enters the leaf, but exists before there is any visible sign of it. Clowes (1961) has stated that, "the interests of anatomy are perhaps best served now by dropping the word (cauline) out of our vocabulary" in the light of the above discussion. It is not correct because the procambial trace to the leaf is not continuous from the stem procambium as it looks apparently, but in reality the trace contains two parts, a cauline part and a foliar part separated by a short region of parenchymatous cells.

The other controversy relates to the directional wave of differentiation of procambium. Many earlier workers assumed a basipetal procambialization from the leaf down the stem. There is no real proof for it, at least for the main trace of the dicots leaves (basipetal wave is observed in some monocots for lateral traces). Evidence for acropetal differentiation includes the fact that the state of differentiation suggests that the youngest tissue is at the top of the strand. As mentioned earlier, the apparently–looking continuity of the procambial strand from the stem into the leaf is not really so. Serial transections of about 5–10 μm thickness reveal a discontinuity at the junction where the leaf joins the stem. From this region of discontinuity, the procambium extends acropetally into the leaf and basipetally into the stem. This region is also marked by the presence of morphologically–distinct type of tracheary elements from those occurring either in the stem or in the leaf, these elements apparently get differentiated from parenchyma cells (without change of size or shape) and not from procambial cells.

When the procambial strands are initiated in the residual meristem, an increasing number of cells become involved in longitudinal divisions to increase the transverse area of the procambial strands (Fig. 119). Thus, the early growth in transactional size of the procambium occurs at the expense of the cells surrounding those that start dividing first. At the end of this stage of growth, one daughter cell of the two resulting from a division may become incorporated in the procambium, and the other may mature as a ground tissue cell. Thus, vascular and non–vascular cells are ontogenetically closely related. The latitudinal growth of the procambium takes place in the above manner resulting in its expansion in the tangential extent. However, latitudinal growth through addition of cells on the periphery overlaps with the increase in the number of cells within the procambium and with the beginning of differentiation of vascular elements. The planes of such divisions vary with plants or in the same plants. In a number of cases the xylic procambial cells may divide in a more orderly manner (often in radial seriation) than the phloic procambial cells.

The procambium forms the base from which primary vascular tissue i.e., primary xylem and primary phloem, differentiate. The first differentiated xylem elements, called **protoxylem** elements, are characterized by annular and helical elements, while the later–formed xylem elements, called **metaxylem** elements, are characterised by helical–reticulate, reticulate and pitted elements. These elements have decreasing length from the first-formed protoxylem to the last-formed metaxylem. Most commonly, the wave of transverse differentiation is along the radial direction with protoxylem facing the centre of the stem axis and metaxylem towards the periphery (Fig. 119). The protophloem differentiates first towards periphery of the procambial strands, while metaphloem differentiates towards the centre. In other words, the wave of differentiation of primary

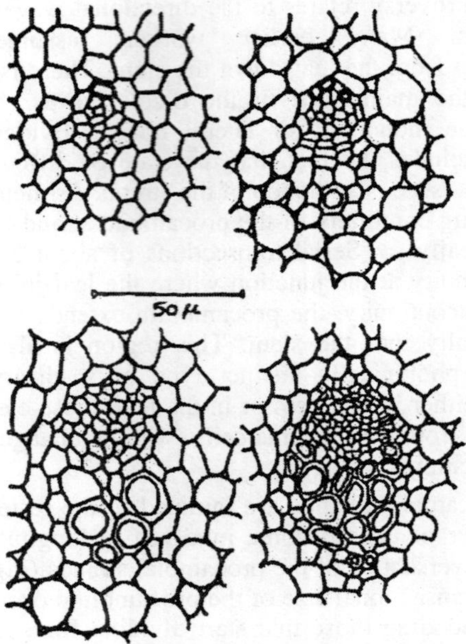

Fig. 119. Progressive stages in the enlargement (due to addition of more procambial cells) and development of a vascular bundle in *Commelina benghalensis.* Phloem development proceeds from periphery to centre while xylem development proceeds from centre to periphery until both meet each other.

xylem is centripetal, while that of the primary phloem is centrifugal. Primary phloem differentiation starts earlier than primary xylem differentiation and the earlier phloem elements are characterized by nacreous walls (shining walls). The metaphloem elements are shorter than those of protophloem cells. The last–formed metaxylem and metaphloem elements are close to each other when the differentiation of primary vascular tissues is completed.

The concepts of protoxylem and metaxylem have always been subjects of dispute to plant morphologists. The originally proposed distinction between the two was the time of their appearance or their differentiation. According to this view, the first–formed xylem elements from procambium are the protoxylem and the later–differentiated ones are metaxylem. But where and when should one draw the line between the two in terms of the time concept? The second view tries to draw morphological boundary between these two based on the types of secondary wall sculpturing. Accordingly, protoxylem elements are characterized by annular and helical thickenings, while the metaxylem elements by reticulate and pitted elements. Also, this definition is not accurate as there are some cases where protoxylem are with pitted elements and metaxylem with annular / helical elements. Esau (1965b), after reviewing these concepts, proposed that protoxylem constituted that portion of the primary xylem that is formed during the elongation of the organ containing the primary xylem, while metaxylem represents those elements formed after the elongation of the organ is completed. The same holds good for protophloem and metaphloem.

The longitudinal course of vascular differentiation in the procambium more or less follows the same wave as that of procambialization. The general features illustrated and observed in many plants are (i) appearance of phloem before the xylem, and (ii) acropetal differentiation of xylem and phloem in the leaf and basipetal in the stem i.e., differentiation is bidirectional.

5.4.4. Development of Root Tissues

Whether one accepts the QC as the promeristem or support the presence of initials for different tissue zones of the root around the periphery of QC (as most people do) or considers both the QC and the dividing cells around it together as promeristem, there is no controversy about the fact the tissue files of developing root body generally emanate from this area (Fig. 120). Root ablation experiments, about which a brief mention was made earlier, carried out in *Arabiopsis,* have clearly demonstrated that positional information, rather than cell lineage relationships, plays the most important role in the determination of cell fate (Scheres and Wolkenfelt 1998). In these experiments, specific cells were ablated with a laser and its effect on neighbouring cells was observed. For instance, when all four QC cells were ablated, they were replaced by the initials. Ablation of a single daughter cell of a cortical initial had no effect on subsequent divisions occurring in that initial, which was in contact with other cortical initials. When all cortical daughter cells bordering a cortical initial were ablated, however, that initial was unable to generate files of parenchymatous and endodermal cortical cells. Apparently all initials depend on positional information from more mature daughter cells within the same cell layer. In other words, the initials of the root apical meristem apparently lack intrinsic tissue pattern–generating information (Van den Berg et al 1997). This is contrary to the traditional view of meristems as autonomous pattern–generating machines.

According to Rost (1994), there are at least five levels of structural organization in the root: individual cells, packets, cell files, tissue cylinders and tissue sector (Fig. 121). Each unit is an independent functional unit and yet each is dependent on the previous lower level and on the next higher level. Hence, any model explaining root development must account for organization at each level. However, the classical view of the root tip is that it consists of three defined regions (Ivanov 1973); the **meristematic zone**, the **elongation zone** and the **maturation zone** (Fig. 122). The first

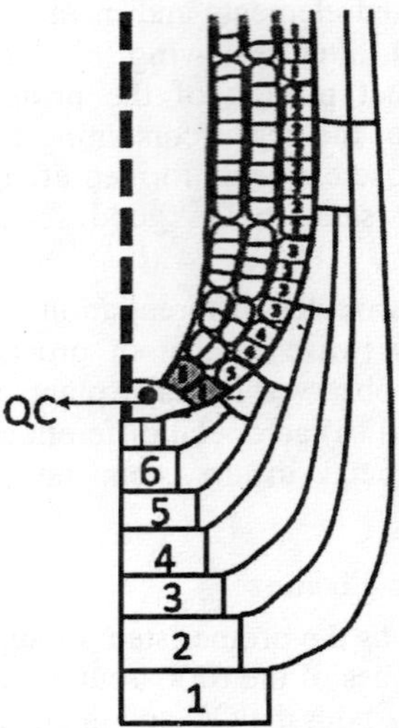

Fig. 120. Diagrammatic representation of the pattern of growth in the root apex of *Petunnia hybrida*; The numbers indicates the sequence of formation of cells via proliferative divisions in both columella and root cortex where cells of common ancestry occur in packets. QC - Quiescent centre; i - initial (Based on Vallade et al 1983)

includes the promeristem and the region of actively diving cells (involved in both formative and proliferative divisions). The elongation zone is characterized by the elongation of the derivative cells from the meristematic zone. This is followed by the maturation zone, where cells generally do not elongate any further. At a level of organization, both the root cap and the root proper may be envisioned as consisting of files of cells emanating from the promeristem. However, the limits of these three zones are not clear–cut in that all the tissue files of a zone do not end at the same level in the body of the root and start the next zone. At the same level of the root, the processes of cell division, cell elongation and maturation overlap not only in different tissue regions, but also in the different cell files of the same tissue region and even in individual cells. The considerable overlap between different files of the root is indicated by Ivanov (1973), who analyzed root development in terms of transition points. According to him, the important factor that differentiates between root–cell files is the position of the transition points between cell division and elongation and between the duration of cell growth at the end of elongation zone and at the start of final maturation (Rost 1994). Hence, analysis of root tip organization and development of different root tissues from it requires visualizing some kind of multilevel mechanism, whereby differentiation events can be regulated within cell files and between different zones. What regulates the differences between adjacent files and what mechanism regulates the spatial modulation of the transition points need to be clearly answered (Rost 1994). For example, transition points can move within a cell–file relative to the growth rate of the root. In fast–growing roots, differentiation events occur farther from the tip than in slow–growing roots.

As indicated earlier, cell division activity is tissue–specific and the boundaries of the meristem are not the same for all tissues. Cell division activity in different tissue layers stops at various distances from the root apical meristems and invariably it extends for greater distance in the dermal layer than in the remaining tissue regions. There are two types of cell divisions within cell files:

1. **Formative division** (T division), which occurs on the periphery of the QC or the initial longitudinal division (radial or periclinal) that starts any new cell file. This division is the first in a file and its derivative cells usually retain the characteristics of the existing cells within the same file. Formative divisions are important in determining pattern formation. The radial formative divisions increase the number of cells in an individual cell layer and thus the diameter of the root.
2. **Proliferative division** or **transverse division,** which occurs within a file. Each file apparently maintains a more or less set number of proliferative divisions (Rost 1994). For instance, the xylem tracheary element precursor in *Vicia faba* divides five

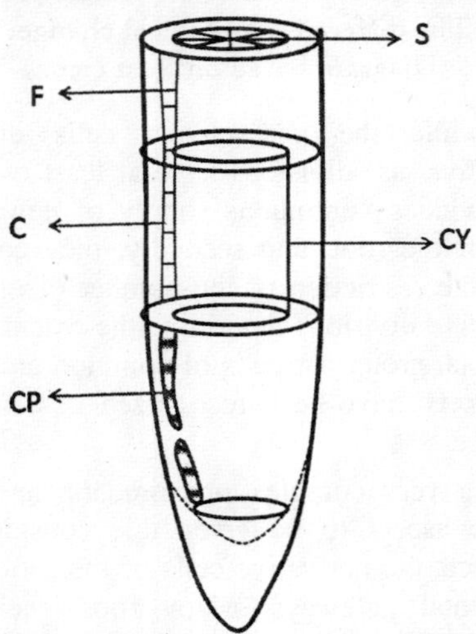

Fig. 121. Diagrammatic representation of the different independent developmental and functional units in root apex. (C -cell, CP – cell packet, F - cell file, S - tissue sector and CY - tissue cylinder) present in a root tip as per the concept of Rost (1994)

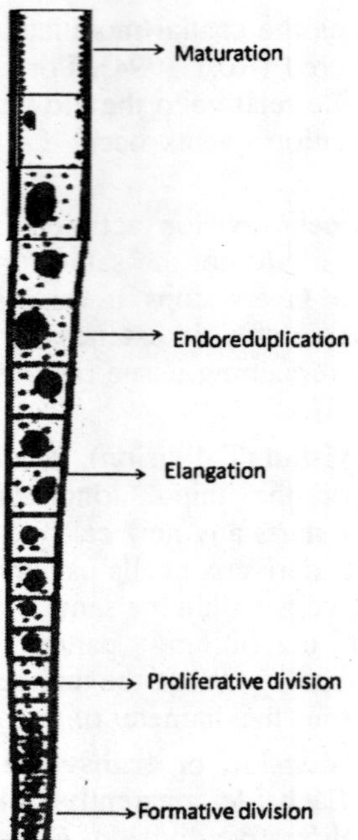

Fig. 122. Diagrammatic representation of a central metaxylem file in a root tip. The different cytological changes are indicated. Diagram based on Rost (1994)

times, while the parenchyma cells divide seven times. Proliferative cell divisions have at least two functions: Firstly, they provide a continuous supply of new cells to ensure the growth of the root, and secondly, they coordinate cell length through the restriction of the number of divisions in each file. Proliferative divisions determine the extent of the meristem. In some roots, groups of cells of common ancestry, referred to as **cell packets**, have been recognized in the files (Fig. 120 and 121).

Root growth is a very complex phenomenon and it is regulated by several genes. One aspect to be given due consideration is the gene expression that directs the derivative cells of the root meristem to follow specific developmental pathways. Many root–specific genes that are transcribed in the cortical cells have been known and identified. One such gene, which has been studied in detail, has been isolated from a cDNA library made to poly (A)–RNA of water–imbibed seeds of *Brassica napus* and this gene is called *Axis–abundant* (*Ax–92*). The transcripts of this gene are detected in the ground meristem and cortex of the seedling root, but are not present in significant amount either in the root cap or in vascular cylinder. *Ax–92* is believed to define the initial cortical cell differentiation events, as its expression is seen in the root cortex of the torpedo embryo of transgenic plants (Deitrich et al 1992). On the contrary, transcripts of one of cDNA clones of maize roots, *ZRP 3* (for *Zea root preferential*) are found in many layers of the inner cortex, while the endodermis and three or four layers of the suberized cells of outer cortex (exodermis) are found to express transcripts of another clone *ZRP4*. The maximum expression of *ZRP3* gene is confined to 2 cm while that of *ZRP4* gene to about 8 cm length of the root tip. A gene homologous to maize *ZRP3* gene is known in bean root and is expressed in the cortical meristem in a decreasing gradient from the tip towards the base. This kind of expression is likely to be related to the gradual differentiation and change in function of cortical cells eliminating the need for a continuous supply of the gene product (Choi et al 1996).

Typically, the cortex matures first and its cells get vacuolated and develop intercellular spaces close to the apex. The prospective metaxylem can also be distinguished close to the apex (especially in taxa without a pith and with a centrally–located metaxylem). The loci of differentiation of primary xylem and primary phloem are along alternate radii in the so–called central core of procambium. Protoxylem differentiates from periphery towards the centre of this core and hence, it is **exarch**. Similar is the wave of differentiation of primary phloem. Hence, metaxylem and metaphloem are more centrally located. If there is pith, it differentiates from the centrally located cells of the core of procambium. Since proliferative divisions in the various tissues or cell files do not stop at exactly the same distance from the root apex, the basal boundary of the meristem or region of cell division is not clearly defined. Hence, the term 'relative meristem height' is preferentially used.

By employing kinematic method by which local rates of cell division and cell expansion can be measured simultaneously, Baskin (2000) has clearly established that different tissues or cell files stop dividing at different distances from the apex, but while dividing, the rate of division of cells in all tissues is the same. In contrast to the constancy of cell division rate, the number of dividing cells in the meristematic region varies widely, indicating that the root must control exit from the cell cycle at the base of the meristem. It is now clearly established that cell division continues well up to the region where cell length increases rapidly. As indicated in Chapter 3, cell division and elongation are often coupled and the zones having these two activities in the developing root may actually constitute one developmental zone (Scheres and Heidstra 2000).

5.5. ORIGIN OF AXILLARY BUDS AND BRANCHES

5.5.1. Structure and Development

Shoots branch at their apices while roots are not (Clowes 1961). In a few lower vascular plants such as *Psilotum, Lycopodium* and *Selaginella,* branching occurs at the shoot apical meristem, without any reference to the leaves. The original apical meristem in these cases undergoes a median division into two equal parts, each of which forms a shoot. This type of branching is described as **dichotomous.** When a branch arises laterally at or near the shoot apex, the branching is called **monopodial.** This is the most common type of branching in seed plants. These branches commonly arise as buds in the axils of leaves, and these buds in their 'nascent' state are called **axillary buds** (Fig. 123). Judged from most studies, the term 'axillary' is somewhat inaccurate, since the buds generally arise on the cauline region or stem, but soon become displaced closer to the leaf base, or even into the leaf itself by subsequent growth adjustments (Garrison 1955). But, because of the long usage, the term 'axillary bud' has been conserved in descriptive botanical literature.

In most seed plants, axillary meristems originate as **detached meristems.** Axillary meristems arise at variable distances from the apical meristem, most commonly in the 'axils' of second or third leaf from the apex. Thus, they are initiated somewhat later than the leaves subtending them and in the meristematic tissue derived from the shoot apical meristem, but separated from it by vacuolated and potentially–differentiated cells. Hence, the pockets of meristematic cells are called detached meristems. In a few species, the axillary meristems appear to be originated from the adaxial (upper) surface of the leaf primordia and hence, appear to be foliar in the origin.

Experimental evidence indicates that a leaf determines its axillary bud (Snow and Snow 1942). For instance, if a leaf primordium is surgically removed before its axillary bud is initiated, the bud will fail to develop. On the contrary, even if a very small portion of the base of the leaf primordium remains, it is enough to trigger the development of the axillary meristem (Snow and Snow 1942). The need of a leaf primordium in the formation of its axillary bud is supported by the *Arabidopsis phabulosa–Id* (*phb–Id*) mutation. In wild *Arabidopsis,* axillary buds normally develop in close association with the adaxial surface of the leaf base, while in the *phb–Id* mutant, abaxial (lower) leaf fate is transformed into adaxial fate, resulting in the development of axillary buds on the underside of the leaf (McConnell and Barton 1998). More likely, the adaxial basal leaf fate plays an important role in promoting axillary meristem development. Transcripts of members of a family of maize genes called *KNOX* and *ROUGH SHEATH (RS)* are expressed in the developing axillary bud (and internode) near the base of the leaves. Thus, *KN1, KNOX* and *RS* genes together predict the sites of axillary bud (and its associated metamer).

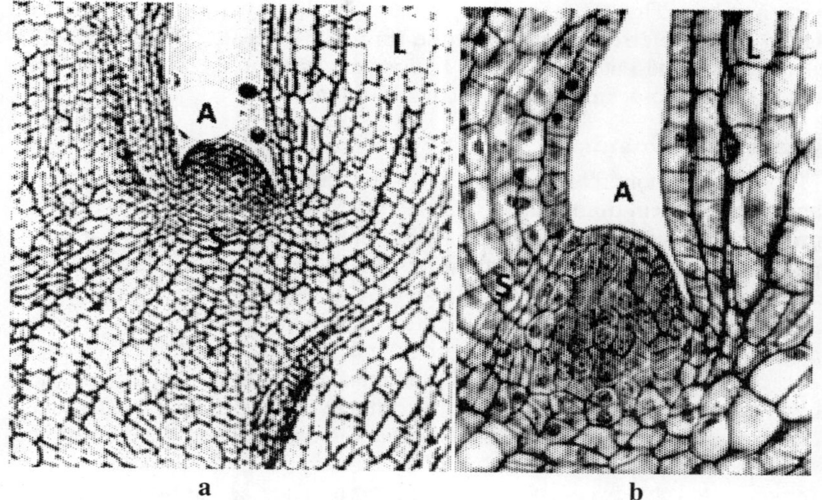

Fig. 123. L. S. of axillary buds (A) of two different plants (a and b) to show their position relative to the stem and leaf (L). The shell zone (S) is also shown (b - Photograph courtesy of Marshall Sundberg, Emporia State University)

Buds do not always develop in the axil of every leaf. In some taxa, buds are absent altogether. This is true of many unbranched taxa. The application of benzyladenine to the shoot apex of five–to–seven day old *Stellaria* seedlings promoted axillary meristem development in the otherwise empty leaf axis, thus stressing the importance of cytokinins in the initiation of axillary buds. Rarely more than one bud may occur in the axil of a leaf. Such buds are called **accessory buds** and these occur in addition to the axillary bud. Invariably both the axillary and accessory buds arise from the same group of detached meristems (Garrison 1955).

Adventitious buds may arise in various parts of the plants, without any relation to the main apical meristem. They may develop naturally or may be induced to develop in wounded areas, callus or chemically treated–regions. Mostly, adventitious buds are exogenous in origin. Generally, prior to adventitious bud formation, there is dedifferentiation of mature living tissues. Axillary buds (as well as adventitious buds) may remain temporarily or permanently dormant or may grow into new shoots soon after their formation.

When a bud is formed, periclinal and anticlinal divisions occur in a variable number of peripheral layers in the cauline region just near the leaf axil. Consequently, the bud meristem becomes elevated above the surface. Sooner or later, the apical meristem of the bud is organized. In a number of plants, ordered cell divisions occur along the basal and lateral limits of the incipient bud. This results in the formation of the zone of parallel curving layers, often referred to as **shell zone** (Fig. 123), because of its shell–like shape (Shah and Patel 1972). The bud is connected to the vascular system of the stem by procambial traces called **bud traces.**

5.5.2. Apical Dominance

A phenomenon related to axillary buds and their development into branches is **apical dominance**. It is a common observation that the actively growing shoot apical bud of herbaceous plants often slows down or even inhibits the development of lateral or axillary buds. This phenomenon is called apical dominance. It is intimately related to branching habit of the plants. A co–ordinated control of apical dominance can be considered as largely responsible for the 23 architectural patterns so far recognized in tropical arborescent taxa. Apical dominance is very strict in tall and unbranched plants and is highly flexible in branched and bushy taxa. Auxin, cytokinin and less characterized factors mediate or control the activity of axillary buds. Soon after auxins were discovered, it was demonstrated that excision of apical bud eliminated apical dominance and triggers the growth of lateral buds, and application of auxins at the excised apical bud region brings back apical dominance. Subsequently, it was shown that application of cytokinins on axillary buds of an apically dominant plant released their inhibition to grow, thus demonstrating the involvement of an interaction between auxins and cytokinins in apical dominance (Cline 1991). Transgenic tobacco plants, incorporated with a chimeric gene that encodes the enzyme isopentenyl monophosphatase, show a several–fold increase in the cytokinin that promoted extensive growth of lateral buds (Li et al 1992). Similarly, transgenic *Petunia* and tobacco plants incorporated with an IAA biosynthetic gene showed absence of branching (Sitbon et al 1992).

Subsequently, new information has come to implicate genetic control in apical dominance. In maize, compared with its closest wild relative teosinte, and the *teosinte branched 1 (tb 1)* mutant that are branched on most nodes, branching is limited to occasional nodes and hence, apical dominance is present. In fact, decreased branching and increased apical dominance has been one of the major traits to be selected during the domestication of maize from its ancestor teosinte (Reinhardt and Kuhlemeier 2002). Some parallels are there between the *TB1* gene that suppresses tiller formation and the *CYCLOIDEA (CYC)* gene of *Antirrhinum majus* that suppresses the growth of floral organs. It is suggested that *TB1* gene acts as a repressor of axillar bud growth in maize (Doebley et al 1997). Conversely, in the tomato mutant *lateral suppressor* (*ls*), the vegetative axillary meristems are suppressed leading to non–formation of lateral branches.

There are two types of branching behaviour in plants; **sylleptic branching** or **syllepsis** and **proleptic branching** or **prolepsis**. In syllepsis.

there is continuous development of a lateral from a terminal meristem to establish a branch, without an evident intervening period of rest of the lateral meristem. In prolepsis, there is a discontinuous development of a lateral from a terminal meristem to establish a branch, with some intervening period of rest of the lateral meristem. However, both prolepsis and syllepsis are alternate fundamental states of the same meristem. There is some information to show that syllepsis (at least in some species) is correlated with rapidity of shoot growth i.e., the greater the rate of extension of a shoot the greater is the likelihood of its lateral meristems developing by syllepsis. This correlation led Tomlinson and Gill (1973) to suggest that the switch from a 'lower' state that determined prolepsis of a lateral meristem to another 'higher' state that determined syllepsis, is conditioned by a 'threshold', which in turn is determined by growth 'vigour' of the parent shoot. It provides an alternative to explanations of the branching solely in terms of apical dominance. Brown et al (1967) suggested that apical dominance should really be thought of as 'apical control'. The effects of apical control are described by the term **'acrotony'** in contrast to **'basitony'**.

5.5.3. Genetic Control of Branching

Genetic studies, carried out in members of Solanaceae, have shown the importance of genes in branching of plants. In general, the apical meristem of the Solanaceae plants forms a terminal flower and the axillary meristems grow out and develop for certain period before terminating their own development and initiating new inflorescences (Reinhardt and Kuhlemeir 2002). This reiterative growth is called **sympodial growth**; the shoot segment that is formed by an individual axiallary meristem before its termination is called a sympodial unit. There are species–specific variations in the sympodial growth in solanaceous members such as tobacco, tomato and petunia, and these variations can be explained by the number of sympodial shoots initiated and the degree to which sympodial units are reduced. Genes regulating sympodial growth have been identified by genetic analysis. In the tomato mutant *self–pruning* (*sp*), the sympodial units are reduced successively. The *SP* gene of tomato is orthologous to *CEN* and *TFL* genes of *Antirrhinum* and *Arabidopsis* respectively.

5.6. ORIGIN AND DEVELOPMENT OF LATERAL ROOTS

Lateral roots are always endogenous in origin and positionally related to pre–existing vascular tissue. The lateral roots of angiosperms arise a little far away from the root apical meristem along the primary root axis through cell divisions in the pericycle that form new meristatic centres (Fig. 124). However, in ferns, the lateral roots arise very close to the root apex and that too from the endodermis and not from pericycle. It should be emphasized that not all pericycle cells, but only those positioned at restricted radial locations opposite or adjacent the protoxylem poles, serve as progenitors of lateral roots. The organization of the lateral root primordium is basically similar to that of the embryo radicle as evident through histological studies (Fig. 124).

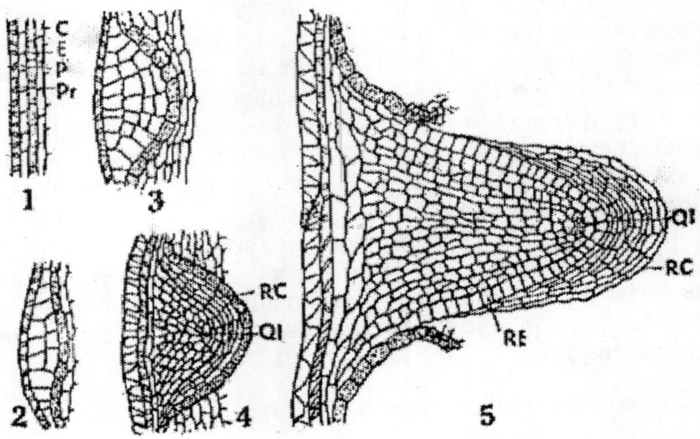

Fig. 124. Various stages in the origin of a lateral root; C - Cortex; E - Endodermis; P - Pericycle; Pr -Protoxylem; RC - Root Cap; RE - Root Epidermis; QI - Quiescent centre initial

The lateral root is initiated by the periclinal division of pericycle cells into an outer and inner layer. Further periclinal divisions produce the root cap, epidermis, cortex, pericycle and stele of the lateral root. The initials that contribute to its apical meristem could not be easily identified by many. However, Swamy and Krishnamurthy (1975b) could trace the origin of QC and dividing cells around it. Soon the primordium pierces the cortex of the parent root and comes out.

Let us now consider the signals that trigger lateral root initiation. It is known for sometime that exogenous application of auxins promotes lateral root development in a number of plants, especially in their cuttings. It was then shown that, in intact roots, the effect of auxin on the development of lateral roots is dependent upon the factors supplied from the older basal part of the root. However, subsequent studies have shown that other than auxins, thiamin, nicotinic acid, adenine and one or a few micronutrients are also needed for lateral root development. Studies carried out by Lin and Raghavan (1991a, b) have shown that the cytodifferentiation pathways of the lateral root initial and the endodermis (in ferns) are due to alterations in gene expression (and possibly in the pericycle in angiosperms). The role of auxin is partly supported by an investigation of *Arabidopsis* mutants such as *superroot (sur)* and *aberrant lateral root formation (alf)*. The former mutant produces a very large number of adventitious roots from parts of seedling other than root due to an elevated level of auxin. The latter group of mutants overproduce lateral roots (*alf 1–1*) or no lateral root (*afl 4–1*) at all. The *alf 3–1* mutant prevents the outgrowth and maturation of lateral root primordia initiated in the pericycle. These mutants are rescued by the application of IAA that perhaps induces divisions in pericycle and maintains its continued cell divisions and cell viability. A number of other genes have also been implicated in lateral root initiation and development. These genes preferentially express themselves in developing lateral roots. They include those genes that code for cell cycle regulatory proteins and cyclins such as *CDC 2* and *CYC* gene products. This has been demonstrated through in situ hybridization and localization of cDNA probes of these genes in the roots of wild–type and transgenic plants (Martinez et al 1992). It was further shown that although the pericycle cells acquire the potential to initiate lateral root development through the activation of *CDC* genes, the primordia formed grow out of the parent root cortex only when they express *CYC* genes.

5.7. ROOT–STEM TRANSITION

The most intriguing aspect that has attracted the attention of many early investigators is the way in which vascular connections are established between the parts of root and stem of a seedling produced from the germating embryo. The vascular tissues are arranged in collateral or bicollateral bundles in the stem, but in the root, phloem and xylem occupy different radii. How these geometrically differently–oriented vascular tissues of root and stem get bridged up? The bridging phenomenon has been studied by many investigators in the seedlings of diverse plants. Unfortunately, almost all the studies have been done on mature seedling, where vascular differentiation has already proceeded to a very great extent. These studies have been summarized by Eames and McDaniels (1947) and Esau (1965b). The former two authors have recognized three major types of transitions in the so–called **transition region** (Fig. 125).

According to them, the length of the transition region varies in different taxa. It is generally short and is about 1–2 mm long. It may lie at the top of the radicle or in the hypocotyl. In some species, the transition zone may appear as a region of depression or swelling. All the three types of transitions recognized explain the continuity between the root and stem vasculatures on the basis of splitting, fusion, torsion etc. of the vascular strands. Although such phenomena are seemingly evident in transverse sections of seedlings at the so–called transition region and thus, represent the condition at a particular stage of development, they are more apparent than real. What actually happens should be studied from the embryo onwards from an ontogenetic perspective.

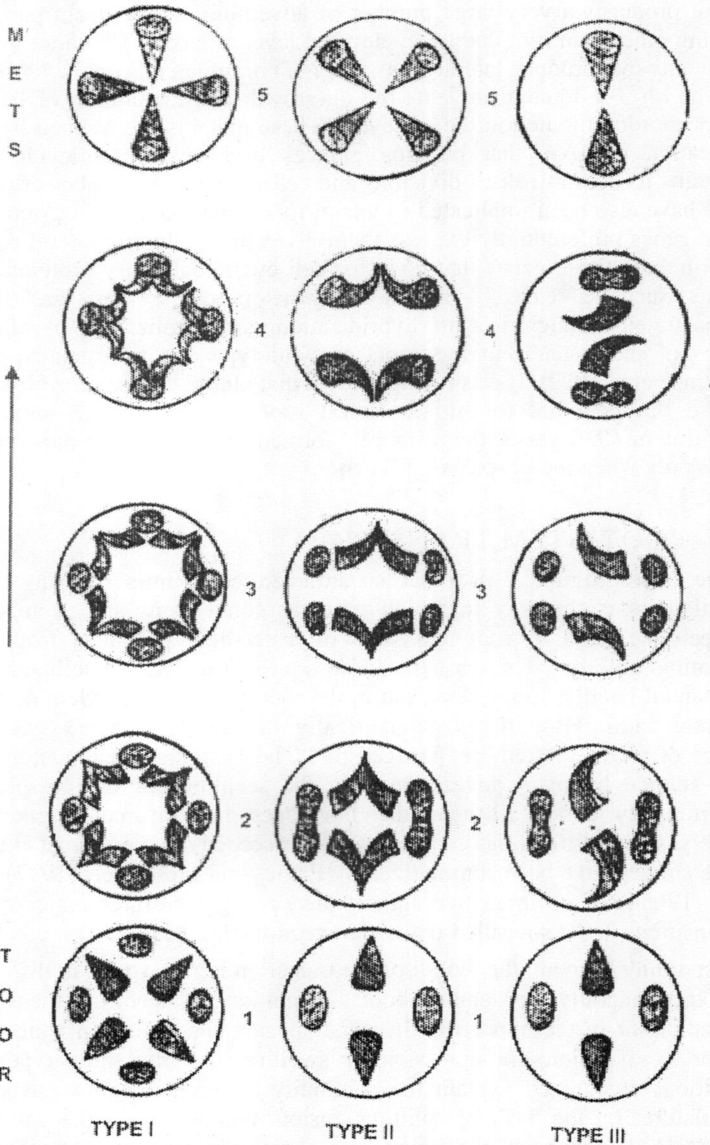

Fig. 125. Diagrammatic representation of three major types of root–stem transition as recognized by Eames and MacDaniels (1947)

The first attempt to study the transition region from an ontogenetic perspective was that of Marsden and Bailey (1955) in *Clerodendrum trichotomum*. In the mature embryos and during their germination, the xylic procambial strands of the root and of hypocotyls not only lie along alternating radii, but also are exarch in the root and endarch in the hypocotyls. This situation remains the same even after the differentiation of the protoxylem. Thus, the hypocotyls and the root are sharply delineated from one another and it is as if the former is directly superimposed over the latter. However, at the time of differentiation of metaxylem, continuity becomes established through the formation of vessels and/or tracheids. Further detailed studies on a variety of dicots and monocots, carried out by late professor B.G.L. Swamy and his students at Chennai, India, revealed that the differentiating metaxylem that bridges up the continuity between the root and the hypocotyl is not derived from procambium, but from differentiated parenchyma cells in the hypocotyl region (Fig. 126). The transition phenomenon in the light of their studies may be summarized as follows:

1. In the mature embryo and during the early stages of its germination, the xylic procambial strands of the root and hypocotyl are discontinuous and the boundary between the two organs is very sharp and abrupt.
2. Such a situation persists even during the differentiation of the proxylem.

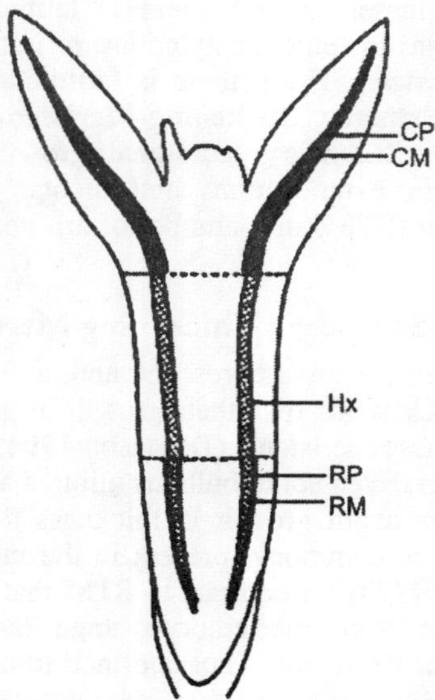

Fig. 126. Diagrammatic representation of the L. S. of a germinating embryo to show the build-up of the root metaxylem (RM) with cotyledon metaxylem (CM) through hypocotyls xylem (HX) that gets differentiated from parenchyma. Root (RP) and cotyledon protophloem (CP) are also shown.

3. The metaxylem differentiation becomes initiated in the overlapping parts of the xylic procambia of the root and stem, thus, establishing continuity. In the monocots, metaxylem is differentiated throughout the transverse area of the transition region to form the nodal plate. In both cases, the metaxylem always differentiats from mature parenchyma.
4. Thus, in the initial stages of the establishment of continuity, there is no evidence in favour of phenomena such as splitting, fusion and torsion of mature vascular strands.

5.8. LATITUDINAL GROWTH IN PLANTS

Latitudinal growth, also called growth in thickness or girth of the axial organs such as stem and root, is often seen after the primary longitudinal growth is over. But, in some monocots and in very few dicots, a type of latitudinal growth is seen almost along with the longitudinal growth. Growth in girth is termed 'primary' if it occurs near the shoot apex. It occurs in diverse ways. First, there is growth due to an increase in the size of the apical meristem itself. This kind of expansion commonly happens when a plant is growing from the embryo stage to the mature vegetative phase (Fig. 89), but it may then be reversed, the apex becoming narrower during reproductive phase. Second, there is latitudinal growth due to expansion in the pith, as in some dicotyledons, or in the cortex as in a few taxa such as *Sempervivum*. Third, there is latitudinal growth due to the activity of a special **Primary Thickening Meristem (PTM)**, as in many monocots (Fig. 127). Secondary latitudinal growth is produced by a **vascular cambium** in gymnosperms and dicots, or by a **Secondary Thickening Meristem (STM)** in a number of arborescent and herbaceous monocotyledons.

5.8.1. Primary and Secondary Thickening Meristems

In many arborescent, semi–arborescent and a few thick herbaceous monocot taxa, the thick stems owe their growth in girth to the activity of PTM, STM or both these meristems (DeMason 1994). These plants lack a true vascular cambium. Even some bulbous alliums and other similar taxa have PTM that brings about growth in thickness to some extent. Thus, PTM may be said to be commonly present in the monocots (Swamy and Krishnamurthy 1974, 1979) (in contrast to STM that is restricted to some monocots only) right from the embryo stage onwards. Meristematic activity occurs all over PTM and is not confined to one or a few layers as in vascular cambium.

PTM occurs in the shoot tip, but in spite of it, the meristem is often classified as a lateral meristem (DeMason 1994). As the name implies, the PTM is a primary meristem. The most important characteristics of taxa with PTM are:

- Several leaf primordia that are produced in quick succession in the shoot apex even when the plant is still a seedling. Hence, the shoot apical meristem appears to be sunken below the surrounding tissues.
- The adult diameter of the stem is often attained even in the seedling itself.

The PTM occurs at the base of all these leaf primordia at progressively declining angles from the youngest to the oldest leaf primordia in such a way as to make the central cylinder of the stem wide and the cortex narrow. In other words, the cell files of PTM are nearly vertically oriented in the distal portion of the stem, but are more horizontally oriented in the

Fig. 127. A - Median L. S. through vegetative apex of *Phormium* showing the apical meristem, successively produced leaf primordia and the abruptly thickened primary stem axis through the activity of primary thickening meristem located at the bases of leaf primordial; B - Details of the primary thickening meristem (Philipson et al, 1971).

lower portion. In some taxa, there is often only a narrow region of tangentially flattened cells that has traditionally been identified as the PTM. The amount of stem thickening that occurs in different species can be correlated to anatomical features of PTM that importantly include: (i) An apical meristem sunken below surrounding stem tissues (ii) The extensive anticlinal files of cells that run from the outer cortex into the central cylinder (iii) A more or less distinct lateral meristem, the PTM, which is located between the cortex and the central cylinder.

To understand the mechanism of stem thickening due to PTM, DeMason (1994) selected *Allium cepa* for study. DeMason calculated the stem height from the base of the vasculature to the centre of the apical meristem, and the maximum stem width. The author argued that if the stem width grows more rapidly than the stem height, one can infer that stem thickening is a significant process. Then, a standard height was ascertained (for *Allium* it was ascertained as 2.5 mm), along with the measuring of the following additional parameters at the same height (i.e., at 2.5 mm) on plants from different age groups: (i) Number of cell files in the cortex (ii) Width of the central cylinder (iii) Angle of the cell files in both the cortex and the central cylinder. The results obtained by DeMason (1994) showed that the number of cells per file in the cortex increased through time until primary tissues became mature at that level, but the width of the central cylinder increased beyond maturation of primary tissues. The first result is evidence that *Allium cepa* indeed has only primary thickening growth. The angle of the cell files in both the cortex and the central cylinder increased rapidly for the first two months and then increased more slowly before finally levelling off. The continued increase in angle of the cell files in the central cylinder explains why the central cylinder increases in width after cell division and why cell enlargement due to primary growth ceases. This distinct reorientation of cells is a unique phenomenon in plant development and plays an important role in the mechanism of primary thickening in monocots.

However, cell division activity also contributes to primary thickening. With correct staining procedures and radiolabelling followed by auto–radiography, a wide meristematic region can be recognized at the top of the stem that narrows basipetally and finally disappears. In serial T. S., this region is a continuous cylinder of dividing cells. The pattern of maturation of growth in the girth of stem is centrifugal in the central cylinder and centripetal in the cortex (DeMason 1994). The sequence of events has been divided into three stages that occur simultaneously. The first event is the initiation of cell files in the crown, which happens immediately adjacent to the base of the apical meristem and must occur by anticlinal cell divisions in the youngest cell files. The second event is elongation of the cell files by both periclinal cell division activity in the PTM and by subsequent cell enlargement along the length of the files. The final (third) event is reorientation of the cell files to a horizontal plane. This further increases the width of the stem. Procambial strands arise both directly from the apical meristem and from the PTM. Swamy and Krishnamurthy (1974, 1979) have also shown that in most monocots, the internode is partly derived from the PTM (also called **pericambium** by some early investigators) and partly from the shoot apical meristem.

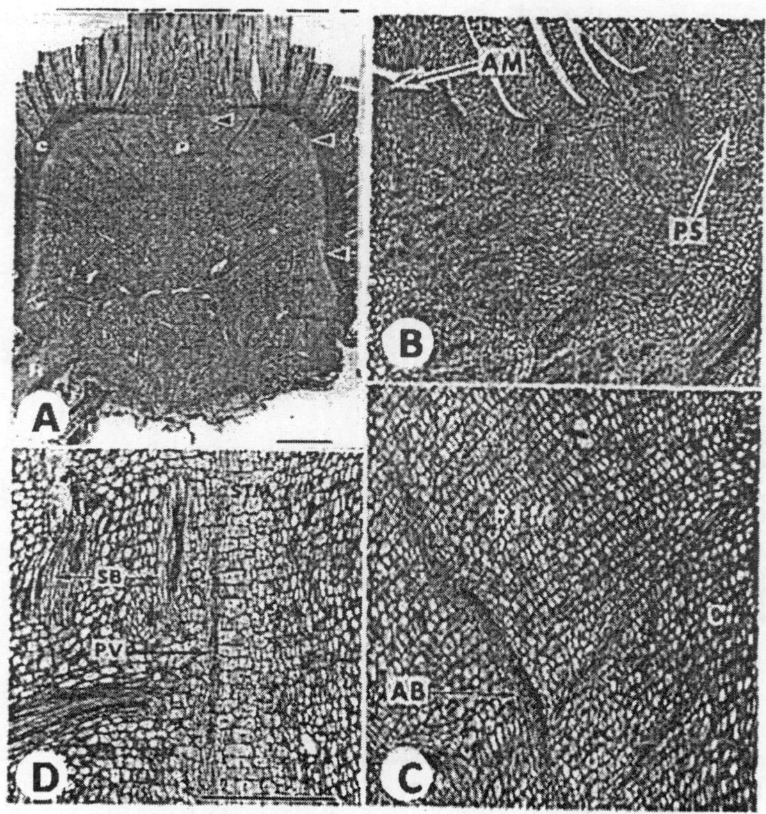

Fig. 128. *Yucca wthipplei*; A - Median L. S. of the apex of one-year-old stem; Arrows indicate primary and secondary thickening meristems; B - L. S. of apical meristem and the subjacent primary thickening meristem; C - Meristematic activity confined to a recognizable zone of tangentially flattered cells; D - Secondary thickening meristem region; AB - Primary axial bundle; AM - Apical meristem; C - Cortex; P - Pith; PS - Procambial strand, PTM - Primary thickening meristem; PV - Secondary provascular strand; SB - Secondary axial bundle, STM - Secondary thickening meristem (photographs supplied by DeMason).

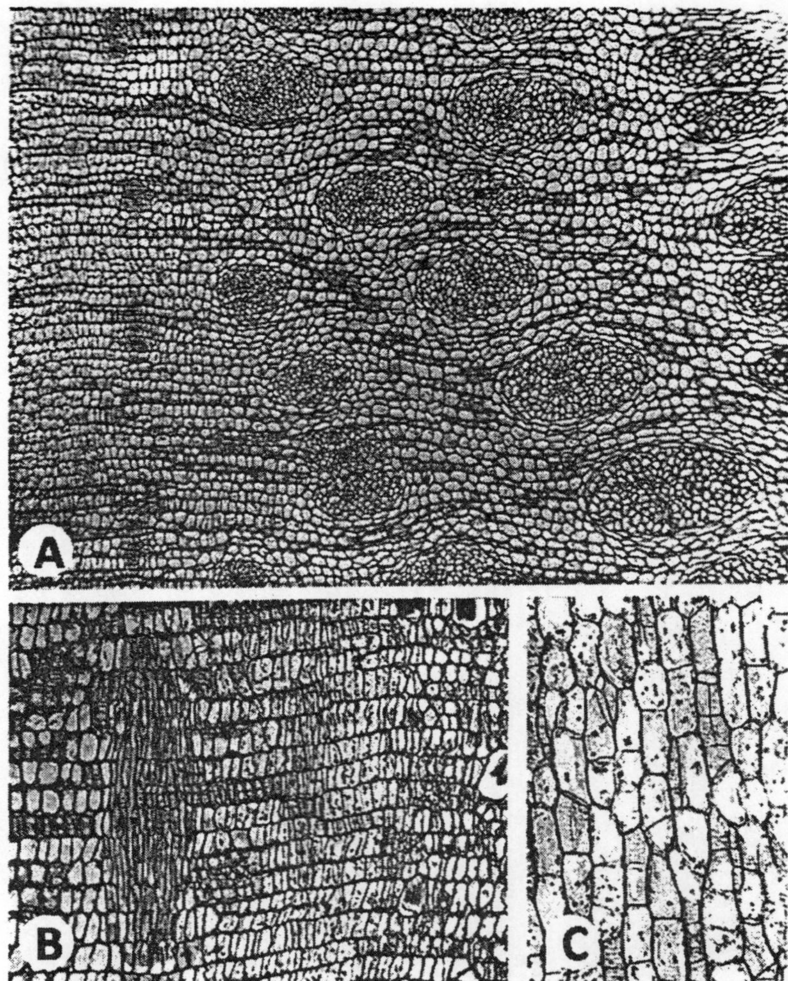

Fig. 129. *Cordylene australis*; A - T. S. of portions of stem showing STM (on the left) and its derivative ground tissue in which vascular bundles are embedded; B - R. L. S. of the same showing the meristem, cortex (on the right) and procambial strands (on the left); C - T. L. S. through the meristem region. Note the absence of distinction into ray and fusiform initials (Philipson et al 1971).

Many arborescent and some semi–arborescent species of monocots show thickening growth by a secondary thickening meristem (STM) (Figs. 128 and 129). This has been wrongly called vascular cambium or cambium by some botanists.

There is functional and developmental relationship between *PTM* and *STM*. To demonstrate this, one has to address the following questions (DeMason 1994): (i) How does the function of the STM relate to the function of the PTM? (ii) Are the STM and PTM longitudinally continuous in an actively growing stem? (iii) At a given level of the stem, does the STM replace the PTM? (iv) How does the *STM* arise for the first time in the stem in relation to the origin of the PTM? If the two meristems are developmentally related, all the criteria should be satisfied. The two meristems should be longitudinally continuous in a growing shoot, the PTM should become the STM at a given level in the stem over time, and the *STM* should arise after and in continuity with the PTM in a seedling. Studies on taxa with STM have shown that all criteria proposed above to demonstrate developmental relatedness have been fulfilled (DeMason 1994). The PTM and STM form a continuum.

The mechanism of stem thickening (due to STM) in *Yucca,* for example, which has secondary thickening, is very similar to that in *Allium*, which has only PTM and has no secondary thickening. Thus, it demonstrates a functional relationship between PTM and STM. The stem thickens by the formation, lengthening and reorientation of cell files, regardless of whether there is secondary thickening or only primary thickening. Another feature that supports the notion that the STM behaves similar to the PTM is that shoot–borne roots are formed by the PTM in *Allium,* but by STM in *Yucca*. A third functional similarity between the two meristems is in the method of vascular tissue development from their derivatives.

The STM produces cells principally towards the inside, although in some cases it supplies some cells of secondary parenchyma toward the exterior. From the inner derivatives, vascular tissue differentiation takes place in the form of distinct vascular bundles or strands (containing xylem and phloem), which are embedded in a parenchymatous matrix. In some, the parenchymatous matrix becomes fully or partially sclerenchymatous (Fig. 129 A). This type of secondary thickening is totally different from the true secondary growth produced through the activity of the vascular cambium.

5.8.2. Vascular Cambium

As briefly indicated in Chapter 1, the vascular cambium is mainly responsible for growth in the girth or latitudinal growth or **secondary growth** of axial plant organs (the stem and the root) in many dicots and gymnosperms. Although a vascular cambium has been reported in pteridophytes (Bhambie 1994), monocots (Refer Rangarajan and Swamy 1980) and in the appendicular organs such as leaves, such reports are erroneous, as the so–called vascular cambium, in these cases, has neither the typical structure nor the characteristic activity and behaviour. The vascular cambium is a lateral meristem, often forming a continuous sheath, and producing **secondary xylem** or **wood** on the inside and **secondary**

phloem or bast on the outside (Fig. 130 A, B). However, there are exceptions. As the volume of secondary xylem goes on increasing, the vascular cambial sheath (or ring) continues to increase in its perimeter through anticlinal divisions of its constituent cells. The duration of production of wood and bast by the vascular cambium decides the degree of growth in girth of the plant axis. In most trees it continues till the death of the plant to result in massive axial organs as in some giant conifers and *Eucalyptus* species, while in herbaceous species, it has highly limited activity. In most taxa, there is only one cambial ring that is active throughout the life of the plant, while in a few, especially in some lianes, more than one cambial ring is produced in succession, each acting for a variable period to produce alternate zones of wood and secondary phloem (Fig. 131). More details on successive vascular cambia and other cambial variants are provided later in this section.

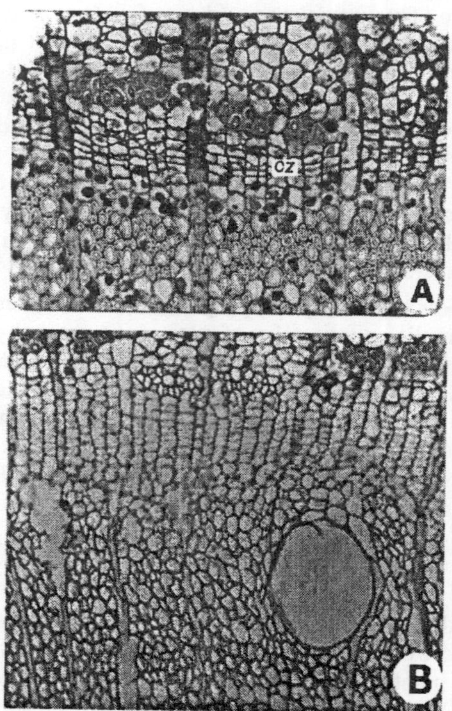

Fig. 130. A - T. S. of stem of *Dalbergia sissoo* showing dormant cambial zone (CZ) with secondary xylem below and secondary phloem (with fibers) above cambial zone; B - T. S. of stem of *Albizia amara* showing actively dividing cambial zone, secondary xylem (below cambium) and secondary phloem (above the cambium). (Photographs courtesy of Dr. N. Venugopal)

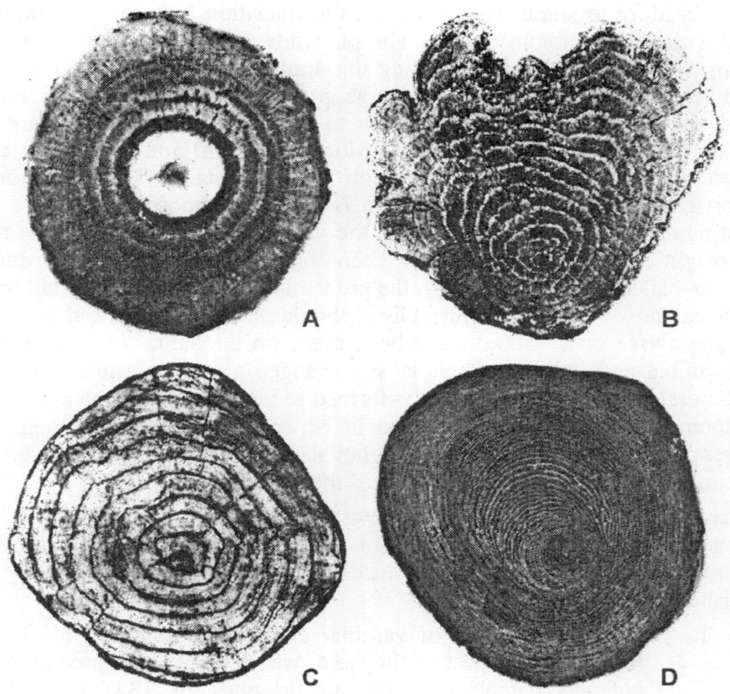

Fig. 131. Successive cambial activity and the resulting alternative zones of secondary xylem and phloem; A - *Cycas circinalis;* B - *Gnetum ula;* C - *Dalbergia lanceolaria;* D - *Avicennia officinalis*

5.8.2.1. *Origin of Vascular Cambium*

The consensus of opinion among plant morphologists is that the vascular cambium originates from the procambium. After the production of primary xylem and primary phloem, the left–over procambium is believed to give rise to the vascular cambium after little or no modification. This assumption has often led to statements as follows: "The procambium and cambium may be looked upon as two developmental stages of the same meristem" (Esau 1965b); "Separate terms like procambium and cambium are needed for convenience only" (Sterling 1946). This assumption that the vascular cambium is related to the procambium developmentally has gained a deep ground among morphologists, although they have realized the following differences between the procambium and the vascular cambium respectively: Gabled cell ends in radial view in contrast to flat ends, shorter cells contrary to longer cells, storied in relation to non–storied arrangement, non–vacuolate protoplast contrary to vacuolated one, and without ray initials in contrast to with ray initials (Fahn et al 1972).

According to some morphologists, the transition between procambium and vascular cambium is very abrupt, while others speak of a gradual transition between the two. During this transition, some procambial cells are believed to grow through intrusive growth to become fusiform initials and some others divide transversely to form ray initials. There are yet others who speak of the presence of long (fusiform) and short (ray) cells even in the procambium (Refer to Soh 1990 for details). The formation of short ray initials is very crucial as it is the most important character that distinguishes the procambium from the vascular cambium. Also, the time of origin of this character has not been convincingly demonstrated during the so–called transition between the procambium and vascular cambium or at least shown to be quite variable. It should also be pointed out here that all the above observations have been made on the stem, where the xylic procambia and phloic procambia are arranged along the same radii in a collateral manner. Since the early–formed secondary xylem and secondary phloem are believed to be formed in between the primary phloem and xylem in the fascicular parts, it is but natural to visualize the origin of vascular cambium from the residues of the procambium.

However, to generalize the above contention with reference to the phenomenon of cambial origin in a wide variety of other situations is unwarranted. The following instances may be mentioned (Swamy and Krishnamurthy 1980b):

1. Procambial origin of vascular cambium is not possible in the interfascicular part of the stem, where there is no procambium, but only parenchyma cells are present (Fig. 132). It has been argued by many morphologists that the formation of interfascicular cambial strip takes place in line with the fascicular cambium and under the latter's influence. However, now enough evidences are available for the fact that the reverse is true.

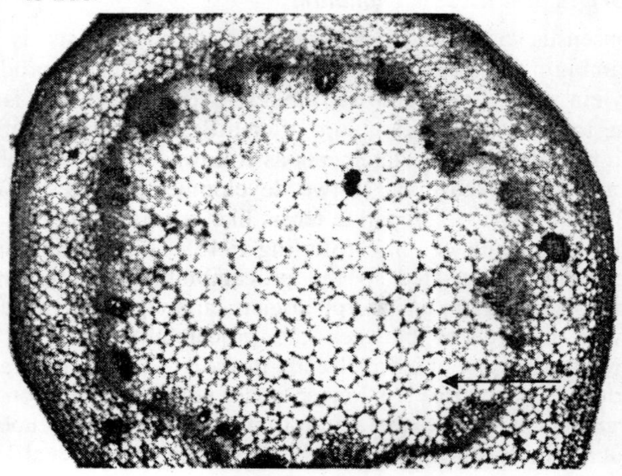

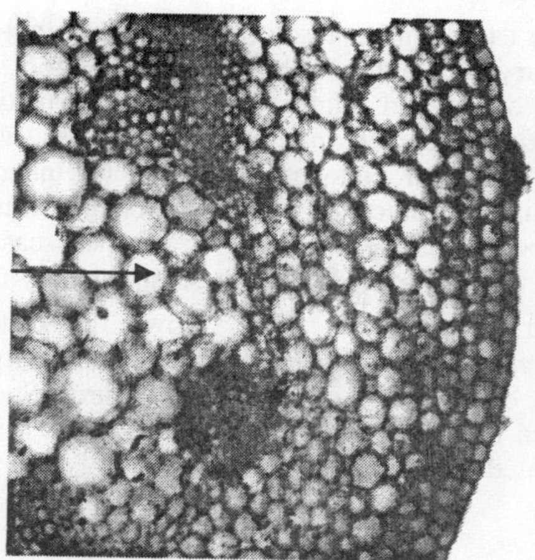

Fig. 132. A - Transverse sections of the stem of *Wedelia* species showing the orgin of interfascicular cambium from parenchyma cells (arrow); B – An enlarged portion of the same.

2. In the root, an essentially similar condition prevails. The vascular cambium arises partly from the conjunctive parenchyma below the phloem groups and partly from the pericyclic parenchyma above the protoxylem groups and the two are subsequently united to form a fluted irregular cambial ring (Fig. 133). Although some (for example, Esau 1965b) argue that the entire vascular cylinder of the root is procambial in origin, and the derivation of not only the primary vascular tissues, but also of the ground tissue (conjunctive parenchyma), pericycle, pith (if present) etc., from a procambium is unwarranted.

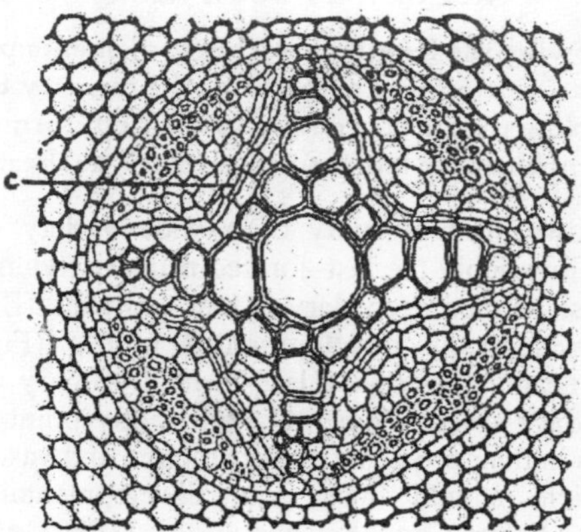

Fig. 133. T. S. of central portion of a dicot root just during the commencement of cambial initiation (C)

3. Vascular cambium can also originate in the plant unconnected to its primary body either in the stem or in the root. The successive cambia of several plants owe their derivation to already differentiated parenchyma outside the primary vascular system, such as those of endodermis, pericycle, inner cortical cells, epidermis, or even of the primary phloem. Medullary cambia of some species arise in the pith. In all these cases, the cambial origin is quite unrelated to the procambium.
4. Vascular cambium is also known to differentiate from parenchyma cells of the callus formed after normal or artificial wounds, or formed in cell or tissue cultures. Such a situation is also seen in grafts. Here, again the origin of vascular cambium is unrelated to procambium

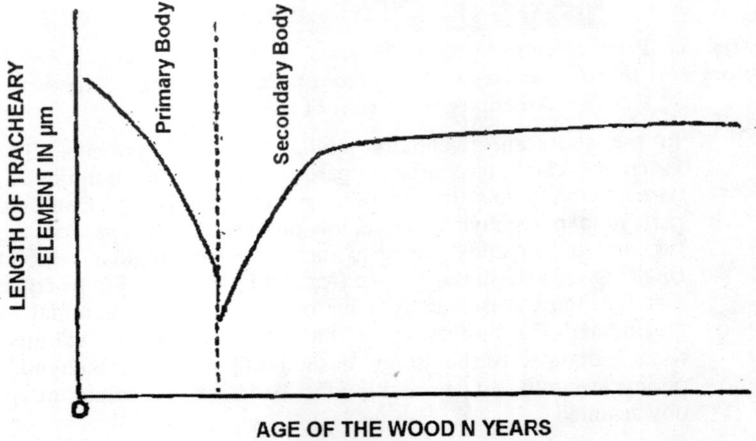

Fig. 134. Graph showing the length of tracheary elements plotted against age of wood. Broken vertical line demarcates the primary body from the secondary body. Note the shorter nature of the first-formed secondary tracheary element than the last formed primary tracheary element.

It has been known from a study of a wide variety of dicots and gymnosperms that invariably the last-formed primary xylem elements are longer than the first-formed secondary xylem elements (Bailey 1944), a phenomenon referred to as **developmental hiatus** (Fig. 134). The difference in the linear values of the last-formed primary xylem and the first-formed secondary xylem elements significantly point out that there must be differences in their precursors (procambium and vascular cambium respectively), which should have no developmental relationship whatsoever. Hence, Swamy and Krishnamurthy (1980) emphasized the following:

1. The procambium is not a prerequisite for the formation of a vascular cambium, even in the stems.
2. In all cases, except in the fascicular parts of the stem, the vascular cambium has been shown to arise from parenchyma. In this region also, it is suggested that the vascular cambium should have an identical origin from parenchyma, which then is stimulated to differentiate into vascular cambium. Parenchyma cells involve only lesser number of changes to become vascular cambial initials than the procambial cells.
3. This also raises very pertinent questions: (i) All groups of vascular plants have procambium (and primary xylem and phloem), but why only gymnosperms and dicots have a vascular cambium? (ii) If procambial origin of vascular cambium is valid, why the procambium of monocots and pteridophytes does not get differentiated into vascular cambium?

As detailed above, the topographic details on the origin of vascular cambium vary. However, the following four types can be recognized in the stem, depending on the activity of this meristematic ring: (i) the primary vascular tissue have the form of an almost continuous vascular cylinder in the internode so that the interfascicular areas are either absent or very narrow. Therefore, the secondary vascular tissues form a continuous cylinder. E.g., *Tilia, Nicotiana*, and *Veronica* (ii) The primary vascular tissues may form a system of strands, but the secondary tissues may so develop that they form a continuous cyclinder. Examples are Conifers, *Sambucus* and *Prunus*. This is the commonest type. (iii) In the interfascicular regions the cambial ring may cut off only parenchyma so that both the primary and secondary vascular tissues have the appearance of a system of strands. E.g., *Aristolochia* and *Vitis* (iv) The secondary growth due to vascular cambium may be confined to the vascular bundle regions, but not to interfascicular region. Examples are *Trifolium* and *Cucurbita*.

5.8.2.2. Structure

The vascular cambium consists of two morphologically–distinct types of initials and in this respect differs from the apical meristem. They are the fusiform and ray initials (Figs. 135 and 136), forming the vertical and horizontal systems of this tissue respectively. The apical meristem can give rise to organs (stem, root, or leaf), while the vascular cambium can give rise to only specific tissues i.e., xylem and phloem. The two types of initials are interwoven as this fulfils the requirement of a thickening meristem (Iqbal 1994). The relative quantum of these two elements within a vascular cambium varies from plant to plant. In some, there may be a greater number of ray initials, and in some, less or even absent (leads to raylessness in wood). Consequently, the number of fusiform initials is either decreased or increased.

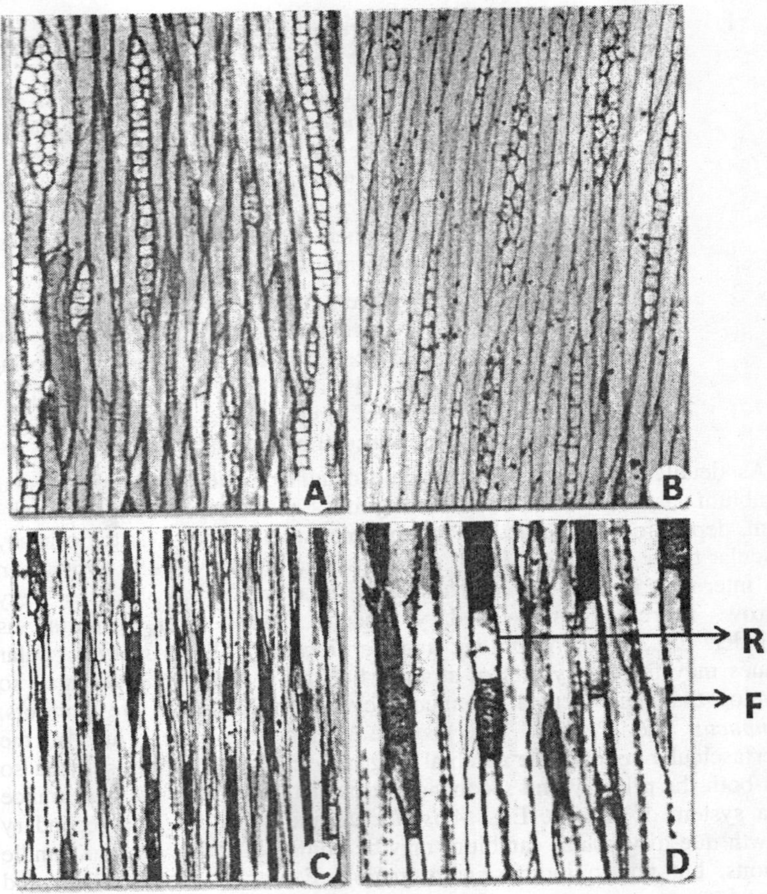

Fig. 135. T. L. S. of non-storeyed vascular cambium; A - *Albizia amara*; B - *Tectona grandis*; C - *Terminalia tomentosa;* D - *Calophyllum inophyllum;* All show the fusiform (f) and ray initials (R). (Photographs courtesy of Dr. N. Venugopal)

The relative arrangement of fusiform and ray initials has led to the categorization of vascular cambium into two distinct types: the **non–storeyed** (also spelled **non–storied**) or **non–stratified** (Fig. 135) and **storeyed** (also spelled **storied**) or **stratified** vascular cambium (Fig. 136). Storeyed vascular cambium is of very rare occurrence and is seen only in some dicotylenous taxa such as some members of Leguminosae, Malvaceae etc. As seen in tangential longitudinal sections, the non–stratified cambium has highly overlapping **fusiform** and **ray initials**, whereas in the stratified cambium the fusiform and often the ray initials also have a more or less distinct tiered arrangement. Moreover, the fusiform initials are relatively shorter (a minimum length of 120 μm) in stratified type. All gymnosperms have only non–stratified cambium, in which the fusiform initials are considerably longer (up to 9000μm) than those of the non–stratified cambia of dicotyledons (upto about 450μm). Of course, cell sizes vary with species, age of the plant, prevailing climatic conditions and habitat. Normally, the maximum length (called genetic length) of the fusiform initials is obtained as the plant ages, invariably around 60 years in large trees and maintains this length till the death of the tree.

The fusiform initial is tangentially flattened, pointed at both ends and possesses 8–32 facets, with an average of 18–21, and each initial will be in contact at least with 14 other cells. The volume is tens or hundreds of times of the volume of an apical meristematic cell. The ratio between length and width of the fusiform initials may be 30:1 to 600:1. The fusiform initials are highly vacuolated (Fig. 137) and this appears to be a means to minimize the amount of cytoplasmic materials synthesized and to reduce energy expenditures at each cell cycle. Thus, the vacuoles economically extend the reach of the cytoplasm. The cells are very similar in ultrastructure to parenchyma cells. The cytoplasm contains larger mitochondria, relatively more differentiated plastids, peroxisomes, ribosomes, lipid bodies, dicytosomes, ER, microtublules, lomasomes and storage substances. The fusiform initials produce all axial components of secondary xylem and phloem by periclinal divisions.

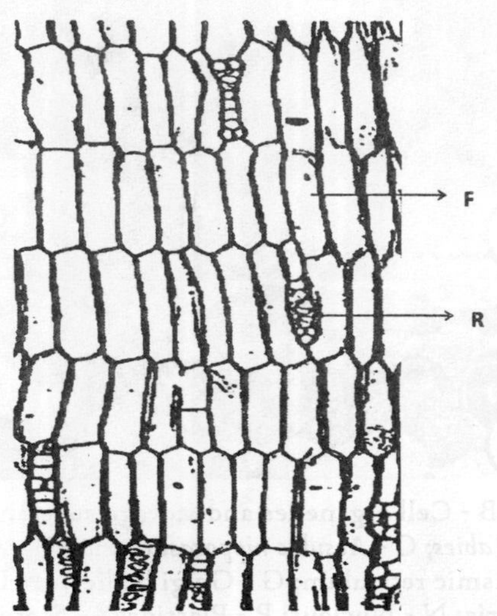

Fig. 136. T. L. S. of storeyed cambium of *Aeschynomene aspera*; F - Fusiform initial; R - Ray initial

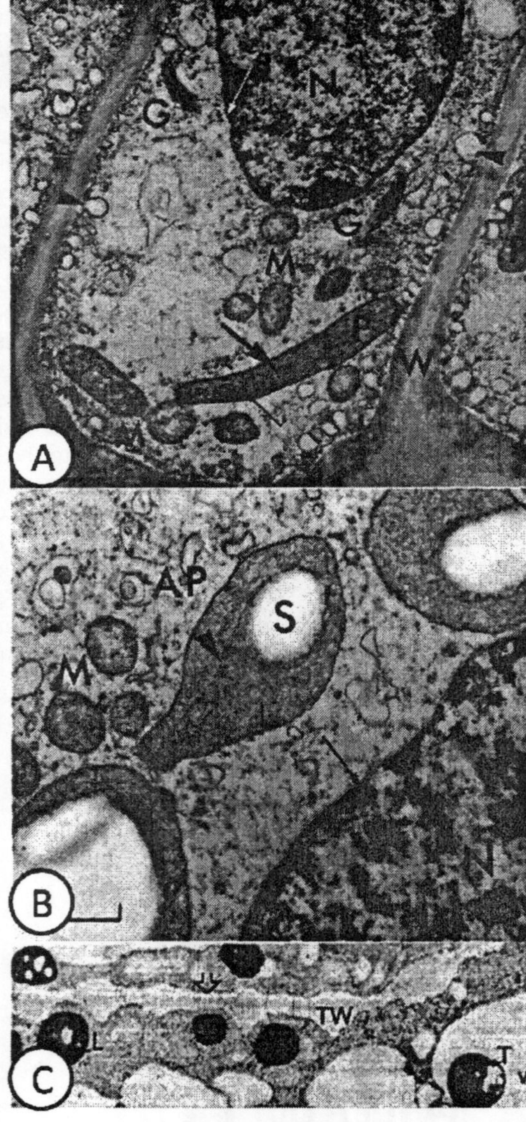

Fig. 137. A and B - Cell organelles and storage substances in cambial cells of *Picea abies;* C - *Aesulus hippocastanum;* AP - Amyloplast; ER - Endoplasmic reticulum; G - Golgi bodies; L - Lipid bodies; M -Mitochondria; N - Nucleus; P - Plastids; S - Starch; T - Tannin bodies; TW - Tangential wall (arrow), V - Vacuole; W - Wall. (A and B - Timell 1973; C - Courtesy of Prof. K. S. Rao)

The ray initials form the horizontal or radial system of the vascular cambium. In T. L. S., they show varied height depending on the species, with cells of varied numbers. The ray initials may be **uniseriate, biseriate** (Fig. 135) or **multiseriate,** and may be **homocellular** or **heterocellular** (with **procumbent** and **erect cells**) (Fig. 138). Their cells have shorter tangential diameters than radial diameters (evident from R. L. S.). Ray initials have well–developed cell–to–cell connections through plasmodesmata and form the main symplastic route for horizontal communication. The number of plamodesmata in the tangential walls between ray initials is specifically high. In many taxa, especially in conifers, special ray initials giving rise to **ray tracheids** are seen at the margins of the ray initials. The ray initials produce the horizontally–aligned rays of wood and phloem.

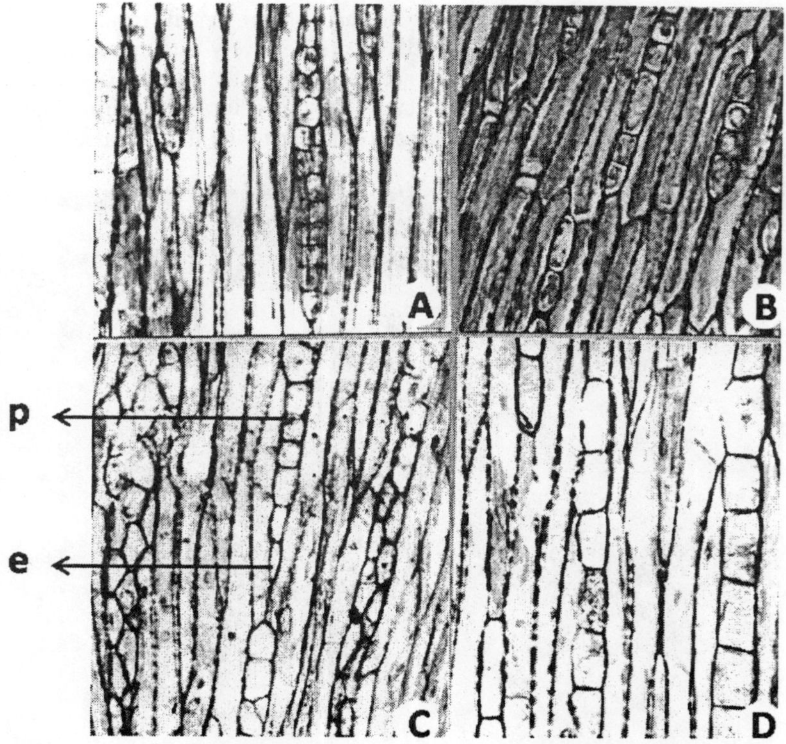

Fig. 138. T.L.S. of the vascular cambium; A - *Albizia amara;* B - *Dalbergia sisso;* C - *Tectona grandis;* D - *Terminalia tomentosa*; ray initials in A and C show both procumbent (p) and erect (e) cells, while those of B and D are with procumbent cells only (Photographs courtesy of Dr. N. Venugopal).

The cell walls of fusiform initial have fibrillar structure and this fibrillar structure seems to play a significant role in regulating the direction of growth in the expanding walls. This wall also accommodates the conflicting demands of cohesion, whereas radial expansion of the cambial cells and their derivatives requires extensibility. Radial walls (R walls) are usually thicker than tangential wall (T walls), especially during dormancy (Fig. 139). The chemical composition of these two walls is found to be different. This difference is due to the facts that T walls are produced afresh at each periclinial division and their expansion is limited, while R walls remain plastic and experience a constant radial extension that requires a loosening of the fibrillar architecture of the wall associated with turgor pressure and hence necessary for cellular readjustments. The cell walls of the cambial zone, thus, exhibit a far greater capacity of radial extension (up to 100 times) than that of tangential extension. The dissimilarity between the two walls is progressively established after the cells have completed their division. Radial walls consist of cellulose skeleton embedded in xyloglucans, arabinogalacturan proteins. T walls have xyloglucans and very little to zero acidic pectins and xylans.

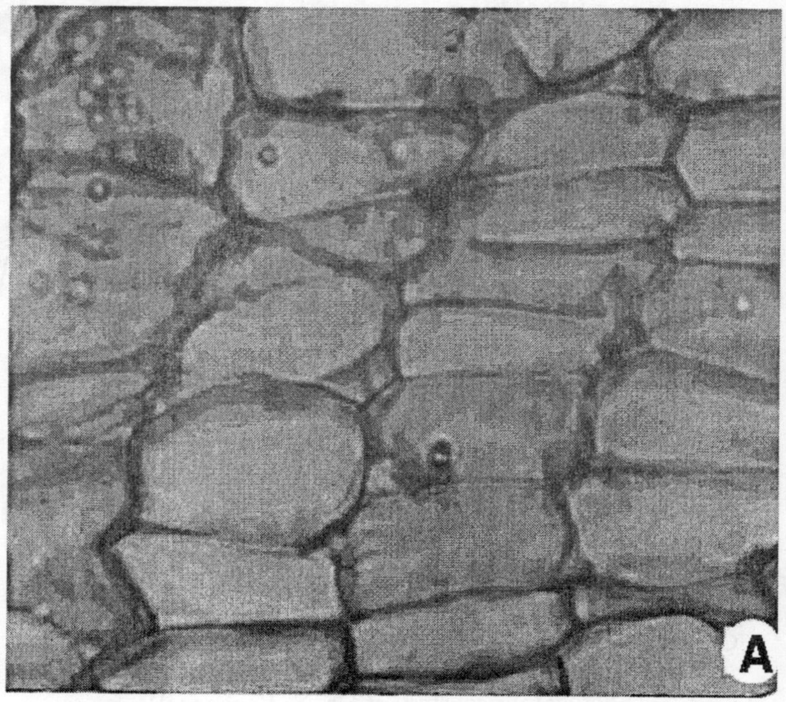

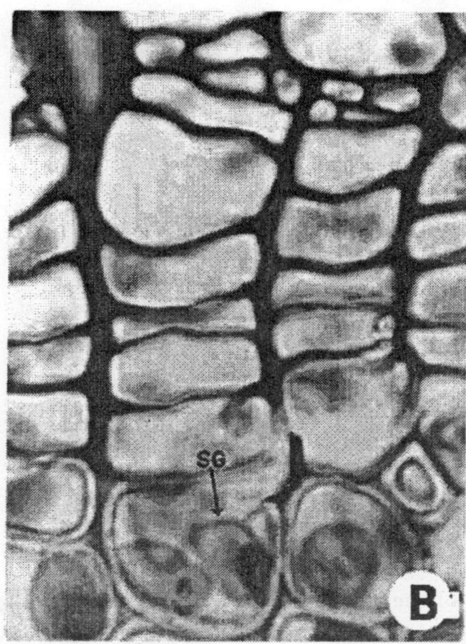

Fig. 139. T. S. of portion of the cambial zone of *Albizia amara* during active (A) and dormant (B) periods; Note that the radial walls are thicker than tangential walls and both walls are thicker in dormant cambium than in active cambium; SG - Starch grains (Photographs courtesy of Dr. N. Venugopal)

One of the greatly–debated aspects relating to vascular cambium is whether the cambial meristem has s single layer of initials or more than one i.e., whether it is a zone. It is very difficult to identify a single initial layer, even during the most dormant condition (Fig.140), and this fact supports the zonal concept.

Even if a single initial layer is assumed to be present, the fact is that its immediate derivatives cannot be distinguished with precision from this initial layer. This debate is made more complicated by the naming and semantic controversies, especially with reference to the application of the term 'cambium' to the component layers. Some apply this term to the initial layer only, while other apply this term to the entire meristematic zone consisting of a single layer and many derivative layers. However, the composition of the meristem, irrespective of being termed as 'cambium' or 'cambial zone', is the same for both these groups. Hence, this terminological discrepancy should not be overemphasized (Iqbal and Ghouse 1990).

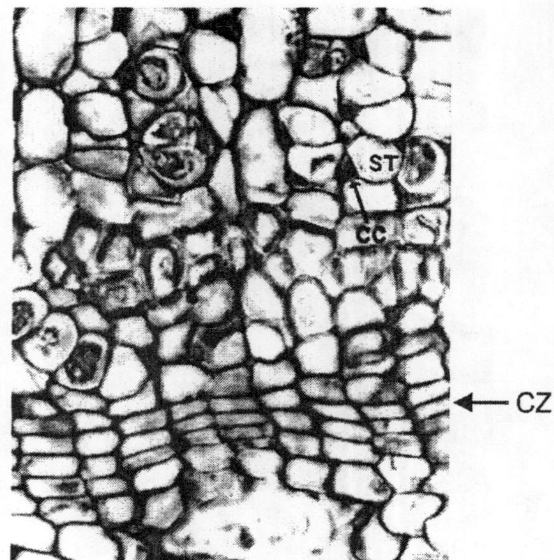

Fig. 140. T. S. of the dormant cambial zone in *Tectona grandis;* Phloem on the outside and xylem on the inside of the cambial zone (CZ); CC - Companion cell; ST - Sieve tube member. (Photograph courtesy of Dr. N. Venugopal)

5.8.2.3. *Cambial Variants*

A typical vascular cambium characteristically forms a single complete ring within the stem and root, as seen in transverse sections, and produces a more or less symmetrical cylinder of secondary tissue, secondary xylem or wood on its inside and secondary phloem or bast on its outside. Departures from the typical cambium and its activity are not uncommon and the secondary growths resulting from such departure were earlier called **anomalous secondary growth** or **thickening** (Philipson 1990). But, since these departures are specific to the taxa, which show them, this terminology has been replaced by the term **cambial variants**. The more common cambial variants and their modes of activity are given below:

1. The cambial ring may be 'normal' (typical), but the tissues formed by it may not be of uniform extent throughout its circumference, often resulting in eccentric or fluted stems. For example, in *Fagus sylvatica,* the secondary xylem produced by the fascicular and interfascicular sectors of the cambium is noticeably different, especially in young axes. The stems of many woody climbers (linanes) have interfascicular and fascicular sectors. The cambial strips of interfascicular sectors produce only parenchyma, while those of the fascicular sectors

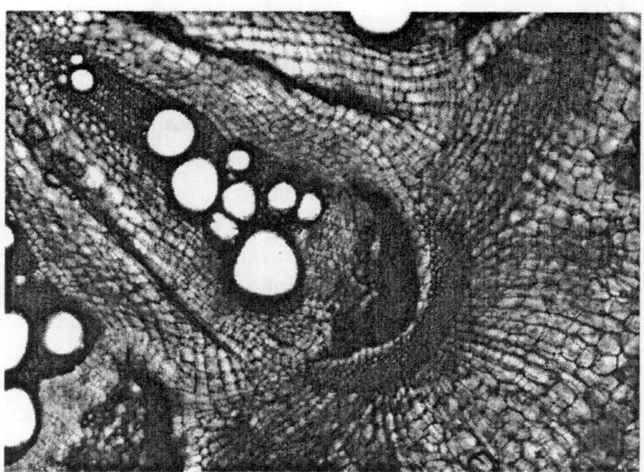

Fig. 141. T. S. of portion stem of *Coscinium;* Secondary growth through cambial activity is restricted only to the fascicular region.

alone produce secondary xylem and phloem (E.g., some species of *Aristolochia, Clematis* or *Coscinium*) (Fig. 141). Subsequently, as the bundles increase in radial extent, the fascicular sectors of secondary xylem become subdivided by arcs of cambium producing only parenchyma instead of tracheary tissue. These parenchyma–producing cambial arcs are ill–defined.

2. **Uneven cambial activity:** Eccentric or fluted stems are sometimes produced by cambial activity taking place at unequal rates on different sectors of the vascular cambium. This asymmetric activity can be related to asymmetry in the arrangement of branches, as can be noticed in some fastigiated tress, with crowded, more or less vertically oriented branches in which the activity of vascular cambium is accentuated towards the outer side of the branches on which leaves and branches are concentrated (Philipson 1990). Uneven activity of cambium is also seen in the flutes or buttresses of some tress. Asymmetric cambial activity is also characteristic of **Reaction Wood** formation in horizontal branches and leaning stems. Asymmetry is very particularly seen in woody climbers with flattened, ribbon–like stems. In these cases the cambial activity is retarded all around except at two opposite sites. These two opposite sites are not the same throughout the length of the stem and their positions are dependent on where the liane adheres to the support and where they are free. In a few other lianes, several

sites of cambial retardation and acceleration alternate around the stem, resulting in an axis with a star–shaped section. The effect of these anomalies is to break up the mass of xylem into more or less discrete strands. It is suggested that the phloem in these stems of lianes is protected within or between xylem and it reduces the chances of disruption of translocated material to the root system. Extreme examples of cambial variants are found in lianes and also they possess a number of distinct morphological strategies. Some of the strategies are given below: The breaking up of the vascular tissue into discrete strands enables short arcs of vascular cambium to cease xylem production (E.g., Members of Acanthaceae, Apocynaceae, Bignoniaceae, Icacinaceae and Passifloraceae). The central mass of xylem gets furrowed and as the stem grows in diameter, the width of the furrows increases through the activity of adjacent arcs of vascular cambium that may also no longer produce xylem (Fig. 142). More furrows may be formed between the original ones. As a result, the core of wood becomes stellate in T. S. (Fig. 143 A), or in extreme cases may break up into a number of irregularly dispersed, separate vascular groups. The furrows in the xylem are filled with a large amount of phloem (Fig. 143 B).

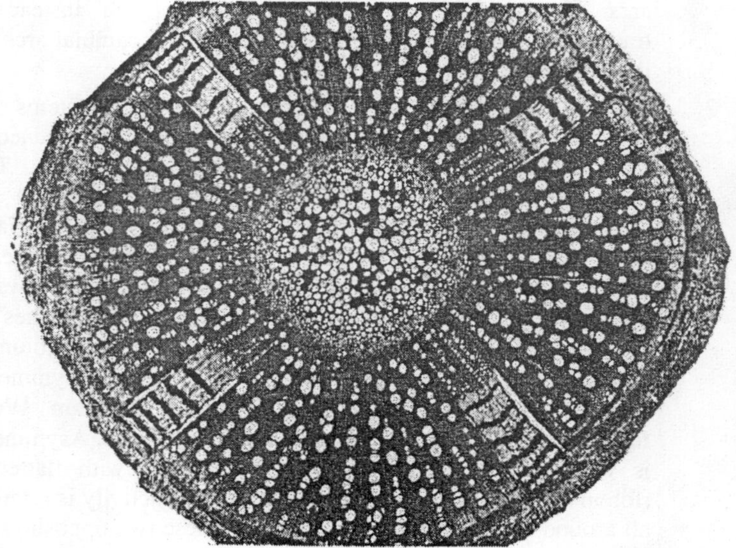

Fig. 142. T. S. of an old stem of *Pyrostegia venusta*. Four sectors of the cambial ring form more phloem than xylem (Philipson et al 1971)

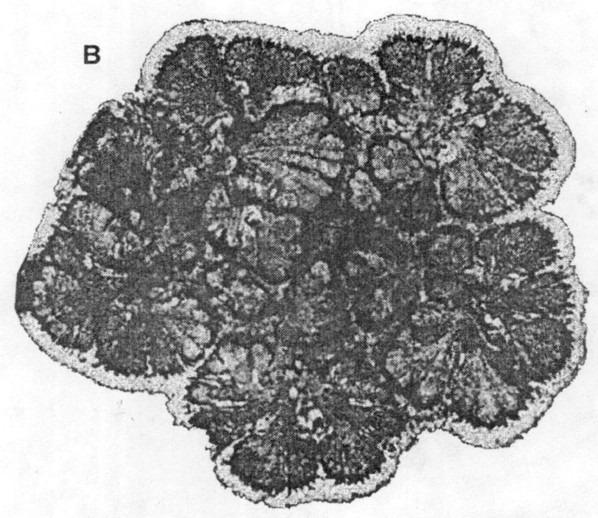

Fig. 143. A - *Passiflora* species deeply furrowed secondary xylem; B - Much dissected stele *Passiflora multiflora* (Photograph Courtesy of Dr. W. L. Stern)

3. **Accessory cambia:** Additional cylinders of secondary vascular tissues may be formed external to the secondary vascular tissues through the formation of circular cambia in the cortex; in *Paulinia pinnata* of Sapindaceae these supernumerary cortical vascular cylinders unite with the central vascular cylinder at the nodes. Thus, the first formed cambium and the additional cambia are continuous. Such supernumerary cambia are produced in large numbers in several lianous taxa of Sapindaceae, Malpighiaceae and Acanthaceae and sporadically in Caesalpiniaceae, Icacinaceae, Convolvulaceae etc., which through their activity produce many dispersed vascular groups in the stems.

   In bicollateral bundles, two strips of cambia arise, one between outer phloem and xylem and the other between inner phloem and xylem (Fig. 144). Two cambial strips are also formed in taxa with intraxylary phloem. The second cambium may develop between the protoxylem and the medullary phloem. Such medullary cambia usually function in a normal manner, producing secondary xylem and phloem but the relative position of these two tissues is reversed. Secondary xylem is added towards the outside and secondary phloem towards the inside (i.e., pith). In some taxa such as *Macropiper*, where medullary

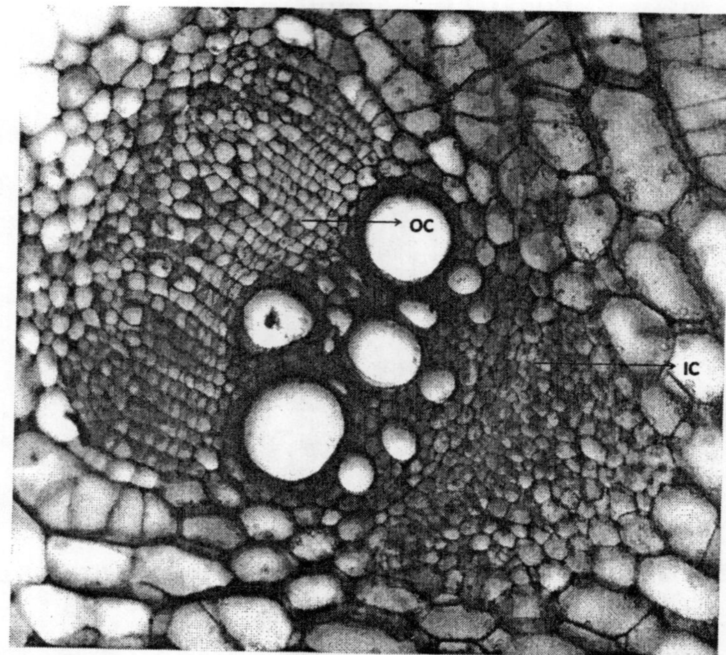

Fig. 144. Additional vascular cambium associated with bicollateral vascular bundles of *Cucurbita* species; IC - Inner cambium; OC - Outer cambium

bundles occur in addition to normal vascular bundles, vascular cambia may arise in the medullary bundles also and cut off secondary tissues. A similar activity may be seen in cortical bundles, as in *Chionanthus*.

4. **Included phloem:** Included phloem means strands of secondary phloem embedded within a mass of secondary xylem (Fig. 145 A and B) and this may be formed in a number of ways. In the *Combretum* or *Thunbergia* type, the internal derivatives of small arcs of cambium differentiate as secondary phloem for a short time before resuming the formation of secondary xylem. The resulting small mass of phloem becomes 'included' inside a mass of xylem tissue. In the *Strychnos* type, xylem production is retarded over small arcs of cambium so that secondary phloem at these sites becomes slightly sunken in grooves on the surface of the secondary xylem. The original circular outline of the cambial sheath is restored by new arcs of cambium that form outside the sunken parts of the secondary phloem. In some members of *Brassicaceae* and *Onagraceae*, the third type of

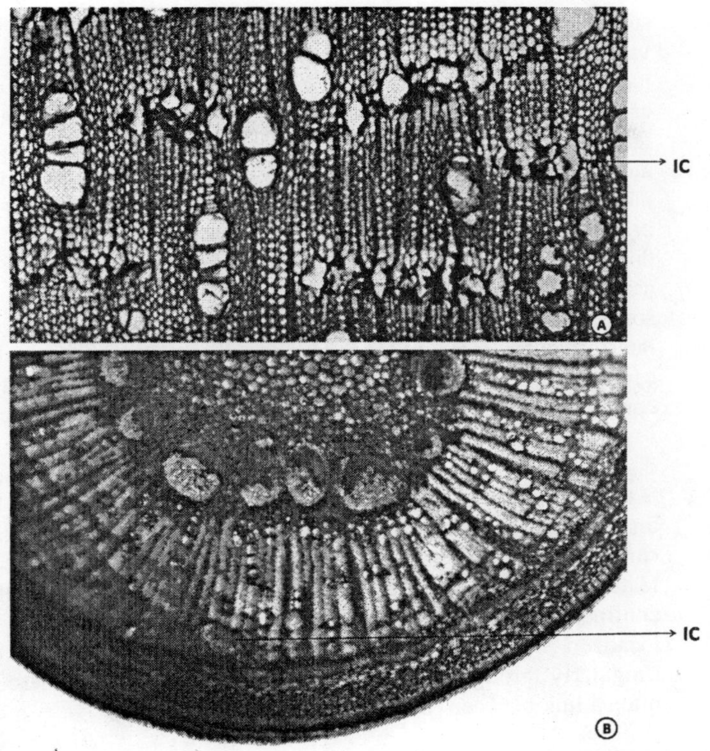

Fig. 145. Included phloem (IC);
A - *Aquilaria agallocha*; B - *Strychnos nux-vomica*

included phloem is seen. Here, certain parenchyma cells of the xylem, already some distance from the cambium, divide to form files of elongated cells that then differentiate into phloem strands. Connections between included phloem and external phloem strands have been shown in many taxa.

5. **Successive cambia:** In some members of a number of families such as Aizoaceae, Amarathaceae, Asteraceae, Avicenniaceae, Buxaceae, Capparidaceae, Caryophyllaceae, Celastraceae, Chenopodiaceae, Connaraceae, Convolvulaceae, Dilleniaceae, Fabaceae, Loranthaceae, Menispermaceae, Molluginaceae, Nyctaginaceae, Phytolaccaceae, Plumbaginaceae, Rhabdodendraceae and Stylidiaceae, the original cambial ring formed soon ceases to function and is replaced by a second cambial ring, which in turn, is replaced by others. Thus, a

succession of cambial rings is formed and these are active for variable periods of time, each producing secondary xylem on the inside and secondary phloem on the outside. Successive cambia are also recorded in species of *Gnetum*. Although many features between successive cambia of these taxa are common, there are some modifications in detail. These pertain to the mode of origin, the extent of the new cambia as complete rings or as sectors, and the type of secondary xylem and phloem laid down by them. In many species with successive cambia, the first cambium appears in the normal position (E.g., *Avicennia*), but in some species no normal cambial sheath develops (i.e., no interfascicular cambium and with highly ephemeral fascicular cambium), and the first complete cambial ring arises in the cortex (E.g., *Heimerliodendron, Bougainvillea*). The secondary vascular tissue derived from successive cambial activity is distinctive as it not only contains included phloem, but in many taxa also includes large amount of parenchyma that may be dispersed in regular rows defining the boundaries of each cambial ring (Fig. 146 A and B). Another commonly–noticed feature of the secondary tissue (as seen in T. S.) is the confinement of conducting tissue (xylem and phloem) to localized bundles dispersed either in concentric circles or irregularly throughout the ground tissue (Fig. 146 B). There is often a link between 'bundles' from adjacent cambia.

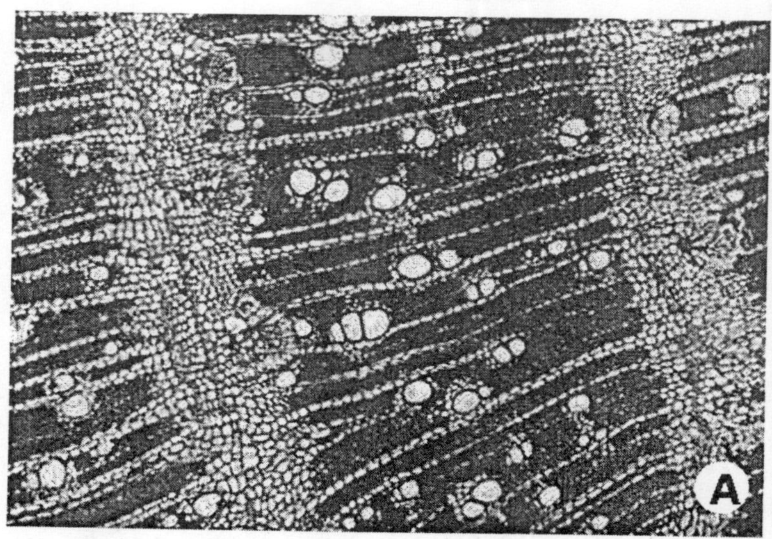

Fig. 146. Successive cambial activity; A - *Avicennia resinifera;* Broad zones of secondary xylem alternate with narrow parenchymatous zones in which strands of phloem and sclerenchyma are present; B - *Heimerliodendron brunonianum* Zones of fibers with embedded vessels alternate with zones of parenchyma in which phloem cells are present along the same radii as xylem (Studholme and Philipson 1966)

5.8.2.4. *Production of Wood and Bast by Vascular Cambium*

The fusiform initials cut off the secondary xylem on the inside and the secondary phloem on the outside through tangential divisions (periclinal divisions). Simultaneously, the ray initials cut off the xylem and the phloem rays. The volume of secondary xylem cut–off by the cambial ring is many times more than that of the secondary phloem. Therefore, the extent of the outward shift of the cambial ring is determined by the volume of the xylem cut–off by the cambial ring. Once the phloem and xylem mother cells are cut off by the cambial cells through periclinal divisions, they differentiate in sequential and overlapping phases. Differentiation occurs in three phases: (i) The phloem and xylem mother cells may divide further (ii) The cells that have ceased dividing usually enlarge in the radial direction (in the case of vessel elements, even in tangential direction) and (iii) Subsequently, phenomena such as secondary wall formation and lignifications in xylem and sieve area development in phloem either continue or commence. These three phases of differentiation often overlap. Laying down of secondary walls in tracheary elements, for instance, frequently begins before the growth of the primary wall ends. However, the expansion of cells often ceases by the time lignifications starts. Complete differentiation of a derivative cell usually takes a few to several weeks from the time of its production. The duration varies with species and cell type. As indicated in Chapter 4, the different cell elements of the secondary xylem and secondary phloem differentiate in relation to relative concentrations and interactions between different growth regulators.

Structure of wood

The components of wood derived from the activity of the vascular cambium belong to either of the following two systems: (i) **Vertical, axial** or **longitudinal system** (ii) **Horizontal, radial** or **transverse system** respectively derived from fusiform and ray initials. The vertical system of xylem consists of the tracheary elements, fibers and axial parenchyma, while the horizontal system is composed of xylem rays. The gymnospermous woods are generally known as **soft woods** in contrast to **hardwoods** of dicots. Soft woods are **homoxylous** and have only tracheids as tracheary elements, while the hard woods are **heteroxylous** and have, in addition or in place of tracheids, vessel elements. Moreover, axial parenchyma cells are more abundant in hardwoods than in soft woods.

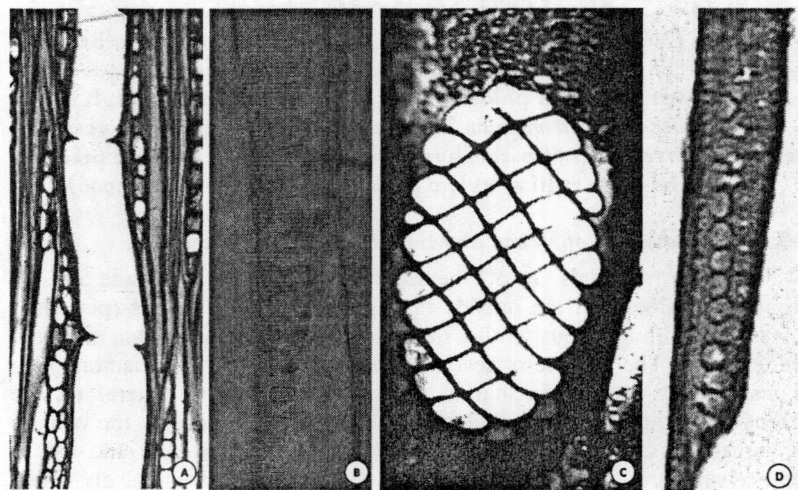

Fig. 147. A - L. S. of vessel elements of *Aquilaria agallocha* with simple perforation plate between them; B - Vessel elements of *Euphorbia* species with scalariform perforation plates; C - Vessel element of *Iryanthera* species with reticulate perforation plate (Photograph P.E. Gasson); D - Portion of vessel element of *Ephedra* species showing foraminate performation plate

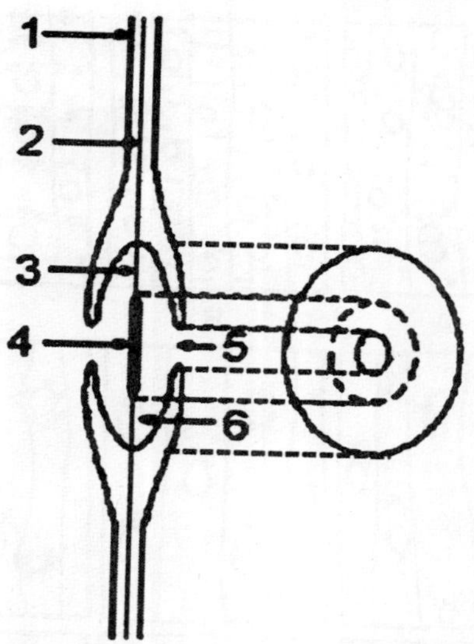

Fig. 148. Structure of bordered pit in sectional and surface views; 1 - Cell wall; 2 - Middle lamella; 3 - Pit membrane; 4 - Torus; 5 - Pit aperture; 6 - Pit chamber

The tracheary elements belong either to tracheids or to vessel elements, the former **imperforate,** while the latter have special openings called **perforations**, usually on either of their end walls. A number of vessel elements placed one over the other form the **vessels**. Tracheids are longer and narrower than vessel elements. The **perforation plates** contain the perforation. The plate may be **simple** (with one perforation) (Fig. 147 A), **scalariform** (with many perforations arranged in the form of steps of a ladder) (Fig. 147 B), **forminate** (with a number of circular perforations in a plate) (Fig. 147 D) or **reticulate** (Fig. 147 C). Both tracheids and vessel elements have various kinds of thickenings on their lateral walls and these may be annular, helical, reticulate or pitted, and the pits are **bordered** (Fig. 148).

As seen in T.S. of wood, vessels may be seen singly (**solitary**) (E.g., *Eucalyptus*), in multiples of two or more (E.g., *Acacia*) or in clusters (E.g., *Pistasia*). The multiple or clustered nature may be in the radial or tangential direction (Fig. 149). The axial parenchyma cells may be absent or almost so (**scanty**) as in some conifers or fairly abundant (up to 50% of the wood) as in many dicots. Abundant axial parenchyma may be **apotracheal** (not associated with vessels) or **paratracheal** (closely associated with vessels). Several subtypes are recognized under both these categories (Fig. 149).

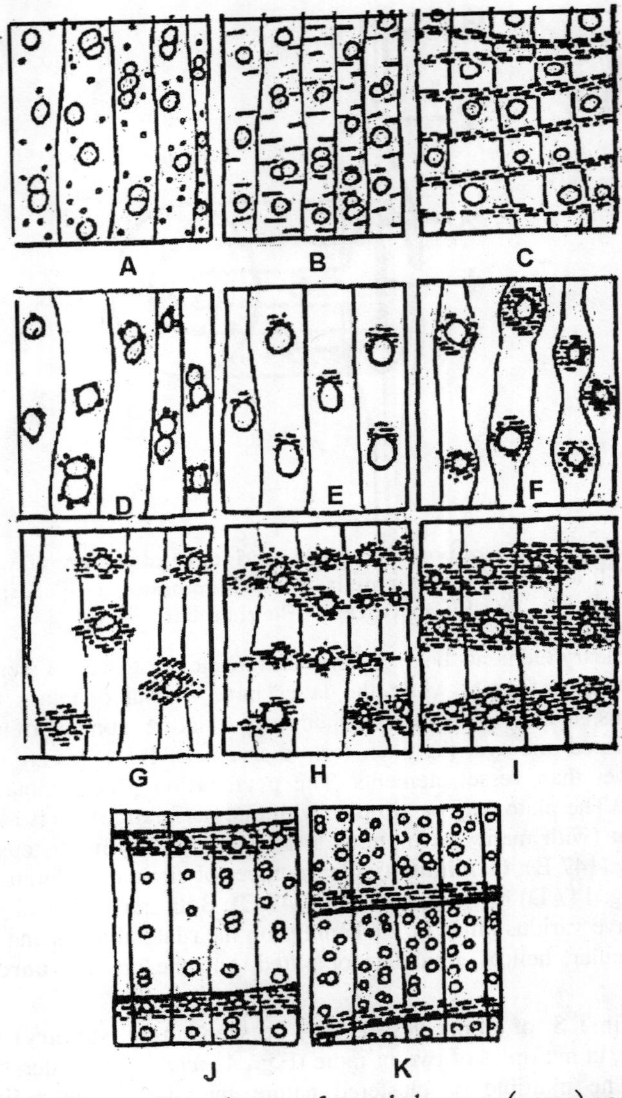

Fig. 149. Types of distribution of vessel elements (pores) and axial parenchyma; Circles represent vessels while dots, horizontal dashes and other shaded regions represent axial parenchyma. Horizontal lines indicate growth rings while vertical lines the rays A–K are parenchyma types. A - Diffuse; B - Diffuse aggregate; C - Apotracheal banded; D - Scanty; E - Paratracheal; F - Vasicentric; G - Vascentric aliform H - Vascicentric aliform–confluent; I - Vasicentric banded; J - Terminal; K - Initial.

The wood fibers are also variable in quantity depending on the taxa and may be almost absent or scanty (as in conifers) to abundant (as in some dicots). They may be **libriform** (thick walled), **septate** (less thick–walled and with thin transverse septa of variable numbers) or **gelatinous**. In some taxa, transitional cells between fibers and tracheids called fiber tracheids are seen (Fig. 68).

The wood rays, as stated earlier, constitute the horizontal systems of the wood. Rays are parenchymatous and vary considerably in their size. Their height is variable and normally several times greater than their width. The height in TLS varies from one to several cells, while width in TLS varies from a single linear row (**uniseriate**) to two rows (**biseriate**) or many rows (**multiseriate**). In dicots, the rays are either **homocellular** (**homogeneous**), where all cells of the ray are alike and radially elongated, or **heterocelluar** (**heterogeneous**), where there are two kinds of cells, **procumbent** cells and **erect** or **upright** (vertically elongated) cells. Both unicellular and multicellular rays may be either homocellular or heterocellular. A few conifers have **ray tracheids** that are intermediate between ray cells and tracheids.

Sap wood and heart wood

When more and more wood is being formed, the first–formed, more centrally–located wood gradually becomes functionless (i.e., do not conduct sap any more) leaving the peripheral wood part alone functional. The functionless part of wood is known as **heart wood** or **duramen** and the functional part is known as **sap wood** or **alburnum** (Fig. 150).The loss of function is also accompanied by the death of the axial and ray parenchyma and a change in wood density and colour. The death of parenchyma is preceded by altered metabolic rates, changes in the activity of many enzymes and changes in moisture content. For example, the enzymes phenylalanine deaminase and laccase show a progressively decreased activity. A clear–cut gradient is often evident in the activity of enzymes from the periphery towards the centre. However, some other enzymes show an increased activity along the same direction. Yet other enzymes show peak activity in the transition region between sap wood and heart wood. As examples for peak activity enzymes, we can cite peroxidase, sucrose oxidase, and tyrosinase. There is also an increased ethylene production in the transition region. The nuclei of wood parenchyma cells change in shape and size, lose vitality, and show a decrease in chromatin content and size. The cells of wood parenchyma lose their storage contents such as starch grains and organelles. In other words, they undergo PCD. The cell walls also become lignified.

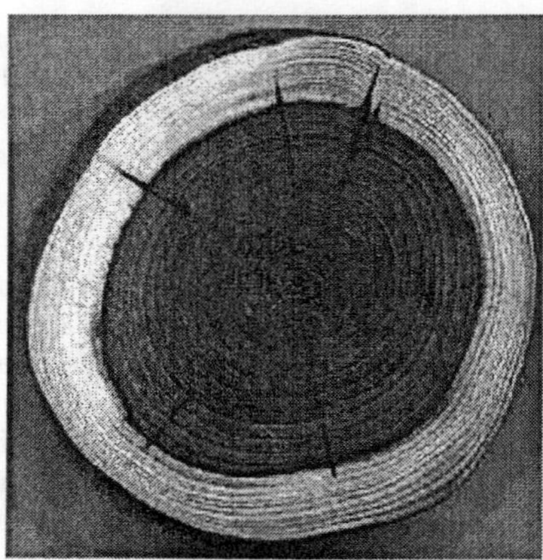

Fig. 150. T. S. of an old stem showing heart wood (coloured) and sap wood (white). Annual rings are also seen.

The amount and rate of heart wood formation varies greatly with species, tree age, rate of growth, environment and many other factors. Again, the age at which heart wood formation begins in a tree also varies greatly with species. In many species of *Eucalyptus,* it is initiated after five years, in *Pinus,* after 15–20 years, in *Fraxinus* after 60–70 years and in *Fagus* after 80–100 years.

The change from sap wood to heart wood may be gradual or it may be quite abrupt. In the first case, it is invariably possible to see a transition zone between the heart wood and sap wood. The transition zone is lighter than heart wood in colour and contains fewer living parenchyma cells than in the sap wood. This zone is very often known in literature as **Dry wood zone.** Rarely, the heart wood may not be formed at all or if formed, its quantum is very negligible. One of the characteristic features by which the heart wood is distinguished from sap wood is the difference in colour. The sap wood is usually white in colour and the heart wood is red, brown, yellowish or even black or in intermediate shade. However, it is not obligatory that the heart wood should have a different colour from that of the sap wood (e.g., *Ilex*). The change in colour of heart wood is mainly due to the deposition of phenolic substances (condensed tannins), but it may also be due to the deposition of other assorted dyestuff, flavonoids, oils, resins, gums, salts of organic acids, aromatic compounds, crystals, alkaloids etc. These substances may be deposited either in the cell lumen or on the cell wall.

The ray cells and axial parenchyma cells, before becoming dead, may intrude into the lumen of the vessel elements through bordered pits and grow as balloon–shaped structures called **tyloses** in the lumen (Fig. 151). In some instances, the tyloses appear collapsed and wrinkled. The nucleus and the cytoplasm also move into the tyloses. The tyloses either partially or fully occlude the lumen of the vessel elements. In some cases, the tyloses may undergo compartmentation inside the vessel lumen. The walls of tyloses may remain thin or get lignified. The tyloses often contain inclusions of various categories including calcium oxalate crystals.

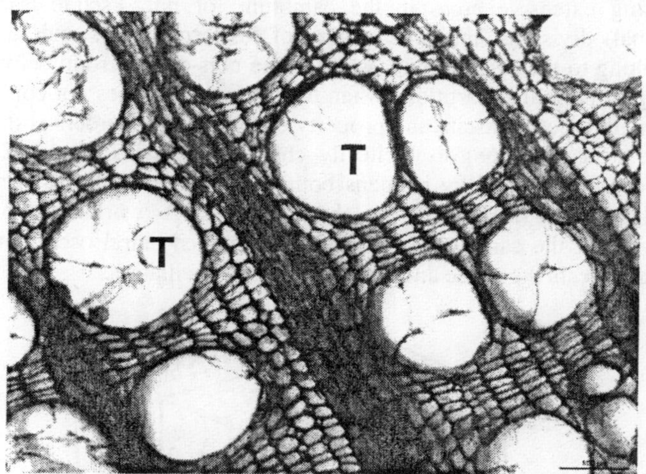

Fig. 151. Tyloses (T) in the vessels of *Gmelina arborea*

What factors are responsible for the sap wood region to become heart wood? Where the controls emanate from? There are two theories on the source of heart wood control. One theory emphasizes that the control emanates at the transition region. The high metabolic activity, including enzyme activity, of parenchyma cells in the transition region supports this theory. The second theory emphasizes that the controlling factors emanate in the cambial zone and these factors are transported through rays to initiate all heart wood changes in the transition region. The constant ratio of distance between the periphery of heart wood and of sap wood throughout the life of the axis is cited as an evidence for this theory.

Spiral grains

A brief mention must be made on **spiral grain** formation in wood of trees. The term 'spiral grain' refers to the alignment of the main conducting and structural elements of wood at an angle to the vertical axis of a tree stem (Harris 1981). The direction of the spiral grain may be to the left or right of the upper extremity of the longitudinal axis of the tree as viewed by an observer on the ground, and is thus described as left or right spiral grain or left or right grain angle. The spiral grains are not a defect in a tree, but rather are a natural phenomenon, which emerge from a variety of normal growth patterns. Although external causes have been implicated in spiral grain formation, evidences are found for its genetic control. Spiral grains are explained as follows (Hejnowicz 1990): Several developmental events in the vascular cambium are **chiral** i.e., they occur in two alternative orientations left (S) and right (Z). The major types of events are classified as whether they are **cyclic** (repetitive) and conform to the domain type (i.e., occupying extensive areas in the cambium) or not. Pseudo–transverse (anticlinal) divisions, intrusive growth of fusiferm initials and fusion of rays belong to the cyclic group, while to the non–cyclic group belongs the twisting of periclinal partitions when their orientation does not conform to the domain type. A fascinating property of the domain borders is that they migrate so as to belong to cyclically–changing domains (Fig. 152). This cyclic change of chirality happens both in time and space resulting in a travelling wave of chirality in the cambium. A working hypothesis proposes that the cambium is a system of coupled chiral oscillators. The coupling affects the phase and the period of the oscillations.

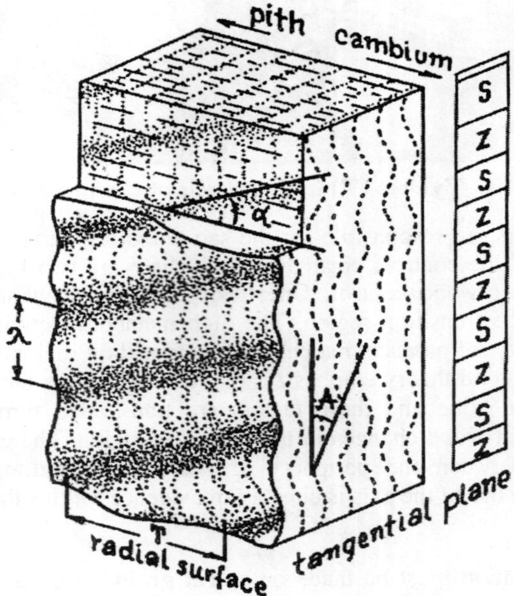

Fig. 152. Schematic diagram of a blocking wavy–grained wood with radial surface split and partially cut; A - Amplitude of the grain undulation or the maximum deviation of the grain from its longitudinal axis; λ - Length of the grain wave; α - Angle between the undulation crest or tough and ray direction; T - Period of grain undulation. Right - domain pattern that existed in the cambium when the wood on the tangential surface was formed (Hejnowicz 1990).

Bast

The bast or secondary phloem consists of **sieve elements (sieve cells** in gymnosperms and **sieve tube members** in angiosperm), **companion cells** (in angiosperms), **Strasburger cells** or **albuminous cells** (in gymnosperms), phloem parenchyma, phloem ray and phloem fibers (**bast fibers**) (Fig. 153).

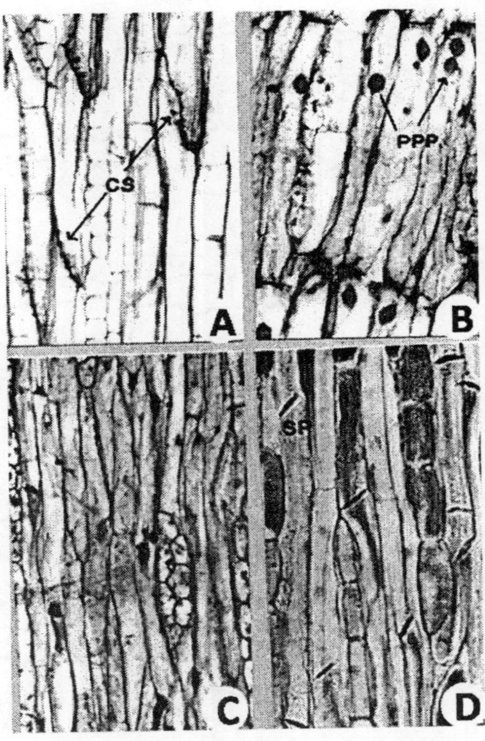

Fig. 153. Structure of bast (secondary phloem) in TLS; A - *Albizia amara;* B - *Dalbergia sissoo;* C - *Tectona grandis;* D - *Terminalia tomentosa;* CS - Compound sieve plate; PPP - Proteins; SP - Simple sieve plate. (Photographs courtesy of Dr. N. Venugopal)

In some cases, there are sclerieds instead of or in addition to fibers, secretory cells or canals, and idioblasts in phloem. All cell types, except the ray, of the phloem are derived from fusiform initials through periclinal divisions and constitute the vertical system (Fig. 153).

5.8.3. Phellogen

A lateral meristem closely related to vascular camnium is the **phellogen** or **cork cambium.** The meristem is also called **subero–phellodermal**

meristem. It is a lateral meristem located on the periphery of the axis of gymnosperms and dicots and has its first initiation, depending on the plant, in the primary phloem, primary cortex, subhypodermal layer or even epidermis. It generates through periclinal divisions on the outside **cork tissue** or **phellem** (also called **secondary suber)** and on the inside secondary cortex, also called **phelloderm** (Fig. 154).

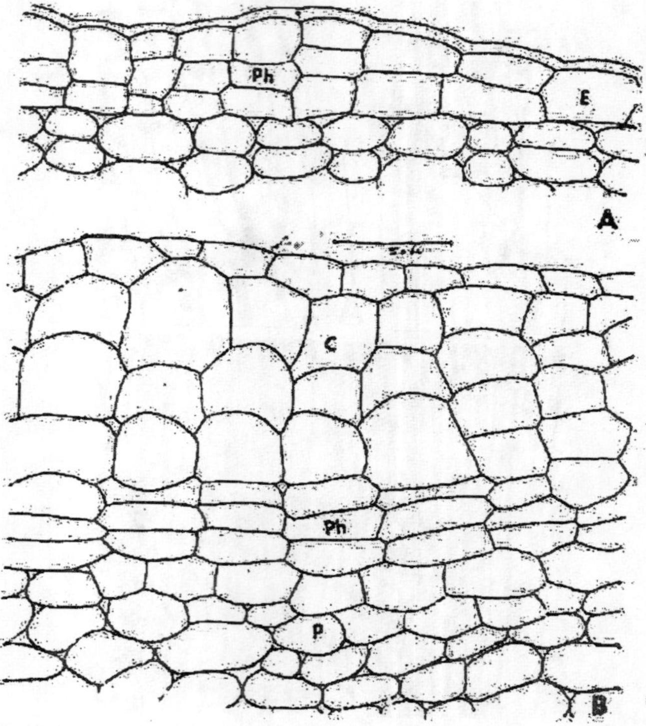

Fig. 154. A - T.S. of the peripheral part of the stem of *Ipomaea* species showing the initiation of phellogen (ph) from epidermal tissue (E); B - T.S. of the peripheral part of the stem of *Nyctanthes* species showing derivatives of phellogen (ph); cork cells (C) towards outside and phelloderm (P) towards inside

At some places, the derivatives of the phellogen on the outside form the **lenticels**, which facilitate gaseous exchange. The entire derivative tissue of the phellogen together constitutes the **periderm**. The phellogen originates soon after the secondary growth is initiated. It can function for sometime or it is often exfoliated and periodically replaced by more internal, newly-formed phellogens. New phellogens arise either in the phloem or in phelloderm.

5.9. ORIGIN AND DEVELOPMENT OF LEAVES

One of the characteristic features of the shoot system of vascular plants is the presence of leaves on the stem. Leaves are the vital components of the shoot system and are actively involved in photosynthesis and transpiration. The leaves are **appendicular** in nature and are produced on the stem. In very rare cases, the stems are leafless. A major question that is often asked is why leaves are formed at all. Not much is known as definite answers to the question, although a few suggestions have been put forward and tested. In this connection, an obvious question to ask is, why stems bear leaves, but roots do not? The major hypothesis proposed is that the surface of a shoot apex is under compression and therefore, tends to produce folds in the form of leaves. But, surgical experiments done on shoot apical meristems in the form of shallow incisions in the sites of the next leaf primordium showed that there is not much compression or tension in the surface of the shoot apex. This problem is yet to be resolved satisfactorily.

5.9.1. Phyllotaxy

The regular order of arrangement of leaves on the stem is called **phyllotaxy** or **phyllotaxis** (Jean 1994). The most common phyllotaxy is **spiral**, with the angle of divergence between successive leaves about 137.5°. The other types of phyllotaxis are **distichous** (with leaves disposed at 90° from each other in pairs at each node), **opposite superposed** (leaves in pairs in each node, with each pair aligned one above the other along the same horizontal axis), **opposite decussate** (each successive pair of leaves at right angles to the previous and next pairs) and **whorled** (with three or more leaves at each node). The majority of phyllotactic patterns noticed in flowering plants are based on spiral (helical) arrangement of leaves. The single spiral that can be drawn through connecting the centres of all leaves in the order of their succession is called the **ontogenetic** or **genetic spiral** (Fig. 155). This spiral was used by the early botanists to determine the numerical value of phyllotaxis. They expressed the phyllotaxis by a fraction in which the numerator is the number of leaves intervening between two vertically superimposed leaves. In the figure 155, the phyllotaxis thus calculated will be 5/18 (see attachment of leaves 14 and 27, or 15 and 28 in figure 155c). The common fractions seen in plants are 1/2, 1/3, 2/5, 3/8 and the less common fractions are 5/13, 8/21 etc. This series of fraction belongs to the so–called **Fibonacci summation series** 0, 1, 1, 2, 3, 5, 8, 13, 21 (Figs. 156, 157 A and B). The numerator and denominator of each succeeding fractions is the sum of the numerators and denominators of the two preceding fractions. Each of these fractions represents the angle intervening between the centres of successive leaf primordia. This angle is approximately 137.5° for all fractions, as mentioned earlier. Hence, with very few exceptions, all genetic spirals are probably alike.

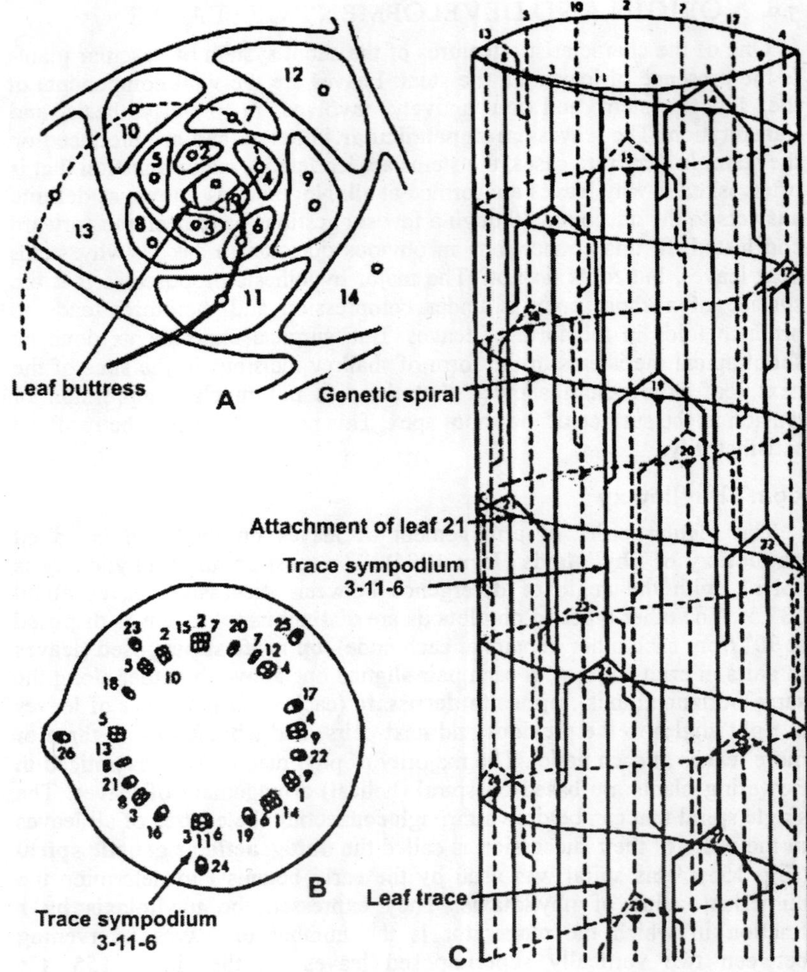

Fig. 155. Diagrams of the primary vascular system of *Hectorella caespitosa* in T.S. (A and B) and in three-dimensional view (C); A - Stem apex (a) and leaves 1–14, in each of which the midrib is indicated by a circle. The two curved lines (one broken, the other double) indicates a pair contact parastictics; B - Stem is cut near node of leaf 23. The traces of cut above the level of their connections with other traces are indicated by blackened phloem regions; C - Diagram showing interconnections of leaf traces (Skipworth 1962).

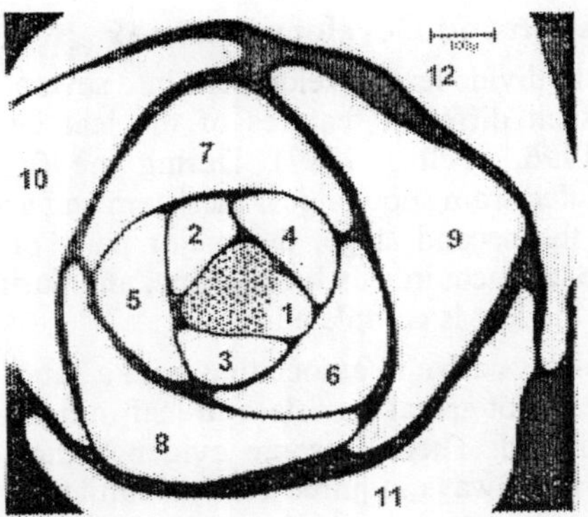

Fig. 156. T. S. near shoot apical meristem of *Saxifraga* (stippled) showing leaves 1–12 cut transversely; it shows 2 + 3 phyllotaxis (Clowes 1961).

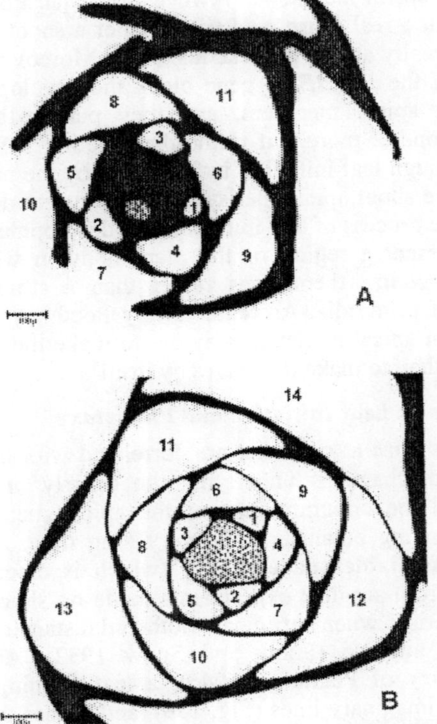

Fig. 157. A - T. S. near shoot apical meristem (stippled) of *Sempervivum tectorum* with a 3 + 5 phyllotaxis; B - T. S. near shoot apical meristem (stippled) of *Saxifraga* species. The youngest leaves have a 3 + 5 phyllotaxy, while the older have a 2 + 3 phyllotaxis (Clowes 1961)

5.9.2. Leaf Initiation and Development

It is convenient to divide leaf development into several stages depending on the time at which different features of the leaf become determined (Sylvester et al 1996, Poethig 1997). During the first stage, the leaf primordium is initiated from shoot apical meristem and acquires its identity as a leaf. During the second stage, the major parts of the leaf become determined and the leaf acquires its basic shape, and during the final phase the histogenesis of the leaf is completed.

Leaves almost always arise in association with a shoot apical meristem. Yet, the role of the shoot apical meristem in leaf initiation has never been very clearly established. There is some evidence that actually a shoot apical meristem is not always required for leaf initiation. In certain plants such as *Begonia* and water cress, leaves or leaf–like appendages can develop in the absence of a shoot apical meristem, and also in calli derived from tissue cultures, leaves sometimes develop in the absence of an associated shoot apical meristem. However, it is questionable (i) whether they are leaves in a real sense, and (ii) whether a shoot apical meristem–like structure is really absent in these instances. Moreover, in *Arabidopsis*, mutant alleles of the *PINHEAD* gene block the development of a fully–functional shoot apical meristem, and they permit the initiation and development of one or more leaf primordia (Poethig 1997). These results suggest that although leaf initiation is facilitated by the unique physiology or structure of the shoot apical meristem, the meristem does not appear to actively direct the process of leaf initiation and development. The meristem is likely to represent a region of the plant body in which leaves may *spontaneously organize* themselves rather than a structural entity that *produces* the leaf primordia. As remarked by Poethig (1997), one of the functions of shoot apical meristem may be 'to make the tissue that makes the leaves rather than to make the leaves by itself'.

5.9.2.1. *Theories on Leaf Initiation and Phyllotaxy*

Leaf primordia arise at sites that are correlated with the phyllotaxis of the shoot. The mechanisms underlying the orderly production of leaf primordia around the circumference of the apical meristem have been highly debated among botanists for a very long time. According to the **Next** (or **first**) **available space theory** (which is essentially based on results obtained from surgical experiments made on shoot apices), a new leaf primordium arises when sufficient width and distance from the summit of the apex are attained (Snow and Snow 1932). According to the **Phyllotactic theory** of Plantefol (1946), a leaf–forming impulse move acropetally along imaginary lines (Fig. 158), and causes the initiation of a leaf at the place of the shoot apical meristem where it ends. In most dicots, there are said to be two **foliar helices** starting with each of the two cotyledons, but the number is not invariable and can change during ontogeny. Also in monocots, the number of foliar helices is variable and can be several hundreds. Plantefol's theory has not been supported outside France. According to the **Physiological field theory** (Wardlaw 1949) or **Repulsion theory** (Richards 1948), as each new leaf is initiated in the meristem, it is surrounded by a physiological field within which the initiation of another new primordium is inhibited until the position for the next leaf primordium comes to lie outside the existing field. In other words, a kind of repulsion spreads from the summit of the apex and from the centres of each of the existing primordial, which prevents the formation of new primordia. At places where this repulsion is below a critical value, a new leaf will arise. The repulsion force can be a growth inhibitor or a competition for growth promoting substances, but there is no information on this point. According to this theory, since the repulsion is emanating from points (i.e., centres of existing primordia and summit of meristem), the leaf about to be formed is also determined at a point in its centre and when the site of the centre is determined, the rest of the leaf base is then decided. In the Next available space theory mentioned above, on the other hand, the leaf primordium must first be determined as a whole and not just at its centre. Also, it is the whole of each existing primordium, which restricts the space available for future leaves.

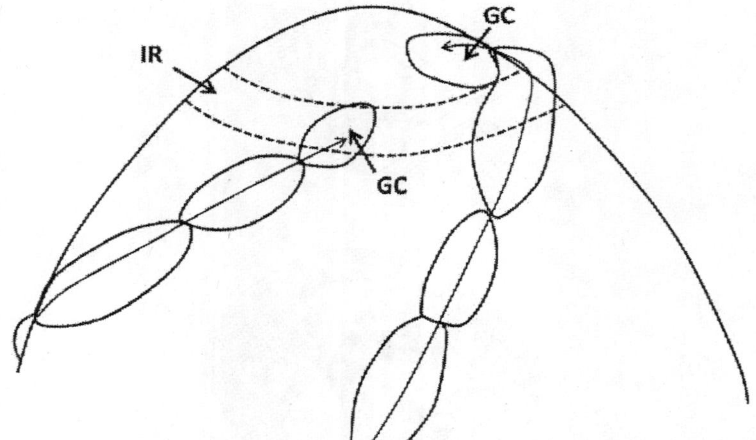

Fig. 158. Vegetative apex interpreted in accordance with the concept of foliar helices and contiguity; Two foliar helices are represented, each one proceeding from a generating centre (GC), moving in the initiating ring (IR) (Plantefol 1947)

Green and his team proposed two biophysical mechanisms for the regulation of leaf initiation. The first model (Green and Lang 1981) suggests that the key event in leaf initiation is the production of a field of cells in which cellulose microfibrils are in a roughly circular arrangement. These circular microfibrils are believed to prevent cells from expanding laterally, thus, forcing the primordium to expand out of the plane of the apical meristem. In a meristem with pre–existing cells, the site of the next leaf primordium is specified by the way in which these pre–existing leaves modify the cellulose pattern within cells of the shoot meristem. Green (1986; see also Selker et al 1992) proposed the **Biophysical forces theory**, according to which biophysical forces in the growing apex determine leaf primordium initiation sites; a new primoridum is initiated when a region of the tunica surface bulges or buckles, a condition brought about in part by a localized reduction in the surface layer's ability to resist pressure from the tissues below (Green 1999); loops of cellulose microfibrils reinforce the walls of these tunica cells and create the favourable conditions for leaf initiation. It is suggested that local stress variations created by buckling trigger the periclinal divisions commonly associated with leaf formation. This hypothesis is supported partly by studies in which localized application of **expansin** (an extracellular protein that increases cell wall extensibility) to the shoot apical meristem of tomato induced the formation of leaf–like outgrowths. It is likely that expansin promotes cell wall extensibility in the outer cell wall layer of the tunica resulting in outward bulging of the tissue (Fleming et al 1997) (Fig. 159).

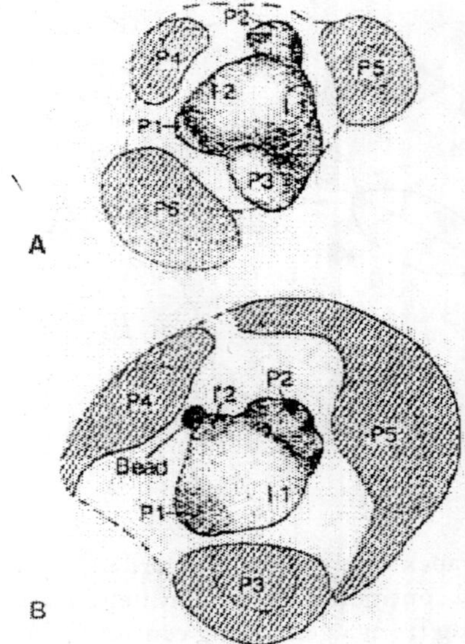

Fig. 159. Induction of leaf primordia on the shoot apical meristem of tomato by expansin; A - Diagram of the meristem, which has generated leaf primordia P6 (oldest) to P1 (youngest); The next leaf primordia will arise at positions I1 and I2; B - Diagram of the meristem five days after placing expansin–loaded beads. A bulge I'2 is present on the meristem near the site of the expansin–soaked beads (Fleming et al 1997)

In situ hybridization studies have indicated that expansin genes are specifically expressed at the site of leaf primordium initiation in tomato and rice plants (Reinhardt et al 1998). The attractive aspects of the biophysical theories are that they provide a mechanism not only for the spatial positioning of leaf primordia, but also for leaf primordial initiation. The latter aspect is not covered by chemical theories i.e., how a leaf is made is not addressed. On the contrary, there is no evidence to demonstrate that the orientation of cellulose within the cells of the shoot apex or in a leaf primordium actually regulates the rate of orientation cell expansion. Moreover, while the second biophysical explanation predicts that the leaf primordium is subjected to compression to start with, most studies have shown that cells in leaf primordial rudiment are either under tension or show no evidence of being under any kind of mechanical stress (Selker et al 1992).

The **Procambial strand theory** of Larson (1983) suggests that the procambial strands decide the location where a new primordium is to be intiated. The phyllotaxy is correlated with the architecture of the vascular system of the stem so that the spatial relation of the leaves to one another is part of the overall pattern in shoot organization. The developmental relationship between the leaves and the leaf traces in the stem suggests that procambial traces associated with the prospective leaf primordial sites apparently provide a transport pathway for auxin or similar substance that promotes leaf primordium initiation. In *Arabidopsis*, the precocious leaf procambial traces is detected as a high density region of expression of the gene *PINHEAD (PNH)*. This expression of *PNH* preceded down–regulating *STM* at the leaf primordial site and hence, may be regarded as an early marker of leaf initiation than the loss of *STM* expression.

The above and other chemical models for leaf inception suggest the presence of promoting and inhibiting chemicals and their interaction, which is controlled by one of several diffusion–reaction mechanisms (Meinhardt 1984, Nelson and Dengler 1997). Several investigators have implicated auxin in the regulation of phyllotaxis (Cleland 2001). If the vegetative shoot apex of tomato is cultured on a synthetic medium containing a specific inhibitor of auxin, leaf production by it is completely suppressed (Reinhardt et al 2000). But, microapplication of IAA to the surface of such an apex restored leaf formation. In intact plants, auxin is supplied by the developing leaves or leaf primordia. A model has been proposed by Stieger et al (2002), according to which auxin efflux carriers control the delivery of auxin to the apical meristem, while influx and efflux carriers regulate its distribution within the meristem. It is further proposed that the efflux carrier plays a role in the redistribution of auxin within the meristem and the influx carrier presumably is needed for correct phyllotaxy. Cytokinin alone or a combination of cytokinin and auxin has been known to induce leaf intiation in a variety of tissues suggesting a role for cytokinnins in leaf initiation process. In this context, it is interesting to note that the *Arabidopsis* cytokinin–resistant mutant *Cyr 1* produces few or no leaves (Deikman and Ulrich 1995). The above described chemical models, although, are very good at generating spatial patterns of leaf initiation and maturation, the existence and identity of these signalling factors have not been proved beyond reasonable doubt. Hence, it is not clear how to evaluate the accuracy or usefulness of these chemical models (Poethig 1997). Whether these chemicals act as promoters or inhibitors (or as both) is also not very clearly established.

Callos et al (1994) found a mutation in *Arabidopsis* called *forever young (fey)* causing anarchic initiation of leaves in a totally different phyllotactic pattern. The *FEY* gene encodes a protein showing a significant homology to **nodulin** (found in root nodules of legumes) and a limited homology to the enzymes reductases and dehydrogenases. Although the role of this protein is not yet clear, it is likely that it regulates leaf formation by modification of a signalling molecule. Studies on *Arabidopsis*, maize and a few ferns indicated the genetic basis for phyllotactic determination and for the determination of the leaf primordium itself. For instance, down–regulation of the *KNOTTED 1 (KN1)* class of homebox–containing plant genes, which were originally identified in maize, provides an early molecular marker of leaf initiation in the shoot apical meristem (Sinha 1999, Brutnel and Langdale 1998). The *KN1* gene in maize is specifically down–regulated at the site of leaf initiation, as shown by in situ hybridization and immunocytochemical methods. The expression of this gene is high in the apical meristem itself. The localization of the KN1 protein is observed to correspond with the boundaries of the *KN1* gene expression. These observations have led to the view that the *KN1* gene may function to maintain the apical meristem in an indeterminate state, and its down–regulation in the site of future leaf primordia may have important consequences in assigning the leaf's fate as a determinate organ. In agreement with clonal and anatomical investigations, the expression of this gene is in the lateral veins of the knotted leaves. Down–regulation of the *KNOTTED1* class genes, *KNAT 1* and *STM 1*, in *Arabidopsis* also marks the sites of leaf primordial initiation. It is suggested that *HBK1* gene, which is found in the shoot apical meristem of *Picea alba*, may also play a role similar to the *KNOTTED* genes in angiosperms.

Phyllotaxy appears to be a non–mutagenic trait. Although many genes with a regulatory function in leaf formation and positioning are known, genetic screens have mostly failed to yield mutants with specific phyllotactic phenotypes i.e., homeotic transformations of one phyllotactic pattern into another.

5.9.2.2. *Early Stages in Leaf Initiation*

Leaf primordia come into being following a change in the orientation of growth and division of cells in both the tunica and the corpus of the flank meristem (peripheral zone) of the shoot apical meristem. The first histological events commonly associated with primordial initiation are changes in the rates and planes of division of the flank meristem cells resulting in the formation of a protrusion called the **leaf buttress** (Figs. 89 and 160) on one side or both sides of the axis depending on phyllotaxy type. This respectively results in an asymmetric or symmetric enlargement.

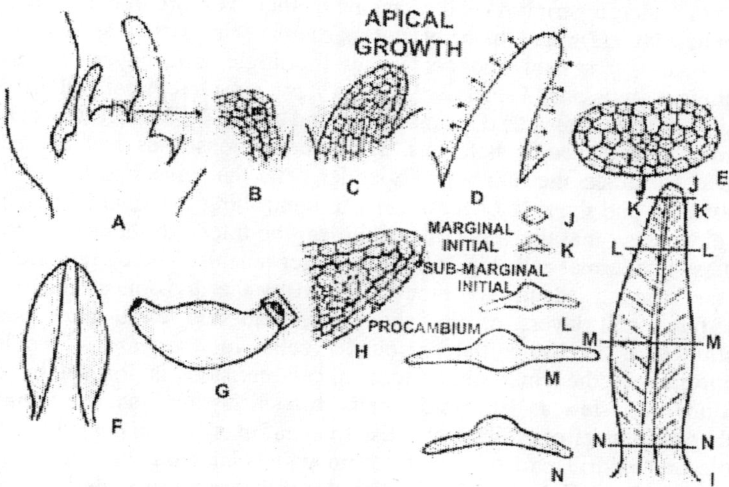

Fig. 160. Diagrammatic figures of development of leaf as per conventional concept; A - L. S. shoot apex; B - Portion marked in A, enlarged to show periclinal division of a cell in the second tunica layer; C - L. S. of older leaf primordium; D - Still older leaf primordium showing directions of growth (arrows); E - T. S. of D; F - Entire older leaf primordium at the time of lamina formation. G - T. S. of F; Dark areas represent areas of marginal meristem activity; H - Enlargement of portion marked in G showing marginal and submarginal initials; J– N represent transections at the levels marked in I.

The period or interval between the initiation of two successive leaf primordia (or pairs or whorls of primordial depending on the type of phyllotaxis) is called a **plastochron** (also spelled **plastochrone**). The changes in the appearance of the shoot apex occurring during one plastochron are **plastochronic changes.** There is a **maximal phase** when the leaf primordium is about to arise on one side of the meristem and a **minimal phase** when there is no primordial initiation (Figs. 89 and 91). The duration of a plastochron depends on the plant and also, in the same plant, depends on environmental conditions (especially light and temperature) or age. A *PLASTOCHRON 1 (PLA1)* gene has been known from rice plant; this gene is implicated in the regulation of duration of vegetative phase by controlling the plastochron intervals. Sometimes, the term **phyllochron** is used to refer specifically the interval between the visible appearance or emergence of successive leaves.

As indicated earlier, the initiation of a leaf primordian is associated with an increase in the frequency of periclinal divisions of the outer corpus cells of the flank meristem at the site of initiation. Invariably the tunica layer does not undergo periclinal cell divisions. If there is more than one tunica, the inner layer(s) may or may not participate in periclinal divisions. Occasionally, as in many monocots, even the single tunica layer undergoes periclinical divisions. Generally, only a few cells are involved in the division to result in a leaf primordium and these cells are often called **founder cells** by some botanists. However, a consensus has not been reached so far on the number of founder cells that give rise to the leaf primordium and the role of these cells in controlling the final leaf shape. The older view that the leaf primordium can be traced to the activity of a single subepidermal cell (Fig.160) is no larger tenable. The current view is that the growing leaf primordium is formed by the division of cells in at least three cell layers of the shoot meristem and each layer being constituted of 50–200 initial or founder cells. In *Arabidopsis,* the leaf primordium in the embryonic shoot apical meristem is formed by the divisions of as few as 10 founder cells in each layer. Thus, the variable number of founder cells first gives rise to a leaf axis (which constitutes the petiole and midrib part of a leaf), from which later on the lamina gets developed. According to some, the founder cells are, in turn, derived from at least a few **pre–founder cells**. The founder cells express *ZWILLE/PINHEAD (ZWI/PIN), PIN–FORMED 1 (PIN1/AUX1)* and *REVOLUTA (REV)* genes. The boundary of the founder cells is characterized by the expression of genes such as *CUC*. The polarized leaf primordium expresses genes such as *AS1, LFY, ANT, CUC/PID/LAS, KANADI/YABBY* and *PHB/PHV/REV*. The growth of the leaf axis in length is more apical to start with, but becomes diffuse later. In some monocots, the length in leaf axis is also contributed by an intercalary growth at the base of the leaf axis. However, some people have questioned the causal relationship between periclinal divisions and initiation of the leaf primordium. The gamma plantlet of wheat studied by Foard (1971) supports a non–causal relationship. In this plantlet, a polarized cell elongation unaccompanied by cell divisions accounts for the initial protrusion of the leaf primordium and hence, a slight shift in the polarity of expansion of a group of founder cells already determined rather than a change in their division plane might be responsible for leaf primordial growth.

5.9.2.3. *Determination of Leaf Primordia*

Most information on leaf determination and commitment of leaf primordia to develop into leaves (i.e., autonomy) come from tissue culture and surgical experiments on fern leaf primordia and shoot apices. Culture of primordia that have already been determined as prospective leaves has been done in *Osmunda cinnamomea* and a few other ferns. These develop into small leaves, but in a developmental pathway similar to fronds on intact plants (Steeves 1959). This proves that such leaf primordia are determined in advance over differentiation, thus proving their autonomy. But, culture of still younger primordia resulted not in leaves but in leafy shoots, and in the latter instance, young leaf primordia acted like apical meristems. It is clear that no irreversible change has occurred in the leaf primordium at the time of its inception, but the primordium acquires the ability to develop into a leaf independent of any control from the rest of the plant later in development. Surgical experiments, in which the centre of the shoot apical meristem or very young primordia from the fern shoot apices is isolated by incisions and allowed to complete development on the plant, have supported the view that the leaf primordium acquires autonomy only some time after its inception. In general, cultured leaf promordia and surgically isolated leaf primordia of angiosperms show a great degree of propensity to reorganize themselves and give rise to leaves, showing that the commitment of the leaf primordia to mature into leaves is more binding in angiosperms than in ferns (Steeves and Sussex 1989). It should, however, be pointed out that leaf primordia are not determined at the level of the whole organ in plants such as species of *Impatiens* in which floral meristem after initiation of flowers reversed to vegetative growth due to altered photoperiodic treatments.

5.9.2.4. *Post–Initiation Development*

Extensive studies have been carried out on a wide variety of species on leaf development and histogenesis after initiation of the primordium. A detailed analysis of leaf development in angiosperms has indicated much difference of opinion and has resulted in two classical views: (i) The final shape with fully differentiated tissues is determined by a series of properly oriented or directed cell divisions closely accompanied by appropriate patterns of cell elongation throughout the initial leaf primordium (i.e., not restricted to marginal meristems). (ii) Subsequent to the cessation of apical growth of the leaf primordium, the lamina is produced by specific **marginal** or by both marginal and **submarginal meristems** along the lateral primordial leaf axis margins (Figs. 160 and 161). The histological appearance of these marginal (and submarginal) meristematic regions is associated with the formation of the lamina. A number of patterns of derivation of different laminal tissues have been recognized (Fig. 162)) in different plants leading to the evolution of different leaf anatomies.

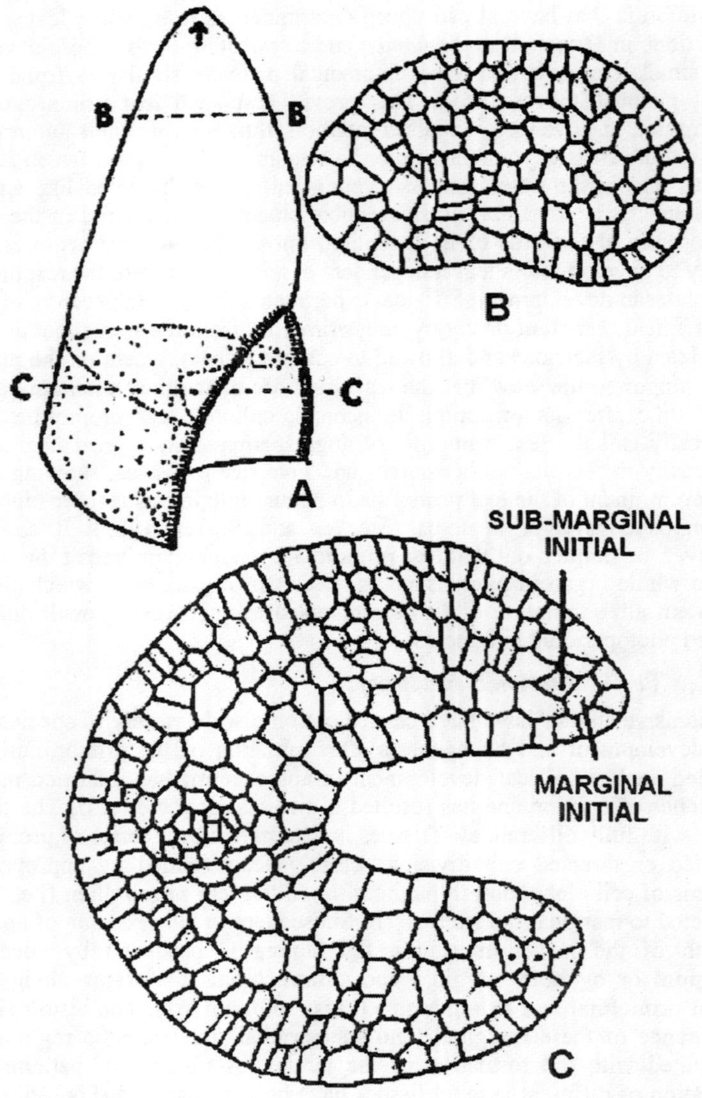

Fig. 161. Some stages in the development of the leaf of *Comelina benghalensis*, a monocot; A - An entire young leaf primordium showing sheathing leaf base and the regions with intercalary growth (stippled) and marginal growth (hatched); Arrow indicates apical growth; B and C - T. S. of A at places marked BB and CC respectively; note the marginal and submarginal initials.

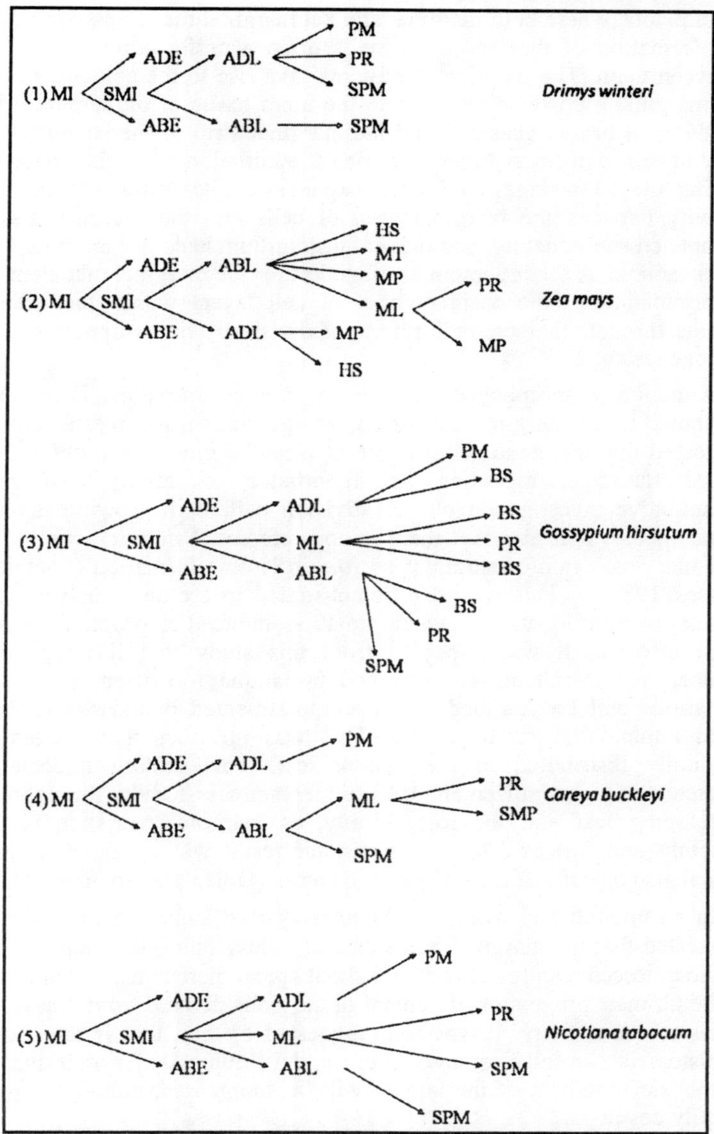

Fig. 162. Different patterns of leaf histogenesis from marginal meristem strip (MI - Marginal initial; SMI - Submarginal initial; ABC - Abaxial epidermis; ABL - Abaxial layer; ADE - Adaxial epidermis; ADL - Adaxial layer; BS - Bundle sheath; HS - Hypodermal sclerenchyma; ML - Middle layer; MP - Mesophyll; MT - Midrib thickening; PM - Palisade mesophyll; PR - Procambium; SPM - Spongy mesophyll;

In plants, where both marginal and submarginal meristems are active in the formation of the lamina, there is often a perfect division of labour between them. The marginal meristems give rise to the epidermis and the submarginal meristems give rise to the inner tissue of the lamina. Merrill (1986 a, b) has emphasized that the marginal strip of meristematic tissue may have a significant function during the initiation of the lamina and not during the later stages of lamina expansion. The distinction in mitotic activity between the marginal strip of cells and the proximal tissue is dampened subsequently, and the leaf primordium becomes more uniformly meristematic. A second important function for the marginal meristem is the determination of the basic number of cell layers within the extending lamina through characteristic planes of division. This is important in leaf histogenesis.

Although a morphogenetic role in lamina formation is frequently attributed to the marginal meristems, a large number of investigators have indicated that this zone is not a site of elevated mitotic activity and have raised questions regarding its importance in lamina development. Quantitative investigation on cell division pattern in growing lamina of some species indicates that there is no preferential mitotic activity in the leaf margins, again questioning its role in lamina formation (Poethig and Sussex 1985 a). This was also demonstrated in a clonal analysis of leaf development in tobacco in which radiation–induced chlorophyll mutations were effected. It was expected from this study that if marginal and submarginal meristems are involved in lamina formation, sectors with mutations will be confined within compartmented boundaries extending from laminal margin to midrib. This was not seen and sectors were randomly distributed in the mature leaf. This pattern indicated the existence of a generalized and diffuse meristematic activity throughout the developing leaf and the role, if any, of marginal meristem is minor (Poethig and Sussex 1985 b). A similar result was obtained through a clonal analysis of leaf development in cotton (Dolan and Poethig 1991).

In an unpublished work, Krishnamurthy (the author of this book) has suggested that the marginal meristems of a developing leaf lamina is akin to the quiescent centres of root and shoot apical meristems, and although it is the ultimate progenitor of laminal tissue, cells derived from it have more meristematic activity. It was also suggested by him that if the marginal meristem is carefully removed from a developing leaf primordium, the further development of the lamina will be suppressed, although a petiole usually develops.

Many investigators advocate the presence of several more or less discrete developmental domains in a leaf during its initiation, along the dorsoventral, centrolateral and proximodistal axes of the developing leaf (Poethig 1997). The dorsiventral symmetry, which is apparent as soon as the leaf emerges from the shoot meristem, is due to the generally flatter adaxial than abaxial side of the primordium. Subsequently, the dorsoventral axis of the leaf is defined by the pattern of cellular differentiation, with some leaf tissues confined to its dorsal and some to its ventral sides. The demarcation of the leaf into central and lateral domains is marked by the differentiation of a distinctive band of cells along the lateral margins of its primordium. Hagemann and Gleissberg (1996) have called this region of the leaf primordium the **'blastozone'**, because the more ususal term 'marginal meristem' implies features that are not observed in all species. Cells in the blastozone expand laterally to form the lamina or pinnae, whereas the central region of the primordium differentiates into the midrib or rachis. The proximodistal domains are defined by the way in which the blastozone develops. Along this axis, a leaf can usually be divided into a distal region, which produces a relatively broad lamina, a proximal region (the petiole), in which the lamina is either absent or reduced and a basal region, which extends around the stem as **phyllopodium** to a greater or lesser extent.

Importance of mutants in understanding leaf development

It is interesting to note that most mutations that affect leaf initiation also affect major aspects of leaf development and morphology. The absence of ordered cell divisions during leaf development has been demonstrated by a study of the *tangled 1 (tan 1)* mutant of maize. A separation between cell divisions and leaf laminal shape during its extended growth has been demonstrated in this mutant. Contrary to the wild plant in which cells divide both transversely and longitudinally as the leaf grows to its final size and shape, randomly oriented cell divisions replace the longitudinal divisions in the mutant leaves. These aberrant cell divisions slow down the rate of leaf growth, but the overall leaf shape is attained at the end (Smith et al 1996). These results indicate that precise, spatially–oriented cell divisions appear to play a relatively minor role in the elongation growth of leaf primordia and in the final shape of the leaf. Two mutations, *lam 1* and *fat,* in *Nicotiana sylvestris* have been shown to modify leaf morphology through their control on the orientation of cell divisions in leaf primordia. In the first mutant, the leaves are bladeless throughout development and anticlinal divisions proceed normally in the L1 and L2 layers of the primordium, but periclinal divisions in L3 layer, responsible for mesophyll development, are arrested. In the second mutant (*fat*) additional periclinal divisons take place in the mesophyll during lamina formation and expansion, producing abnormally thicker lamina. Hence, it is likely that *FAT* gene functions as a negative regulator of periclinal divisions (McHale 1992). In *Arabidopsis,* certain mutant leaves are narrower and thicker or shorter and rounder than the wild–type leaves and an analysis of these has provided evidence for the existence of genes that control polarity–specific cell elongation along the directions of width and length of the leaf respectively (Tsuge et al 1996, Bohmert et al 1998). *ARGONAUTE* (AGO) gene is one among these genes and this controls the lateral expansion of leaves through its unidentified soluble protein product (Bohmert et al

1998). Another gene is *ROTUNDIFOLIA (ROT)* and this is involved in the longitudinal growth of leaves. It belongs to a class of steroid–producing cytochrome P–450 (Kim et al 1998). The actual mode of action of these protein products is not clear.

The most important feature of a leaf is its dorsiventral character with growth retardation on the abaxial or ventral side in comparison to the adaxial or dorsal side assumed to occur very early in leaf ontogeny. A simple model for dorsiventrality has been developed after studying *phantastica* (*phan*) series of mutations of *Antirrhinum majus*. The mutant plants produce a great variety of leaf morphologies among which the most characteristic is the needle–like morphology lacking dorsiventrality, which has an apparently normal initiation, but during subsequent development, concentric layers of cells, normally belonging to the ventral side, fill the leaf. The above results support the view that an inductive influence (dorsalization) of the *PHAN* gene product is needed to specify the identity of dorsal cell types of the developing leaf (such as dorsal epidermis, palisade and spongy mesophyll) (Waites et al 1998). The *phabulosa* (*phb)* mutant of *Arabidopsis* shows the transformation of abaxial leaf fates into adaxial leaf fates and the development of radially symmetrical, laminaless leaves, thus emphasizing the view that juxtaposition of adaxial and abaxial cell fates is essential for lamina growth (McConnell and Barton 1998). A mutation in the *LEAFBLADELESS* (*LBL*) gene of *Zea mays* produces diverse leaf morphologies and an extreme manifestation of this is the production of radially symmetrical structures (Timmermans et al 1998) and hence, like the *PHAN* and *PHB* genes, the *LBL* gene also appears to be necessary for the adaxial cell identity maintenance in the developing leaves.

Dominant mutations of the *KN1* gene in maize also affect development of the leaf blade. These lead to chaotic growth or knobs, which are generally confined to the lateral veins in discrete regions of the leaf and which can be traced to localized sites of extra cell divisions and expansions in the cell layers of the leaf. A few more morphological abnormalities may also be noticed among different stocks of *KN1* mutant. The expression of the *KN1* gene in transgenic tobacco plants causes alterations in leaf size and shape. Since the gene is believed to function in maintaining cells in the indeterminate pathway, over–expression of the gene resulting in ectopic shoot formation on the leaf surface, in addition to leaf lobbing, is seen as expected. The leaf also changes from dorsiventral to a radially symmetrical one that is characteristic of a stem. Yet, the nature of the information that regulates leaf shape remains elusive to a large extent even today.

It is pertinent to mention here the slightly variable abnormal characters found in the leaves of *rs* mutants of maize. It induces proliferation of a sheath–like tissue at the base of the leaf blade, and defects in vein morphology, proliferating mesophyll cells and abnormal cell shape. Similar to *KN1* gene, both *RS* and the *KNOX* genes (another member of the maize homebox gene family) are not expressed in the leaf primordia, but in the shoot apical meristem preceding leaf initiation and subsequently as a ring at each leaf insertion (Jackson et al 1994). Interestingly, it has been found that *KNOX* gene is initially repressed in the leaf founder cells flanking the shoot apical meristem of both the *rs* mutant and wild–type maize. Thus, it is clear that the acquisition of leaf cell fate depends upon the down–regulation of *KNOX*–like genes and the *RS* gene also plays a role in this event. Based on all results obtained from the analysis of mutants, it appears that a general strategy of possible action is that the fate of cells of the apical meristem may be defined by the independent or interactive functions of *KN1, KNOX, RS* and other similar genes. The activity of these genes is not exclusive to leaf initiation, since similar changes in their expression patterns are noticed during floral meristem formation.

Development of lobed simple leaves and pinnately compound leaves

Changes in leaf shape are initiated during early stages of development. Hagemann (1973) proposed the concept of **meristem fractionation**, in which localized zones of meristematic activity become separated by regions of vacuolated cells, to describe the mechanism giving rise to lamina lobes and leaflets. For example, the mitotic index of incipient leaf lobes within the lamina margin of *Tropaeolum peregrinum* is significantly higher than in adjacent regions. After lobe formation mitotic index becomes more uniform throughout the lamina. However, some other botanists believe that lobes and leaflet primordia (in pinnately compound leaves) are produced in close proximation from 'generative centres' located at distal or proximal regions of the incipient blade. As per this model, the shape changes associated with leaflet formation can be recognized in a uniformly meristematic zone before 'fractionation' takes place. Whereas lobes of leaves are usually formed as part of an already defined lamina margin, leaflets are usually formed directly on the axis of the leaf primordium. Leaflet primordia and whole leaf primordia have similar forms initially and then undergo the same process of lamina formation. In compound leaves, the pattern of lamina formation is identical to that of simple leaves, but the timing is delayed.

5.9.2.5. *Differentiation of Leaf Tissues*

The differentiation of cell types of the leaf is essentially related to the adaptations of plants to various habitats and environmental factors, especially light and water (Dengler 1994). Whatever be the leaf shape and size, it has basically an epidermal tissue, mesophyll and vasculature. The stages in the differentiation of the different laminal tissues are illustrated in figure 163. Epidermis is the ultimate extension of the tunica layer of the shoot apical meristem, but whether all or only a few selected primordial cells of L1 layer of the flank meristem are involved in the formation of the primordial epidermis. A fate–mapping study of maize leaf with reference to cell lineage of the epidermis was carried out by Carioli et al (1994) and this study showed that at least four cells of the leaf primordium, all derived from one progenitor cell, are involved in the determinative decisions leading to the formation of the epidermis in a region of the leaf bordered by two lateral veins. This study shows that the leaf epidermis is constructed of a number of clonal compartments, each separated by a vein, and cells belonging to one domain do not spill over to the next domain during development.

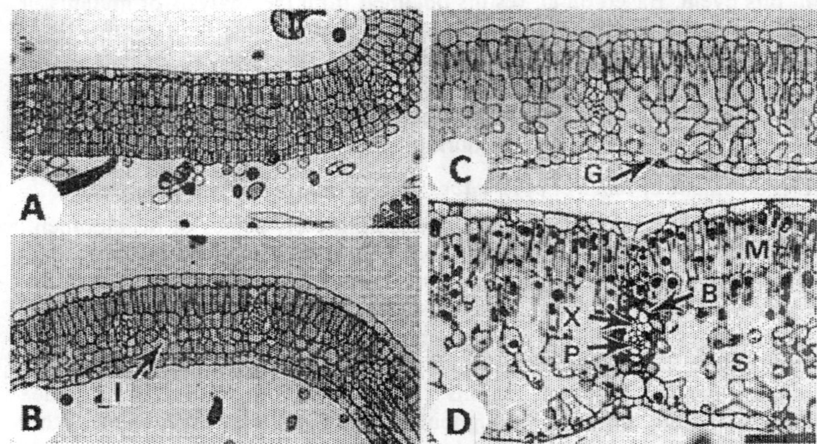

Fig. 163. T. S. of developing leaves of *Fagus grandifolia* illustrating the differentiation of different laminal tissues (Dengler et al 1975)

The ground tissue of the leaf (i.e., mesophyll) differentiates well before the emergence of leaf. The greatest changes in cell cross–sectional area, intercellular space enlargement, and plastid growth occur just before the leaf blade emerges.

5.9.2.6. *Evolution of Leaf*

The phylogenetic mapping of leaves among land plants indicates that leafy shoot systems have evolved at least five times over the course of evolution of Embryophyta (which included Bryophyta with so–called 'leaves' and Lycopsida having microphylls). These analyses have also shown that among Euphyllophytes (all megaphyllous taxa that includes Pteridophytes other than Lycophyta and seed plants) the leaves of seed plants and pteridophytes are not homologous.

Molecular developmental data may throw useful light on the origin of leaves, as for example, may help to assess the antecedent structure (a sterilized sporangium) of leaves of Lycopsida. Unfortunately, there have thus far been no studies on the molecular basis of the evolution of leaf development over the course of land plant diversification. Analyses of homologues of genes, involved in diverse stages of leaf development in some angiospermous taxa, and discussed in the earlier pages of this chapter in other land plants, may provide insights into the evolutionary origins of leaves and leaf–like structures. However, the study of the genetic basis of compound leaf development, particularly the study of the expression of *KNOX* genes, have provided mechanistic support for the leaf–shoot continuum model, the hypothesis that leaves are, at least in Euphyllophytes, derived from stem like (axially indeterminate) organs (Bharathan and Sinha 2001). Compound leaves are known to have evolved many times in angiosperms from ancestral simple leaves. The expression patterns of *KNOX* gene in various *Lepidium* species that have either compound or simple leaves have been studied and this study stressed the importance of the evolution of morphology (of leaf type) in a phylogenetic context. *KNOX* gene expression certainly plays a role in leaf morphogenesis, but the manner in which the expression of this gene has been co–opted to form either compound or simple leaves can differ among plant species. Simple and compound leaves have evolved independently several times via alterations in different components of leaf molecular programmes. There is a fundamental difference between *KNOX* genes expression in the compound leaves of ferns and that of seed plants (Refer Friedman et al 2004). While *KNOX* expression is shown at the initial leaf primordium in extant seed plants (including angiosperms and gymnosperms such as *Picea* and *Welwitschia*), expression is not down–regulated in the fern *Anogramma chaeophylla.* It is an observation that emphasizes the independent origin of leaves in ferns and seed plants.

Pre–Fertilization Reproductive Growth and Development 6

6.1. INTRODUCTION

Vegetative growth often culminates in the production of reproductive structures, the flowers in angiosperms and the cones or strobili in gymnosperms. This requires the transformation of vegetative shoot apical meristem into a **floral meristem** or an **inflorescence meristem** depending on whether the plant has a single flower or an inflorescence. Although what causes the making of a flower is still unclear at the physiological and molecular levels, there is sufficient advancement in our understanding of the changes leading to floral initiation and flowering. By the word 'flowering', the changes associated with the formation of the primordium of a single flower or of an inflorescence are denoted. However, it should be kept in mind that there are basic structural and anatomical differences between a flower and an inflorescence, especially the fact that in the former the vegetative meristem is transformed into a determinate floral meristem and in the latter it is indeterminate to some extent and for some time (although sooner or later it also becomes determinate). It should also be kept in mind that flowering is not an abrupt developmental decision of the plant, but it is the culmination of a series of closely integrated developmental events. Flowering, thus, involves the following events:

1. Conversion of vegetative apex into a floral or inflorescence apex
2. Floral initiation and differentiation of floral primordium
3. Differentiation of individual floral parts
4. Floral maturation, especially of anthers and ovary and the structures contained in them
5. Anthesis and opening of flowers and receptivity for pollen receipt.

The first event is often referred to as **floral evocation**. Evocation not only involves physiological and molecular changes in the concerned transitional vegetative apex, but also in the leaves, especially the **flag leaf** that is located below the flower or inflorescence.

6.2. FLOWERING

The process of flowering is treated under various co–ordinated sub–events that together lead to flowering.

6.2.1. Acquisition of Floral Competence

The appearance of flowers is a useful marker of reproductive competence of an angiospermous plant. It is generally believed that a plant must attain a certain stage of vegetative development, often known as **ripeness–to–flower,** before it can flower. In this connection, it is believed by some that there exists a packet of cells in the embryonic shoot apex itself and that it is passively carried forward in the adult plant to produce flowers. Evidence for this is shown by clonal analysis in maize plant. But, most others support the view that the vegetative shoot apex is determined florally only after producing a considerable number of leaves and vegetative nodes. A juvenile–to–adult transition is often a prerequisite to flowering, especially in woody taxa. The probable unifying idea for the transition from vegetative to reproductive phase is that there should be complete transformation in the activity of the shoot apical meristem from the one it had during vegetative phase of development.

Molecular studies have revealed that the control of floral meristem identity is regulated by genes, called **floral meristem–identity genes.** These are the earliest acting genes in reproductive development. The acquisition of inflorescence or floral identity on the meristem consists of a series of gene–controlled steps and the mutation in one of the genes will cause arrest in the transition or deviation at the earliest point when the gene product is required. A detailed analysis of these genes has been made on plants such as *Antirrhinum majus* and *Arabidopsis thaliana.* This analysis has shown an interaction and integration of (i) *chromatin regulators* (ii) MADS-box genes and (iii) floral identify genes (Krishnamurthy and Bir Bahadur, 2014). The vegetative programme initiated in the embryo or seedling of *Arabidopsis* normally switches on to an inflorescence programme only if the wild chromatin regulator *EMBRYONIC FLOWER 2 (EMF 2)* gene is activated, while the mutant *emf* seedling bypasses the rosette vegetative stage to directly produce an inflorescence. The other wild genes that function as chromatin regulators are *SPLAYED (SYD), Terminal Flower2 (TFL2), FIE, CURLY LEAF (CLF)* etc. Again, in this species, *LEAFY (LFY)* is a key gene required for the formation of flowers on the inflorescence axis; *lfy* mutants are similar to wild plants throughout development, but never produce complete flowers on the inflorescence. Thus, the *LFY* gene product is a factor that promotes the switching on of meristem to a floral one. The *lfy* mutant's homologue in *Antirrhinum* is *floricaula (flo)* and there is about 70% identity between the predicted protein products of *LFY* and *FLO* genes. *ZFL* in maize and *RFL* in rice are the homologues of FLY. This idea of genes affecting the change from inflorescence to floral meristems is also true in case of *apetala 1 (ap 1), cauliflower (cal)* and *usual floral organs (ufo)* mutants of *Arabadopsis*; the *squamosa (sqa)* and *fimbriata (fim)* mutants of *Antirrhinum majus*; and the *falsiflora (fa)* and *anantha (an)* mutants of tomato. However, we do not have any clear–cut idea about how the protein products of these genes control meristem identity at the cellular and molecular levels, but it is likely that signal transduction pathways that activate certain meristem identity genes are controlled by phytochrome and GA. It is also likely that the control of an indeterminate inflorescence meristem and a determinate floral meristem is closely linked in a complicated way with endogenous growth regulators, environmental factors and elementary chemical modifications of DNA (See Krishnamurthy and Bir Bahadur, 2014 for more details on the other genes).

At the time of floral evocation, cells of the shoot apical meristem commit themselves, often irrevocably, to produce a flower or an inflorescence. This is evident from in vitro culture of 'transitional' meristems or parts thereof (even surface layer) on a suitable culture medium, when they produce flowers or inflorescences.

6.2.2. Physiology of Floral Evocation

A number of plants flower once the ripeness–to–flower period is completed, without needing any further treatment. In such plants, the events associated with floral evocation often overlap with the final stages of vegetative maturation. However, in some species floral evocation needs the presence of specific environmental conditions, the most critical of which are day length and temperature. In many plants, flowering is coordinated under natural conditions year after year by the more or less reliable seasonal fluctuations in day length and temperature cues.

6.2.2.1. *Photoperiodic Control*

Photoperiodically–sensitive plants are often grouped under three major response groups:

1. **Short–day plants (SDPs),** which flower only when they are exposed to light periods of less than a critical length in a 24–hour light–dark cycle. As the day–length in excess of the threshold keep the SDPs in the vegetative state, these plants need to be appropriately labelled as **long–night plants.**
2. **Long–day plants (LDPs)** or **short–night plants,** which flower only when the day–length exceeds a certain critical duration.
3. **Day–neutral plants (DNPs),** which are far more than the other two categories. They flower regardless of photoperiod, but only after a specific ripeness–to–flower vegetative phase.

Photoperiodic control on flowering is genetic. Single–gene mutants regulating flowering time have been isolated in a few plants such as *Arabidopsis*, maize, pea, sorghum and tobacco. In pea plant, photoperiodic control of flowering is probably controlled by genes such as *STERILE NODES (SN), DAY NEUTRAL (DNE),* and *PHOTOPERIOC RESPONSE (PPD),* which are complementary to one another. Mutations in recessive alleles of any of these three genes result in an early–flowering day–neutral condition. In the facultative LDP *Arabidopsis* more than 30 mutants that regulate phototropic effects have been isolated. One such is the *constans (co)* mutant, where the activity of the product of this gene is a limiting factor of flowering. *CO* gene expression acts in a pathway that promotes the transcription of floral meristem–identity genes such as *LFY* and *TFL* in response to an inductive LD treatment. *CO* and *LUMINIDEPENDENS (LD)* genes have already been cloned. The other important genes are *FLOWERING LOCUST (FT)* and *SUPRRESSOR OF CONTANSI (SOC1).* The homologue of *FT* in rice is *Hd3a* and of *CO* is *Hd1*.

The primary effect of a favourable light–dark cycle in the control of flowering is a change in the plant that enables it to flower subsequently, even if there are unfavourable photoperiodic conditions at that time. This change in state is called **photoperiodic induction.** There is an enormous body of evidence to implicate the leaves in the sensing of day–length and masking of all leaves of a plant affects the flowering. Even if a single leaf is exposed to correct photoperiod, it is sufficient to induce flowering. It is believed that the photoperiodically–induced leaves synthesize a floral stimulus that gets transmitted to the apex and the synthesis of this stimulating substance is influenced by the absolute length of dark period. This is evident from night–break experiments in which light is provided for short durations in the time. These experiments subsequently led to the discovery that **phytochromes** of leaves are involved in sensing photoperiods. Photoperiodic signalling pathways are mediated by both phytochrome A and B as indicated by investigations made on phytochrome–deficient mutants of *Arabidopsis* such as *hy 3* (*phy B*), *fhy 2* (*phy A*) and *pef 1*.

The transmissible floral stimulus (produced in the leaf and transported to the apex) produced by the action of phytochromes has been called **'florigen'**, but all the attempts, so far made, to chemically characterize or identify this substance, have failed. Initially, florigen was considered as a hormone or as a mixture of hormones, especially involving GA. Flowering due to application of GA and the study of GA–deficient mutant, *ga 1* (deficient in producing *ent*–kaurene, a precursor in GA biosynthesis) of *Arabidopsis* provided strong evidence for the role of GA in flowering in several LDPs, but in none of the SDPs under LD condition. This has been suggested as strong evidence against GA being the florigen, atleast in SDS. Studies made on a late–flowering *indeterminate (id)* mutant of maize have suggested the presence of a gene involved in the production of a transmissible floral stimulus. It has also been suggested that inhibitory substances produced in the leaves under non–inductive conditions might be important in photoperiodic control of flowering and this may even negate the promotive roles of favourable photoperiods. This 'inhibitory hypothesis' has been supported by grafting experiments, for example, in species of *Nicotiana.* Like the florigen, the chemical nature of the inhibitory substance is unknown. The inhibitory substance is called **antiflorigen.**

6.2.2.2. *Temperature Control*

It has been commonly observed that many annuals and biennials have a chilling requirement for flowering. This low–temperature treatment is often referred to as **vernalization**. It is conceivable that low temperature, like photoperiodism, exerts a specific effect on apical meristem, although low temperature does not mimic photoperiodism. Dividing cells are necessary for thermal induction. GA can substitute cold treatment since chilling increases the level of endogenous GA. Most of the work on the role of vernalization in flowering has been on winter wheat, rye and *Hyoscyamus niger.* In the last instance, after vernalization, the plant must be maintained under LD conditions. Cold treatment is believed to produce a stimulating substance named **vernalin,** and like florigen, the exact nature of vernalin is yet to be deciphered, at though some believe that it may be GA.

6.2.3. Structural Changes and Molecular Control in Floral Evocation

It is evident from what has been detailed so far is that morphological and functional changes in the shoot apical meristem mark floral evocation. The attempts made so far to interpret, define and describe floral evocation have all centred on the shoot apical meristem employing diverse techniques such as light and electron microscopy, histochemistry, autoradiography and biochemical and molecular methods (Nougarède 1967).

During floral evocation, the apex increases in diameter, and becomes more conspicuous and often dome–shaped. The tunica–corpus organization is changed into a **mantle–core** organization due to the redistribution of mitotic activity of the cells. Mantle cells become more meristematic than the core cells, which look more parenchymatous (Bernier 1988). It is from the mantle cells that floral primordia arise. Mantle also contains more rapidly–cycling cells than the core whose slowly–cycling cells often are considered as an initial reservoir that replenishes the mantle, if necessary.

An important question regarding change in the pattern of mitotic division to result in mantle–core organization is whether it is a cause for floral evocation or whether it is part of the floral evocation and development event. Increased mitotic activity was observed to be synchronous with the arrival of the floral stimulus in the meristem. In *Sinapsis alba* (a plant that requires one LD cycle in about 18 hours after floral induction), for instance, there is a shortening of the **G2** phase of cycling cells and by the recruitment of non–cycling **G2** cells into mitosis, a mosaic of rapidly cycling cells and relatively non–cycling cells at the floral meristem results. This is followed by a shortening of mitotic cycle duration of the rapidly–cycling cells and entry into **S** phase by some of the non–cycling **G1** cells, and thus, causing a second wave of mitotic activity in about 62 hours. The net result is the synchronization of mitotic activity that appears to be an important feature of floral evocation. In *Silene coeli-rosa* (a plant that requires at least seven LD cycles to flower), there is shortening of **G1** and **G2** phases during photoinduction resulting in the accumulation of **G2** cells mainly in the mantle of the meristem. This is followed by mitotic synchrony at eight to nine days under SD conditions. Here also, the net result is the attainment of synchrony in mitosis. However, in this taxon a significant increase in the mitotic index is observed in the shoot apex as early as 0.5–1 h after induction began due to abbreviated **S** phases, which hastens the mantle cells through synthesis of DNA. An increase in DNA replication rate, thus, appears to be the earliest cytological event of photoinduction and floral evocation.

Some of the early cytotochemical studies indicate that changes in RNA, proteins and enzymes are implicated in floral evocation, thus emphasizing the fact that transition to flowering is associated with a general increase in macromolecular synthesis in the prospective floral meristem. Subsequent histochemical studies, combined with in situ hybridization, indicated a characteristic pattern of accumulation of transcripts of genes for some ribosomal proteins such as RPL 2 and RPL 18 and a down–regulation of arginine decarboxylase (ADC) at the distal end of the meristem, while cells at the base of the meristem showed increased expression of transcripts.

There is an increase in the amount of transcriptionally-active dispersed chromatin in the nuclei of meristem cells, as demonstrated by ultrastructural studies made in *Sinapsis alba* (Bernier et al 1993). Similarly, the restriction of RNA synthesis in the rib meristem region of the meristem indicates that the activation of cells in this region is one of the first events in floral evocation. This, in fact, coincides with or just precedes the arrival of floral stimulus both in *Sinapsis alba* and *Pharbitis nil*. Perhaps the majority of this RNA is mRNA, although there is growing evidence for the synthesis and regulatory functions for microRNAs (miRNAs); GA and *ARGONAUTE (AGO1)* are essential factors in miRNA-mediated (miRNA 159 and 172) pathways (Kidner and Martienssen 2005). Many experiments involving the use of metabolic inhibitors have also shown the importance of RNA synthesis (perhaps mRNA) during floral evocation, emphasizing the idea of floral–specific mRNA, both in terms of its quantity and quality. It is not yet clearly known whether the changing pattern of protein synthesis is important in photoperiodic floral induction, especially in the light of the fact that there is both the appearance of new proteins and disappearance of extant proteins. Although the demonstration of new RNA and proteins in vernalization–induced floral evocation has not been done so far, it is most likely that such an increase would be present. A comparison of mRNA population from vegetative and floral tissues of *Nicotiana tabacum* showed the existence of a complex gene expression phenomenon during floral evocation in this plant also. Based on this study, it has been suggested that a few thousand genes might be required to specify floral evocation and development of floral organs.

Apical meristems of *Sinapsis alba* and tobacco harvested at a different time after exposure to correct photoinductive cycles were used to make cDNA libraries. In *Sinapsis alba* transcripts of three genes, *Sa MADS–A* (necessary for floral meristem specification), *Sa MAD–B* (essential for differentiation of special cell types in the flower) and *Sa MAD–D* begin to accumulate rapidly and specifically in the apical meristem of photoinduced plants. *Sa MAD–A* mRNA initially appears in the central zone of the meristem, where the first signs of floral evocation are seen, as mentioned earlier. Thus, in floral evocation, expression of *MADS–Box* genes is the earliest molecular signal. Similarly, *FA2* (for 'floral apex') gene is critical in tobacco for transition of vegetative meristem into floral meristem. Besides these, genes isolated from apices of photoinduced plants, genes up–regulated in mutant flower arrested at early stages of development may be involved in floral evocation. Such marker genes have been demonstrated in tomato. However, it is to be mentioned that investigations of genes in floral evocation are, on the whole, too meagre to make any serious conclusion on the molecular mechanism of floral induction, and hence more research is required.

6.2.4. Formation of Floral Organs

6.2.4.1. General Organization and Variations in Floral Structure

The formation of floral organs on the floral meristem takes place immediately after floral evocation and sometimes in an overlapping manner with it. A typical angiosperm flower has four compressed whorls (in some taxa spirally arranged) of modified leaves; **sepals, petals, stamens** and **carpels**. The former two whorls are sterile, while the latter two fertile. All these four whorls are borne on a **receptacle, thalamus** or **torus.** The receptacle is a modified vegetative shoot apical region with nodes and internodes, which are highly telescoped between the appendages. The aggregation of sepals make up the **calyx**, of petals make up the **corolla**, of stamens make up the **androecium,** and of carpels make up the **gynoecium.** The stamens represent the male reproductive unit, the carpels represent the female reproductive unit, and stamens and carpels together are concerned with sexual reproduction. If the floral organs are arranged in whorls, members of successive whorls alternate in position with those whorls directly above and below. If they are spirally arranged, the phyllotactic spiral is usually the same as that of the leaves. The four floral whorls are generally produced in acropetal succession by the floral meristem, with the carpels maturing at the last and their production is precisely determined according to the blueprint specific for each plant species (Fig. 164). In some species it is basipetal with the carpels arising first and sepals the last. Some people use the words 'centripetal' instead of 'acropetal', implying the initiation of floral whorls beginning at the periphery and proceeding inwards, and 'centrifugal' for basipetal wave (proceeding from centre to periphery). In flowers possessing a large number of appendages of any one morphological category, say many petals, these two waves of differentiation may be evident even within the whorls of the same category. However, these directional waves are not always maintained in some flowers. Development of stamens conforms to the centripetal direction in some families such as Annonaceae, Magnoliaceae and Myrtaceae. The direction is reversed in other families such as Cactaceae, Malvaceae and Hypericaceae. The differential time factor in the ontogenetic appearance of the whorl is, however, only a transitory phase confined to the earlier stages of development and the order of maturation need not necessarily synchronize with that of initiation (Swamy and Krishnamurthy, 1980a).

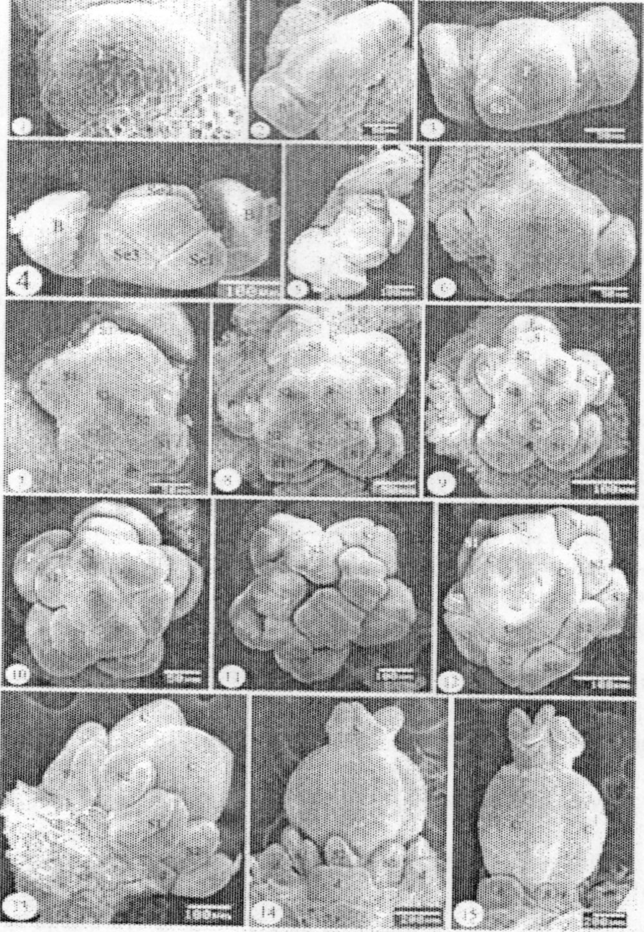

Fig. 164. SEM photographs of floral organogenesis in *Jatropha curcas*; 1 - Inflorescence (I) primordium; 2 - Initiation of two bract (B) primordia; 3 - Origin of one sepal primordium (Se1); 4 and 5 - Initiation of all five petal primordia (P); 7 - Initiation of the outer whorl of stamen primordia (S); 8 - Initiation of the inner whorl of stamen primordia (S Series); 9 and 10 - Inner and outer whorls of stamen primordial; 11 - Primordia (C); 13 - Stamens and carpels; 14 - Carpel closing; 15 - Female flower with carpels, but stamens degenerate (Liu et al 2008) (Reproduced with permission from China Academic Journal Electronic Publishing House)

There is a great variation in the number of whorls in a flower as well as in the number of organs in each whorl, and also the extent of fusion of organs of the same whorl or of different whorls. There is also great variation in the elaboration of organs in size and form. These variations result in an almost endless variety of flower types such as simple and complicated, primitive and advanced, and symmetric and asymmetric (Swamy and Krishnamurthy 1980a).

The flower is an exclusive feature of the angiosperms. Detailed phylogenetic analysis based on morphological, anatomical and ontogenetic studies have led to a standard definition of a flower as a modified vegetative shoot (which is highly telescoped and with determinate growth). This concept assumes that the floral organs are homologous to leaves. The vascular system of the flower is similar to that of a vegetative leafy axis, but specialized flowers have modified vasculature due to various types of fusion, cohesion and adnation between different floral parts, and/or amplification or reduction of vasculature (Swamy and Krishnamurthy 1980a). In some cases, phylogenetic abortion of floral parts is also the causative factor.

The gynoecium of flower is the most important organ as a similar organ is not found in any other vascular plant group. Although the overall morphologies of angiosperm gynoecia are highly variable, almost without exception the gynoecium occupies the central position. Pistil (often used interchangeably with the word 'gynoecium') development is initiated by the formation of carpel primordia at the centre of floral meristem. A characteristic feature of early pistil development is the occurrence of carpel fusion in multicarpellary, syncarpous pistils. Although fusion is an obvious requirement for the formation of compound pistils from two or more carpels, it also occurs in plants with simple pistils in which the fusion of the carpels' margin is necessary to form the closed pistil. Two types of fusion are recognized as participating in the formation of pistils (Verbeka 1992). Carpels are said to be **congenitally fused** when a compound pistil is directly produced as a single structure on the floral meristem. The compound pistil consists of multiple carpels inferred from the number and locations of vascular traces and rows of ovules or from other morphological and anatomical features. This type of fusion is also referred to as 'phylogenetic fusion' because progressive fusion of the carpels is believed to have occurred during evolution from progenitors with less fused or unfused carpels. The contrasting process of **post-genital fusion** (also called **ontogenetic fusion)** occurs when usually separate carpels meet and fuse to form a single structure (e.g., tomato). Post–genital fusion is commonly followed by redifferentiation of the contacting epidermal cells into normal parenchyma cells undistinguishable from other cells of the ground tissue of the ovary. Normally, the pistils either show congenital fusion or post–genital fusion depending on the species, but in plants such as *Arabidopsis,* the pistil shows both congenital and ontogenetic fusions.

The details of periclinal chimeras in relation to leaf development can be recalled from Chapter 5. Three meristem layers i.e., L1, L2 and L3 respectively give rise to the epidermis, outer layers and inner layers of the floral organs. The only part of the flower that has an exclusive origin from a single meristem layer is the style, i.e., L2 layer. However, the determinative decision that causes cells of these three meristem layers to opt for forming particular organs of the flower is not rigidly fixed in some plants. Hence, the genesis of floral organs depends neither upon specific cell lineages within the meristem nor on the position of individual cells or cell layers.

6.2.4.2. *Interaction between Different Floral Organs*

Different whorls of floral organs may influence each other's differentiation in the floral meristem, suggesting an interaction between them. Even when tested in the most favourable culture medium, the sepals suppressed the further growth of the floral buds of *Aquilegia formosa.* But, on surgical removal of sepals, the buds resumed normal growth; perhaps, inhibitory substances produced by sepals limited the growth of other organs. However, how such inhibitory substances in the growing buds of intact plants are overcome is not clear. Quite contrary results were obtained in cultured floral meristems of tobacco plant, where existing floral organs promoted the differentiation of yet to be formed floral organs. The suppression or removal of sepal primordia arrested the formation of petals, and the suppression of petal primordia inhibited the initiation of stamens and so on. These observations favour a gene activation based on the **operon model** in such a way that inducing substances released by earlier–formed floral organs may govern key development decisions leading to the formation of succeeding organs. However. studies made in *Arabidopsis* have shown that during floral ontogeny, inductive influences from adjacent whorls are not required for appropriate organogenesis in the floral meristem, but positional information bequeathed to the organ primordia is likely to coordinate the development of floral organs. In an experiment involving a genetic knockout of petals and stamens in the flower of transgenic *Arabidopsis* and tobacco, without actually surgically removing them, but engineered by ablation with the *DIPHTHERIA TOXIN A (DTA)* gene fused to the promoter of a petal and stamen–specific *APS* gene, sepal and carpel formation were found to proceed normally in the absence of the other two whorls.

The results of some surgical experiments have emphasized the presence of progressive loss of the flexibility of some floral organ primordia as the meristem advances through sequence of developmental stages. In *Portulaca grandiflora,* a longitudinally–bisected floral meristem, left to regenerate, failed to produce new members of the whorls of floral organs already present at the time of bisection, but produced only the subsequently–to–be–formed floral whorls. A similar observation was made on bisected floral meristems, cultured in vitro, of tobacco and *Aquilegia formosa.*

6.2.4.3. *Genetic Control of Floral Organ Differentiation*

Floral organ identity

The major question in floral organ differentiation is how cells of the floral meristem faithfully establish their position, generation after generation, and subsequently develop into the characteristic floral organs, particularly the male and female sexual organs made up of the appropriate cell types. Most of the previous attempts have been focused on sex determination i.e., to study the cytological and genetical basis of stamen and gynoecium differentiation in flowers. In hybrids between two species of *Nicotiana,* stamen primordia quickly deviated from normal developmental pattern and exhibited characters of the pistil like stigma tips and this has been attributed to cytoplasmic genetic elements of the female parent involved in hybridization. Very fruitful genetic studies on sex determination have been carried out in maize florets. In this plant, many mutations leading to selective suppression of either stamens or carpels are known. Mutated genes such as *ts2* and *tunicate (tu)* permit the development of pistillate florets instead of staminate ones in tassel. However, most studies on sex determination have been carried out on plants with unisexual flowers, especially in cucurbits. Culture conditions can transform stamen primordia into carpels, as the former appear to be undetermined. IAA can increase this transformation. However, no hormone treatment is found to transform carpel primordia into stamens. There appears to be a general antagonism between stamens and carpels.

In the last three decades, the differentiation of floral organs is being studied from a genetic perspective. The experimental approach used for this was to generate mutants that interfere with the ordered development of the different floral parts on the meristem. Because such mutations cause certain organs or a series of organs to be substituted by other organs not normally found in that position of the floral receptacle, they represent **homeotic** transformations; this implying that what is newly regenerated does not conform to what is being replaced. Thus, mutant flowers reveal the subtle influence of floral–organ identity genes whose protein products are essential for the regular patterned formation of organs on the floral meristem. Some of these genes have been cloned, especially those of *Arabidopsis* and *Antirrhinum*. The most important floral homeotic mutants of *Arabidopsis* and their phenotypic expressions are shown in Table 3. On the basis of the phenotypes of the mutant flowers, a canonical working model, called **ABC model,** was limitally used to explain how the four whorls of floral organs are specified. This model invokes three classes of homeotic genes: A, B and C, each affecting two different whorls on the floral meristem. The class A genes (*AP2* and possibly *LUG*) are active in the first (calyx) and second (corolla) floral whorls, class B genes (*AP3* and *PI*) are active in the second and third (androecium) floral whorls, and class C genes (*AG*) function in the third and fourth (gynoecium) floral whorls.

Another important postulate of the ABC model is that a combination of gene products present in certain whorls determines the identity of the floral organ differentiating in those whorls on the meristem, while the gene products present in certain other whorls independently determine the identify of the floral organ differentiating in these whorls. A and C genes act alone and respectively control sepal and carpel development, B controls petal development in combination with gene A in the same whorl and C promotes stamen development in combination with gene B in the third whorl (Fig. 165). This model has to be necessarily plastic in the sense that the products of A and C genes cross–regulate each other in a mutually antagonistic way. This necessarily means that any region of the floral meristem, where *AP2* gene is expressed, will not have a functional *AG* gene in the first and second whorls. Similarly, when the *AG* gene is expressed in the region of the fourth whorl, the *AP2* gene will not function in the third and fourth whorls. But, when the activity of A is knocked off by a mutation in the *AP2* gene, the activity of C will be abnormally high in the first and second whorls. When the activity of C is removed by a mutated *AG* gene, then the activity of A dominates in the third and fourth whorls. Thus, as per this model, it is easy to understand why the lack of A, B and C activities results in mutant phenotypes represented respectively by *ap2, ap3/pi* and *ag* (Fig. 165). This model is confirmed in the double mutant *ap3* or *ap2* and *pi* or *ap2* in which A and B classes of genes are knocked out and only the activity of C gene is evident in all four whorls, producing flowers entirely consisting of carpels. In the same say, when both B and C genes are eliminated, as in *ag/pi* or *ag/ap3* double mutants, the flower formed contains only sepals, specified by A–type genes. Triple mutant *ap2/ag/pi* flowers consist of only leaf–like appendages, as the activity of all three genes is absent.

Table 3. Floral homeotic mutants of *Arabidopsis* (Data from various sources)

Serial No.	Name of mutant	Phenotypic expression
1.	*agamous (ag)*	An outer whorl of four sepals, two inner whorls of 10 petals (5 + 5), and a variable number of intermediate organs.
2.	*apetala1 (ap1)*	Sepals are transformed into bracts and flowers are formed in the axils of the bract-like organs.
3.	*apetala2 (ap2)*	A gynoecium-like outer whorl, second and third whorls of stamens, and a fourth whorl of carpels.
4.	*apetala3 (ap3)*	The second whorl organs are transformed into sepals, and the third whorl organs are transformed into carpelloid stamens or into normal carpels.

5.	*floral (flo)*	Carpels are absent in the fourth whorl or are replaced by stamens or by stamen–carpel intermediate organs
6.	*leunig (lug)*	Sepals are petaloid, staminoid or frequently carpelloid; petals with stamen characteristics; reduced number of stamens and carpels
7.	*pistillata (pi)*	The second whorl of organs is transformed into sepals, the third are absent or carpel–like; the fourth whorl is of carpels of abnormal size.
8.	*unusual floral organ(ufo)*	The outer whorl of sepals is followed by second and third whorls of sepals, petals, stamens or carpels; the third whorl carpels are fused with those of the fourth whorl.

In situ hybridization techniques showed that *AG* mRNA is expressed almost exclusively in the stamen and carpel domains of the meristem. Similarly, the expression of the *AP3* and *PI* genes is restricted to petal and stamen domains of the meristem of wild–type plants. Transcripts of the *AG, AP3* and *PI* genes are initially detected in alien locations on the meristem, but they are subsequently detected in organ–specific locations. This is in agreement with the view that these genes are important in the early determination of the three inner whorls of the flower. The prediction from the model that *AP2* gene product function to negatively regulate *AG* gene activity is also indicated by the observation that *AG* gene transcripts are present in the primordia of all four sets of floral organs of an *ap2* mutant (Refer to Raghavan 2000).

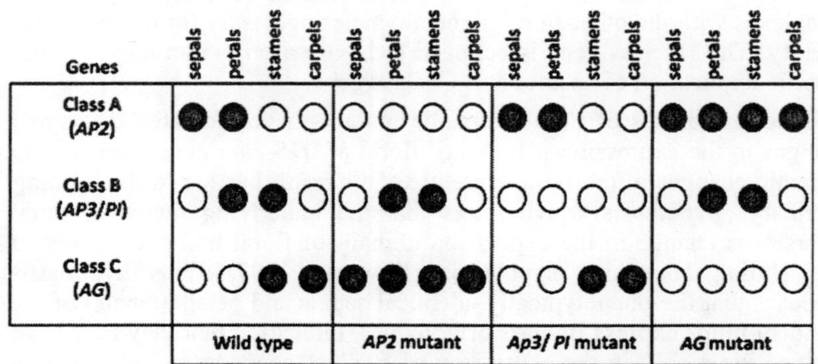

Fig. 165. Diagrammatic representation of the operations of the genes in wild type and three homeotic mutants of *Arabidopsis*; the genes active in each floral whorl are indicated by the darkened circles. Hollow circles indicate that the particular gene is absent in respective floral organs. (Diagram based on Dr. Raghavan 2000)

Transgene technology has been successfully employed to show that each of the A, B and C classes of genes is sufficient to specify a particular organ on the floral apex. If the *AG* gene is over–expressed in *Arabidopsis* under the control of the CaMV255 promoter, the resultant effect is to swap the *AP2* gene products with high levels of *AG* gene products, and the resultant floral phenotype on the transgenic plant is closely similar to that of the *ap2* mutant. This is considered as direct evidence to show that the function of the wild-type *AG* gene is to inhibit *AP2* function. The effective impact of the ectopically expressed *AP3* gene is evident in the flower of transgenic *Arabidopsis* by a partial conversion of carpels to stamens, but the sepals are unaffected. On the contrary, when plants, expressing both the *AP3* and *PI* genes, are produced by crossing progenies harbouring chimeric constructs of the genes, flowers with two outer whorls of petals and two inner whorls of stamens are produced. Here, it is the expression of a combination of organ–identity genes that provides a strong evidence for the interaction of class B genes with class A and C genes in specifying petals and stamens, as predicted by the model.

It is not known whether the ABC model, which was first described in eudicots, is applicable to flowers including of lower eudicots such as Ranunculaceae, where all the floral organs, especially perianth lobes, are produced in spirals and not in whorls. Also, the ABC model has not taken into account several additional observations made on the functions of three classes of genes resulting from the interactions between them; nor has the model fully incorporated the activities of newly–discovered floral–organ identity mutants. Some of the refinements to the model are described by Ma (1997). The ABC model has been updated as ABC+DE model to include the D–class ovule–specific genes and the E-class genes that are expressed in the three inner floral whorls and form quaternary protein complexes with the other floral homeotic genes necessary for correct organ identity. The D–class gene is believed to have arisen via an angiosperm–specific duplication of an ancestral C–class gene.

Genetic analyses of floral homeotic mutants have indicated that small changes in the expression patterns of floral *MADS–Box* genes can lead to dramatic changes in the floral structure. This result has led to the "sliding boundary" hypothesis which states that the underlying factor in floral diversity is change in the expression domain of floral homeotic genes or the 'sliding boundary' hypothesis (Bowman 1997). This hypothesis proposes that the phenotypically–identical sepals and petals (tepals) of the monocot tulips express B–class orthologues, indicating that they may have evolved via a shift in the expression of B–class genes to include the first whorl in addition to the second and third whorls. However, this is not happening with the expression of B–class genes in the tepals of lily and asparagus and also in Ranunculaceae, thus, not supporting the sliding boundary hypothesis for all taxa with tepals. Several meristem–identity genes such as *AP1, CAL, CLV, LFY* and *UFO* of *Arabidopsis* also interact in a complex way mediating between floral meristem determination and the activation of floral organs on the meristem. (See Krishnamurthy 2014).

Homeotic mutations in *Antirrhinum* and a few other taxa have also been studied. *Antirrhinum majus* has been studied for a long time. It has large, zygomorphic flowers with five sepals and petals (gamopetalous), four stamens (two short and two long), a staminode, and a gynoecium. A large number of mutations affecting various aspects of floral organs have been reported in this species. These include homeotic mutants such as *deficiens (def), globosa (glo)* and *sepaloidea (sep),* which have altered first three whorls of the flower to result in sepal, sepal and carpel instead of sepal, petal and stamen respectively. These mutants have phenotypes similar to *ap3* and *pi* mutants of *Arabidopsis*. Thus, their genes belong to the B class, affecting the second and third whorls on the floral meristem. Similarly, the *plena (ple)* mutant of *Antirrhinum majus* is similar to the *ag* mutant of *Arabidopsis,* and it affects the third and fourth whorls of floral meristem. Its genes belong to C class. The first and second whorls of the floral meristem are affected by semi–dominant mutations called *ovulata* (*ovu*). This has flowers with carpels in place of sepals, and stamens in place of petals (thus similar to the *ap2* mutant of class A gene). Thus, it is probable that the same molecular mechanisms that operate in *Arabidopsis* also operate in *Antirrhinum majus.* This is shown by studies on the localization of cloned genes. In the wild–type flower, transcripts of *DEF* and *GLO* genes are expressed in those organs that are homeotic (petals and stamens). *PLE* transcript gets expressed in the stamens and carpels of wild–type flower, thus showing its homology to *AG* gene of *Arabidopsis.* But, in the *ovu* mutant, *PLE* transcripts are no longer restricted to the inner two whorls. They are also expressed ectopically in the outer two whorls. Since *PLE* is a class C gene, its activation causes the formation of sex organs in place of sterile organs in the *ovu* mutant.

An *AG*–like gene cloned from *Brassica napus* has been shown to express in transgenic tobacco and to convert the sepals into carpels and petals into stamens, thus, resembling the *ap2* mutant phenotype. The presumption that an *AG* gene function is crucial for normal stamen and carpel development in tomato is also indicated by the fact that transgenic tomato plants that express antisense RNA of tomato *AG* gene have petaloid organs instead of stamens, and pseudocarpels instead of carpels. On the contrary, when sense RNA is expressed in the transgenic plant, flowers with fruit–like tissues in the place of sepals, and staminodes in place of petals are formed.

In *Petunia hybrida* a potential class B gene, called *FBP* (for '*floral binding protein*'), has been found. Its transcripts are expressed only in the petals and stamens of wild–type flowers. If a chimeric *FBP* gene is introduced into transgenic *Petunia hybrida* in the sense-orientation to cause a defect in the function of class B gene, as expected, the flowers show homeotic conversion of petals into sepals, and stamens into carpels. *Petunia hybrida* also has a spontaneously occurring *green petal* (*gp*) mutant in which sepals are formed in the place of petals, but unlike in the *ap3* mutant of *Arabidopsis,* a simultaneous conversion of stamens into carpels does not occur. In another spontaneous mutant *blind* (*bl*) of this plant, there is a partial conversion of petals into stamen tissues, and sepals into carpelloid tissue. A *Petunia* gene that shares sequence homology with the *Antirrhinum majus DEF* gene has also been reported. This can restore the wild-type phenotype to some extent in the *gp* mutant under transgenic condition. This indicates that the inserted gene is vital for petal development in *Petunia hybrida.* The petal–forming function of the transgene was verified by its ectopic expression in the wild–type *Petunia hybrida* that resulted in the formation of sepals in the second whorl, as the mutant phenotype. Similarly, an ectypic expression of another *Petunia* gene, which shows sequence homology to the *AG* gene of *Arabidopsis* and the *PLE* gene of *Antirrhinum majus,* leads to the recapitulation of the *bl* mutant genotype in transgenic *Petunia hybrida.* These findings and the exclusive expression of the gene in the stamens and carpels of wild–type flowers clearly show that the *AG–DEF* homologue of *Petunia* is a class C gene. The AG–like genes found in tobacco and the *DEF*–like gene found in potato also get expressed in the stamens and carpels, as expected from the *AG* gene, or in the petals and stamens, as expected from the *DEF* gene.

The MADS-box *AG* gene of *Arabidopsis* and the *DEF* gene of *A. majus* have been cloned. The functional motif shared by these genes has been given the name *MADS-Box.* An important role for a network of *MADS-Box* genes in specifying floral organ identity in *Arabidopsis* and *Antirrhinum majus* has been clearly established by demonstrating that the deduced protein products of the *AP1, AP3, PI, PLE* and *GLO* genes show striking sequence similarity with the proteins of the *AG* and *DEF* genes. *MADS-Box* proteins are, in reality, transcription factors. How a limited number of transcription factors interact to control floral organ development is not yet clear. It is believed that a sequence–specific binding of the transcription factors to DNA might lead to transcription of different genes in each whorl of the flower. However, because of the overlapping in the activities of the major floral–organ–identity genes, the molecular basis of floral development appears to be more complicated than is implied by the above mentioned simple statement (Raghavan 2000; Krishnamurthy and Bir Bahadur, 2014). We also do not know as to how the transcription factors select the genes for binding and whether these genes make the floral protein products for the formation of each type of floral organ or whether some intermediate products are involved. We still have a long way to go to fully understand the genetic basis of floral organ formation, as also to understand the evolution of flower and floral organs (Refer to Box. 6.1.)

Floral organ number

There is very little molecular data regarding genes whose expression controls the correct number of units of each organ to be initiated on the floral meristem. Mutation in *Arabidopsis* genes such as *CLV, PAN* (for '*perianthia*'), *WIG* and *FON* (for 'floral organ number') causes an increase in the number of units present in some whorls, while mutations in *REVOLUTA (REV), SUPERMAN (SUP)* and *TOUSLED (TSL)* genes lead to reduced organ numbers in the whorls due to the formation of incomplete or aborted structures or due to the occasional loss of floral organs. In *fasciata (fas)* mutant, there are fewer petals and stamens and more sepals in their respective whorls than the wild type flower, while the carpel number is unaffected. Unlike in *clv* and *wig* mutants where the change in organ number is correlated with a great increase in the size of the floral meristem, *pan* mutants go through ontogenetic changes through normal development, independent of changes in size, cell number or cell pattern of the floral meristem. The *PAN* gene can, thus, be considered to act directly in the process by which cells assess their position within the floral meristem; probably it is the gene that controls floral organ number in the floral meristem.

Symmetry of flower

Mutant genes of *Antirrhinum majus* also help us to understand the molecular basis of floral symmetry. Wild–type flowers have dorsiventral symmetry. In this flower, the symmetry is the result of unequally–sized floral organs, especially of the dorsal and lateral petals, and of the number of fertile stamens. Mutants such as *cyc* have radially symmetrical flowers and hence, lack the genetic functions associated with the lateral and dorsal domains of petals and the staminal regions. This is evident by the RNA expression pattern of *CYC* gene in wild–type flowers, whose transcripts are restricted to the above indicated domains or regions. Thus, *CYC* gene promotes the larger size of dorsal petals and retards dorsal stamen growth, and this confuses the mechanism of action of this *CYC* in dorsiventrality. It is suggested that it acts together with *DICHOTOMA* (*DICH*) gene to result in dorsiventral flowers.

BOX 6.1. EVOLUTION OF FLOWERS AND FLORAL ORGANS

Angiosperms or flowering plants represent the most recently evolved land–plant group, but their origin and early evolutionary history are less known than the other vascular plants for various reasons. The most important reason is the total uncertainty about the identity of the closest seed plant relative of angiosperms. Recent molecular phylogenetic studies have shown that extant gymnosperms may be sister group to the flowering plants and establishing angiosperm out–groups, based on characteristic polarities and homologies of angiosperm features such as carpel, four–sporangiate anthers and second integument, is still critical. These studies have also indicated *Ranales* as the earliest angiosperm group with monotypic *Amborella* that is probably sister to all angiosperms, and Nymphaeales that is sister to all remaining flowering plants. Hence, reconstruction of the morphology of the earlier flower has often been derived from comparison among these most extant primitive angiosperm lineages. It is assumed that such a flower should be hermaphrodite without any sepal–petal dichotomy (only perianth) in the helically arranged floral leafy whorls. There is also a general indication that sepals, petals and perianth have had separate evolutionary origins.

Molecular genetic studies have been brought into the phylogenetic derivation of the flower, as these have provided some clues to the diversification of floral morphology. Although such studies turn to the role of the homeotic transcription factors, *MADS-Box* genes, it is the floral identity genes that have received the most attention in evolutionary (and developmental) studies, as their functions are mainly conserved across a breadth of taxonomic groups and hence can be used to identify widely conserved organ identity programmes (Friedman *et al.* 2004). Genetic studies are consistent with not only the dependent origins of petals, but also with the independent origins of sepals, petals and tepals.

It has been suggested that alterations in expression patterns of the *MADS-Box* and other floral development genes resulted in the angiosperm flower. There are varied expression profiles for *LFY* homologies in diverse species. Also, recent studies of the proteins from many plant species eludicate how there are changes in the conserved DNA-binding domain, over evolutionary time. Both these could contribute to the likely diverse functions (Maizel et al 2005). Investigation on LEAFY (floral transcription factor) and B-class MADS-Box genes in gymnosperms have led to the proposal of the mostly male theory of floral origin (Frohlich 2003), which states that the male and female reproductive units were combined in the angiosperm flower by the emergence of ectopic ovules on a subset of microsporophylls in originally unisexual cones of a gymnosperm ancestor. Another hypothesis for the origin of flowers has suggested a homeotic transformation of reproductive organs to either male or female from unisexual gymnosperm cones, leading to the evolution of an ancestral hermaphrodite flower (Theissen et al 2002).

Further studies are needed towards floral genes of angiosperms and their gymnosperm homologues to test either of the above two hypotheses.

Refer to Friedman et al (2004) for more information.

6.3. OVULE AND FEMALE GAMETOPHYTE

The structural unit in which the development of the female gametophyte takes place is the **megasporangium**. Although this term is generally used for the **megaspore**–bearing units of tracheophyta (i.e., vascular plants), the units in the seed plants are called **ovules** as they possess a **nucellus** in which the megaspore mother cell is differentiated and which is covered by a jacket called the **integument(s).** Unlike in the non–seed–bearing Tracheophyta, fertilization of the **egg** (developed inside the female gametophyte) and consequent development of the embryo is initiated while the seed (fertilized ovule) is still attached to the parent sporophyte. Typical ovules are absent in Balanophoraceae and Loranthaceae and in the latter the female gametophyte develops in the **mamelon**, which is a conical projection developed at the base of the ovary (Swamy and Krishnamurthy 1980a).

6.3.1. Configuration

The ovule primordium arises as a tiny protuberance on the placenta of the ovary. While the primordium is increasing in size, a small annular tissue thickening appears at the point of attachment of the primordium. The thickened belt grows at a relatively faster rate, soon enclosing the blob of the primordial tissue within (Figs. 166 and 167).

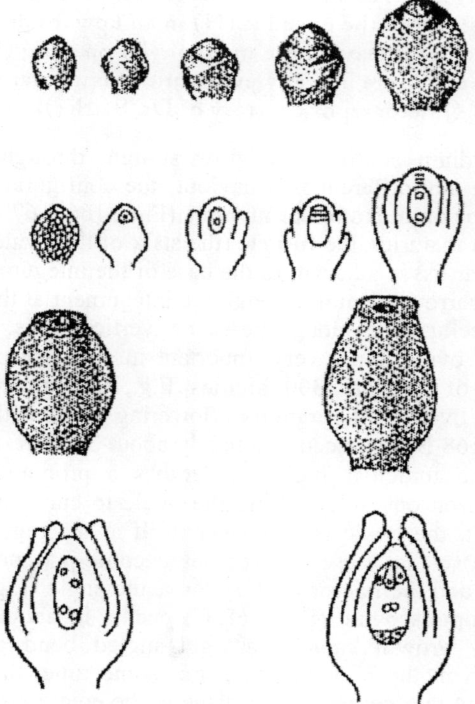

Fig. 166. Diagrammatic representation of the development of an orthropous ovule; for each stage of the entire ovule development, a corresponding L. S. of the ovule of that stage is shown.

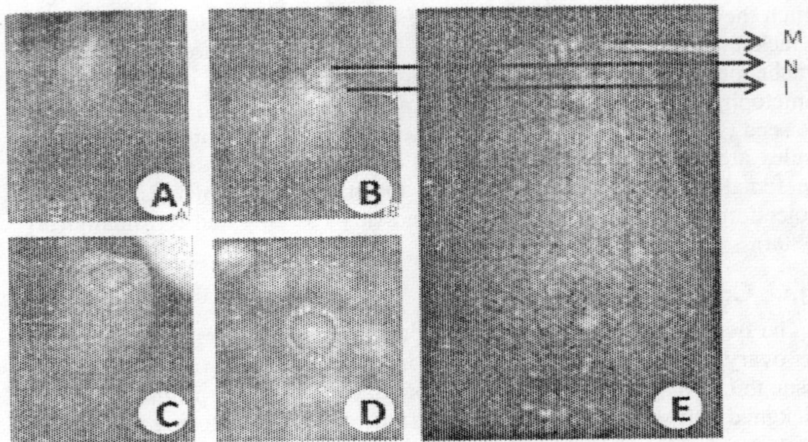

Fig. 167. Photographs of development of entire ovule of *Myrica gale*; Integument (I) arises as an annular outgrowth from the base of the ovule and grows around the nucellus (N) in an upward direction until it completely envelops it leaving the microphylar opening (M) at the tip. A to C and E are lateral views while D is a polar view (Photographs courtesy of Dr. Sattler).

If the primordium continues to grow straight throughout its course without showing any differential behaviour, the configuration of such an ovule is said to be **orthotropus** or **atropus** (Figs. 166, 167 and 168 A). In such an ovule, at maturity, the **funicle** (the stalk of the ovule), the **chalaza** (part of the ovular tissue adjacent to the base of the integument(s)) and the **micropyle** (the narrow opening through the integument at the extreme free end of the nucellus) lie along the same vertical axis. This type of configuration of ovule is not very important in angiosperms as it occurs only in about 20 of the nearly 300 families (E.g., **Polygonaceae**). The most common type of ovule configuration in flowering plants is the **anatrapous** condition (Fig. 168 B) that occurs in nearly about 200 out of 300 families. In this type, the funicular region undergoes a pronounced degree of localized growth on one side making the ovule to curve on the opposite side. As a result, the ovule is bent over itself at the region of chalaza–funicle combine so that these two regions become juxtaposed. Thus, the funicle, chalaza and the micropyle form an acute angle (E.g., Lamiaceae). The **campylotropous** ovule (Fig. 168 C) occurs in about five families. Here, the early growth causing a right–angled bend is much more pronounced than in the previous type. For some time, the nucellus and integuments grow in a perpendicular plane to the erect axis of the funicle. At the end of this phase, new loci of active division become initiated in the sub–terminal part of the ovule on the anti–funicular side, as a result of which the micropylar part curves downward. But it never becomes juxtaposed as close to the funicle as in the anatropus ovule and the attachment point of the funicle, chalasa and the micropyle form an obtuse angle (E.g., Fabaceae). **Hemitropons** (E.g., Primulaceae and 12 other families) (Fig. 168 E) and **amphitropons** (E.g., Crossosmataceae and three other families) (Fig. 168 D) ovules are intermediate between orthotropons and anatropous ovules. It must, however, be noted that the ovular configurations described above are for ovules just before fertilization. In a few cases, the configuration may change during post–fertilization stages. For example, taxa of Leguminosae, Alimataceae, and some families of Centrospermales.

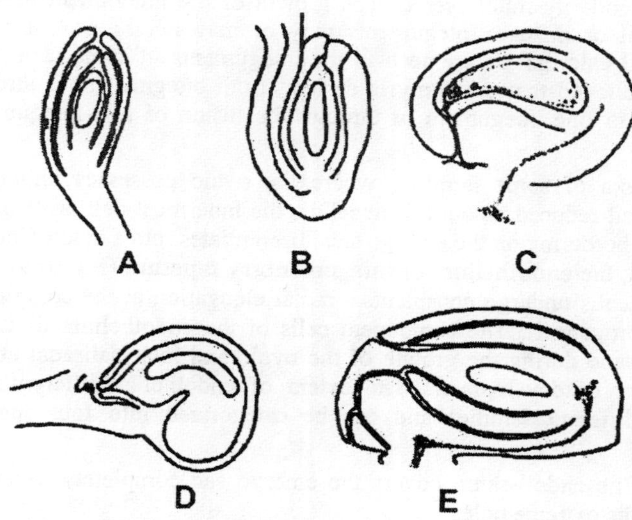

Fig. 168. Types of ovule configuration; all in L. S.; A - Orthotropous; B - Anatropous; C - Campylotropous; D - Amphitropous; E - Hemi–anatropous

6.3.2. Structure of Ovule

6.3.2.1. *Micropyle*

The **micropyle** may be formed by the only integument in **unitegmic ovules**, by the inner or outer or both integuments in **bitegmic ovules**. If both integuments form the micropyle, the bitegmic ovule has an **endostome** (covered by the inner integument) and an **exostome** (covered by the outer integument) (Fig. 168). The micropyle may be just an opening or a canal, and may be angled, curved or zigzag in contour. In a number of taxa, the integumentary cells lining the micropyle are different from other integumentary cells. The micropyle directs the pollen tube growth, but it may also have other important physiological functions as well. In many cases, the micropyle is occluded after fertilization.

6.3.2.2. Integuments

The ovules, depending on the species, are unitegmic or bitegmic. The former condition is seen only in some dicot families, and the latter is seen in both dicots and monocots. Although specific to each family, some 20 families have both unitegmic and bitegmic taxa. In bitegmic integuments, the inner always arises first followed by the outer. In bitegmic ovules, the outer integument is derived from subdermal initial while inner integument is derived from dermal initial, or both the integuments are dermal in origin. In unitegmic ovules, it is always dermal in origin. Integuments may become secondarily multilayered. The growth of the integument may be apical, basal or diffuse. Integuments may or may not be vascularized. Stomata and chloroplasts are seen in the integuments of some taxa. It is generally believed that unitegmy is derived from bitegmy either through elimination of one integument or through the fusion of two integuments into one.

In the taxa of some families, where the ovule possesses unitegmic condition and reduced amount of nucellus, the innermost cell layer of the integument bordering on the embryo sac differentiates into a cytologically–distinct unit, the **endothelium** or **integumentary tapetum** (Fig. 169). The constituent cells undergo conspicuous radial elongation at the completion of megasporogenesis. The component cells of the endothelium divide in anticlinal plane during the growth of the ovule and look palisadal at any given stages of development. The extent of endothelial differentiation varies in different families and can be categorized into four specific patterns:

1. The endothelium covers the embryo sac completely except at the extreme poles
2. The endothelium is limited to the antipodal half or three–fourths of the embryo sac.
3. The endothelium is confined to half to three–fourths of micropylar region of the embryo sac.
4. The endothelium is confined to the middle part of the embryo sac.

It is generally believed that the endothelium provides nutrition to the embryo sac and also protects the embryo sac. Although it is likely to serve a nutritive role, the nature of the nutritive substance provided or how it is provided is not clear. The cells have dense cytoplasm and prominent nucleus (or more than one nucleus) that may be polupAloidal or polytenic. The walls often have transfer cell morphology. The endothelial layer enlarge radially (with or without accompanying cell divisions) and occludes the embryo sac in incompatible crosses and in otherwise sterile ovules. This process is called **somatoplastic sterility**.

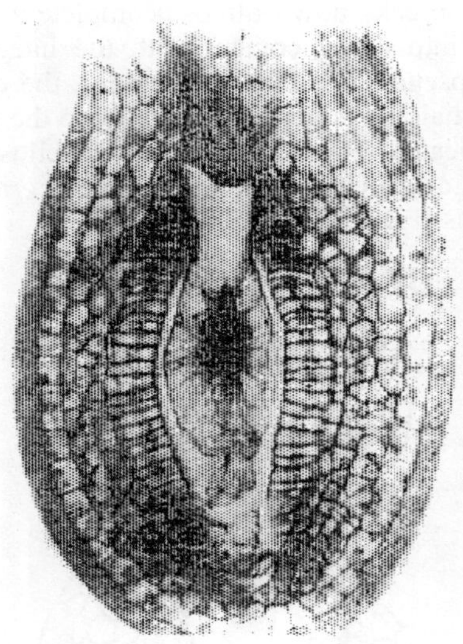

Fig. 169. L. S. of ovule of *Downingia bacigalupii* just after fertilization showing zygote, darkly-coloured pollen tube, the primary endosperm nucleus and a prominent endothelial layer around the embryo sac (Photograph courtesy of Dr. D. A. Kaplan)

6.3.2.3. *Nucellus*

The tissue of the ovule enclosed within the integument(s) is the **nucellus.** Depending on the extent of this tissue, two extreme conditions are recognized:

1. **Tenuinucellate,** where the nucellus is represented by a few cells either arranged in a linear fashion or as a mass (e.g., Scrophulariaceae and Campanulaceae)
2. **Crassinucellate,** where the nucellus is of considerable volume (e.g., Burseaceae, Cucurbitaceae and Malvaceae).

The absence or formation, respectively, of a parietal cell is usually associated with two types of nucellus. It is the repeated division products of the parietal cell that contributes to the massive crassinucellate condition. In view of this, the megaspore mother cell is towards the nucellar epidermis or deeply embedded (Fig. 170).

Cells located at the base of the nucellus also becomes meristematic in some cases and adds to the crassinucellate condition. In some cases, the nucellar epidermis at the micropylar end divides repeatedly periclinally to contribute to the mass of the nucellus. This type of nucellus has been named as **pseudocrassinucellate** by Davis (1966). In tenuinucellate ovules, the nucellus breaks down almost completely and the developing embryo sac comes into direct contact with the integument(s), while in crassinucellate and pseudocrassinucellate ovules, the disintegration of the nucellus is often partial and is generally limited to the apical region where the embryo sac is located. Thus, a part of the nucellus persists even up to the mature seed stage as **perisperm** (e.g., *Piper nigrum*) in many ovules with massive nucellus (Fig. 222).

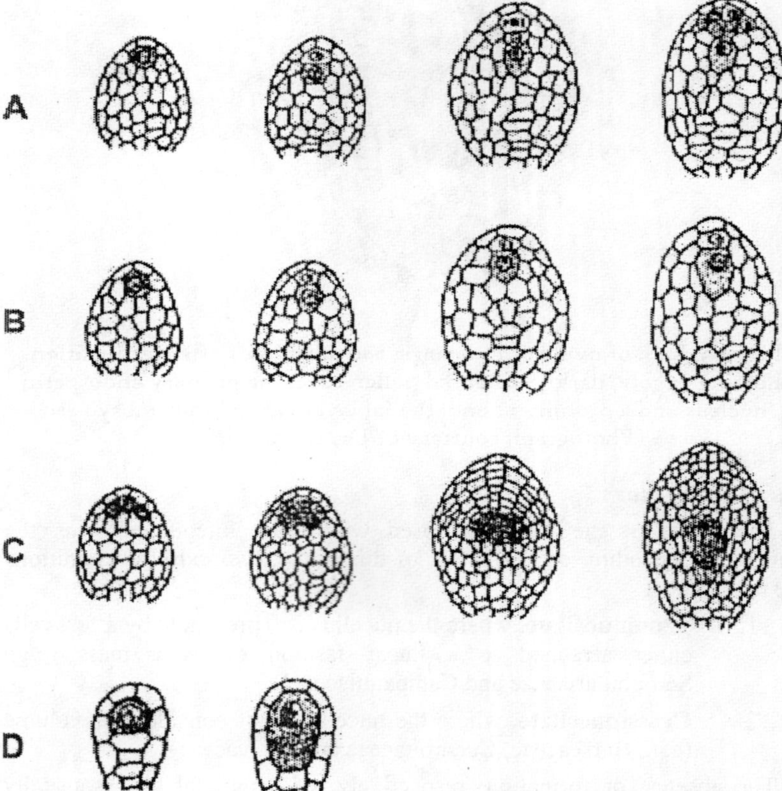

Fig. 170. Different types of nucellus based on ontogeny and quantum of cells of the nucellus; A - Crassinucellare ovule with parietal cell contributing to nucellar cell increase; B - Crassinucellate ovule where cells are not contributed by the parietal cell, but by the divisions of the existing cells; C - Pseudocrassinucellate ovule with epidermis contributing cells to nucellus; D - Tenuinucellate ovule where the nucellus is with retain meagre cells. The megaspore mother cell lies immediately inside the nucellualr epidermis. (Swamy and Krishnamurthy 1980a)

6.3.2.4. *Chalaza*

The part of the ovule that is subjacent to the base of the integument(s) is designated as the **chalaza.** It is the region of the ovule from where the integuments originate and where there is no distinction into nucellus and integument(s). Chalaza indicates a pole of the ovule, which appears to be the seat of important metabolic activities during successive developmental stages, both during pre- and post–fertilization stages. Hence, it is difficult to delimit the boundaries of chalaza either morphologically or physiologically. Attention has already been drawn to the role played by the chalaza in the establishment of the different configurations of ovules, in the origin of integuments and in the organization of an intercalary meristem to increase the nucellar volume. It is to be pointed out here that the ovular traces terminate in the chalaza and further ramification of the integumentary vasculature, if present, is also initiated in the chalaza both before and after fertilization. More pronounced growth of the embryo sac also takes place only in the chalazal direction.

The chalaza serves as an important topocentre of the ovule. As indicated above, it is from the chalaza that the integuments take origin. It is also the topocentre for the initiation of the aril in arillate seeds. Recent molecular–biological data have indicated that the ovule develops and matures by maintaining a distinct proximal–distal axis (Grossnicklaus and Schneitz 1998) and this axis is characterized by a three–tiered arrangement of pattern elements: Distal (or nucellar), Chalazal (or central) and Proximal (funicular) domains (Fig. 171). Recent data further suggest that the chalazal domain can be subdivided into two subdomains (Baker et al 1997), based on its role in the production of either inner or outer integument. About a dozen genes are already known to affect the integument in *Arabidopsis* by operating at the chalazal domain (Table 4).

Some ovules exhibit the differentiation of a histologically–distinctive pad of cells in the chalazal region immediately outside the nucellus and it is called **hypostase** (e.g., members of Malvaceae, Onagraceae) (Fig. 172). The other terms used for hypostase are **postament**, **podium** and **pseudochalaza**. The constituent cells may be thick–walled or thin–walled depending on the taxa. Hypostase has been believed to play a role in the supply of nutrients to the growing embryo sac, in stabilizing the moisture status of the ovule and in controlling the chalazal–ward elongation of the embryo sac. In some, an **epistase** is present on the micropylar part of the nucellus (e.g., *Costus*).

Table 4. Chalaza as a topocentre of operation of genes involved in ovule or seed development. All genes in the mutant form affect the development (Data from different sources).

Sr. No.	Wild-type genes whose mutant forms are known to be involved	Effect seen in mutants
1.	*INNER NO OUTER INTEGUMENT (INO)*	Only inner integument is formed; no outer integument.
2.	*BELL 1 (BEL 1).*	Both the integuments are not organized. Inner integument is totally absent, while the outer is represented by an amorphous entity or collar-like structure.
3.	*SUPERMAN (SUP)* (Also known as *FLO10* gene)	Development of outer integument is affected; grows asymmetrically around the ovule.
4.	*SHORT INTEGUMENT1 (SIN1)* (maternal gene)	Integument development is affected; no clear distinction between inner and outer integuments; they are extremely short. The gene also causes defects in embryo such as funnel shaped cotyledon or masses of unorganized tissue.
5.	*AINTEGUMENTA (ANT)*	It lacks integuments. Also, there is no embryo sac formation.
6.	*HUELLENLOS (HLL)*	It lacks integuments. Also, there is no embryo sac formation.
7.	*ABERRANT TESTA SHAPE (ATS)*	It produces a single fused integument and hence, no distinction between two integuments.
8.	*UNICORN (UCN)*	It produces supernumerary integuments.
9.	*BLASIG (BAG)*	It affects cell division and cell shape in integuments.
10.	*STRUBBELIG*	It affects cell division and cell shape in integuments.
11.	*FBP7, FBP11* (MADS-Box genes) (in Barley and *Petunia*)	They affect ovule determination and development in carpels; there is no endospern development

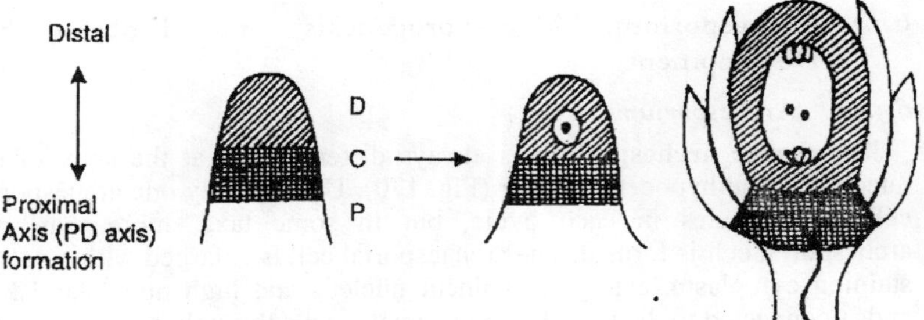

Fig. 171. Diagrammatic representation of the proximo–distal polarity in a developing ovule showing the three pattern elements. The proximal (P) domain forms the funicle, the central or chalazal (C) domain forms the chalaza–integument complex and the distal (D) domain forms the nucellus and embryo sac. Chalazal domain may possibly have two subdomains (Modified from Grossniklaus and Schneitz 1998).

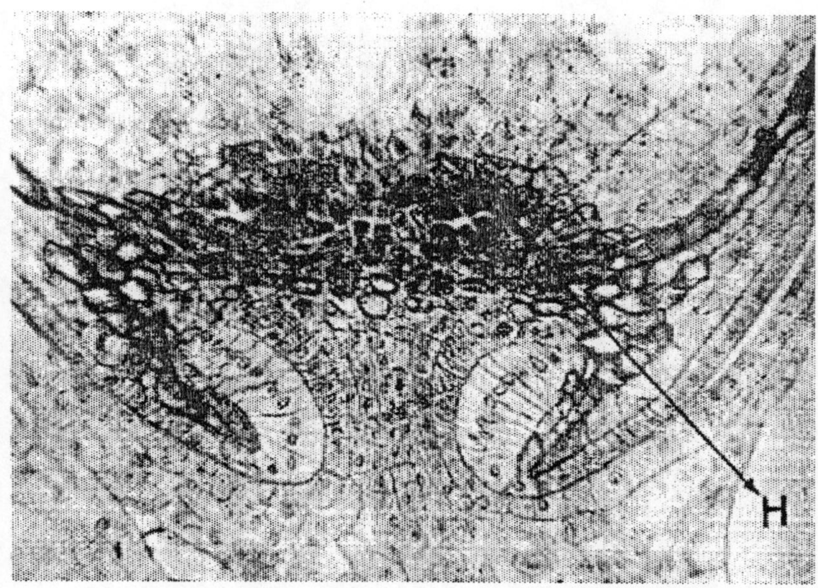

Fig. 172. Median L. S. of the chalazal part of the ovule of *Helianthemum* showing hypostase (H) (Kapil and Maheshwari 1964) (Reproduced with permission from International Society of Plant Morphologists, New Delhi)

6.3.3. Archesporium, Megasporogenesis and Embryo Sac Development

6.3.3.1. *Archesporium*

The female **archesporium** is always differentiated at the apex of the nucellus in the hypodermal layer (Fig. 170). Usually only one archesporial cell differentiates in each ovule, but in some taxa, more than one archesporial cell is formed. The archaesporial cell is enlarged with a deeply staining cytoplasm, a large prominent nucleus and high nucleolar RNA, and is connected to the neighbouring somatic cells through plasmodesmata.

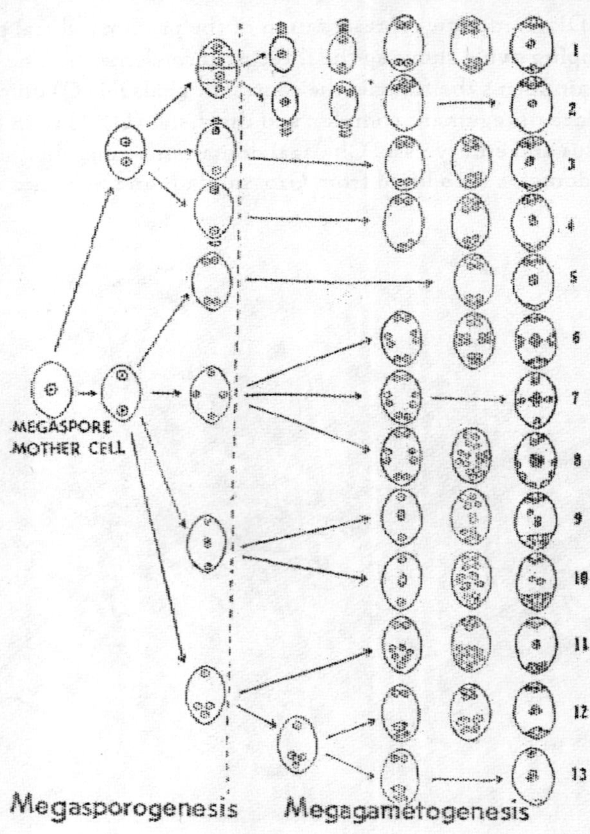

Fig. 173. Schematic representation of different types of embryo sac development in angiosperms; (1 and 2 - Monosporic types; 3 and 4 - Bisporic types; 5–13 - Tetrasporic types); 1 - Polygonum; 2 – oenothera; 3 - Allium; 4 - Endymon; 5 - Aoxa; 6 - Penaea; 7 - Plumbago; 8 – Peperomia; 9 and 10 - Chrysanthemum cinerarifolium; 11 - Drusa; 12 – Fritillaria; 13 - Plumbagella types.

In crassinucleate ovules, the archaesporial cell cuts off a **primary parietal cell** towards the micropylar end and an inner **sporogenous cell**. The derivatives of parietal cell add cells to the nucellar volume. In some tenuinucellate ovules, the parietal cell does not divide further and hence, the sporogenous cell is located very near the surface. In crassinucellate and psuedocrassinucellate ovules, the sporogenous cell is located deep in the nucleus. The sporogenous cell enlarges and becomes the **megaspore mother cell** (Figs. 170 and 173).

6.3.3.2. *Megasporogenesis*

The megaspore mother cell or **megasporocyte** somewhat elongates parallel to the long axis of the nucleus. Its nucleus is positioned nearer to the micropylar pole of the cell. Just before the onset of meiosis, the plasmodesmatal connections that the cell had with the neighbouring cells are lost and a special wall made of callose (α β, 1–3 glucan) is deposited around the cell. If callose wall is not deposited, the megasporocyte will not enter into meiotic cell division (Krishnamurthy 1976). The deposition of callose seems to have an intrinsic relationship with the differentiation of a functional megaspore and the type of embryo sac ontogeny. There is unequal deposition of callose on the wall of the megaspore mother cell and this is, in fact, related to the polarization of the cell organelles and the cell as a whole. Callose is laid down first and then becomes thickest on the side of the wall where the functional megaspore is destined to develop i.e., at the chalazal pole in **monosporic**, micropylar pole in **biosporic** and all around the cell in **tetrasporic** types of embryo sac development. Significant ultrastructural changes and regrouping of organelles occur in the megaspore mother cell as the hallmark of changeover from the diploid to the haploid state. The plastids and mitochondria are cycled through a stage of dedifferentiation, which begins at the early meiotic prophase, and then through a stage of redifferentiation at the tetrad stage. During dedifferentiation, there is a loss of contents from the plastids and the spherical or rod–like conformation of mitochondria that is accompanied by the appearance of electron–dense moieties. The organelles become highly dispersed and are far away from one another. At tetrad stage, the organelles regain their normal configuration. There is an array of new ribosomes. This is followed by a steadily increasing number of small nucleoli, which begin to bud off from the nucleolus of the nucleus of megaspore mother cell. Polysome initiation and appearance of paracrystalline protein inclusions happen towards the end of meiosis. The concentration of cytoplasmic and nucleolar RNA decreases during the onset of meiosis, probably due to the rapid increase in the size of the megaspore mother cell (three to four times more the size of archesporial cell), but much of the fall results from a prophase diminution or almost total loss of ribosome content.

The behaviour of megaspore mother cell and the method of meiotic cell division depends upon whether the embryo sac development belongs to the monosporic, bisporic or tetrasporic types (Fig. 173). In the monosporic type, where only one of the four megaspores (especially megaspore nuclei) gives rise to the embryo sac or female gametophyte or megagametophytes, the nucleus of the megaspore mother cell divides meiotically **(heterotypic division)** *in situ,* followed by the deposition of transverse wall resulting in a smaller micropyle cell and a larger chalazal cell. It must be emphasized here that it is during this division that the two resultant daughter nuclei attain the haploid chromosome constitution and thereby acquire the requisite for the sexual cell. Both the dyad cells again divide (**homotypic division**) by transverse wall to result in four, linearly arranged megaspores that collectively constitute the **megaspore tetrad.** The dividing walls are richly callosic, richer than the lateral walls. The three micropylar megaspores are generally smaller and degenerate early due to PCD, as indicated in Chapter 4 of this book. The uninucleate chalazal megaspore, which has maintained a relatively larger size throughout megasporogenesis, undergoes further overall enlargement. During this process, the wall becomes greatly depleted of its callose and becomes thinner (with callose persisting). A condition where the micropylar megaspore of the tetrad becomes functional is indeed rare, but has been reported, raising the question: What mechanism determines the selection of the functional megaspore from a group of four megaspores?

In the bisporic development, a wall is formed between the **dyad** nuclei following the first division of the megaspore mother cell i.e., at the end of meiosis I. The micropylar or chalazal dyad cell degenerates sooner or later through PCD, and the persistent dyad is involved in further development. The persistent dyad's nucleus completes the second meiotic division (homotypic or mitotic division) and results in two megaspore nuclei (haploid) (Fig. 173).

In the tetrasporic development, the meiotic division in the megaspore mother cell results in four haploid megaspore nuclei all inside the same cell (**coenomegaspore**) (Fig. 173). Thus, in monosporic, bisporic and tetraporic developments essentially one, two or four megaspore nuclei (megaspores) that respectively take part in the formation of the embryo sac. All stages leading to the formation of the four megaspores from a megaspore mother cell are described as megasporogenesis.

6.3.3.3. Megagametogenesis

The formation of the embryo sac from one, two or four megaspores is designated as **megagametogenesis.** Further classification of the monosporic, bisporic and tetraporic embryo sac types is based on the number of nuclei or cells contributing to the mature organization of the embryo sac, number of further nuclear divisions, nuclear fusions, loss of nuclei, type of mature organization of the embryo sac etc. The monosporic category is classified further into **Polygonum** and **Oenothera** types, bisporic into **Allium, Endymion** and **Schizandra** types and tetrasporic into **Adoxa, Plumbago, Penaea, Peperomia, Drusa, Fritillaria,**

Plumbagella and **Chrysanthemum cinerarifolium** types (Fig. 173) (Swamy and Krishnamurthy 1980a). A modified classification of embryo sac types has been provided by Swamy and Krishnamurthy (1975a and 1980a), where importance has been given to the time of chromosome reduction.

6.3.3.4. *Organisation of the Mature Embryo Sac*

Once the required numbers of nuclei are formed depending on the type of embryo sac ontogeny, their organization sets in. Invariably, three nuclei get organized into cells at the micropylar end as an **egg apparatus,** with an **egg cell** in the centre and two **synergids** one on either side of the egg cell. Two nuclei move to the middle of the embryo sac and get organized as part of the **central cell.** These two nuclei are called **polar nuclei.** Three nuclei organize themselves as the **antipodal cells** at the chalazal end of the embryo sac. The egg has its nucleus positioned towards its chalazal end and a large vacuole occupies its micropylar end (Figs. 174 and 175). The egg cell has a wall only at its micropylar facet or the wall may be totally absent all around it in some taxa. Whether a thin wall is formed at the chalazal part of the egg and then gets disappeared or whether from the beginning there is no wall in this region is a matter of dispute, since both conditions have been recorded in the literature. The absence of a wall at the chalazal region of the egg is apparently necessary for the ultimate transfer of male gamete from the synergid to the egg. The orientation of the endoplasmic reticulum (ER) in egg cell is parallel to the plasma membrane over a large area of the cell, but forms a network at the micropylar end.

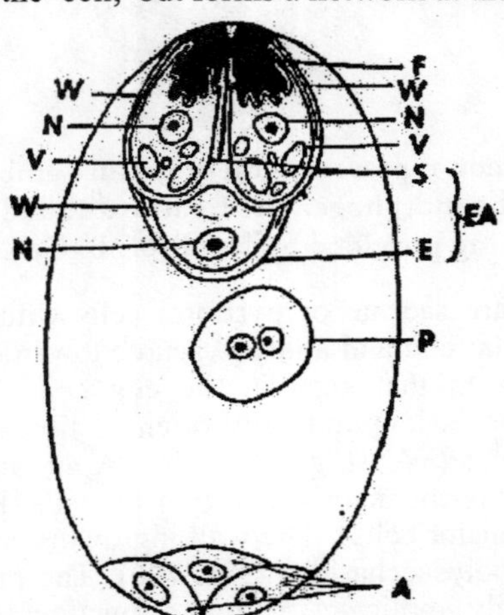

Fig. 174. Diagrammatic representation of a typical mature embryo sac; A - Antipodals; E - Egg; EA - Egg apparatus; F - Filiform apparatus; N - Nucleus; P - Secondary embryo sac nucleus; S - Synergid; V - Vacuole; W - Wall

Organelles such as mitochondria (more in the micropylar region), Golgi bodies and plastids are associated with ER and all of these are concentrated around the nucleus, and ribosomes are also abundant. Starch may be present. There is a prominent nucleus in the egg. A substantial amount of DNA is detected in the cytoplasm, but its exact location is not clearly known. Histones are also present in the cytoplasm. In *Plumbago and Plumbagella types,* where synergids are absent, the egg itself has a **filiform apparatus** (FA) (Refer to Swamy and Krishnamurthy 1980a).

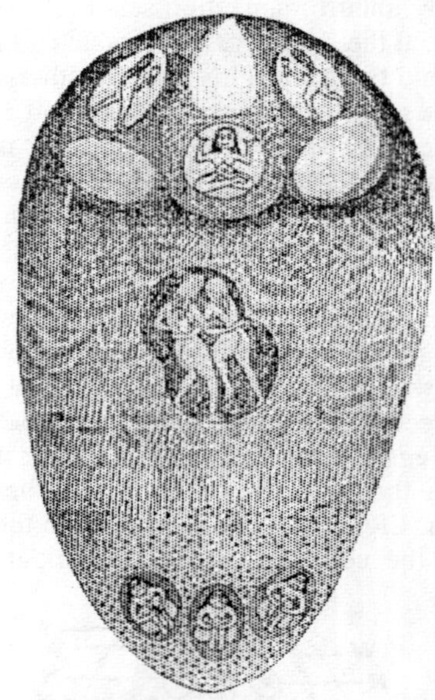

Fig. 175. Cartoon representation of mature embryo sac showing differences in the morphogenetic fields exhibited by the embryo sac (Drawing provided by late Prof. B. G. L. Swamy)

The synergids are saccate or pyriform cells with their nuclei situated near their micropylar end and a large vacuole towards their chalazal end, a condition opposite to that seen in the egg cell. The apical region of synergids often shows longitudinally oriented fibre–like structures called the filiform apparatus (FA) (Fig. 174). The FA, in fact, is a special type of wall ingrowths and is characteristic of transfer cells (Refer to Chapter 4 for more details on transfer cells). The wall ingrowths are finger or plate–like and are made of polysaccharides (Fig. 196). The presence of FA in the synergids necessarily implicates them in a functional role as transfer cells, but the direction of translocation, whether into or out of these cells is not clear. But, it is more probable that it is out of these cells that all substances, especially chemotropic substances, are transferred to direct the pollen tube. Some believe that the FA of the two synergids does different functions.

The FA of the degenerated synergid controls the pollen tube growth through the transfer of chemotropic substances out of it, while FA of the other synergid transfers materials into the embryo sac from other regions (especially micropylar) of the ovule. The material released by FA is synthesized within the synergids. It consists of a homogeneous osmiphilic substance indicating its non–cellulosic composition; cell ablation studies in synergids of *Torenia fournieri* demonstrate their control over both pollen tube guidance and reception (Higassiyama 2001). The wall ingrowths have a central core with tightly packed microfibrils of cellulose having an outer sheath of non–microfibrillar material. Synergids are connected to the central cell through plasmodesmal connections. The maximum concentration of ER in the synergids is towards their micropylar end and its density is decreasing towards the base. Dictyosomes are seen in great density in the middle region of the cell, while spherosomes are spread over evenly throughout the cell. The structural pattern described above is taken to indicate that the synergids before fertilization are in a more metabolically active state than the egg (Swamy and Krishnamurthy 1980a).

Normally, there are three antipodal cells or nuclei located at the chalazal end of the embryo sac. Only one is seen in Plumbago and Plumbagella types of embeyo sac. In some plants, antipodal cells are with more than one nucleus. Syncytial antipodals are also seen in some cases. In a few plants, the antipodal nuclei undergo endoreduplication and reach up to 1024 C, frequently displaying polytene chromosomes without distinct bands. Nucleolar DNA amplification is also seen in some taxa. In some species, especially grasses, the antipodal cells increase in number. In the grass *Sasa paniculata,* up to 300 antipodal cells have been reported. In yet other taxa, the antipodal nuclei or cells are highly ephemeral and disappear sooner or later. The antipodal cells often have been suggested to play some role in the control of endosperm development, at least in some grasses, although considered for a long time as inert structures not doing any important function. Recent researches indicate the involvement of antipodals in the secretion, absorption and transport of nutrients to the developing central cell and egg before and after fertilization. This is evident from the presence of active nucleoli, a ribosome–polysome pattern, active ER, transfer cell morphology of walls (restricted in some to most–chalazally located cells, but may be seen in cells towards the central cell as in *Spinacea),* plasmodesmata (between antipodals and between antipodals and central cell) etc. Degeneration of antipodals starts from the micropylar end onwards. Antipodal haustoria develop in some taxa (Swamy and Krishnamurthy 1980a).

The central cell is the largest cell of the embryo sac. It normally has two nuclei, the polar nuclei (a misnomer according to some, as they do not occur at the poles of the embryo sac) remaining juxtaposed or exhibiting various degrees of fusion or even complete fusion to form a diploid **secondary nucleus**. Only one nucleus is present in Oenothera type of embryo sac while more than two occur in Penaea, Plumbago and Peperomia types. The central cell is highly vacuolated and its cytoplasm consists of all types of organelles, with a high proportion of dictyosomes, which are in an active state. Around the polar nuclei or secondary nucleus an increasing ER system is developed. There is often a tremendous increase in surface area of the nuclear envelops. Starch is stored abundantly in some taxa, while lipid accumulation is seen in some others. Nucleoli are invariably large. There is abundant cytoplasmic RNA, presumably mostly ribosomal. Free ribosomes and many mitochondria are present, especially near the micropylar end. The vacuole of central cell constitutes an important reservoir of sugars, amino acids and inorganic salts. The vacuolation is considered to be looked upon as a preparation for the rapid growth in length of the embryo sac during post–fertilization stages. The walls of central cells, in some taxa, show transfer cell morphology (Swamy and Krishnamurthy 1980a).

6.3.4. Gene Expression during Embryo Sac Development

Gene expression during megagametogenesis has a dual function: (i) Orchestrating the gametophytic programme during the division of the megaspore nucleus, and (ii) Assigning the characteristic fates to different cells formed (Raghavan 2000). There is a likelihood of differential gene expression in the different cells of the embryo sac, although derived from a single source, as indicated by differences in mRNA profiles and ribosomal population of these cells. For example, more mRNA and ribosomes are especially characteristic of the egg apparatus and central cell. Genetic control of embryo sac organization was first revealed by the presence of the *gametophytic factor1* (*gf1*) mutant in Arabidopsis. Only after mid-1990s, a large number of gametophytic mutants have been isolated (Brukhin et al. 2005), which are deficient in one or more of the developmental processes of the embryo sac such as megaspore specification, initiation of megagametogenesis, mitosis, establishment of polarity, migration of polar nuclei and their fusion, cellularization of embryo sac, antipodal cell death and PCD of synergids. Defective organization of embryo sac is revealed by ***bel1***, ***sin***, ***ant*** and ***prolifera*** (***prl***) mutants of *Arabidopsis*. In the first three mutants, embryo sac development is affected either through the absence of meiotic division or by the failure of functional megaspore to complete its nuclear divisions to form the embryo sac. The *PRL* gene is expressed throughout the plant body, especially in dividing cells, but its effect in the arrest of embryo sac development at four–nucleate stage is very significant in producing a defective embryo sac. The other embryo sac–specific (but not affecting sporophytic tissues of the ovule) mutants of *Arabidopsis* are generally classified under the phenotypic classes *megasporogenesis defective* (*msd*) and *embryo sac defective* (emd) (Schneitz et al 1997).

6.4. ANTHER AND MALE GAMETOPHYTE

6.4.1. Introduction

The **anther** is the male sexual counterpart of the megasporangium. The term **microsporangium** is often used as a synonym for the anther, but the former term has a much wider connotation as it represents the homologue of the microspore–producing structures of non–angiospermous tracheophyta groups, especially the pteridophytes. Though there are a number of similar developmental features between the anther of angiosperms and the microsporangium of other tracheophytes, the gametophyte organization and behaviour present significant divergences. The male gametophytic cycle in the angiosperms is extremely abbreviated in time and space, and the male gametophytic body is often composed of just two cells– the **vegetative** and **generative cells** (Swamy and Krishnamurthy 1980a). Anther and male gametophyte development involoves very precisely controlled cellular processes and a diverse range of genes and their interactions (see details in Scott et al. 2004; Ma 2005; Yang et al. 2007).

6.4.2. Structure and Development of Anther

6.4.2.1. *Structure of Anther*

A typical anther is tetrasporangiate, although bisporangiate (E.g., Malvaceae), unisporangiate (E.g., *Arceuthobium*) and octosporangiate (E.g., Bixaceae) anthers are also known in angiosperms. There is a central column of sterile tissue called the **connective** through which vasculature extends from the filament or anther stalk. In tetrasporangiate anthers, at maturity, the four sporangia coalesce to form two thecae or bags containing the pollen (male gametophyte). The sporangia are usually surrounded by an epidermal layer on the outside, followed by wall layers on the inside. The latter is made up of an **endothecial layer** (sometimes more than one), one to a few **middle layers** and a **tapetal layer** covering the sporangial locule (Fig. 176).

4.2.2. *Development of Anther Wall*

The anther primordium, as seen in transection, appears somewhat squarish to rectangular in outline. The cells are homogeneous and parenchymatous with a covering layer of epidermis. The **primary archesporium** differentiates as a single or a group of two to few adjacently–placed cells in the hypodermal layer at the four corners of the anther primordium (Fig. 177). In fact, these cells extend vertically from base to apex of the sporangium (Fig. 177 A). The archesporial cells are distinct from the rest of the cells by their larger size, and greater avidity for nuclear and cytoplasmic stains. The archesporial cells divide by periclinal wall to form an outer **primary parietal layer** and an inner **primary sporogeneous layer**. Generally both the layers undergo further periclinal

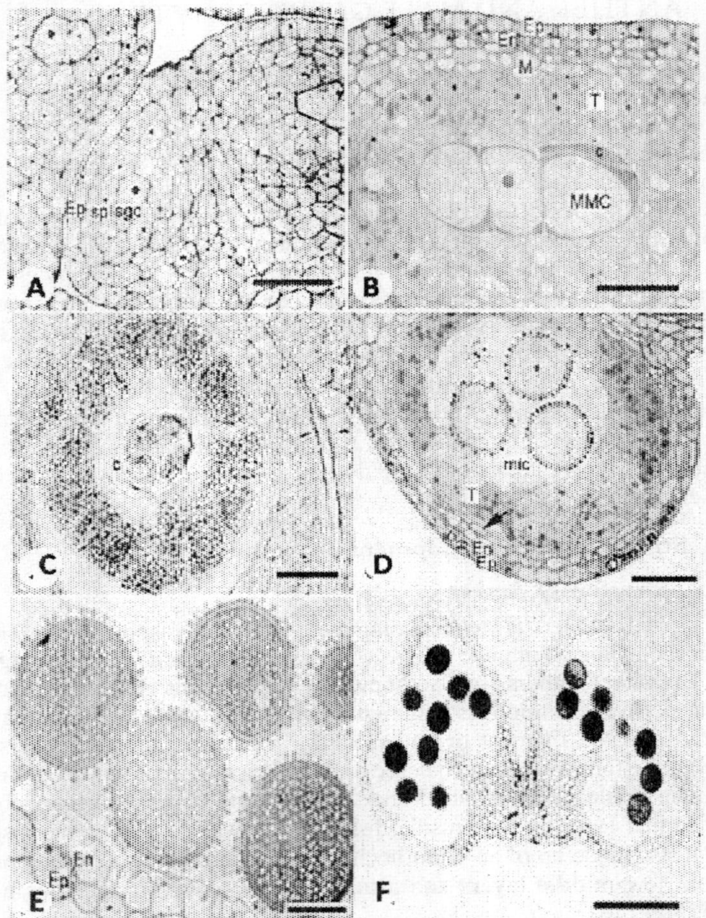

Fig. 176. Photographs of stages in anther, microspore and male gametophyte development in *Jatropha curcas*; A - Sporgeneous cell (spc) and secondary parietal layers (spl) beneath epidermis (Ep); B - Microspore mother cells (MMC) enclosed by callose (c); anther wall consists of epidermis (Ep), endothecium (En), middle layers (M) and tapetum (T); C - TEM photograph of tetrahedral microspore tetrad enclosed by callose (c); note the prominent tapetal layer around the locule; D - Free microspores (mic) with prominent, centrally placed nucleus; anther wall consists of epidermis (Ep), Endothecium (En), prominent tapetum (T); E - Mature anther T. S. with epidermis (Ep), prominent endothecial cells with thickening bands (En) and microspores; F - T. S. mature anther with four microsporangia fused to form two thecae at the time of dehiscence (Liu et al 2007) (Reproduced with permission from Torrey Botanical Society)

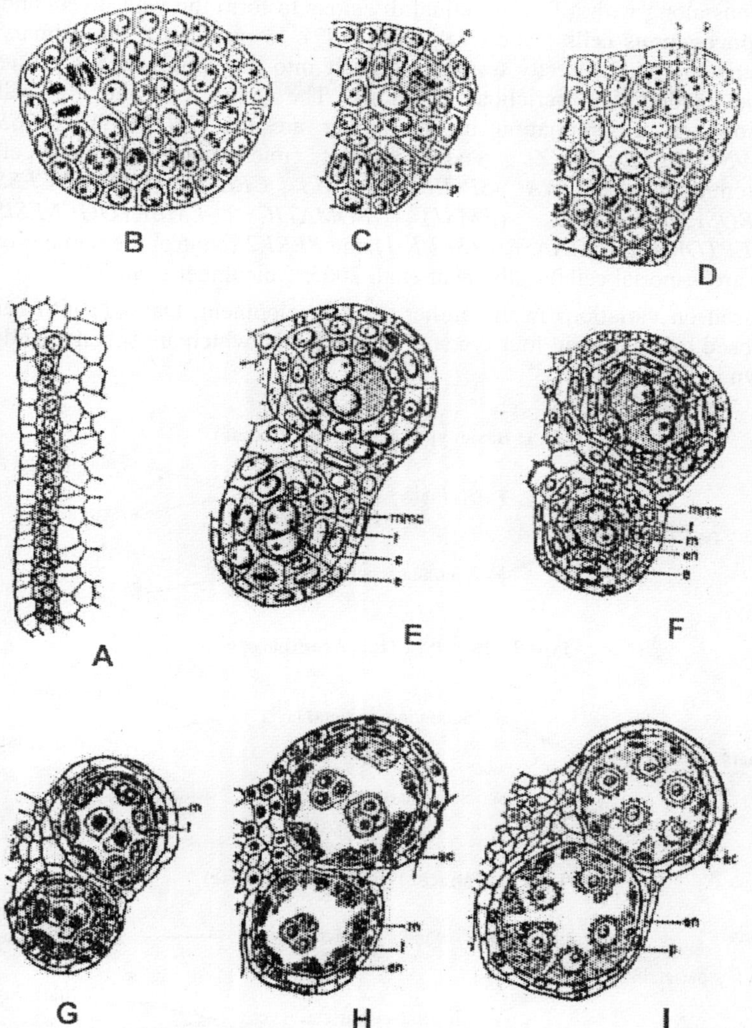

Fig. 177. Sketches of anther, microspore and male gametophyte development; A - L. S. of thecal portion of a very young anther of *Cananga odorata* showing hypodermal archesporial cells; B–I - Successive stages of anther development in *Podolepis jaceoides*; Arrows in E and F indicate derivation of inner tapetum from the connective tissue cells. a - Archesporial cell; c - Hypodermal layer; e - Epidermis; ec - Epidermal cuticle; en - Endothecium; m - Middle layers; mmc - Microspore mother cells; p (in C, D) - parietal cell; p (in I) - Disintegrated portion of tapetum; s - Sporogenous cells; t - Tapetum (Davis 1961).

divisions along with a few anticlinal divisions to form the wall layers and the **sporogenous cells** respectively (Fig. 177). In rare cases, the primary sporogenous layer directly transforms itself into the sporogeneous tissue without undergoing periclinal divisions. The genes required for cell division and differentiation in the anther are the *SPOROCYTELESS (SPL)/NOZZELE (NZZ)* gene (controls microspore mother cell differentiation), *EXTRA SPOROGENOUS CELLS (EXS)/EXCESS MICROSPOROCYTE1 (EMS1) SOMATIC EMBRYOGENESIS RECEPTOR-LIKE KINASE 1 (SERK 1)* and *SERK2* (control the number of male archesporial cells) (Alberecht et al. 2005; Colcombet et al. 2005)

Based on variations in the anther wall development, Davis (1966) has proposed the following four types of anther walls, which are schematically shown in Fig. 178.

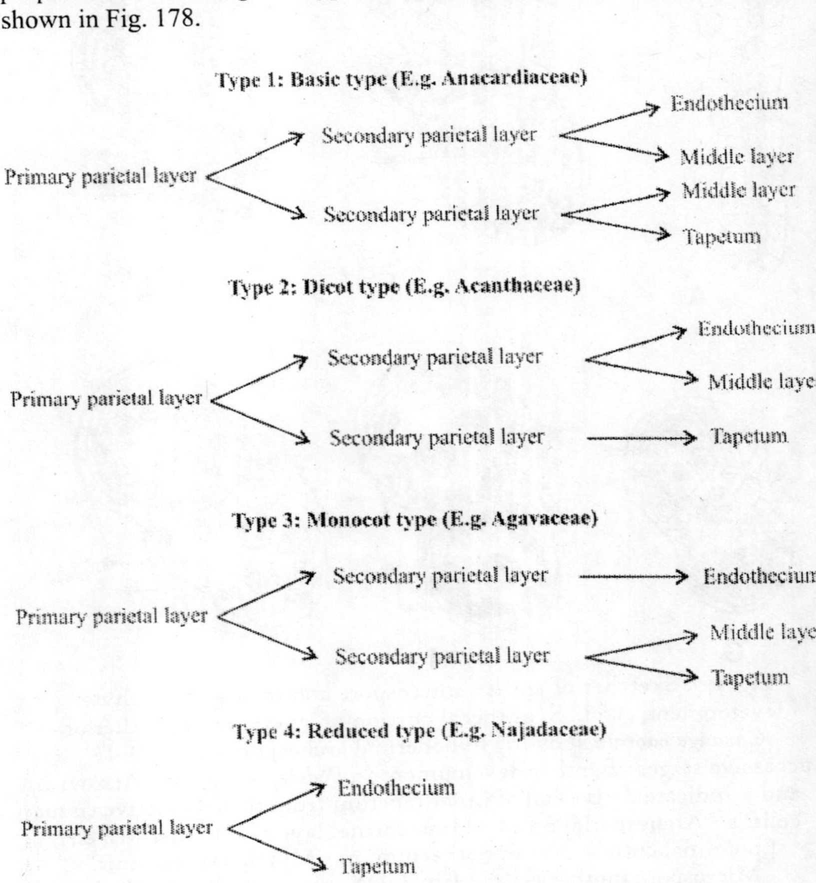

Fig. 178. Types of anther wall development (Davis 1966)

Endothecium

The endothecial cells are often radially elongated and develop thickenings in the inner tangential walls and rarely on radial walls too at anther maturity (Fig. 179). The morphological patterns of thickening conform to varieties of banded type. Although the thickenings absorb aniline dyes readily, the thickening material is not callose but α–cellulose; in some it may get slightly lignified. In the anthers of aquatic taxa, saprophytes, cleistogamous flowers and extreme parasites, endothecial differentiation is absent. Transcriptional activity is required for the differentiation of the endothecium and this is evident in the localization of poly(A)–RNA in rice anthers by in situ hybridization using [^{3}H] poly(U) as a probe (Raghavan 2000). Just before meiosis, although poly(A)–RNA concentration sharply decreases in epidermis and middle layers, a considerable amount of poly(A)–RNA is retained in the endothecium. Low levels of poly(A)–RNA are also found in the endothecium even after the completion of meiosis in the microspore mother cells (Raghavan 2000). The histone *H3* gene also is shown to activate the differentiation of endothecium in both rice and wheat plants; this gene is especially expressed in the endothecium of anthers of wild–type and transgenic rice. However, the mechanism by which this gene helps in endothecial differentiation is not clear yet.

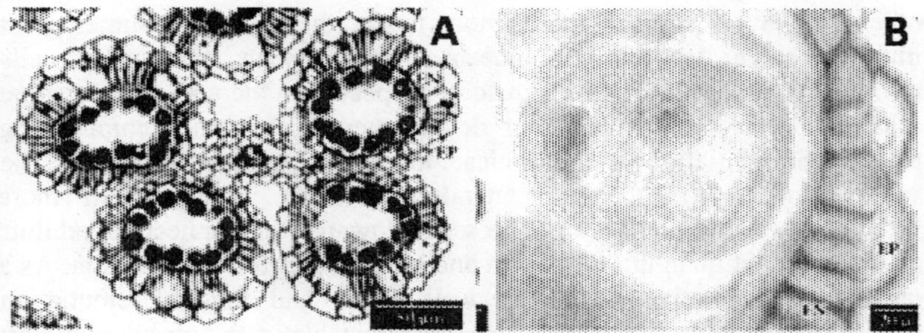

Fig. 179. A - T. S. of anther of *Olyra humilis* showing prominent endothecium and cellular tapetum; B - T. S. of portion of the anther of *Eragrostis solida* showing prominent endothecium (EN), epidermis (EP) and two–celled pollen (Nakamura et al 2010) (Reproduced with permission from *Instituto Internacional de Ecologia* São Carlos SP - Brasil)

A very critical analysis of anther–specifc cDNA clones in tobacco anthers supports the contention that the whole programme of cytodifferentiation of the anther wall, disintegration of middle layers and anther dehiscence is a cascade of temporal and spatial gene expression events orchestrated in the anther wall (Raghavan 2000). The cDNA clones implicated in anther wall differentiation are *TA56* encoding a thiol endopeptidase, and *TA20* encoding an unknown protein. By following the expression of these two clones by in situ hybridization, it was shown that *TA56* transcripts accumulated in the anther wall in the prospective stomium region at a very early stage of development. As the anther matures, there is an appreciable decrease in the intensity of hybridization signals in the cells in and around the stomium at the same time as the cells around the connective acquire the signal. *TA20* transcripts are initially expressed in all layers of the anther wall, but as the anthers mature they are found concentrated in the connective cells around the vascular bundle. Selective expression of *TA56* gene transcripts in the stomial region suggests a role for endopeptidase in anther dehiscence (Koltunow et al 1990). When the stomium region alone is ablated with a cytotoxic gene fused to the *TA56* gene promoter, the anther development is normal, but fails to dehisce (Beals and Goldberg 1996), again supporting the above finding. Transcripts of anther– specific cDNA clones isolated from tomato anthers are also expressed in the wall layers of anther, particularly in epidermis and endothecium. The proteins encoded by these genes show homology to Kunitz trypsin inhibitor (KTI) and pectinase enzyme. Genes encoding a number of other enzymes are also expressed in the anther wall layers of the plants (Raghavan 1997).

Anther tapetum

The innermost layer of the anther wall is the **tapetum**. It borders on the mass of sporogenous cells. Because of its strategic position between the other wall layers and the sporogenous tissue, the tapetum assumes great importance. The tapetum is unique in several respects, although its cells resemble the sporogenous cells in some respects and the wall cells in some other respects. The sequence of developmental changes accompanying tapetal differentiation is very clear towards the outer sector of the sporangium as the thecum generally increases in dimension more prominently along this sector. The sector towards the connective exhibits an insignificant quantum of growth and the least degree of distortion. As a result, the distinctiveness of all wall layers and their differentiation (especially tapetum) are more clearly recognizable in the protuberant part of the anther thecum. It is again along this sector that the origin of the tapetum from the innermost wall layer is clearly demonstrated. The differentiation of the tapetum towards the connective side, however, is not in any way related to the derivatives of parietal layer. The cells of the connective tissue that borders the sporogenous cells undergo appropriate histological changes and align themselves in line with the tapetum of parietal origin (Fig. 180). Thus, the anther tapetum is **dual** in origin. Although the dual origin of tapetum is eventually lost in many plants, some histological distinctions are still retained in a number of species. There are differences in cell size, cell shape, number of cell layers, nuclear number, time of differentiation etc. between the proximal and distal tapeta.

The tapetal cells are richly protoplasmic with prominent and metabolically active nuclei. Nuclei increase in number (2–8) in some taxa (E.g., *Antigonon*, *Mimusops*), while in others they become polyploidal (due to endomitosis or due to secondary nuclear fusion) (E.g., *Curcubita*), or polytenic (upto 16 times increase in DNA content). Anomalous chromosome behaviour generally accompanies the nuclear division. Raphides, rhomboidal crystals, mitochondria, rough ER, Golgi complex, membrane–bound ribosomes, plastids and a few other organelles are reported in the cytoplasm of tapetal cells in some taxa. In addition, the cytoplasm contains innumerable granular bodies, and lipid membrane–bound, electron–dense organelles known as **pro–ubisch bodies, pro–sphaeroids** or **proorbiscules**. The shape of these bodies may vary greatly. For example, they maybe granular, rod–like, stellate, circular, perforated disc–like or compound multiperforate plate–like. They accumulate below the plasmolemma and then begin to disappear from the inside of the cell and are seen immediately thereafter in the plasma membrane. From here, they are believed to be transported first to the inner tangential walls of the tapetal cells and then to the exine of the microspores, where they attain **sporopollenin** nature. Hence, the **ubisch bodies** are to be considered to represent a transport form of sporopollenin (Fig. 185).

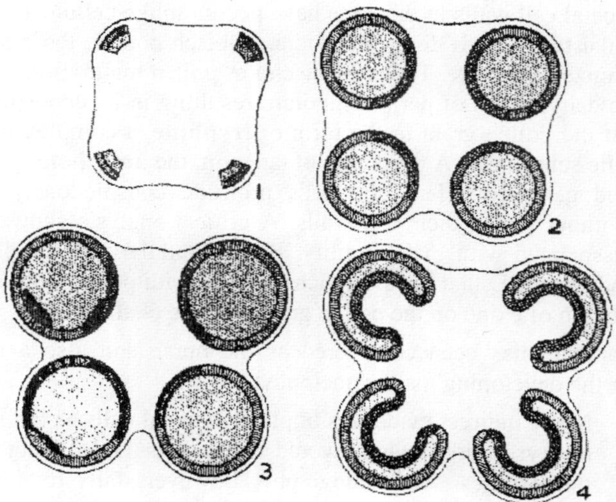

Fig. 180. Dual origin of anther tapetum; Single hatched portion of the anther tapetum is of parietal origin while double hatched portion is derived from the connective tissue (Periasamy and Swamy 1966).

Two distinct types of anther tapetum are known. In the first type, the tapetal cells retain their walls and eventually perish as such and *in situ*, and without change in shape (by Programmed Cell Death or PCD). This type is known as the **glandular, secretory** or **parietal** tapetum (e.g., *Helleborus*, *Tarenna*) (Fig. 179). In the second type, the cells lose their inner tangential and radial walls due to hydrolytic enzymes produced by the tapetal cells themselves and the protoplasts of different cells coalesce to form a mass of viscous consistency. It flows into and fills the sporangial cavity all around the developing microspore mother cells. This type is called the **periplasmodial tapetum** (e.g., *Commelina*, *Theriophonum*) (Fig. 181).

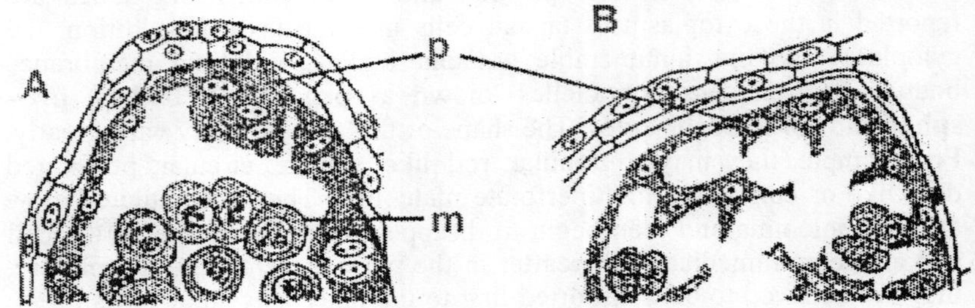

Fig. 181. Periplasmodial tapetum; A and B - Two stage in the organization of periplasmodial tapetum (p) of *Commelina subulata*; m - Microspore mother cell (Chikkanniah 1963)

The tapetal cell walls in this type have pectin unlike cellulosic cell walls of glandular tapetum. It does not produce Ubisch bodies. Periplasmodium is an organized structure. Towards the end of pollen maturation, there is an excessive dehydration of periplasmodium resulting in its deposition on the surface of the pollen grain in the form of **tryphine**, a complex mixture of hydrophilic substances. A third type of tapetum, the **amoeboid** type, is also recognized generally. Here, the cells protrude conspicuously onto the locule, without losing their cell walls. A critical analysis shows that this type is associated with male sterility. Both glandular and periplasmodial tapeta undergo PCD that occurs synchronously with pollen mitoic division and formation of exine on the pollen grains. (Love et al. 2008).

The tapetum has been considered as the nurse and also a regulating tissue for the developing male gametophyte.

1. Many indirect evidences implicate tapetal cells as the source for deoxyribosides, which would be used for DNA synthesis in the microspores. Autoradiography, however, fails to demonstrate the transfer of DNA components from tapetum to microspores.

2. Circumstantial evidences are there to indicate that carbohydrates and pollen reserves of mature pollen may result, at least partially, from the transfer of soluble sugars and peptides or amino acids from the tapetum. In many taxa, there is a close correlation between tapetal breakdown and the appearance of reserves in pollen.

3. As indicated earlier, pollen wall compounds or their precursors are supplied by the tapetum from elaeoplasts and tapetosomes (**Ubisch bodies** for sporopollenin, **pollenkitt** and **tryphine**) (Hsieh and Huang 2009; Parish and Li 2010). These also include a variety of proteins and incompatibility/compatiability–controlling substances. These processes involve a variety of genes and their interactions (see details Zhang et al. 2006; Yang et al. 2007; Wilson and Zhang 2009; Parish and Li 2010; Xu et al. 2010; Phan 2011).
4. Tapetum controls male sterility or fertility through its effect on **callase** (β, 1–3 glucanase) production at the right time when the callose sheath around the microspore tetrad needs to be dissolved to release the microspores. Precocious or delayed production of callase by tapetum promotes male sterility.

In situ hybridization with [^{3}H] poly(U) has revealed that mRNA accumulation is one of the metabolic activities that prepares tapetal cells for their function. Commensurate with the high metabolic activity, the tapetal cells express a variety of genes. At least five tapetum–specific m–RNAs and two mRNAs that are also seen in other anther tissues (TA series mRNAs) and genes were established by in situ hybridization and by the use of chimeric gene constructs in transgenic plants (Koltunow et al. 1990). These mRNAs get accumulated and decayed in the same temporal sequence during tapetum ontogeny and have been identified from a cDNA library of tobacco anthers. One of them is *TA29* whose product is a glycine–rich cell wall protein that may be involved in exine formation. An *Arabidopsis* gene, *MALE STERILITY2 (MS2)* is expressed in the tapetum of homologous and non–homologous hosts. The sequence similarity of the gene product to a protein that converts fatty acids to fatty alcohols has linked the gene to pollen exine formation (Aarts et al 1997). Anther tapetum–specific gene whose protein products show amino acid sequence similarity to cell wall proteins has been identified from tomato anthers. Other tapetum–expressed genes from the anthers of *Brassica napus*, flower buds of *Antirrhinum majus*, floral apex of *Sinapsis alba*, tassels of maize, and microsporocytes of *Lilium henryi* are known to encode small secretory proteins that are like lipid transfer proteins (Raghavan 2000). Some tapetal sequences got from anther cDNA libraries of *Brassica napus* and *Arabidopsis* specify β, 1–3 glucanase, which is required for the release of microspores from tetrads. Genes that encode proteinase inhibitors of β, 1–3 glucanase action have been isolated from tomato anthers. Thus, the timing of prevention or promotion of callose breakdown (around microspore treated) through this enzyme is controlled by tapetal gene activity. The other genes involved in tapetal development and function are *ABORTED MICROSPORES (AMS)*, the rice orthologue of *AMS* called *TAPETUM*

DEGENERATION RETARDATION, TAPETAL DETERMINATION1 (TRD1), DYSFUNCTIONAL TAPETUM1 (DYT1), its rice orthologue *UNDEVELOPED TAPETUM, DEFECTIVE IN TAPETAL DEVELOPMENT AND FUNCTION1 (TDF1), MYB 80* etc (see Krishnamurthy 2014a).

6.4.3. Microsporogenesis and Microgametogenesis

6.4.3.1. *Microsporogenesis*

The sporogenous cells either directly or after some divisions give rise to microspore mother cell(s). The microspore mother cells possess thin cellulosic cell walls. Plasmodesmal connections are largely concentrated on the outer walls of the peripherally–positioned cells and establish connections directly with the tapetal cells. Dictyosomes and plastids without starch grains are characteristically present in the cells. Most of the DNA is synthesized in microspore mother cells during the premeiotic interphase, but a meagre percentage of DNA is synthesized during zygotene–pachytene stage, which is necessary for the ordered progress of meiosis. Similarly, active RNA and protein synthesis takes place during premeiotic stage, but a fall in RNA and protein synthesis is seen during meiotic prophase. There is also a fall in ribosmal population after the initiation of meiosis, but the population is restored after homotypic division. There is also a reorganization of mitochondria and plastids in the haploid microspore as they are partly degraded during meiosis. Similarly, the plastids show reduction in lamellar system and lose starch, if present, but redifferentiate the lamella system and amyloplastids at tetrad stage. Just at the time of onset of meiosis, a callose wall is deposited inside the original cellulosic wall. In fact, callose deposition is considered as a prerequisite for meiotic induction and continuance (Krishmamurthy 1976) (Refer to Chapter 3). Any irregularity in callose deposition or metabolism results in male sterility. Callose deposition starts on the walls of microspore mother cells close to the tapetum and gradually extends to the more centrally located cells of the anther. Initially, the callose wall is incomplete leaving many gaps in the wall through which massive cytoplasmic channels between neighbouring microspore mother cells (but not with tapetum) are established. These channels reach their maximum development during zygotene–pachytene stage. These channels are very important for establishing synchroneity in meiotic division in the entire anther locule. Callose is highly impervious to most molecules and so far only about six molecules are known to be allowed through it. Thus, it is a highly isolating and insulating material. The cytoplasmic channels are sealed off towards the end of metaphase I in plants with **successive division** and at the anaphase II in plants with **simultaneous division**.

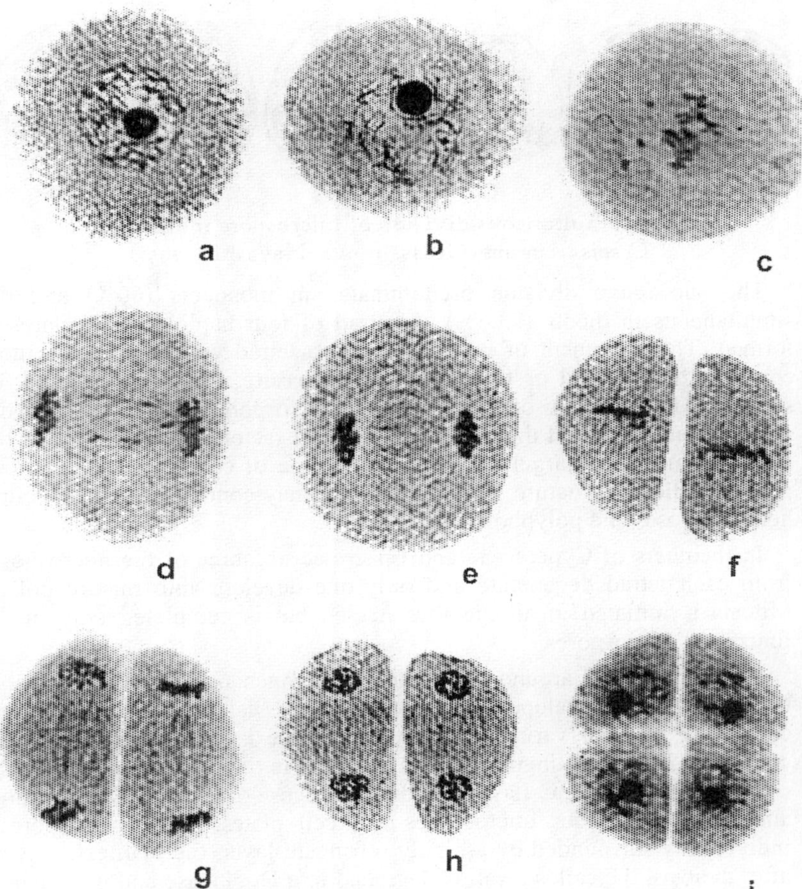

Fig. 182. Successive division of microspore mother cell of *Brachiaria* (Fuzinatto et al 2007) (Reproduced with permission from Genetics and Molecular Research, Ribeirão Preto, SP, Brasil)

Two types of meiotic division are known in the microspore mother cells, either of which results in the formation of four haploid microspores. In one type, a cell wall is promptly laid between the daughter nuclei at the end of each division (Fig. 182) and is called the **successive division**. During this division, the cell plate is formed first at the centre of the cell, progressing towards the periphery (centrifugal). In the second type, known as **simultaneous division**, the separation of all four microspores is effected at the end of the second division. The separating furrows are first initiated at the periphery and then they proceed towards the centre (centripetal) (Fig. 183).

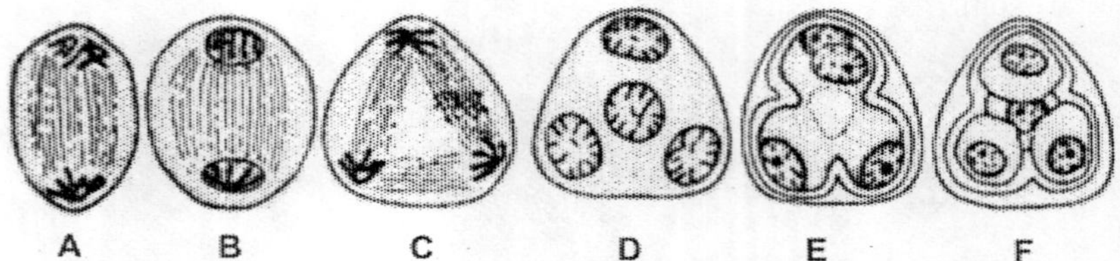

Fig. 183. Simultaneous division of microspore mother cell of *Crepis tectorum* (Greassimova-Navashina 1951)

The successive division predominates in monocots (60%) and the simultaneous in dicots (93.6%). A tetrad of four haploid microspores is formed. The alignment of microspores in a tetrad varies, from the most common **tetrahedral** or **tetragonal** to **decussate**, **linear** or inverted T–shaped. The position of the dividing microspore mother cell in the sporogenous mass and the impact of physical factors operating within the sporangium appear largely to control the type of configuration. The cell plate is callosic in nature, but, in addition, may contain proteins, pectins, hemicelluloses and polyphosphates.

In members of Cyperaceae and Epacridaceae, three of the microspores from each tetrad degenerate and only one develops into mature pollen. Mitosis is initiated in all the four nuclei, but is completed only in the functional microspores.

The callose wall around the tetrad is heterogeneous. The outer layer is the most well–developed and often forms lateral expansions with concentric rings of variable density. The three layers following this are concentric layers, distinguished by a variation in their relative densities and constitute a small zone. Layer 5 is the least dense of all layers. It surrounds and isolates the four microspores and cell plates. Each microspore is individually surrounded by a highly refringent layer, the **primexine**. Soon after meiosis, the callose wall is degraded into D–glucose and polymers of D–glucose of different lengths, which serve as a base material for exine formation and pollen nutrition. As a result of callose degradation, the individual microspores are separated out of the tetrad configuration. This degradation is caused by an enzyme β–1, 3 glucanase (also called callase), a product of tapetum. This enzyme is present in low quantities even during meiosis in the tapetal cells, but increases suddenly during late tetrad stage to cause the separation of microspores. In *Arabidopsis* anthers, a candidate gene called *QUARTET* (*QRT*) is required for the separation of microspores from the tetrad. The mutation in this gene affects the release of microspores. There is a patchy formation of callose between the microspores of this tetrad (Preuss et al 1994). In the *qrt* mutants the microspores fail to separate due to the fusion of the exine layers of adjacent microspores resulting from a failure of pectin degradation. So, *QRT* gene controls pectin degradation. Mutation in *AMS* and *DYT1* genes causes microspore degeneration.

In some taxa, failure of microspores to separate out from the tetrads results in the formation of compound pollen grains. These may be dyads or tetrads (E.g., some Ericaceae) or polyads of 8–16 grains (E.g., Mimosaceae, Orchidaceae) called **massulae** (Fig. 184). An extreme case of adherence, where pollen grain mass appears as a tissue, is seen in **pollinia**, as are noticed in some Asclepiadaceae and Orchidaceae.

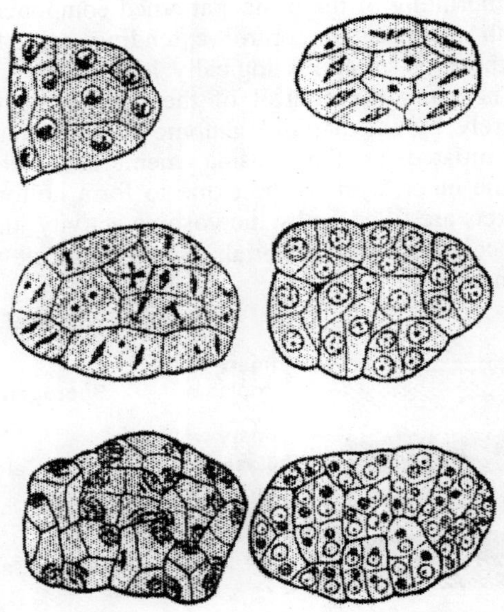

Fig. 184. Developmental stages in the Massulae of *Peristylus spiralis* (Drawing supplied by late Prof. B.G.L. Swamy)

6.4.3.2. *Microgametogenesis*

The newly–released microspores increase in volume many times due to the intake of locular fluid. Up to this time the microspores bear no **exine** (the outer wall of pollen), but contain a thin layer of **primexine** between the original callose wall and plasmalemma (except at the germinal pore regions where it is absent or thin). The primexine is distinguishable from the callose by its electron opacity. It has a matrix (presumably made up of cellulose), and radially–directed rods, the **probaculae**, and **profoot** layer. The chemical nature of the last two is not clear. The deposition of sporopollenin layer begins immediately after release from the tetrad (Fig. 185). The characteristic pattern of the sporoderm is determined by features already imprinted in the primexine during the period of enclosure in the tetrad (Blackmore et al. 2007). Others have provided evidences that the initial exine pattern laid down in the microspores is controlled by the plasma membrane. Callose has been implicated in causing this imprinting by acting as a template. The mutant of *DEFECTIVE IN EXINE PATTERN FORMATION (DEX1)* in *Arabidopsis* shows delay in the edevelopment of primexine and abnormal deposition of sporopollenin (Paxson-Sowders et al. 1997). Intine formation begins in the vicinity of the germinal apertures and spreads throughout to form a continuous layer and this growth is said to be associated with dictyosome activity. Although other studies seem to suggest that the imprinting of the basic–patterned component of the exine is associated with cytoplasmic control extending back to the diploid microsporocyte, these observations logically lead to the conclusion that exine formation is under the control of the haploid genome. Intine is programmed entirely by the haploid genome of the microspore. Intine organization is initiated at the plasma membrane, which separates irregularly from the inner layer of the exine to form hollow pockets and these hollow pockets are filled up by dictyosome activity, through fibrillar materials of pectocellulosic wall materials. Some implicate the role of ER in intine formation also.

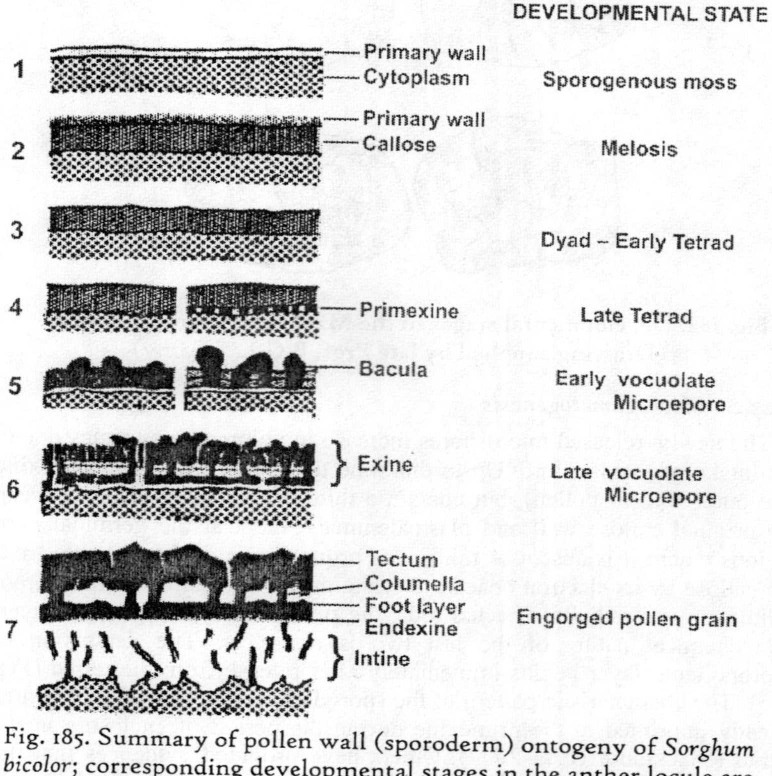

Fig. 185. Summary of pollen wall (sporoderm) ontogeny of *Sorghum bicolor*; corresponding developmental stages in the anther locule are also mentioned opposite to each figure (Christensen et al 1972, Swamy and Krishnamurthy 1980a)

The time span between the microspore and its asymmetric division varies in diverse taxa. Under typical conditions, the microspore nucleus occupies a central position and the cytoplasm shows a number of minute vacuoles spread evenly (Figs. 186 and 33). Prior to division, the microspore nucleus moves towards a side that is generally away from the furrow (future pore). Most of the plastids and mitochondria are displaced into that part of the cytoplasm, which is away from the nucleus. Active ribosomal RNA synthesis takes place during interphase in the microspore nucleus. Usually a conspicuous vacuole appears on the side opposite to the region where the nucleus moves. The nucleus then divides and a callose wall is formed between the two daughter nuclei. This wall is always curved leading to the formation of a small lens–shaped daughter cell (appearing spindle–shaped in cross–sectional view) called the **generative cell**. The sister cell is conspicuously larger and is called the **vegetative cell** (Fig. 33). Thus, the division is asymmetric, as already discussed in Chapter 3. The callose wall, wherever formed, is highly transitional and is retained only for about 10–20 h. In the *sidecar pollen* (*scp*) mutant of *Arabidopsis*, some of the pollen grains produce an extra vegetative cell even before the generative cell is formed. This gene probably regulates the asymmetric division by controlling the distribution of cytoplasm.

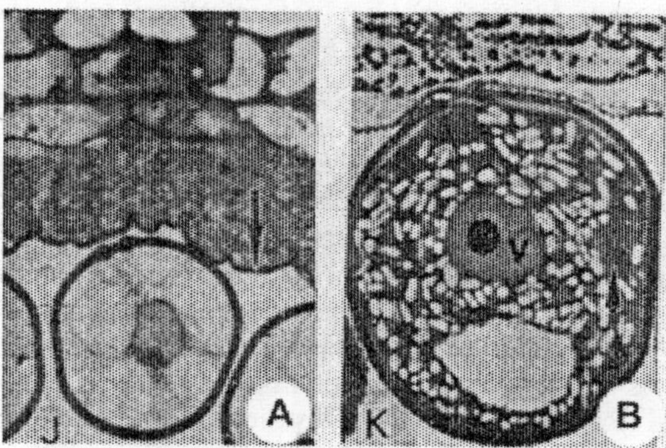

Fig. 186. A - T. S. of anther of *Sorghum bicolor* with microspore in mid–vacuolate stage. The microspore has a thick wall and a well–developed pore adjacent to tapetum. The nucleus is at the centre surrounded by vacuoles. B - Pollen enlarged. The vegetative nucleus (V) and generative cell (Arrow) are shown. Numerous starch grains are present. (Photographs courtesy of Dr. Christensen).

The generative cell soon separates itself off from the microspore wall and becomes embedded freely in the cytoplasm of the vegetative cell (Fig. 33), and by which time its callosic wall is also lost. This may or may not be accompanied by a change in its shape. This separation is effected by the growth of callose in between the plasma membrane of the generative cell and the intine until the generative cell is pinched off from the intine. At this stage, the pollen is said to be mature i.e., ready for pollination. The new location of the generative cell obviously provides a new environment for cell–to–cell interactions. At this stage the generative cell is surrounded by a double membrane envelope, by a distinct cellulose wall or by a transient callose wall. The cytology of the two cells of the **male gametophyte** (pollen) varies pronouncedly between themselves. The generative cell is less dense due to very poor or no RNA and proteins. Minute vacuoles filled with water and fatty substances are also present. DNA content of its nucleus is very high (rises to 2C level) and the nucleolus is not very conspicuous. Axial microtubules have also been recorded, which are implicated in maintaining the shape of the generative cell. However, there has been some disagreement concerning the generative cell organelle composition. Mitochondria, dictyosomes, lipid bodies and ER are reported. Plastids are observed in very few taxa. However, there is generally a scarcity of organelles in the generative cell. The vegetative cell shows dense cytoplasm due to a greater amount of RNA and proteins. The usually lobed nucleus is poorer in DNA (at the 1C level) and possesses a relatively large nucleolus. The pollen grains may have starch as one of the storage products at maturity in many angiosperms (e.g., Alimataceae, Malvaceae), but not in others (e.g., Acanthaceae, Liliaceae).

The division of the generative cell leading to the formation of two male gametes or sperms takes place either in the pollen grain itself (in 25% of angiosperms) or in the pollen tube (in 75% of angiosperms) and consequently the pollen grains are liberated either at the two or three–celled stage. The nucleus of the generative cell undergoes DNA synthesis, but there is no appreciable synthesis of RNA and proteins in the cell. Development of the pollen grain, thus, involves switching off of DNA synthesis in the vegetative cell without interfering with transcription. Transcription is slowed down without impairing DNA synthesis in the generative cell (Raghavan 2000). The division of the generative cell in the pollen grain is associated with normal mitosis (Fig. 33) followed by cytokinesis through cell plate formation or through a process of furrowing. The mechanism involved in the division of the generative cell in the pollen tube is not clear as yet because of spatial restraints. It appears to be normal mitosis (Raghavan 2000). In any case, the cytoplasmic organelles associated with the generative cell are reported to be present in the sperms and also at the same quantity. It is now generally accepted that maternal inheritance of plastids is the norm of flowering plants. The use of DAP1 and fluorescence microscopy indicated the absence of plastid DNA and this has been confirmed in the generative cell or sperm cells of pollen grains of about 82% of the species surveyed, whereas about 18% are potentially capable of biparental plastid inheritance.

6.4.3.3. *Sporoderm*

The sporoderm of pollen is made of two layers, an outer **exine** and an inner **intine**; exine is further divisible into layers, strata (Figs. 187 and 188) and further stratified sub layers.

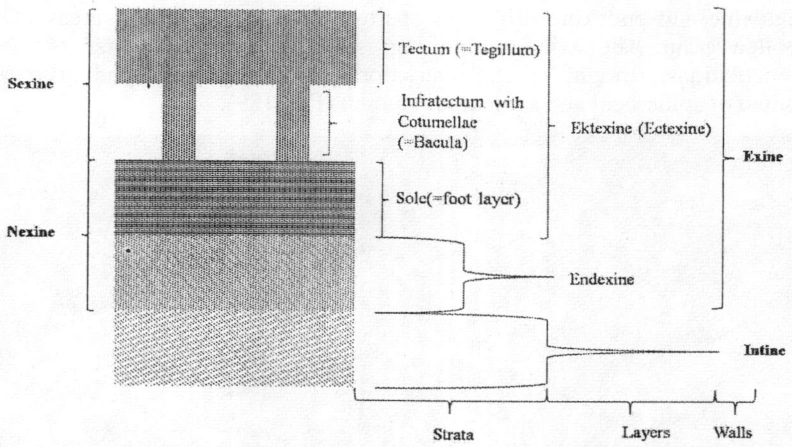

Fig. 187. Sporoderm structure and terminology as per the Association of the Palynological Society of France (1975)

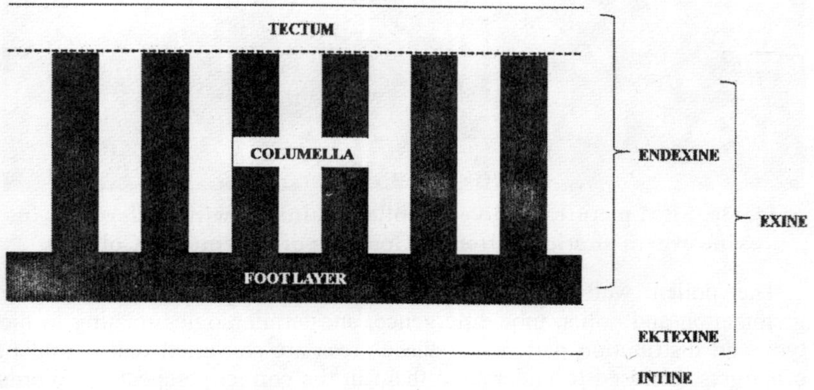

Fig. 188. Sporoderm structure and terminology as per Walker (Based on Walker 1974)

The intine consists of polysaccharides such as pectins, hemicelluloses, cellulose and callose together with proteins, the last being involved in incompatibility operating at the stylar level. The exine is made up of **sporopollenin**, a highly polymeric molecule consisting of carotenes and carotenoid esters resistant to microbial and chemical degradation. The exine is generally further divisible into **ektexine** and **endexine**; the ektexine exhibits three minor zones respectively called **tectum** or **tegillum**, **infratectum with columella** or **bacula** and **foot layer** or **sole**. The characteristic pattern of the sporoderm is organized by the variation in the relative thickness, extent and design of these strata. The amount of ektexine and endexine differs in apertural and non–apertural areas of a pollen grain. The exine sculpturing is in the form of coarse or fine reticulations, straight or wavy striations, zones of ridges and grooves, pitted or spinescent and sometimes smooth (Fig. 189).

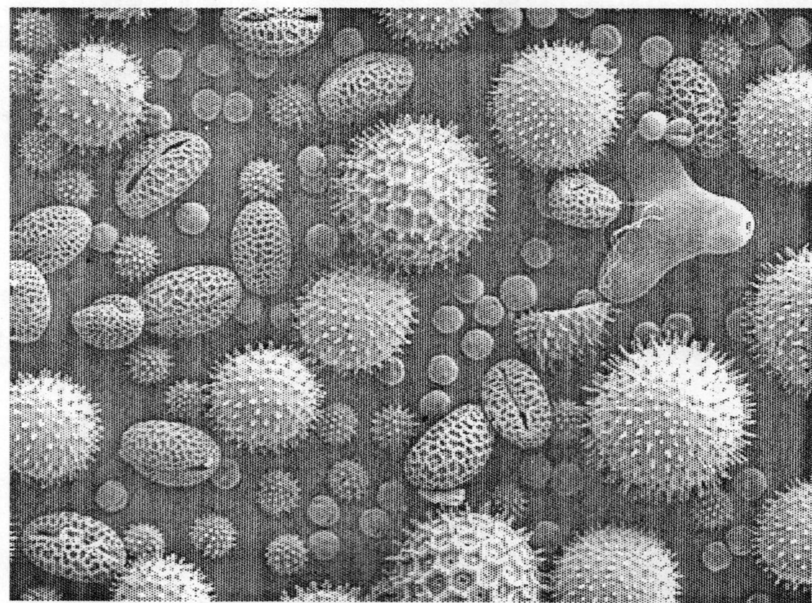

Fig. 189. SEM pictures of diverse pollen grains showing differences in exine ornamentation patterns (Courtesy of Dartmouth College)

The pollen wall may or may not contain specialized areas for germination and pollen tube emergence, and terminologies relating to the type and distribution patterns of these areas are so varied that a student often gets confused to understand them in the correct perspective. Words such as **germpores, colpa (mono, tri, poly and acolpate types), aperture, colpora and pseudocolpa** occur in pollen literature. The pollinial sporoderm is very interesting. The pollinial wall is acetolysis–resistant and hence, sporopollenin in nature, except for a specific region (i.e., **germ furrow**) through which the pollen tubes emerge during germination. There has been a controversy as to whether the exine patterning is controlled by the sporophytic genome or the gametophytic genome, but some recent evidence indicates that the control is done by the microspore mother cell.

Two types of pollen wall proteins have been discovered, emphasizing the fact that the pollen wall is a living, physiologically active system. One type is found in the cellulosic intine in the form of radially arranged tubules or tangentially–oriented leaflets. This is especially concentrated at the germinal aperture areas. The other type of pollen wall protein is found in the spaces and cavities of the exine. The first type of protein is synthesized by the microspore cytoplasm, while the latter, as indicated earlier, is from the tapetum. These proteins are considered to assist in pollen germination, nutrition and tube penetration in the stigma and also in the expression of incompatibility relationships. A part of these proteins are enzymes, mostly hydrolytic, such as esterases (predominantly in the exine), acid phosphatases (predominantly in the intine), amylase, proteases, chitinases, and ribonucleases. Allergic proteins are also present in the sporoderm.

6.4.3.4. *Genetic Control of Pollen Development*

The results of studies using cytochemical, autoradiographic, biochemical and molecular studies on RNA, and protein synthesis of developing pollen grains indicate that pollen development is controlled by a temporal and spatial programme of differential gene expression. The period leading up to the first pollen mitosis is marked by major contribution of rRNA in the total RNA synthesized. This is consistent with the idea that among the multiple copies of 5s RNA genes that control pollen development, some are turned off following the peak synthesis, whereas a few persist for an additional period. Quantitative variations in mRNA populations have also been noted in pollen development. A use of mRNA accumulation in nature pollen is their serving as templates for the first proteins of germinating pollen. Both qualitative and quantitative differences are detected in the protein synthetic activities of pollen grains at different stages of their development. The proteins include lysine– and arginine–rich histones, which begin to accumulate in the generative cell nucleus and sperm nucleus. Their syntheses are linked to the transcription of the haploid pollen genome. Extensins and arabinogalactan–rich proteins, which are important in pollen tube wall formation and growth and in the **recognition–rejection reaction**, are also synthesized.

Genes involved in pollen grain development have been isolated and characterized, especially in *Arabidopsis, Zea mays*, *Tradescantia*, tomato, *Oenothera*, *Brassica napus*, *Brassica oleracea,* tobacco, *Lilium*, cotton and *Petunia*. The isolated genes turned out to be members of small families present in one or few copies in the genome, and none belonged to large multigene families (Raghavan 2000). Much weight has been given to the concept that pollen development involves the timely and sequential expression of genes of both gametophytic and sporophytic origin. Hence, genes of two distinct sets are shown to be activated in a temporal fashion during pollen development. Transcripts of one set, commonly called **early genes**, become active as soon as the microspores get released from the microspore tetrad, but these have only a short period of activity. The **late genes** become active after pollen mitosis and continue up to pollen maturity. The nature of the proteins encoded by early genes appears to be important in coding for cytoskeletal proteins such as actins and tubulins, while the late gene–encoded proteins are needed for pollen maturation and early pollen tube growth (Stinson et al 1987). Non–expression of any of these genes results in male sterility. The proteins encoded include pectin lyases, pectin esterase, polygalacturonase, protein kinases, ascorbate oxidase, thioredoxins, actin–depolymerizing factors, Zinc finger class proteins, RNA helicases, pollen allergins, ATPase, ubiquitin, osmotin, stress proteins, PR– proteins, malate synthase, super oxide dismutase etc.

The genetic basis of phenotypic differences between vegetative and generative cells has been analysed by the use of mutants, transgenic molecular markers and in situ hybridization with cloned genes. An associated asymmetry in gene expression is detected along with asymmetric division that results in these two types of cells. Mitotic division is not a prerequisite for expression of a vegetative cell–specific gene, but a symmetric alignment of this division silences gene expression in the generative cell. Twell (1995) has shown that ablation of the vegetative cell of transgenic tobacco pollen by the expression of the cytotoxic DTA gene linked to the tomato pollen–specific gene inactivates the generative cell and prevents its function (Raghavan 2000). How the vegetative cell controls the generative cell's activity is not clear.

The molecular mechanisms underlying the production of male gametes are beginning to be unraveled (see Kim et al. 2008; Brownfield et al. 2009; Ron et al. 2010; Zheng et al. 2011). The genes involoved include *DUO POLLEN1 (DUO1), APC/C, CYCB1; 1, PLA2* paralogues etc.

6.4.4. Anther Dehiscence

Simultaneously with the maturation of the microsporangium, the wall layers aligned in the groove between the adaxial and abaxial pairs of sporangia fail to undergo histological modification. This linear strip of tissue is the **stomium** that predetermines the locus of anther dehiscence. With the continued bulging out of the distal anther wall, the stomium appears as if seated in the furrow. The stomium cells form the weak zone of the anther wall. A few cell layers of parenchyma are present in inward continuation of the stomium. This septum–like tissue separates the adaxial and abaxial sporangia. The resorption of septum towards maturity causes the merger of the sporangia on each side of the anther. About this time, the stomium cells slightly elongate in the radial direction due to the pressure exerted by the anther wall on either side. The mechanical action of the endothecium causes an evagination of the anther wall along the stomial line. Finally, the anther dehisces to release the pollen. In some taxa, anther dehiscence takes place by terminal pores or valves. Dehiscence of anther is absent in aquatic taxa and in cleristogamous taxa. Here, the pollen are released often due to the disintegration of the anther wall. The molecular aspects relating to anther wall in general, and stomium and endothecium in particular have already been discussed.

The events involving dehiscence of the anther may be expected to get mediated by structural features of the filament since the former is dependent upon the latter for the transport of water and nutrients. Dehiscence of anther is largely a desiccatory process. Therefore, any histological feature promoting a rapid loss of water from the stamen or a disruption of water supply to anther may facilitate anther dehiscence. The filament seems to be well–adapted to promote rapid loss of water once the necessary developmental stage is reached, with open stomata, a weakly developed cuticle, a prominent intercellular space system and xylem lacunae. The desiccatory dehiscence process has cohesion and hygroscopic mechanisms. Hygroscopic mechanisms depend entirely upon volumetric changes in the cell walls, whereas cohesion mechanisms involve volumetric changes in the cell lumen, with the cell walls undergoing largely passive deformation. The latter mechanism primarily involves cohesive forces between water molecules in the cell lumen. Dehiscence occurs when the cohesive forces are exceeded. In hygroscopic mechanism, adhesive forces are prominent rather than cohesive forces. Although most workers now accept a cohesive mechanism, both forces are important in anther dehiscence.

In general, the angiosperm pollen grains are partly dehydrated at the time of dispersal, with a water content of 10–30%. Further water loss invariably occurs during pollen transport. Then the level of water content is comparable to that found in dry seeds. The pollen grains are also metabolically poor. The dehydration causes disorganization of the membranes of the vegetative cell. The plasmalemma is not distinguishable as a continuous membrane, and neither mitochondria nor plastids have well–developed envelops. The dehydration is also manifest in their walls. Apparently this dehydration helps the dispersal capability of pollen. Because of low water content (water stress) proline accumulates in the pollen in large amounts.

6.5. DOUBLE FERTILIZATION

Fusion of the male and female gametes is **fertilization**. In the gymnosperms, of the two sperms contained in a pollen tube, one fuses with the egg and the other degenerates. This process is distinguished as **single fertilization**. On the contrary, in angiosperms, both the sperms are directly concerned in the sexual process, one sperm fusing with the egg and the other with the polar nuclei or secondary nucleus. The first act is **syngamy** and the second is **triple fusion**; the former resulting in **zygote** (diploid) and the latter in **primary endosperm nucleus (PEN)** (usually triploid). As both the sperms in the pollen tube take part in sexual fusion, this process is distinguished as **double fertilization**. It should be emphasized that double fertilization is an exclusive feature of angiosperms. Double fertilization provides not only the essential stimulus for the development of embryo and endosperm, but also for the continued growth of the seed and fruit. Events happening in the stigma, style, pollen on the stigma, pollen tube in the style and ovule, and in the components of the embryo sac are all important for effective double fertilization. Events happening to the pollen or pollen tube in the stigma, style and somatic tissues of the ovule embody **pollen–pistil interactions** or **progamic phase** (Raghavan 2000).

6.5.1. Stigmatic Environment

6.5.1.1. Structure of Stigma

Since, after pollination, the pollen grains are lodged on the stigma, the stigmatic environment plays a vital role in the further behaviour of the lodged pollens. Pollen adhesion to the stigmatic surface is largely determined by the extent of wetness and/or surface features of stigma, and also on the sculpture of the pollen exine. Stigmas, at the time of pollination, have been classified into **dry** and **wet** types. The wet stigmas have the receptive surface covered with a secretion called **stigmatic exudate** produced by the subepidermal cells of the stigma, while in the dry stigmas the stigmatic surface is covered by papillae having a cuticle, which in turn is covered by a proteinaceous hydrated layer called **pellicle** (Fig. 190).

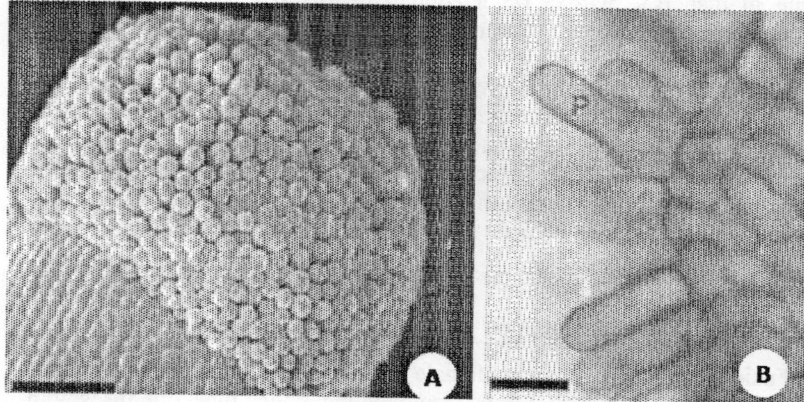

Fig. 190. *Qualea multiflora*: SEM picture of stigma surface (A) and a section showing papillae with pellicle (B) (Carmo-Oliveira and de Morretes, 2009) (Reproduced with permission from Sociedade Botânica do Brazil)

The wet stigma plays an important role in pollen germination in plants with two–celled pollen, while dry stigma in plants with three–called pollen. The pellicle, in addition to proteins, contains water, lipids, amino acids, phenolics, sugars, Ca^{2+} etc. It also shows high esterase, acid phosphatase and other enzyme activities. The origin of the pellicle is not very clear, but, since the cuticle is discontinuous, the pellicle may have been extruded to the surface through these discontinuities. The pellicle is physiologically and functionally comparable to the stigmatic exudate. The pellicle-containing dry stigmas are further classified into sub–types. Dry stigmas are seen in grasses, Acanthaceae, Asteraceae, Brassicaceae, Amaranthaceae, Bromeliaceae, Oxalidaceae etc. Wet stigmas with stigmatic exudates are characteristic of plants of Apiaceae, Liliaceae, Fabaceae etc. The exudate covers the stigmatic surface and is either produced by the stigmatic tissue as in *Petunia* or by the **stylar canal cells** bordering the hollow style, and then gets extruded onto the stigmatic surface. The exudate is secreted by the ER of the concerned cells. The chemical composition of the exudates is variable depending on the plant. It has a varying proportion of lipids, carbohydrates, minerals, phenolic compounds and proteins/peptides/amino acids. If largely lipoidal, it normally does not help in the nutrition of the pollen (e.g., *Petunia*). If aqueous with carbohydrates and small amounts of proteins, it is probably nutritive to the pollen. The lipoidal material helps in trapping the pollen and protects the stigma from disiccation and wetting. Phenolics are presumed to help in the protection of stigma from microbes and pests, and also in the promotion or inhibition of pollen germination. There is either one spurt of stigmatic exudation or two depending on the plant.

The time interval between pollination and pollen germination on the stigma varies in different species. In general, the pollen of herbaceous taxa germinate relatively earlier than in arborescent species. Also, there is a corresponding temporal relationship in the rate of pollen tube growth in the two habits, and, therefore, of the time between pollination and fertilization, shorter in herbs and prolonged (over days or even months) in trees. For instance, the period between capture and germination of pollen is less than five minutes in many grasses, Brassicaceae and Asteraceae. The time interval between pollination and pollen germination also depends on the inherent degree of viability of pollen, external factors affecting pollen viability and compatibility or otherwise of the stigma. In compatible stigmas, the pollen germinates, while in incompatible ones the pollen does not germinate. It was already mentioned that the pollen grains are dehydrated during and after release, but once they land on the stigma, they get hydrated so that the well–known volumetric changes or **harmomegathy** can result. The water is supplied by the stigmatic exudates of wet stigmas or by the hydrated pellicles of dry stigmas. The rapidity of pollen hydration depends upon the nature of stigma. In dry stigmas, hydration is gradual and controlled by the water potential of the stigmatic papillae, while in wet types, hydration is very rapid. For example, in rye, within three minutes after lodging on the stigma, the pollen may take up $6X10^{-8}$ cm^3 of water indicating a flow of $3.5X10^{-10}$ cm^3/sec^{-1} at the interface. Therefore, the wall of the stigma must present a low resistance to water flow and the plasmalemma must be unusually permeable.

6.5.1.2. *Comptibility and Incompatibility Reactions*

When the pollen takes up water from the stigma, extremely rapid changes occur in the pollen, especially in the vegetative cell and also in pollen wall. The pollen wall proteins are released onto the stigma, first exine proteins followed by intine proteins. The range of proteins released is wide and 26 different proteins have been separated in rye and about 40 different proteins in *Brassica oleracea*, when compared to control pollen not kept in contact with stigma. Since some of these proteins synthesized by pollen are highly phosphorylated, it is likely that protein phosphorylation can account for signal transduction in a compatible pollination. Initially, these proteins do not bind to the stigma, but soon bind themselves to the stigmatic surface establishing a close interaction. The stigmatic surface receives different kinds of pollen grains and hence, there must exist a physiological mechanism to ensure that only intraspecific and compatible pollen grains are allowed to germinate and produce an effective pollen tube. This physiological mechanism is referred to by the term **recognition**. The favourable reaction is known as **compatibility** and a rejection is called **incompatibility**. Both compatibility and incompatibility may be **self** or **cross**. In self–incompatibility, pollen from the same flower or from another flower of the same plant is generally avoided. Cross–incompatibility refers to avoiding pollen from another species or from the flower of another plant of the same species. Recognition of the pollen appears to take place during the pollen–stigma interaction and the further behaviour of the pollen depends on this recognition. Compatibility or otherwise has been shown to be mainly the result of interaction of the proteins released by the pollen with the proteins of the stigma present on the pellicle or stigmatic exudates, and this reaction is similar to the antigen–antibody reaction. Both the intine and exine proteins released are specific recognition factors, which interact with the receptor sites or on the stigma papillae. Ca^{2+} is also involved in pollen recognition–rejection reaction by serving in cell signalling. Hence, compatible pollen grains produce transient Ca^{2+} peaks in the stigmatic papillae (for example, *Brassica napus*) adjacent to the applied pollen grains. But, how this system is activated to promote hydration and germination of pollen is not clear. Thus, in instances of 'acceptance', the pollen is allowed to germinate to produce a pollen tube, which penetrates the stigma cuticle and brings about a recognition response, whereas 'rejection' is indicated by the failure of the pollen to germinate or by the production of abnormal tubes, which are often unable to enter into the stigma. In some cases of incompatibility, pollen tube may pass the stigma, but are arrested in the style.

The genetic basis of pollen recognition and rejection has been elucidated to some extent. Mutants of *Arabidopsis* have contributed new information on this aspect. Pollen–stigma interactions are shown to be regulated by specific components of the pollen wall tryphine. The *pop1* (for 'defective in pollen–pistil interaction') mutant is male–sterile and its pollen grains fail to get hydrated on the stigma and germinate. However, under *in vitro* conditions they are able to germinate, and hence failure to germinate is not due to loss of viability / fertility. The mutant grains lack long–chain lipids as well as tryphine. *CERI* locus mutants of *Arabidopsis* also have pollen that do not hydrate on the stigma and lack tryphine with the normal complement of lipids. Information on recognition is also provided by flavonoid deficient *Petunia hybrida*. The pollen grains fail to germinate or produce only a protuberance *in vitro*, but addition of kaempferol, a flavonoid, added to the wild-types stigma along with this pollen promotes germination.

6.5.2. Pollen Germination and Pollen Tube Growth

Immediately after hydration, the plasmalemma of the vegetative cell of pollen grain once again becomes resolvable. The mitochondria assume a normal appearance and profiles of ER appear. After hydration, vacuolation begins and a normal tonoplast surrounds each vacuole. Protoplasmic streaming is revived. The time to undergo these changes may be two minutes to several months depending on the degree of dehydration of pollen.

Pollen germination involves nuclear movements towards a germinal area and the formation of a pollen tube, and therefore, the nature and mode of action of cytoskeletal elements that provide the motive force for germination are of great importance (Tiwari and Polito 1990a, b). There are changes in actin configuration during pollen germination as revealed by immunofluorescence studies. The loose network of actin microfibrils occurring throughout the vegetative cells is soon replaced by an entangled web of fibrils converging towards the aperture from where the pollen tube is to emerge (See Cai et al 2005). The contribution of Microtubules (MTs) to pollen germination depends on their association with actin microfilaments (MFs). The metabolic changes associated with pollen germination include efflux of metabolites, increased respiration, increased rates of RNA and protein synthesis etc. Mature pollen in many species contain stored mRNAs that code for the first proteins needed for germination and pollen tube growth, but stored mRNAs are not enough to code for all of the essential proteins required for continued growth of pollen tube. In some, during germination, rRNA and tRNA are produced. A battery of genes is essential for pollen tube growth. These include the genes that encode polygalacturonase, pectin lyase, and pectin esterase, all being cell wall degrading enzymic proteins. Other enzymic proteins encoded are ascorbic acid oxidase and receptor kinase. GA has also been implicated in pollen germination and tube growth.

A part of the intine of the pollen confronting the germinal area protrudes and grows out in the form of a pollen tube. Local secretion of hydrolytic enzymes is involved in the dissolution of the wall confronting the pore. Most generally, a single tube arises from a pollen grain, irrespective of the number of germinal areas. There are, however, instances (e.g., Malvaceae, Cucurbitaceae taxa) where more than one tube emerges from a corresponding number of germinal areas. Upto 14 tubes have been reported in species of *Malva*. Branching of the single tube is also seen in some taxa (e.g., *Casuarina*). In all cases, however, only one tube or branch carries the two sperms and eventually reaches the ovule.

The growth of the pollen tube, as already described in Chapter 3, is restricted to its tip region of about 4–7µm. The wall of the tube is made up of cellulose embedded in a non–crystalline polysaccharide matrix, probably pectin. Callose is absent at the tube tip, but forms a layer inside the tube wall a little behind the tip. Some authors have reported an outer pectic layer, a middle cellulosic layer and an inner callose layer in the proximal part of the tube below the tube tip. Immunofluorescence studies using pectin monoclonal antibodies have reported two patterns of pectin deposition, one as periodic annular deposits found coating the pollen tube walls of species with solid styles, and the other as a more uniform sheath as in pollen tubes of species with hollow styles. Pollen tube tip is richer in esterified pectins than the proximal parts.

The pollen tube grows penetrating the stigmatic cuticle through the production of cutinase, passes the stigma and enters the style. Studies made on a few plants indicate that pollen tubes do not penetrate the cuticle of stigmatic papillae from which the surface exudate has been removed enzymatically. Hence, it is inferred that the pollen grain contributes a precursor of the cutinase enzyme, which then gets activated by a factor present in the stigmatic exudate. When stigma cells are ablated by the introduction of a stigma–specific gene fused to the cytotoxic *BARNASE* gene, a few pollen grains germinate on the stigmatic surface, but the pollen tubes do not penetrate the style. This block to pollen tube to penetrate the style becomes fully restored by the application of an exudate of the wild–type stigma. The essential factors in the exudate necessary for pollen tube penetration appear to be lipids (cis–unsaturated triglycerides) and in the presence of these lipids, pollen tubes even penetrate the leaves. Thus, lipids on the pollen tube wall and stigmatic exudate or pellicle have an important role in pollen–pistil interaction and in directing pollen tube growth. Studies made also indicate that components of the pollen recognition system are present even in floral organs, but are segregated to the stigmatic surface by the action of genes such as *FIDDLE HEAD* (*FDH*) of *Arabidopsis*.

There is great disparity in the recorded rates of pollen tube growth in plants: maize 3 µm sec^{-1}, rye more than 2 µm sec^{-1}, *Lilium* 0.6 µm sec^{-1} and *Crocus* 0.07–0.1 µm sec^{-1}. Growth of pollen tube is very rapid in some Asteraceae and Brassicaceae. The distance that the tube has to travel depends on the plant, for example, only a few mm in sugar beet and about 45cm in *Zea mays*.

6.5.3. Stylar Environment

The further growth of the pollen tube depends on the histological structure of the style. In styles with a centrally located canal (**hollow** or **open type** of style) the bordering glandular–appearing surface layer of cells, called **canal cells,** act as **transmitting tissue** in aiding the growth of the pollen tube ectotropically along their surface (Fig. 191 A, B) (E.g., some members of Liliaceae). In *Lilium*, for instance, the cuticle of the canal cells is ruptured during the secretion of style–based stigmatic exudate and hence, this ectotropic growth is very smooth. The canal cells have 8–14 µm thick **secretory zone** on the side facing the canal. This zone in *Lilium* has three regions. The canal cells are often multinucleate or polyploidal. Their transverse walls are thin and are traversed by abundant plasmodesmata, while the longitudinal walls are thick. The cells are surrounded by an intercellular substance of different electron density than the cell wall. This intercellular substance, a secretion product of the transmitting tissue, is not comparable to the middle lamilla, but is more complex and comparable to secretion fluid formed in the stylar canal. The intercellular substance is reported to contain certain proteins. In styles, which are **solid** (or **closed**) (E.g., Malvaceae, Solanaceae), the pollen tube passes between the cells (all the cells are considered here as components of the transmitting tissue) (Fig. 195 B) in the intercellular substance (i.e., pectic layer), although in cotton the pollen tube grows through the cell wall layer and not through the middle lamella. A third category of style has been recognized and has been named **half–closed**. This is intermediate between open and closed styles and is found in members of Cactaceae and in *Artabotrys* of Annonaceae. This type of style possesses a rudimentary type of transmitting tissue. Whatever be the type of style, the styles are often rich in boron and supply this to the growing tube. In fact, after pollination, the canal in open styles and the intercellular spaces in solid styles are filled with the secretion produced by the transmitting tissue, which serves as nutrients and regulatory materials for the growing pollen tubes. The starch present in the styles of many taxa is hydrolysed and this hydrolysed product forms part of the secretion. Pollen tubes obviously use this for the synthesis of their own walls.

The intercellular substance and secretion found in the canal are also very important in controlling compatibility or otherwise of the growing tube. This secretion is a heterogeneous mixture of sugars, proteins, glycoproteins, lipids and minerals. In this mixture, especially arabinogalactan–rich proteins and extensin–like (hydroxyproline rich proteins) proteins are important. The latter are often specific to the mature styles, as shown by immunofluorescene histochemistry, in the extracellular matrix to the total exclusion of the cell wall. These glycoproteins are also found in the cytoplasm and cell walls of compatible pollen tubes. Hence, these proteins help in the growth, nutrition and guidance of pollen tubes besides controlling recognition–rejection of pollen tubes in the stylar tissue. If an incompatible tube is found, it is arrested by the recognition–rejection reaction. Some experiments show that pollen tube growth through the style is mediated not only by the stylar matrix and its chemicals, but also by mechanical or electrical signals that interact with the stylar matrix (Raghavan 2000). As per a model for tube direction, pollen tube growth through style is reckoned as an authentic directional cell migration along the transmitting tissue by the interaction with a substrate adhesion molecule present in the stylar matrix. This substance is shown to be a homologue of human **vitronectin**. The other popular mechanism proposed is the attraction of pollen tube towards the ovule by a chemotropic substance of ovular origin.

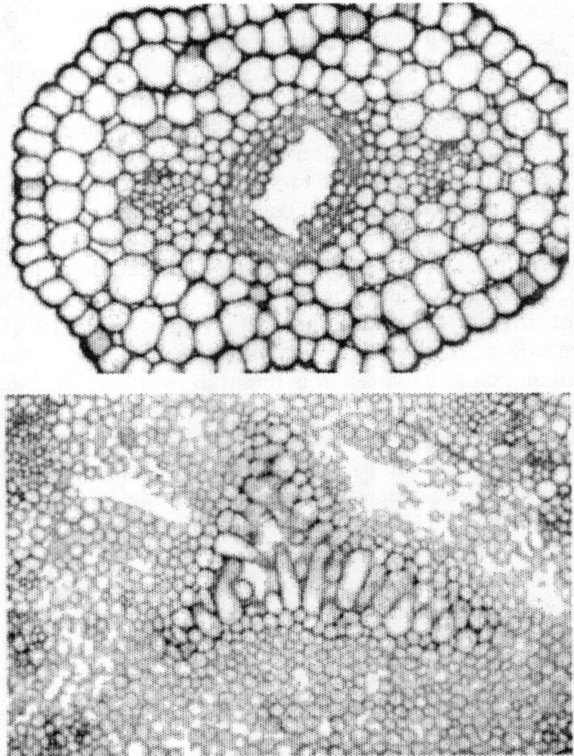

Fig. 191. A - T. S. of hollow or open type style showing canal cells or transmitting tissue lining the canal, over the surface of which pollen tubes grow towards the ovule; B - T. S. of solid or closed type of style showing loosely-packed transmitting-tissue cells in between which the pollen tubes grow towards the ovule (Leins and Erbar 2010)

6.5.4. Growth of Pollen Tube into the Ovule

After reaching the locule part of the gynoecium, the pollen tube continues its course generally on the surface of the placenta and reaches the funicular region. From here, the tube tip grows into the micropyle. Whether such is the situation in instances where the ovules are free and the locule large, and whether tubes can grow through free space in the ovary is yet to be studied with greater care. But, in many taxa the growth of the pollen tube inside the ovary towards the micropyle of ovules is guided by a structure called **obturator**. The obturator may be placental in origin as in some Euphoriaceae, funicular in origin (most common) as in Anacardiaceae, Lamiaceae etc., stylar canal in origin as in Thymeliaceae or ovary wall origin as in *Ottelia* (Fig. 192 A and B). Both placental and funicular obturators occur in Rutaceae. The obturator cells are often secretory in nature and this secretion is believed to attract the pollen tubes towards them, thus, enabling them to guide the tubes.

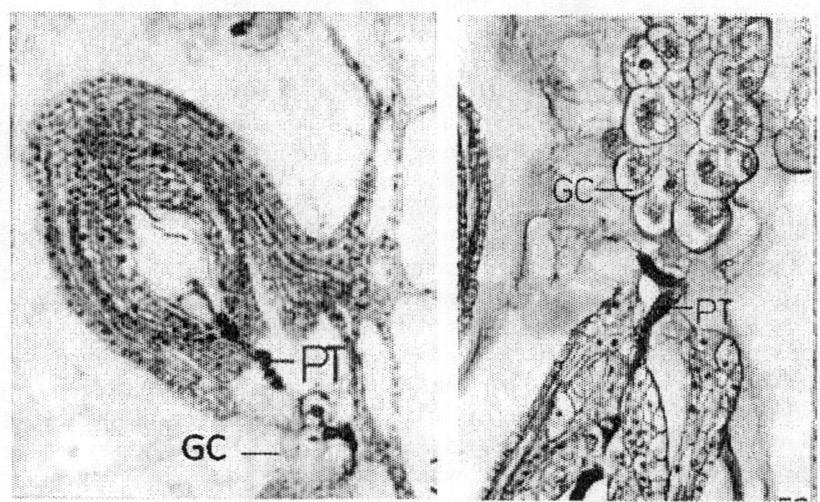

Fig. 192. L. S. of portion of an ovary of *Ottelia alismoides* showing the guidance of pollen tube (PT) by the ovary wall obturator (GC)

In the large majority of taxa, the pollen tube enters the ovule through the micropyle, and this process is referred to as **porogamy** (Figs. 192 and 193 A). In a few taxa belonging to Casurianaceae and Betulaceae, the pollen tube passes through the raphe along the vascular trace and reaches the chalazal region of the ovule (Fig. 193 B). Here, the tip of the tube undergoes proliferation into blind branches, which spread in haphazard directions. One of the branches carrying the sperms enters the nucellus, reaches the micropylar end and enters into the embryo sac. This process is called **chalazogamy**. In porogamous taxa, where there are nucellar layers in between the micropyle and the embryo sac (i.e., in crassinucellate and pseudocrassinucellate taxa), the pollen tube has to pass through the nucellus. In taxa such as cotton, a column of nuclear cells extending from the embryo sac into the micropyle degenerates just before the arrival of the pollen tube. The pollen tube grows between the radial walls of degenerated nucellar cells, crushing some of them, but not entering them. In some, the nucellar cell walls on pollen tube path become mucilaginous making a convenient passage for the penetration of the pollen tube. This becomes mucilaginous from the synergid region towards the micropyle. Basically, pollen tube growth in the nucellar region is intercellular, but growth of tube between the cell wall and plasma membrane has also been reported.

There are two prevailing opinions concerning the mechanism involved in the entry of the pollen tube into the embryo sac. Some believe that there is secretion of enzymes by the pollen tube itself and these enzymes dissolve the embryo sac membrane at the point of contact. Yet others conclude on the basis of electron microscopic studies that pore formation on the embryo sac sheath is caused by the secretions of one of the two synergids, which receives the pollen tube (Fig. 194). They also mention that the filiform apparatus of the synergid transports the secretion product towards the micropylar tip of the egg apparatus where the pollen tube establishes the contact with the embryo sac. Through the pore thus formed, the pollen tube contacts one of the two synergids and this synergid receives the pollen tube (Fig. 195). This synergid shows signs of degeneration long before the time of actual entry of the tube. The filiform apparatus of the degenerated synergid seems to control and regulate the molarity of the synergid. Probably, after degeneration, this synergid has a lower molarity than the persistent synergid. This molarity is also lower than the molarity of intercellular spaces of the bordering nucellar tissue. This reduction in molarity has something to do with the pollen tube growth, but the actual relation is not known. The tip of the pollen tube enters this synergid through the filiform apparatus end (Fig. 195 A) and abruptly ceases growth the moment it comes in contact with the synergid cytoplasm. What causes this sudden cessation of pollen tube growth is not yet known. A subterminal opening is usually formed in the pollen tube and its contents are released into the synergid cytoplasm (Fig. 195 B). Occasionally, the contents are released by the rupture of the tube tip. The release of contents (Fig. 195 C) is supposed to be caused by low oxygen tension in the embryo sac and the lowering of oxygen tension was increased with the greater proximity of the approaching pollen tube. This is probably due to the fact that the tube grows aerobically up to the stylar base, but anaerobically in the ovary. Soon after discharge of the contents, the pore in the tube is closed by the formation of a callose plug. The other synergid, which has remained unaffected, usually appears to undergo a slight enlargement, persists for some time and then finally degenerates through PCD.

Many investigators have postulated that chemotropic responses of the pollen tube guide its unerring entry into the embryo sac. The nature and source of such a chemotropic substance are strongly debated. A number of

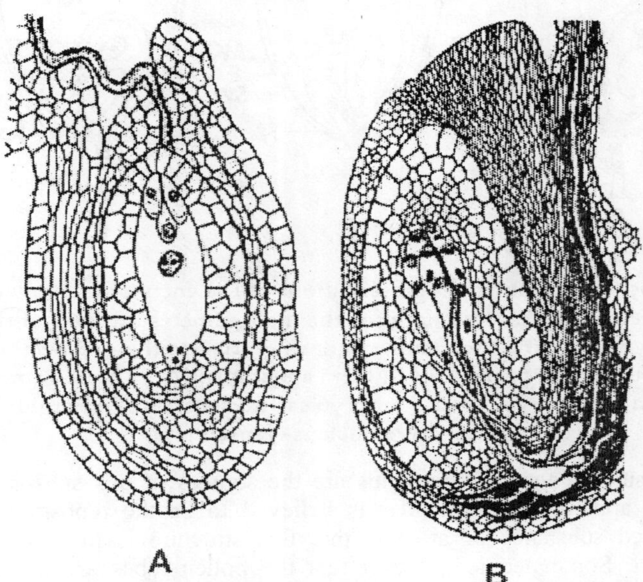

Fig. 193. A - L. S. of ovule of *Ottelia alismoides* at the time of fertilization. Pollen tube enters into the ovule through micropyle (Porogamy); B - L. S. of ovule of *Casuarina* at the time of fertilization. The pollen tube transverse along the raphe, branches freely in the chalazal part of the ovule and one of the branches that carries the male gametes grows vertically up and enters into one of the many embryo sacs swollen at both ends (chalazogamy). (Drawing supplied by late Prof. B.G.L. Swamy)

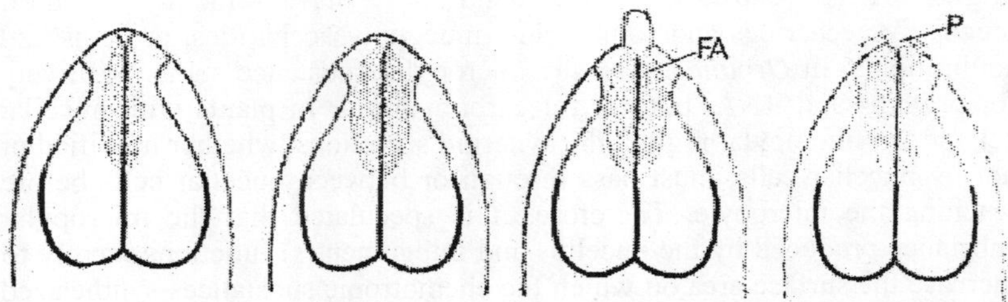

Fig. 194. Formation of filifrom apparatus (FA) with synergids and of the pore (p) at the tips of the embryo sac probably through the secretions of the synergids; the egg is not shown. (Van der Pluijm 1964)

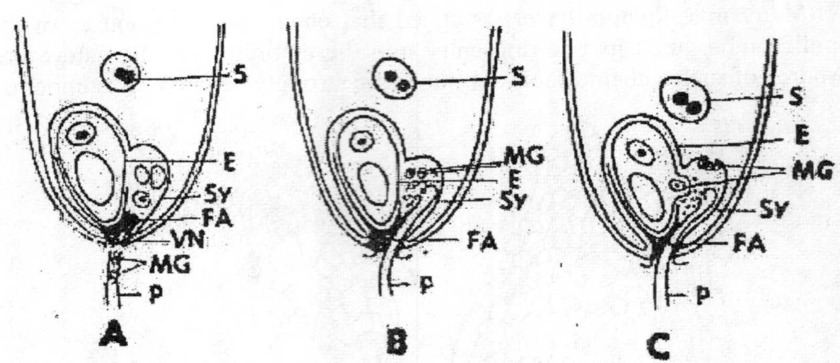

Fig. 195. Diagrammatic representations of the entry of the pollen tube into the embryo sac; discharge of the male gametes into one of the two synergids and their subsequent migration to their respective destinations; E - Egg; FA - Filliform apparatus; MG - Male gametes; P - Pollen tube; S - Secondary embryo sac nucleus; Sy - Synergid; VN - Vegetative nucleus (Jensen 1973)

people suggest that the synergids are the sources of this substance. The filiform apparatus in particular is believed to be the repository of the concerned substance or at least that this structure is involved in this transport. Some people question that the pollen tubes are attracted by chemotropic stimuli produced from filiform apparatus. They cite as evidence the fact that pollen tubes enter into aborted or already fertilized ovules also. A few others attribute this attraction to the micropylar substances, which, in turn, are got from the degenerating nucellus, integument(s) and/or the synergid. While secretion of substances by synergids is apparently a universal phenomenon, retention of these or any other material in the micropyle is rare. In *Paspalum* and a few other plants such as *Ottelia* (Fig. 196), the micropyle is filled with a Periodic Acid–Schiff (PAS)–positive carbohydrate substance of integumental and nucellar origin. This substance is found to have mainly neutral mucopolysaccharides and some acidic mucopolysaccharides, proteins and acidic lipids. In *Ornithogalum*, the micropylar substance stains positively for proteins and RNA, but not for carbohydrates. In plants with nucellar cells in the micropylar region, the synergid secretions, whether modified or not by nucellar cells, must pass through or between nucellar cells before reaching the micropyle. Therefore, it is speculated that the micropylar substance produced by the nucellus and integument(s) functions merely to increase the surface area on which the chemotropic substances synthesized by the synergid may be exposed. The thin sheet or hymen of exudates in the micropyle ensures the localization of chemotropic substances. This serves as a viable system for pollen tube attraction. Which of these chemicals, carbohydrates, proteins or lipids are chemotropic substances is not yet clear. There are evidences for and against calcium as an elicitor of chemotropic activity. Amino acids and amino mixtures coupled with sugars have also been suggested as possible chemotropic agents. Another hypothesis suggests that any growth factor can act as a chemotropic factor by causing a shift in the angle made by the centre of the growing tube tip, with respect to the rest of the tube. The tube would then continue to grow straight in the changed direction without the presence of a concentration gradient of the chemotropic factor until another shift in the direction

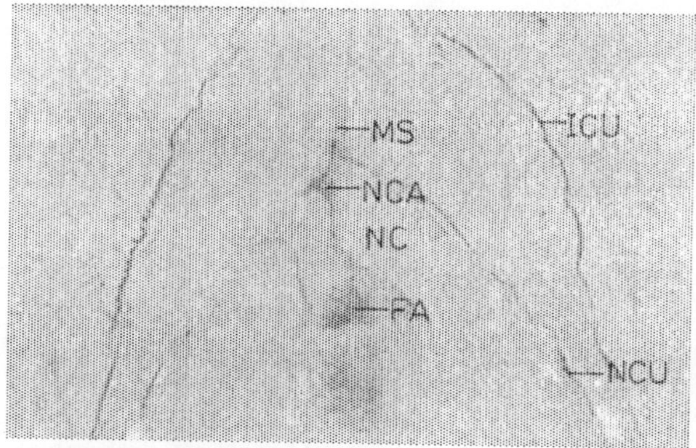

Fig. 196. L. S. of ovule of *Ottelia alismoides* just before fertilization, stained with PAS reaction for the demonstration of insoluble polysaccharides in the stylar canal, produced by the synergid and probably involved in the attraction of pollen tubes; FA - Filiform apparatus; ICU - Cuticle of the integument; MS - Micropylar carbohydrate substance; NC - Nucellar cap; NCA - Nucellar canal; NCU - Nucellar cuticle

occurs. According to this hypothesis the factors responsible for assembling the various components of the tube wall at the pollen tube tip can act as chemotropic factors.

6.5.5. Male Gametes and Double Fertilization

In many taxa, the vegetative nucleus stays in the pollen grain itself and does not move into the pollen tube, while the generative cell (in two–celled pollen) or the two sperms derived from it (in three– celled pollen) move into the pollen tube. It is believed that MFs of the tube cytoplasm are responsible for making them move into the tube and also in the tube. However, there are a number of examples where the vegetative nucleus enters the pollen tube and travels to varying distances when the tube grows. DNA synthesis is absent in the nucleus of the vegetative cell, but RNA and protein synthesis continues unaltered. The vegetative nucleus often lies in the neighbourhood of the sperms or generative cell (Fig. 197).

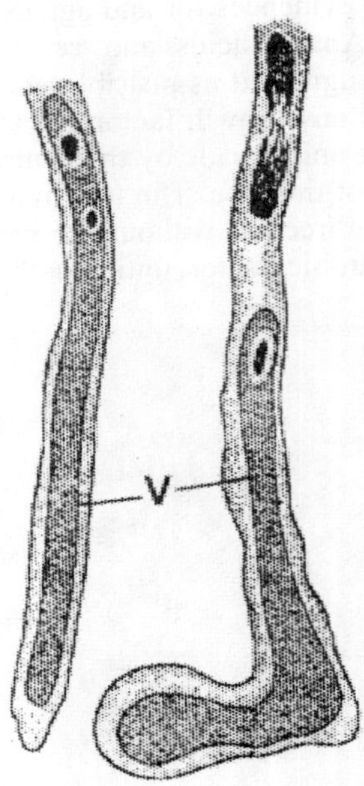

Fig. 197. Tips of two pollen tubes of *Cymbidium bicolor* showing the highly hypertrophied vegetative nucleus (V) and dividing generative cell (Drawing supplied by late Prof. B.G.L. Swamy)

In either case, it undergoes degeneration sooner or later. It is suggested that it undergoes PCD. In some taxa, they become lobed, enlarged in size, and hypertrophied and polyploidal. The exact function of the vegetative cell is not clear.

In the growing pollen tube, the two sperm cells derived from the division of the generative cell are always in close proximity and they are often connected through plasma canals. Both the sperms are enclosed by the plasma membrane of the tube. The functional complex of the two sperm cells of common origin within the vegetative cell in the pollen grain or pollen tube is designated as '**male germ unit**', although it is not universal in flowering plants. In some plants, the connection between two sperms of a pair and between one sperm and the vegetative cell is ephemeral or absent, especially in the pollen grain, but appears to get stabilized in the pollen tube. This unit condition might help in easy sperm delivery. The sperm that embraces the vegetative cell nucleus is larger of the two sperms and is rich in mitochondria and poor in plastsids. The smaller sperm is not associated with the vegetative cell. It has plastids, but is poor in mitochondria (Russell 1984). This sperm dimorphism may have far–reaching implications in double fertilization process of angiosperms.

Pollen tube releases its contents into one of the two synergids through a subterminal pore (Fig. 195). Almost immediately after this, pores are formed between this synergid and the egg, and between this synergid and the central cell. The contents released into the synergid by the pollen tube include the two sperms and **X–bodies**. The latter are DNA–positive and are suggested to represent remnants of the vegetative nucleus of the synergid itself or supernumerary sperm cells.

The sperms released into the synergid cytoplasm are definite cells with their own hyaline cytoplasm and limiting membrane. They often remain attached to one another for sometime before being separated. From now on the sperm cells show repeated change of shape, a feature which had led to the inference that they reach their destinations by autonomous movements, and this movement is believed to be directed by MFs. For a long time, the mechanism of syngamy and triple fusion has been under discussion in the light of the **mitotic hypothesis** and in connection with cell cycle events (Batygina and Vasilyeva 1998). In the course of their development, the two sperm cells derived from the generative cell fail to complete their mitotic cycle and remain temporarily blocked in one of the final steps of telophase, arbitrarily designated as **GO**. Hence, the sperm nucleus exhibits telophasic state and this state is continued till the time of fusion. The male nucleus, destined to fuse with the egg, reaches its destination earlier than its counterpart, probably because of the shorter distance it has to cover, but syngamy is completed much later than triple fusion. Several investigators using light microscope report that the cytoplasm of the sperm also enters the egg cell and the central cell, and thus, conceive the fusion at two levels: plasmogamy and karyogamy (e.g., *Plumbago*). On the other hand, there are reports to the effect that only the nucleus of the sperm enters the egg cell (and central cell) (for e.g. in *Spinacea*). In *Spinacea*, since the plasma membrane of the degenerated synergid is lost previously, the sperm cell membrane comes into contact with the plasmalemma of the egg cell, followed by the fusion of the two membranes facilitating the entry of one sperm nucleus into the egg cell. The same changes happen also in the central cell membrane region facilitating the entry of the second sperm nucleus into the central cell. Subsequently, the plasma membranes are closed again. The egg nucleus, on completing a full mitotic cycle during its formation, progresses to the **GO** period in which it stays till fertilization. The egg nucleus and its nucleolus, and the sperm nucleus and its nucleolus undergo slight enlargement by the time of syngamy. The DNA content of both is low. A difference in the electric charge of these two nuclei is said to result in mutual attraction. At the time of contact, the nuclei of both egg and sperm cells are usually in the same period of the cell cycle i.e., **GO**.

Syngamy is a slow process. The time taken for complete fusion varies in different taxa depending on the type of fusion. Three types of nuclear fusion have been recognized and these three differ from one another by the speed of fusion of male and female nuclei (Gerassimova-Navashina 1960,

1969; Batygina and Vasilyeva 1998) (Fig. 198): (i) **Pre–mitotic type**, where the male nucleus fuses with the female even while they are in the telophasic structure i.e., the gamete nuclei unite before the onset of mitotis, and the sperm nucleus completes its mitotic cycle after its fusion with the egg nucleus. (ii) **Post–mitotic type**, where the fusion between male and female nuclei is delayed till the male nucleus completes its telophase and enters into the next mitotic prophase, while in contact with the female nucleus. (iii) **Intermediate type**, which is in between the first two types. It is characterized by karyogamy taking place during mitosis. Thus, the type of fertilization is determined depending on karyogamy occurring before, after or during mitosis (i.e., it is correlated only in the final period, M period, of cell cycle) respectively. Moreover, depending on species, pre–mitotic type includes three subtypes of karyogamy i.e., **G1**, **S**, and **G2** karyogamies, while in post–mitotic and Intermediate type only one type of karyogamy is possible i.e., **M** type of karyogamy (Batygina and Vasilyeva 1998) (Fig. 199). The product of syngamy is the diploid zygote, which is the starting point for the next sporophytic generation.

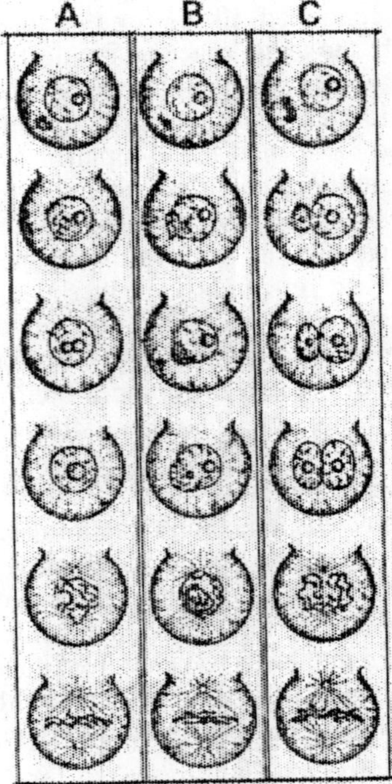

Fig. 198. Three types of syngametic fusions A. pre–mitotic type. B. Intermediate type C. post–mitotic type (Gerassimova–Navashina 1960)

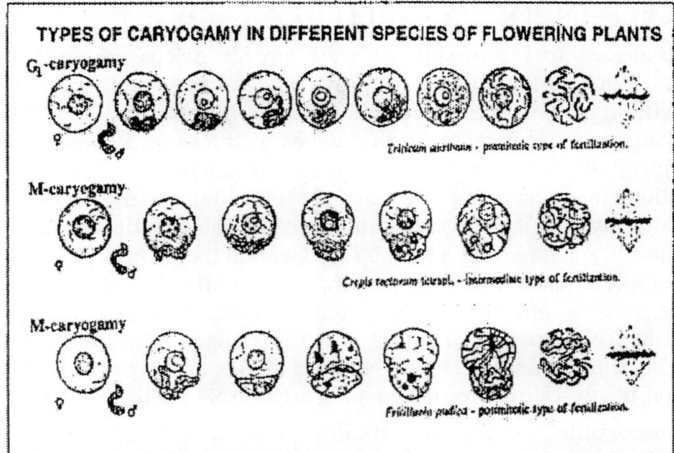

Fig. 199. Types of Karyogamy (between male gametes and eggs) in different species of flowering plants (Gerassimova–Navashina and Batygina 1958)

The sperm that is involved in triple fusion loses its plasma sheath before contacting the two polar nuclei or secondary nucleus (Fig. 200) located in the central cell. Wherever the two polar nuclei have not fused, their fusion is hastened by the presence of the male gamete. Triple fusion is completed much earlier than syngamy, although the reverse is true in some species of *Spinacea*. The resultant product, normally triploid, is the primary endosperm nucleus (PEN), which produces the **endosperm.**

Fig. 200. Stage just preceding triple fusion in *Trillium;* the male gamete has lost its extracellular matrix and is lying adjacent to the two polar nuclei with which it will fuse (Swamy 1948) (Reproduced with permission from Lierre, Belgium)

6.6. SELF-INCOMPATIBILITY

Some aspects of incompatibility were already discussed earlier in this chapter. More information is provided in this section on self–incompatibility. Some plants are self–fertile in that the pollen of the same flower can cause fertilization in the ovules of the same flower. However, the majority of flowering plants have a mechanism to prevent self–fertilization, to reject self-pollen and to force outcrossing through self–incompatibility, an elegant genetic mechanism. The genetic system of self–incompatibility is based on a single locus, called the *STERILITY* (*S*) locus, with multiple alleles. These alleles are very tightly linked and hence, in breeding programmes they are inherited together, as if they were a single gene. Pollen germination or tube growth is stopped when the pollen and stigma, on which they fall, have the same alleles at the S–locus, but does not stop if the two structures have non–matching S– alleles.

Homomorphic[1] self–incompatibility is divided into two types: **gametophytic** and **sporophytic**. In the gametophytic type, the self–incompatibility phenotype is determined by the genotype of the pollen grain (a gametophyte), while in the sporophytic type, pollen rejection is caused by the genotypes of the pollen parent plant (a sporophyte) and not by the pollen genotype. Gametophytic type is the most common and occurs in a vast majority of taxa studied (E.g., Liliaceae, Solanaceae, Fabaceae and Rosaceae), while the sporophyte type is known only in members of Brassicaceae, Asteraceae, Convolvulaceae and a few other families. Generally, in the gametophytic type, pollen rejection mechanisms operate in the style leading to the blocking of pollen tube growth, while in the sporophytic type rejection takes place in the stigma resulting in the failure of pollen to germinate or the failure of pollen tube to penetrate the stigma.

In compatible crosses, pollen tube tip is successively sealed off from the rest of the tube through successive callose plugs, keeping the tip free of callose. In gametophytic incompatibility, the first step is the irregular deposition of callose in the pollen tube, often leading to the plugging of the entire tip of the pollen tube and even the pollen grain itself. In sporophytic incompatibility, callose deposition occurs at the tip of the stigmatic papillae between the cell wall and plasma membrane, and this callose deposition is triggered by tryphine proteins of the pollen wall. It should, however, be emphasized that callose deposition is only one of the many mechanisms that together promote self–incompatibility.

---

[1] Incompatibility is also classified into **homomorphic** and **heteromorphic** types based on whether the flowers involved in crossing are morphologically similar or different; heteromorphic type has flowers which differ in the lengths of their stamens and styles (i.e., short stamen–flowers have long styles and vice versa) (distylous and tristylous flowers respectively have flowers with styles and filaments of two or three lengths respectively).

In both types of self–incompatibility systems, the acquisition of competence for self–pollen rejection is known to be developmentally regulated. This competence is acquired only about one or two days before anthesis and in open flowers through stigma or style–related glycoproteins called **S–proteins**, which are expression products of S–gene. In gametophytic type, the S–gene product is S–RNase (for Stylar RNase). In some taxa with gametophytic incompatibility, cytosolic Ca^{2+} fluxes act as a second messenger and control self–incompatibility by preventing pollen tube growth. Sporophytic incompatibility implicates stigma–specific proteins called **S-locus–specific glycoproteins (SLGs)**. The mechanism that activates self–pollen rejection probably lies in another gene, *S–LOCUS RECEPTOR KINASE (SRK)*, which is genetically linked to *SLG* gene at the S–locus.

Post–Fertilization Reproductive Growth and Development 7

7.1. INTRODUCTION

Double fertilization induces sudden and rapid changes in the ovary and the ovules contained in it. The ovary develops into the **fruit** and hence, the fruit may be defined as a fertilized ovary. It can be recalled that the ovary, style and stigma together makeup the gynoecium or pistil, which is equivalent to an appendicular, leaf–like structure i.e., the carpel. The carpels are exclusive structures of flowering plants. On the contrary, in gymnosperms, there are no carpels and hence, the ovules develop after fertilization into naked **seeds**. The fruit provides a suitable environment for seed development and maturation, and often serves as a mechanism for dispersal of mature seeds. Invariably the development of the fruit takes place concurrently with that of its seeds. A seed may be defined as a fertilized ovule. During fruit development, the style and stigma normally abscise off, as also the other floral parts such as stamens, petals and sepals do. But, in some taxa not only one or more of these floral parts persist, but also grow to variable extent along with the fruit or seed. They often influence the development of the fruit in a substantial way.

7.2. FRUIT DEVELOPMENT

7.2.1. Fruit Types

Fig. 201. A and B - Aggregate fruits; C - Multiple fruits; A - *Polyalthia*; B - *Annona*; C - *Morinda*

In order to understand the details of fruit development, one must know about the various kinds of fruits that are known from angiosperms. There is a great morphological diversity of fruit types. Basically, fruits are divided into **simple, aggregate** and **multiple** fruits. If only one fruit is produced per flower, irrespective whether the flower has monocarpellary or syncarpous pistils, the fruit is called a simple fruit. If a flower produces more than one fruit, the fruit type is called an aggregate fruit. In aggregate fruit, each free carpel of the pistil develops into a separate fruit (as in *Polyalthea* or *Michelia*) or as in cases such as *Annona*, where all the fruits developing from separate carpels subsequently fuse to form a single fruit (Fig. 201 A and B). If several flowers together form a single fruit, it is called a **collective** or multiple fruit. Invariably, the whole inflorescence (with many flowers) forms a single fruit involving the fusion of fruits of all flowers into one unit as in jack fruit, pineapple or *Morinda* (Fig. 201 C).

7.2.1.1. Simple Fruits

Simple fruits are classified into **fleshy** and **dry** depending upon their condition at maturity i.e., whether, the fruit wall or **pericarp** becomes fleshy or dry at maturity.

Fleshy fruits

There are several categories of fleshy fruits:

1. **Berry:** This type of fruit is developed from a superior ovary. The fleshy pericarp is usually divided into a thin, papery **epicarp** or **exocarp** making up the outermost zone of the fruit and an inner, massive, fleshy mesocarp. In the grape fruit, the whole mesocarp is fleshy, while in tomato, in addition to fleshy mesocarp, the placenta develops to an exaggerated proportion (Fig. 202 A). The seeds are, thus, eventually embedded in the juicy placental mass. In contrast, the septal tissue of tomato fruit does not divide much and so does not increase in size conspicuously. Date fruit is a one–seeded berry.
2. **Pepo**: It is a berry developed from a three–carpeled, syncarpous, inferior ovary with parietal placentation. The epicarp is thick, the mesocarp is moderately thick, while the placenta are highly proliferative and thick. Pepo type of berry is seen in most cucurbits (Fig. 202 B and C).
3. **Hesperidium**: This type of fruit is related to a berry and develops from a multicarpellary, syncarpous, superior ovary with axile placentation. The pericarp is differentiated into three zones: a leathery, thick, outer exocarp (epicarp) with oil glands (seen as spots on the surface), a mesocarp with a very thin, papery and often white tissue or with a loose, spongy tissue depending on the species, and an inner thick zone of endocarp giving rise to juice sacs that develop as multicellular hairs. The

basal part of the hair is differentiated as a stalk (e.g., orange, lemon and other citrus fruits) (Fig. 202 D).

4. **Drupe**: In this type, the pericarp is differentiated into three distinct regions: an outer skin or epicarp, a thick and fleshy mesocarp and a hard and stony **endocarp** (e.g., mango and coconut) (Fig. 202 E).

5. **Pome**: This type of fruit develops from an inferior ovary. The torus becomes fleshy and proliferative and forms a major part of the fruit, while the real fruit with a thick, somewhat stony envelope and enclosing seeds is seated deep in the middle of this false fruit or **pseudocarp** (e.g., apple and pear) (Fig. 202. F and G).

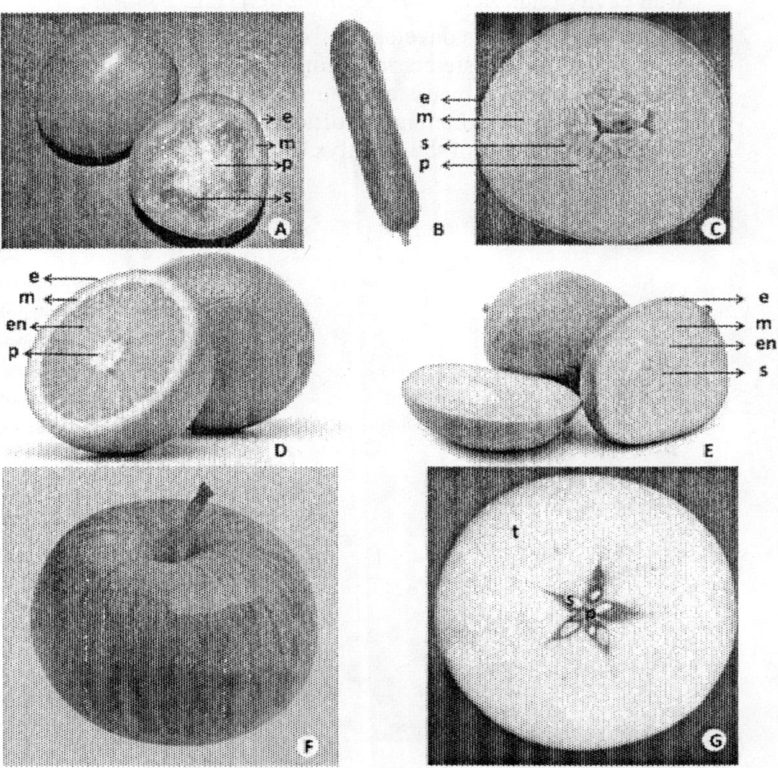

Fig. 202. Simple fleshy fruits; A - Tomato, a berry, entire and in sectional view; B and C - *Cucumis*, a pepo, whole fruit and T. S. respectively; D - Orange, a hesperidium, entire and in sectional view; E - Mango, a drupe, entire and in sectional view; F and G - Apple, a pome, entire and in sectional view respectively; e - Epicarp; en - Endocarp; m - Mesocarp; p - Plancenta; s - Seed; t - torus

Dry fruits

The dry fruits may also develop either from a monocarpellary ovary or from a multicarpellary syncarpous ovary. The dry fruits dehisce (in various ways) to release the seeds (**dry dehiscent** fruits) or remain **indehiscent**. In the case of indehiscence, the seeds are released after the decay of the fruit wall.

A. Dry indehiscent fruits - These are further categorized into many subcategories:

1. **Achene**: An achene has the simplest pericarp structure. This type of fruit develops from superior, one–celled, one–ovuled ovary. The seed is free from the fruit wall, but fills the whole fruit cavity (e.g., *Achyranthes*, *Mirabilis*) (Fig. 203 A).
2. **Cypsela**: An achene developing from an inferior ovary is called a cypsela (e.g., Asteraceae members). Although it develops from a bicarpellary ovary, at maturity, it has only one locule and one seed. The fruit often has a tuft of pappus hairs at its one end, which are products of the epicalyx (Fig. 203 B).

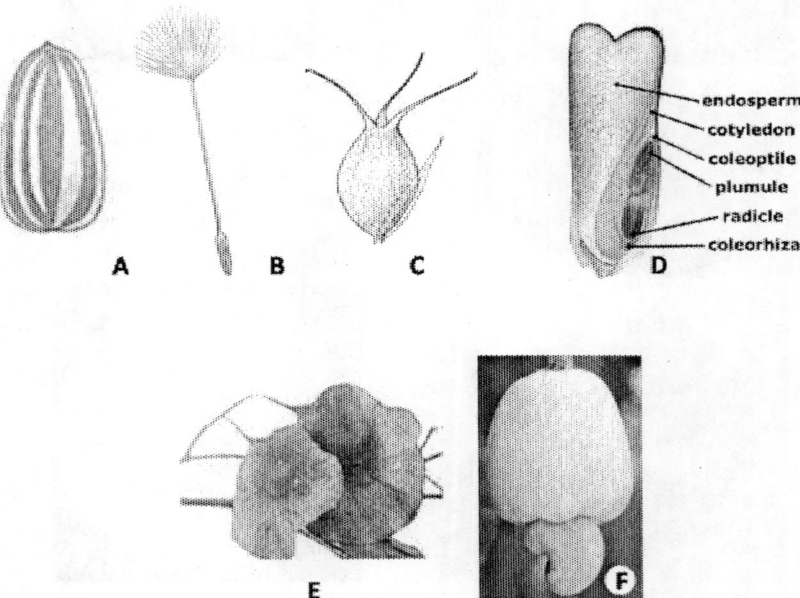

Fig. 203. Simple dry indehiscent fruits; A - *Mirabilis*, an achene; B - Dandelion, a cypsela; C - *Amaranthus*, a utricle; D - L.S. of Maize, a caryopsis; E - *Pterocarpus*, a samara; F - Cashew nut, a nut; a pseudocarp developed from pedicel is also shown.

3. **Utricle**: An achene with a small seed occupying only the basal part of the fruit chamber is an utricle (e.g., **Amaranthus**) (Fig. 203 C).
4. **Caryopsis**: A modified achene in which the seed coat is fused with the pericarp is called caryopsis or grain. Most cereal grains belong to this type of fruit (Fig. 203 D).
5. **Samara**: It is a winged achene (e.g., *Pterocarpus*) (Fig. 203 E).
6. **Nut**: A large achene with a very hard, woody pericarp is a nut (e.g., cashew nut) (Fig. 203 F).

B. Dry dehiscent fruits – They may develop from monocarpellary or multicarpellary and syncarpous ovaries. The following are the various kinds of dry dehiscent fruits:

1. **Legume** or **Pod**: It develops from a monocarpellary ovary with marginal plancentation. It dehisces longitudinally into two equal valves (e.g., most legumes) (Fig. 204 A).
2. **Follicle**: The fruit opens up along one of the margins (e.g., *Calotropis*) (Fig. 204 B).
3. **Lomentum**: It is a fruit type that develops from a monocarpellary ovary and it breaks into several one–seeded bits along transversely constricted regions of the fruit (e.g., *Acacia*, *Mimosa*) (Fig. 204 C).

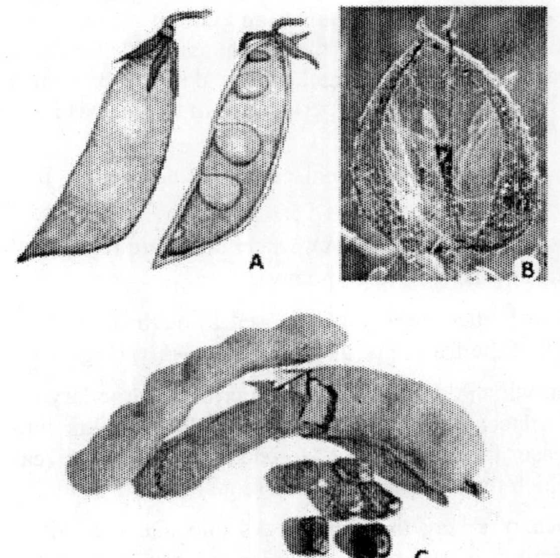

Fig. 204. Simple dry dehiscent fruits; A - Legume (Pea); B - Follicle (*Pergularia*); C - Lomentum (Tamarind)

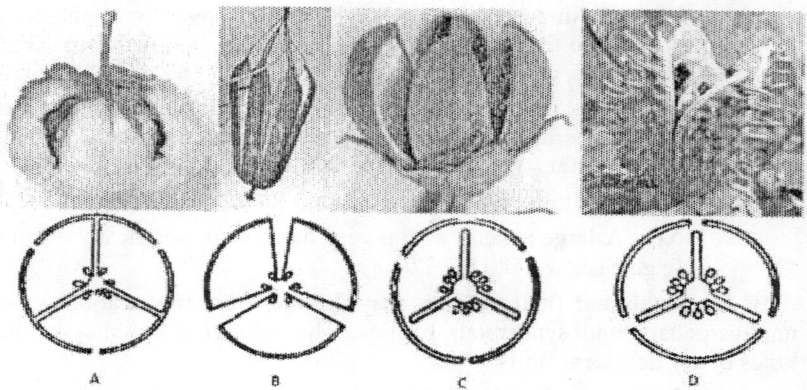

Fig. 205. Simple dry dehiscent fruits of Capsule type; A - Loculicidal (cotton); B - Septisidal (*Aristalcholia*); C - Locuticidally septifragal (*Lagerstroemia*); D - Septicidally septifrgal (*Datura*); upper row shows dehiscing fruits, while lower rows the diagrammatic representation of the type of dehiscence.

4. **Capsule**: It develops from a syncarpous and multilocular ovary with axile placentation (Fig. 7.5). It may split longitudinally along the middle of each locule into as many valves as there are carpels, as in lady's finger and cotton (**loculicidal capsule**), or longitudinally along the septa as in *Aristolochia* (**septicidal capsule**), or **septifragal**, where dehiscence results in the total breaking away of the pericarp from the fruit as in *Lagerstroemia* and *Datura*.

There are other types of dry dehiscent fruits developing from syncarpous ovaries:

- **Porous capsule** (Fig. 206 A), where seeds are released through pores at the tips of the fruit (e.g., poppy).
- **Pyxidium**, where seeds are released through the opening of a lid at the apex of the fruit (e.g., *Portulaca, Lecythis*) (Fig. 206 B).
- **Siliqua**, which develops from a superior, bicarpellary ovary with free central placentation and where the pericarp splits into two halves from base upwards leaving a central false septum (called **replum**) with which the seeds are attached (e.g., mustard) (Fig. 206 C).
- **Schizocarp**, where the fruit breaks into one–seeded portions called **mericarps** (e.g., *Abutilon*, coriander and castor) (Fig. 206 D).

Fig. 206. Other capsule type fruits. A. Porous capsule (poppy) B. Pyxidium (*Lecythis*) C. Siliqua (mustard), D. Schizocarp (*Abutilon*).

7.2.1.2. *Aggregate Fruits*

The aggregate fruits (Fig. 207) may be aggregate of follicles (e.g., *Michelia*), achenes (e.g., *Thalictrum*), berries (e.g., *Annona*) or drupes (e.g., *Raspberry*).

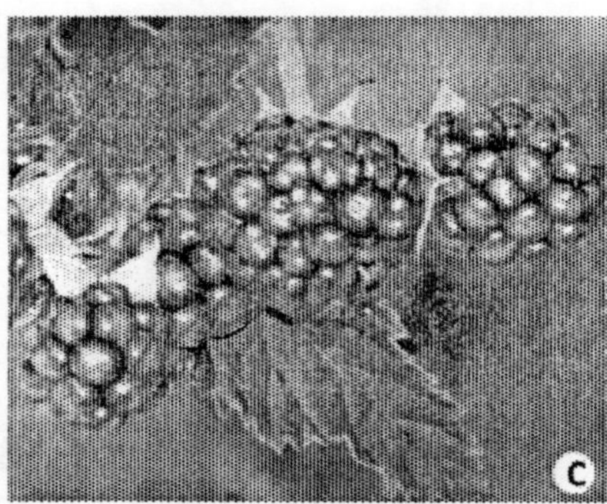

Fig. 207. Aggregate fruits; A - Aggregate of follicle (*Michelia*); B - Aggregate of achenes (*Thalictrum*); C - Aggregate of drupes (Raspberry)

7.2.2. Basics of Fruit Development

Fruits develop from a gynoecium, particularly from the ovary part. Hence, fruit development would be more correctly defined as the further differentiation of a pre–existing organ. In pseudocarpic fruits, organs other than the gynoecium, such as the receptacle, bract, floral tube or pedicel/peduncle also participate in the formation of the fruit. When an ovary develops into a fruit, the ovary wall becomes the pericarp. The outermost part of the ovary becomes the epicarp (or exocarp), while depending on the fruit type, both mesocarp and endocarp, or mesocarp alone, are derived from the inner wall tissue of the ovary. Depending on whether the fruit is a dry or a fleshy fruit, the pericarp tissue becomes entirely dry or fleshy to various extents respectively. The placentum and/or septa in fleshy fruits also become fleshy to various extents with or without accompanying cell proliferation and enlargement.

In most plants, the early fruit development can be divided into three phases. The earlier phase involves the development of the ovary and the decision to abort or to proceed with further cell division and fruit development, which is generally referred to as **fruit set.** In the second phase, there is fruit growth primarily due to the cell division. The third phase begins after cell division ceases. During this phase, fruit growth continues, mostly by cell expansion, until the fruit reaches the final size. This growth phase is the most visible and physiologically most significant because of the strong sink activity exercised by the expanding cells. Different fruit types display variations of this general developmental programme. For instance, in some cases such as avocado, cell division in the pericarp continues until shortly before ripening (Gillespy et al 1993).

7.2.3. Biochemical factors in fruit development

Normal development of fruit, however, is dependent not only on double fertilization, but also on additional growth regulating substances that emanate from the endosperm and embryo of the seed and also from the sources such as persistent floral parts, if any, flag leaf and vegetative leaves. These substances are necessary not only for the ordered development of the fruit, but also for preventing abscission and premature fruit fall. It has been clearly demonstrated that the developing seed is an essential prerequisite for fruit growth. The removal of a few developing seeds results in misshapen fruits. Such manipulations also affect the morphology of structures associated with the fruit. In strawberry, where the thalamus is a major constituent of the fruit, the removal of achenes in certain longitudinal or transverse rows cause the development of long and narrow fruits and of short, thick fruits respectively (Fig. 208). The failure of one or more seeds to develop in multiovulate ovaries leads to the formation of fruits with greatly altered symmetry and shape. The pericarp tissue of the fruits, in such cases, is developed only around the persistent seeds. The importance of seeds in fruit growth is also substantiated by the fact that an early degeneration of embryo due to natural causes affects the maturation of fruits.

It has also been established that the developing seeds exert their impact on fruit development through chemical substances of hormonal nature, especially through GA. This is proved by the fact that the extracts of seeds bring about the growth of the pericarp tissue, even in unpollinated and unfertilized ovaries. The active chemical principle contained in the seeds appears to be similar in most fruits. For example, extracts of immature corn kernels and of the seeds of apple have been shown to promote the growth of tomato ovary. There is also a close relationship between the ultimate size of the fruit and the number of seeds it contains. The growth regulators present in the seeds belong to the categories of auxins, gibberellins, kinins and growth inhibitors. These substances have been extracted from a variety of seed sources. In strawberry, for instance, free auxins have been isolated from the achenes, but none are present in the receptacles. The amount of growth regulators fluctuate greatly during the various stages of development. The peak gibberellin production in *Pharbitis nil* is coincident with the period of greatest seed development and this peak is reached in 13 days after anthesis. Similarly, the level of auxin reaches the maximum mid–time between pollination and fruit maturation. Generally, kinins occur in greatest concentration in very young seeds and fruits. The role of gibberellins in fruit development deserves special mention. There are enough evidences to show a direct correlation between growth rate in the developing fruit and the endogenous GA level in fruits such as apricot, peach, bean, grape and orange. GAs are found instead in *Pharbitis nil.* Previous studies support the contention that particular endogenous GAs in a given type of fruit are involved directly in the growth of the fruit, while the other GAs may be implicated otherwise in fruit maturation. However, the exact function of gibberellins in fruit development is not clear. A decrease in endogenous GAs appears to be associated with the ripening of fleshy fruits.

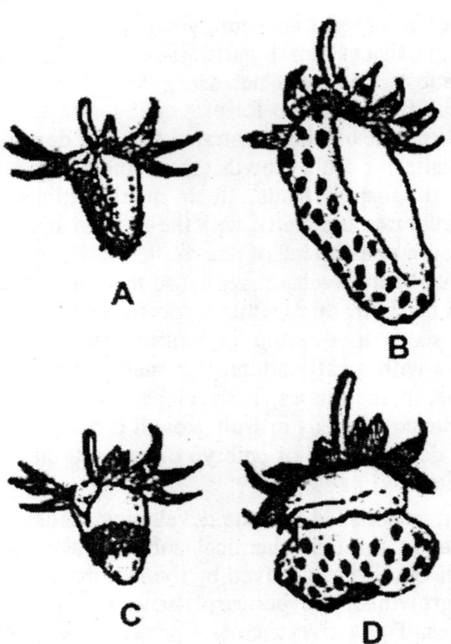

Fig. 208. Diagrammatic representation of the importance of seeds in strawberry fruit development; A and B - A few vertical rows of seeds have been removed (A) to result in an elongated fruit (B); C and D - A few horizontal rows of developing seeds have been removed (C) to result in a broad fruit (D) (Nitsch 1950)

Some studies have been done on chloroplast changes in the fruit. In tomato fruit, there is a significant decrease in the transcript levels of the gene for *RbcS* in cells of immature fruits compared with leaves. The levels of *RbcS* and *Cab* mRNA are also considerably lower than those of *RbcL* and *psbA* throughout the fruit development (Piechulla et al 1986). The role of *RbcS* gene in pericarp development was found out by assessing the pattern of gene expression of individual members of the gene family, composed of five genes *(RbcS1, RbcS2, RbcS3A, RbcS3B* and *RbcS3C)*. Within this domain, only *RbcS1* and *RbcS2* are expressed in the cells of the pericarp. The transcripts of these genes accumulate during early fruit development, but decrease rapidly during fruit maturity.

7.2.4. Physical Factors in Fruit Development

Many fruits show a pattern of growth that follow a single sigmoidal growth curve. There is an initial phase of slow growth characterized by active cell division, followed by a period of major increase in size, fresh and dry weight, dominated by cell expansion and a final phase of declining growth rate during which ripening is usually initiated. In apple, there are about two million cells at anthesis and about 40 million at harvest. In some fruits, such as avocado, in which cell division continues throughout the life of the fruit and in which growth continues all the time, the fruit remains attached to the tree. Certain fruits show a more complex growth pattern with double or even triple sigmoidal growth patterns due to the presence of two or three rapid growth phases interrupted by one or two periods of little or no growth (E.g., figs, grapes). The two rapid growth phases are designated as period I and III, while interrupting slow growth phase is designated as period II. In grapes, about 40% of the total growth is seen in period III, and in peach, about 80% of the total growth is seen in period III. Studies on the development of gourd fruits reveal that the differences in morphological form in the various races are due to the differential growth rates in diverse dimensions. In the long, narrow type of fruits, the length increases faster than width, but at a relative constant rate, while in the broader fruit, the width increases faster than length, again at a constant relative rate. In bottle gourds, which have races producing miniature as well as giant–sized fruit, the relative growth rate remains the same and the shape or size of the two kinds at maturity differ because of a difference in the duration of growth rather than rate. Differences in fruit shape depend mainly on the relative number of cells lying along the different axes and therefore, a genetic control of the planes of cell division must be involved. The orientation of mitotic spindles and the consequent plane of cell plate formation bears a direct relationship with the direction of fruit growth. The differences in fruit size are generally caused by the variations in both cell number and size. The innermost tissues of the fruit suffer cessation of cell division sooner than the peripheral tissues, but they show a more rapid increase in cell volume during mitosis.

There are also differences in growth rate in different parts of the same fruit. For instance, the apple fruit is a product of the combined growth of the fleshy receptacle and the fertilized ovary. Intensive cell division ensues following double fertilization as a result of which a substantial increase in fruit diameter is reached 5–6 days after pollination. Although cell division occurs throughout the developing fruit, the duration of both cell division and cell enlargement varies in different regions of the same fruit. The thick endocarpic part develops rather early at a rapid rate reaching maximum development long before any other region of the fruit. This part is formed from the layer just below the inner epidermis of the ovary. The five or six layers of derivative cells become thick–walled. As a result of continued growth of the surrounding tissues, the endocarp may often become split. In contrast, the mesocarp continues growth for a further period of two months. During earlier stages of development, these parenchymatous tissues grow in size both due to cell division and cell enlargement, but later on, only cell enlargement predominates. The growth of mesocarp ceases at the end of about 150 days after pollination.

7.2.5. Fruit Ripening

Fruit ripening involves a series of changes in colour, texture (which depends on the degree of contact between the cells and on the nature of cell walls), taste, flavour and aroma (due to production of volatile compounds) occurring in the fully–developed fruit (Fig. 209). Thus, the structure and composition of the already developed parts of the fruit is altered such that it becomes acceptable to be eaten either by humans or by other animals, or become easy to get decayed by microbes to release the seeds. Ripening usually occurs while the fruits are still attached to the plants, the prominent exception being avocado, where ripening starts only after separation from the plant.

Fig. 209. Three stages in ripening of banana fruit; Note the change in colouration of the fruit with ripening.

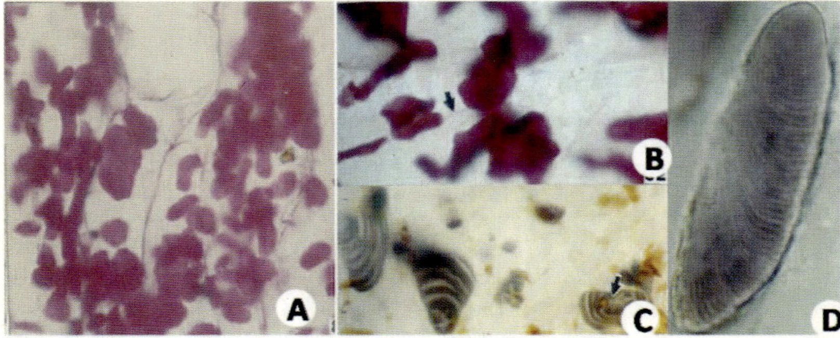

Fig. 210. A - Starch grains in unripe banana; B and C - Degradation of starch grains in ripening banana due to enzyme action; D - A single starch grain enlarged to show degradation in a portion on the left hand side of the grain where the amylopectin layers are no longer evident (Photographs courtesy of Dr. C. Santhakumari)

The onset of ripening is initiated usually by sugar accumulation invariably due to starch hydrolysis in many fruits such as banana (Fig. 210) and apple, loss of acidity, drying up of latex (Fig. 211), softening and colouring of skin (usually a change from green to yellow, red or other colours). Physiologically, ripening is usually marked by an upsurge in the rate of respiration, called **climacteric respiration** (roughly two to ten times the level of the minimum respiration period value depending on the fruit). Only after climacteric respiration period, fruits of most species reach their optimum eating quality.

Treatment of unripe fruits with ethylene promotes both climacteric respiration and other ripening–associated changes. Similarly, there is retardation of ripening by inhibitors of ethylene biosynthesis. Hence, it is clear that ethylene is a major regulator of fruit ripening. Moreover, ethylene is produced by all fruits during as well as preceding the ripening. Peak production of ethylene proceeds (e.g., banana), coincides with (e.g. mango, pear, avocado) or immediately after (e.g., apple, tomato) the peak in respiration. There are some fruits, which are **non–climacteric**, for example, the fruits of chilli, citrus, cherry and some varieties of grapes. Also, these fruits do not show an accelerated production of ethylene i.e., ethylene production in these fruits is uniform, although very low.

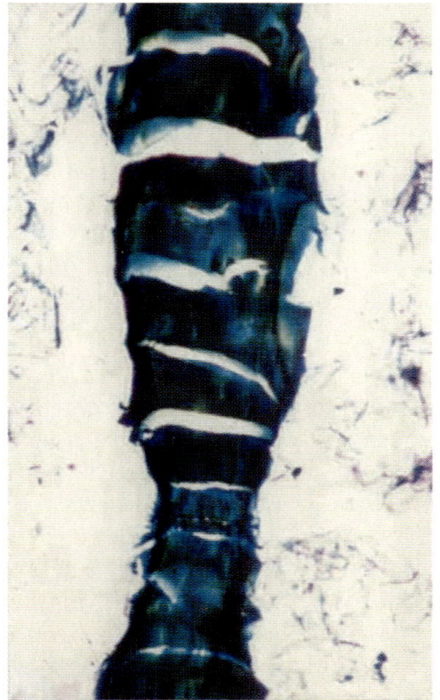

Fig. 211. Phenolic–containing laticifers in ripe banana fruit; Compare the laticifers of unripe fruit shown in Fig. 86C (Photograph courtesy of Dr. C. Santhakumari)

The process of ripening also involves other complex changes in the metabolism of fruits in which both anabolic and catabolic processes take place. These pertain to the production of new pigments and new volatile compounds and degradative processes such as those involved in the conversion of chloroplasts into chromoplasts, and in changes in the adhesion between cells leading to fruit softening. A lot of structural changes in plastids of ripening fruits have been recorded through electron microscopic studies. Cell wall changes are more important as they make the fruit soft in texture and such changes are often very complex. These metabolic changes are related, in climacteric fruits, to the changes in ethylene production and respiration. Cell enlargement, presence of polyploid nuclei (up to 64–ploid), increase in water content and accumulation of organic substance such as sugars and organic acids are some of the common changes in the ripening fruits. Accumulation of tannins (condensed phenol), leucoanthocyanins (responsible for astringent taste) etc. are the other changes. There is a decrease in acidity associated with an increase in sugar content in many ripening fruits. The change in flavour and aroma during ripening is due to the release of a diverse array of volatile compounds in varying quantities. These volatile compounds include organic acids, alcohols, esters, carbonyl compounds, hydrocarbons, terpenoids, phenolics etc.

The changes, both physical and chemical, during ripening of mature fruit have led to a controversy whether ripening is a degradative process or an active metabolic process. In support of the latter view, earlier studies cite the increase in protein content, retention of protein and RNA–synthesizing capacity, and the prevention of ripening through inhibitors of protein and RNA synthesis. In strawberry, striking changes in mRNA subsets, including the appearance of new mRNAs in ripening fruits, a decrease in other mRNAs with advancing ripening and the total disappearance of certain other mRNAs in over–ripened fruits are seen, again indicating the strong metabolic activity during ripening.

Some information is available on the genetic control on fruit ripening process, especially related to chloroplast–chromoplast transition and the cell wall softening process. In tomato, during ripening, the transcript levels of various genes coding for protein of photosystems I and II, and stroma decrease to extremely low or zero levels, with the exception of *psbA* transcripts in the chromoplasts, which is at least 20–fold higher than the transcript level of other photosynthetic genes. The *PSY* gene, responsible for the enzyme phytoene synthase and essential for carotenoid synthesis in tomato, has been shown to be expressed during ripening of the fruit. Gene expression related to cellulase, pectin methyl esterase (PME) and polygalacturonase (PG) has been studied in avocado and tomato fruits through the investigation of the levels of translatable mRNA. A nearly 50–fold increase in cellulase transcripts in avocado fruits during ripening was reported. Gene activity expressing PME, responsible for dissolution of middle lamella pectins, increases 2 to 3–fold during the beginning of ripening in tomato, but decreases during ripening. The tomato mutants such as *never ripe (nr), non–ripening (nor), and ripening inhibitor (rin)* are good examples in which fruits do not undergo normal ripening, and the ripening–associated chlorophyll degradation, carotenoid synthesis, ethylene production, and appearance of polygalacturonase activity are negatively regulated to varying degrees. The pattern of expression of the *PME* gene in the fruits of *nor* and *nr* mutants is similar to that of wild–type fruits, whereas it is down–regulated in the *rin* mutant. Fruits of plants transformed with an antisense *PME* gene are found to contain reduced levels of *PME,* yet they ripen normally and display other ripening characteristics (Raghavan 2000).

In tomato, the initiation of fruit ripening signals the accumulation of PG in the pericarp, where it accounts for 3–5% of the total soluble proteins of the fruit. It is not expressed in the fruit locules where the seeds are located (Tieman and Handa 1989). The developmental regulation of PG gene expression plays a major role in the ripening of tomato fruit. Transcripts of PG gene show more than 2000–fold increase in mature ripe fruits and it is significant that they are almost non–existent in the immature green fruits. The expression of PG gene appears to be regulated at the transcriptional level, with contributions from post–transcriptional processes such as mRNA stability or transport. In the mutants already mentioned, there is a greatly reduced level of PG and its gene transcripts. These results have also been broadly obtained in transgenic experiments. Although PG expression in the transgenic *rin* (mutant) fruit results in cell wall degradation, there is no significant effect on fruit softening or colour development. This questions the generally held view that cell wall pectin degradation primarily causes fruit ripening. The explanation for this may come from the identification of a gene for expansin from tomato fruits. The transcripts of expansin gene are expressed in abundance in ripe fruits, especially during the period of softening and are present only at barely detectable levels in *nor* and *rin* mutants. Hence, it is likely that expansin, by loosening bonds between cellulose and hemicellulose, may play a much more important role in ripening than pectin degradation. Additional genes called *E8* and *E4* (ethylene–related genes) need to be expressed for fruit ripening in tomato. *E8* gene is activated during normal fruit ripening and also in unripe fruits exposed to ethylene. The action of this gene is mediated by a DNA–binding component that specifically interacts with the base–pair sequences flanking the gene. This is based on the finding that DNA–binding activity is low in unripe fruit and increases during ripening. *E4* gene transcripts are also abundant in ripening fruits. As in *E8*, the same DNA–binding protein probably coordinates the *E4* gene.

Not much information is available on genes regulating fruit ripening in other taxa. Limited data indicate the expression of genes such as those of cytochrome P–450 from avocado, pectin lyases, endochitinase, β–1,3 glucanase, ascorbate peroxidase, metallothianein, ACC–oxidase and a senescence–related protein from banana, pectin lyases and HSP from strawberries, chitinase and thaumatin–like protein from grape and endo–β–1,4 glucanases from *Prunus*.

7.2.6. Histology of Fruits

As already indicated, the fruits are either fleshy or dry at maturity. Depending upon the type of fleshy fruit, the pericarp as a whole or a major part of it remains essentially parenchymatous. The fleshy zone of the fruit may be homogeneously parenchymatous as in tomato or it may be interspersed with varying proportion of sclerenchyma (sclereids or fibres) (as in pear, some varieties of apple), tanniferous cells (as in banana), laticifers (banana and lady's finger), secretory canals or cavities (as in *Citrus*) etc. The pericarp of dry fruits is made up of sclerenchymatous or sclereified cells of diverse morphological shapes, although initially many of them are parenchymatous.

The outermost layer of the fruit is usually the epidermis, derived from the epidermis of the ovary. This layer may be smooth, often covered by epicuticular wax and a cuticle, or only by the cuticle. In some fruits, the epidermal cell may show unicellular, multicellular or both unicellular and multicellular hairs or trichomes. The trichomes may be glandular or non–glandular of diverse categories depending on the species. The fruit epidermal cells may resemble the epidermal cells of the stem and leaves in form and wall structure. Many fruits have stomata, which are often functional, at least during the major part of fruit development. The ordinary epidermal cells are usually isodiametric with straight or variously sinuate walls. Many fruits have a corky tissue instead of an epidermis. Such a tissue has suberized cells; lenticels may be present in such a tissue (E.g., pear).

7.2.7. Dehiscence of Fruit and Release of Seeds

Next to pollen transport, seed dispersal is the most important single factor promoting gene flow in populations. As indicated earlier, dry dehiscent fruits release seeds through dehiscence of pericarp, while in dry indehiscent fruits, seed are released only after degradation of pericarp tissue, usually by animals or microbes. In the latter, there are no special structures that are involved in fruit dehiscence or are organized. In some instances, especially in fleshy fruits, the pericarp is eaten by a variety of animals (including humans) and then the seeds are released.

In dry dehiscent fruits, special modifications of the entire fruit wall or part of it are involved in dehiscence in characteristic ways. Anatomic structure - basic and secondary, gross and minute- is important in dehiscence. The form of vascular skeleton is usually less important in dehiscence. The type, amount and distribution of fibrous, stony and soft tissues determine the locus or the position of dehiscence.

In the legume fruit, the pericarp usually consists of epidermis, the mesocarp of relatively thick parenchyma and an endocarp of stone cells on the inside of which there occur either a thin–walled inner epidermis alone or a few parenchyma cells. The relative thickness of the different tissues and their histology are the chief factors that cause a range of structural variability in legumes. The dehiscence is brought about by the twisting of the two valves of the fruit and this again is dependent on the anisotropic shrinkage of the thickened walls of the pericarp and the orientation of microfibrils in their walls. The walls are swollen in a direction at right angle to the long axis of the microfibrils and consequently the greater shrinkage during drying coincides with this direction.

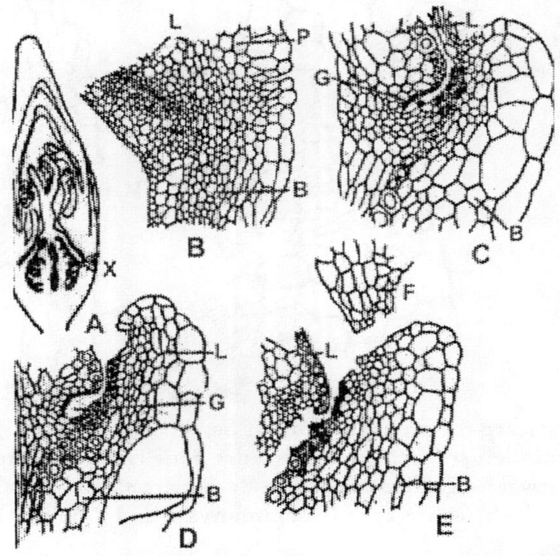

Fig. 212. Progressive anatomical changes associated with dehiscence of the fruit in *Portulaca;* Some parenchymatous cells at the junction (X) of the lid (L) with hypanthium (B) become lignified; Dehiscence takes place along the thin–walled cells just above this lignified zone; P - Perianth (Subramanyam and Raju 1953)

In capsules that dehisce through a lid, the line of dehiscence on the fruit wall is marked in diverse ways depending on the species. In *Portulaca* (Fig. 212), some parenchymatous cells at the junction of the lid and hypanthium gradually become lignified. Simultaneously lignification is also evident in the outer epidermal cells of the lid. Lignification of the hypanthial part gradually extends downwards. Dehiscence takes place along the thin–walled cells just above this lignified group of cells. Splitting is effected by the pressure exerted by the fully–developed seeds and also by the deepening of the groove into the zone of thin–walled cells. In

Sesuvium (Fig. 213), a band of small–celled zone appears in the middle region of the ovary during the earlier stages of fruit development. The adjacent walls of the inner and outer epidermal cells above and below this zone become slightly lignified. Towards the maturity, some of the inner epidermal cells of the lid part develop fibrous thickenings. Dehiscence of the fruit takes place along the small–celled zone.

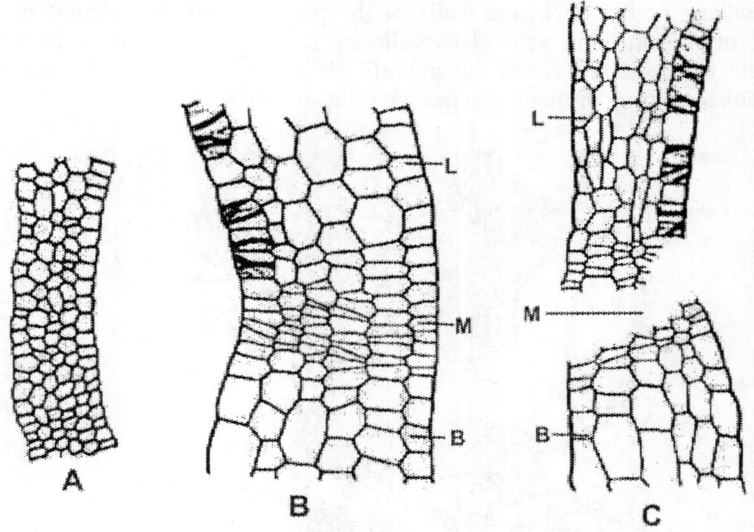

Fig. 213. Pericarp region of the fruit of *Sesuvium* at successive stages leading to fruit dehiscence; The dehiscence zone is parenchymatous and is present between lignified zones; B - Basal part of ovary wall; L - Lid; M - Dehiscence zone (Subramanyam and Raju 1953)

7.3. SEED DEVELOPMENT

7.3.1. Introduction

Although a seed was defined as a fertilized ovule earlier, such a rigid definition is not often followed and the term is used in a loose sense. As Eames (1961) had rightly pointed out, the definition that the seed is a fertilized ovule is inadequate as the term is applied even to the immature fertilized ovule. The presence of a mature embryo cannot be strictly used as a criterion for the definition of the seed as many already–released 'seeds' often contain immature embryos. Therefore, the term is used here in the sense of a product of a fertilized ovule containing an embryo at some stage of development.

As briefly indicated earlier, the seed is one of the diagnostic characteristics of the phanerogams. It is a dispersal unit i.e., a **diaspore,** and it is resistant to extremes of climate and without undergoing much morphological change. It can persist over considerable periods of time. At the same time, the seed, as said earlier, secures the safe spread and survival of a species. Much of the changes that happen during development are the products of post–fertilization development and are meant for the important function mentioned above. These changes also account for the variations noticed in seed size, shape and weight. The seeds vary from very tiny dust–like particle (each weighing around 20 μg fresh weight) to very large ones as in some palms (each weighing more than 6 kg).

Double fertilization results in the production of zygote and PEN. It also triggers activities in certain maternal tissues of the ovule such as the chalaza and the integument, and in some species the nucellus also. Seed formation, hence, requires the co–ordinated development of both products of double fertilization, and also of the maternal tissues of the ovule. The seed is genetically a **triantic entity** (Evenari 1984, Krishnamurthy 2003). It is the only structure in a plant that has this distinction and hence, makes it very unique. Its sporophytic maternal tissues such as the chalaza, seed coat and nucellus (wherever present) have a 2n genetic makeup with maternal (m) and paternal (p) ploidy ratio of being 2m: 0p; the zygote/embryo, resulting from syngamy, also has a 2n genetic makeup but exhibiting 1m: 1p ploidy ratio; the endosperm, a product of triple fusion, in most angiosperms, has a 3n genetic makeup with a ploidy ratio of 2m: 1p. If these ratios between m and p are altered in the three components of the seed, its development is disturbed or abnormal. In addition, certain component cells of the endosperm (mainly endosperm haustorial cells) and of embryo (suspensor cells) exhibit endopolyploidy or polyteny.

7.3.2. Role of Chalaza in Seed Development

The behaviour of the chalazal tissue constitutes an important aspect of seed development. As already indicated in the previous chapter, the chalaza is an exclusively sporophytic maternal component of the seed. Chalaza is the epicentre of seed initiation and development not only from a topographical perspective, but also from a functional standpoint. It may also be noted that the chalaza functions as a topocentre for seed development due to the fact that it serves as a topocentre for ovule development as well. As mentioned earlier, the topographical relationship between the chalaza, the funicle and the micropyle varies in the different configurations of the ovule. This relationship is liable to further alteration during post–fertilization development. It should also be noted that the ovule generally does not undergo a *pari passu* growth during the formation of seed. The relatively small ratio of the chalaza in relation to the remainder of the ovule at maturity (Fig. 214 A and D) becomes completely reversed in the mature seed (Fig. 214 C, F and G). The major trends of behaviour of the chalaza during post–fertilization development may be broadly given as follows (Periasamy 1962):

1. **Normal chalazal development:** The chalaza continues to occupy a comparatively small portion of the seed, so that there is no appreciable alteration of the size relationship that existed between the chalaza and the remainder of the ovule prior to fertilization (Fig. 214 B, E). This developmental trend is seen in members of Scrophulariaceae, Passifloraceae, and some Acanthaceae etc.
2. **Massive chalazal or pachychalazal development:** The chalaza shows an overall increase in size so that the earlier size relationship is significantly altered in the seed. In such cases, a major part of the seed volume is occupied by the chalaza as compared to the case during pre–fertilization development (Fig. 214 C and F). Such a growth is seen in species of *Degeneria, Myristica* and many Rubiaceae.
3. **Perichalazal:** Here, the chalaza, instead of showing an overall growth, increases only along a certain direction so that the seed as a whole loses the radial symmetry of the ovule stage and attains a bilateral symmetry (Fig. 214 G). Perichalazal type of seed development is seen in members of Annonaceae, Vitaceae, some Menispermaceae etc.

However, it should be cautioned that it is not always possible to assess the individual share contributed by the above types of growth in any given seed since there is practical difficulty in demarcating the limits of the chalazal tissues from that of the integuments. This difficulty is further accentuated in unitegmic ovules, where normal chalazal and massive chalaza types are met with. Whatever the type of chalazal growth, this tissue forms an appreciable part of the mature seed in a vast majority of plants.

The chalaza not only acts as a contributor to the overall increase in the size of seed, but also functions as a controlling centre of seed development (especially of seed coat, endosperm and embryo) (Krishnamurthy 2003). It is interesting to note that the chalazal control on the embryo and endosperm is effected in spite of the absence of any direct or symplastic connection between them (Evenari 1984, Thorne 1985). The lack of symplastic connection, in fact, prevents the transmission of viruses from maternal to embryo tissues. As indicated earlier, many genes operate in the chalaza–integument complex area of the ovule or the developing seed. The *SHORT INTEGUMENT* (*SIN*) mutant causes defects in embryo development, and the affected embryos either possess funnel–shaped cotyledons or masses of unorganized tissue, thus affecting pattern formation in the embryo. This is an example for a maternal gene affecting embryo development irrespective of the latter's genotype. *AINTEGUMENTA* (*ANT*) and *HUELLENLOS* (*HLL*) mutants lack integuments and do not show seed development. *FBP7* and *FBP11* genes

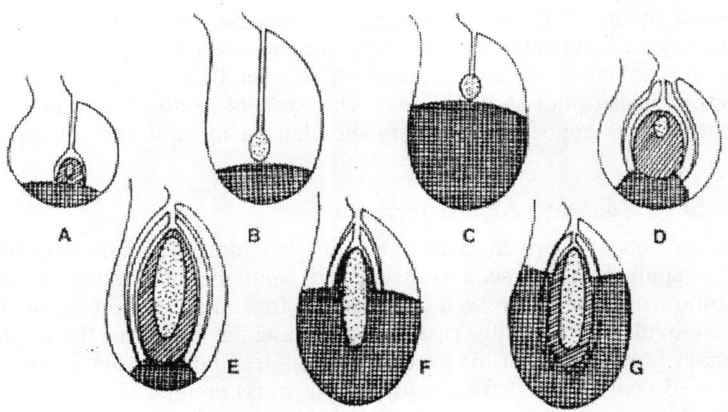

Fig. 214. Behaviour of chalaza during seed development; A and D - Uni and Bi-tegmic ovules respectively before fertilization; B and E - Normal chalazal type; There is no appreciable change in the proportion of the chalaza in the seed as compared to the ovule; C and F - Massive chalazal type; Chalaza shows an overall increase of size in seed; G - Perichalazal type where chalaza increase only along a certain direction so that the seed becomes bilaterally symmetrical; The seed is stippled, nucellus cross-hatched and chalaza netted (Periasamy 1962)

in barley and an ovule-specific *MADS* gene in *Petunia* are additional examples of sporophytic maternal genes, operating in the chalaza–integument domain. These are shown to affect endosperm development. The chalaza epicentre also controls the orientation of developing embryos along the chalaza–micropylar polar axis through genes operating in this domain.

The chalaza is also a functional centre by serving as a seat of very high and regulatory physiological activity during seed development. How do nutrients move from the parent plant to the developing seed? Several accommodations have been made for this purpose in the chalaza, especially at the chalaza–endosperm interface. Unloading of nutrients from phloem terminals takes place in the chalaza, where these nutrients get assimilated in the form of simple sugars, amino acids, fatty acid etc. Then, they are transported from chalaza to the basal endosperm region apoplastically. To facilitate this apoplastic transport, not only the endosperms cells, but also the chalazal cells in the chalaza–endosperm interface are provided with transfer cell morphology (Charlton et al 1995, Howell 1998). In fact, a cDNA called *Bet 1* was found and this may serve as a marker for these cells during this functional apoplastic transfer between chalaza and endosperm. *Bet 1* encodes a small cell wall protein (probably related to expansins) that may be unique to transfer cells. This may be necessary for cell wall extension at the places of cell wall ingrowths. Water transport is controlled at the chalaza at the time of initiation of seed desiccation. Preliminary data indicate the possibility that the chalazal domain may also serves as the centre of emanation of signals for PCD, especially in the endosperm haustorium and seed coat. The wave of death of seed coat and endosperm cells appears to be from the chalaza towards the micropylar pole.

7.3.3. Seed Coat and Accessory Structure

The seed coat is derived mainly from the integument(s). The term **testa** is often applied to the seed coat derived from the only integument in unitegmic ovules or to the seed coat derived from the outer integument in bitegmic ovules. In the latter case, the inner integument forms the **tegmen** that again forms a part of the seed coat. Non–technical people denote the entire seed coat (derived from both integuments) as testa. Externally the seed coat varies in colour and features of the seed surface that may be smooth, wrinkled, striate, ribbed, furrowed, reticulate, tuberculate, alveolate, hairy, spinescent, gummy etc. These have been nicely brought out by scanning electron microscopy.

The role played by the different cell layers of the integument(s) varies from plant to plant. The integuments and their cell layers may be said to be **proliferative** if they undergo periclinal cell divisions to increase the number of cell layers, **degenerative** if they undergo degeneration to various degrees instead of undergoing periclinal cell divisions, **proliferative and degenerative** if they undergo periclinal divisions during

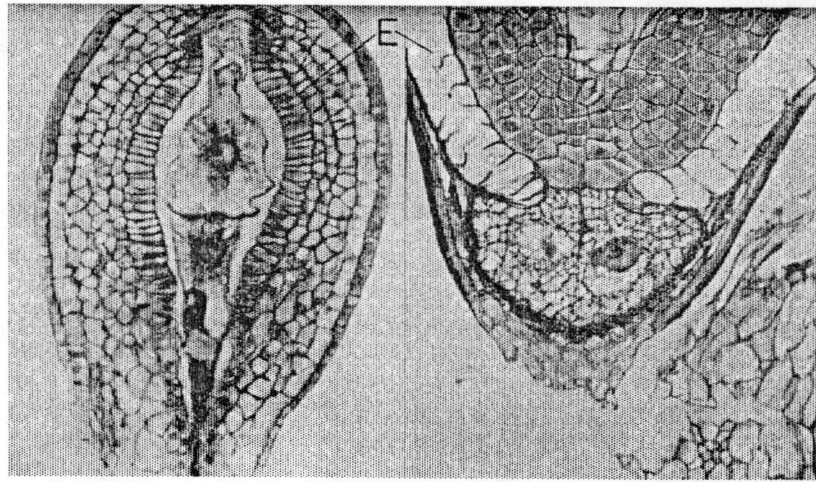

Fig. 215. L. S. of developing seeds of *Downingia* at the two–celled endosperm stage (A) and a very late stage with several endosperm cells and chalazal haustorium (B); Note the prominent endothelium (E) at both the stages (Photograph courtesy of Dr. D.A. Kaplan)

early seed development, but get degenerated to varying extent during later stages of seed development, or, as in rare instances, integuments are neither proliferative or degenerative in terms of increase in cell layers during seed development. In many instances of bitegmic ovules, degeneration is most commonly noticed only in the entire tegmen, in the inner layers of the only integument in unitegmic ovule or in the inner integument and middle layers of outer integument in bitegmic ovules. In ovules having endothelial development, the endothelial layer not only persists as a part of the seed coat, but also generally undergoes peculiar histological modifications (Fig. 215 A and B). Both integuments persist as conspicuous elements of mature seed in some members of Rosaceae, Rutaceae, Euphorbiaceae etc.

Just as there is considerable variability in reference to the persistence or otherwise of the different cell layers of the integuments, the histological structure of the persisting cell layers also varies in diverse species. The simplest type is known in the Orchidaceae where the seed coat is constituted of only the outermost reticulate layer of transparent cells of the outer integument; air–filled space is seen between this layer and the embryo (Fig. 216). Single–layered seed coat is also known in some saprophytic and parasitic species of genera such as *Monotropa* and *Orobanche*. One to three (rarely only one) layers alone are present in the testa of many Apiaceae members. In *Zea mays* the integuments are entirely absorbed.

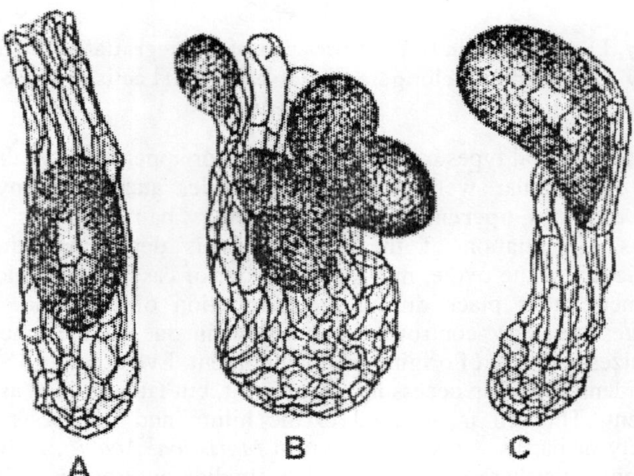

Fig. 216. Seeds of three different races of *Spiranthes cernua*, an orchid, respectively showing a single sexual embryo (A), many agamaspermic embryos of integumental origin (B) and with one agamospermous embryo (C); Note the single layered, transparent seed coat (Drawings supplied by late Prof. B. G. L. Swamy)

In seeds with fleshy or very hard testa, the histology of the component cells is complicated and diverse. The outermost layer of hard seeds is usually made up of thick–walled cells, often covered by a thick cuticle. These thick–walled cells are sclereids. In many hard seeds (such as those in legumes), there is a palisade of **malpighian cells** (sclereids) below the epidermis. In some cases, the seed coat swells in contact with water indicating the presence of mucilaginous substances in the superficial cell layer(s) (e.g., *Linum*). In fleshy seeds, the juicy part is contributed largely by the outer cell layers of the outer (or the only) integument. In pomegranate, the epidermal cells undergo an exaggerated degree of radial elongation (Fig. 217). It also makes up the edible part of the seed.

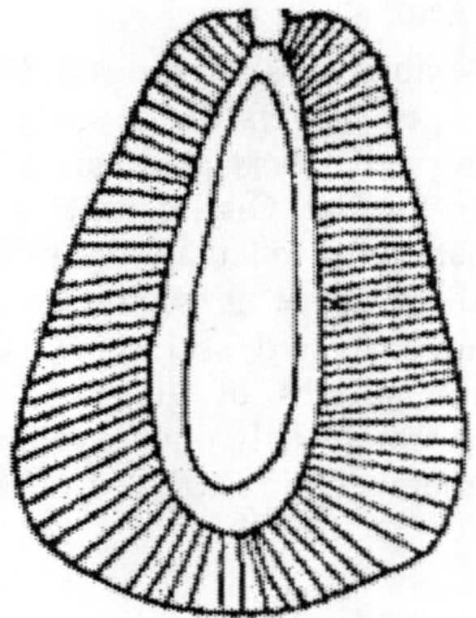

Fig. 217. Diagrammatic L. S. of the seed of pomegranate showing the very much radially elongated, juicy epidermal cells of the outer integument

There are several types of local outgrowths or appendages in the seed in the form of glandular swellings, hooks or prickles, ancillary pulpy or scaly **aril**–like envelops, **operculum, caruncle**, wings, hairy tufts etc. Although the locus of initiation of these structures is determined during the differentiation of the ovule, in a large number of cases their rapid and full development takes place during the maturation of the seed. Aril–like appendages are very controversial outgrowths due to their heterogenic shapes, sizes and way of origin and development. Even by late 18th century aril was identified as an accessory integument, but later dubbed as the third integument. The aril is attached to the hilum and it invests the seed completely or partly. Arils are known in *Myristica, Acacia*, *Polygala* etc. Apart from arils there are several other smaller outgrowths, which were given different names such as **strophilum, epiphysis, spermotylium, arillode, arilloid, false aril, hilar aril, funicular aril** etc. and most of these terms have been rejected. The true aril arises, as in the seed of *Trianthema,* from the base of the ovule covering the seed completely at maturity (Fig. 218).

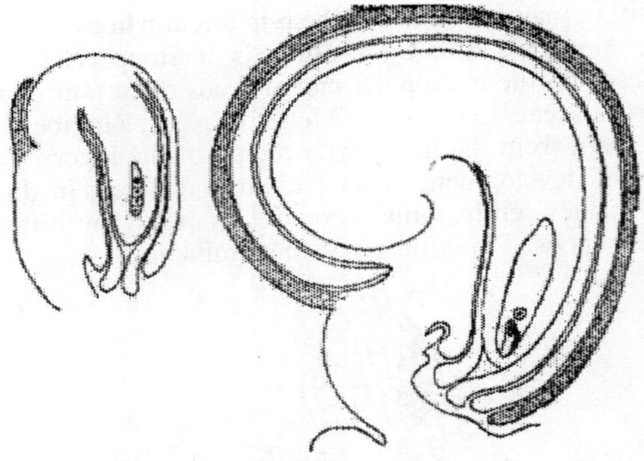

Fig. 218. L. S. of the developing seeds of *Trianthema monogyna* showing the development of aril (stippled) (Bhargava 1935)

In taxa, such as *Myristica,* the structure differentiates as an annular protuberance from the funicle and envelops the seed as an orange–coloured membranous structure known as the mace of commerce. This is called funicular aril. A similar origin is known for the aril of *Acrotrema,* where the aril covers a major part of the seed in the form of a skirt. The aril of *Pithecellobium* is fleshy and edible. The so–called **middle integument** or arillode of the Annonaceae seeds (e.g., *Cananga odorata*) (Fig. 219) arises, unlike an integument, in the transmedian longitudinal plane of the ovule as two groups of meristematic cells on either side of the vascular trace. In the early stages, the meristematic activity is limited to the chalazal zone and as development procedes, the meristems widen out towards the raphe and antiraphe sides so as to establish connection with the inner face of the outer integument on either side of the vascular trace. In the micropylar part, this structure develops mainly as lateral extensions of the outer integument. The extensions meet along the transmedian line and fuse together. Corner (1976) refuses to accept a morphological subdivision of arils and arillodes since too many intermediate forms are known within the arillate families. He prefers to use these terms as general names for the 'pulpy structure which grows from some part of the ovule or funicle after fertilization' that 'invests part or the whole seed'. Other recent authors also doubt the usefulness of a morphological distinction and follow a similar point of view.

It is preferable to retain the terms **caruncle** and **strophiole** for purely descriptive purposes because of their long usage. It is preferable to apply the term 'caruncle' only to the small, disc–like appendages, the attachment and growth of which is limited to the **exostome** rim. The micropyle may be discernible in its centre (Fig. 220). In some, the location of the caruncle is on the funicular side. Its function is believed to be to help in water absorption during seed germination. Since the caruncle is often coloured and/or sugary in many taxa, it may help in **anemochory** (dispersal of fruit or seed by animals). The **operculum** is a stopper–like or plug–like structure present at the micropylar end of seeds of certain families such as Araceae, Droseraceae, Lemnaceae, Melastomaceae, Zingiberaceae etc. It is normally derived from the micropylar region of the inner integument, but the pathway of development of operculum is different in different cases: from nucellar apex, **endostome**, exotome, or partly by hilar region. The operculum is supposed to influence seed germination.

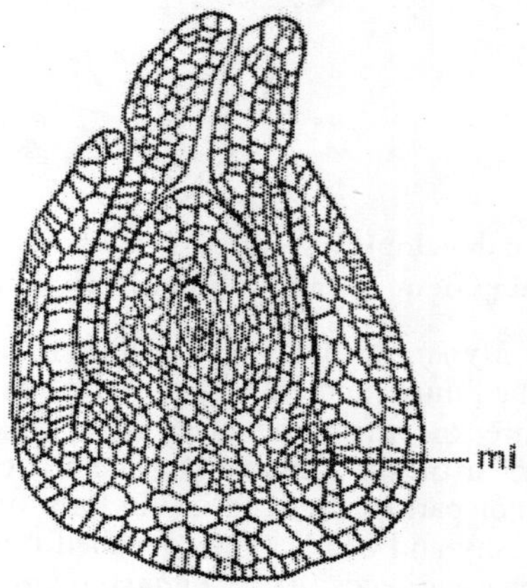

Fig. 219. Transmedian section of the ovule of *Cananga odorata* just after the initiation of the so–called middle integument; Mi - Middle integument (Periasamy and Swamy 1961)

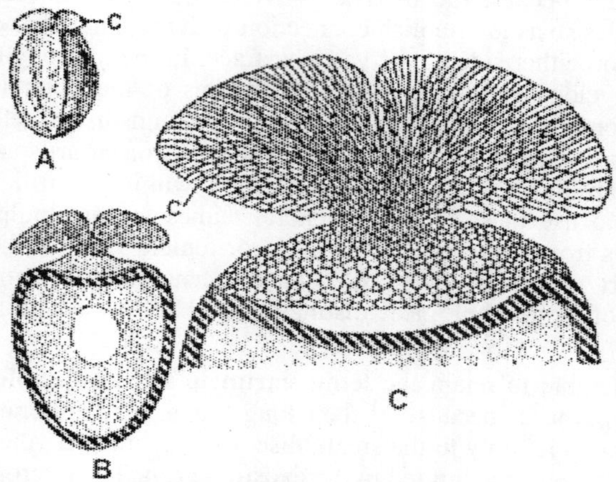

Fig. 220. Caruncle (C) of *Euphorbia dulcis*. The rim of the integument forms this structure; A - Entire seed; B and C – L. S. of the same (Kapil 1961)

Many seeds have wings or **alae**. The wing is a modification of the outer integument. The perichalazal locus of the integument widens out in the form of a membranous structure. In some taxa such as *Cercidiphyllum*, the chalazal and raphe tissues grow out into the triangular wing (Fig. 221). Rarely, the funicle is also involved in the wing formation. A correlation exists between the structure of wings and its function in the seed dispersal. It has an optimum strength and a minimum biomass.

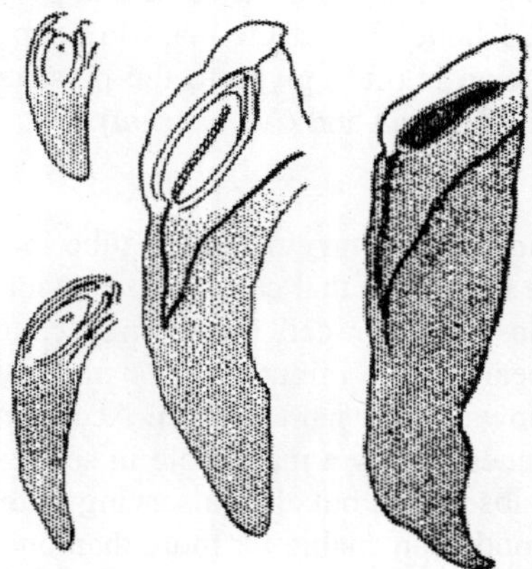

Fig. 221. Development of wing (stippled structure) in the seeds of *Cercidiphyllum* (Swamy and Bailey 1949)

7.3.4. Perisperm

Normally, the nucellus is almost completely lost by the time the embryo sac matures or during early post–fertilization development in unitegmic, tenuinucellate ovules. But in a few unitegmic ovules and many bitegmic crassinucellate or pseudocrassinucellate ovules, the nucellus may even persist in the seed as a fairly conspicuous tissue. This nucellar tissue in the seed is called **perisperm** (Fig. 222).

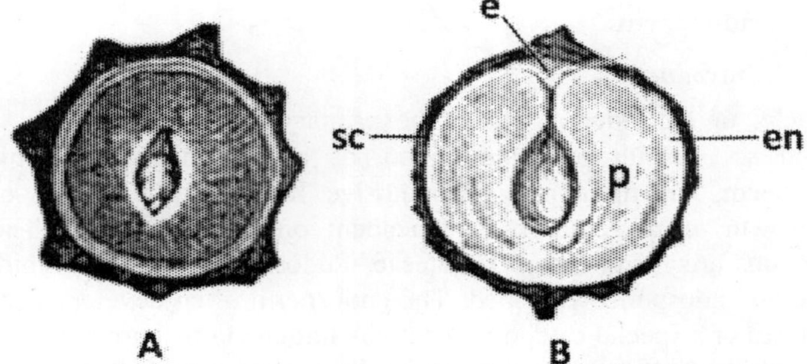

Fig. 222. T. S. (A) and L. S. (B) of seeds of *Piper nigrum* showing perisperm (p); e - Micropylar region; en - Endosperm; sc - Seed coat.

Perisperm is present in members of families such as Amaranthaceae, Cannaceae, Capparaceae, Chenopodiaceae, Phytolaccaceae, Piperaceae, Portulacaceae and Zingiberaceae, either as a single layer or as more than one layer. Perisperm is often studded with food materials of various kinds such as starch, proteins and lipids and also secondary metabolites, and does the function of accessory nutrition. Some cells of the perisperm are multinucleate, either due to cell fusion or due to polyploid nuclei formed after nuclear fusion. About 95% of the mass in pepper seed is perisperm. The term **chalazosperm** is often given to the persistent nucellar tissue in the chalazal zone (as in *Canna* and *Cyanastrum*).

7.3.5. Micropyle

The micropyle aids in the entry of pollen tube into the ovule. In some taxa, the cells of the micropyle that come into contact with the pollen tube are destroyed, while the outer cell layers may continue to persist in a shrivelled state, appearing as a minute scar or pinhead in the mature seed. The scar is often known by the name **hilum**. According to some botanists, the micropyle continues to play a major role in seed germination, not only by helping in water absorption but also in serving as the point of exit of the radicle. But, the second is untenable for more than one reasons. Even in the classical example of the bean, the diameter of the micropyle is many times smaller than that of the radicle. Secondly, the histology of the bordering tissue of the micropyle clearly shows that the cells are conspicuously thick–walled and compact, thereby ruling out any possibility of distension of the pore diameter when the seed comes in contact with moisture or water. Thirdly, careful observations indicate that the radicle pushes itself out of the seed coat at the locus, which is in no way related to the micropyle and this process is associated with the rupture of the seed coat at the point of exit. Thus, the micropyle ceases its function after the pollen tube entry.

7.3.6. Endosperm

7.3.6.1. Introduction

Nuclei or cells derived from the primary endosperm nucleus (PEN) formed as a result of triple fusion, as already noted, constitute the **endosperm**. This tissue is largely nutritive. In the largest majority of taxa, the growth of the embryo is dependent on the endosperm. The only exceptions are members of Trapaceae, Podostemaceae and Orchidaceae, where no endosperm is formed. The endosperm often develops accessory structures of a special category known as **haustoria** that are believed to be more efficient means of absorbing nutrients from surrounding tissues and also serve as seats of synthesis and transport of nutrients to the embryo.

The endosperm may persist in the mature seeds (**endospermous** or **albuminous seeds**) or not; wherever it persists, it is a repository of stored food, and the germination of the seed appears to be dependent on this tissue. In all cereals, the endosperm constitutes the edible part.

7.3.6.2. *Types of Development*

The position of PEN inside the central cell, prior to its division, varies considerably. It either remains at the place of its formation, moves to the centre of the embryo sac or to the antipodal pole depending on the species. Three types of endosperm development are known: *ab initio* **cellular,** *ab initio* **nuclear** and **helobial**. In dicotyledons, depending on the taxon, the first two types are known; reports of the third type are erroneous. In the monocotyledons, all the three types are recorded, although cellular type is found only in Araceae and Lemnaceae. In the first type, the division of PEN is promptly followed by a cell wall, so also in all the subsequent divisions (Fig. 223 A). In the second type, the division of PEN as well as the subsequent divisions of the resultant two daughter nuclei are not followed by cell wall, so that several free endosperm nuclei are formed (Fig 223 B). However, subsequently, the wall formation is initiated between the free nuclei to form cells. In the helobial type, the PEN always moves to the antipodal pole and divides there to form a large micropylar cell or chamber and a small chalazal cell or chamber (Fig. 223 C). In the micropylar cell, the subsequent divisions are free nuclear, while in the chalazal chamber the nucleus may not divide at all, or may divide a few times to result in some free nuclei or the nucleus may undergo endomitosis, polyploidy or polyteny. In the micropylar chamber, cell wall formation may happen at late stages of ontogeny.

In taxa with cellular endosperm, early cell divisions are followed by transverse walls in elongated embryo sacs. Later on, division in vertical and oblique planes occurs, almost at the end of the initial transverse divisions. In some plants, it is possible to discern tiered arrangement of endosperm cells even after a few vertical or oblique divisions are completed (Fig. 224 A–E). This indicates the operation of laws governing cell division. The largest bulk of endosperm is formed by the derivatives of cells that are located in the middle part of the embryo sac so that the shape of the endosperm mass tends to conform to a sphere or spheroid. If the embryo sac is spherical or ovoid, the phase of transverse division is abbreviated and oblique or vertical divisions set in early. There are a few instances where the PEN division is followed by a vertical wall (Fig. 224 F and H) (e.g., *Adoxa*); the subsequent one or two divisions are also vertical. Later transverse or oblique divisions follow.

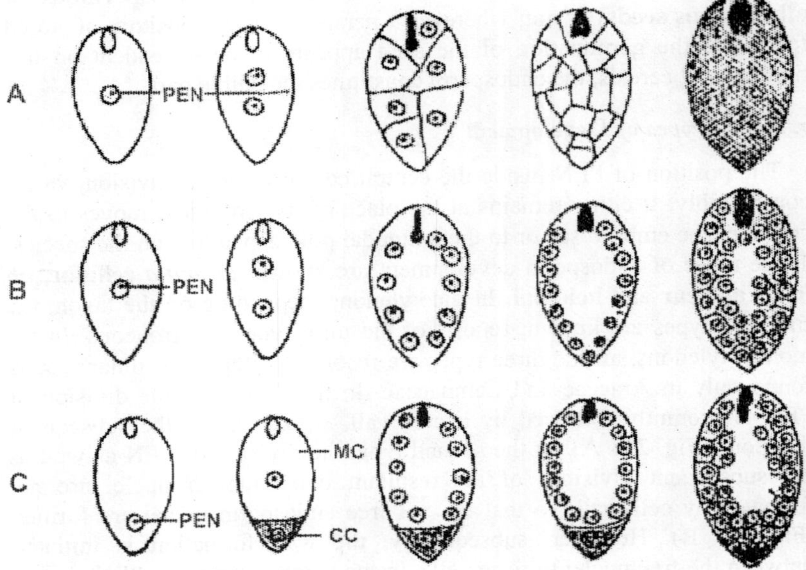

Fig. 223. Major types of endosperm development; A - Cellular; B - Nuclear; C - Helobial; CC - Chalazal chamber; MC - Micropylar chamber; PEN - Primary endosperm nucleus (Swamy and Krishnamurthy 1980a)

In taxa with nuclear endosperm, the free nuclear phase may last to a variable time in ontogeny. In *Coffea*, for instance, walls appear soon after four nuclei are formed. However, in many plants, wall formation is delayed until a large number of free nuclei are formed. There are also instances where the nuclear endosperm never forms cells (e.g., *Limnanthes*, *Oxyspora*). The initial few free-nuclear divisions are synchronous and normally, these continue to happen until the nuclei become embedded in a peripheral sheath of endosperm. It should be kept in mind that the embryo sac rapidly enlarges during post–fertilization development and although this enlargement is overall, a more pronounced growth takes place towards the chalaza in several taxa. Hand in hand with this enlargement, the free nuclei together with the associated cytoplasmic matrix continue to grow similarly. This process is generally accompanied by a faster rate of mitosis at the growing locus. Thus, there is an increasing gradient in reference to the density of distribution of endosperm nuclei from the micropylar to the chalazal end. Cytoplasmic nodules or vesicles, called **endosperm nodules,** are known in the nuclear endosperm of Cucurbitaceae, *Musa*, coconut etc. These are aggregates of nuclei ensheathed by a common mass of cytoplasm; the function of these nodules is not clear. In embryo sacs of relatively small dimensions possessing a spherical shape, the wave of cellularization in nuclear endosperm spreads from the periphery to the centre (Fig 225 A). In elongated embryo sacs, the initiation of this wave is seen at either poles spreading towards the centre (Fig. 225 B). In those embryo sacs, which show more pronounced growth in the chalazal pole, the wave of cell formation starts in the micropylar region (Fig. 225 C). Cell formation may not be completed fully throughout the embryo sac in several taxa, either due to the slow rate of cell wall formation or due to the very fast growth of the embryo sac cavity during post–fertilization development.

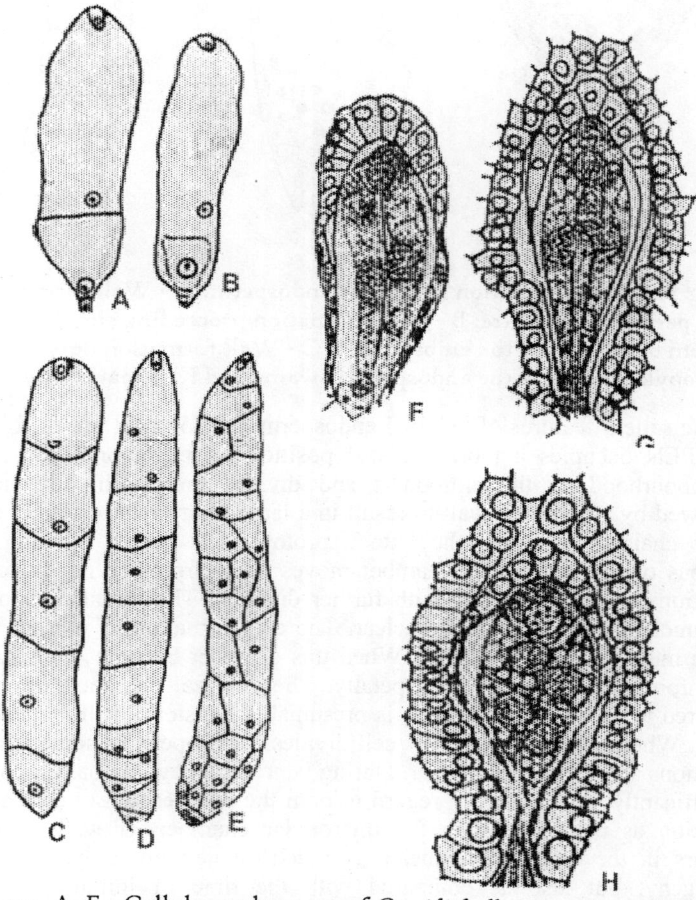

Fig. 224. A–E - Cellular endosperm of *Cercidiphyllum japonicum* showing initial transverse division followed by vertical divisions (Swamy and Bailey 1949); F–H - Cellular endosperm of *Adoxa moschatellina* showing first and second vertical divisions followed by transverse divisions (Legerberg 1909).

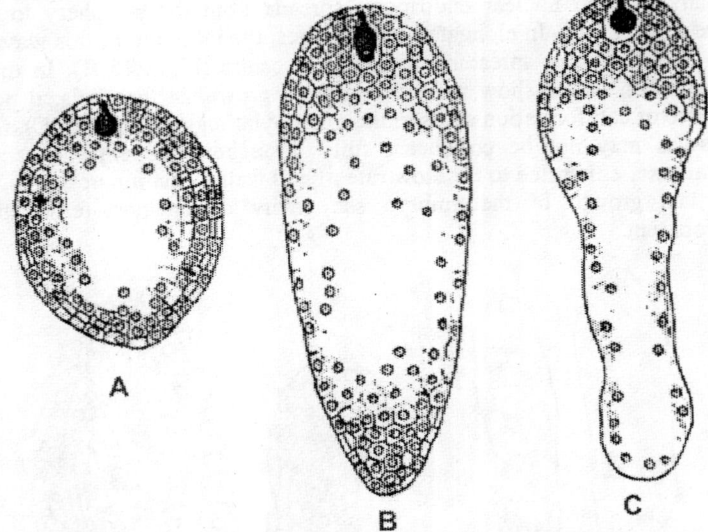

Fig. 225. Wall formation in nuclear endosperm; A - Wall formation from periphery to centre; B - Wall formation proceeding simultaneously from both poles of the embryo sacs; C - Wall formation only in the micropylar region of the endosperm (Swamy and Krishnamurthy 1980a)

The salient features of helobial endosperm development are as follows: The PEN occupies a more chalazal position in the embryo sac in the neighbourhood of the antipodals and divides there. This division is followed by a transvers wall to result in a large micropylar chamber and a small chalazal chamber; the latter's protoplast is deeply staining. The nucleus of the micropylar chamber moves away from the newly formed partition before proceeding with further division. The initial divisions of this nucleus are always free nuclear. The cell formation, if present, takes place much later (Fig. 223 C). When this chamber becomes cellular, cell wall formation proceeds centripetally. The chalazal chamber, often also referred to as **basal apparatus,** is presumably haustorial in function (Fig. 235). When the nucleus of this cell divides, as happens in some taxa, the divisions are again free nuclear. During later ontogeny, the basal apparatus significantly lags behind in regard to both the number and rate of nuclear division as compared with the micropylar chamber. If wall formation occurs in the basal apparatus, it is much belated not only in its own ontogeny, but also as compared with the time of initiation of wall formation in the micropylar chamber.

7.3.6.3. Endosperm Haustoria

In a number of plants special structures are organized in the endosperm either in one or both the poles of the embryo sac. These structures present morphological and histological distinctiveness from the remainder cells of the tissue and are believed to aid in the nutrition of the developing embryo; hence they are called **haustoria**.

In the cellular endosperm, the simplest type of haustorial organization is exhibited by a single cell, which is cut off during the first division of the PEN. In *Opilia amentacea,* the PEN cuts off the tubular chalazal part of the embryo sac from the dilated micropylar part. Once formed, the chalazal tubular cell grows deep into the pedicel of the flower; its nucleus becomes hypertrophied. The micropylar cell builds the entire bulk of endosperm through repeated cell divisions (Fig. 226). In *Cansjera rheedii,* this chalazal tubular haustorial cell gives rise to a number of tenuous branches which ramify deeper through intercellular route. In addition, the lowermost endosperm cells, immediately around the primary haustorial tube, also elongate into tubular, branched, haustorial structures (Fig. 227). A simple, chalazal, two–celled haustorial tube is seen in species of *Magnolia*. A chalazal, intracellular haustorial organization is seen in *Nothapodytes foetida.* Here, the initial growth of the haustorial cell is, however, intercellular. The chalazally located cells are penetrated by the haustorium, and the protoplasts of the haustorial cell and the adjacent cells freely establish connections through pores. This process is repeated in several chalazal cells, and the nucleus of the original haustorial cell becomes greatly hypertrophied with great distortion in shape. Its major lobes are directed towards the direction of haustorial spread (Fig. 228). In *Impatiens,* the micropylar endosperm cell enclosing the zygote undergoes one or two divisions in the same plane, and its terminal cell forms the initial for the origin of a micropyle haustorium. This structure pushes itself out of the micropyle through elongation, reaches the placenta and there undergoes conspicuous dilation, putting forth numerous branches that ramify in the funicular tissue (Fig. 229). In many gamopetalous taxa, haustoria are produced by both micropylar–located and chalazal–located cells and the initials for these are cut off during the first two divisions of the PEN (e.g., Scrophulariaceae). The relative complexity of both these haustoria varies depending on the taxa. For instance, in species of *Veronica,* micropylar haustorium is more aggressive than chalazal (Fig. 230), while in *Bignonia* the reverse is true (Fig. 231).

In nuclear endosperm, various degrees of haustorial complexity are noticed. In the simplest case, as exhibited by *Stemona tuberosa,* there is a greater density of nuclear distribution in the chalazal pole of the endosperm that is enclosed in the dense cytoplasm. When these nuclei divide, adjacently located spindles often fuse to result in polyploidal nuclei (Fig. 232). This region serves a haustorial function. An analogous situation is exhibited by *Hypericum mysorence*, where the chalazal endosperm nuclei aggregate themselves in a dense mass of cytoplasm to form a morphologically distinct **cyst**, which is haustorial in function (Fig. 233). In Leguminosae, Cucurbitaceae, Proteaceae and some Euphorbiaceae (E.g.,

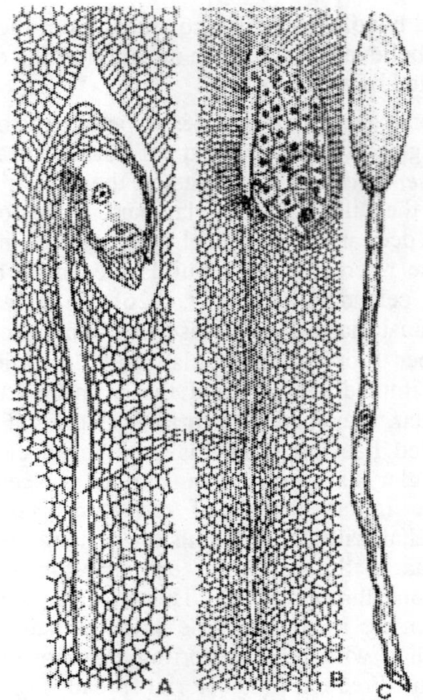

Fig. 226. A and B - Organization of endoperm haustorium in *Opilia amentacea*; chalazal tubular cell organizes itself into a haustorium (EH), while the micropylar cell forms a mass of endosperm cells; C - Whole mount of the endosperm and its haustorium (Swamy and Dayanand Rao 1963)

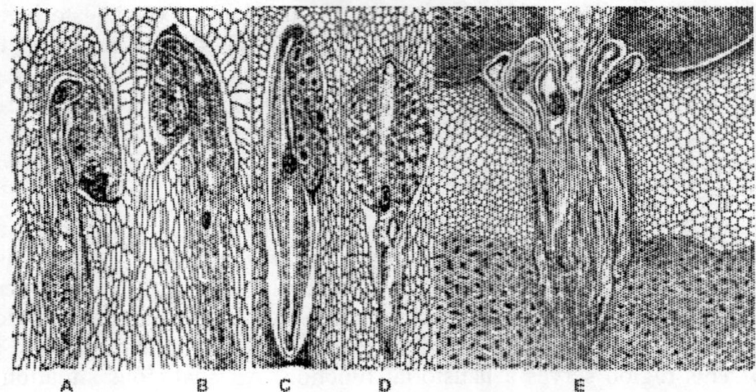

Fig. 227. A–D - Various stages in the development of endosperm and its chalazal haustorium in *Cansjera rheedi*; E - L. S. of a portion of the basal part of the ovule showing secondary endosperm haustoria developing from the basal endosperm cells (Swamy 1960)

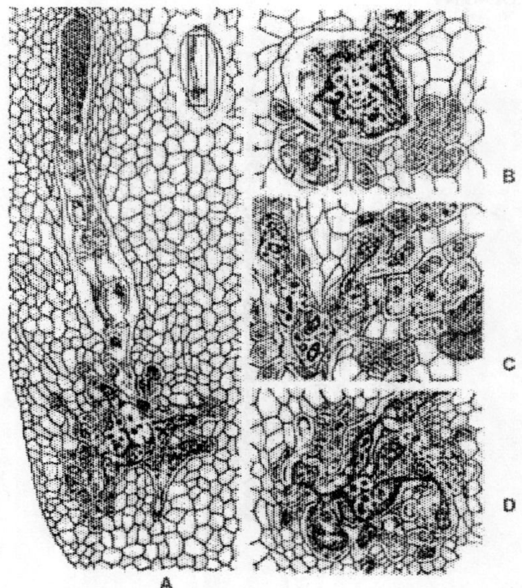

Fig. 228. Endosperm haustorium of *Nothapodytes foetida;* The chalazal haustorium, to start with, is intercellular but becomes highly complicated later on due to fusion products of the haustorium and the protoplasts of the invaded cells. The nuclei also becomes hypertrophied and distorted in shape (Swamy and Ganapathy 1957).

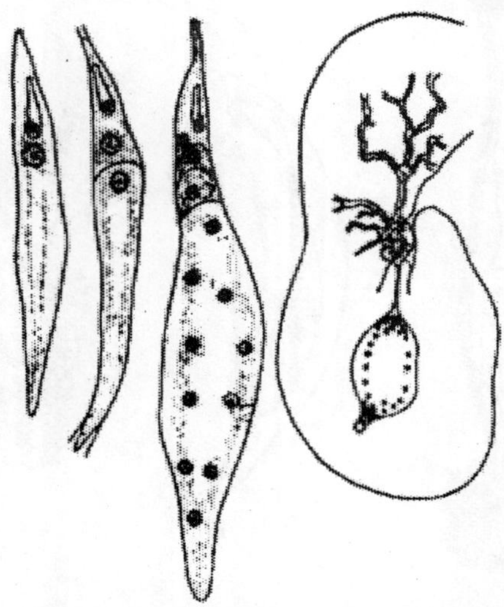

Fig. 229. Development of endosperm and branched micropylar endosperm haustoria in *Impatiens leschenaultia* (Narayana 1965)

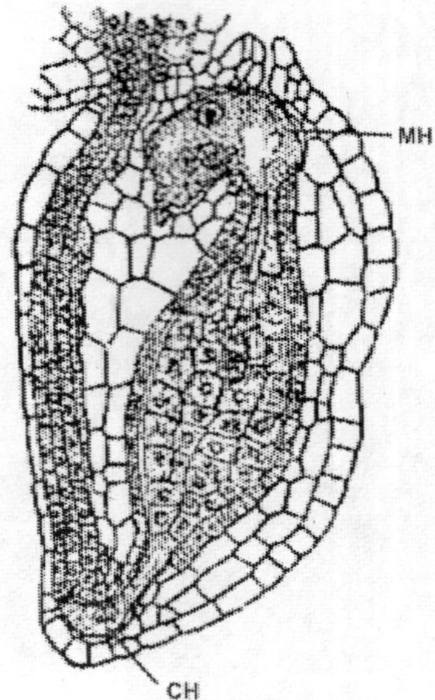

Fig. 230. Micropylar (MH) and chlazal (CH) endosperm hanstoria of *Veronica* (Arekal 1966)

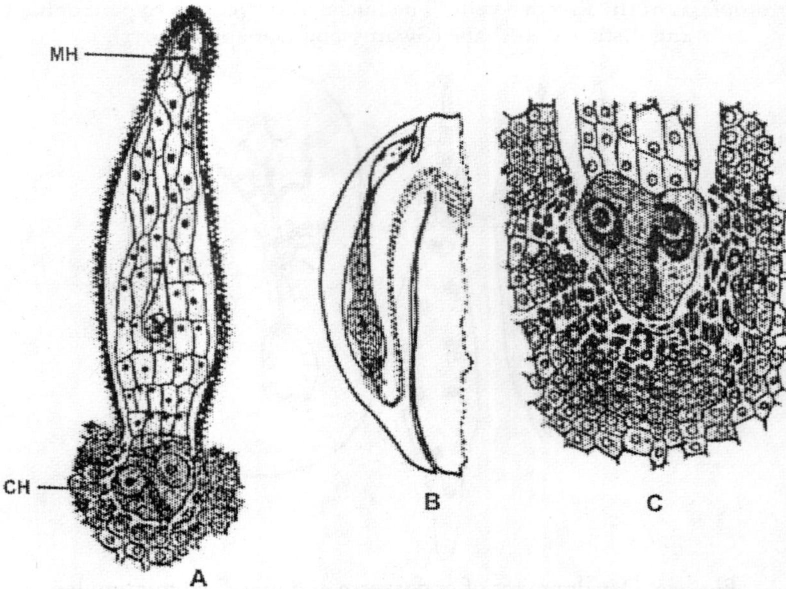

Fig. 231. Micropylar (MH) and Chalazal (CH) endosperm haustoria of *Bignonia megapatomica* (Swamy 1941)

Croton) the chalazal, coenocytic haustorial part of the endosperm is characteristically developed (Fig. 234). It is sharply delimited from the micropylar part of the endosperm, which by this time becomes cellular, and exhibits a variety of shapes such as saccate, narrow or broad tubular, curved tubular, vermiform etc. The length of these haustorial tubes in Cucurbitaceae varies from a few millimetres to 1.6 cm. The most impressive display of coenocytic haustorium is reported in Proteaceae. For example, in *Grevillea robusta,* the coenocytic, haustorial, chalazal, vermiform appendage contains several nuclei (Fig. 234).

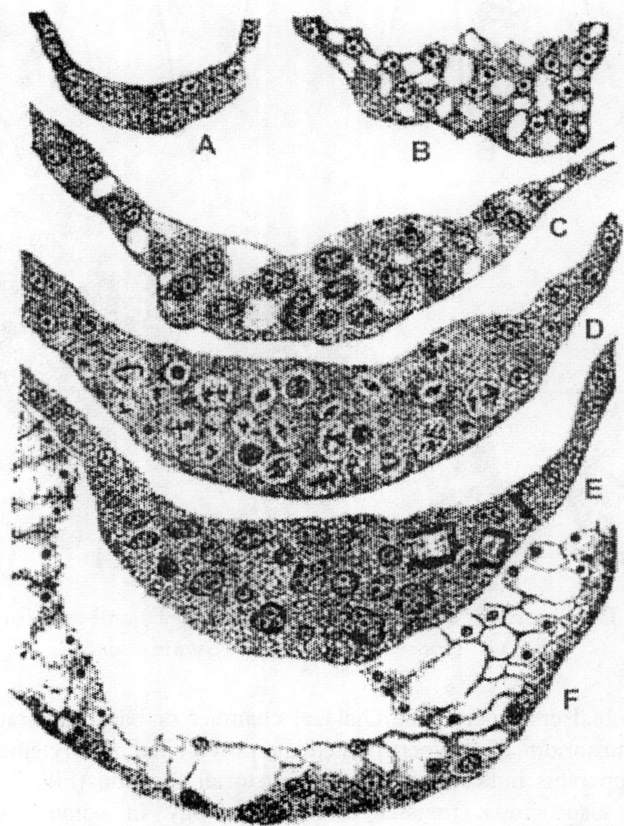

Fig. 232. Chalazal region of the nuclear endosperm of *Stemona tuberosa*; A–C - Some of the nuclei grouped in a pocket of cytoplasm; D – Division of nuclei; E and F – Fusion of spindles lying close to one another resulting in polyploidy nuclei (Swamy 1964)

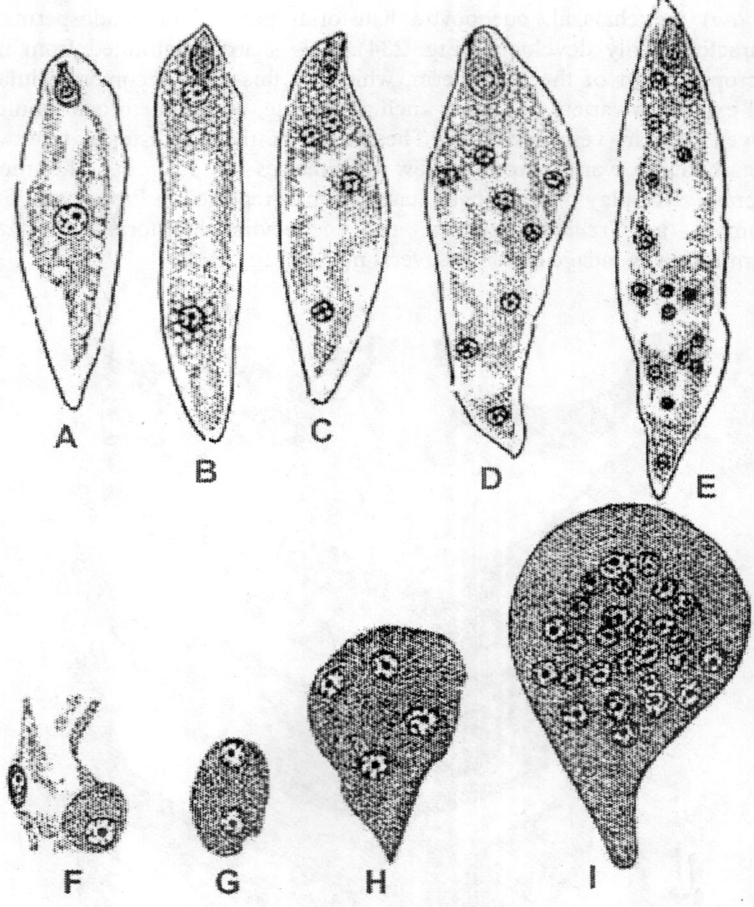

Fig. 233. Development of nuclear endosperm (A–E) and cyst formation (F–I) in *Hypericum mysorense* (Swamy 1946)

In helobial endosperm, the chalazal chamber or basal apparatus itself acts as haustorium. The hypertrophied and polyploidal or polytenic nucleus of this apparatus indicates its strong haustorial function (Fig. 235). This chamber also shows transfer cell morphology in some taxa (e.g., *Haemanthus*).

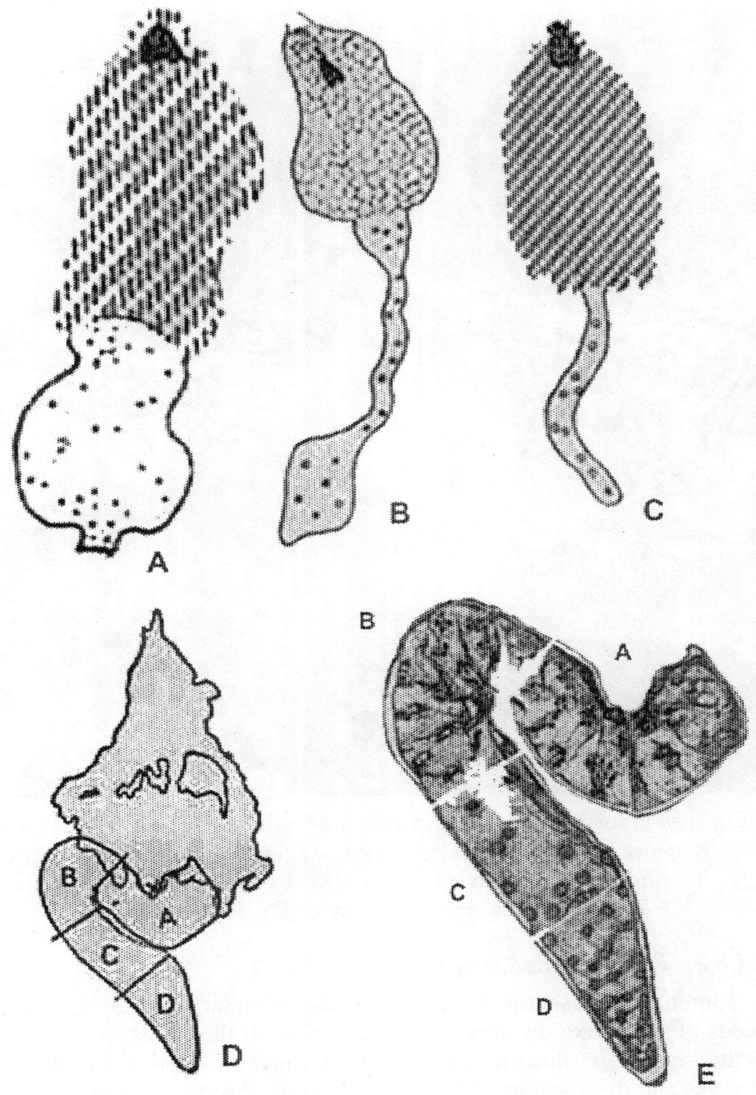

Fig. 234. Haustorial structures associated with nuclear endosperm; A - *Mimosa pudica*; B – *Crotalaria laburnifolia* C–*Coccinia grandis*. In these figures (A–C), embryo is shown as black; D - *Grevillea robusta*. Four regions of the haustorium shown here are enlarged in E. (D and E - Kausik, 1942)

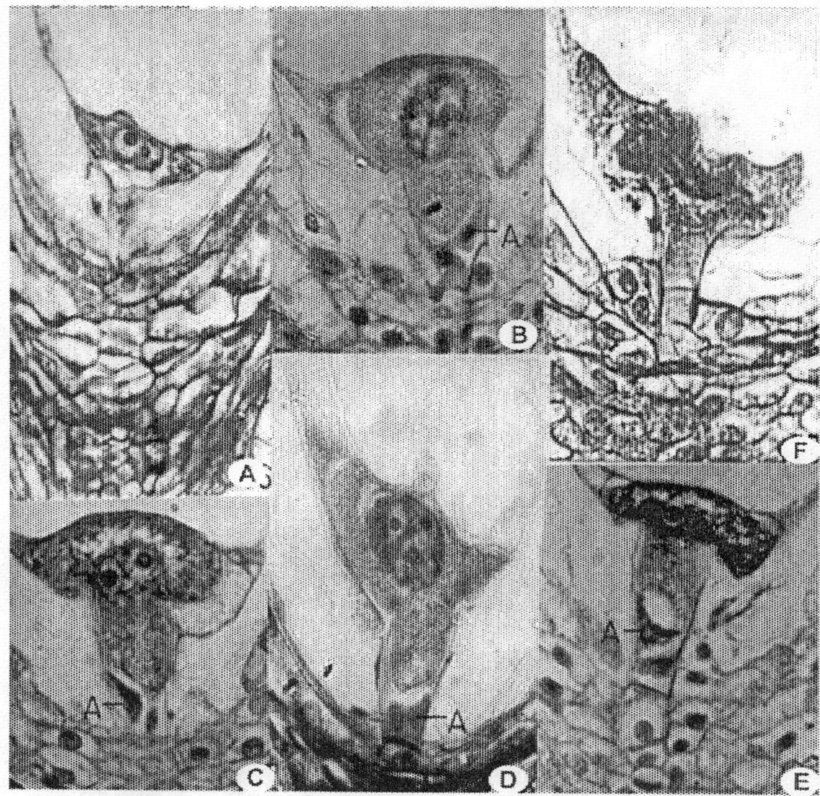

Fig. 235. Haustorial chalazal chamber of helobial endosperm of *Ottelia alismoides* in various developmental stages; Note a highly enlarged nucleus. Antipodals (A) are also shown in some figures (Photographs courtesy of Dr. R. Indra)

7.3.6.4. *Ruminate Endosperm*

Ruminate endosperm is one of the many peculiar features of the mature seeds of some taxa. In such seeds, the surface of the endosperm, instead of being smooth, is thrown into irregular fissures or depressions in varying degrees and this contour being closely paralleled by the irregular surface of the seed coat (Periasamy 1962). Rumination is reported in about 30 of the dicot and four of the monocot families. The cause of rumination rests with either the seed coat or the endosperm and is due to localized meristematic activity during post-fertilization stages resulting in ingrowths or infoldings of the seed coat and ridges and furrows in the endosperm. These formations often involve (i) differential radial elongation of the component cells of any one layer (or the only layer) of the seed coat, or (ii) the participation of the entire seed coat. To the first belong certain taxa of Scrophulariaceae (Fig. 236 A and B), where the persistent endothelial cells undergo unequal radial elongation. In *Passiflora calcarata* the innermost layer of the outer integument shows unequal radial elongation of the cells. In the second category, the ingrowths are formed either by the excessive elongation of the constituent cells as in Annonaceae, Aristolochiaceae etc. (Fig. 236 E and F) or by the localized meristematic activity as is seen in Myristicaceae, Degeneriaceae, Vitaceae, Palmae, Ebenaceae, Menispermaceae etc. (Fig. 236 C and D).

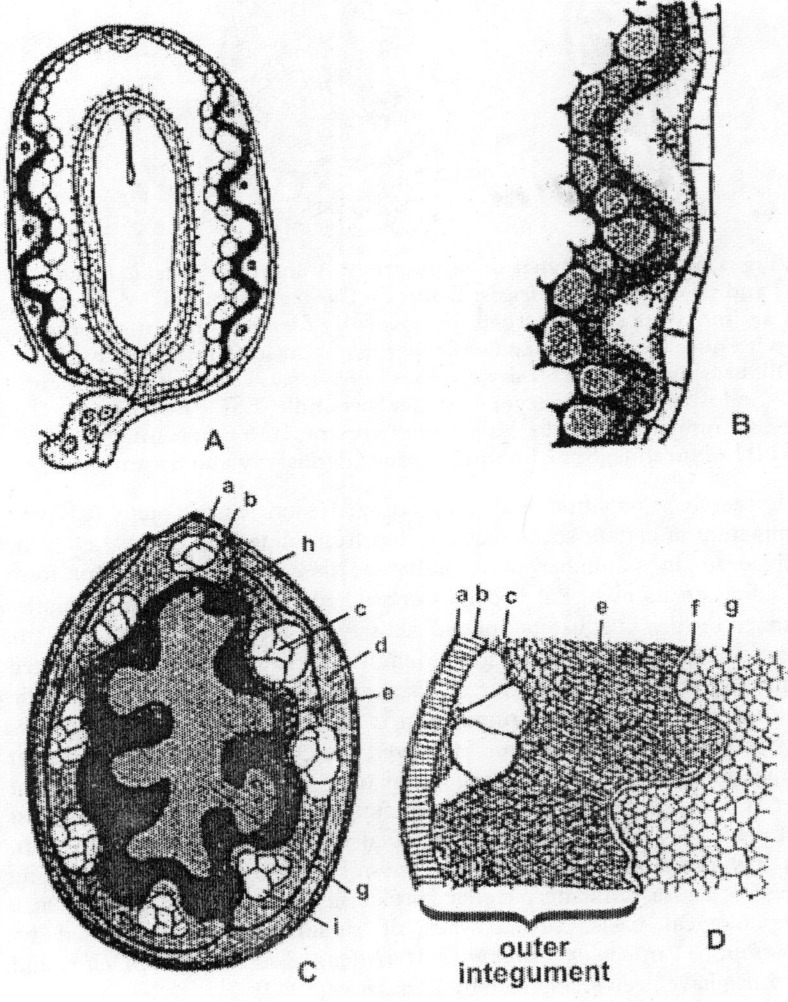

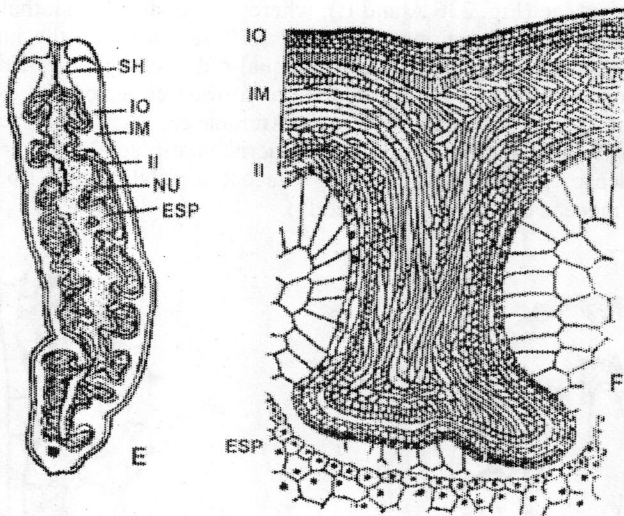

Fig. 236. Different types of rumination: A and B - *Vandellia hirsuta*; C and D - *Degeneria vitiensis;* E and F - *Cananga odorata*; F - A Portion of an immature seed enlarged to show fully formed rumination process with nucellar remains and endosperm; a - Cuticle; b - Epidermis; c - Oil–bearing cells; d - Fleshy coat; e - Stony coat; f - Inner integument; g - Endosperm; h - Embryo; i - Vascular bundle; ESP - Endosperm; II - Inner integument; IM - Middle integument; IO - Outer integument; NU - Nucellus; SH - Subhilar tissue (Periasamy and Swamy 1961)

Excessive elongation and the localized meristematic activity cause rumination in certain seeds that develop from unitegmic ovules, as in the Rubiaceae. In a number of Acanthaceae the seed coat does not form infolding on its own, but becomes consumed and pushed in an irregular manner by the unequal peripheral activity of the enlarging endosperm. Seed–coat activity leading to formation of rumination is usually associated with seeds that exhibit massive chalazal and perichalazal mode of growth. Consequently, the field of rumination is altered depending on the degree and nature of chalazal growth. However, the chalazal region is devoid of rumination processes, while these rumination processes are confined to that part of the seed coat constituted by the integuments. It is also to be noted that the seed–coat activity does not actually produce ruminate endosperm, but only a ruminate nucellus in the case of bitegmic ovules with persisting nucellus, and a ruminate nucellar surface in the unitegmic ovules. On a morphological basis, seven types of rumination as represented by *Passiflora, Verbascum, Annona, Myristica, Coccoloba, Spigelia* and *Elytraria* have been recognized by Periasamy (1962).

7.3.6.5. *Endosperm Cytology*

Although triploidy is the common nuclear status in the endosperm, there are many taxa where the degree of ploidy varies with the initial number of polar nuclei that make up the secondary embryo sac nucleus. In Oenotheraceae, for instance, the endosperm is diploid as there is only one polar nucleus in the embryo sac. In several 16–nucleate embryo sacs, the secondary nucleus varies from 4n–8n, and in all these, the PEN has one ploidy higher due to fusion of a sperm nucleus. In nuclear endosperms, as also in the haustorial structures in all three basic categories of endosperm development, the nuclei undergo conspicuous morphological and cytological changes indicating intense physiological activity. The most common feature concerns the increase in the size of the nuclei, for instance, in maize endosperm some nuclei undergo 1000–fold increase in size in 24 days subsequent to pollination. This enlargement is accompanied or caused by endomitosis. In *Gagea,* the enlargement of the nucleus is accompanied by an increase in chromosome number. In *Pedicularis,* the chalazal haustorial nucleus shows 96–ploidy and the micropylar haustorial nucleus 192–384–ploidy. The highest recorded ploidy is 24576 in *Arum maculatum.* The increase in nuclear size or ploidy level is often accompanied by an increase in the number of nucleoli as well. After a few days of endosperm development, there is neither normal mitosis nor endomitosis, but there is a selective replication of nucleoli. This represents the amplification of ribosomal genes involved in the increased protein–synthetic efficiency of endosperm cells. The polyteny due to metaphase mitotic inhibition, observed in endoserm nuclei, is different from the one that occurs in the salivary glands where polyteny takes place in the interphase, corresponds to what is called **dispersion endomitosis** described by cytologists such as Nagl. However, in *Borassus* endosperm nuclei interphase endoreduplication occurs.

Special mention must be made about the levels of DNA methylation of endosperm nuclei; they have reduced methylation when compared to the nuclei of embryos or vegetative tissues (Hsieh et al. 2009; Gehrig et al. 2006, 2009). The functional significance of this is not known but it is likely that it generates small interfering RNAs (siRNAs).

7.3.6.6. *Endosperm–Maternal Tissue Balance*

The endosperm – maternal tissue relationship is a deciding factor in seed set. The effects of the maternal tissue on endosperm (and embryo) are called **sporophytic maternal effect**, a study of which is essentially required in the light of statements such as the following: "..... development from ovule to mature seed appears to be largely independent of informational cues from the maternal plant .... and is therefore primarily under genetic control" (Sussex and Dale 1979). Our knowledge and understanding of the individual and synergistic roles of these three components (endosperm, embryo and maternal tissues) in influencing seed development and maturation are still very poor. In addition, the effects of the gametophytic elements (embryo sac components) on embryo, endosperm and maternal diploid tissues of the ovule are also not very clear. This effect is called **gametophytic maternal effect**. In other words, the effects of the 'legacy of the mother plant' need to be properly understood.

In addition to the maintenance of chromosome number in the maternal tissues, which is less than that of the endosperm cells, a nutritional balance should also exist between them, In an interesting theory called **somatoplastic sterility,** Brink and Cooper (1939) suggested that in a normal balance between endosperm and maternal tissue, the endosperm successfully transfers nutrients from the ovule to the immature embryo. If the endosperm lacks its normal vigour because of inbreeding, incongruous interspecific matings, improper chromosomal balance or chemical treatments, then the surrounding maternal tissues (integuments, nucellus) become hyperplastic at the expense of the endosperm. In other words, in the normal seed, the size of the maternal tissue, immediately lining the endosperm, is merely accommodated to that of the rapidly expanding endosperm to which it is subordinated, and in abnormal instances, this nicely adjusted balance is upset very early in the ontogeny of the seed.

7.3.6.7. Functions of Endosperm

Endosperm may be said to have two major functions: (i) Regulatory (ii) Nutritive. The first implies that endosperm regulates and controls the orderly development of the embryo, right from the zygote to maturity, as well as embryo size (Hong et al 1996). The second implies that the endosperm is the source of nutrients for the growing embryo up to seed maturity (Lopes and Larkins 1993) and in many cases (in endospermous or albuminous seeds) for the germinating embryo until the seedling becomes autonomous and takes care of its nutritional requirement. The regulatory role is not easily realized when compared to the nutritive role of endosperm.

The evidences for the regulatory role of endosperm come from various sources. It is known that the vacuole of the central cell or endosperm chamber constitutes an important reservoir of sugars, amino acids and inorganic salts and their concentrations in the chalazal and micropylar poles are totally different, thus establishing a physicochemical gradient with different pH at different regions. This gradient appears to be different in the pre–fertilization stage, immediately after fertilization, and considerably after fertilization. Naturally, this difference in gradient affects the osmoticum of the egg cell or zygote and embryo. In other words, the chalazal and micropylar regions of the central cell or endosperm have different 'topophysic effects' (positional effects) (Evenari 1984), emphasizing the existence of different morphogenetic fields in the embryo sac in pre–fertilization as well as post–fertilization stages. There is growing evidence that the endosperm is a rich source of growth substances, which play an important regulatory role in the development of young embryos. IAA has been isolated from immature corn kernels, while there is an interaction between tryptophan, auxin and nicotinic acid in the endosperm of maize. Indole pyruvic acid has also been reported from maize endosperm. The endosperm of apple accumulates auxin–like substances that are not IAA. Coconut milk has an **'embryo factor'**, as well as 1, 3 – diphenyl urea, both of which play a regulatory role.

The changes in symmetry from filamentous to globular, and from globular to heart–shaped and torpedo phases are likely to be regulated by the endosperm, as has been shown by Krishnamurthy (1988). The transition between these phases is marked by specific changes in the endosperm; particularly it can be demonstrated in nuclear endosperms and helobial ensosperms: the free nuclear phase, the cell-wall formation phase and the stage at which storage products are deposited. During the free–nuclear phase, the development of the globular embryo phase is completed. At the initiation time of wall formation, the hypophysis and epiphysis are differentiated, and also the cotyledonary development is inititated. When the storage products are deposited in the endosperm, a rapid development of cotyledon(s) occurs. In cases of seed failure, if wall formation fails or delayed between endosperm nuclei, the globular embryo correspondingly stops to transform itself into the heart–shaped stage or proceeds to heart–shaped stage after delay, as in colchicine treated *Argemone mexicana.* In citrus–crosses also, nuclear endosperm does not become cellular, with corresponding absence of transition of globular embryo into cordate embryo.

The nutritive role of the endosperm to the growing embryo is very well documented. The endosperm is the major source where the zygote and the developing embryo look for the nutrient substances. The transfer of nutrients from endosperm to the embryo has been strengthened by the isolation of two genes from maize endosperm that are specifically expressed during its early stages of development: *BET* (for 'basal endosperm transfer layer') whose expression is restricted to the basal endosperm transfer cells with extensive wall ingrowths (Hueros et al 1995), and *ESR* (for 'embryo surrounding region') whose expression is seen in the endosperm in the immediate vicinity of the growing embryo. The *ESR* gene probably plays a direct role in the nutrition of the embryo by exporting gene products into the embryo (Opsahi–Ferstad et al 1997). The nutritive role of endosperm is also evident from a study of interspecific and intergeneric crosses, where hybrid embryos abort due to faulty endosperm development, and also from embryo culture experiments. Mid–way during embryogenesis, the endosperm gradually becomes committed to the synthesis and accumulation of storage products. Not much is yet known on how the metabolism of the endosperm, programmed for a modest synthesis of house–keeping proteins, gets deflected to a large–scale synthesis of storage substances. Starch, lipids and a complex of proteins are the main storage reserves of the endosperm. The main storage proteins are generally classified under **albumins, globulins, prolamins** and **glutelins**. The principal storage proteins of the endosperm of maize, barley, sorghum and a few other cereals are of the prolamin type, whereas those of wheat are glutelins and prolamins. In rice, glutelins are the major proteins, although prolamins and globulins also form significant percentage. Generally, the cereal endosperm proteins are named after the generic name of the cereal from which they are known such as *Zein* (prolamins from *Zea mays*), *Hordein* (prolamins from *Hordeum vulgare*) and *Oryzenin* (glutelins from *Oryza sativa*). The appearance of protein bodies in the vacuoles begins after the nuclear endosperm becomes cellular. ER is likely the location where storage protein molecules aggregate to form protein bodies and the latter are then transported to the vacuole by Golgi bodies. Subsequently, there is an increase in size and number of protein bodies. These protein bodies are changed into a translocation–competent form before being transported to storage sites. Where the endosperm contains more than one type of protein body, what mechanism enables them to be sequestered into distinct protein bodies in the endosperm cells is not very clear. Different domains of the rough ER, called protein body ER and the cisternal type ER, are the sites of synthesis of prolamins and glutelins respectively in rice (and perhaps in other cereals as well). The ER membranes are unique with respect to the type of storage proteins synthesized. However, the different proteins and their fractions accumulate at different times during caryopsis development.

7.3.6.8. *Genetic Control of Endosperm Development*

Evidences are available to the fact that endosperm development is under genetic control. One line of evidence for this is the analysis of mutants that impair endosperm development. These mutants include both embryo sac-specific mutants and endoperm-specific mutants. The former operate through meternal gametophytic effects to affect very early endosperm development. Examples for this category are the *fertilization independent endosperm (fie)* mutant and *fertilization independent seed (fis)* mutants like *fis1, fis2* and *fis3* of *Arabidopsis*. These can form 'endosperm' in the absence of fertilization (see full literature in Krishnamurthy 2014b). The *Arabidopsis* mutant *medea* (*mea*) also displays gametophytic maternal control on endosperm (and embryo) through its protein product, a part of **polycomb** group complex that suppresses cell proliferation. Endosperm–specifc mutants are known in cereals, particularly in maize, and these include mutants such as *brittle (bt), floury (fl), opaque (op), shrunker (sh), sugary (su),* and *waxy (wx)*. These affect the nature of the storage reserves in the endosperm. The *defective kernel (dek)* mutants show lesions in the endospem development (and embryogenesis) and the phenotypic effect of this mutation is grains with the collapsed or shrunken endosperm with soft and fluid consistency. Detailed studies made on *miniature seed (mn)*, a *dek* mutant, have shown that there is an impairment of movement of photosynthates to the endosperm and this causes a physical destruction of cells at the placento–chalazal pole, seen as a gap between the pedical and the endosperm. This has been attributed to the presence of low levels of an *MN* gene–encoded, endosperm–specifc cell wall invertase in the mutant endosperm and the consequent osmotic imbalance due to failure of sucrose mobilization, particularly in the chalazal region. The *shrunken endosperm expressing xenia (sex)* mutations of barley, though analogous to *dek* mutations, show profoundly different effects, which range from the total lack of aleurone cells to the complete loss of starchy endosperm traits, even cell types that are unknown in wild–type, are developed.

7·4· EMBRYO

7·4·1· Introduction

The embryo is the miniature sporophyte that is produced as the end–product of zygotic ontogeny as the fertilized ovule becomes the seed. The embryo is the most important and integral component of a seed. A seed without an embryo serves no purpose. With inputs from diverse disciplines, the subject of angiosperm embryo has now attained a fascinating facelift. Technical improvements in the study of embryogenesis have also contributed substaintially. Particular attention has been paid to the isolation of viable male and female gemates and development of an in vitro fertilization system and to study the zygote and its immediate division products (Kranz et al 1999; Kranz and Lorz 1993; Hoshino et al 2004; Okomoto and Kranz 2005). The seemingly simple process of embryogenesis has now turned out to be very complex and full of exciting problems. The zygote is the starting point for development of the embryo. Embryogenesis involves an array of developmental episodes that occur in an ordered sequence and result finally in the mature embryo in which the structural and functional organization of the adult plant becomes imprinted. Development entails the interplay of a series of integrated and ontogenetically–coordinated factors. Cell multiplication is progressively associated with the origin of new localized centres of growth, which, in turn, cause development of specific parts and organs of the embryo. Although a number of morphologically distinguishable contours such as the filamentous, globular, heart–shaped, torpedo and mature embryos form at different stages of embryogenesis, the entire gamut of events is to be looked upon as a continuous process in which any given stage is intimately related to the previous stage and also to the stage that follows. This relationship is not to be conceived strictly in terms of cell lineage and tier systems, but in terms of cells that have specific positions in the multiplying mass of cells. During successive developmental stages, different chemical and structural components are fabricated under the direction of a coordinated set of genes and the cells are organized into patterns distinctive of the different histogenic regions of the embryo. The interaction of structural, physiological, biochemical and genetical factors or controls during embryogenesis is highly complicated and involves mechanisms not yet fully delineated. The zygote and its products possess an internal environment of their own, which again interacts with external environment present in the embryo sac or endosperm. The modus operandi of these factors is not yet clear. Whatever is known is outlined below.

7.4.2. Zygote

As indicated in the previous chapter, the zygote is the fusion product of one of the male gametes, carried by the pollen tube, and the egg cell (Fig. 237). The egg cell which is comparatively poor in cytoplasmic organization and metabolically fairly quiescent prior to syngamy undergoes sudden transformations, once fertilization gets effected. Many properties of the egg change during the first few hours.

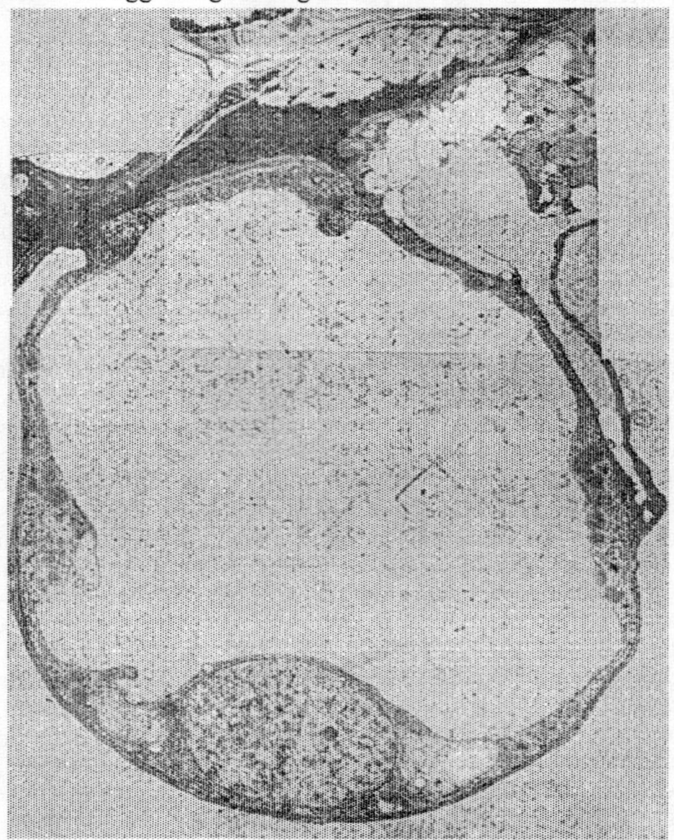

Fig. 237. Electron micrograph of the zygote of *Quercus*; The nucleus is located at the chalazal end, while a large vacuole occupies the micropylar part (Photograph courtesy of Dr. Mogensen)

The mechanism that triggers the egg cell to develop into the zygote cannot be fertilization alone or pollination alone, but something more than that. Dramatic changes occur in the number and position of organelles of the egg when it becomes a zygote. These changes include regrouping and increase in the quantum of the ER, increase in starch grains in plastids, elaboration of additional cristae in mitochondria, generation of new populations of ribosomes and polysomes, and increase in dictyosome activity. All organelles clump around the nucleus, which itself shifts toward the chalazal pole. However, the most notable change is the formation of a wall around the egg (within 36–50 h, depending on the species), which was devoid of a wall at its chalazal end prior to the fertilization. The size of egg changes immediately after (8–10 h) fertilization in some species. In cotton, for example, the zygote decreases nearly to one half of its original size. This shrinkage is apparently characteristic of most zygotes that are to proceed with normal development, as hybrid zygotes that are destined to abort do not shrink. Zygote shrinkage is believed to be related to the osmotic gradient caused by the initial rapid growth of the central cell and the young endosperm contained in it that would cause water to move from the vacuole of the zygote to the central cell or endosperm. This growth results in the creation of a sudden disequilibrium between the zygote and the endosperm around it. However, the shrinkage of zygote is only a momentary phenomenon and it is soon overcome by the zygote enlargement prior to its division (Krishnamurthy 1994).

It is debated whether the zygote has sufficient machinery of its own to synthesize its basic needs such as proteins, or it obtains it from its surroundings. Many believe that the first proteins synthesized by the zygote are coded by the stores of mRNA bequeathed from the egg cell i.e., no RNA is synthesized by the zygotic machinery, but the mRNA involved in protein synthesis is a leftover from the egg, a phenomenon very common in animal zygotes. However, some believe that the zygote is capable of active synthesis. Such synthetic activity of zygotes has been demonstrated in cotton, tomato and tobacco as evidenced by the formation of a new population of ribosomes. It is likely that both situations may prevail depending on the species.

In the vast majority of angiosperms, the zygote embarks on division immediately after syngamy, but in some instances it experiences a period of rest before initiating division (up to seven days in tobacco and up to several months in *Pistacia).*The dormancy may be due to a delayed fusion of male and female nuclei, or due to a delay in the formation of a complete wall around the zygote. The post–fertilization enclosure of the zygote with an intact wall all around, providing the necessary insulation and isolation, is a necessity because of its different genetic makeup. It also forms a pre–requisite for its embarking on division. For this purpose, a callose wall is reported to be laid down around the zygote during the initial period after fertilization, as callose is a highly insulating substance.

One of the characteristic features of the zygote (and the embryo derived from it) is its polarity. Although the basic polarity of the zygote is inherited from its precursor egg cell, syngamy has further accentuated this basic polarity. The polarity of the egg and the zygote, in turn, is largely due to the polar electric gradient that already exists in the embryo sac. Only sexually derived zygotes (and embryos derived from them) are polarized structures, while adventive embryos and embryoids derived from tissue cultures are not polarized. The evidence for the polarized nature of the sexually–formed zygote is shown by its cytological organization and also by its asymmetric division to produce two unequal cells, which have entirely different fates. In the two–celled proembryo, the smaller terminal cell is endowed with most of the cytoplasm that the zygote had, while the larger basal cell inherits the large vacuole of the zygote. Mutant embryos of *Arabidopsis* that lack the cortical and preprophase band of microtubules do differentiate normally with spatially positioned embryonic organs. This indicates that an ordered arrangement of microtubules is not the main determinant of the form of developing embryos (Traas et al 1995).

7·4·3. Embryogenesis

Wardlaw (1955) considered the zygote and the embryo derived from it as a complex, specific, **diffusion reaction system** that functions in conformity with the laws of physical chemistry and mathematics (Refer to Chapter 4). It is also a gene–determined reaction system operating under the sustaining environmental conditions prevailing at the micropylar milieu of the embryo sac or endosperm. The division of the zygote is asymmetric and results in two unequal daughter cells, a smaller terminal cell and a larger basal cell. The embryo is formed by the subsequent divisions of these two cells and these divisions are temporally regulated by gene expressions and DNA replication. There is considerable variation in the extent of the respective contributions of the two cells to the embryo. However, on the basis of this, five major types and many subtypes of embryogeny are recognized (Johansen 1950, Creté 1963). A number of embryologists believe that the outstanding feature of embryogenesis is the orderly and almost predictable sequence of cell divisions during embryogeny in any given species. Such a belief enabled Souèges (1937) to establish his **'Laws of embryonomy'**, which were followed faithfully by many subsequent investigators. Embryonomy takes into account the diverse relationships expressed by certain laws and enables a definition of the embryonic types, which form the fundamental units of embryogenic classification. The four main laws of embryonomy are as follows:

1. **Law of origins:** As embryogeny enables us to establish the lineage of the cells in an embryo, it is possible to define quite precisely the origin of all cells in a species with respect to any particular stage in the development.
2. **Law of numbers:** The number of cells produced in the different generations varies with the species and depends on differences in the rate of division in the cells of the same generation.
3. **Law of dispositions:** In normal embryonic development, the derivative cells are formed by divisions in definite and predetermined planes and their positions correspond to the role, which they are destined to play.
4. **Law of destinations:** In any species, when the development is normal, the cells always give rise to well–determined regions in the body of the embryo.

This classic approach of Souèges (1937), often called the **'Cell lineage concept'** or **'Mosaic theory'** (Street 1976) has been stated by Johansen (1950) as follows: "In order to elucidate the laws of embryonomy with utmost precision and clarity and in as complete a manner as possible, the employment of formulae has become necessary. These formulae indicate precisely the relationships, which the individual cells present to one another during the course of their segmentations, with respect to their origin, to their number, to their disposition and to their eventual destination or their histogenic function. Each and every cell has a reason for its existence, its origin can be demonstrated, its destination determined, and its position is invariably the same. A superfluous cell would seriously upset the harmonious balance". Therefore, according to the cell lineage concept, (i) great significance is to be attached to the order and planes of divisions in the young embryo; (ii) during these divisions, the individual cells of the embryo inherit different cytoplasmic potentialities from different derivatives of the zygotic division; and (iii) these differences determine the exact role they and their daughter cells play in organizing the embryo and its parts. The cell lineage concept gained great support for many years and contributed much to the recognition of the major types of embryogenesis and the variations under each type. As a consequence, many embryologists tended to assess the type of embryo development from the first few divisions in embryogenesis and to allocate the ontogeny into one of the recognized types without looking beyond the globular stage.

The concept of cell lineage and its fixity has been questioned in the last three decades. It is questioned whether a reliable system of classifying angiosperm embryo ontogeny can be formulated on the basis of successive steps in the early division patterns and a dogmatic emphasis on the destiny of cells formed. Even within the same taxon, distinct variations were observed in the early ontogeny of the embryo and cell lineages (Periasamy 1977). **'The regulative theory of embryo organization'** (in contrast to 'mosaic theory'), proposed by those who worked on somatic embryogenesis, postulates that segmentation patterns observed during early embryogeny are controlled solely by physical factors such as surface tensions (and has no taxonomic significance). It further emphasizes that during early development of the embryo, its constituent cells do not inherit distinct and specific cytoplasmic potentialities, but remain undetermined and uncommitted. An embryogenetic 'field' may exist in combination with a position effect i.e., a given cell acts in the embryo in relation to the surrounding cells. In other words, it is not the cell or cell group itself that determines the future histogenic region of the embryo it gives rise to, but only the position that the cell or cell group occupies in the developing young embryo [the so–called 'positional information' or 'Wolpert model' (Wolpert 1970, 1971, 1981)]. Therefore, according to this school, allocation of parts of the mature embryo to initials at the 8–celled to 16–celled stage of the young embryo appears to be based on topographical correspondence of the initials with the parts of the mature embryo, rather than on the actual proof of derivation of parts from the initials. Thus, the end product (mature embryo) is more important than the means of achieving the end product. In the embryo of *Arabidopsis,* Scheres et al (1994) used a transgenic marker consisting of the maize transposable element *ac* incorporating a CaMV35S promoter–GUS gene fusion. Through this method, they traced the hypocotyl, the root and root meristem back to compartments in the four tiers of cells in the heart–shaped embryo. There were no restricted cell lineages for the root and hypocotyl.

The dicot embryo experiences four distinct morphological phases during its ontogeny: **filamentous**, **globular**, **cordate** and **torpedo** phases (Figs. 238 and 239). The filamentous phase results from the variable number of transverse divisions in the zygote. The globular phase is initiated by two successive vertical divisions at right angles to each other in one, two or (rarely) more of the more basal cells of the filamentous embryo (Periasamy 1977). The globular embryo comprises of a variable number of cells depending on the plant, but in all cases it comes to initiate the **epiphysis** at the prospective shoot pole and the **hypophysis** at the prospective root pole (Refer to Chapter 5). The differentiation of these two poles is a very important morphogenetic step because the further orderly development of the embryo depends on whether these two poles are formed or not (Swamy and Krishnamurthy 1975b, 1977). In the absence of the differentiation of epiphysis and hypophysis, no transition from globular to cordate phase takes place. If epiphysis does not differentiate, the cotyledons do not develop (Krishnamurthy 1988). For instance, the globular embryo does not undergo transition to the cordate stage and does not have a normal shoot apes i.e., the epiphysis, in the reduced tomato mutant. Despite the ontogenetic similarity between the dicot and monocot embryos up to the octant stage, the form of mature embryo is distinctly dissimilar in the two groups of plants, as the monocot embryos have only one cotyledon, while the dicots have two. Hence, there are no cordate and torpedo phases in monocots. All the morphological phases of embryogeny also involve conspicuous changes in symmetry of the embryo. From filamentous to globular phases, the embryo is typically radially symmetrical, and from globular to heart–shaped stage, it becomes bilaterally symmetrical (because of two cotyledons) in the case of dicots. In monocots, the embryo has only one cotyledon and hence, it becomes unilaterally symmetrical from the radial symmetry of the globular phase. As explained already, this symmetry change in the embryo is controlled by the surrounding endosperm.

As embryogenesis proceeds from zygote onwards, there is a synthesis of DNA with progressive cell divisions. Virtually, all cells of early stage embryos undergo a regular cycle of DNA synthesis followed by division of the nucleus. Hence, all newly–formed cells of the young embryos have the 2C level of DNA. In the embryos of many legumes and certain other plants, the cells of developing cotyledons continue to synthesize DNA by endoreduplication after divisions resulting in a progressive increase in the DNA content of the cells of fully–grown cotyledons to a level as high as 64C in species of *Pisum* and 256C in *Phaseolus vulgaris*. It is not very clearly known whether the increase in DNA content represents endoreduplication of the entire genome or the selective amplification of certain sequences such as the storage protein genes, but there is more likelihood for the first possibility only.

7·4·4· Histological Differentiation

During embryogenesis, histological differentiation proceeds at specific regions of the developing embryo. Many early embryologists believed that histogenesis starts only during late embryogenesis and becomes morphologically obvious only after the late globular or heart–shaped stage. Consequently, a number of histologists began their histogenetic studies only from late globular or cordate stage and were under the impression that the histological ontogeny up to this stage is of no consequence. Also, by 'histological differentiation', they meant only the origin of the different meristems such as protoderm, ground meristem, procambium, and root and shoot apical meristems. However, it should not be forgotten that this morphological differentiation is preceded by distinct biochemical and histochemical changes, which can be demonstrated by special staining techniques, electron microscopy, autoradiography etc. Therefore, any attempt to deal with differentiation in the embryo should encompass not only the temporal and morphological aspects, but also the physical, physiological, biochemical and genetic bases of the differentiation phenomenon.

The **protoderm** is perhaps the first to differentiate in the growing embryo. Its time of origin varies with taxa, but it is generally differentiated in the octant stage (Fig. 239). Once the protoderm is differentiated, its cells divide only anticlinally. The differentiation of the protoderm is related to the cutting off of an internal cell in the embryo. Probably because of this relationship, Periasamy (1977) attached great importance to the cutting off of the internal cells in the proembryo and attempted to classify angiosperm embryogeny on this basis. According to him, the increasing exposure of newly–formed cells from the zygote to the internal environment of the future embryo, rather than to the external environment of the endosperm, makes the segmentation of the first internal cell (which leads to the differentiation of the protoderm), and this is an important morphogenetic event. The differentiation of the protoderm further increases exposure of the embryonal cells to its internal environment. The time of differentiation of the protoderm appears to be dependent on the regulatory effect of the extra–embryonal environment, especially the one that prevails at the micropylar pole of the embryo sac, but studies on somatic embryogeny indicates that the basic control over its differentiation lies in the embryo itself.

Almost concurrent with the differentiation of the protoderm, the **hypophysis** and **epiphysis** become initiated at the opposite poles of the globular embryo (Fig. 238). As indicated earlier in Chapter 5, this differentiation implies that polarized gradients of the zygote persist in the developing embryo. The hypophysis refers to a group of cells that become differentiated in the prospective root pole, while the epiphysis is a similar zone in the prospective shoot pole. The hyphophysis later forms an integral part of the root apical meristem and constitutes the base for the organization of the quiescent centre (Swamy and Krishnamurthy 1975b). Similarly, the epiphysis becomes integrated into the shoot apical meristem as the prospective, relatively quiescent central mother cells zone (Swamy and Krishnamurthy 1977). The cells of hypophysis and epiphysis have identical histology, cytochemistry and ultrastructure. They are larger than their adjacent cells in the embryo, less densely cytoplasmic, more vacuolated, and poorer in RNA and protein content. They are large nucleated and relatively quiescent or almost inactive mitotically. Experiments in which an exogenous supply of DNA analogues, such as ^{3}H–thimidine, was used, demonstrated the mitotic quiescence in these cells. The hypophysis and epiphysis remain more or less unaffected through further embryogeny, both with respect to the number of constituent cells and their histology and ultrastructure. In a few plants there may be a slight increase in the number of constituent cells and the newly added cells possess characteristics identical to the earlier cells. As already discussed, there is no doubt that the quiescent centre of the root is organized by the hypophysis. Claims regarding the disappearance of this zone during later stages of embryogeny are likely to be based on faulty observations made on non–median longitudinal sections of the embryo.

The shoot apical meristem of the embryo is organized around the epiphysis, which serves as its nucleus. Similarly, the radicular meristem is organized with the hypophysis as its epicentre (Figs. 238 and 239).

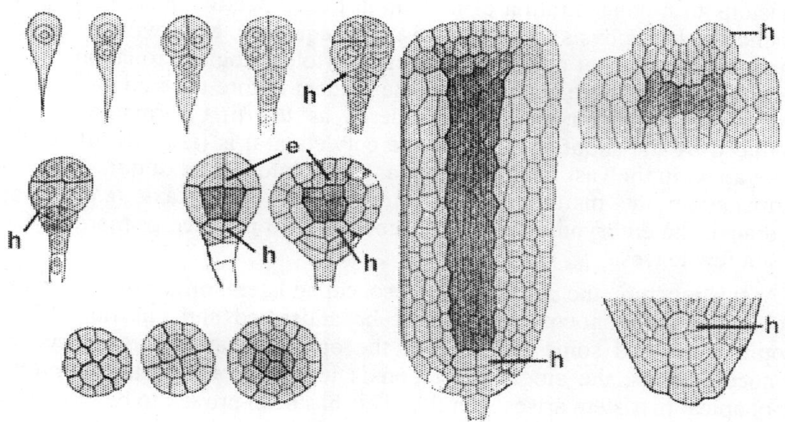

Fig. 238. Various stages in the development of the dicot embryo, as represented by *Sphenoclea zeylanica*; Stippled cells represent epicotyl (including epiphysis); Single–hatched and cross–hatched cells represent the plerome; E - Epiphysis; h - Hypophysis; p - Leaf primordium. (Swamy and Padmanabhan 1961)

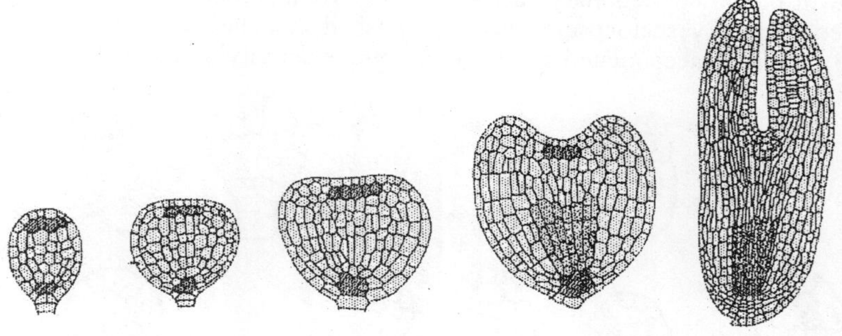

Fig. 239. Developmental stages from late globular to mature embryo of *Catharanthus roseus*; The epiphysis and hypophysis are shown in the respective poles in the first four stages, while the plerome in shown in the last two stages (Philip 1972)

The epiphysis becomes the central mother cell zone of the shoot apical meristem (or the apical zone of the French school), while the hypophysis organizes the quiescent centre of the root apical meristem. In other words, the differentiation of the root and shoot apices begins with the organization of hypophysis and epiphysis respectively. Although the embryonic apices are very small, their origin and differentiation can easily be discerned and so it is wrong to assume that these two meristems have an undistinguishable origin. During the organization of the two meristems in the developing embryo, a change from the random distribution of cell divisions to a concentration of mitosis in the respective poles takes place around the hypophysis and epiphysis. Consequently, even within the very small embryonic root and shoot apices, a cytohistological zonation can be identified. Since these changes become evident before the initiation of the cotyledons, the latter can be considered as the first formative organs produced by the shoot apex. Once the cotyledon(s) is (are) identified, the shoot apex, in the vast majority of taxa, lapses into a state of quiescence or dormancy in the mature embryo. However, in some taxa (e.g., many legumes), the embryonic shoot apex continues to produce primordia of at least a few leaves.

At this juncture, the question of the so–called lateral origin of embryonic shoot apex of monocotyledons must be addressed and clarified. Many people were, and some still are, of the opinion that in the embryo of monocotyledons, the single cotyledon is terminal, while the embryonic shoot apical meristem arises laterally. This has been proved to be wrong by detailed studies made on several monocotyledons by Professor B. G. L. Swamy and his school at Chennai (Swamy and Krishnamurthy 1980a) and by Dr. Barbara Haccius and her students in Germany. Their work has proved beyond doubt that both the embryonic shoot apex and the cotyledon of monocots have their origin in the terminal position in late filamentous and early globular embryonic stages, but because of the more pronounced growth of the cotyledonary sector, the embryonic shoot apical sector (or epicotyledonary sector) appears to be pushed to a lateral position (Fig. 240). This is accentuated by the poor mitotic activity in the cells of the

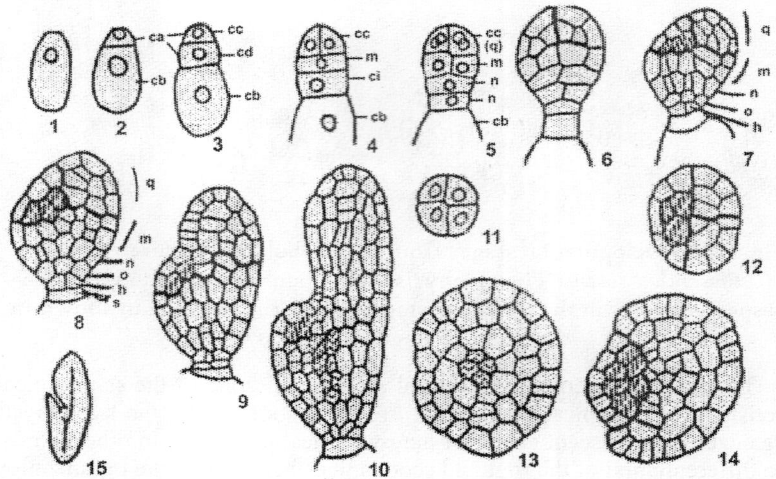

Fig. 240. Development of the monoct embryo as represented by *Ottelia alismoides* showing the different tiers; 11–14 represent transections of 5, 7, 10 (basal part) and 10 (epicotylary) respectively; Prospective epicotylary region is shown in the central part of figures 10–13 (Swamy 1962)

shoot apical sector. Hence, the epicotyledonary sector of the embryo of monocotyledons is as terminal as that of the dicotyledons embryo.

The **ground meristem** and the provascular meristem are blocked out almost simultaneously during embryogeny (Fig. 238 and 241). Both are derived from the cells of the central core of the late globular or early heart–shaped stage by way of cell enlargement, decreased stainability, and increased vacuolation. Depending on the taxon, the ground meristem may produce only a cortex or both cortex and pith. The former situation occurs in *Phlox drummondii, Dianthus chinensis, Sphenoclea zeylanica, Downingia bacigalupii* and *Stellaria media.* In all these cases, the peripheral layer of cells of the central meristematic core in the late globular stage is the progenitor of the cortex of the mature embryo. During elongation of the embryo, these cells divide periclincally to form an embryonic cortex of 3–4 concentric layers. In *Juglans regia, Pisum sativum, Nerium oleander* (Fig. 241) and *Sesamum indicum* both pith and cortex develop from the ground meristem. The first evidence of pith is the presence of enlarged, lightly–staining cells in the central region of the globular embryo; by continued divisions, these cells differentiate as the pith of the hypocotyl part of the mature embryo.

Our knowledge regarding **procambialization** and **vascular differentiation** in the embryo is meagre. This lack perhaps stems from a lack of understanding among past workers of the actual features of the procambium and also from a lack of biochemical markers for identifying the procambium and its activities in embryo. Many past workers designated the whole central core of the cells of the embryonic axis as procambium. The region was also called 'plerome' in accordance with the histogen concept of Hanstein, "stelar promeristem", "provascular meristem", "residual meristem", "prodesmogen" and "procambium in incipient stage". Considering the entire zone as procambium implies that the procambium gives rise not only to vascular tissue, but also to non–vascular elements such as pith (when present), conjunctive parenchyma, pericycle and endodermis. This confusion caused problems in interpreting the wave of procambialization in the embryo and the temporal origin of the procambium. Esau (1965b) admitted that the central core of cells of the embryonic axis initially retain the characteristics of the early embryonic cells and only gradually assume the characteristics of the procambium. Not all cells of the central core become transformed into the procambium that differentiates from it. The important characteristics of the procambium are as follows:

1. Presence of a single cell type
2. Small and isodiametric cells in transverse sections
3. Cell division in all directions
4. Absence of intrusive growth

5. Cell walls of uniform thickness
6. Presence of plastids and many small vacuoles
7. Basophilic nature and intense staining with pyronin for RNA.

Only a few of these traits are shared by all the cells of the central core of the embryonic axis.

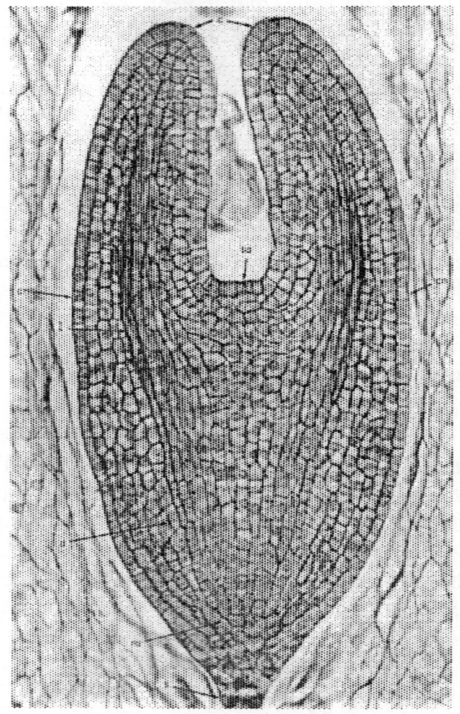

Fig. 241. Median L. S. of the mature embryo of *Nerium oleander* c. cotyledon; cn - Cotyledonary mode; d - Result of recent periclinal division in cortex; l - Laticifer; p – Pith; pr - Procambium; ra - Root apex; s - Suspensor; sa - Shoot apex (Mahlberg 1960)

It is true that in most embryos, the blocking out of the so–called provascular cells takes place in the central region of the late globular or early heart–shaped embryos, but it is wrong to confuse it with procambial differentiation as claimed by a number of investigators. The blocking out takes place as a continuous structure not only in the embryonic axis, but also in the central core of the cotyledons (Fig. 241), and this trait has given rise to confusion in interpreting the longitudinal wave of procambialization also. Consequently, we find such statements as the following: "The procambium of the cotyledons, hypocotyl and radicle is one continuous tissue system". Keeping this in mind, the following account on what actually happens is given. Procambialization of the embryo is noticed earliest in the first appendicular structure, the cotyledon. The linear derivatives of the subapical group of initials form the locus of differentiation of the median trace of the cotyledon. The blue print for procambial development in the radicle is laid down much later i.e., after the pre–radicular meristematic centre has become active. The medianly aligned linear derivatives of this meristem differentiate in part into the procambial system, in which the metaxylem locus becomes histologically distinctive (Fig. 242). When procambial initiation has commenced in the cotyledon and the root regions, the hypocotyl part exhibits a conspicuous developmental lag, although having a central core of cells. This situation emphasizes that: (i) the root and hypocotyl are organs that are ontogenetically derived independently; (ii) the procambial system of the cotyledonary procambium has no connection with that of the radicle or of the hypocotyl, as the latter exhibits a pronounced chronological lag in reference to procambialization. The initiation of the pre–procambium is probably under the control of the parent plant and the control may come through hormones. Although cytological evidence does not permit an early detection of pre–procambium, cytochemical markers have been devised for an early identification of these cells. The marker is carboxyl esterase, an enzyme associated with cell wall metabolism, which is active in cells destined to differentiate into vascular elements. Not all cells of the central core of embryonic axis and cotyledons show the activity of this marker enzyme, but only those that are truely procambial (Krishnamurthy 1994).

The procambial cells that differentiate in the embryo are determined and committed to develop into vascular elements i.e., the programme persists irrespective of any subsequent programming events. On the contrary, the cells of the procortex are not rigid, but are plastic. We do not have adequate information on what causes the fixity of the procambial commitment, but the available evidence points to the involvement of specific genes. Information is deficient even on differentiation of vascular tissues from procambium in the embryo and this is particularly true of the phloem. The differentiation of the first tracheary cells from the procambium strictly follows the corresponding temporal sequence described above for procambium. The earliest protoxylem and protophloem appear in the cotyledon and the radicle, when only the procambial initiation is detected in the hypocotyls–proper. The origin of the procambium in the cotyledon and radicle is traceable to the respective meristems, but the origin of the hypocotyl procambium is not related to any apical meristematic unit, but has it from its own constituent cells at appropriate loci. Vascularization of this procambium is also correspondingly delayed. Phloem differentiation generally precedes that of xylem, and this too, is discontinuous as is for procambium or xylem.

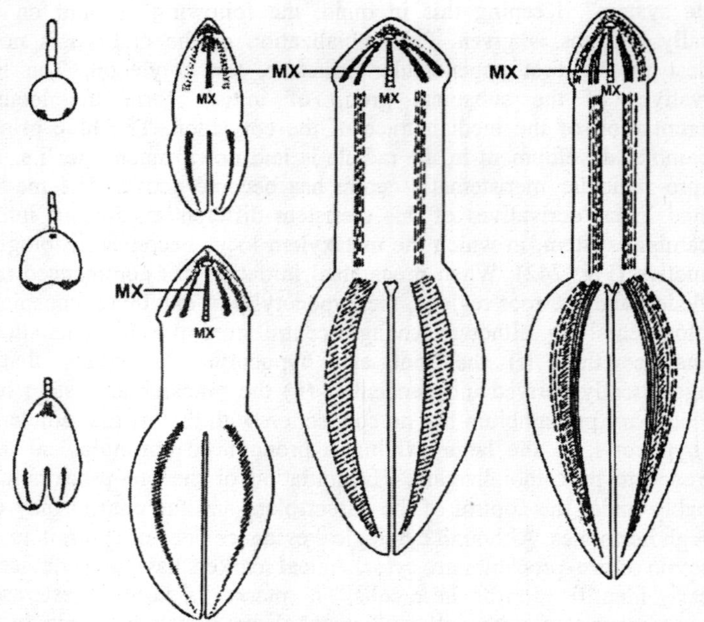

Fig. 242. Stages in the development of dicot embryo; The locations of meristematic regions (stippled), procambialization (hatched) and vascularization (broken lines) are shown; mx - Metaxylem (Drawing supplied by late Prof. B.G.L. Swamy)

7.4.5. Genetic Control of Embryogenesis

7.4.5.1. *Maternal Genetic Control*

An important question that needs to be resolved concerns the importance of parental conflict and infanticide during early embryogenesis (and endosperm development), as controlled by maternal genes (Viella-Caldazia wt al. 2000). There are evidences to show the presence of gametophytic maternal effect on embryo (and endosperm) at least for three days after fertilization (Grimanelli et al. 2005; see also Kohler et al. 2005), although there are also evidences to indicate that the required switch from maternal to embryonic control of development occurs after fertilization at the zygote itself in angiosperms i.e., at a very much earlier developmental stage of embryo than in animals (Kranz et al 1999; Okamoto and Kranz 2005). The *fertilization independent endosperm* (*fie*) mutant and also *fertilization independent seed* (*fis*) mutants such as *fis1*, *fis2* and *fis3* in *Arabidopsis* can form 'endosperm' in the absence of fertilization (Grossniklaus et al 1998b, Ohad et al 1996, Chaudary et al 1997), and this can indirectly affect the 'zygote' if formed. In plants heterozygous for *fie*, *fis1*, *fis2*, and *fis3* loci, the secondary nucleus starts dividing in the absence of pollination. The *sin1* mutant of *Arabidopsis* is also a maternal sporophytic gene mutant. Thus, these gene products are likely to be involved in the activation of reproductive programme in response to fertilization. All four loci are not or very poorly transmitted through the female gametophyte. If fertilized, seeds derived from female gametophytes carrying the mutant alleles abort irrespective of the paternal contribution. Two other *Arabidopsis* mutants, *emb173* and *medea* (*mea*) (Grossniklaus et al 1998b, Grossniklaus and Viella–Caldaza 1998) display gametophytic maternal control on seed development, if the genes are inherited through the female gametophyte. The gametophytic maternal effect of *mea* results in aberrant growth regulation during embryogenesis. The embryos derived from *mea* eggs show excessive growth and die during the desiccation phase of seed. Thus, these mutants display a gametophytic maternal effect, where the genotype of the embryo sac controls the subsequent sporophytic generation. (Pagnussat et al. 2005).

A gametophytic maternal effect can be mediated by several different mechanisms (Grossniklaus and Schneitz 1998): (i) It may be caused by haplo–insufficiency (resulting in gene dosage alternations in the endosperm, which inherits two mutant alleles from the mother, but only one paternal wild type from pollen in outcross; (ii) The mutation probably affects a stored factor present in the cytoplasm of the egg and/or central cell that is required for seed development after fertilization; (iii). The mutant may disrupt an imprinted gene **(genomic imprinting)** that is only expressed from the maternally inherited copy; or (iv). The mutation in genes that are expressed during female gemetophyte ontogeny, but whose expression products are needed for embryo and endosperm development soon after double fertilization.

7.4.5.2. *Post–Fertilization Non-maternal Genetic Control*

Development of the embryo is also shown to have a genetic basis, other than the maternal. A lot of concerted gene action and diverse patterns of gene expression are needed to build the tissues of the embryo. Changes in the pattern of protein accumulation and synthesis in developing embryos are the immediate consequences of gene expression. Cells in different locations of the developing embryo attain different gene–based commitments. Biochemical changes are triggered at these locations at appropriate times due to sequential evocation of specific genes or **chreode** *sensu* Waddington (1957), which can be analysed by the specific proteins synthesized and the specific mRNA sequences employed. A developing embryo synthesizes a number of proteins (including enzymes). On an average, about 20,000 genes (Fig. 243) were known, even 20 years before, to be operating during the whole gamut of embryogenesis (Goldberg et al 1989b, 1994). This is evident from studies on *Gossypium, Arabidopsis, Nicotiana, Zea, Pisum,* Soya bean and a few other taxa. Of these, one fourth to one fifth genes are unique to the embryo and other genes are shared by other organs of the plant. Of these embryo–specific genes, 10–12% (40–50) are believed to be master regulator genes. Some of the genes involved in embryogenesis, hence, have a house–keeping role i.e., fulfilling structural, metabolic and synthetic genes, while others encode storage proteins and organ–specific and tissue–specific proteins.

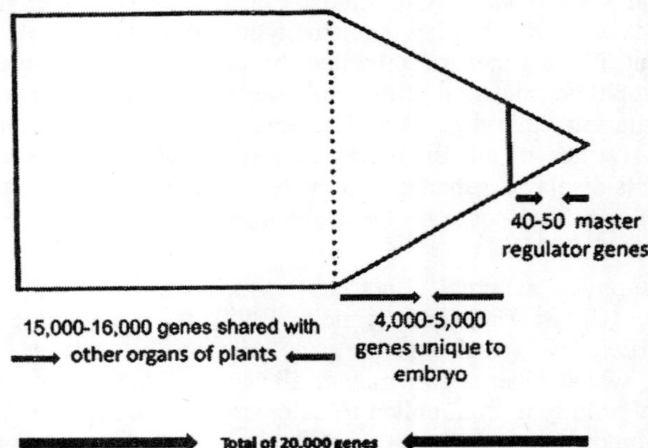

Fig. 243. Diagrammatic representation of the total number of genes shared by the embryo with other organs, total number of genes exclusive to the embryo and the total number of master regulatory genes. The number of genes is approximate and may vary in different taxa.

The presence of approximately the same number of different mRNAs of 14,000 to 18,000 in embryos of two morphologically–distinct age groups supports the notion that there are no changes in the absolute number of structural genes expressed during embryogenesis. New ideas have been obtained regarding the modulation of the appearance and disappearance of mRNAs and proteins during embryo development in cotton (Dure 1985) (Fig. 244). As shown in this figure, there are separate subsets of mRNAs that are expressed during embryo development and germination. These include mRNA groups encoding a large set of proteins for embryo development and germination, including actin and tubulin. These form group 1. Embryo–specific proteins that are present throughout embryogenesis and during early seed germination form group 2 genes but these are turned off after embryogenesis. The proteins of group 3 genes are active during early embryogeny only. The proteins of group 4 genes accumulate in embryos of all developing stages except at the early stages of embryogenesis and become abundant during late embryogenesis. Proteins of group 5A are active from mid embryogenesis to mature seed stage while those of group 5B are LEA protans abundant in late embryogeny. Proteins of group 6 genes are believed to be involved in embryo desiccation and dormancy. Proteins of group 7 genes are enzymes involved in germination. A two dimensional SDS–PAGE of proteins synthesized by maize embryos of different ages indicate that, of more than 25 polypeptides found, a subset of two polypeptides is found to be synthesized throughout embryogenesis, another subset of 14 polypeptides is synthesized during the period of morphological differentiation of embryos, and a third, smaller subset is synthesized when the embryos lapse into quiescence or dormancy (Boothe and Walden 1990).

Embryo–specific expression of cotyledon–abundant and embryo axis–abundant genes isolated from a cDNA library made to poly (A)–RNA of seedlings of *Brassica napus* was also studied. These two sets of genes are expressed during critical stages of embryogenesis, with transcripts of the cotyledon–abundant genes reaching maximum levels in embryos about to dessicate (Harada et al 1988). The expression of numerous developmental genes isolated from the sporophytic organs of plants has been followed in embryos of wild–type and transgenic plants. These include genes encoding homebox proteins, cell cycle control protein kinases, extensins, histones, chloroplast proteins, lipid transfer proteins, phytochrome, actin, MADS–Box proteins etc. during some stage of embryogenesis. Most mRNA sequences transcribed during embryogenesis persist as stored mRNAs in seeds and serve as templates for the synthesis of proteins of early germination stages (Raghavan 2000).

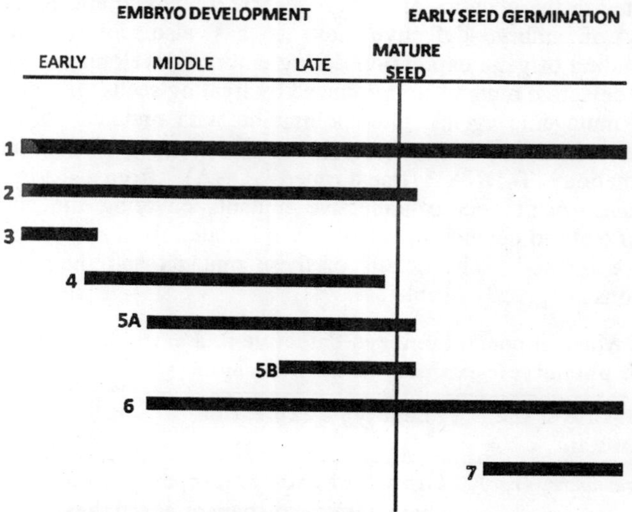

Fig. 244. Diagrammatic representation of the groups of abundant mRNAs represented in developing and germinating embryos of cotton (Dure 1985)

It should be emphasized that gene expression during embryogenesis is more cryptic and appears to be more fundamental than that determined by plane of cell division or cell lineage (Raghavan 2000). A good strategy to identify genes and characterize their expression in transgenic embryos of transgenic plants by a promoter trap method promises to provide markers for restricted cell types of proembryos or globular embryos before spatial patterns of gene expressions are activated. Lu et al (1996) have isolated and characterized a cDNA clone called *ATML 1* (for *'Arabidopsis thaliana meristem L1 layer'*) and this encodes a new homeodomain protein. After the asymmetric division of the zygote of this plant, transcripts of this clone are first expressed in the apical cell of the two-celled proembryo and subsequently get concentrated in the cells derived from this apical cell up to the globular embryo stage. At the globular embryo stage, the expression of the transcripts of this clone is restricted to the protoderm cells and disappears from the inner cells. Thus, the expression of *ATML 1* gene transcripts is considered to be the first molecular marker of the apico–basal and radial pattern formation in the developing embryo. *KT 1* gene forms a good marker of polarity in the globular embryos of soya beans as the gene gets transcribed in a small number of cells at the micropyler pole of the embryo. *STM* gene forms a marker for the earliest stage of differentiation of the shoot apical meristem in *Arabidopsis.*

A procedure for isolating the basal and apical cells of 2-celled proembryo of maize and the subsequent use of these cells as starting points for detecting genes that are up- or down-regulated in these two cells have contributed to the genetics of embryogenesis (Okamoto and Kranz 2005). A study of embryo–defective mutants has also contributed to our understanding of gene expression during embryo development. Generally, embryo–defective mutants are produced by treating seeds with chemical or physical mutagenic agents, after pollination with mutated pollen grains, and after insertional mutagenesis with transposable elements or transformation T–DNA (transformed DNA) from *Agrobacterium tumefaciens*. A number of defective mutants covering the entire time course of embryo development have been isolated from *Arabidopsis* using the above methods. The details of these mutants and their phenotypic expressions are given in Table 5.

Table 5. Most important embryo-defective and embryo-pattern mutants and their phenotypic expressions in *Arabidopsis*

Sr. No.	Name of the mutant	Phenotypic expression
1.	*raspberry (ras)*	Growth of embryo–proper inhibited with characteristic suspensor anomalies.
2.	*suspensor (sus)*	Same effect as above.
3.	*twin 1 (twn 1)*	Same effect as of the first mutant.

4.	*twin 2 (twn 2)*	Apical cell of two-celled proembryo arrested, but suspenser is proliferated.
5.	*leafy cotyledon (lec)*	Cotyledons become leafy both morphologically and anatomically; embryo maturation features (such as desiccation and dormancy) not seen.
6.	*cup-shaped cotyledon (cuc)*	Embryo is with fused cotyledons.
7.	*gurke (gu)*	Embryo lacks both shoot apical meristem and cotyledons.
8.	*fackel (fc)*	Cotyledons are attached to the root, and no hypocotyl present.
9.	*pastaccino (pas)*	Embryo is with misshaped cotyledons and with a short hypocotyl.
10.	*mp*	Embryo is without hypocotyl or radicle.
11.	*gn*	Embryo is reduced to a cellular mass.
12.	*wus*	Embryo lacks shoot apical meristem.
13.	*stm*	Embryo lacks shoot apical meristem.
14.	*rootless (rtl)*	Embryo lacks radicle.
15.	*knolle (kn)*	Embryo lacks protoderm.
16.	*keule (keu)*	Embryo with abnormal epidermal cells.
17.	*fist*	Radicle pattern in embryo is affected.
18–22.	*fs, hydra (hyd), knopf (knf), mickey (mic)*	Embryo shape is affected and produces abnormal seedlings.

The *raspberry* and *sus* mutants have arrested embryo–proper, but the suspensor undergoes abnormal enlargement and divisions to express embryo–like behaviour, while the *twin* mutant has a good number of twin seedlings (one from embryo–proper and one from the suspensor), emphasizing the fact that *RASPBERRY*, *SUSPENSOR* and *TWIN* genes are required to maintain the normal potential in embryo–proper and suspensor, and their identity through their effect on signalling mechanism. *LEC* gene is required for normal cotyledon development and for the exhibition of features of embryo maturation such as desiccation, dormancy and accumulation of storage proteins.

A number of mutants that cause abnormalities in cell lineage patterns have been described in *Arabidopsis*. These affect apico–basal pattern, radial pattern and other pattern–related aspects of the developing embryo. The genes responsible are referred to as **pattern genes**. The genes that express specifically in the apical cell of the 2-celled proemberyo are *MP, BDL, WOX2* and *AtML 1,* while those restricted to basal cell are *PIN7,*

WOX8 and *WOX9*. Mutants that lack apico–basal polarity and pattern include *gurke (gr), fackel (fc), pasticcino (pas), mp, gn, wus, stm, rootless (rtl)* etc. The mutants that affect radial pattern are *knolle (kn)* and *keule* (keu). Shape mutants such as *fs, hydra (hyd), knopf (knf)* and *mickey (mic)* produce very abnormal seedlings. Histological studies on pattern mutants reveal that anarchic cell divisions in zygote or proembryo cause problems in embryo patterns. For instance, *gn* mutant has symmetric division of the zygote and this is followed by several random divisions of the terminal cell alone resulting in a defective embryo (Mayer et al 1993); this gene has its effect through *PIN1* location. The *fs* gene does not affect the asymmetric division of the zygote, but subsequent divisions in the apical cell are largely irregular to result in a radially–enlarged embryo with a telescoped apico–basal axis and an incipient radicle. In *fist* mutant, protoderm is not formed, while *mp* mutant is with abnormal lower part in the embryo with proliferative hypophysis. The *kn* and *keu* mutant embryos show chaotic divisions in the octant embryo and have large multinucleate cells with incompletely–formed cell walls. Thus, it is very clear that the body plan is specified by a hierarchy of genes acting at much–defined stages of development and embryo patterns are not dictated by cell size or shape or orientations of planes of cell divisions, but probably by the relative positions that cells occupy. It is evident from the cloning and characterization of pattern genes that the protein products of these pattern genes function as components of a network of ubiquitous proteins that affect common cellular functions. These include other intracellular transport proteins associated with golgi bodies, vesicle traffic directing proteins during cell division, transcription factors etc.

Some details are available on the genes affecting storage protein synthesis and accumulation in endosperm and embryo. The major work relates to zein genes in endosperm. There are five molecular mass classes of this protein: 10, 15, 19, 22 and 28 kDa; α–zeins (19 and 22), β–zein (15), γ–zein (28) and δ–zein (10). The op2 mutation affects zein synthesis resulting in a soft, floury endosperm. The mutation also inhibits 22 kDa zein, but enhances the synthesis of the lysine–rich EF–protein. The programming of storage protein gene expression during endosperm development in rice, wheat, and barley are similar to *Zea mays* (in zein synthesis) in many ways. The cloning and the sequencing of genes involved in cereal endosperm proteins have led to the recognition of a seven–nucleotide sequence (TGTAAAG) called the '–300 bp element' or the 'endosperm box'. This sequence is believed to programme the endosperm–specific gene expression in cereals. The mRNAs that encode storage proteins of embryos increase from very low levels, become dominant species in the embryo cotyledons during peak–protein accumulation period and again declines to very low levels at embryo maturity. Thus, there is a mechanism that limits gene transcription coincident with embryo maturation. It is also evident that storage protein synthesis is generally regulated at the transcriptional level, but there also appears to be some post–transcriptional regulation for some storage proteins (E.g., conglycin). There are evidences for the fact that the genes encoding storage proteins are targets of the growth–hormone action that leads to an increased transcription and eventually to increased protein accumulation. Cloning and sequencing studies made on several legume storage proteins indicate the presence of conserved upstream sequences, known as the 'legumin box' and the 'vicilin box' in the genes determining tissue specificity. The legumin box is a 28–bp DNA element identified in the promoter regions of most globulin genes and consists of motifs known as RY–repeats (5'–CATGCAT–3'; 5'–CATGCAC–3'; and 5'–CATGCATG–3'), while vicilin box is a highly conserved C–rich region approximately 120–bp from the transcription–start site of the vicilin genes. These two boxes are absent in cruciferin and napin genes.

Cell fate in developing embryo is also dependent on positional signal molecules like auxin (Jürgens 2001; Frimml and Wisiniewska 2005) and auxin-induced genes, either directly or indirectly, like *GNOM/EMB 30 WOX9, 8, 2, YODA* etc. (see Okamoto and Kranz 2005).

7.4.6. Embryo Maturation

Defining a mature embryo is made difficult by the varying connotations of the word 'mature'. This term has been used to describe the final stage attained by a seminal embryo i.e., an embryo with a root meristem at one pole, and a shoot meristem and one or two cotyledons at the other pole connected by the hypocotyls–root axis. This word is also used to describe an embryo characterized by the cessation of cell division and, to some extent, cell expansion. Also, 'maturity of an embryo is reckoned in terms of its ability to give rise to a normal seedling plant upon germination' (Raghavan 2000). The maturity of an embryo is also difficult to define from its metabolic status as subtle physiological or biochemical changes continue to occur in the embryo even after the attainment of a morphologically mature structure. 'Mature' seminal embryos are, to some extent, independent and are able to grow when supplied with a limited diet. Some authors have even considered the weight of the embryo to identify its maturity, for example, in cotton, embryos weighing more than 100 mg (final weight 120 mg) are deemed to be mature. However, maturity may safely be associated with a combination of most, if not all, of these characteristics.

Embryo maturity generally synchronizes with **seed ripening**, but in several species an approximately ripe seed may contain an immature embryo that continues to grow to maturity intraseminally after the seed is released. These intraseminal changes in the embryo, collectively called **after–ripening** (e.g., coconut), make the embryo fit enough for germination. The after–ripening period varies from 8 d to nearly 12–18 m. Generally, the embryos of the after–ripening category do not have a dormant period. In all the above, the factors that promote after–ripening are intraseminal.

7.4.7. Nutrition of Embryo

The nutritive requirements of the growing embryo are provided by the sporophytic maternal tissues of the ovule or developing seed and the endosperm. Studies have revealed that the embryo, during its development, is heterotrophic (perhaps with the exception of taxa with **chloroembryos**), particularly the zygote. The stage of embryo development, when autotrophy is attained, varies from species to species. The structure of the embryo sac after the zygote is formed indicates the importance of wall projections (transfer–cell morphology) in the transport of metabolites from the surrounding cells. For instance, these projections start developing from the micropylar end of the embryo sac wall, close to the zygote, and are retained till the heart–shaped stage is attained in the embryo. ^{14}C–labelled photosynthates were traced in the ovule using autoradiography studies (Chamberlain et al 1993) and these studies showed that wall ingrowths are intimately involved in the absorption of nutrients (Fig. 245). The label was also initially seen in the hypostase, before it was finally seen in the zygote. This study shows that the zygote obtains its nutrients through two major transport pathways: one from the outer integument to the base of the zygote and the central cell, and the second through the hypostase into the central cell. Nutrition to the growing embryo is also provided through the suspensor whose cells also have transfer–cell morphology.

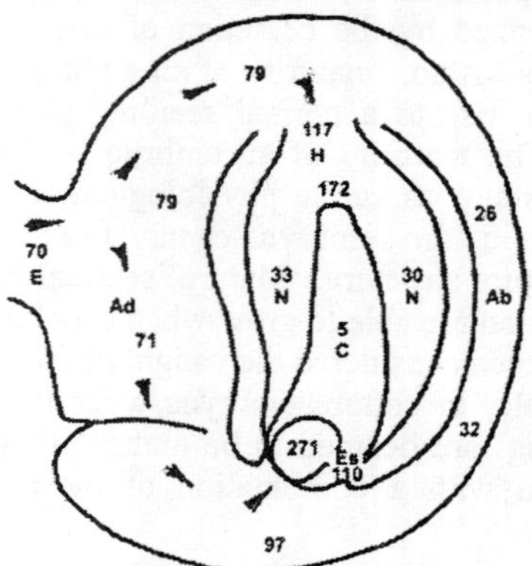

Fig. 245. Diagrammatic L. S. of the soya bean ovule at zygote stage showing the pathways (arrowheads) and patterns of accumulation of labelled photosynthates in specific regions; Numbers indicate the numbers of autoradiographic silver grains per 1000 μm^2; Ab - Abfunicular outer integument; Ad - Adfunicular outer integument; C - Embryo sac; Es - Endosperm; F - Funicle; H - Hypostase; N - Nucellus (Chamberlain et al 1993)

7.4.8. Synthesis of Storage Substances

Mature embryos of many taxa are rich in storage materials such as proteins, carbohydrates or lipids found in their cotyledons. Most, if not all, of these metabolites are synthesized when cell division almost stops and cell expansion commences. However, the synthetic process is built up progressively and culminates when the seminal embryos reach maturity.

Compared with other storage products, embryo proteins have been the subject of intense study mainly by those interested in gene action and expression during embryo development. The total protein fraction of an embryo broadly falls into two categories: heterogeneous metabolic proteins that include enzymes and ribosomal, chromosomal and membrane proteins, and more homogeneous storage proteins (including lectins). The recognition by Osborne (1924) of the four types of proteins (albumins, globulins, prolamins and glutelins) based on solubility classes is still valid. Albumins are soluble in water, globulins in salt solutions, prolamins in alcoholic solutions and glutelins in dilute acid or alkali. Globulins and albumins are the major embryo–storage proteins, with globulins dominating in many species. In legumes, the major storage proteins are the globulins, whereas prolamins and glutelins predominate in the cereals. The two major components of globulins are the 7–8 S and 11–12 S proteins. The globulins are named after the plants or families from which they were first described such as vicilin, convicilin, legumin, conglycinin, glycinin, phaseolin, cruciferin and helianthinin. Napin is one of the about 25 albumins known. Storage proteins appear to originate both in the vacuolar system and the plastids in cereals. In legumes, the vacuoles, rough ER, and dictyosomes are involved in the formation of storage–protein bodies indicating that protein–body formation in embryo cells shares at least some of the features of this process described for endosperm earlier.

7.4.9. Embryo Suspensor

As already mentioned, the asymmetric, transverse division of the zygote produces a small apical cell and a large basal cell. The small apical cell gives rise to the embryo–proper, while the large basal cell usually organizes the embryo suspensor and the future root cap. Although there are several instances of embryos which do not possess a morphologically–distinct suspensor, there are an equal number of instances where the organization and elaboration of the suspensor result in the production of highly impressive morphological formations. Such structures are presumed to function as haustoria.

In a number of species, the free cell of the suspensor alone takes on the haustorial function, the remaining cells perhaps acting as a channel of transport. Under these conditions, the concerned suspensor cell undergoes conspicuous hypertrophy and acquires typical histological features of a haustorial cell (Fig. 246 A). In some plants, the haustorial cell grows out of micropyle and becomes embedded in the placentum (Fig. 246 B and C). The most impressive morphological forms of suspensor organization are seen in the Leguminosae (*sensu lato*) and Orchidaceae (Figs. 247 and 248).

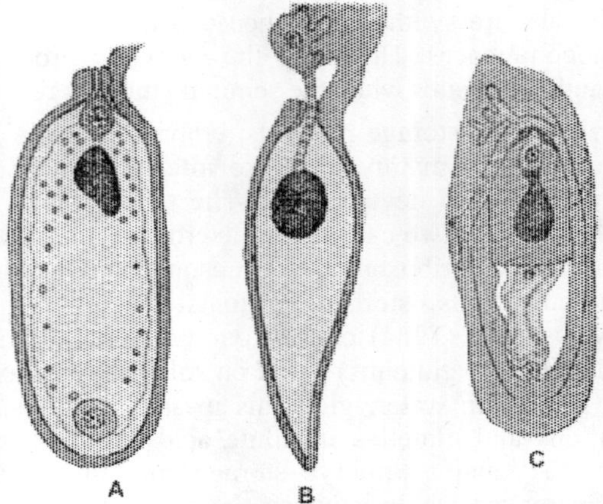

Fig. 246. L. S. of developing seeds of *Blyxa* (A), *Habenaria* (B) and *Rosularia pallida* (C) showing the nature of suspensor. (All drawings were prepared by Late Prof. B.G.L. Swamy)

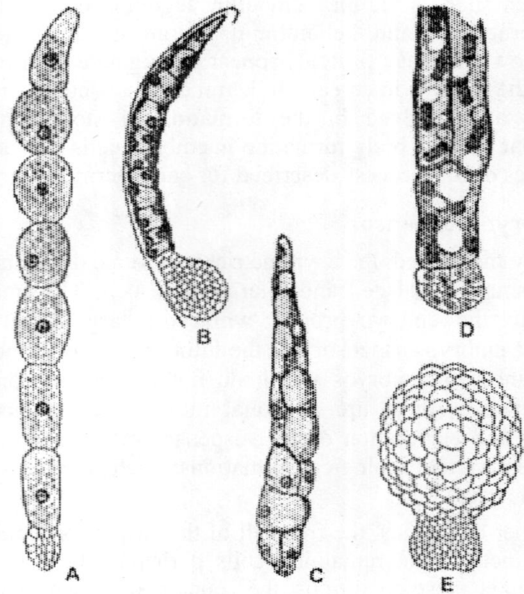

Fig. 247. Variation in the organization of suspensor in Leguminosae; A - *Ononis;* B - *Lupinus;* C - *Cicer;* D – *Pisum;* E - *Cytisus*

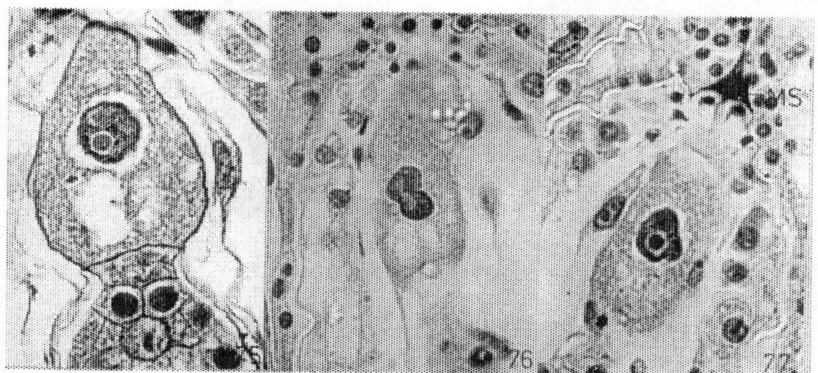

Fig. 248. Terminal haustorial cell of the suspensor of the embryo of *Ottelia alismoides* showing the hypertrophied nucleus (Photographs courtesy of Dr. R. Indra)

The suspensor, thus, is now shown through electron microscopic, histochemical and biochemical studies, to be not only a locus of synthesis of nutrients and growth regulators, but also a transfer region of food materials from other parts to the embryo. The suspensor nuclei of many species are hypertrophied and polyploidal or polytenic (Figs. 26 and 248). A single suspensor cell may also have more than one nucleus in many taxa (as many as 70 nuclei in *Cuscuta reflexa*). Endoreduplication, coupled with polyteny, is a way of life for the nuclei of suspensor cells. The DNA content of nuclei of suspensor cells of different species of *Phaseolus* ranges from 16 C to 8192 C (in *P. coccineus*). In *P. coccineus* polyteny associated with puffing of chromosomes (as in giant chromosomes of dipteran larvae) is seen indicating selective amplification of certain genes and underreplication of certain other genes. Once the embryo approaches maturity, the suspensor cells undergo PCD.

7.5. SEEDLING DEVELOPMENT

The young sporophytic plant coming out of the seed at the time of germination is called a **seedling**. It is also sometimes called a **sapling**. Seedling–like structures produced under in vitro culture conditions from embryoids or callus are often called **plantlets**. The production of a seedling during the germination of a seed involves a complicated series of morphological, physiological and biochemical processes or events.

7.5.1. Viability of Embryos

Embryo viability may be defined as the ability of the mature embryo to germinate and produce a seedling under favourable conditions. Embryos vary appreciably in their viability. In some taxa, the embryo loses its viability immediately after the seed becomes mature, while in others, it can germinate even several years after seed maturation (up to 250 y in some cases). However, the normal viability period in most species ranges from 5–30 y. It can be prolonged in many cases under proper storage conditions.

The gradual loss of viability is often referred to as **ageing**. A number of changes take place slowly in a mature embryo that lead finally to the total loss of viability. The following changes are more important:

1. Damage to and disintegration of membrane–lined organelles such as plastids
2. Dispersion of ribosomes
3. Disappearance of dictyosomes
4. Rupture of lipid and protein bodies
5. Accumulation of toxic and/or mutagenic compounds resulting in chromosomal aberrations and inhibition of growth, as seen in many cereals
6. Accumulation of phenolics not noticed in viable embryos, as seen in cereals
7. Increase in acid phosphatase activity in the cytoplasm
8. Absence of further cell division
9. Loss of mitochondrial function leading to the loss of an ATP–generating system
10. Reduction in or total loss of protein synthesis due to structurally and biochemically–defective ribosomal and post–ribosomal fractions.

Thus, the loss of viability of embryos is caused by general damage to macromolecules including nucleic acids, membrane compounds and enzymes. Ageing embryos undergo autolysis in their cells, although at a slow pace. Autolysis is initiated by the appearance of **cytolyosomes** that arise by the fusion of segments of ER. This results in the fine compartmentalization of the cytoplasm and finally in the death of cells. A similar phenomenon is observed in ageing embryo suspensors.

7.5.2. Germination

A seed containing the 'mature' embryo remains in a state of suspended animation till it germinates to produce a seedling. This suspended animation is due to **quiescence, after–ripening** or **dormancy**. The concept of after–ripening had already been discussed. Most seed physiologists consider quiescence of seeds as an externally–controlled phenomenon and dormancy of seeds as an internally–controlled phenomenon. Quiescent seeds germinate when they are provided with appropriate environmental conditions. The onset of germination involves profound structural and biochemical changes in the embryo and in the extra–embryonal tissues of the seed (Bewley 1997). Most, if not all, of these changes are triggered when seeds begin to imbibe water. It is the first step in seed germination.

Reactivation of embryonal growth during germination universally first occurs in the radicle and then is followed in the shoot. Nucleic acid synthesis, particularly RNA synthesis, has proved to be a useful marker for the onset of biochemical changes in the embryo just as imbibition of water is the first morphological marker. DNA synthesis has been studied by a variety of methods including cytochemical and autoradiographic techniques. DNA synthesis is indicated by evidences such as chromatin dispersion, nucleolar vacuolation, increase in the transcriptional activity of the chromatin, loss of histones from DNA and an increase in the activity of endogenous RNA polymerase. These changes imply that the cells of the embryo, which are arrested in the **G1** phase when the seeds mature, proceed to complete the cell cycle swiftly with the onset of imbibition of water by the seed. Moreover, a progressive loss of DNA integrity and also DNA fragmentation occurs in the embryos of dry seeds. DNA synthesis is necessary to repair the damaged strands, to reinstate the functional integrity of the genome and to reduce the lag time for germination (Raghavan 2000). This necessitates the initiation of activity of a DNA ligase that takes care of this repair, and this enzyme activity is shown just after water is imbibed. This repair removes the barriers for full transcription of the genome and allows for successful germination. However, it should be kept in mind that the onset of DNA synthesis and genome replication are not essential components of early germinational growth of the embryo as protrusion of the radicle outside the seeds ordinarily occurs in many seeds even before the initiation of DNA synthesis in the meristem. But, it should also be kept in mind that for the successive continuance of germination, DNA synthesis is a must.

The preparatory events culminating in seed germination are dominated by RNA and protein syntheses. RNA and protein syntheses begin long before the initiation of DNA synthesis, in many cases just after 30 minutes of water intake by the seed. Among the two, protein synthesis precedes RNA synthesis, probably from stored or leftover mRNA (often called long–lived mRNA), using pre–formed ribosomes. The first RNA species to be synthesized appears to be mRNA, and that too by the chromatin and not by the nucleolus. Ribosomal RNA (rRNA) is synthesized only subsequently.

The active mitotic divisions in the embryo during germination require the engenderment of several proteins, mainly enzymes, involved in growth and development. The initial demands are met with by the activation of pre–existing enzymes or by proteins synthesized by reliance on stored mRNAs and subsequently by those made on newly synthesized mRNA templates. Given the vast varieties of proteins required for seed germination, only a few are synthesized on stored templates. Protease, carboxypeptidase, malate synthase and isocitratase are some enzymes synthesized from pre–existing mRNA. Soon there is a slowing down of the use of stored mRNAs. There is also an increased recruitment of newly–

synthesized mRNAs for translation. A number of new polypeptides are synthesized shortly after imbibition of water by the seed as a consequence of new transcription. The appearance of unique mRNA populations in germinating embryos is to be considered as diagnostic of a differential gene expression program underlying germination. Although the distinction between proteins made on stored mRNAs and newly–synthesized mRNAs may seem arbitrary during the early phase of imbibition of the seed, it appears to be important in the progress of germination and the formation of the seedling that depends on new gene functions (Raghavan 2000).

Details on storage reserves in developing seeds were given earlier. These food reserves are mobilized during seed germination. The reserves are commonly found as storage forms of fats, proteins and starch in the embryos of dicots and in the endosperm of monocots. Their digestion into simple molecules is facilitated by hydrolytic enzymes such as lipases, proteases and amylases. Starch mobilization studies during germination are carried out in cereals that have endosperms of the most specialized type. These starchy reserves are found in the inner cells of the grain that are surrounded by 1–4 layers of aleurone cells. While the starch–filled inner cells are dead at maturity, the living aleurone cells, with the help of gibberellic acid (GA), hydrolyze the starch. GA regulates α–amylase genes and the α–amylase produced de novo by aleurone layer(s) is secreted into the starch–filled cells. The products of starch hydrolysis are translocated to the embryo to pomote its germination. The aleurone cells, induced by GA, produce several iso–forms of α–amylase, which can be categorized into a high iso–electric point group and a low iso–electric point group. GA also induces de novo synthesis of a host of other enzymes such as proteases, glucanases, phosphatases, deoxyribonuclease (DNase) and ribonuclease (RNase) by the aleurone cells during germination. A number of experiments have indicated that this stimulation of α–amylase activity is due to the production of new mRNA. The mRNA encoding the enzyme is made during the first few hours after the aleurone cells encounter the hormone (Raghavan 2000). There are evidences to show the transcriptional control by GA dedicated to the production of α–amylase in the aleurone cells. Aleurone cells also showed an increase in total α–amylase mRNA abundance in the presence of the hormone (GA), but not in its absence. Endosperm tissues treated with GA show increased activity of wall–degrading enzymes such as α–arabinofuranosidase and (1→3, 1→4) β–glucanase, which help in the dissolution of the endosperm and in the easy availability of nutrient materials to the germinating embryo. There are also evidences for the production of α–amylase by the scutellum in some cereals such as barley.

Growth Movements 8

8.1. INTRODUCTION

In the late 19th century, Charles Darwin and Francis Darwin (1880) recognized that plants, although rooted to the soil, exhibit a range of fascinating movements that often can be interpreted as adaptations to their environments. Of course, growth movements were known to people even earlier to Darwin, for example, to Sachs (1832–1897) who pioneered modern scientific methods to study plant movements. Sachs invented the **clinostat** (also spelled as **Klinostat**), which he used to study phototropism and gravitropism. Subsequent to these great people, several others contributed substantially to our knowledge on growth movements. This chapter summarizes the available knowledge on plant growth movements.

8.2. SOME BASIC CONCEPTS

Most plant movements fall into two natural categories:

1. **Tropisms or Tropistic movements**, in which the direction of environmental stimulus determines the direction of movement.
2. **Nastic movements** that are induced by external stimuli (sometimes interacting with an internal timing mechanism), but in nastic movements the direction of stimulus does not determine the direction of movement.

Both tropistic and nastic movements are often the result of differential and irreversible growths, but they can also be caused by differential and reversible uptake of water into special cells, called **motor cells**, that together form part of a pulvinus. Stems growing away from gravity or towards a light source are good examples of tropistic movement, while leaf movements during dusk and dawn, and stomatal opening and closing are examples of nastic movements. Some movements are examples of **taxis**, in which an organism or a structure such as a pollen tube moves towards or away from some stimulus such as light, warmth, cold or a chemical gradient. Most taxic movements are exhibited by lower plants.

The energy and mass parameters of the environment always influence plants. Changes in one or more of these parameters often induce either tropistic or nastic responses, or both. An environmental change that induces a plant to exhibit some kind of movement is called a **stimulus**. Often, a stimulus induces in the plant a process that continues even after the stimulus no longer exists in its initial form. This is the case with daily movements in leaves that change their course in response to the rising or setting of the Sun. Similarly, a stem continues to bend towards the light or away from the gravity for some time even after the light or gravitational stimulus has been removed (Salisbury and Ross 2004).

The stimulus always acts on some machinery that is part of the plant, and the plant part that receives or perceives the stimulus is called the **receptor**. Once the stimulus has been perceived by the receptor, it is changed (i.e., **transduced**) into some other form, often called a **signal**. The signal then leads to some **motor response** (growth or pulvinar action) that forms the actual cause of the plant movement (Hart 1990, Haupt and Feinleib 1979). Haupt and Feinleib (1979) have introduced and discussed the following three steps in plant movements (Salisbury and Ross 2004):

1. **Perception:** How does a plant or its part detect the stimulus emanating from the environment that causes the correct response? For example, what pigment absorbs the light involved in the expression of phototropic movement, or what in the cells or tissues of roots responds to gravitational stimulus? Where in the plant is the mechanism of perception located? And, what is that mechanism? All the above questions are difficult to answer as the plant organs such as leaves, stems and roots, unlike animal eyes and ears, are not known to be specialized to respond only to one stimulus.

2. **Transduction:** How does the mechanism of perception transduce the perceived stimulus to target cells in the concerned plant organ where movement occurs? What signal does it send? What biochemical or biophysical changes occur in response to such a stimulus? Researchers have suggested the active involvement of electrical and/or chemical messengers such as growth regulators in transduction.

3. **Response:** What actually happens during growth movement? How is the response carried out? Any hypothesis put forward in this connection to explain the mechanism of perception and transduction must also account for the observed motor responses during movement.

Two generalizations have emerged from the work on tropistic and nastic movements conducted during the past 3–4 decades: (i) Similar mechanisms within a plant often cause different responses. For example, the turgor changes in cell volume caused by K^+ in and out of stomatal cells during opening and closing of stomata as well as in leaf movements on touch in sensitive plants such as *Mimosa*. (ii) Different mechanisms can produce similar responses in different or even in the same plant, for example, the fact that different pigment systems are involved in the phototropism of different organisms.

8.3. NASTIC MOVEMENTS

Leaves, leaflets of compound leaves and floral leaves (especially petals) often exhibit nastic movements. An upward bending of these organs is called **hyponasty** and a downward bending of these organs is called **epinasty**. Often, leaf movements are caused by pulvini at the base of the petiole, lamina or leaflet, but they also occur in many plants without a pulvinus. Epinasty occurs, for instance, when cells at the tip of petiole or lamina, especially in the main veins, elongate irreversibly more than those on the bottom, while the opposite is true for hyponasty.

8.3.1. Nyctinasty

In many species, the leaf moves from a nearly horizontal position during the day to a nearly vertical position at night and this phenomenon has been recognized for a very long time. Typical nyctinastic movements are rhythmic events controlled by interactions between the environment and the **biological clock**. Bicompound pinnate leaves (e.g., *Albizia*, *Mimosa* and *Samanea*) (Fig. 249) are convenient for study of nyctinasty. These compound leaves often exhibit striking **sleep movements**. In *Albizia*, at night, the tips of opposite leaflets fold together to close, rise upward, and point toward the distal end of the rachilla. In *Samanea*, they fold down instead of up (Fig. 250).

In both cases, cells of pulvinus that swell during leaf opening are called **extensors**, whereas those that shrink are known as **flexors**. The pulvinus of *Samanea* is a straight cylinder during the day when the leaflets are open, but becomes a curved cylinder or hook–like at night when the leaflets are closed. Studies carried out in *Mimosa pudica* have shown that the leaflet closing is accompanied by loss of K^+ from ventral cells and absorption of K^+ by dorsal cells, although the ventral cells do not lose the K^+ to be absorbed by the dorsal cells. The reverse happens during leaflet opening.

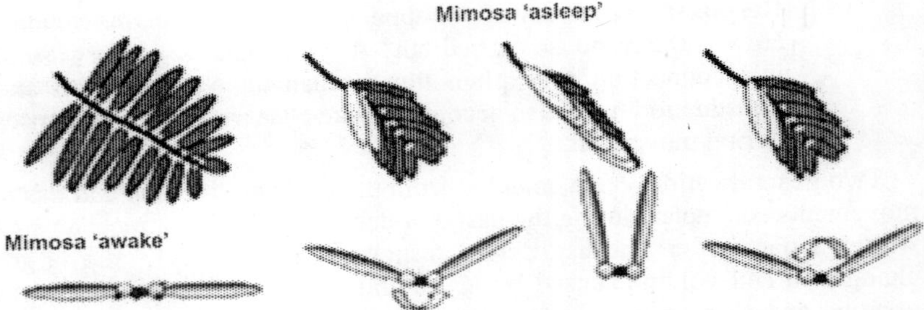

Fig. 249. Stages in sleep movements in *Mimosa* species in whole leaflets and their representation in transections of leaflets (Source: foothillsfancies.blogspot.com)

Fig. 250. Sleep movement of *Samanea* leaf; A - At noon; B - In the evening; Note the straight and bent shapes of the pair of secondary terminal pulvini

Changes in Cl^- are parallel to those changes observed in K^+. The same changes were recorded in *Samanea* also. Thus, in all three taxa, water movement occurs in response to the osmotic driving force of ion transport mentioned above, similar to what happens during stomatal opening and closing. In some manner, changes from light to dark or dark to light, mediated via phytohormone and a blue–absorbing pigment, activate these mechanisms, possibly through effects on the phosphatidylinositol cycle. It is interesting to note that light acts in opposite ways in the extensor and flexor cells. For example, the transition from white light to darkness activates the H^+ pump in flexor cells and inactivates the H^+ pump in extensor cells. Light may also indirectly act by resetting the biological clock. In experiments conducted with leaflet pairs from detached pinnae of *Mimosa,* it was evident that during day time closed leaflet pairs got opened up in response to blue or far–red light, but not to orange or red light. During night, they did not respond to light, but instead opened in response to applied IAA, NAA or 2.4–D. In some legume taxa, intense sunlight makes leaflets to fold in the upward direction, but at night in the downward direction.

8.3.2. Hydronasty

It is also known as **hygronasty**. **Hydronasty** involves a folding or rolling up of leaves, which occurs in response to water stress instead of light. These events reduce exposure of the leaf surface to dry air, supplementing stomatal closure to reduce transpirational loss of water and to minimize the dangers of photoinhibition. Folding and rolling movements are caused by loss of turgor in thin–walled motor cells called **bulliform cells** (Fig. 66). As their turgor pressure decreases, constant turgor in cells on the lower side of the leaf causes the folding of the leaf.

8.3.3. Thigmonasty

This refers to nastic movements resulting from touch. Thigmonasty is of widespread occurrence, although of especial occurrence in certain members of Mimosaceae. The most notable examples are species of *Mimosa*. The leaflets and leaves of this sensitive plant rapidly fold together upon being touched, shaken, heated, rapidly cooled, or subjected to an electric shock. Even when only one leaflet is touched or otherwise stimulated, the stimulus is transmitted throughout the plant making all leaflets to fold. The advantage of this movement to the plant is uncertain, but one suggestion is that the folding leaflets startle away insects before they begin to eat the foliage. The folding is caused by water transport out of motor cells in pulvini, an occurrence associated with an efflux of K^+(Salisbury and Ross 2004).

With reference to the transmission of signals in *Mimosa*, there is evidence for two distinct mechanisms, one *electrical* and the other *chemical*. The electrical response, first proposed in India by Bose in 1907 and then by Houwink (1935) in Holland, is as follows: An **action potential** is a change in voltage that forms a characteristic peak when plotted as a function of time. In *Mimosa,* the action potentials are similar to those occurring in animal nerve cells, but are comparatively much slower, and are caused by fluxes of specific ions across cell membranes. In *Mimosa,* they travel through parenchyma cells (connected by plasmodesmata) of xylem and phloem at velocities of up to about 2 cms^{-1}. The action potential will not pass from one leaflet to another through a pulvinus, unless the chemical response is also simultaneously elicited, in which case several leaflets may fold up. The chemical response was first reported by the Italian scientist Ricca in 1916 and was caused by a substance that moved through the xylem vessels along with the transpirational stream. This substance was originally called **Ricca's factor** after its discoverer, but now is identified as **turgorin**. Its movement elicits electrical responses that travel ahead of it from one leaflet to another in vascular parenchyma cells.

The other well–known examples of thigmonasty are the carnivorous plants such as the venus–fly trap and similar taxa. Here also, an action potential is obviously useful to the plant as an insect touches one or more sensory epidermal glandular hairs (Fig. 251). Action potentials move from the hairs into the leaf lobe tissues and cause the leaf to snap shut immediately. In some taxa, the hairs themselves bend down trapping the insect, again through action potentials in the hairs themselves. This usually traps the insect. The insect is then digested by enzymes secreted from the leaf providing a nitrogen and phosphate source for the plant. The rapid closing of leaves is an example of acid growth caused by rapid pumping of hydrogen ions into the walls of cells on the outside (lower side) of the leaf trap in response to the action potentials from the trigger hairs on the inside (upper side). This has been demonstrated very elegantly by India ink marking experiments by Williams and Bennett (1982), although later work had doubted the acid–growth hypothesis.

8.3.4. Thigmo–Morphogenesis and Seismo–Morphogenesis

The developmental effects of mechanical stimulation, especially of rubbing, on plants were started to be studied in the early 1970s. Most vascular plants studied responded to mechanical rubbing with slower stem elongation and increased stem diameter to result in short, stocky plants, sometimes only 40–60% as tall as controls (Fig. 252).

These and similar responses to mechanical stress were called by Jaffe (1973) **thigmomorphogenesis**. Although this is related to thigmonasty described earlier, rapid plant movements were not involved. Flexing and wind influence also cause thigmomorphogenetic responses. Similarly, shaking the plant periodically (in a shaker table, for example) for a few seconds to minutes once to many times a day causes a similar developmental effect, although there is no physical contact of the plant with the applied force. This phenomenon is called **seismomorphogenesis**. The thick, shorter plants produced under both conditions are not easily damaged by natural mechanical stress factors in comparison to their taller counterparts. What causes these developmental responses to mechanical stresses? A change in growth–regulator pattern was suspected, probably through increased ethylene, inhibition of auxin transport, disappearance of GA activity, and/or involvement of calcium. How a change in growth–regulator balance might come about is not known, but a change in membrane permeability that allowsions such as K^+ to pass rapidly from the symplast into the apoplast following mechanical stimulation may be the cause. This change in membrane permeability may, in turn, affect amounts of growth regulators at subcellular sites at which they act. In *Arabidopis*, it was found that mRNA increased 10–100% 30 minutes after the plant was mechanically stressed by spraying with water or by other means, and this plant becomes 50% less tall than the unstressed control plant. The mRNA was found to produce at least four and probably five genes that were activated by mechanical stress. One of those genes coded for calmodulin and two others coded for two closely–related proteins. Calcium levels increased in stressed cells, and so it is possible that calcium–calmodulin is a secondary messenger in the gene activation.

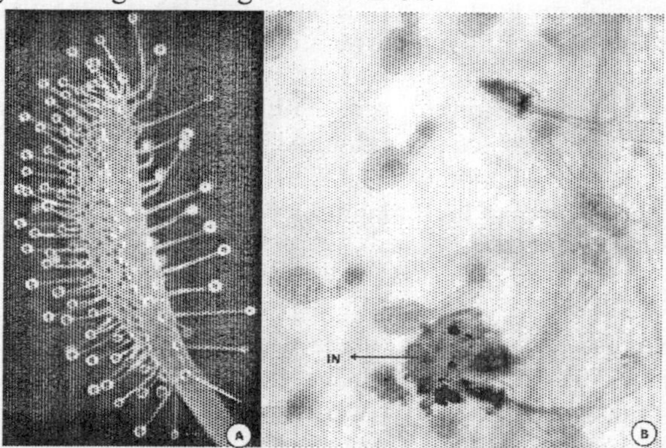

Fig. 251. Thigmonastic movement involved in the capturing of an insect (IN) by the glandular hairs of *Drosera* species

Fig. 252. The effect of number of daily rubs of the stem (once up and once down, between thumb and forefinger, with moderate pressure) on the growth of young bean plants. From left to right, the numbers of stimuli were 0, 2, 5, 6, 20 or 30 (Jaffe 1976)

8.4. TROPISMS

8.4.1. Introduction

As already indicated, tropisms involve differential (or unequal) growth (usually cell elongation) that occurs on different sides of an organ and at a particular direction in response to the direction of an environmental stimulus. For instance, roots grow downward and stems grow upward in response to gravity i.e., **gravitropism**. Earlier gravitropism was known by the term **geotropism** that means tropistic response to the Earth. Because this response is actually to a gravitational force, gravitropism is a more preferable term than geotropism. Stems and leaves frequently orient themselves with respect to light rays i.e., **phototropism**. Earlier it was known by the term **heliotropism** that means tropistic response to the Sun. Roots seldom exhibit phototropism. **Thigmotropism** is a response to contact with a solid object, as exhibited by climbing plants that grow around a pole or stem of another plant. There are other kinds of tropisms as well. All tropisms involve the three earlier postulated steps of perception, transduction and response.

8.4.2. Phototropism

Detailed studies seem to have started in right earnest by Darwin and Darwin (1880). They found that canary grass coleoptiles would not bend toward dim light if the tips were cut off or shaded. The use of coleoptiles as experimental materials was subsequently exploited by Frits Went, which led to the discovery of auxins. The Darwins also found that oak coleoptiles also bent towards light when their tips were covered i.e., some phototropic sensitivity occurs below the tips, but the **tip response** is about 1000 times more sensitive than the **base response**. Thus, if dim light is used, most response in the form of curvature towards light is localized in the tip of the coleoptile, while if higher light levels are used, bending begins almost simultaneously along the entire length of the coleoptile.

8.4.2.1. Perception

Dose–response relations

In this context, one of the primary aspects of importance is the question of **reciprocity** i.e., whether the response is proportional to the duration of exposure, to its **energy** or **photon flux** (energy or photons per unit area per unit time = **irradiance** or **fluence rate**, often loosely called **intensity**), or to the total dose, that is, the product of exposure duration times flux [called **fluence** (flux x time = fluence or flux = fluence / time)]? If the response is to total dose then flux and duration bear reciprocal relationship to each other, so the **law of reciprocity** is said to hold (Salisbury and Ross 2004). The degrees of phototropic curvature as a function of fluence are shown in Fig. 253. Such curves are called **dose–response curves** or **fluence–response curves**. In the Fig. 254, the first, ascending part of the curve is the same for all three photon fluxes (irradiances), so reciprocity, as mentioned above, holds over this low–fluence range. This part of the curve is called **first–positive curvature**. At higher irradiances (part of the curve labelled as C), meaning short exposure times, first–positive curvature is followed by decreasing curvatures with increasing fluence producing a bell–shaped curve. In a few monocots, such as oat coleoptiles, this trend continues until the organs bend *away* from the light, a response called **first–negative curvature.** With further increase in fluence, a **second–positive curvature** and even a **third–positive curvature** become evident. At the intermediate irradiances used, labelled as B in the Fig. 254, the descending part of the curve is greatly reduced, but second–positive curvature is still apparent. At the lowest light levels (longest exposure times, labelled as A in the Fig. 254), only a shoulder in the curve suggests a second–positive curvature. Thus, it is very clear that reciprocity holds only for the first–positive part of dose–response curves. When 24 different irradiance levels were used and dose–response curves analyzed, it was evident that the second–positive curvature occurs at the same time at all irradiance levels (about 40 min after the beginning of irradiance). Thus, as the irradiance levels decrease, first–positive curvature is delayed (reciprocity holds), but second–positive curvature comes at the same time regardless of irradiance level, so eventually the two come together, eliminating first–negative curvature.

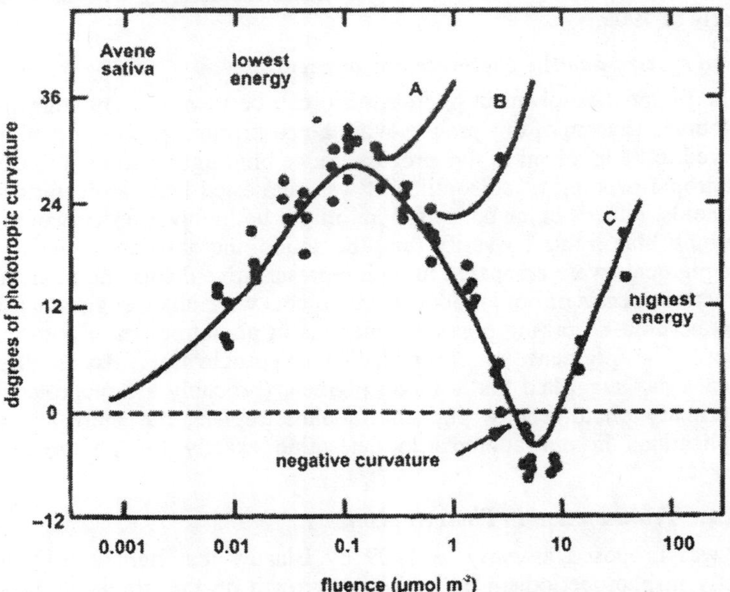

Fig. 253. Phototropic response of *Avena sativa* coleoptiles as caused by increasing fluences of unilateral blue light (436.8 nm) following red light; Note first positive, decreasing, and second positive curvatures. Light energy was held constant, but exposure times were varied to provide the total doses shown in abscissa. The irradiances were 0.014, 0.14 and 1.4μ mol m^{-2} S^{-1} for lines respectively labelled as A, B, and C; Points marked are actual data, but apply only to curve C (Zimmermann and Briggs 1963)

Based on theory and some experiments, it was suggested that light, in fact, has two effects in phototropism: (i) Light acts as a trigger for the bending response. (ii) Light decreases the sensitivity of the organ to subsequent light. This latter effect is non–directional and is referred to as a **tonic effect** (when two exposures are given, the tonic effect caused by the first exposure gradually decays away, so after about 20–25 min the exposure to the second irradiance is the same as if the first exposure had not been given; recovery has occurred). Hence, it was also concluded that the curvatures beyond the first–positive response are not really independent phenomena, but are the result of the desensitization of the first positive curvature system (Salisbury and Ross 2004).

Blue light is involved in coleoptile bending, while red light changes the sensitivity of the tissue to the blue light that actually causes the bending. So the phytochrome pigment system (activated by red light) plays a vital role in determining the sensitivity of the coleoptiles to the blue light that causes bending. The fact that blue light causes the bending was discovered as early as 1909.

Action spectra and the photoreceptor pigment

The pigment involved in phototropism can be identified by comparing the action spectrum of this process with the absorption spectra of pigments believed to be involved in this process. Since blue light is known to cause phototropic bending of coleoptiles, it was suggested that carotenoids and flavonoids, either one or both of them, might be involved in curvature by absorbing blue light. To verify this, the action and absorption spectra of these pigments were compared in both representative dicots and monocots. The action spectra of both monocots and dicots were alike, suggesting that the same blue–absorbing pigment functions in phototropism in both plant groups. The pigment has been called **cryptochrome**. Accumulating evidence has suggested that a flavin pigment (probably a flavoprotein) is the primary photoreceptor pigment in phototropism, but there are still complications in our attempts to determine exactly the photoreceptor pigment.

8.4.2.2. *Transduction in Phototropism*

It was proposed as early as 1909 by Blaauw that light acted rather directly in photoperiodism by inhibiting growth on the irradiated side of coleoptiles or stem through destruction of auxin. But, in 1926, it was demonstrated by Cholodny and Went independently that auxin is not destroyed but apparently migrates from the irradiated to the shaded side of a unilaterally–irradiated coleptile tip (Fig. 254).

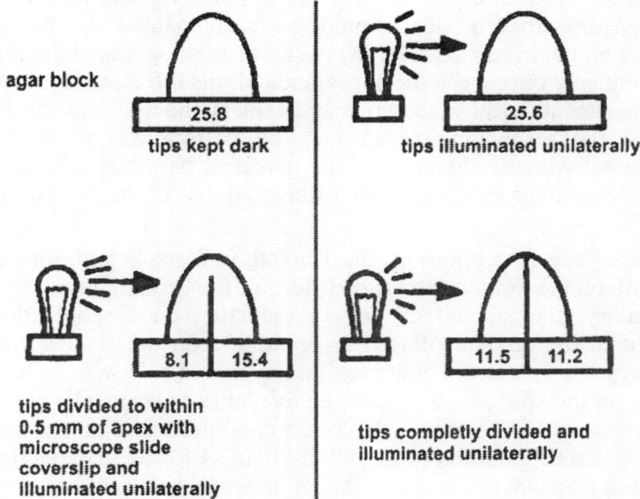

Fig. 254. Experiments to demonstrate that unilateral illumination of the coleoptiles tips of maize results in the transport of auxin from the lighted to the shaded side of the tips; Numbers on the agar blocks represent the degree of curvature caused by the application of the blocks unilaterally to decapitated oat coleoptile stumps. In the partially split tips, part of the auxin was laterally transported above the dividing barrier, but in the completely split tips this was not seen (Briggs 1963).

This figure shows the results of an experiment in which the amount of auxin transported into agar blocks at the base of a coleoptile tip is indicated by the degrees of curvature elicited when the agar blocks were placed on decapitated coleoptiles. This is the basis for the standard Avena curvature test for auxin. Coleoptile tips exposed to light exported as much auxin as those kept in the dark, suggesting that auxin destruction by light did not occur. When the agar and part of the coleoptile were divided with a thin piece of mica and the tip was exposed to light from one side, the amount of curvature was twice as great for the block under the shaded side as for that under the irradiated side. Division of the entire tip with the mica prevented lateral auxin transport in the tip, so auxin amounts collected from both sides were the same.

The **Cholodny–Went model** suggests that light illuminated on one side of the coleoptiles in some way causes a transport of auxin towards the shaded side, accounting for the basic transduction mechanism involved in phototropism. This was accepted in the next few decades, but was questioned again in 1980, for example, by Firn and Digby (1980). In the light of all information available till about the middle of 1980s, Briggs and Baskin (1988) enlisted the following four criteria to test the Cholodny-

Went model: (i) In an organ curving due to phototropism, acceleration of growth on the shaded side should be accompanied by the growth retardation on the irradiated side. (ii) A lateral auxin–gradient development must accompany or precede the appearance of this differential growth. (iii) It must be unequivocally shown that auxin is indeed a factor limiting growth in the responding organ i.e., the coleoptile. (iv) It must be demonstrated without doubt that the auxin differential established is sufficient to account for the growth differential that is observed on the two sides.

Regarding the first criterion, the Cholodny–Went hypothesis suggests that growth on the lighted and shaded sides of the bending organ should be compensatory: the amount of growth reduction on the lighted side is equivalent to the amount of growth promotion on the shaded side. The Blaauw hypothesis stresses inhibition on the lighted side with no (or some) inhibition on the shaded side. Based on available experimental results, the conclusion seems to be that under some conditions and in some species there is a general inhibition of growth by light as it was first suggested by Blaauw, but that superimposed on this or under some conditions, instead of this, there is a compensatory growth possibly caused by transport of auxin (or an inhibitor?), as suggested by the Cholodny–Went model. Regarding the second and third criteria, several studies on transport of radiolabelled IAA have demonstrated that both the direction and rate of its transport in the coleoptiles is definitely influenced by light, though the results obtained were often controversial. But, in monocot coleoptiles, evidence for the Cholodny–Went model is quite strong. There is good evidence that a different mechanism acts in at least some dicots, where no auxin gradient across a phototropically stimulated hypocotyl exists. It was, however, demonstrated that a gradient of an inhibitor, **xanthoxin**, existed across the hypocotyl, with 69%–70% of this inhibitor on the lighted side was demonstrated. Gradients in three such inhibitors, which were isolated and characterized, were also shown in phototropically–stimulated radish hypocotyl. It was subsequently suggested that an inhibitor (or two unidentified inhibitors in some) is found along with the auxin in greater amount on the illuminated side. Thus, in spite of evidences in favour of Cholodny–Went hypothesis mentioned for the monocots, the results obtained also strongly support an inhibitor model for dicots, as suggested by Blaauw even for grass coleoptiles.

8.4.3. Solar Tracking

Solar tracking is found in many plants. In this process, the flat leaf blade remains nearly at right angles to the Sun throughout the day, maximizing the light harvested by the leaf. This was studied in detail in desert plants such as desert lupine, as also in some plants of other ecosystems. Solar tracking is a true tropistic movement, since the orientation of the leaf blade is determined by the direction of the Sun's rays, but it is neither positive nor negative as is stem phototropism. It is **diaphototropism** (or **diaheliotropism**) as the leaf orients itself at right angles to the light rays. Leaf blade orientation is controlled by motor cells in a pulvinus where the lamina joins the petiole. The movement of water in and out of these motor cells is reversible and is controlled by osmotic changes effected by ions such as K^+. At sunset, the lamina is almost vertical and its tip faces the Western horizon where the Sun sets, but after one or two hours, the lamina assumes a resting position at right angles to the petiole, only to start moving again one to two hours before sunrise. At sunrise, the lamina faces the eastern horizon. In other words, the leaf orientation keeps track of the Sun and its movement. If plants are left in total darkness for several days, they continue to reorient their leaves each 24 hours to the point where the sun first appeared the last time it irradiated the plants (Salisbury and Ross 2004). Solar tracking is also the most sensitive to blue light, but more details on action spectra have not been obtained. Cells along the major veins in the flattened lamina are said to detect the direction of solar rays and send a message to the motor cells in the pulvinus leading to the reorientation of the lamina. There is some evidence that auxin may transmit the message. However, the situation is not that simple because the message can be either positive or negative, depending on the direction of the light with respect to the polarity of the vein cells. If light rays fall on a major vein cells at right angles, the message is neutral. If the rays are directed toward the tip of the leaf, the exported message is positive, so the cells in the sector of the pulvinus, associated with the vein, absorb water, causing the lamina to rise toward the beam of rays. On the other hand, if the rays are directed towards the base of the vein, a negative signal is sent from the major vein to the same sector of motor cells at its base, causing them to lose water and the lamina to decline toward the beam of rays.

8.4.4. Skototropism

When a seed of a climber or liana such as *Monstera gigantea* in the tropical rain forest germinates on the floor of the fairly dark forest, how does the developing seedling find a support to aid its climb into the light? Such a seedling grows towards the darkness of the nearest support trunk. It can be considered akin to negative phototropism, with the response towards the darkness rather than away from the light. Experiments done on seedlings showed that the seedlings invariably grew towards a dark spot, rather than away from a lamp. The phenomenon was named **skototropism**, i.e., a directional response to the darkness. Once the seedling reaches the tree trunk, it becomes positively phototropic.

8.4.5. Gravitropism

Growth movements towards or away from the Earth's gravitational pull is called (respectively **positive** and **negative**) **gravitropism.** Roots are usually positively gravitropic and primary roots are generally oriented more vertically than secondary (lateral) roots. The secondary roots sometimes grow at, rather constant, nearly horizontal angle. Tertiary roots and roots of higher order may hardly be gravitropic and may grow in random directions. This is necessary for the root system to explore the soil more thoroughly. Stems and flower stalks are typically **negatively gravitropic**, but the response is highly variable, depending on the plant. The main stem or tree trunk grows 180° away from the Earth's centre of gravity, but branches, petioles, rhizomes, stolons etc. are usually more horizontal. Vertical growth of main stem and tap root due to gravitropism is called **orthogravitropism,** while more horizontal growth is **diagravitropism**. The growth of an organ at any determined angle (α) to the vertical ($0° < \alpha < 180°$) is called **plagiotropism** or **plagiogravitropism** (meaning 'oblique'). Leaves and branches of a number of plants are plagiotropic. Organs with no response to gravity are **agravitropic**.

8.4.5.1. Gravitropism in Roots

Most studies on gravitropism have been done on roots. The downward growth of primary roots of most plants results from three events: (a) Perception of gravity (b) Transformation of gravity into a physiological stimulus (c) A stimulus–directed response characterized by differential growth that produces downward curvature (Moore 1994).

Perception of gravity

Charles Darwin and his son Francis (1880) were among the first botanists to study root gravitropism. One of their earliest experiments and also those of Ciesielski in Poland made in 1872 demonstrated that removing the root cap rendered roots non–responsive to gravity. These experiments suggest that the root cap is necessary for gravi–irresponsiveness.

In *Zea mays* and in many other plants, the primary root has at the base of the root cap a **calyptrogen** or root cap meristem (Refer to Chapter 5). Cell divisions in the calyptrogen continually push cells into the cap, where they sequentially differentiate into columella, which forms the central part of the cap, and into peripheral cells. Each columella cell contains 15–30 amyloplasts (each with two or more starch grains) that sediment to the lower sides of the cells (Fig. 255).

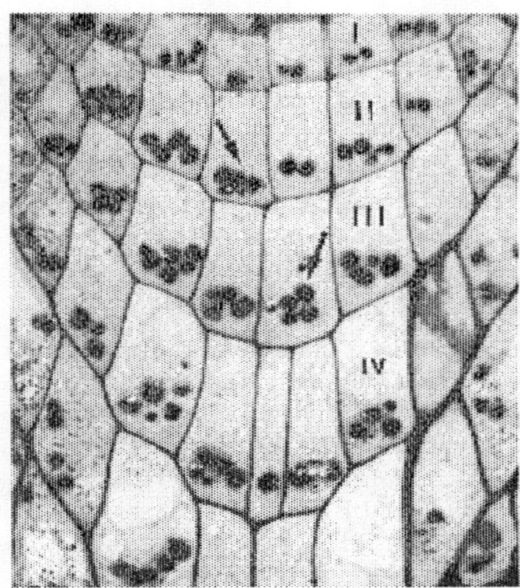

Fig. 255. Root cap of the Cress root showing Statoliths (indicated by arrows) (Volkmann and Tewinkel 1997)

The sedimentation of amyloplasts in columella cells and their likely interactions with other cell structures such as the ER and/or plasmalemma are believed by many to underlie how roots perceive gravity (Moore 1994). This concept is often called the **Statolith theory.** As per this concept, the amyloplasts are the statoliths that are located in **statocytes** (cells with statoliths). This conclusion is based on the following observations:

1. Amyloplasts gets sedimented in columella cells just before the beginning of root gravicurvature (Audus 1979).
2. It has long been known that the so–called **presentation time** (i.e., the minimum time required to elicit a gravitropic response) is closely correlated with the rate at which amyloplasts settle at the base of columella cells (about 12–20 s in *Lepidium*).
3. Removing the root cap makes roots non–receptive to gravity (Jackson and Barlow 1981). Replacing the detached root cap once again restores graviresponsiveness of the root and this recovery by decapped roots usually occurs within 24 h after decapping and positively correlates with sedimentation of amyloplasts in their tip cells (Hillman and Wilkins 1982). However, decapped roots regain graviresponsiveness long before a new root cap is regenerated from calyptrogens, thus, indicating that a root cap per se is not necessary for graviresponsiveness.

4. Treatments that dissolve starch from amyloplasts make root non–responsive to gravity. These roots regain graviresponsiveness when they regenerate starch in their tips (Iversen 1969).
5. If roots (or coleoptiles) are treated with gibberellins and kinetin at elevated temperatures, all starch in the amyloplasts disappears, as does the response to gravity.
6. Maize mutants with smaller amyloplasts than the wild type are less sensitive to gravity.

Although the above correlations are convincing, correlation is not causality (Moore 1994). In fact, several recent observations have questioned the importance of columella amyloplasts in gravity perception in roots. For instance, it has been shown that, sedimentation of amyloplasts in columella cells does not ensure that a root responds to gravity (Ransom and Moore 1983, Olsen and Iversen 1980, Moore 1985 a, b). It has also been reported that the roots of TC7 *Arabidopsis* mutant contain no starch or amyloplasts, but yet respond to gravity (Caspar and Pickard 1989), indicating that neither amyloplasts nor starch are important for root graviresponsiveness and hence, in perception of gravity. However, other workers reported that gravitropic response was present in these mutants, but was significantly slower due to the fact that starchless plastids settle in the statocytes more slowly than amyloplasts of the wild type. Hence, starch may not be necessary for graviperception, but might aid perception when present, and settling of plastids may be necessary. Saether and Iversen (1991), however, claim that plastids in roots of TC7 mutant do contain starches and sediment in response to gravity.

To settle, a statolith must have a density significantly greater than the medium in which it is suspended (Salisbury and Ross 2004). Starch has a density of 1.3, which is greater than that of the cytosol that is close to 1.0. Densities of other organelles in the cell are often very close to that of the cytosol, so they are not expected to settle. There is evidence that some organelles besides amyloplasts or starchless plastids do settle, but the rates do not correlate with presentation time. Though an organelle such as a ribosome is dense enough, it will not settle if it is too small. It will remain in suspension because of its Brownian movement. Thus, amyloplasts or starchless plastids are the most likely candidates for the role of statoliths.

Recent evidence suggest that (a) integrin–like proteins present in the cell or extracellular matrix junction may be required for gravisensing by roots (Wayne 1992), and (b) roots respond to gravity by sensing gravitational pressures exerted by the protoplast and not by sedimentation of amyloplasts (Lee et al 1992).

Response to gravity and differential growth

Although it is likely that the root cap is the region of perception of gravity, the downward curvature characteristic of root gravitropism occurs in the elongating zone, which is 2–4 mm behind the cap. If the roots are oriented horizontally, growth along the lower side of the root is reduced causing the root to curve down, usually within 10–30 min after reorientation. Some signal that inhibits growth presumably moves from the cap to the elongating zone of the root. Some studies have suggested that this signal is an inhibitor. The following observations support this suggestion:

1. Decapped roots elongate faster than intact roots and fixing a second cap to an intact root decreases elongation.
2. Inserting impermeable barriers such as a mica plate between the root cap and the elongating zone causes roots to curve away from the side of the root into which the mica strip was inserted.
3. Removing half of the root cap causes roots to curve toward that side of the root with the half cap, irrespective of the root's orientation.

Some studies originally suggested that this inhibitor might be ABA (Pilet and Chanson 1981); however, more recent observations, as indicated below, show that this suggestion was wrong (Moore 1994):

1. ABA initially stimulates root growth over a wide range of concentrations, but it inhibits growth only at high concentrations after prolonged periods (Mulkey and Evans 1981).
2. Mutants unable to synthesis ABA and also chemically–treated roots having undetectable amounts of ABA are strongly graviresponsive.
3. If gravicurvature of root is due to a gradient of ABA, one would expect gravicurvature to be reduced or eliminated when roots are submerged in a solution of ABA. Such treatments do not significantly reduce gravitropism.

Hence, ABA is probably not the inhibitor that controls root gravitropism. Rather the following observations suggest that IAA acts as the inhibitor in gravitropism. Probably IAA acts through its promotion of acid growth (Moore 1994):

1. Exogenously applied IAA inhibits root growth at concentrations 100–1000 times lower than those at which ABA is effective.
2. IAA applied to the upper side of the elongating zone of horizontally–oriented roots moves to the lower side of the elongating zone.
3. Artificially induced asymmetric distribution of IAA across root tips induces gravitropic–like curvature of the root.

4. Gravicurvature of the root is preceded by an asymmetric efflux of IAA that is strongest along the upper rapidly growing side of the elongating zone. Inhibitors of auxin–induced proton secretion (E.g., orthovanadate) and auxin transport abolish asymmetric IAA efflux and gravicurvature.

Therefore, auxin probably controls the differential growth that causes gravitropism by acting as an inhibitor. However, the importance of other substances acting as inhibitors cannot be completely ruled out.

Transforming the gravitational signal

Our poor understanding of how roots perceive gravity has also hindered our understanding of how roots transform gravity into a physiological response. Perception of gravity by roots, as discussed earlier, is presumably dependent on movement or transduction of one or more statoliths, which, in turn, initiates the physiological changes that finally cause gravicurvature. It has been suggested that transduction of the gravitational signal involves amyloplasts contacting the plasmalemma or distal complex of the ER of the columella cells (Hensel 1984). These interactions then somehow influence production and/or transport of growth regulators responsible for gravicurvature. Although impressive, these models are probably not generally valid for root gravitropism, since (i) sedimented amyloplasts, seldom, touch the plasmalemma of columella cells, (ii) columella cells of most roots examined so far do not have a 'distal complex' of ER, and (iii) as already stated, the role of amyloplasts as statoliths is questionable.

Not all models for transmission of the gravitational signals are based on amyloplast movement and the resulting cellular interactions. As indicated earlier, physical pressure exerted by the weight of the protoplast rather than movement or pressure exerted by the amyloplasts per se can affect gravitropism by altering transport properties of membranes. However, it was stated by Audus (1962) that this force would be insignificant as compared to the osmotic force of the cytoplasm, and hence, it could not be vital for sensing gravity. Wilkins (1978) considered that electrical charges on amyloplasts could induce an electrical polarity that, in turn, could affect permeability and transport across the plasma membrane. Interestingly, amyloplasts have an electrical potential of approximately -19mV. This hypothesis cannot explain graviresponsiveness by starchless (and amyloplastless) mutants of *Arabidopsis*.

Transduction of the gravitational signal into differential growth probably involves some type of communication between the cap and elongation zone, since the signal is perceived in the former and the gravicurvature takes place in the latter. Studies have indicated that a root's perception of and response to gravity are closely linked by free and mobile calcium (Ca^{2+}), as shown by the following examples (Moore 1994):

1. Application of Ca^{2+} asymmetrically to root tips results in curvature toward the side of the root to which Ca^{2+} was applied (Lee et al 1983 a, b).
2. Calcium–deficient roots do not show graviresponsiveness. Also, chelating Ca^{2+} in root tips with EDTA makes roots non–responsive to gravity. Graviresponsiveness is once again seen when EDTA is replaced with Ca^{2+} (Lee et al 1983 a, b).
3. Calcium moves downward across tips of horizontally–oriented roots, but such a polar transport is not seen across decapped roots (Lee et al 1983 a). Such a downward transport of Ca^{2+} is not a consequence of gravicurvature because a gradient of endogenous Ca^{2+} forms even before the onset of gravicurvature (Moore 1994).
4. Inhibitors of Ca^{2+} translocation such as 2, 3, 5–triiodobenzoic acid (TIBA) also disrupt auxin transport and gravitropism (Lee et al 1984)

The above observations indicate that Ca^{2+} moves to the lower side of tips of horizontally–oriented roots, and the resulting gradient of Ca^{2+} somehow influences gravicurvature. The nature of this influence is unknown, but probably involves activation of calmodulin, a small calcium–binding protein that activates many enzymes vital to cell function. This conclusion is based on the observations mentioned below (Bjorkman and Leopold 1987, Hasenstein and Evans 1988, Moore 1994):

1. Calmodulin is abundant in root tips, and inhibitors of calmodulin suppress gravitropism.
2. Aluminium (Al^{3+}) binds to calmodulin and inhibits its activation by calcium. Correspondingly, unilateral application of Al^{3+} to root tips has an effect opposite to that of Ca^{2+} i.e., roots grow away from the Al^{3+}.
3. Applying Al^{3+} unilaterally to a root cap stimulates movement of auxin to the concave side of the elongating zone of a curving root.
4. In corn roots requiring red light for gravisensitivity, mRNA coding for calmodulin increases within 5 min after exposure to red light.

Calmodulin may activate Ca^{2+} or auxin 'pumps' in the plasma membrane, which can produce apoplastic gradients in the root tip that affect gravicurvature (Moore 1994). We do not know what controls the accumulation of Ca^{2+} and/or other effectors along the lower side of the root tip during gravitropic response. Electrical asymmetries may be involved and these asymmetries may redistribute calcium (Ca^{2+}) and other ions across root tips.

Peripheral and epidermal cells of root tips produce large amounts of mucigel, a hydrated acidic polysaccharide containing sugars, organic acids, vitamins, enzymes, and amino acids. Mucigel has several important functions (such as protection, lubrication, absorption of water and nutrients, and attracting symbiotic microbes), one of which may be to provide a route for movement of gravitropic effectors. This conclusion is based on and/or is consistent with the following observations (Moore 1994):

1. Calcium is abundant in mucigel, especially on lower side of horizontally–oriented roots. This calcium enhances auxin transport.
2. Calcium applied to mucigel along the upper surface of root can be collected from the mucigel covering the lower surface of roots. This movement is likely to be apoplastic.
3. Application of mucigel asymmetrically to root tips induces gravitropic–like curvature; roots that naturally lack mucigel are non–responsive to gravity; continuous washing of mucigel from roots significantly reduces gravicurvature; adding mucigel to roots lacking it promotes gravitropic response in such roots.
4. The onset of graviresponsiveness by decapped roots is preceded by production and accumulation of mucigel at their tips.

Taken together, all these results suggest that gravitropic effectors moving from the root cap to the elongating zone move at least partially through the apoplast. This is consistent with Clarkson's (1984) suggestion that calcium moves out of cells into the apoplast during gravitropism.

Several reports suggest that growth and gravitropism by roots are controlled by the outer–cell layers of the root (Iwami and Masuda 1974, Nelson and Evans 1986). This implies that gravitropic signal movement between columella and elongating zone of the root might take place through the apoplastic domains of these outer layers. However, this observation is proved to be not applicable to all plants. Perhaps the stelar region acts as a conduit for the transport of gravitropic signal from root tip to the zone of elongation.

Calcium and auxin

A model for root gravitropism: Moore (1994) suggested the following model for root gravitropism based on an analysis of all the previous literature:

1. In vertically–oriented roots, auxin moves through the stele towards the cap, where it is distributed to the cortical and epidermal tissues. This acropetal flow of auxin through the stele accounts for the large amounts of IAA present in root caps. In tips of vertically–oriented roots, pumps and symmetrical electrical currents distribute Ca^{2+} symmetrically in the apoplast.

2. If the roots are horizontally oriented, electrical asymmetries favouring movement of Ca^{2+} back and downward through the apoplast are induced resulting in Ca^{2+} accumulation along the lower side of the root tip.
3. This accumulation of Ca^{2+} along the lower side of the root tip activates calmodulin, which, in turn, activates pumps that redirect auxin towards the lower side of the root tip. Calcium may also sensitize epidermal, cortical, and/or endodermal cells to auxin.
4. The increased amount of auxin along the lower side of the root inhibits cellular elongation, thereby causing downward curvature of the root.

One important detail complicates our understanding of the gravitropic response of roots. In certain cultivars of maize, radish and a few other species, the nature of the gravitropic response depends on light. Roots of seedlings, growing in the dark, are diagravitropic, but when exposed to light these roots become positively orthogravitropic and bend downward. Red wavelengths are most effective and the phytochrome system has been implicated.

Gravitropism in non–graviresponsive roots

Almost all research regarding root gravitropism has centred on the vertically–growing primary or tap roots of a few cultivated species. However, most roots, particularly lateral and secondary ones, do not grow straight down, but grow at various angles relative to gravity, in order to more efficiently explore the soil for water and minerals (Moore 1994). Columella cells of such non–graviresponsive roots also contain sedimented amyloplasts as in graviresponsive roots. Hence, they should be perceiving gravity if amyloplasts are, in fact, doing this. Tips of lateral roots placed asymmetrically on detipped primary roots induce gravitropic–like curvature, suggesting that these roots can produce gravitropic effectors. Then why they do not respond to gravity and do not grow vertically? It is suggested by Moore (1994) that these roots do not respond to gravity because they do not establish significant gradients of gravitropic effectors, probably due to reduced columella tissue. In fact, if the columella tissues of these lateral roots are of sufficient volume, they grow vertically down. Hence, there is a distinct correlation between the size of columella and the degree of graviresponse.

8.4.5.2. *Gravitropic Response in Stems and Coleoptiles*

Studies on gravitropic responses of horizontally–oriented coleoptiles, hypocotyls and stems provide results strikingly similar to those observed with high irradiances in phototropic bending. The top (upper) surface of these structures ceases growth almost immediately, while growth of the bottom (lower) surface speeds up or sometimes continues at about the same rate or even slower. Most plant physiologists have assumed that stem bending occurs because auxin moves to the lower tissues, promoting their growth, as per Cholodny–Went model. But, there are complications that cannot be explained by this model.

The response and the mechanism of bending

As known even in late 1880s, if a growing stem is turned on its side and restrained in such a way that it cannot bend upward, it does bend some hours to days later after the restraint is removed, often to more than 90° and within 1 to 10 seconds (Salisbury and Ross 2004). Cells on the bottom of such a restrained stem elongate, but not as much as if the stems were not restrained. Because the stem is held straight and cells are not dividing, cells located at the top side are stretched. During restraining, the diameters of bottom cells increase, whereas those of top cells decrease; upon release from restraint, top cells shorten and become thicker, while those on the bottom elongate somewhat and become thinner. This results in the rapid bending.

In a horizontal stem, cell growth on top does not merely virtually stop; those cells will not grow even if they are stretched by elongation of bottom tissues. In such a stem, pressure in bottom tissues increases while growth continues, and growth stops in top cells in spite of decreasing pressure. Top cells return nearly to their original dimensions when the stem is released from restraint resulting in bending.

Perception

In stems, coleoptiles and hypocotyls, the sites of gravity perception and response are the same. Some people have incorrectly considered the stem or coleptile tip to be the site of gravity perception in the stem, as the root cap is the site of perception in roots (Salisbury and Ross 2004). Even after much stem portion above its rapidly growing region has been removed, that region continues to bend upward in response to gravity when the stem is kept horizontally. Like in roots, amyloplasts are thought to be the statoliths in stems. In several angiosperms, amyloplasts are confined to one or two inner layers of the cortex (wrongly considered by some as endodermis) called the **starch sheath,** just outside the vascular bundles. Again, like in roots, the evidence that these amyloplasts in stems play the role of statoliths is only strongly suggestive but not conclusive. Amyloplasts have been demonstrated to settle in the starch-sheath cells of stems, and all gravitropically responsive stems so far studied do have amyloplasts.

Transduction

The application of the Cholodny–Went hypothesis to explain gravitropism in stems and coleoptiles has been questioned by several workers. Considering the rapid cessation of growth on the upper side of a horizontally–laid stem compared with the normal growth rates on the bottom side of it, it is difficult to imagine that changes in auxin concentration can occur rapidly enough or reach values capable of accounting for these differences. Growth rates on the upper and bottom sides of a horizontal stem can differ by a factor of 10 or more, especially when parts of the upper surface do not grow at all or even show considerable shrinkage. Even if we assume that auxin concentration is in complete control, upper cells will have to be almost completely depleted of auxin within a very short time after the stem gets horizontally disposed. Many workers have found only minimal or even no gradients in auxin concentration or auxin transport across gravitropically stimulated stems, especially in dicots.

Even though some workers have questioned the Cholodny–Went hypothesis, others have defended it. Yet others have suggested a modified and elegant Cholodny–Went model (Salisbury and Ross 2004). It has been proposed that dicot stem tissues differ in their sensitivity to auxin. The epidermis is considered to be more auxin–responsive in contrast to subepidermal tissues that are less responsive. During gravitropism, auxin transport might be limited in distance, with auxin moving out of responsive epidermal cells on the upper side of the stem into the subjacent cortex and then from non-responsive cortex of the bottom side into epidermis on the lower side. In such a case, it is technically very difficult to measure auxin gradients. Such gradients may not appear at all when a horizontal stem is split into upper and lower halves and auxin level is measured in each half stem.

Even if auxin gradients do not always account for gravitropic bending of stems, it is clear that such a bending does not occur when there is insufficient auxin. Brauner and Hager (1958), in their experiment, removed the tips of sunflower hypocotyls (likely to be a source of auxin) and then, after four days, gravistimulated the hypocotyls by disposing them in a horizontal position. Bending did not occur. However, if the hypocotyls were returned to the vertical position and the cut–stump region supplied with IAA, the hypocotyls bent in the predicted direction. This phenomenon was called by them **gravitropic memory** or **mneme**. The gravitropic bending of maize roots was eliminated by treatment with several auxin transport inhibitors, thus providing strong support for the Cholodny–Went hypothesis.

There have been studies on stem gravitropism with growth regulators other than auxin. GAs, for example, occur in high concentrations on the bottom of horizontally–disposed and gravitropically–stimulated stems. Sometimes, significant gradients were absent until after some bending has occurred. However, such gradients in GAs were recorded in leaf sheath pulvini of cereal grasses during bending. In some experiments, ethylene inhibitors were shown to inhibit gravitropic bending proving its importance in this process. But, ethylene has been shown to have no role in the gravitropic bending of some other plants. Ethylene probably acts by inhibiting growth on the top of the stem placed horizontally. Yet when ethylene is measured in the tissues, it is found to increase at about the same rate as gravitropic bending, but in the *bottom* tissues instead of in the top tissues. Calcium ions (Ca^{2+}) also apparently play an important role in shoot gravitropism. Ca^{2+} concentrations have been observed to be higher on the tops of horizontally–disposed stems and Ca^{2+} is known to inhibit cell elongation, perhaps by overcoming auxin effects.

Much more seems to be involved in transduction in gravitropic stem bending than a simple transport of auxin. To explain this observation, an alternative **changing sensitivity hypothesis** has been proposed. According to this hypothesis, gravistimulation can lead to a change in tissue sensitivity to auxin. If tissues on the lower side of a horizontally–oriented stem become more sensitive to auxin than those on the upper side, the lower side will then respond more to the auxin present in the stem, leading to more growth and upward bending whether or not auxin concentrations in the stem is changed. Most plant physiologists, who have worked on stem gravitropism, have failed to consider this possibility. Brauner and Hager's (1958) work on gravitropic memory was the starting point for this hypothesis. According to these workers, sensitivity to auxin changed during gravistimulation of decapitated or cold–treated sunflower hypocotyls, and hence, their tissues responded differently to the auxin that was applied after they were returned to the vertical posture. These authors found several pieces of evidence suggesting that sensitivity to auxin did change in response to gravitropism (Salisbury and Ross 1992). The 'changing sensitivity to auxin' hypothesis can be in addition to or instead of an auxin–transport mechanism to explain gravitropism in stems.

It is also necessary here to explain the results obtained in clinostat experiments. These simulate weightlessness for a plant by rotating it slowly about a horizontal axis. If the graviperception time exceeds the rotation time then the vector of gravitational stimulation would add algebraically to zero, as though the plant being so rotated were truly weightless (Salisbury and Ross 2004). Now, space exploration programmes provide orbiting laboratories, and hence, it is possible at present to compare the early clinostat–based results with those obtained by allowing plants to grow in true microgravity conditions (milligravity condition will in fact be more appropriate) (Refer to Box 8.1). The most notable response of plants on a clinostat is epinasty of leaves. It was suggested that this response may not be due to weightlessness, but due to mechanical stresses, which in turn lead to the production of ethylene and leaf epinasty. This explanation for clinostat–induced epinasty is not correct as evident from data obtained in space–shuttle laboratory experiments, where epinasty was seen, but there was obviously no mechanical stress.

BOX 8.1. PLANT EXPERIMENTS IN SPACE

Space provides a unique environment that can be used for plant experimentation and such experiments are important because plants are vital to life on the Earth and may also become vital for future space missions. There are at least two broad areas where such experiments have been undertaken. One relates to gravitropism and other to the interplay between cell cycle events and environment. In gravitational research, space provides an environment in which gravity can be manipulated below the Earth norm of 1 to a state comparable to weightlessness. Controlled experiments on seeds, seedlings and herbaceous plants such *Tradescantia*, oats, sunflower, wheat and daylily as well as embryoids have been conducted in space for the past almost 65 yrs, using both manned and unmanned spacecrafts of both US and former USSR. Two specific questions were considered in such experiments on gravitropism: (i) What are the fundamental effects, if any, of the near zero G space environment on plants? (ii) How can this zero G be used to understand aspects of more general importance to plant biology, such as metabolism, growth and development. The results of near zero gravity experiments on gravitropism have been summarized in this chapter.

The other aspect studied was the chromosomal and nuclear abnormalities frequently encountered in cells of various plants exposed to space due to direct and indirect effects. These effects are not due to weightlessness alone, but are due to a combination of factors including "the biological status of systems and the way in which they experience multiple stresses" such as radiations, weightlessness, vibration, and acceleration. The extent, to which space–specific changes become manifest, is dependent on the extent of pre–existing stresses in the system. It is also evident from space experiments that the more completely developed a system, the less likely it is to show cell-related abnormalities during growth; the less morphologically complex, the more are the abnormalities.

Plants in space grow initially in all directions as there is no real upside or downward direction, but they soon get adapted to grow in a more stable direction. Because of the weightless conditions, plant experiments are stored in sealed containers (Fig. 256), as otherwise soil and water will float around freely. These containers also provide the plants with right levels of gases, water, light and temperature.

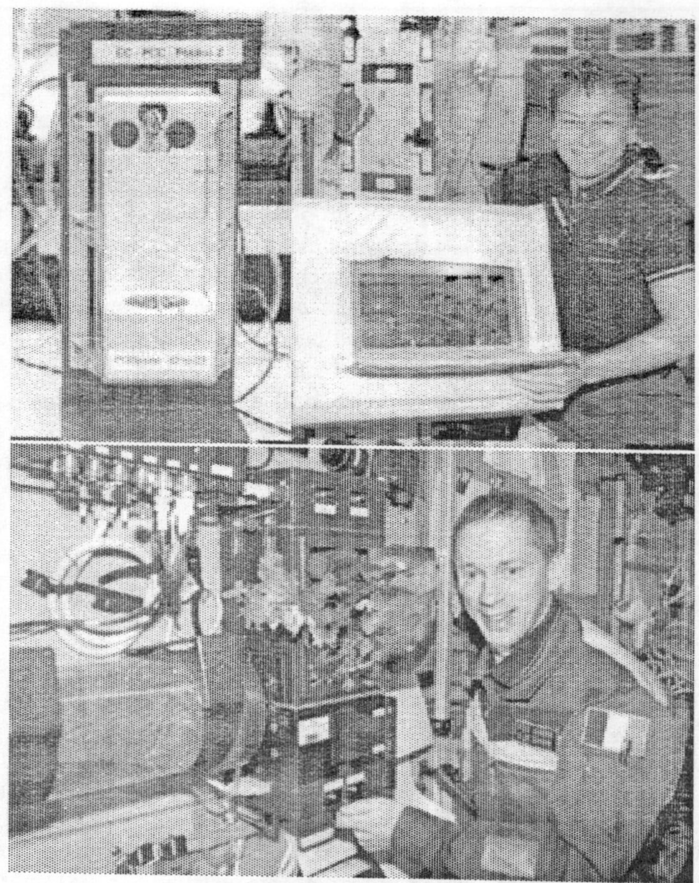

Fig. 256. European Space Agency pictures
See Dutcher et al 1994, Krikorian 1998 for more information

Attention should also be focused on observations made on false pulvini of grasses during stem gravitropism. In the Panicoid grass stems, the gravity–sensing tissue is located in a swollen basal internodal region near the node. It can respond to gravity if it is not fully mature. In some grasses, the base of the leaf sheath can also sense gravity. In the Festucoid grasses, only the leaf–sheath base possesses this gravisensing tissue. This tissue is often called a pulvinus, although it is only a **false pulvinus** as it acts by differential growth and not by reversible changes in cell volume, as do the true pulvini that control nyctinastic leaf movements. The false pulvinus of grass is a very specialized organ of meristematic and maturing cells that both detects and responds to gravity (Dayanandan et al 1977). As its cells mature, they become parenchyma, collenchymas and a few vascular tissue cells. Response to gravity consists of differential cell elongation; the lowermost cells undergo elongation, while the uppermost cells do not elongate (Fig. 257). Continued differential growth of the pulvinus can bring the shoot to a more or less vertical position, usually after 48–72 h. Starch statoliths sediment within 2 min of horizontal disposition of the stem, and a curvature response is seen within 15–30 min of gravity stimulation. If the starch is hydrolysed in barley nodes through the application of α–amylase or kept in the dark, which results in starch loss, no graviresponse is seen. Reconstitution of the starch by treatment with 0.1 M sucrose restores the upward–bending response in these false pulvini. As with stems and roots, there are many complications and few positive answers relating to the graviresponse of false pulvini of grasses. For example, IAA, GA, and ethylene are all known to accumulate more in the lower half of the pulvinus following gravistimulation, but ethylene first appears only about 5.5 h after the bending begins. There were evidences reported for changes in tissue sensitivity to auxin in the false pulvinus of oats, again supporting the 'changing sensitivity hypothesis'.

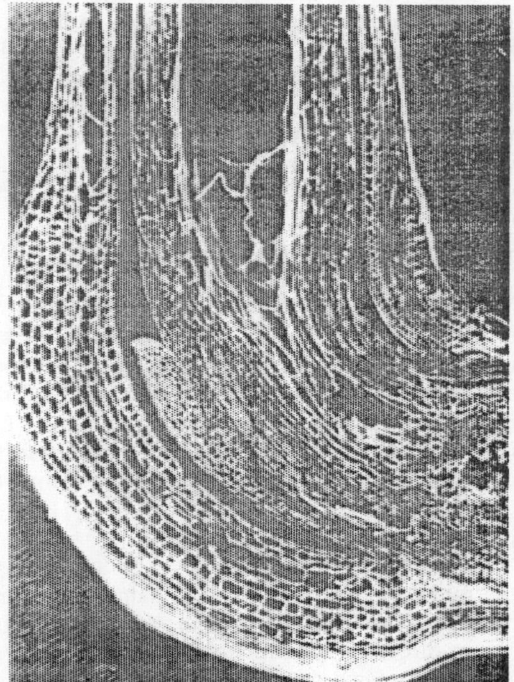

Fig. 257. The gravitropic response of a grass pulvinus (SEM picture); Note the large cells on the bottom (convex side) compared with the much smaller cells on the top (concave side) (Photograph courtesy of Dr. P. Dayanandan)

8.5. OTHER TROPISMS

Several other types of tropisms have been described in the past, but no detailed study has been made on most of them. **Hydrotropism** or **hygrotropism** refers to growth of plants, particularly of their rhizoids or roots (and their root hairs), towards water or moisture. The nature of the root system, the volume it covers and its deep or surface spread are all essentially decided by the locus and extent of water availability in the soil. The *ageotropum* mutant of pea was studied with reference to hydrotropism. The roots of this mutant do not respond either to gravity or to light, but do respond to a humidity (moisture) gradient. The roots often grew up out of the soil, but if they encountered relative humidity in the atmosphere that was below 80–85%, they turned and grew back into the soil (Salisbury and Ross 2004). Removal of the root cap eliminated most of the hydrotropic response, although the rate of root growth was not affected, suggesting that the hygroscopic responses are localized in the root cap.

Chemotropism refers to the response in terms of growth towards chemicals. **Electrotropism** is the response in the form of movement toward electric currents. **Thigmotropism** refers to touch movements. Many plant organs, especially tendrils, respond to touch. Tendrils bend towards the point of contact, wrapping around a support. In the case of curling of tendrils around a solid object, a distinction has to be made between thigmonastic coiling, which happens when the direction of movement is independent of the direction of the stimulus, and thigmotropic coiling, which happens when the movement occurs in response to the directed stimulus. In pea tendril, the coiling involves an initial rapid reaction just behind the tip of the tendril, followed by a slow continuous process of bending along the entire length of the tendril. Although the coiling of the tendril is definitely due to turgor–based cell changes, it is also associated with a pronounced drop in the level of ATP. Thus, it provides the impression that coiling is related to an energy source supplied by photosynthesis. Two additional interesting observations are that coiling is enhanced by light and light effect can be duplicated by ATP in the dark. From these results it has been argued that an ATPase–sensitive contractile protein acting as an ion–pump could be a factor in the coiling of tendrils.

8.6. CIRCUMNUTATION

Ever since the middle of the 19th century, plant biologists were aware that axial plant organs such as stems and roots rarely grow straight, although their mean direction is maintained straight and vertical for long intervals of time. In other words, they believed that their instantaneous growth direction oscillates and describes an ellipse slowly about their mean vertical direction, as seen from above. The axis of the ellipse can vary: At one extreme, the ellipse approximates a line and at the other, a circle (Brown 1991). The direction of oscillation may be clockwise or

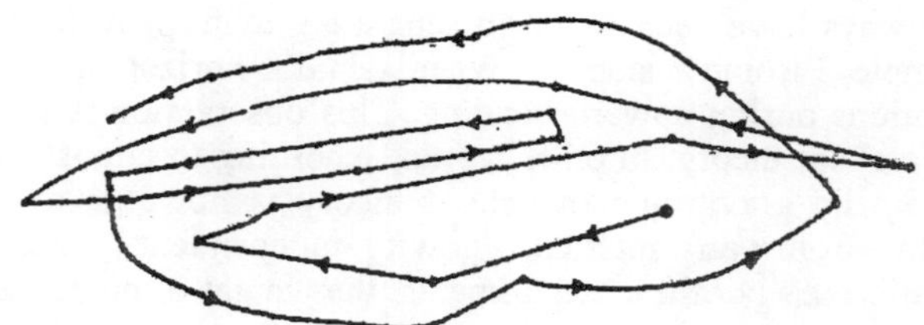

Fig. 258. Drawing from the book "The power of movement in plants" written by Darwin to illustrate the oscillatory movement during circumnutation.

anticlockwise. As the axial organ grows in length, its tip advances and traces a helical path in three dimensions. The oscillation frequencies are in the range of about 50 μHz, with periods of about 20–300 minutes (for example, less than 2 h for a sunflower seedling). Darwin (1880) particularly had drawn attention to the stem tip tracing a more or less ellipse as it grows in length (Fig. 258). They called this phenomenon **circumnutation** (originally called **revolving nutation**). Its utility for climbing plants is obvious because the stem is likely to encounter some kind of support as it circumnutates. Any function it may serve in non–climbing stems is not yet evident, but Darwin suggested that virtually all movements that we have discussed earlier are modifications of circumnutation. Circumnuitation is of widespread occurrence, but it occurs only during organ growth and when growth stops, circumuntation also stops. Hence, it is absolutely growth dependent.

The Darwins (1880) also suggested that circumnutation occurs in response to some internal rhythmic control i.e., it is an endogenously–controlled nastic movement (an internal oscillator model). This model persisted for a very long time until about 1967 when Israelsson and Johnsson proposed the **gravitropic offshoot model** (Johnsson 1979). They provided this mathematical model based on the assumption that circumnutation is really a matter of gravitropic overshoot, in which case, it would be a tropistic response driven and regulated by a sequence of self–sustaining gravitropic reactions with overshoots. As a stem reaches the vertical posture in response to gravity, it may continue to grow more on what was the lower side passing the vertical before an opposite gravitropic response takes over (Brown 1991). This overshooting leads to an oscillation. The mathematical model shows that it is not difficult to convert a back and forth waving to an elliptical or even circular motion. If the overshoot model is correct, circumnutation should occur neither on a clinostat, nor in an orbiting space laboratory. Although the effect can be secondary, many species do stop circumnutating on a clinostat; others do not. In a space experiment, where gravity is eliminated, sunflower seedlings continued to nutate for many hours in microgravity (less than 10⁻

[3] g). Other ways have been found to separate gravitropism from nutation. For example, some stems, when laid horizontally, continue circumnutations during upward bending. This observation is incompatible with an overshoot theory. In other words, according to clinostat and space experiments, the gravitropic overshoot theory is not able to rule out a contribution from an internal, gravity–independent mechanism to circumnutation. As per these experiments, the weight of evidence seems to be against an exclusive overshoot mechanism. However, such a mechanism can function in some species and not in others.

The relationship between circumnutation and gravity needs further discussion in the light of studies made during post–space orbital experimentation period (Brown 1993). The main question addressed is whether gravitropic response controls circumnutation or circumnutation induces gravitational responses. In other words, which of these is the cause and which of these is the effect? All proponents of gravitropic model believe that gravitation is the cause of circumnutation (Kiss 2006). Agravitropic mutants of *Pharbitis nil* and *Arabidopsis thaliana* are defective, but not entirely lacking in circumnutation (Hatakeda et al 2003, Kitazawa et al 2005) emphasizing the presence of some sort of relationship between gravitropism and circumnutation. This group has shown that the *SCR* (for *scarecrow*) gene of *Arabidopsis* regulated both circumnutation and gravitropism. The unpublished work of Krishnamurthy and Karthikeyan on tension wood also shows that the two are related, but it also shows that circumnutation causes gravitropic response and not vice versa. The results obtained by Krishnamurthy and Senthil Kumar (1992) on embryo development in *Polygala* species also show that circumnutation is initiated as early as embryogenesis (Fig. 259) and this endogenously–

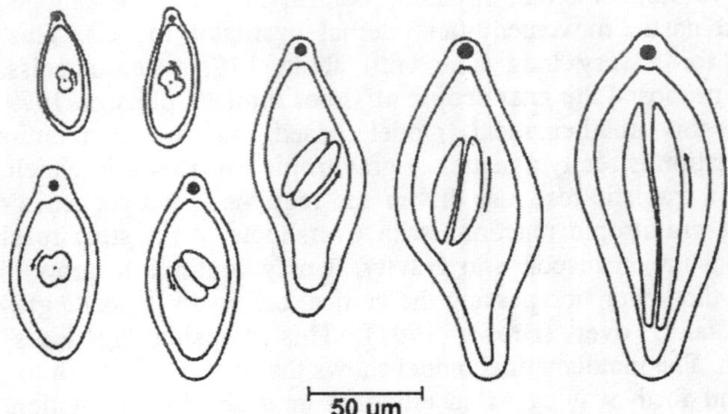

Fig. 259. T. S. of the developing seeds of *Poygala* species at successive stages to show rotation of the embryo (arrows). The position of the raphe vascular bundle (dark dot) is used as the marker to find out the direction of rotation of embryo (Krishnamurthy and Senthil Kumar 1992).

controlled phenomenon continues until the death of the plant. It is also vital in controlling the phyllotactic spiral. The above result also emphasizes that the above gravitropic response is only incidental to circumnutation and not its causative factor.

8.7. REACTION WOOD

As a branch of plageotropic tree grows, it may be expected to bend downward because of its increased weight (due to production of more and more twigs and leaves). Such a phenomenon is sometimes observed, but it is resisted by the formation of **reaction wood (RW)** (Krishnamurthy 2007). This wood has an increased xylem (wood) production on either the upper or lower side of the branch by more rapid division of the vascular cambium on that side. In the conifers, reaction wood is formed on the lower side of leaning branches and bent stems (Fig. 260a) and this wood by expansion (compressive strain) pushes such branches more upright, maintaining a more constant branch angle. The tracheids of this reaction wood are rounded in transactional plane and have intercellular space (Fig. 260b). The tracheid walls become abnormally thick and contain more lignin (40% of the total mass) than normal wood (NW) (30%) and less cellulose (40%) than usual (50%). Such a wood is called **compression wood (CW)**. In dicots, RW is formed on the upper side of the leaning branches and bent stems (Fig. 261) and contracts (tensile strain) to pull the branch toward the trunk by tension to enable the branches to maintain a constant angle. This wood is called **tension wood (TW)**.

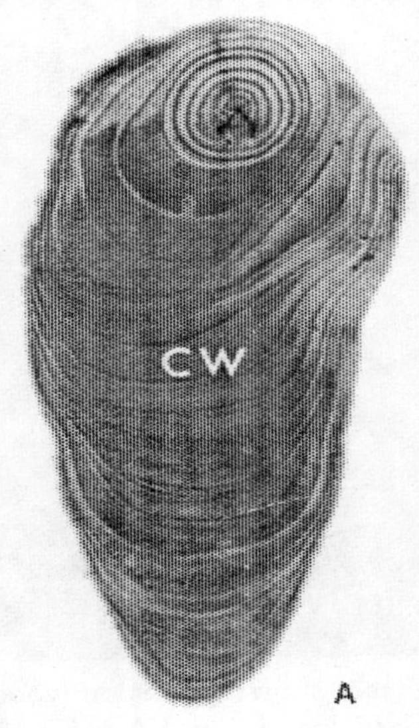

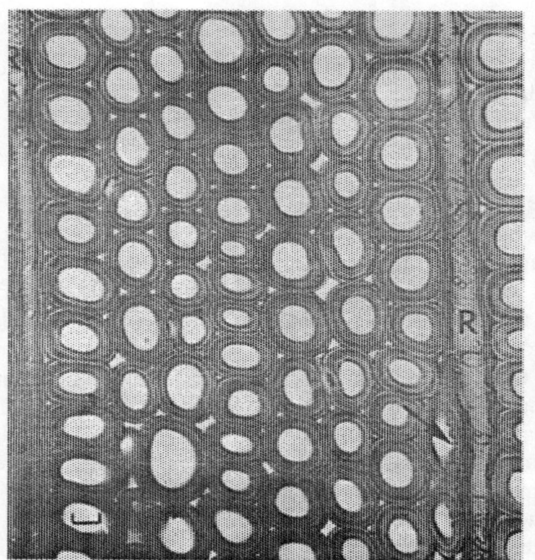

Fig. 260. Compression wood (CW) of *Picea abies*;
A - T. S. of whole wood; B - A portion of the wood enlarged to show circular tracheids with intercellular spaces; R - Ray (Timell 1973)

Fig. 261. T. S. of a dicot stem showing tension wood (Reproduced with permission from Keslick and Son, West Chester, PA, USA)

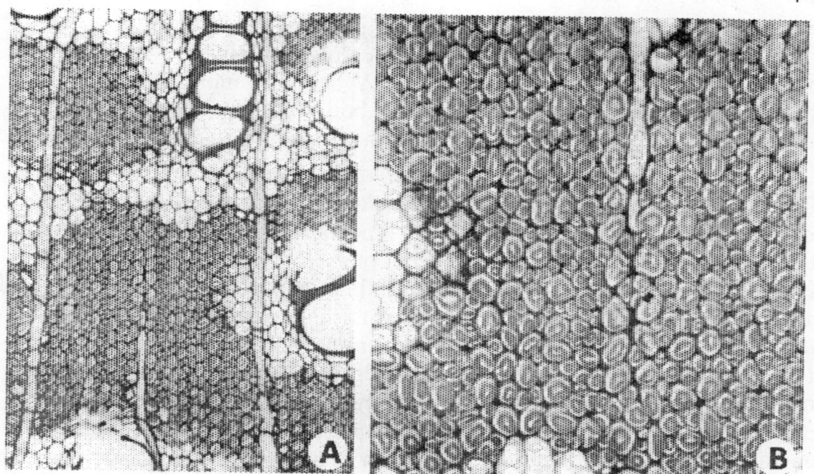

Fig. 262. T. S. of tension wood of *Cassia* species showing gelatinous fibres (A); Enlarged portion of the same (B)

The RC produces high levels of strain and stress. Strains are changes in length per unit length (cm/cm), while stresses are forces per unit area (km/cm^2, or Pascals, 1 Pascal = 1 Newton/m^2). Stress causes strain and vice versa, and the two are related. The mechanical action of RC results from anatomical and structural characteristics. Some of these characteristics are macroscopic, but others are evident only microscopically or submicroscopically. Not only these characteristics of RW follow a continuum from the NW, but also the strain generated by its cells. RW is, thus, only quantitatively different from NW. The TW contains more fibres called **gelatinous fibres (G–fibres)** (Fig. 68 D and 262) and fewer tracheary elements than NW. The G–fibres have secondary walls that are thicker than those in NW because they form an additional, often convoluted, cellulose–rich layer (called **sg–layer**) (along with carboxylated acidic polysaccharides).

In contrast to conifer CW, TW in dicots does not become unusually rich in lignin (15% as compared to 20% in TW), but have more cellulose (65%) when compared to NW (55%). Within TW, the level of tensile stress is related to the proportion of G–fibres in the tissue. The CW tracheids have an unusually high (45°) microfibril angle (MFA), whereas the Sg layer of TW fibres has an unusually low (5°) MFA and the MFs run almost parallel to the fibre axis. TW develops tensile strains 2–3 times those of NW, while CW develops compressive stresses of <100 kg/cm^2 compared to the tensile stresses of 40 kg/cm^2 in NW. Ten fasciclin–like arabinogalactan–rich proteins are specifically expressed in TW, but not in the cambial zone. These as well as extensins (hydroxyproline–rich proteins) are localized immunocytochemically close to the inner part of Sg layer of TW fibres

(Hariharan and Krishnamurthy 1995, Lafarguette et al 2004, Jothi et al 2010). It is suggested that these structural, stress proteins are linked to the specific mechanical properties of TW. TW becomes more pronounced at greater leans (for example from 10°–20°), both through more G–fibres and through more highly–developed G–fibres in terms of Sg–layer thickness.

RW can be a gravitropic response and a mechanism to maintain plageotropic growth, or a response to tensions and pressures resulting from bending, or both. On balance, gravity appears to be more important than other factors in the formation of RW, but other forces or factors seem to be involved (Timell 1986). The important functional aspect of RW is the strain it develops and bending movements that these stains generate. In fact, all types of wood seem to generate strains, but not such large ones as in RW. Of the proposed strain generating mechanisms (Bamber 2001), none seems adequate to explain all the strain phenomena. MFA seems to be a key in explaining the direction, compression or tension, of the strains generated. It seems unlikely that MFA can be varied independently of cell length, so the developmental regulation of the direction of growth strains may be complex. A combination of several mechanisms seems most likely to explain strain phenomenon (for example, as per the **unified hypothesis** of Okuyama et al 1994 a, b), for instance, some combination of the swelling of the secondary walls during lignification and the contraction of cellulose microfibrils (change in MFA) during their crystallization. Evidence for this is provided by the upregulation of cellulose synthase genes and of genes encoding enzymes involved in lighin biosynthesis during RW differentiation. CW has a β, 1–3 glucan called **laricinan** in more amount than in NW and this has been linked to a strain generating mechanism.

Several phytohormones are shown to be involved in the regulation of RW development and activity, especially TW formation. Expression of members of the AUX or IAA gene family is regulated during TW formation. Redistribution of IAA and other hormones, especially ethylene, may explain these results, but the cause of hormone movement may prove complex. When young plants of Arizona Cypress were treated with ethylene, branches bend upward (hyponasty). Other treatments that cause an increase in endogenous ethylene also induce branch hyponasty. Furthermore, when ethylene was removed in the air from around the plant through the use of mercuric perchlorate, branches grew downward. Thus, ethylene might be involved in plageotropic growth of branches, or ethylene could reflect changes in auxin or other compounds. ACC oxidese gene is also shown to be regulated in TW.

Rhythmic Phenomena and Growth Periodicity 9

9.1. INTRODUCTION

Almost nothing is really constant in this world and in its environment. Several kinds of environmental changes, both stochastic and deterministic, have happened during the geological history of the Earth in diverse regions of the world, both on long–term and on short–term bases. Long–term environmental changes in any region of the world are implied by the term **climate**, while short–term changes are denoted by the term **weather**.

Weather changes are related to wind velocity, temperature, light level and humidity in time intervals from minutes to perhaps a few hours. All these changes are superimposed on a daily (**diurnal**) cycle. The term 'diurnal' has often been used in literature synonymously with the term 'daily', but it must be kept in mind that it also implies 'daytime' or the light span, just as 'nocturnal' implies night time or the dark span. Weather cycles normally last for several days. These weather cycles are invariably superimposed on the **annual cycle** of seasons. Weather trends are also related to **tidal (lunar) cycles** that influence the tidal–zone environment and possibly other ecosystems also. Similarly, weather trends may also be related to the 11 year **sunspot cycles** (Salisbury and Ross 2004).

Long–term climatic changes occur over periods of centuries to millennia. It is greatly advantageous for any organism, be it a plant, animal or microbe, to anticipate and adjust itself to these long–term and short–term environmental changes i.e., the climatic and weather cycles respectively. Three of them, related to the mechanics of our solar system, are regular enough so that this anticipation and adjustment should be possible: the daily, the tidal, and the annual cycles. For any plant (or animal or microbe) to anticipate and get prepared for these regular changes in its environment, it needs a **clock** or **oscillator** and various other associated mechanisms. Thus, a clock or oscillator can be considered as an internal processor of temporal inputs from the environment. The output from the processor regulates the timing of developmental events within the plant (Salisbury and Ross 2004).

The timing system or clock should have at least two major sets of characteristics (see Salisbury and Ross 2004):

1. It should not be unduly affected by capricious environmental factors operating around the plant, particularly by those factors that cannot be accurately predicted such as temperature, light intensity, wind velocity, humidity etc. Even if a biological clock (or oscillator) is not very sensitive to all the above factors, which itself would be very impressive, it would be much more impressive if such a clock could run with the accuracy achieved by man-made mechanical and electronic clocks (Salisbury and Ross 2004). If such type of accuracy and perfection are absent, the biological clock may soon get out of phase with the reenvironment, and hence, will be of no use to the plant. However, the inherent accuracy of the biological clock can be obtained by an alternative method i.e., it can be regularly reset or resynchronized by some dependable feature of the plant's environment. It would not matter much if this clock loses or gains an hour or so a day and go out of phase provided that it gets reset each day at sunrise or sunset or both. It is likely that the biological clock may possess one status suited to day time and another to night time and thus, resetting at dawn and dusk

may keep the clock in tune with the prevailing environment. It is conceivable that a clock may simply be driven by changing characters of the environment and in this case, the plant would need only to track the environmental changes, "without actually keeping time or anticipating future environmental events" (Salisbury and Ross 2004).

2. There must exist coupling mechanisms that enable the plant to take advantage of the biological clock's time–keeping. For example, a plant can save its energy that may enable it to direct and concentrate its available resources to photosynthesis during day time and to other metabolic activities during night time. Many similar manifestations of biological time–keeping are seen in plants.

It is sometimes easy to recognize the clock–controlled activity in plants, but there are also highly sophisticated clock mechanisms that cannot be easily recognized. The following two important questions can be asked: What is the mechanism of these biological clocks? How do they work? At the moment, not much is known on the clock mechanism. However, the available information is summarized below.

9.2. SOME BASIC CONCEPTS

One of the oldest and the most well–known daily changes observed is the **sleep movement** of leaves of some plants, with leaves drooping down after dusk and becoming erect during dawn. It was wondered even about 300 years back whether these sleep movements were driven by environmental changes i.e., the daily light–dark cycle, or whether they might be controlled by some time–keeping system within the plants. If leaves move only in response to environmental changes, timing is **exogenous**, and if the response is to an internal clock, timing is **endogenous**. It was shown even by 1729 that the leaf movements are under endogenous control, which was confirmed in 1759 and subsequently by Pfeffer between 1875 and 1915. However, the major breakthrough came in 1920s when Rose Stoppel in Hamburg, Germany studied the leaf movements in a dark room at constant temperature. The maximum vertical position of the leaves was seen by her at the same time each day. From this investigation onwards, the basic concepts on rhythmic phenomena and growth periodicity started emerging.

A **rhythm** is the repeated occurrence of some event or function, and the **period** of the rhythm is the time interval between recurrences of a recognizable point on the cycle of this event (for example, the occurrence of the maximum vertical position of the leaf in sleeping movement). Since the period of rhythm is almost exactly 24 hours, Stoppel reasoned that a biological clock could probably not be so accurate and hence, some factor in the environment must be resetting this clock on a daily basis. Since in her experiment the plants were kept in dark and at constant temperature, the critical environmental factor could not be day–light or temperature. She called this critical factor **'factor x'**, as she was not sure about its identity. Later Bünning and Stern in 1930 identified this 'factor x' as red light, which Stoppel used at night while watering the plants. When red light was stopped, Stoppel's prediction became true i.e., the maximum vertical position of the leaf came about an hour and a half later each day, so the leaf movement cycle soon became out of phase with day and night outside the dark room. Thus, the concept of **free–running period** came into existence, and this is the period of a rhythm that is under constant environmental conditions. Since the free–running period was greater than 24 hours, there appeared to be no alternative to an endogenous clock that gets reset or **entrained** every time to the natural cycle of day and night, probably by dawn or dusk or both. Thus, the free–running internal clock betrayed its inaccuracy and displayed its endogenous nature. By late 1950s, it was suggested that rhythms with a free–running period of approximately, but not exactly, 24 hours should be called **circadian** (*circa* = approximately; *dies* = day).

There are three rhythm characteristics, which are as follows:

1. **Period** - It is the time interval between comparable points on the repeating cycles. The term '**phase**' is used in a specialized sense to indicate any point on a cycle that is recognizable by its relationship to the rest of the cycle. It may be specified as the time after an initial reference time, such as mid–day or mid–night of the first cycle, or it may be given in degrees if we assume that one cycle (one period) equals 360°. If the cycle or period is described by a sinusoidal function, then the maximal phase of the cycle is called the **acrophase**. Thus, the period refers to the time interval between acrophases.
2. **Range** - It is the difference between the maximal and the minimal values, and the **amplitude** equals half of the range, or the extent to which the observed response varies from the **mean**.
3. **Pattern** of the cycle - Many rhythms follow a sinusoidal curve, but many variations from this pattern are seen. For example, a sharp maximum may be followed by a broad minimum, or the slope of the curve approaching the maximum may be steep, while that approaching the minimum may not be so (Salisbury and Ross 2004).

When plants are exposed to an environment that fluctuates according to some period and the rhythms exhibit the same period, they are said to be **entrained** to the environment, as opposed to free–running. This entrainment to the environment can be brought about by many factors such as light and dark conditions separated by dawn and dusk. Such entraining cycle is known as **synchronizer cycle** or **Zeitgeber** (time giver). The term entrainment is used when the Zeitgeber is a fluctuating environment with several regular cycles. If an environmental stimulus is given only once (for example, a single flash of light) and the phase of the rhythm is shifted in response to it, then the rhythm is said to have been **phase-shifted** or rephased.

9.3. RHYTHMIC RESPONSES TO ENVIRONMENT

A lot of detail has accumulated on this topic and hence only a few effects of light, temperature and applied chemicals will be dealt with here.

9.3.1. Light

Light, especially weak red light is a subtle entraining factor and therefore, is of great importance as a potential Zeitgeber. According to one approach, it became readily apparent that the rhythms could be entrained to some light–dark cycle other than a 24–hour cycle. The rhythms can be entrained even by shorter cycles of 20–21 h and in rare instances even of 10–16 h or a longer cycle of 28–38 h. Another approach is to first allow a rhythm to become strongly established by a cycling environment and to let it run free in continuous darkness. Then, a brief interruption of light is given at various times during the free–running rhythm. When a flash of light is given during subjective day, there is virtually no effect upon the rhythm i.e., if light comes during the phases typical of a day in a natural cycling environment, the following phases of the cycle are not much affected. However, when the light interruption comes during early subjective night the rhythm is typically delayed (i.e., the next peak comes later than otherwise expected). It appears as if the flash of light acts as dusk, but by coming later, it causes a delay. As the light flash is given later and later during subjective night, the extent of the delay increases until a point is reached at which the flash of light suddenly results in an *advance* of the rhythm, rather than in a delay (that is, the next peak comes earlier than expected). The flash of light is then acting as dawn rather than as dusk (Fig. 263).

In light entrainment, some photoreceptor pigment is involved in absorbing light and thereby getting changed in such a way that it can lead to an advance or a delay of the biological clock. For understanding photobiochemical basis of this response, this photoreceptor pigment must be identified through a determination of its action spectrum for the response. Light of carefully–controlled spectral quality must be given at various times to test its effectiveness on phase–shifting or entrainment. Some plants show a strong response in blue light; some plants in red light (E.g., dark–grown bean plant), which is reversed on exposure to far–red light. Furthermore, in the bean plant, rhythmic leaf movements often continued for several days when the plants are maintained at a constant

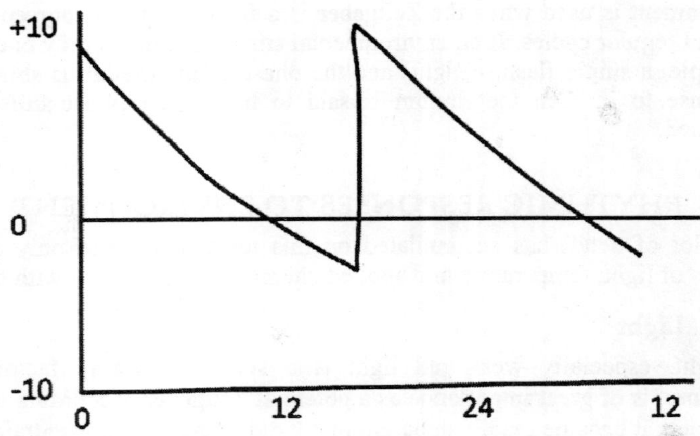

Fig. 263. Phase–shift in petal movements of *Kalanchoe* species following 2 h exposure to red light given at various times during an extended period of continuous darkness (Graph made from data of Zimmer 1962).

temperature and under continuous light, provided that the light is rich in red wave lengths, but contain none of the far–red component of the spectrum. These observations implicate the phytochrome system in entrainment. Additional evidences for the implication of phytochrome system in other plant taxa were subsequently provided by other investigators. Phytochrome system definitely interacts with the biological clock at least in some plant taxa, but there are several complications that need to be resolved before this interaction is fully understood.

9.3.2. Temperature

Studies on the effects of temperature on biological clock were initiated in 1930 by Bünning and Stern. They reported that the effect of temperature on the response of the biological clock of the plant was unexpectedly small. However the leaf–movement period of his experimental bean plants was very sensitive to temperature. It was subsequently proved that the biological clock would be of little value to the plants, if the rate at which it ran was strongly dependent upon temperature, as are most metabolic functions (for instance, the Q10 value concept which is a measure of the rate of change of the biological or chemical system as a consequence of increasing the temperature by 10°C). Thus, the temperature insensitivity of the biological clock was discovered and was strongly appreciated.

9.3.3. Chemicals

The effects of chemicals on biological clock were also not sufficiently understood. If one can find some chemical that will either inhibit or promote biological time measurement, then its mode of action may be tested in order to get insight into the operation of the biological clock. Many chemicals have been investigated, but the obtained results must be interpreted carefully. A given chemical often influences the amplitude of the rhythm without influencing its mechanism, as indicated by the period. There are some instances where the amplitude can be completely damped out and yet when the inhibitor chemical is removed, the cycle proves to be in phase with that of control plants having the same period and to which no inhibitor was applied (Salisbury and Ross 2004). It was shown by Bünning and his associates that several chemicals applied to bean plants appeared to act directly on the clock itself rather than on its manifestations. In these instances, the period was lengthened to some extent, but the amplitude was unaffected or only slightly decreased.

9.4. CLOCK MECHANISM

What is the clock? Where is the clock? Can there be more than one clock or mechanism in the same or in different plants? These are some of the most important questions on biological clock that are yet to be convincingly answered. We know more about the location and number of clocks than what we know about how they work. A plant can have more than one clock. For example, the bean plant has circadian rhythms in stomatal opening and leaf movement with different free–running periods. This is really possible only if there are different clocks (although it is still possible that these two rhythms may have different free–running periods resulting from cell–specific modification of the same clock components). So far, there is no evidence that the clock has different mechanisms in different organisms, but more studies need to be made on this issue. A model for circadian rhythms in plants that has input pathways, such as light and temperature, interacting with one or more central oscillators (clocks) is shown in Fig. 264.

The central oscillator (s) is then involved in generating output rhythms via a range of cellular or systemic signalling pathways (Kreps and Kay 1997). Phototransduction is both an input and output pathway, while the output pathways regulated by the clock include stomatal and leaf movements. That the clock is located in the membranes is indicated by a number of evidences obtained through the application of inhibitors and through studies on mutants. In plants, the phytochrome system apparently plays an important role in the rephasing of clocks and it is considered to interact with membranes. Several clock mechanisms involving membranes have been proposed. Some *substance X*, possibly K^+, is implicated in models explaining the clock mechanism. Several circadian rhythm mutants have been isolated from *Arabidopsis*. The *timing of cab-1* (*toc 1*) mutation defines a product that is required for normal cycling of several rhythms, such as the gene–expression rhythms of the chlorophyll a or b binding

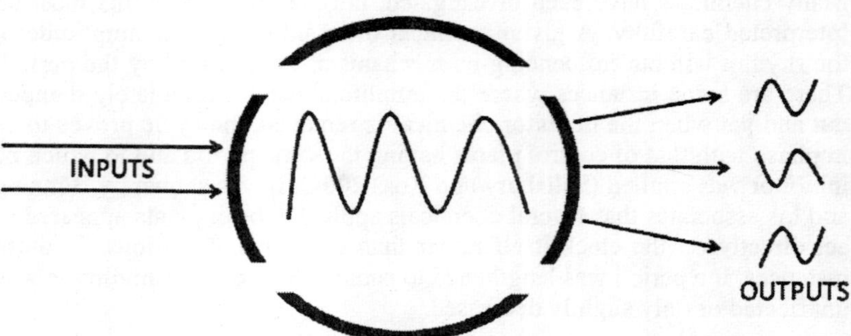

Fig. 264. A general model for organization of the circadian clock system in plants (Based on Kreps and Kay 1997)

protein 2 gene (*CAB2*) and the *cold, circadian rhythm* RNA–binding protein 1 and 2 genes (*CCR1* and *CCR2*) as well as a leaf movement rhythm. The effects of *toc1* on multiple rhythms suggest that the *TOC* protein may be a clock component (Kreps and Kay 1997).

9.5. USE OF CLOCKS

It was mentioned already at the beginning of this chapter that one can deduce the likely existence of a biological clock based on the ways it can be used in a cycling environment. Plant rhythms in metabolic events adjust the plant to prevailing light and temperature conditions. Rhythms in floral anthesis must be closely coupled to the time memory of honey bees and other pollinators. Photoperiodic phenomenon confers upon the plant showing it the ability to occupy a particular ecological niche in seasonal as contrasted to daily time. Several plant species that flower due to photoperiodic induction may do so at different times and in sequence throughout the flowering season, thus providing a constant source of nectar or pollen for the pollinators.

What are the functions of leaf movements? Pfeffer concluded a long time back that leaf movements have no (evolutionary) selection value. Charles Darwin suggested that leaf positions may play a role in the heat transfer between a plant and its environment. A horizontal leaf is in a good position to capture sunlight during the day, but avoids this posture at night to reduce heat radiation. Vertical leaves in a plant community may radiate more heat to each other keeping each other warm. However, recent studies indicate that the temperature difference between vertical and horizontal leaves is small, often less than 1°C. Transpiration can be reduced in leaves that fold up at night. According to Bünning, a horizontally– disposed leaf may absorb moon–light, which in turn may affect the photoperiodic response, but a vertical leaf at night can be a protective device to ensure successful time–keeping in photoperiodism.

9.6. PHOTOPERIODISM

9.6.1. Discovery

W. W. Garner and H. A. Allard working in the USDA laboratory at Beltsville, Maryland in 1910s made two important observations, which they themselves could not explain initially, but which heralded the concept of **photoperiodism**. *Maryland Mammoth* variety of tobacco grew at that latitude to a height of 3–5 m during summer, but never flowered. It flowered profusely when only about 1m tall after being transplanted in the winter green–house. They had also observed that all the individuals of a given cultivar of soya bean would flower at about the same time in summer regardless of their height and the time of their planting. They suspected that some common factor of the environment was responsible for flowering of these two species. To know this factor they tested various environmental factors such as light levels, temperatures, soil moisture, and soil nutrient conditions that might differ between the summer fields and the winter greenhouse. But no combination of these tested factors resulted in flowering of the tobacco plants. Garner and Allard then realized that day length varied throughout the season and was a function of latitude and hence, they tested whether this could control flowering and to their pleasant surprise it did. They found that when the days were shorter than some other days with maximum length, both the tobacco and soya bean plants began to flower. Garner and Allard found that some plants such as spring barley and spinach flower in response to days that are **longer** than some critical day length. They called these plants **long–day plants** (LD plants). They also observed that plants such as tobacco and soya bean flower when the day length is less than some other days with maximum length, hence they called these plants **short-day plants (SD** plants). Some species they studied showed no response to day length. They called these plants **day–neutral plants** (also known as **day length–indifferent plants)** **(DN** plants). These discoveries were published in 1920 and the phenomenon was named **photoperiodism** (This name was suggested by A. O. Cook, a colleague of Garner and Allard at Beltsville). It is now clear that plants are capable of measuring the length of the light or the dark period or both, and hence, there is a clear–cut demonstration of biological clock measurement. Photoperiodism also demonstrates the synchronization concerned with reproduction of plants with seasonal time.

9.6.2. Some General Principles

In the non-mountainous regions of the equator, sunrise and sunset occur at the same time on each day throughout the year, so the lengths of the day and night remain constant. Exactly in the south and north poles, the Sun remains above the horizon for six months each year and below the horizon for the other six months. Again, the day and the night are equal and each is six months long. As one travels from the equator towards either of the poles, the days become longer in summer and shorter in winter. In these intermediate places, the rate at which day length changes varies during the year. During the approach of summer and winter solstices, when days are respectively the longest and the shortest, there is little change from day to day. But, during spring and autumn, the rate of change is much more rapid, as days become longer during spring, and shorter during autumn. Thus, a plant (also other organisms) can detect the season by measuring day and night lengths and how much these lengths change, but since the absolute day lengths at any time of year depend very strongly upon latitude, plants must be 'calibrated' to their location (Salisbury and Ross 2004). This situation has resulted in the evolution of different ecotypes for the same species of plant growing at different regions, which flower at different times of the year as they have to respond to different day lengths.

One must first realize that there are no broad generalities, and no sweeping laws to help us understand the photoperiodic response as photoperiodism is a widespread phenomenon in nature and virtually every aspect of plant development is influenced by the photoperiod. Each species and often each cultivar or variety within a species seems to have its own features of photoperiodic response; probably no two respond exactly alike. Yet some generalizations can be made.

9.6.3. Aspects of a Plant's Life Cycle Controlled by Photoperiod

Several aspects of a plant's life cycle are controlled by photoperiod. Some of these aspects are described below:

9.6.3.1. Seed Germination

The germination of seeds of certain plants depends on the photoperiod received by the parent plant. Mature seeds of some plants are also influenced in their germination by photoperiod to which they are subjected. There are long–day, short–day and day–neutral seeds for germination.

9.6.3.2. Features of the Vegetative Shoot

Stem elongation in response to long days is perhaps the most–widespread photoperiodic effect. This elongation (called **bolting**) can take place before or after flowering. The related effects are a suppression of branching by long days (it is promoted by short days) in several species and on **tillering** in grasses. Many temperate grasses such as barley tiller more under short days, while tropical and subtropical grasses such as like rice tiller more under long days. Some consider that stem elongation induced by long days may not always be a true photoperiodic phenomenon and may be independent of photoperiodism. This may be a species–dependent feature. Many features of leaves are strongly influenced by day length as follows: Long days promote leaf expansion and stomatal density and decrease leaf succulence, content of organic acids and chlorophyll amount.

Rooting of stem cuttings is promoted by long days both when applied to the mother plant and to the cutting itself. The production of potato tubers, cassava and dahlia root tubers, radish etc. is induced by short days, while onion bulb formation is promoted by long days. Short days promote abscission and dormancy in temperate species. The spring juvenation and bud breaking in woody perennials are influenced by day length, especially long days. The colour changes preceding the senescence and leaf fall during autumn and after the bud break during spring are influenced by photoperiod.

9.6.3.3. *Reproduction*

Different types of vegetative reproduction are also influenced by photoperiod. For instance, long days promote runner formation in strawberry and foliar plantlet formation is *Bryophyllum*. Timothy grass forms viviparous plantlets under short photoperiods.

As indicated already, day length is very important in floral induction and flowering in angiosperms and reproductive organ formation in conifers. In fact, flowering is the most–studied phenomenon through photoperiodic treatments. The further development of the various floral organs often is also strongly influenced by photoperiod (Vince–Prue 1975). Most typically, the same day lengths that lead to floral induction also lead to an increased rate of further floral development. But some plants have a photoperiodic requirement for floral initiation, but are day–neutral for floral development or have the opposite. Some require SD for initiation and LD for further development or the opposite.

Sex expression in many species is often strongly influenced by photoperiod, but no simple relationship exists between the two: either femaleness or maleness can be promoted by either short days or long days, depending on species or cultivars. Many studies have been made on the effects of photoperiods (and of other factors) on seed filling, and thus, yield of economically–important food plants. In some plants such as soya bean, this has been definitely demonstrated, but in others the situation is usually complicated by a number of other factors.

9.6.4. Types of Response to Photoperiod

Fig. 265 summarizes some possible effects of day length in each 24–hour cycle on relative flowering. In truly day–neutral plants, which are probably rare, flowering is independent of day length and hence, is represented by an almost horizontal line (1) in the figure. Flowering of LD plants is promoted by LDs and hence, their curves (5) slope upward to the right. Flowering of SD plants take a longer time on LDs and so their curves (6) slope downward to the right. It must also be emphasized that there is both a **facultative** and an **absolute response** to photoperiod. Curves representing plants with an absolute day–length requirement cross the abscissa at the day length called the **critical day**. Hence, there is also a **critical night,** which denotes that night length that must be exceeded for flowering of SD plants or inhibition of flowering of LD plants. In fact, the facultative or the absolute response of most, if not all, plants depends on its age, history, growing conditions and other such factors. One other problem crops up. Often SD plants do not flower if the days are too short. Note that this minimum light requirement in SD plants is represented by a curve that slopes upward to the right in Fig. 265. Such a curve really expresses the response characteristic of LD plants and so, SD plants act as LD plants when the days are extremely short. This brief day–length is required for metabolic activities other than just photosynthesis. The photoperiodic response seems to require some minimum amount of P_{fr} each day and this

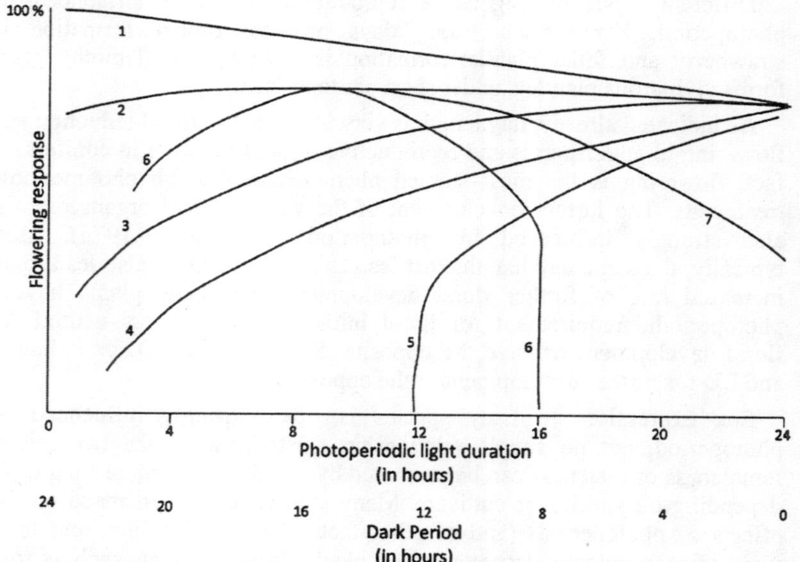

Fig. 265. Diagram illustrating flowering responses to various photoperiodic treatments; 1 - A true day–neutral plant; 2, 3, and 4 - Three different plants whose flowering has been promoted quantitatively by increasing day lengths; 5 - Absolute day–length plant that flowers only when days are longer than 12 h; 6 - Qualitative short–day plant that flowers only when days are shorter than 15.7 h; 7 - Plant that flower on any day length, but better under short days (adopted from Salisbury and Ross 2005)

minimum requirement does not seem to be related to time measurement or to the specific metabolic events that occur during light or darkness. SD plants such as *Xanthium* and *Kalanchoe* also require CO_2 during the light period for proper photoperiodic induction.

The SD and LD responses are completely opposite to one another. The curve 6 in Fig. 265 represents a classical SD plant such as *Xanthium,* while curve 5 represents a classical LD plant such as *Hyoscyamus*. Both flower when the days are about 12.5–15.7 h long (or nights 11.5–8.3 h).

Some plants require only one inductive photoperiodic cycle (SD or LD) for flowering, while others require more than one cycle. All these plants require specific critical day lengths. The three response types observed by Garner and Allard form the foundation for classification of photoperiod response types. Since their time, some other response categories have been discovered. To produce flowers, a few species require LDs followed by SDs, as occurs in late summer and fall (Salisbury and Ross 2004). These plants, when maintained either under continuous LDs or continuous SDs, remain vegetative. These are called **long–short–day plants** (LSD plants). In contrast, **Short-long-day plants** (SLD plants) require SDs followed by LDs, as in spring. There are a few plants that flower only on **intermediate day lengths** and remain in the vegetative state when days are either too short or too long. They have counterparts that remain vegetative on intermediate day length, flowering only on longer or shorter days.

Responses of plants to photoperiodism also depend on the latter's interesting interactions with temperature. A **vernalization** requirement is often followed by a requirement for LDs, as seen during the LDs of late–spring following winter. In other plants, a given response type is noticed at one temperature, but not at another. It may, for example, show a SD response at temperatures above 20°C, but remains essentially day–neutral at cooler temperatures. Poincettia is an absolute SD plant at higher temperature, while morning glory is an absolute LD plant at lower temperature. They are day–neutral only at the intermediate temperatures. Thus, specific day–length requirements at specific temperatures appear to be quite common among plants. Vernalization and a given photoperiod treatment are sometimes interchangeable. For example, in a variety of *Campanula medium,* vernalization is fully replaced by SDs, but LDs are required following either treatment. Indeed, many SLD plants flower after a cold treatment followed by long days.

There is some data to indicate a genetic basis for photoperiodism. Day–length sensitivity is governed either by a single gene, as in *Maryland Mammoth* tobacco and a mutant of *Arabidopsis*, or by several genes as in sorghum and wheat. Response to day length may be either a dominant or recessive feature, again depending on the species. Both single or multiple and dominant or recessive genes can govern photoperiod (as well as vernalization). In wheat, genes controlling photoperiod and vernalization sensitivity are independent of each other, probably controlling different component processes of flowering.

Thus, it can be concluded that although the basic response types are SD, LD and DN plants, there is a wide diversity of photoperiodic response types and much plasticity of response centred around these three types.

9.6.5. Ripeness to Respond or Competence

We have discussed briefly the concept of ripeness to flower in Chapter 6. Only a few plants respond to photoperiod when they are small seedlings. For example, *Pharbitis nil* responds to SDs in the cotyledonary stage itself and species of *Chenopodium* respond and flower even when they are minute seedlings. Most plants such as *Xanthium* must attain a somewhat larger size and in these seedling plants with the cotyledons do not respond. *Hyoscyamus* must be 10–30 days old before responding to LDs. Certain monocarpic bamboo and palm species and several polycarpic tree species do not flower until they are 5–40 y or more old, but it is not known whether or not they can then respond to photoperiod. The condition a plant must achieve before it flowers in response to the environment was called by Klebs (1910) as *Blüreife*, which translates as **ripeness to flower**, but a correct descriptive term is **ripeness to respond**. In many species, the number of required photoperiodic cycles for inducing flowering decreases as the plant gets older. In other words, the 'ripeness to respond' increases with the age of the plant. Often, the plant finally flowers independently of the photoperiod, and it becomes DN. However, in some plants, it becomes more difficult to respond to photoperiod as they become increasingly older, as the younger leaves and recently produced leaves of these plants have declining sensitivity to photoperiodic treatments. Thus, it is evident that individual leaves of plants must also reach a ripeness to respond. In some species, the leaf shows maximum sensitivity when it is first mature (i.e., fully expanded), but in yet others such as *Xanthium,* it is the half–expanded leaf, the one growing most rapidly, that is most sensitive. Ripeness to respond can also be identified by the ability of vegetative meristem transforming itself into a floral meristem i.e., **evocation** (Refer to Chapter 6). If evocation occurs, the meristem is said to be **competent**. The concept of ripeness to respond is also identical to that of **juvenility**, which is defined as the condition of a (woody) plant before it is mature enough to flower (Refer to Chapter 5). Another related concept is **minimum leaf number**, which is the minimum number of leaves the plant produces from seedling to the first flower under the most ideal conditions for flowering.

9.6.6. Role of Dark Period

Even by 1930s, it was asked: which is more important, day length or night length, in photoperiodism? In one experimental approach, days and nights were varied to give cycles that did not equal 24 hours. In such an approach, the critical night, but not the critical day, that remained constant, indicated the importance of the dark period in response. In another approach, days were interrupted with darkness, or nights with light. Interruption of the day with darkness had little or no effect, but interruption of night with light inhibited flowering of SD plants, but promoted flowering of LD plants. This discovery was termed as the **night–break phenomenon**. Thus, it became evident that a period of light during the dark period nullifies the effect of darkness in photoperiodic induction. It was also evident that within rough limits, the total quantity of light energy (i.e., fluence) proves to be the determining factor in the night break, but not the level of light used (i.e., irradiance). It was also proved that light applied even at very low irradiances during the entire dark period is effective (especially in inhibiting flowering of SD plants) and the most effective time when this light is important is often equivalent to the critical night. The most effective wavelength is red light followed by far red. Thus, phytochrome apparently detects the light. The effectiveness of phytochrome depends upon the time of irradiation.

However, there are lot of complications, and the situation is not as simple as it appears. LD plants are often less sensitive and somewhat more quantitative in their response to a night break than are SD plants. Moreover, whereas red light is most effective in a night break with SD plants or a brief night break with LD plants, a mixture of red and far red wave lengths can be more effective with LD plants when it is applied as a long night break, as a day–length extension, or during the continuous light that best induces flowering in most LD plants (Salisbury and Ross 2004). Studies made on *Pharbitis nil* have shown that P_{fr} produced by red light at the proper time during darkness is highly effective at inhibiting induction, whereas the prositive role of P_{fr} produced by red light does not seem to be related to specific events in the induction cycle. Based on these observations, it has been postulated that phytochrome must exist in two forms (two forms of P_{fr}) in SD plants. One of them is necessary for induction of SD plants and is highly stable in darkness, while the other is highly labile in darkness, but inhibits induction of SD plants when it is produced at the proper time. Antibody studies have also confirmed the presence of at least two distinct types of phytochromes, but whether these are similar to the two mentioned above is not clear. It will be considerably helpful if we can identify and measure the various forms of phytochromes in leaves at different times during photoperiodic induction.

9.6.7. Time Measurement in Photoperiodism

The main characteristic feature of photoperiodism is measurement of seasonal time by detecting the lengths of day and night. But, how is time measured by plants in photoperiodism? It is to be emphasized here that photoperiodism is similar to such other examples of biological time measurement as circadian rhythms. Photoperiodism was related to the biological clock by Bünning as early as 1937 and this is agreed upon by many other plant physiologists subsequently. It is, however, questioned whether photoperiodic time measurement is akin to time measurement done by an hour glass (where only one interval of time can be measured) or akin to time measurement done by a pendulum that is an oscillator (where long–term interval of time can be measured). Some features of photoperiodic induction look functioning like an hour glass system; as example for this apparent similarity, we can cite the synthesis of a flower–inducing hormone during floral evocation and the shift from the predominant P_{fr} to Pr. However, elements of hour– glass timing play a relatively small role in photoperiodic timing and photoperiodic response is not as simple as that of an hour glass system. Bünning (1937) suggested that plants may use their oscillating circadian clocks in photoperiodic time measurement and hence it is a rhythmic phenomenon. In some SD plants, at least, this timer is strongly rephased at dawn, can become suspended after some hours in the light, and is then rephased at dusk to a night phase. It must also be emphasized that plants respond sharply to the light changes at dawn and dusk, "ignoring" the changes in irradiance during both day and night, although there may be slight responses to moonlight. Thus, it may be concluded as follows: There are elements of hour–glass time measurement in photoperiodism, but these play a relatively smaller role in timing. On the contrary, oscillating–timer mechanism is very important, at least in SD plants.

9.6.8. Chemical Basis of Photoperiodism

This topic was briefly introduced in an earlier chapter. It was also briefly mentioned that a chemical stimulus emanating from the leaf and transported to the meristem was responsible for photoperiodic induction and it was labelled as 'florigen' by Mikhail Chailakhyan of Soviet Union in the 1930s. Grafting experiments conducted by him not only enabled him to predict the presence of this hypothetical florigen, but also provided two important bits of information: (i) Florigen moves only through a living tissue union between the two graft partners and probably only through phloem tissue. (ii) Florigen frequently seems to move with the assimilate stream, but it can also move by other mechanisms, in addition to the assimilate stream. Grafting experiments done by others have also indicated that although florigen is produced in response to widely different environmental conditions, it is the same compound or at least physiologically equivalent whether it be SD or LD plants in many, if not all, angiosperms. Thus, florigen is a promotary substance. Failure to isolate florigen so far has led to scepticism about its existence.

There is also a strong evidence for the involvement of inhibitory substances or processes in the photoperiodism. Indeed, both promoters and inhibitors must influence flowering. If a plant is induced by exposing only one leaf to suitable photoperiodic conditions, flowering is frequently inhibited by the non–induced leaves. The presence of leaves on a receptor plant in a graft experiment also reduces receptor flowering. These inhibitory effects are often caused by influences on the translocation of assimilates. Attempts to extract promoters and inhibitors have also been disappointing so far.

Thus, it can be concluded that there are circumstantial evidences to show that floral initiation is controlled or strongly influenced by hormones; one or more positively–acting florigens and one or more negatively–acting inhibitors are likely to be involved. These substances remain to be correctly identified and characterized. Recently several studies have highlighted a role for a product of *Arabidopsis FLOWERING LOCUS T (FT)* gene as a component of florigen (Corbesier and Cou pand 2006), particularly FT protein; the homologue of *FT* called *TWIN SISTER OF ST (TSF)* protein in *Brassica napus* (Givavaliso *et al.* 2006) and the orthologue of *FT* called *SINGLE-FLOWER TRUSS (SFT)* protein in tomato (Lifschitz and Eshed 2006).

Several studies have been carried out to record the responses to applied plant growth hormones or growth regulators. Auxins, ethylene, gibberellins, cytokinins, ABA, sterols, γ tocopherol and other growth regulators have been applied to understand the chemical basis of photoperiodic induction. The results of these studies are as follows: The flowering response is often influenced by applied plant growth regulators and other compounds, but a few patterns that can be tentatively discerned have many exceptions. Several compounds, including sucrose and reduced N-compounds, can cause flower formation, but there is no convincing evidence that florigen is or is not one of the well–known plant hormones, although a role for gibberellins seems increasingly likely.

9.7. THERMOPERIODIC AND OTHER TEMPERATURE–RELATED EVENTS

9.7.1. Temperature Effects on Plant Growth

Plant growth is highly sensitive to temperature. Often, a change of a few degrees of temperature leads to a significant change in plant growth rate. Each plant species requires at any given stage of its life cycle and in any given set of environmental conditions, a **minimum temperature** below which it does not grow, an **optimum temperature** (or range of temperatures) at which it grows at a maximum rate, and a **maximum temperature** above which it does not grow and may even die. Alpine and arctic species have low minima, optima and maxima, while tropical species have much higher **cardinal temperatures**. Often different tissues within the same plant have different cardinal temperatures. Temperature changes can induce nastic movements in plant parts such as the petal, and these are referred to by the term **thermonasty**. Temperature also plays important roles in seed germination, floral evocation, induction or breaking of seed or bud dormancy etc. Temperature often affects plant growth through its effect on enzyme action either through denaturation of enzymes or by increasing the kinetic energy of the reacting molecules. However, there are positive responses in plants if the temperature is increased from the minimum to the optimum.

9.7.2. Thermoperiodism

Went (1957) introduced the term **'thermoperiodism'**, a phenomenon in which growth or development is promoted by alternating day and night temperatures. Potato tuber formation and tomato fruit set are promoted by low night temperatures. Stem extension and floral evocation were also due to thermoperiodic responses. However, fluctuating day and night temperatures are not necessary for the optimum growth and development of many other plants. For example, plants such as sugar beet, oats, beans and pea grow as well at optimum constant temperature as they do under varying day and night temperatures.

Many thermoperiodic responses interact with the light environment, typically via photoperodism and probably via balances in the phytochrome system. Different tissues within the same plant have different cardinal temperatures. For proper growth of the entire plant the temperature range during the day should include near–optimal temperatures for the growth of all necessary tissues. The cardinal temperatures for roots and shoots of some plants are different and therefore, if roots and shoots are kept at the same temperature optimal growth may not be obtained.

9.7.3. Vernalization

This process has been studied very extensively. Although known even by the middle of 19th century, it has not attracted much attention until the early 20th century when Gustav Gassner in Germany and Lysenko in former Soviet Union renewed the interest on this topic through their research. Lysenko in fact coined the term **vernalization**. Vernalization from Latin can be translated into 'springization' in English, implying that winter cultivars are converted to spring or summer cultivars by cold treatment. It should, however, be pointed out that the term 'vernalization' has been widely misused in the past. Any plant response to cold, for instance, has sometimes been wrongly referred to as vernalization. It is better to restrict the term vernalization to low temperature promotion of flowering, but not to denote **stratification** or **prechilling.**

9.7.3.1. *Response Types*

There are numerous vernalization responses, depending not only on species, but also frequently on varieties and cultivars within the same species. There are, for example, delayed and non–delayed responses. Most plants that have been studied give delayed response after cold treatment, while a few plants such as brussels sprout form flowers during cold treatment itself or immediately after it. Another, but often the best, way to classify the response types is according to the *age* at which the plant becomes sensitive to cold. Many winter annuals, including some cereals, often respond to low temperatures as seedlings or even as seeds, provided that sufficient oxygen and moisture are present. Since cold requirement is a **quantitative** or **facultative** one i.e., low temperatures result in faster flowering, but not a **qualitative** or **absolute** one i.e., flowering absolutely depends upon cold, there is another basis for classification. Most winter annuals are delayed and quantitative in their response, although some have absolute cold requirement for flowering. The biennials such as beets, cabbages and carrots normally germinate in the spring, leaves die back in the autumn to protect the apical meristem, form new leaves in the second spring and rapidly elongate their flowering shoot (bolting). These plants have an absolute cold requirement and must experience several days to several weeks of temperatures slightly above the freezing point to subsequently flower. Many species of cold–requiring plants cannot be classified in the categories of winter annuals or biennials. These plants, after cold treatment, must have a short–day or a long–day requirement for flowering. In all the above cases, we are not dealing with an endogenous clock system, but with a complex system in which a plant responds to one season by becoming prepared for the next season.

9.7.3.2. *Locus of Response*

It is the vegetative bud, presumably the meristem part of it, which normally responds to cold treatment and become vernalized. Only if the buds are subjected to cold treatment will the plants subsequently flower. Embryos or even isolated meristems from rye seeds have also been vernalized and if a vernalized meristem is transplanted to stock, the latter will ultimately flower. However, if a meristem from an unvernalized plant is grafted onto a vernalized plant after removal of the vernalized meristem, the growing transplanted meristem remains vegetative. Hence, it has been suggested that vernalization requires dividing cells or DNA replication in non–dividing cells. It is likely that the zonal cytology of vegetative shoot meristem needs to be transformed into a meristem with mantle–core arrangement and this transition is promoted by cold treatment (and also by photoperiodism).

9.7.3.3. *Experiments in Vernalilzation*

Vernalization proceeds at a maximum rate over a fairly wide range of cool temperatures depending somewhat upon the species. Vernalization can even occur at a few degrees below freezing point. Thus, there is typically a fairly broad optimum–temperature range (about 0°C–10°C), but some positive effects have also been observed at temperatures even as high as 18°C–22°C in some species. What is the most effective vernalization time? Experiments carried out to resolve this issue have indicated that the minimum lengths of time of vernalization treatment for any observable effect vary from 4 d to 8 wk, depending on the species. Saturation times have been reported to vary from 3 wk for winter wheat to 3 mo for henbane.

If vernalization treatment in a plant is immediately followed by exposure to high temperatures, the plant often does not flower. This reversal effect is called **devernalization.** For a very striking devernalization, the temperatures applied must be about 30°C or higher for winter rye, and also they must be applied for a few days and within 4 or 5 d (somewhat longer for other species) after vernalization. After devernalization, most species can again be revernalized with another appropriate cold treatment.

9.7.3.4. *Role of Growth Regulators*

As stated earlier, if the apical meristem itself responds to vernalization treatments, a translocated flowering stimulus or growth regulator does not seem necessary. Since in most plants, the vernalization effect is not translocated from one meristem to another, either within the same plant or when a vernalized bud or plant is grafted to a non–vernalized plant, it is likely that there is no growth regulator involved in vernalization or at least there is no transport of it from one place to another in the plant. There are, however, exceptions, but the exceptions are limited to a few taxa. It was proposed by Georg Melchers of Germany that a hypothetical vernalization stimulus or hormone, named by him as **vernalin**, is present. (Melchers 1937). However, the isolation and identification of vernalin have been in vain. But results with gibberellin treatment show that its properties are similar to those expected for the hypothetical vernalin. Since gibberellins clearly seem to be involved in vernalization, the question naturally arises: Is gibberellin equivalent to vernalin? For various reasons, researchers do not consider the two to be equivalent. Chailakhyan (1968) suggested that there are two substances involved in flower formation, especially during vernalization, one is a gibberellin or gibberellin–like substance and the other is a substance called **anthesin**. It was further suggested that plants that require low temperatures (or LDs or both) may be lacking enough amount of gibberellins until they have been exposed to the inducing cold environment. On the contrary, SD plants may contain enough of gibberellins, but may lack anthesin. Chailakhyan stated that vernalization–requiring plants produce vernalin in response to cold treatment, but the vernalin is subsequently converted during long days to gibberellins, at least in those plants requiring long days following the cold. This two–substance concept is supported by Melchers (1937) through his experiments.

9.7.3.5. *Genetic and Epigenetic Stabilization of Vernalized Meristem*

Genetics studies on vernalization in spring and winter varities of diploid wheat and barely have revealed two genes that act antogonistically.

Dominant alleles of *Vrn-Arn1* from wheat and *Vrn-H1* from barely reduce the requirement for vernalization and promote a spring growth habit, while dominant alleles of *Vrn-Arn2* and *Vrn-H2* increase the vernalization requirement and promote the winter growth habit. A few recent studies have highlighted the relevance of epigenetic regulation to the maintenance of shoot apical meristem identity. This type of regulation may help to stabilize gene expression in the face of fluctuating physiological and environmental conditions such as the ones met with in vernalization. Vernalization is a classic example of a developmental state that is imprinted to the shoot apical meristem. Grafting experiments have demonstrated that exposure to extended periods of cold imprints the meristem to a state where it becomes more receptive to floral induction signals than to other signals. This imprinted state reflects epigenetic modification of specific genes such as *FLOWERING LOCUS C (FLC)* that regulate flowering, and this occurs both at the level of DNA methylation and chromatin modification with the latter involving both histone modification and association with stabilizing polycomb group proteins (Sung and Amansino 2005, Veit 2006).

9.8. DORMANCY

9.8.1. Concept and Definitions

Only a few plants can effectively function when subjected to temperatures near to the freezing point of water. Under these circumstances, the following question naturally arises: How do plants in temperate and polar regions live in temperatures close to or below the freezing point for several weeks or months in a year? Invariably these plants become dormant or quiescent, and in this state they retain life but carry out very little metabolic activity. To aid in this, they invariably shed their leaves, and their buds and seeds become metabolically inactive and remain dormant through certain changes in their cells that allow them to resist very cold temperatures (Salisbury and Ross 2004). The same changes are also noticed in buds and seeds of some evergreen trees during the winter season. The buds or seeds accumulate or sum the periods of exposure to cold and, thus, they measure the length of the cold winter period and anticipate spring to renew growth and development. Thus, there is a periodic alternation between dormant and active periods of growth in these plants. In the tropical environment, dormancy is often induced by higher temperature (rather than cold temperature as described above). Under very hot temperature regime, the plants cease active growth and lose leaves due to abscission, and their buds enter into a state of dormancy. When normal temperature regimes return, the dormancy of buds is released and new flush of leaves ensues.

Germination (of seeds), as already indicated in an earlier chapter of this book, refers to those events that commence with imbibition of water and terminate when a seedling from the embryo emerges out of the seed coat. A seed may remain **viable** (alive), but yet may be unable to germinate for several reasons that may be *external* (environmental) or *internal* (in the seed). Internal reasons may include the immaturity of the embryo making it incapable of germination as in members of Orchidaceae and Orobanchaceae, while external reasons include lack of moisture or the presence of unfavourable temperature and light conditions in the environment. To distinguish between these two different situations i.e., internally and externally imposed conditions, seed physiologists have used two different terms: **Quiescence** that denotes a condition of a seed when it is unable to germinate only because suitable external conditions are not available, and **dormancy** that indicates a condition of a seed when it fails to germinate because of internal conditions, even though external conditions are favourable for germination. However, there are problems with this categorization. One major problem is that quiescence and dormancy are not mutually exclusive. Further we cannot draw a clear line between internal and external conditions, as one influences the other. In view of these problems, one can define dormancy as the condition of a seed (or any other diaspore) when it fails to germinate even though (i) there is sufficient external moisture, (ii) the seed (or other diaspore) is exposed to appropriate atmospheric conditions found in well–aerated soil or at the Earth's surface, and (iii) temperature is within the range, usually associated with normal physiological activity (10°C–37°C) (Salisbury and Ross 2004). Accordingly, quiescence is the condition of a seed or diaspore when it fails to germinate unless the foregoing conditions are available. Furthermore, the concept of dormancy is also associated with the idea of induction. In almost all cases, germination does not occur during the treatment that breaks dormancy, but only a little later i.e., after sufficient induction time.

9.8.2. Classification of Dormancy

It is suggested by some plant physiologists that one should use the term '**ectodormancy**' or '**ecodormancy**' in the sense of dormancy, and '**endodormancy**' in the sense of quiescence. The term '**paradormancy**' as a form of dormancy is also proposed. In this form, dormancy is controlled by plant parts other than the one actually undergoing dormancy, for example, the control of bud dormancy by nearby leaves. Others like Bewley (1997) classify seed dormancy in a different way as follows: (i) **Seed–coat dormancy** when the seed is prevented from germination by the constraints of the seed coat. It is also called **coat–enhanced dormancy**. Embryos isolated from such seeds are, however, not dormant. (ii) **Embryo dormancy**, where the embryo itself is dormant. Attention should be drawn here on **exegetic** dormancy, which refers to the dormancy of embryos during their development (Evenari 1984). The embryo suddenly stops developing at a particular stage of development for a variable time period, and then restarts its development. In some species, this dormancy may extend for as long as eight years before the seed reaches a 'germinable' condition. Even as early as 1946, this type of dormancy was found in about 46 angiospermous genera. *Ilex* is one of the most familiar examples, of which about a dozen species have been reported to show exegetic dormancy. The reasons for the presence of this type of dormancy are not clearly understood and it is likely that the reason may differ from species to species. The available data indicate that the controlling factor(s) lies in the tissue immediately around the developing embryo. Nitrogen in an insoluble form and growth inhibitors are two of the possible factors. Also, it is not clear as to how this inhibition is overcome, but low temperature treatment, exposure to dark conditions, avoidance of light etc. are known to break this dormancy. Red–far red mechanism has also been suggested, but conclusive proof is yet to be obtained. However, it is interesting to mention that the ultrastructure of exegetic embryos is similar to that of mature dormant embryos.

9.8.3. Induction and Overcoming of Seed Dormancy

9.8.3.1. Germination vs. Dormancy

The lifespan of some seeds due to prolonged dormancy is indeed great, in some cases exceeding the human life span. *Mimosa glomerata*, for example, has longevity of 221 years, in comparison to a normal life span of 10–50 yr. Many seeds lose their viability (Refer to Box 9.1) within weeks and hence, germinate immediately after release. Invariably they cannot also be stored for a long time under typical seed storage conditions and hence, they are often called **recalcitrant** in contrast to **orthodox** seeds that can be stored for a long time without losing viability. Many seeds are said to exhibit **primary dormancy** as they are dispersed from the parent plant in a dormant state. Primary dormancy is initiated as part of the seed development process itself and there is a great variation among seeds with respect to the period of primary dormancy shown by them i.e., from almost no dormancy period (these include the viviparous taxa) to long dormancy periods (in years). Before considering the various aspects of dormancy, its breaking and the resultant completion of germination, it is ideal to ponder why so little progress has been made so far in understanding the dormancy. One of the contributing factors is that we are still not fully aware of the defining events in germination, and without this 'baseline' data, we cannot understand fully the observations on dormant seeds or buds that block their germination. However, studying seed germination has also been difficult because populations of seed do not complete this process synchronously; release from dormancy is also often erratic because the threshold stimulus required to promote germination varies widely among individual seeds of the same species (Bewley 1997). Although a 'biotime' concept has been introduced incorporating a mathematical model to characterize and predict germination of seeds with respect to dormancy and the factors governing it, the concept has posed additional complexities in studying seed dormancy.

Box 9.1. SEED VIABILITY TESTING METHODS

A seed viability test is any technique that is used to determine whether a seed (in some instances dry fruits) appears to be dead or alive. Such a test enables the estimation of proportion of live seeds (or fruits) in a population. Seed viability, thus, is the state of a seed being alive when given a specific and effective viability test. It should, however, be pointed out that most seed viability tests are destructive i.e., they destroy the property they set out to measure. There are two general test types: Germination tests and Rapid viability tests. Germination tests may involve tests on seeds that are pre–treated or not pre–treated with a germination stimulant (which may help in germination of a non–dormant seed or in breaking the dormancy). The advantage of germination test is that it is a direct measure of germination, while the disadvantage is that it often takes a long time for estimation. The rapid viability tests include cut test, X–ray test, tetrazolium (TZ) test, and Excised Embryo (EE) test. In the cut test, a seed is cut open, examined and identified based on whether its embryo is living or dead from the appearance of its tissues. The cut method is quick, cheap and useful in checking maturity of embryos. Its disadvantages are that it is an indirect measure of germination and it is subjective (not objective). The X-ray test determines the percentage of 'filled' (i.e., potentially viable) seeds in a seed population. Although quick, non–destructive and well–suited to wild species, it is an indirect method involving expensive equipment and does not reveal whether the seed tissues are dead or alive. The TZ test is the most commonly used test. Here the selected seeds (or their dissected out embryos) are treated in TZ solution (0.1–1.0 per cent colourless solution at 30°C for a variable period of 1–3 d depending on species) and then the seeds or embryos are categorized according to the colour, intensity and staining patterns, based on which the seeds are identified as viable or non–viable. The basic principle of the test is that only living cells (of embryo) have dehydrogenases, which convert the colourless, soluble 2, 3, 5, triphenyl tetrazolium chloride into an insoluble, red product 2, 3, 5 triphenyl formazan. TZ test yields a quick and dependable method that is particularly suited to dormant seeds, although it is also an indirect and labour–intensive test, often not capable of differentiating dormant and non–dormant seeds or embryos. The EE test employs the extraction of the embryo out of the seed. Seeds with well–developed and normal–looking embryos are often viable. The embryos are then grown (for 7–14 d) on suitable media; whichever embryos survive and grow (and germinate) are to be considered as viable. Although the EE test method is indirect, labour–intensive and not suited to small seeds, it yields quick results and is well–suited to deeply dormant species (ISTA 1999 and Gosling 2004).

9.8.3.2. Dormancy Factors and Overcoming Them

One of the easiest examples of dormancy to understand is the plants with the presence of a hard seed coat that prevents absorption of water and/or oxygen, as in many legumes. In a few species, water and oxygen are unable to penetrate the seeds because entry is blocked by **strophiolar plug** present in a small cleft (strophiolar cleft) present in the seed coat. Vigourous shaking of seeds sometimes removes this plug, allowing water and oxygen to enter into the seed. This treatment is known as **impaction**. Breaking the seed coat barrier through an abrasive action is called **scarification**. In nature, this abrasive process is taken care of by microbes, or when seeds pass through the digestive tract of birds and other animals, or are exposed to alternating temperatures, or by flowing water or by moving rock and soil particles. In laboratories, it can be done through knives, files or sand papers, through treatment with alcohol or other fat–solvents or through acid treatments. In nature, scarification process can happen at the required time, when other conditions required for germination are favourable.

In a number of seeds, even when other conditions such as temperature are ideal, and ample moisture and oxygen are available, germination does not happen if the seeds are still surrounded by fruit tissues or when they are still inside the pulpy fruit. This is seen, for example, in tomato seeds. In this fruit, the osmotic potential of the fruit juice is too negative to allow germination. Moreover, specific germination inhibitors such as ABA may be present. It is also likely that fruit tissues may filter the required wavelength of light that is necessary for seed germination. In some taxa, the seeds themselves contain germination inhibitors. The chemical inhibitors present in the seed must then be leached out before they can germinate. Rainfall or moving water can enable this process of leaching in nature. One of the inhibitors of germination is sodium chloride. Other inhibitors are organic compounds and these include cyanide–releasing compounds, ammonia–releasing substances, aldehydes, essential oils, alkaloids, phenolics and unsaturated lactones such as coumarins and parasorbic acid. Germination inhibitors may sometimes be produced from leaves, roots and other plant parts of the same plant or from other adjacently growing plants. Inhibitory chemicals of such adjacently growing plants are called **allelopathics**. Other plants and organisms may also serve as sources of germination promoters.

9.8.3.3. Role of ABA and GA

The role of ABA in seed dormancy needs to be discussed. ABA in the endosperm of alfalfa has been shown to prevent the embryo from germination. Seeds rarely germinate when they are developing and if there is precocious germination from such seeds, it is shown to be associated with deficiencies in ABA synthesis, availability or sensitivity. Hence, ABA accumulation in seed is low during its early developmental stages, but becomes the greatest during mid–development only to decline as the seed matures. The above results are confirmed through studies made on mutants deficient in ABA or responsiveness, and also through studies involving exogenous application of ABA. The mutants include maize *viviparous (vp)* (at least 10 mutants are known), tomato *sitiens* (*sit*), and *Arabidopsis ABA–deficient* (*aba*) and *ABA–insensitive* (*abi*) mutants. Also, there are *reduced dormancy* (*dro*) mutants of *Arabidopsis* in which neither ABA synthesis nor sensitivity is altered. The presence and differential sensitivity of embryos to ABA may be important in the maintenance of mature seeds in a dormant state. The possibility for the existence of 'dormancy genes', encoding proteins that act to suppress germination, has also been suggested, similar to ABA's control on their expression.

GAs do not appear to be involved in the control of dormancy per se, but rather are important in the promotion and maintenance of germination (Bewley 1997) i.e., they act after the ABA-mediated inhibition of germination has been overcome. Thus, the activities of ABA and GA seem to be linked in seed dormancy, breaking of dormancy and the subsequent germination of seeds.

9.8.3.4. *Role of Temperature*

In many plants, seeds need to be exposed to low temperature and oxygen under moist conditions for weeks or even for months to enable germination. Rarely, moist seeds respond to high temperatures, and several seeds respond best by germinating when daily temperatures alternate between high and low. The long practice of layering the seeds during winter in flats containing moist sand and peat by lay agriculturalists is called **stratification**. As seeds in the flat need to be cooled before germination, a more popular and appropriate descriptive term than stratification is **pre–chilling.** It is a method of **after-ripening,** a term used by some authors in reference to any change that goes on within a seed or bud during the breakdown of dormancy. In a more restricted sense, others use the term 'after-ripening' to maturation changes occurring in the embryo during storage. The problems of after–ripening of seeds are basically concerned with the physiological growth of the embryo. This is evident from the fact that only embryos, excised from fully after–ripened seeds, are able to grow in cultures and germinate, and not those obtained from non–ripened seeds. A variety of cellular and biochemical parameters associated with after–ripening has been followed in seeds of various species and marked differences in cell number, dry weight, oxygen uptake, nitrogen and phosphorus contents, amino acids, organic acids, enzymes, fats, proteins and carbohydrates between embryos of after–ripened and non–after–ripened seeds have been reported.

In nature, low temperature requirement protects seeds from precocious germination during autumn or during unseasonal warm period in winter. We have to understand the chemical changes that go on within the seed during pre–chilling so as to allow it subsequently to germinate when the right conditions come. Pre–chilling definitely affects different phases of embryo metabolism, but it is not clear whether any of these phases is directly responsible for releasing seeds from dormancy. Pre–chilling also enhances nucleic acid (both RNA and DNA) synthesis. It is likely that stored fat massively gets degraded, protein content rises, and starch disappears in the seeds during the cold. It is also likely that pre–chilling removes inhibitors and/or promotes the accumulation growth promoters (that make embryos to germinate) such as gibberellins or cytokinins (auxins have little effect on germination according to some). This is evident not only from the detection of GA in pre–chilled embryo, but also from GA–inhibitor treatment experiments made on pre–chilled seeds where germination is prevented. In many cases, gibberellins are substituted for all or part of the cold treatment, just as they often do in vernalization. Cytokinin effects are usually less dramatic and are much less widespread. It is likely, according to some, that ethylene and possibly IAA are also needed for pre–chilled embryos of at least some taxa to germinate. Thus, both inhibitor disappearance and hormone accumulation in whole seeds have been observed. However, there are numerous complexities that are to be resolved.

9.8.3.5. Role of Light

Dormancy of seeds of a wide variety of taxa is overcome by light. Seeds in which dormancy is broken by light are said to be positively **photoblastic**. A brief exposure of fully water–imbibed seeds to red light induces germination. Red–light treatment followed by far–red light is mutually reversible, indicating the operation of phytochrome, which is apparently present in dormant seeds in the red absorbing form, **pr**. Seed dormancy is broken when **pr** phytochrome is changed to the far red form, **pfr** by red light and is re–imposed when **pfr** is reconverted to **pr**. So, there is a rehydration of **pfr** phytochrome during dormancy breakdown. With the cloning of phytochrome gene, it has been found that light–stable *PHY B* molecular species of **pfr** is responsible for inducing dark germination of *Arabidopsis* seeds. *PHY A* has been assigned a secondary role in germination, specifically in the light–promoted breaking of dormancy.

9.8.3.6. Molecular Basis

The molecular basis for the breaking of seed dormancy still remains to be deciphered partially because some reports seem to contradict each other. For example, compounds that inhibit respiration can often break seed dormancy. Where does the seed dormancy mechanism lie? Seed coat, endosperm or even the embryo may be the likely controlling site of dormancy, but there is good evidence that the embryo itself responds to cold; yet, there is a little direct evidence that inhibitors in seed coats are affected, although they are often present.

9.8.3.7. *Conclusion*

Seed dormancy is generally an undesirable characteristic in agricultural crops where rapid germination and growth are required. However, some degree of dormancy is advantageous, at least during seed development. This is particularly true of plants where seeds are of value as food or for other requirements, as dormancy prevents germination of these seeds, while still on the parent plant or fruit.

It is important to know how dormancy is induced in seeds. At present, very little is known about the nature of the mechanism that pushes the embryo into dormancy. The embryo all along has been undergoing active growth and differentiation, and how it suddenly lapses into a state of quiescence or dormancy is a great puzzle. In this connection, we can speak of a developmental mode, a dormant mode and a germination mode in the embryo. Available information indicate that entry of the embryo into this metabolically inactive state is presaged by desiccation and drying, resulting in the loss of nearly 90% of water from its cells. At this stage, the developmental mode of the embryo is temporarily suspended. Experimental evidence has indicated that dehydration enables an embryo not only to enter first into a dormancy mode from a developmental mode, but also makes it to get ready for a germination mode once the conditions favouring the germination become available. This change is preceded or accompanied by the synthesis of many specific proteins associated with the germination mode and the inhibition of synthesis of key proteins associated with the developmental mode. This transition from developmental to germination mode with an intermediate dormancy mode thus regulates the activation and deactivation of specific genes. Molecular studies on seed dormancy are still in their infancy and only in the past decade some attempts have been made to isolate and characterize genes that are differently expressed in the embryos of dormant and non–dormant seeds. The fact that the onset and maintenance of dormancy is based on differential gene activation has come from a critical analysis of protein and RNA changes during dormancy. It is more likely that LEA (for 'late embryogenesis abundant') genes and their protein products are vital in inducing dormancy. LEA genes are shown to encode proteins such as RAB 17 (for 'responsive to ABA') and RAB 21, which induce tolerance in embryos through ABA to innate dehydration and desiccation. It is also to be mentioned here that the most thoroughly studied hormonal change in seeds as they sense desiccation and lapse into dormancy is the accumulation of ABA.

9.8.4. Bud Dormancy

9.8.4.1. *Factors Involved*

In temperate regions, seed and bud dormancy have much in common, while in tropical environment bud dormancy is induced by higher temperature and thus has a different mechanism from seed dormancy. In buds, whether temperate or tropical, inducing dormancy is as critical as breaking dormancy. Let us first consider bud dormancy in temperate taxa. Bud dormancy almost always develops before autumn colour–change and the senescence of leaves. Buds of many taxa do not grow in mid–summer, sometimes exhibiting a little growth again in late summer before entering into deep dormancy in autumn. Bud dormancy is induced in many taxa by low temperatures, but there is also response to photoperiodism particularly if temperatures remain high. For example, SD treatment on several deciduous species studied in Maryland, USA resulted in the formation of a dormant terminal bud and cessation of internodal elongation and leaf expansion, although the leaves were often retained. Long nights, each interrupted by an interval of light, had the same effect as LDs.

It is interesting to note that different ecotypes within a species have quite different dormancy responses and thus, emphasizing the genetic basis of bud dormancy. Frequent withholding of water accelerates bud dormancy. Similar is the result of restriction of mineral nutrients, particularly nitrogen and changing of day length. Partial bud dormancy often precedes true dormancy and the former can be reversed by moderate temperatures and LDs, but with the gradual approach of true dormancy, attempts to reverse the bud to active growth fail. Then, true dormancy requires special treatments to overcome.

The morphology of the bud is very important in the dormancy phenomenon. A dormant bud typically has greatly–telescoped internodes and specially–modified leaves called **bud scales**. These scales often are very thick and prevent desiccation. They also briefly insulate against heat loss, and restrict movement of oxygen to the bud meristem. They may also respond to ambient light and/or perform other functions. In a sense, bud scales are analogous to the seed coat.

The hormonal factors involved in bud dormancy are not known clearly, but ABA has been implicated in the response in tree taxa. It has been reported in one investigation that if ABA is fed to the trees through the leaves, one can induce bud dormancy, but this experiment could not be repeated with the same result by others. Hence, it is impossible to conclude that ABA is definitely involved in bud dormancy.

9.8.4.2. Overcoming Bud Dormancy

Bud dormancy can be overcome by treatments involving specific temperatures, day lengths, or both. LDs overcome bud dormancy, even in leafless taxa. Apparently the bud scales themselves respond to LDs when dormancy is being overcome and the leaves are not involved. It is also possible that enough light penetrates the bud scale to bring about the response within the [illegible]dial leaf tissues covered by the bud scales. It is likely that both the SD [illegible]uction and the LD breaking of bud dormancy are due to phytochromes, but the situation is not very clear. In some studies involving SD induction, red light is most effective in the night interruption, and the effect is reversed somewhat by a subsequent exposure to far–red light. However, this reversal has failed in several other taxa.

Bud dormancy of many taxa can be overcome by exposure for days to months at temperatures below 10°C. For fruit trees, 5°C–7°C is more effective than 0°C. The effects of chilling upon the breaking of dormancy are not translocated from bud to bud within the plant, but are localized within the individual buds. Several chemicals are known to break bud dormancy and these include ethylene chlorohydrin (applied in vapour form). Warm water immersion (40°C–55°C) of the plant part and gibberellin application is also known to be effective in removing bud dormancy in some taxa

9.8.4.3. Importance of Bud Dormancy and Break

Dormancy of vegetative buds and the subsequent bud breath in temperate taxa and flushing of new leaves (actually a rapid expansion of preformed bud primordia) in tropical species have great relevance to the growth periodicity and phenology exhibited by vegetative shoot. As already indicated, the shoot system is a collection of metamerically constructed modules with apical meristems (terminal and axillary) of varying age and activity. Depending on both the season of the year and internal correlations within the system, these meristems are either actively growing and producing additional metamers, or they are contained within a dormant bud together with a set of partially–developed metamers, awaiting a stimulus to resume growth. When growth does resume (bud break or flushing of leaves), the major part of submodule elongation takes place in the internodal portion of each metamer. There is correlation between the size of structural organization of a plant (i.e., whether unicellular, a few–celled, or multicellular and with a complicated plant form) and the period of growth rhythm. At the higher levels of whole plant, there are often periods when the meristems are either active or inactive.

The alternation of dormant and active periods of growth in a plant is called **rhythmic growth**, in contrast to **continuous growth** where the shoots have no marked (endogenous) periodicity in growth. In rhythmic growth, shoots have a marked endogenous periodicity of growth. It can be regarded as synonymous with **episodic growth**, a term usually used with reference to temperate trees or with **intermittent growth**, a term usually used with reference to tropical trees. The word 'rhythmic' may imply a regular seasonal cycle, though this should not be intended in its usage. The morphological indication of rhythmic growth in the mature shoot system is a more or less pronounced segmentation of the stem axis, referred to as **articulate growth**, with a series of short and telescoped internodes or small leaves. The shoot system is also often characterized by rhythmic branching called **prolepsis** or **proleptic** branching. In prolepsis, there is a discontinuous development of a lateral meristem from a terminal meristem to establish a branch, with some intervening period of rest of the lateral meristem. Continuous growth is again endogenous and is equivalent to **evergrowing** or **non–articulate** growth. Invariably, continuous growth is characterized by **syllepsis** or **sylleptic branching**. In syllepsis, there is a continuous development of a lateral meristem from a terminal meristem to establish a branch, without an evident intervening period of rest of the lateral meristem. There are no basic, but only gradual, differences between rhythmic (determinate) and continuous (indeterminate) shoot growth. Rhythmic or continuous growth may result from differences in the sensitivity of apical meristems to feedback inhibition of a certain intensity that could induce formation of a rest bud or simply slow shoot growth in continuously growing trees.

The dormancy or otherwise of a vegetative bud is fully reflected in and decided by the structural changes that the shoot apical meristem undergoes during periods of unfavourable growth. Conifer shoot apices that have been studied with regard to seasonal variations in structure reveal two basic types of structural changes. For some species such as *Pinus ponderosa*, *Abies concolor*, *Cephalotaxus drupacea* etc. the basic zonation pattern of the meristem does not change, but the height of the apical dome above the youngest node is greater during active growth than during rest or dormancy. Because of this difference, the zones are differentially distributed in the two kinds of apices with regard to the youngest node. The rib meristem occurs below this node in resting apices and partly above this node in active apices. A loss of zonation and the appearance of tunica–corpus–like structure have been reported for the dormant shoot meristems in *Tsuga heterophylla* and *Picea mariana*, which form the second basic type.

9.8.5. Dormancy of Underground Storage Organs

In many cases, temperature variations not only induce the formation of underground storage organs such as bulbs, corms and tubers, but in some other species, dormancy of these organs is also broken and subsequent growth influenced by storage temperature. In some, day length is also involved in their formation or dormancy.

9.8.5.1. *Potato Tubers*

A lot of studies have been done on potato tubers. Four steps have been identified in tuber formation: (i) Stolon induction and initiation from the axil of an aerial leaf (ii) Elongation and branching of stolon towards the soil (iii) Cessation of elongation growth of the stolon (once it reaches inside the soil) (iv) Tuber induction and initiation during which there is radial growth of the stolon tip to form a tuber. All these four distinct steps have been studied in relation to temperature, day length and hormonal treatments. Stolon initiation occurs over a wide range of temperatures and day lengths, but the development of stolons into tubers usually requires more specific conditions. High levels of gibberellins are required for stolon initiation; cytokinin levels should not be too high. LDs favour stolon elongation, but SDs cause cessation of stolon growth. Ethylene stops stolon elongation, but it also stops tuber formation. Tuber formation is a response to lowered gibberellins and ethylene (negative influences). There is a **tuber–inducing substance** that forms in the leaves of some cultivars in response to SDs. Tuberization is optimal at night temperatures of about 12°C. An interaction between photoperiodism and temperature is common in tuberization, as it is in vernalization and dormancy. In sensitive cultivars, on LDs, no tuber will form at any soil temperatures unless the shoots are exposed to low temperatures. Hence, the leaves must detect both the photoperiod and the temperature, and must transmit the tuber–inducing substance to the stolon. This substance is proved to be a complex molecule similar to jasmonic acid.

Tuber dormancy can be induced or prolonged by spraying such growth regulators as maleic hydrazide or chloropropham on the foliage. Potato tuber dormancy enables the tuber to keep all its buds (eyes) dormant and cutting the tuber to pieces often eliminates this dormancy. For breaking dormancy, certainly no cold requirement is needed. It is possible to break potato dormancy by those chemical treatments that are effective in breaking bud dormancy. These include chemicals like 2–chloroethanol, and gibberellins, and hot–water treatment.

9.8.5.2. Other Underground Organs

There are other kinds of underground storage organs in plants such as bulbs, corms and rhizomes. There is scarcity of detailed investigations on the dormancy and quiescence of these organs, although a few studies have been carried out in Holland and in USA. The studies by the Dutch bulb industry are concerned with the determination of optimum temperature requirements for the storage of bulbs and the duration of storage that will enable them to remain dormant or quiescent under storage and to result in the development of leaves, flowers and stems of desirable quality and at the desirable time when sown in field. Another goal for such a research is to shorten the time of flowering, a process known as **forcing**.

Dutch studies further indicated that bulbs must attain a critical size, which often takes two to three years in order to begin to respond to specific storage temperatures by forming floral primordia, either preceded by or simultaneously with the formation of leaf primordia. Temperatures that induce flowering are relatively high when compared to those that effect vernalization of seeds and whole plants, although the response in both is identical. In *Iris*, although floral primordia actually appear during winter (with temperature 9°C–13°C), a high temperature (20°C–30°C) pre–treatment is necessary if floral evocation is to occur at all. This is a typical example of induction similar to vernalization, but here the response is to high temperature instead of low temperatures. Studies made on

Pediomelum subacaule in USA have revealed that these plants with underground dormant buds at the tips of tuberous roots were never truly dormant, but only quiescent as they would grow at any time if temperatures were right and enough soil moisture was available.

9.8.6. Root Dormancy

Much attention has not been paid so far on root dormancy obviously because of its non–commercial importance, underground location and also because of difficulties in studying it. Under very cold or hot conditions, the root tip does not grow continuously or at least at the same rate as it was under optimum temperature conditions. This is especially true for perennial plants (Kozlowski and Pallardi 1997). In *Abies procera*, for example, the roots show periodic deceleration of growth and have periods of dormancy (Wilcox 1954). Dormancy is preceded by lignification of cell walls and by a deposition of suberin, a double process called **metacutization,** in the cortex and root cap throughout a layer of cells that is continuous with the endodermis and completely covers the apical meristem. The latter, thus, becomes sealed off by a protective layer on all sides except toward the base of the root. Externally such root tips appear brown. When growth is resumed, the brown covering is broken or sloughed off and the root tip pushes beyond it. Studies on excised roots indicate that roots may have a growth rhythm not dependent on seasonal changes, but determined by internal factors (Street and Roberts 1952). These conclusions may not be fully correct because such changes invariably occur during very cold seasons in temperate trees and very hot months in tropical trees.

The root apical meristem also undergoes characteristic structural changes during dormancy. The actively dividing promeristem cells stop their division and become low in metabolic activity. When normal conditions resume, new meristem cells are produced by the quiescent centre (QC) that is obviously not affected by dormancy. It has been reported that the QC resumes division during recovery from a period of dormancy induced by cold (Barlow and Rothfelder 1985). The QC in the long roots of *Euphorbia esula* apparently undergoes a seasonal fluctuation in cell production (Raju et al 1976). At the height of the growing season, they exhibit a well–developed QC, but during recovery of growth early in the growing season, a QC is not discernible, since its cells are now possibly involved in reconstructing a functional root apical meristem. However, it should be mentioned that the root tip, unlike a shoot tip, lacks a prolonged dormant phase. This may be evidence to the fact that, unlike the dormant buds of prospective shoot modules, the lateral root primordia are not subjected to multiple correlative influences from elsewhere in the root system. Instead, they are subjected primarily to the influence of the position at which the primordia are initiated and the position at which they would emerge later as young lateral roots.

9.8.7. Periodicity in Latitudinal Growth

Within the limits of genome action, the cambial meristem involved in latitudinal growth (growth in girth of plant axial organs) operates under the influence of internal physiological processes and external environmental factors. The cambium is dormant during times of cold and heat stress that are seen respectively in temperate and tropical taxa. The activation of cambium, a determinate process, ensues when the stress is removed during favourable growing season. Thus, cambial activity is periodic like apical (longitudinal) growth, rather than continuous, which is seen mostly in many evergreen taxa. Normally, species that show continuous longitudinal growth also have continuous latitudinal growth and similarly species that show rhythmic longitudinal growth also exhibit rhythmic latitudinal growth. In temperate species, the radial growth due to cambial activity usually occurs in spring and summer, which is followed by a period of dormancy in winter that extends up to the next spring. Even in the tropical species that apparently have a continuous longitudinal growth, there is some degree of unequal cambial activity, at least during some months in a year, when there is cold or heat stress to some extent. In taxa with rhythmic cambial activity, the number of times the cambium becomes dormant or active is one each in a year, but in certain taxa more than one dormant or active period may be seen within a year depending on changes of prevailing environmental conditions.

The cambium becomes dormant with the approach of winter in temperate taxa and remains dormant throughout the winter. The active cambium becomes gradually restive with low metabolic activity and then becomes dormant without any metabolic activity. The cambial zone becomes three to four–layered during dormancy. The four–layered zone is very characteristic of many conifers and has been historically known as **Sanio's four** named after Sanio, a great wood anatomist. Of these three to four layers, one represents the cambial initial layer, one or two xylic derivatives on the inside and one or two phloic derivative on the outer (phloem) side. A similar dormant cambium is also seen in dicots, although not known by 'Sanio's four' (Figs. 139 B and 140). The cambial cells lose their division activity, both periclinal and anticlinal during dormancy. There is no metabolic activity in the dormant cells. The radial cell walls become much thicker and become prominently beaded (Fig. 266). The cambial cells and cells around the dormant cambium store a number of chemical substances, including protein bodies, starch, lipids, phytoferritin secondary metabolities, crystals etc. The cells contain fewer or no vacuoles; if fewer, the vacuoles are very small. Cytoplasmic streaming is lost. The cytoplasmic organelles become less abundant or poor. The ER becomes smooth and only in rare instances ER remains rough. The plasma membrane is thrown into folds. The ribosomes are free. The dictysomes become fewer and poorly active. The mitochondria become fewer and occur in chains or groups; occasionally they are present singly. The water content of the cells also becomes highly reduced. Growth inhibitors accumulate.

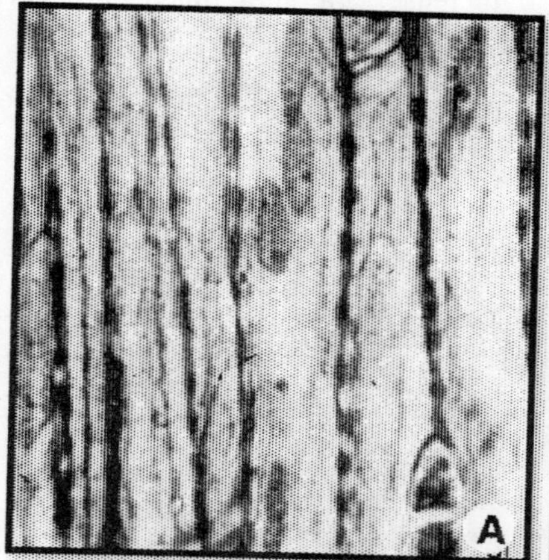

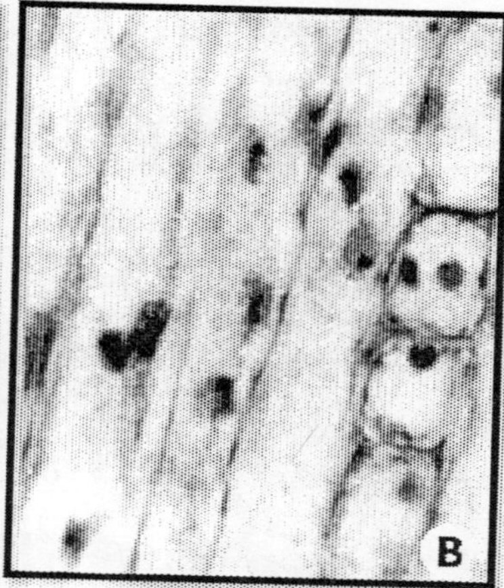

Fig. 266. T.L.S. of portion of dormant cambial region of *Albizia lebbeck* (A) and *Dalbergia sissoo* (B) showing the beaded nature of the radial walls (Photographs courtesy of Dr. N. Venugopal)

There is close relationship between bud break and cambial reactivation in termperate trees and between flushing of new leaves and cambial reactivation in tropical trees. Consequently the number of times that cambium gets reactivated depends on the number of bud breaks or new leaf flushings in a year. Hence, extension growth precedes radial growth. Most workers agree that auxins produced in the breaking buds stimulate cambial reactivation through onset of cell divisions. Generally, the cambium becomes active in diffuse–porous species up to a week before a visible bud break, while in ring–porous species, activity starts 1–9 d before and in conifers 5–15 d before visible bud break. In tropical trees, however, cambial growth gets reactivated generally after visible bud emergence and leaf flushing. During reactivation, the level of growth promoters increases in cambial cells. The initiation of periclinal divisions (Fig. 139A) is usually preceded by cell swelling, with the water content drastically increasing. The growth expansion is predominantly radial and a broad zone of cambium with several rows of cell is resulted (Fig. 267 A and B). The cytoplasm becomes less dense. Vacuoles fuse to form one or two large vacuoles around which cytoplasmic streaming is initiated. Most or all storage products get exhausted and used up in the formation of more derivative cells. High metabolic activity ensues. The radial walls become thinner and their beaded appearance becomes less apparent. The cell organelles increase in number. The ER becomes rough. Dictyosome activity and the number of coated vesicles increase, as also ribosomes. The mitochondria become highly dispersed. In a few cases the number of nuclei in the fusiform initials is reported to increase, although this is highly contested by others.

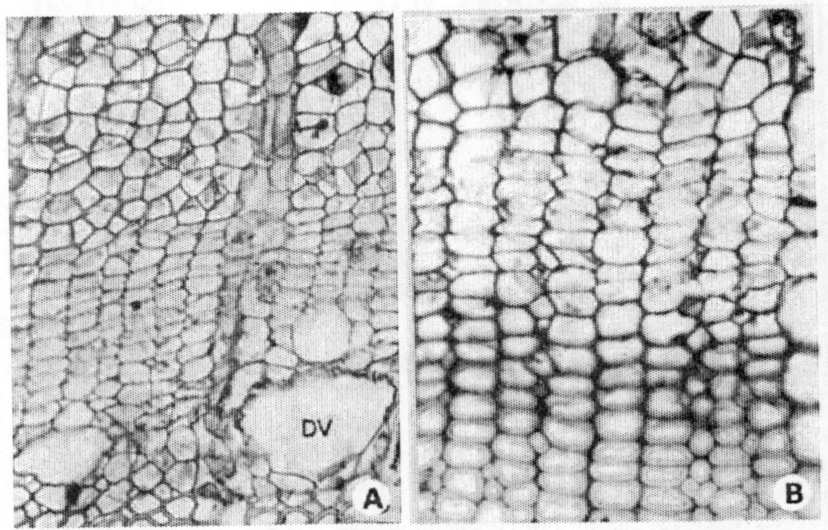

Fig. 267. A - T.S. of active cambial zone and the derivative secondary phloem and secondary xylem in *Tectona grandis*; DV - Developing vessel element; B - Same of *Mangifera indica* (Photographs courtesy of Dr. N. Venugopal)

Cambial reactivation in tropical trees (Figs. 268 and 269) may happen on different months only once in a year and this happens on different months, depending on the taxon (Table 6), More than one cambial reactivation from dormancy per year is noticed in a number of taxa. For example, in *Holarrhena floribunda,* first cambial reactivation happens in February–March and the second in September, in *Ricinodendron heredelotii,* first reactivation in February and the second in December, and in *Psidium guajava* first reactivation happens in March and the second in July. It is of interest to mention here that in the same species growing in different areas, cambial reactivation may commence in different months: *Bombax buonopozense* in February-March and in May, *Ricinodendron heudelotii* in February or March and in December, *Tectona grandis* in early March and early June to early July. There is also variation in the production period of wood and secondary phloem i.e., their production from vascular cambium need not always be simultaneous (Figs. 268 and 269) (Venugopal and Krishnamurthy 1987, 1989), as generally believed.

Cambial dormancy and reactivation are closely accompanied by dormancy and reactivation of the cork cambium or phellogen. Consequently, there is a corresponding active or dormant period of production of secondary cortex (or phelloderm) and cork (or phellem) in a year.

Table 6. Period of cambial reactivation in representative taxa

Sr. No.	Taxa	Period of cambial reactivation
1.	*Some Nigerian taxa*	January–March
2.	*Dalbergia sissoo*	February–March
3.	*Bombax species*	February–March
4.	*Sterculia tragacantha*	late February
5.	*Tectona grandis*	early March
6.	*Cedrela toona*	March
7.	*Ricinodendron headelotci*	March
8.	*Gmelina arborea*	March
9.	*Prospis spicigera*	early April
10.	*Ficus rumphii*	late April
11.	*Bombax buonopozense*	May
12.	*Melia azedarach*	late May
13.	*Streblus asper*	late May
14.	*Polyalthia longifolia*	May to June
15.	*Mimuşops elengi*	early June
16.	*Delonix regia*	early June
17.	*Acacia catechu,*	early June to early July
18.	*Albizia lebbeck,*	early June to early July
19.	*Bombax malabaricum,*	early June to early July
20.	*Shorea rolusta,*	early June to early July
21.	*Syzygium cumini,*	early June to early July
22.	*Tectona grandis*	early June to early July
23.	*Terminalia tomentosa*	early June to early July

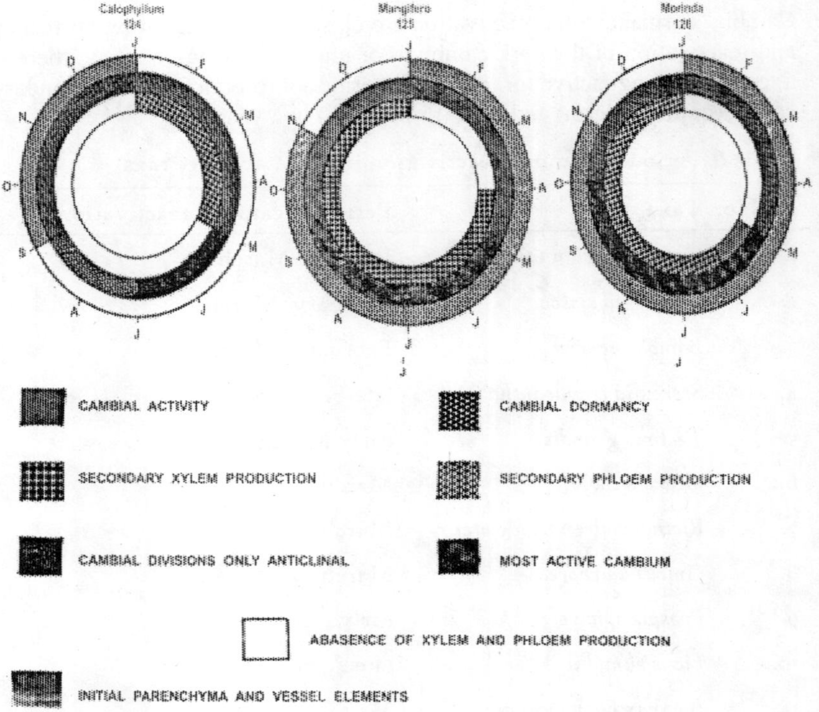

Calophyllum

Mangifera

Morinda

CAMBIAL ACTIVITY

CAMBIAL DORMANCY

SECONDARY XYLEM PRODUCTION

SECONDARY PHLOEM PRODUCTION

CAMBIAL DIVISIONS ONLY ANTICLINAL

MOST ACTIVE CAMBIUM

ABASENCE OF XYLEM AND PHLOEM PRODUCTION

INITIAL PARENCHYMA AND VESSEL ELEMENTS

Fig. 268. Annual cycle of changes in the vascular cambium and the secondary xylem and phloem produced from it in three different taxa. The months are indicated by their first letters.

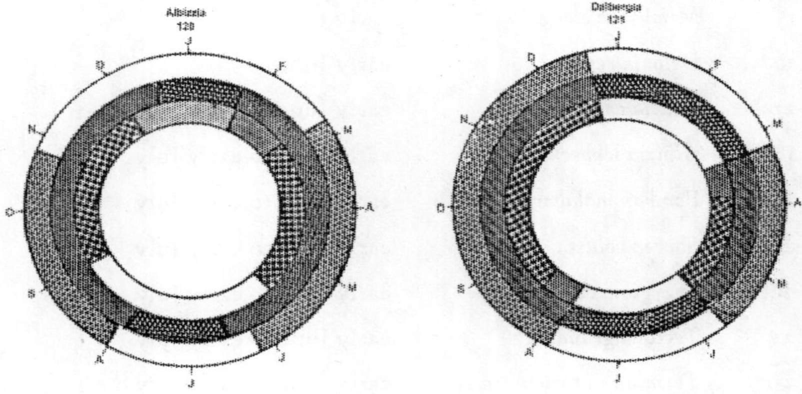

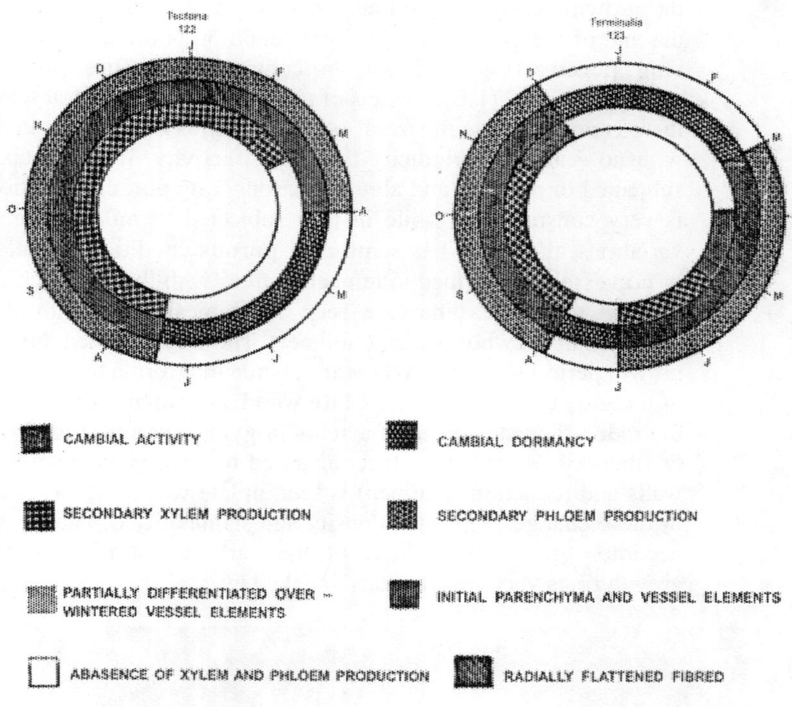

Fig. 269. Annual cycle of changes in the vascular cambium and the secondary xylem and phloem produced from it in four different taxa. The months are indicated by their first letters.

9.8.8. Growth Rings

It was indicated earlier that the activity of the vascular cambium is not uniform throughout its period of existence and this is reflected in the characteristics of the differentiating xylem and phloem. One major result of this non–uniform activity of vascular cambium in a year is the production of **growth rings** (Fig. 270).

The differences noticed in the elements produced in the xylem (wood) during each period of activity are responsible for the demarcation of growth rings. These anatomical differences can be summarized for both tropical and temperate woody plants as follows (Fahn and Werker 1990, Carlquist 2001).

1. Periodical changes in pore (vessel) diameter; the largest usually found at the beginning of the growth season, sometimes in the middle of a growth ring, and the smallest usually at the end of

the growth season. Sometimes vessels are altogether absent at the end of the growth season. This condition is often termed as **ring porous** (Fig. 271 A), in contrast to **diffuse porous** condition (Fig. 271 B) (vessels of almost uniform size or at least there is no gradation in size of pores), which is common in trees with no seasonal periodicity in cambial activity. In many taxa subjected to distinct (and abrupt) seasons, ring porous condition is very conspicuous, while in taxa subjected to mild seasonal variations, the wood has **semi–ring porous** condition. Smaller or no vessels are formed in temperate trees with the approach of winter, while the same are seen in tropical trees with the approach of very hot season (summer). The wood formed during growth period is called **early wood,** while that formed at the end of the growth season is called **late wood**. A similar reduction in the radial diameter of the tracheids in gymnosperms (Fig. 272) or fibers of dicotyledons (accompanied by greater thickness of walls and reduction in lumen) is seen in late woods. As a result of these changes, often the density and stainability of late wood becomes greater than those of the early wood making the growth rings very conspicuous to naked eye.

Fig. 270. T.S. of portion of a trunk of a tree showing growth rings

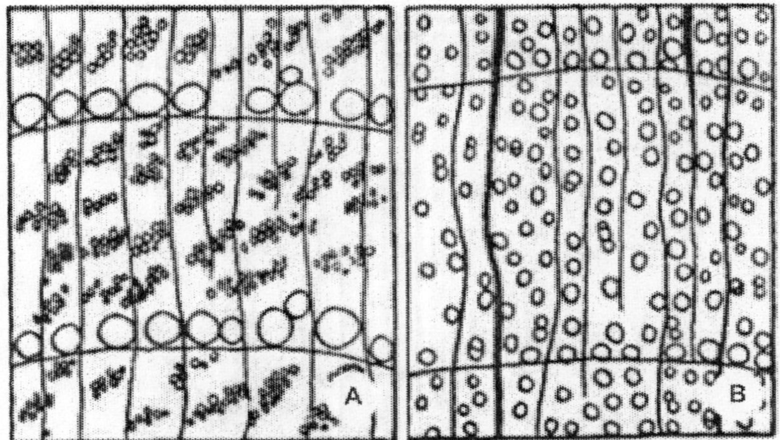

Fig. 271. Diagrammatic representation of ring porous (A) and diffuse porous (B) woods

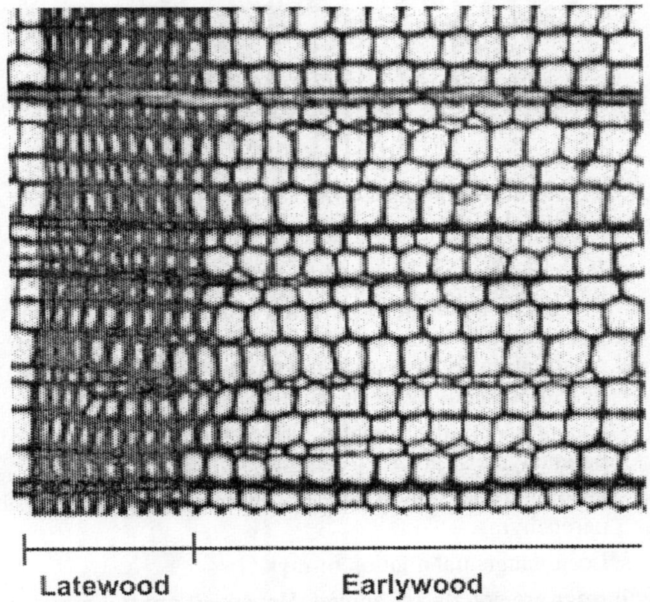

Fig. 272. T.S. of a portion of a conifer wood showing earlier wood and late wood, together making one annual ring

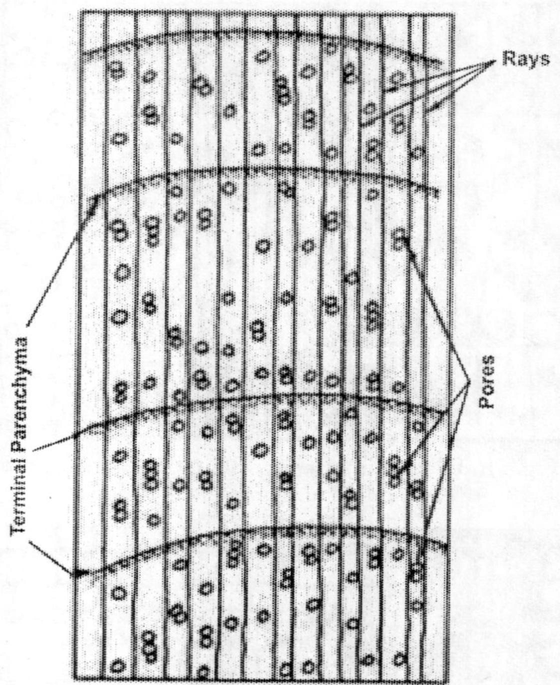

Fig. 273. Diagrammatic representation of a portion of a dicot wood showing terminal parenchyma formed at the end of each growth ring

2. Periodic changes in the pattern of vessel arrangement.
3. Presence of **marginal parenchyma** either in the form of **initial parenchyma** (formed at the very beginning of cambial reactivation) or **terminal parenchyma** (formed at the very end of cambial activity) (Fig. 273).
4. Changes in the length of elements, especially of axial parenchyma strands.
5. Periodicity in the width of alternating bands of fibers and parenchyma.
6. Local tangential dilation of rays.

Growth rings are not always annual. Hence, all **annual rings** are growth rings, but not all growth rings are annual. **False rings**, complete or incomplete, or two rings in a year, may be produced by drought, frost, fire, flooding, and defoliation caused by animals, or a sporadic flush of extension growth of single branches (Fahn and Werker 1990). Growth rings may be discontinuous or missing at certain heights of trees, especially in old or slow–growing trees and in those with restricted photosynthetic activity. In has also been suggested that this phenomenon may also be due to hindered transport of auxin and, possibly, other growth hormones following limitation in bud growth and development of the crown. Growth inhibitors may also be involved in this. False growth rings in gymnosperms are due to thick–walled tracheids appearing in the middle of a growth ring, and in some dicotyledons due to the change in thickness of fiber walls. It should also be mentioned here that the presence of growth rings does not always denote a period of total absence of cambial inactivity.

In the tropics, growth rings are less reliable in denoting annual rings than in temperate areas. Growth rings in tropical trees are sometimes irregular, and in young trees may not be annual and can be totally absent. Although tropical plants invariably grow throughout the year because of the favourable conditions, alterations in radial growth are not uncommon. There are fluctuations in temperature, day length, rain fall and relative humidity, all of which either alone or in varied combinations result in differential rate of photosynthesis and differences in flower and fruit bearing. These changes result in interruption of cambial activity in some trees, which include both deciduous and evergreen trees. Especially in the latter, even if cambial activity is seen to be continuous, variations in its rate give rise to growth rings (Fahn and Werker 1990). 85% of species studied in Malaysia showed distinct growth rings, in Brazil rain forests 35% have clearly–defined growth rings and 22% have poorly–defined growth rings, while in the same country out of trees growing in seasonal climate 60% had distinct, 25% had less defined and 11% had no growth rings. In India 25% trees had well–defined growth rings. The African continent also had trees with distinct, less distinct or no growth rings (Fahn and Werker 1990).

A seperate branch of science called **dendrochronology/ dendroclimatology** has emerged that deals with growth rings in relation to past climatic changes.

Alternate Strategies in Growth and Development 10

10.1. INTRODUCTION

It was remarked earlier in this book that most plants (barring a few algae) have an alternation of generations in their life cycles involving diploid and haploid generations and that the two cardinal points in the life cycle are meiosis, whereby chromosome number is reduced to the haploid condition and gametic fusion resulting in a diploid zygote. While gametic fusion provides a scope for genetic recombination, meiosis helps in segregation of genes to offsprings. This life cycle pattern is faithfully followed in most of the plants, but these plants have the potential for alternative developmental strategies that include the latent capacity of their cells, tissues and organs to develop into whole, new plants bypassing either one or both of these cardinal requirements (meiosis and gametic fusion). Plant regeneration from isolated parts of plants such as bulbs, corms, rhizomes, tubers, stem cuttings etc. is often easily obtained in agricultural and horticultural practices and constitutes the basis of vegetative reproduction or propagation. The ultimate basis of this type of reproduction is the dedifferentiation or redifferentiation ability and totipotency of cells of plants, which may result either in whole plant regeneration as in in vitro cultures or may proliferate to various degrees as in many abnormal growths depending on the availability of the required nutritional inputs.

The ideas on dedifferentiation and totipotency can be traced to the prophecy of the German botanist Haberland at the beginning of 1900s (Refer to Box 10.1). He suggested that it would be possible to grow cells of plants under artificial conditions to produce new plants, although he himself could not achieve it in spite of his best efforts. For a further period of 50 yr from his time, efforts in this direction by others failed to achieve this goal of demonstrating totipotency until Steward et al (1958) obtained whole plants from isolated single cells of carrot. This chapter provides details on these alternative strategies in growth and development exhibited by plants themselves on their own after inducement from biotic and abiotic stresses, and also the details on human ingenuity of demonstrating these innate abilities of plants to bypass meiosis and gametic fusion.

BOX 10.1. THE BEGINNINGS OF PLANT CELL AND TISSUE CULTURE

The history of the concept of plant cell and tissue culture may be said to have begun in 1838–1839 when Schleiden (1838) and Schwann (1839) independently stated the 'Cell theory', in which there was an implicit postulation that the cell is autonomous and even totipotent. But, the above two authors did not have any experimental evidence for this implicit postulation. Meanwhile wound healing and the accompanying callus formation were observed by other botanists and these results were discussed and summarized by Sachs (1880–1882). Among the major conclusions reached by Sachs, the foremost was that plants contain organ–forming substances in their cells or tissues and these substances are distributed in a polar manner. It was also shown that there was a transformation of quiescent cells into dividing cells at the wounded site.

The first approach towards actual tissue culture was made by Rechinger (1893) when he tried to experimentally determine the limit of plant–tissue proliferation by using isolated buds, root slices, stem bits etc. grown on the surface of moistened sand. He neither used nutrient media nor aseptic conditions. The principles of plant cell or tissue cultures were very clearly formulated in 1902 by Goltlieb Haberlandt (1854–1945) an Austrian botanist. Contrary to Rechinger who sought to determine the limits of divisibility, Haberlandt began directly from cell theory and assumed that there were no limits of divisibility of plant materials. Hence, he started working with single cells and Knop's nutrient solution, sucrose, the amino acid asparagine and peptone. The cells were active even after 20–27 d. Haberlandt conducted many experiments whose results were carefully interpreted by him. Using thin slices of potato tuber, he could demonstrate cell division, almost without exception. Although he was not entirely successful, he foresaw the use of cell culture as an elegant and excellent means of studying physiological and morphological problems. Harberlandt's original idea presented in 1902 was called totipotentiality of plant cells: "Theoretically all plant cells are able to give rise to a complete plant". It is a pity that Haberlandt did not pursue further his cell culture studies. However, his unsuccessful attempts, by modern standards, to culture isolated vegetative cells from higher plants (as he selected more specializated and differentiated leaf mesophyll cells and starch containing cells, monocot cells that are mostly recalcitrant, and he used simple nutrient materials) stimulated later investigators to grow artificially cultured cells, tissues and organs.

Goltilieb Haberlandt

See Noe (1934), Gautheret (1983) and Krikorian and Berquam (1969) for more information.

10.2. APOMIXIS

In angiosperms, there is a full range of reproductive phenomena where the usual, normal sexual method of reproduction is replaced by alternative asexual methods. This process is called **apomixis**, a term coined as early as 1908 by Winkler. The existing definitions of apomixis vary greatly. According to Maheshwari (1950), the definition of apomixis is as follows: "Apomixis can be defined as the substitution for sexual reproduction (**amphimixis**) of an asexual process in which the fusion of nuclei does not take place". Grossniklaus et al (1998) have defined apomixis as follows: "Apomixis is a natural process that allows clonal reproduction through seed, resulting in offsprings that are genetically identical to the mother plant". In certain other classifications of apomixis, only the cases of incorrect spore formation, called **aneuspory**, have been taken into consideration and certain other phenomena of disruption of the sexual process such as **pseudogamy**, **androgenesis**, **semigamy** etc. have not been represented. The definition by Maheshwari (1950) has been followed here.

Apomixis has been reported in the representatives of more than 300 genera and 80 families of angiosperms. The highest rate of apomixis is seen in grasses (about 56 genera) and Asteraceae (about 40 genera), which are known for their fast–spreading nature as compared to their sexual relatives. Hence, apomicts are generally cosmopolitan, they lack ecological localization, and very often are highly polymorphic. Apomicts have high viability and ability to compete with other plants and they very often supercede them. A number of alien invasive taxa and weeds are apomicts.

At one end of the apomictic spectrum stand taxa in which there is a partial suppression of sexual process leading to the origin of the sporophyte from haploid female gametophyte, and at the other end are instances of the new sporophyte arising entirely through vegetative reproductive means. All those instances wherein the megasporangial cells, sporophytic or gametophytic, are involved in the production of a new sporophyte are generally treated under the category **agamospermy**. The phenomenon of development of more than one embryo per seed is called **polyembryony**. Although polyembryony is seen sporadically along with normal sexual cycle, it appears to be especially associated with agamospermy where the process of multiple embryo production is denoted by the term **adventive embryony**.

The varieties of agamospermic and polyembryonic phenomena are so great that there are not only confusions in vast terminologies used in apomixis by different botanists, but also it is often difficult to classify them under distinct categories to the satisfaction of all.

10.2.1. Agamospermy

Agamospermy is categorized into **non–recurrent** and **recurrent** categories. Further development under the first category is dependent on pollination. The pollen germinates on the stigma and the pollen eventually discharges the two male gametes into one of the two synergids. The sperms (male gametes), however, fail to take part in double fertilization. The egg independently develops into a haploid embryo, triggered by pollination. In recurrent agamospermy, the resultant product is a diploid embryo. Based on the source of the initial cell (for the embryo), recurrent agamospermy is divided into two subtypes: **Gonial Apospory** and **Somatic Apospory**.

10.2.1.1. Gonial Apospory

In this subcategory, either meiosis in the megaspore mother cell is defective and is accompanied by abnormalities or the reduction process (i.e., meiosis) itself is suppressed. In the first case, the dividing cell or its products promptly degenerate(s), while in the second the nucleus of the cell directly enters into the mitosis to result in the formation of unreduced (i.e., diploid) nuclei. In some species of *Erigeron*, the nuclei that are produced during the first division are unaccompanied by a wall separating

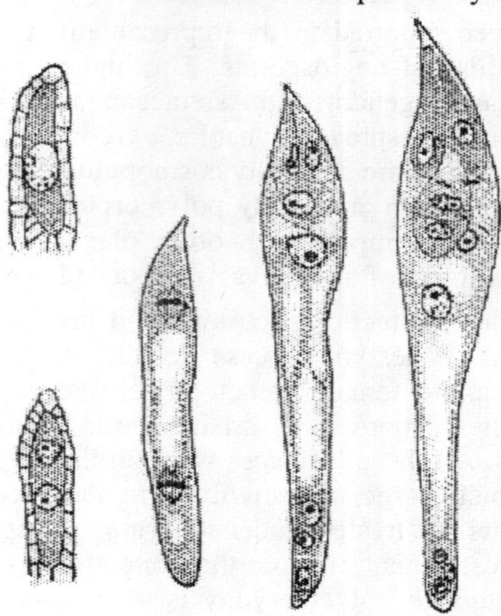

Fig. 274. Gonial apospory in *Erigeron ramosus* (Holmgren 1919)

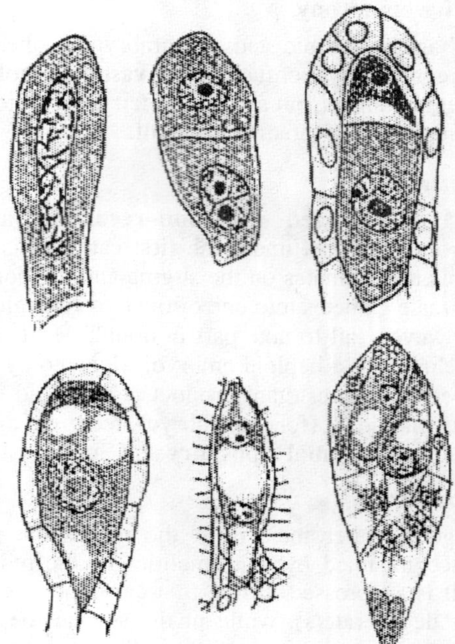

Fig. 275. Gonial apospory in *Taraxacum albidum* (Osawa 1913)

the two daughter nuclei. These nuclei move to opposite poles of the cell, thereby indicating the initiation of polarity. Two more in situ divisions lead to the formation of an 8–nucleate structure, in spite of the absence of chromosome reduction (Fig. 274). In *Taraxacum*, the first mitotic division in the 'embryo sac mother cell' is followed by a wall separating the daughter nuclei. The micropylar cell degenerates, but the chalazal cell continues development to result in an 8–nucleate 'embryo sac'. In this 'embryo sac' also, all the components are diploid and unreduced (Fig. 275).

10.2.1.2. Somatic Apospory

In some angiosperms, where the development of the haploid embryo sac is prevented, a somatic cell of nucellus, situated often close to the defunct megaspore mother cell, begins to grow considerably in size associated with the gradual vacuolation of the cytoplasm. These histological changes invariably happen before its nucleus divides. Such a cell functions as the initial for the aposporic embryo sac. The location of such a cell is highly variable, but it is often in the micropylar half of the ovule, and rarely it is located in the chalazal half. In all these cases, the megaspore mother cell or its immediate division products degenerate(s) before the onset of division in the somatic embryo sac initial. However, in some exceptional cases, both the aposporous (diploid) and normal (haploid) embryo sacs undergo development to varying extent of maturity. Although the pathway of **somatic apospory** is clearly followed in ovules with a single archesporial cell, the ovules with many archesporial cells present a somewhat confused

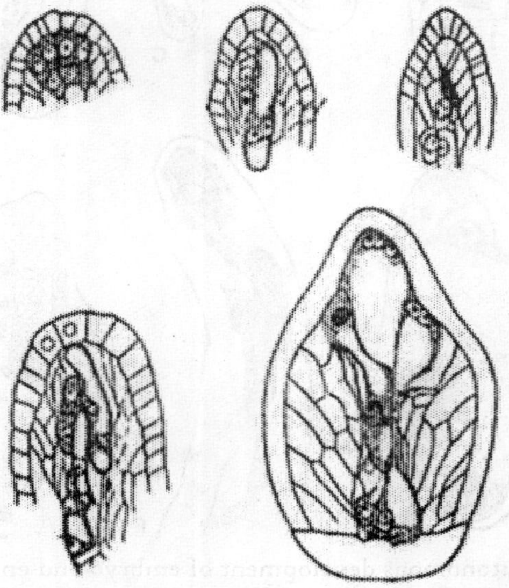

Fig. 276. Somatic aposory in *Atraphaxis frutescens* (Edman 1931)

picture. For instance, the young nucelleus of *Atraphaxis* possesses an archesporial complex of several cells in its hypodermal locus. Of these, only one or two centrally–placed archesporial cells generally function as true archesporium, while the rest pass through a phase of active multiplication prior to the initiation of meiosis in the true archesporial cell(s). The true archesporial cells soon degenerate, while some of the remaining cells of the archesporial complex function as the initials of the aposporous embryo sac initials (Fig. 276).

In a number of agamospermous plants, the embryo sacs, in spite of the organizational similarity to the sexually produced ones, fail to function further and these plants then adopt the asexual methods of propagation. There are, however, instances of agamospermous plants that produce viable seeds with or without fertilization. In the former case, which is very rare, the zygote is triploid.

In some taxa such as *Aerva*, even pollination is not necessary and the egg develops into an embryo autonomously. This process is known as **parthenogenesis**. In *Aerva*, the embryo sac is unreduced with a diploid egg and a tetraploid secondary nucleus. The endosperm is produced autonomously from the tetraploid secondary nucleus, with the embryo being diploid (Fig. 277).

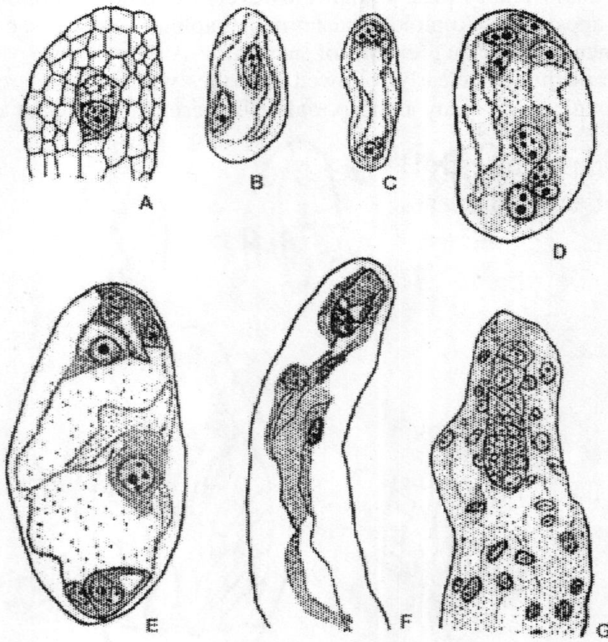

Fig. 277. Autonomous development of embryo and endosperm in *Aerva tomentosa* (Sachar and Murgai 1959)

In spite of the fact that agamospermous embryo sacs exhibit verisimilitudes to their sexually produced counterparts, the agamospermous embryo sac exhibits significant ontogenetic deviations and a wide variety of mature organizations. In many cases, the daughter nuclei, following the first mitotic division in the aposporic initial, never separate from one another, but stay in situ for a significant period of time, thereby suggesting the absence of the operation of polarity. Even in those instances where a larger number of nuclei are formed, a typical linear elongation of the cell is absent, more often assuming irregular shapes. The loci of the component nuclei are also greatly disturbed. All these clearly speak of a basic disturbance in the biochemical diffusion reaction system, in the sense of Wardlaw, leading to a lack of polarity and right organization.

10.2.2. Molecular Basis of Apomixis

Grossniklauss and colleagues have reviewed the molecular aspects of apomixis (Grossniklauss et al 1998). In Chinese chive, according to these authors, there is no omission of any developmental step leading to embryo sac formation. Instead, an exquisitely–timed round of DNA replication is inserted prior to meiotic prophase, resulting in an unreduced embryo sac. The apomixis in *Hieracium* is dominant (i.e., decided by dominant alleles), but further segregation of the progeny into distinct classes of apomictic penetrance have been observed. Cytological studies have shown that the apomictic mechanism varies between different progeny groups. These studies suggest that additional unlinked modified genes may be necessary for higher penetrance of the trait and the successful production of an otherwise viable seed.

Introgression of *Tripsacum* apomixis into maize has also been studied by the above investigators. Using molecular markers, comparative mapping between *Tripsacum* and maize has defined genomic regions in maize that are syntenic to the regions containing the locus that controls apomixis in *Tripsacum*. These regions are now being analysed for possible candidate apomixis genes.

A genomic region that is tightly associated with apomixis in *Pennisetum* has been reported, but surprisingly a homologous genomic region cannot be found in sexual relatives. At present, it is unclear whether the presence of this extra–apomixis–linked genomic DNA is a general characteristic of apomicts, because the mapping strategies used by other researchers have failed to detect it. The origin of this 'exogenous' DNA is a topic of extensive discussion. One possibility raised is that this DNA may have been derived from a parasitic or symbiotic organism, providing an alternative to the hybridization and mutation theories of apomixis.

Grossniklaus et al (1998a) have given information on the identification of ovule–specific genes and their promoters in *Arabidopsis* that may be used in directing candidate genes to alter, influence or stimulate apomixis. There are also reports of genes that are differentially expressed between isogenic lines of wheat displaying haploid parthenogenesis, and those displaying normal sexual development.

Attention is drawn here to the *file* and *fis* genes of *Arabidopsis* discussed in Chapter 7; these genes cause fertilization independent endosperm development.

10.3. POLYEMBRYONY

The phenomenon of development of more than one embryo in a developing seed is known as **polyembryony**. Barring a few habitually polyembryonate plants such as citrus, mango etc., polyembryony is generally an abnormal phenomenon. The increase in the number of embryos may be due to the following:

1. The proliferation of the zygote or its early derivatives (the so-called **cleavage polyembryony**) (E.g., many conifers) (Fig. 278) The cleavage type is of rare occurrence in angiosperms and even in the recorded instances, it is difficult to decide whether the phenomenon stands for a favourable comparison with that in the gymnosperms. In certain orchids such as *Erythronium* and

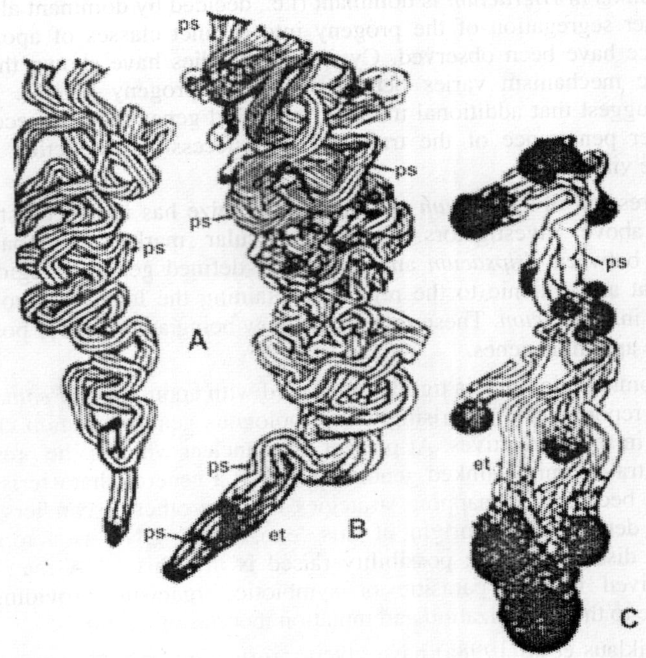

Fig. 278. A & B - Polyemryony in *Podocarpus urbani*; C - Polyembryony in *Torreya californica*; Et - Embryonal tubes giving rise to suspensors; Ps - Pseudosuspensors (A and B - Buchholz 1941; C - Buchholz 1940)

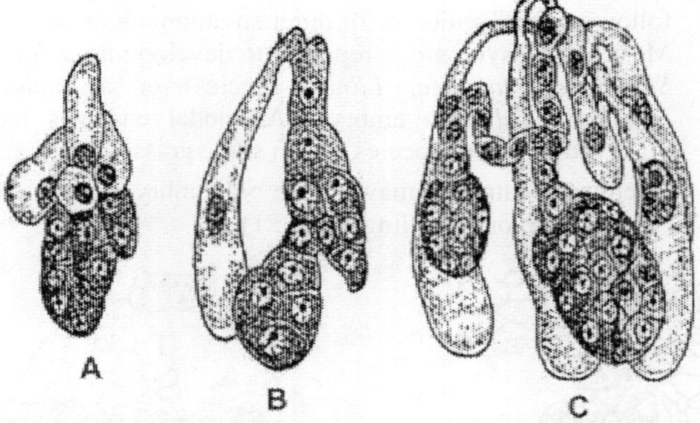

Fig. 279. Cleavage polyembryony in *Eulophia epidendraea* (Drawing provided by late Prof. B.G.L. Swamy)

Linocharis, the zygote or its derivatives divide to form a variable number of irregular groups of cells, each of which independently develops into an embryo. In *Eulophia,* the polyembryonate condition results from one of the following methods (Fig. 279): The zygote divides to form a mass of cells with some of those lying at the chalasal end giving rise to additional embryos, or the filamentous embryo branches out and one embryo develops from each of these branch tips, or buds are produced from the embryonal body that develop into additional embryos. In *Zygophyllum fabago* (Zygophyllaceae) and *Exocarpus* species (Santalaceae) suspensor polyembryony is common.

2. The development of more than one embryo from independent sources of ovular origin, other than from the embryo sac (the so–called **adventive embryony**) i.e., from the cells of the nucellus or integument (usually from the micropylar half of the embryo sac or very rarely from the chalazal half) - These may be produced either in addition to the zygotic embryo or instead of zygotic embryo, usually as a result of pollination stimulus (E.g., *Citrus, Eugenia, Mangifera*) (Fig. 280).
3. Presence of more than one embryo sac within the same ovule (the so–called **macrosporocytic polyembryony**) (e.g., *Casuarina*) (Fig. 193B)
4. By the fusion of two or more nucelli or ovules, as reported in *Nechamandra alternifolia*
5. Gametophytic cells other than the egg giving rise to additional embryos (the so–called **gametophytic polyembryony**), either

following fertilization or through an autonomous development - Most often a synergid is reported to develop into a diploid (E.g., *Sagittaria graminea, Lilium* species) or a haploid (E.g., *Phaseolus vulgaris*) embryo. Antipodal embryos have been observed in *Ulmus* species and in some grasses.

6. Chemical treatments may induce polyembryony and production of more than one seedling (Fig.281)

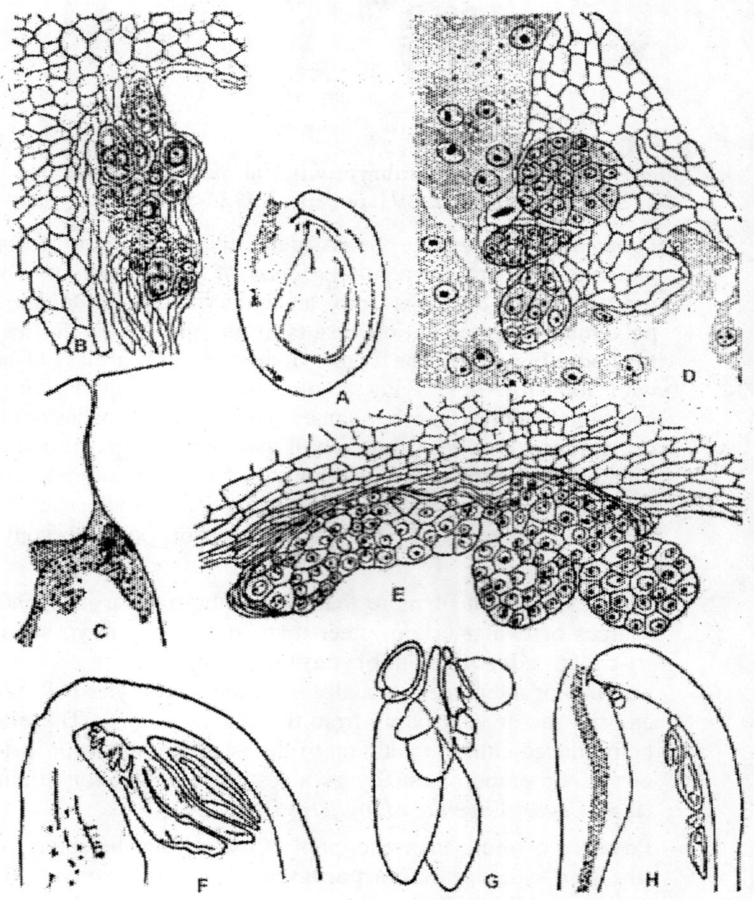

Fig. 280. Adventives embryony in *Mangifera*; A, B and F- *Mangifera indica* var. *olour*; C and D - *Mangifera odorata*; E - *Mangifera indica* var. *Higgins*; G- *Mangifera indica var. Cambodia, nine embayos dissected out from a seed.* H - *Mangifera indica* var. *pabo* showing eight embryos (Sachar and Chopra 1957)

Polyembryony is also classified into **true** and **false** categories. According to this classification, the production of many embryos within, or by projection into, a single embryo sac is called true polyembryony. In false polyembryony, multiple embryos are derived from different embryo sacs. However, according to other botanists, the term 'true polyembryony' should be applied to embryos derived from one or more embryo sacs present in the same nucellus.

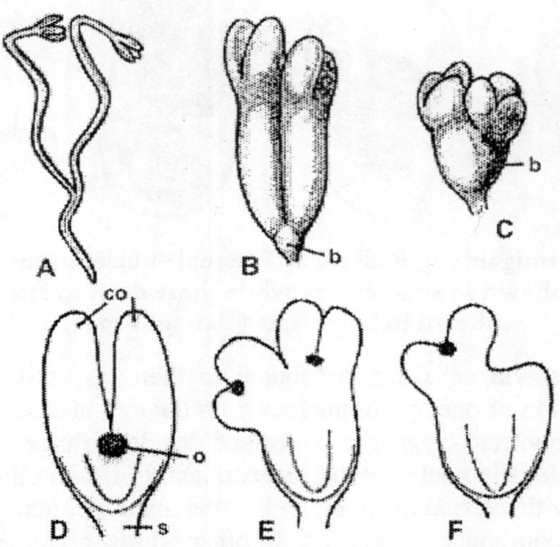

Fig. 281. Chemically induced polyembryony in *Eranthis hiemalis*; The treatment of young embryo with 0.1% 2, 4-dichlorophenoxyacetic acid gives rise to polyemryony; a - Shoot apex; b - Basal region; co - Cotyledons (Haccius 1955)

Many suggestions have been provided to explain the causes of polyembryony, but none has been validated satisfactorily. Effects of recessive genes, production of **necrohormones**, hybridization in some cases, chemical effects etc. have been suggested as the causative factors for polyembryony. The involved chemicals, if any, may be induced by pollination/pollen /pollen tubes/triple fusion.

10.4. HEMIGAMY, ANDROGENESIS, SINGLE FERTILIZATION AND POLYSPERMY

10.4.1. Hemigamy

Hemigamy, originally known by the term **semigamy**, was first discovered by Battaglia in *Rudbeckia speciosa* and *Rudbeckia lacinata* in 1945. Here, normal pollination takes place, pollen tubes enter the embryo

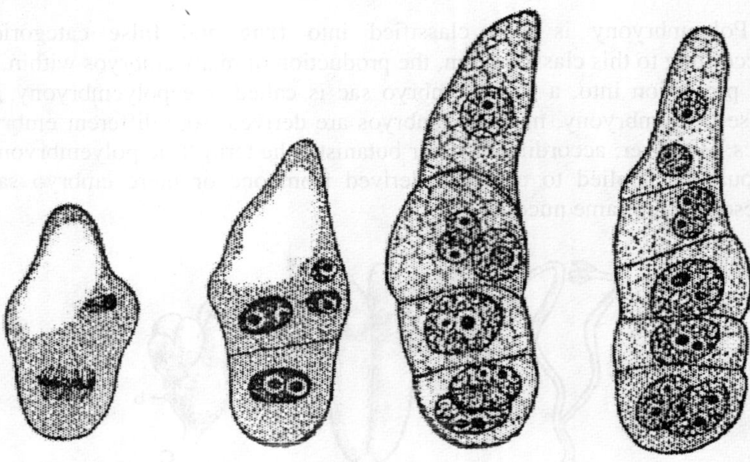

Fig. 282. Hemigamy in *Rudheckia*; The male nuclei in the developing embryo are shown in smaller size while those derived from the egg are shown in larger size (Battaglia 1947)

sac and the sperm cells are distributed to their respective destinations. However, fusion of one sperm nucleus with the egg nucleus does not take place. The binucleate 'zygote' thus formed develops further in which both the male and female nuclei divide independently of each other resulting in an embryo with cells containing either the male or female nucleus or occasionally both nuclei (Fig. 282). In other words, a chimeral or mosaic embryo is developed. Hemigamy is now known in other species of *Rudbeckia*, and species of *Cooperia, Zephyranthes, Arabidopsis* etc. Whether hemigamous embryos are functional in establishing a new plant is not known.

10.4.2. Androgenesis

Androgensis is derived from hemigamy. Here, the male gamete enters the egg cell. Soon after the egg nucleus degenerates, the male nucleus in the egg cytoplasm divides and redivides to form an embryo of male origin. This phenomenon is seen in some interspecific hybrids of *Nicotiana* and in *Crepis tactorum* pollinated with gamma ray–treated pollen.

10. 4.3. Single Fertilization

It refers to the condition where, of the two male gametes released by a pollen tube, only one male gamete that enters the egg or the central cell fuses, while the other does not fuse, to result either in the formation of the zygote or PEN only.

10.4.4. Heterofertilization

It refers to a condition where a male nucleus carried by one pollen tube fuses with the egg to form the zygote, while the other male nucleus carried by a different pollen tube fuses with the polar nuclei or secondary nucleus to form the PEN. Thus, the embryo and endosperm have different origins.

10.4.5. Polyspermy

It refers to a condition where more than two sperms (from the same pollen tube or from more than one pollen tube) enter the embryo sac and fuse with the egg or secondary nucleus to result in a polyploidal zygote or PEN or both.

10.5. INCOMPATIBILITY BARRIERS AND PROMOTION OF FERTILIZATION

One of the major hurdles in reproductive development in plants is the failure of fertilization for various reasons, the most important being the operation of incompatibility factors at various points and at various levels. But, efforts at overcoming these hurdles have provided us not only knowledge on the various aspects controlling fertilization, but also on the (alternative) strategies that plants follow to avoid such barriers during reproductive growth and development. This section deals with control on fertilization, the strategies adopted by the plants and human interventions on this aspect.

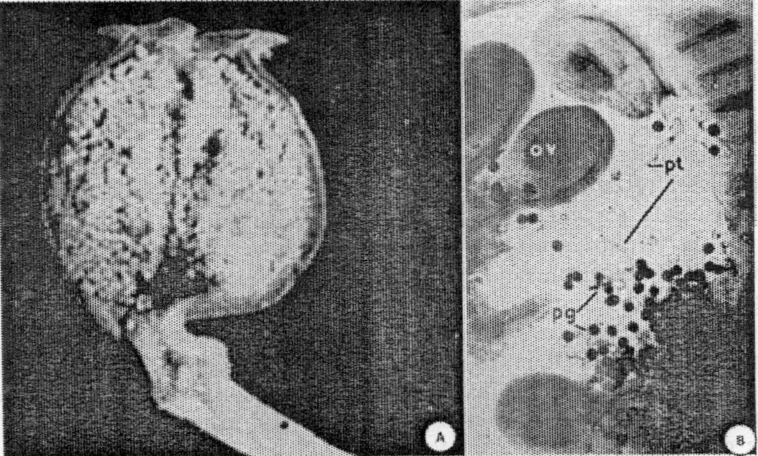

Fig. 283. Intra-ovarian pollination: A - *Papaver rhoeas*, 6-day old ovary, into which pollen suspension has been injected through a syringe, cut open longitudinally to show developing seeds; B - *Eschscholzia californica*, T. S. of portion of ovary 3 d after injection of pollen to show ovules (ov); pg - Pollen grains; p - Pollen tubes; (Maheshwari, 1965)

In those instances where failure of fertilization is due to the short life span of the flower, it has been possible to prolong the life span by preventing the abscission of the flower through the application of appropriate hormones. The first level where incompatibility operates is the stigma. The stigma may either prevent the attachment of the incompatible pollen to its surface (due often to the non–secretion of stigmatic exudates or pellicle) or arrest the germination of the deposited pollen (either by not providing for its nutrient requirements, or by releasing the inhibitory incompatibility proteins and rejecting the pollen or its emerging pollen tube). It is now possible to apply to the stigmatic surface an artificial media containing, for example, boric acid, NAA or extracts of compatible stigma, suited for the germination of the desired pollen. The stigma that is incompatible is chopped off and the pollen lodged on the upper part of the style to facilitate its germination in some instances. Grafting of a compatible style to the lower part of the incompatible stigma allow the pollen to germinate and enable the pollen tube to grow towards its destination.

Wherever incompatibility operates at the second level (i.e., the style), the stigma can be grafted to the top of the ovary after removal of the incompatible style, or the pollen can be placed on the surface of the ovary whose style and stigma have been removed. In some instances, **intra–ovarian pollination** has been carried out, where the pollen grains are either introduced into the ovarian cavity through syringe (Fig. 283) or directly lodged on to the exposed placental tissue. This procedure totally avoids the hostile stigmatic and stylar pathways. Much more promising is the simultaneous cultivation of both ovules and pollen on an artificial medium, thus totally eliminating the gynoecial tissue. This process is called **test–tube fertilization** (Fig. 284).

Clearly, *in vitro* fertilization techniques can be used to overcome incompatibility barriers. Hence, in recent years, there had been attempts to independently isolate male gametes from pollen tubes and eggs from embryo sacs (or their protoplasts) and to fuse them under in vitro conditions (Kranz et al 1999). The fusion between two gametes can be promoted by electrofusion, calcium or polyethyleneglycol (PEG). Also, egg cells without fertilization have been activated to initiate development in vitro. The access to isolated angiosperm gametes and the possibility of handling them allow detailed studies of surface biology of generative cells. One of the open questions that need to be answered is whether fertilization in angiosperms occurs in a non–gametic specific manner as somatic cells fuse or whether specific receptors are involved in this process. In vitro fertilization technology is the ideal methodology to find an answer to the above question. (Okamoto and Kranz 2005).

There are also post–fertilization hurdles that prevent the zygote or proembryo to develop into a mature embryo. One way in which it is done is through the overgrowth of ovular tissues (nucellus or integument) into

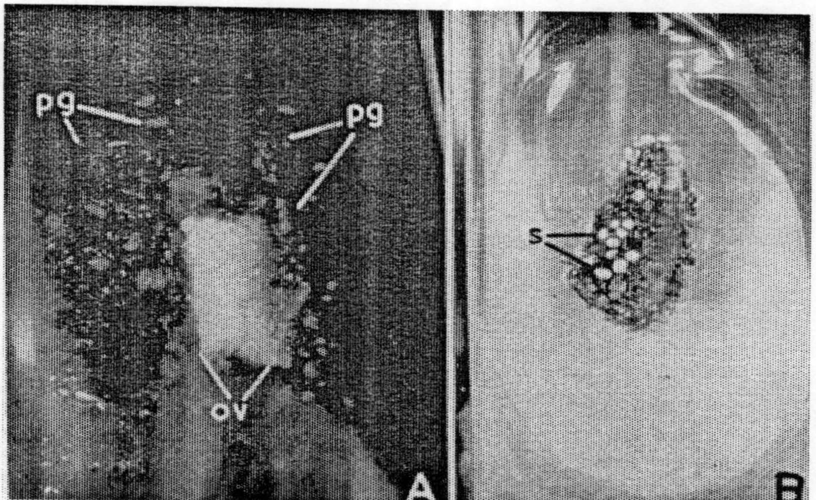

Fig. 284. Test–tube fertilization in *Papaver somniferum;* A - Ovules (ov) attached to a portion of placenta inoculated on Nitsch's basic medium along with pollen grains (pg) on the day of anthesis; B - An older culture showing developing seeds (S) (Maheshwari 1965)

the embryo sac cavity crushing the embryo and endosperm at early stages of seed development. This process is called **somatoplastic sterility**. Such zygotes or proembryos can be excised and grown under artificial conditions, the process being known as **embryo rescue**. Embryo rescue technique is successfully employed to enable many hybrid zygotes/ proembryos to grow to maturity. Wherever endosperm failure is the reason for zygote or proembryo mortality, the same technique can be followed. An embryo undergoes a gradual transition from the total dependence of the zygote to the relative autonomy of the mature embryo, especially for its nutritional and regulatory requirements. This is reflected in the degree of abnormality shown by growing embryos; the degree of abnormality decreases with increasing age of the embryo cultivated.

10.6. PARTHENOCARPY

Parthenacarpy may broadly be defined as the development of the ovary into fruit, without fertilization of its ovule(s). This definition that emphasizes the importance of fertilization, is half–way between that of Noll (1902) who introduced the term parthenocarpy to indicate fruit development without pollination or other stimuli, and of Winkler (1908) who defined it as the development of fruits without seeds or with embryoless seeds (aborted seeds). Thus, there are four different situations under which parthenocarpy develops:

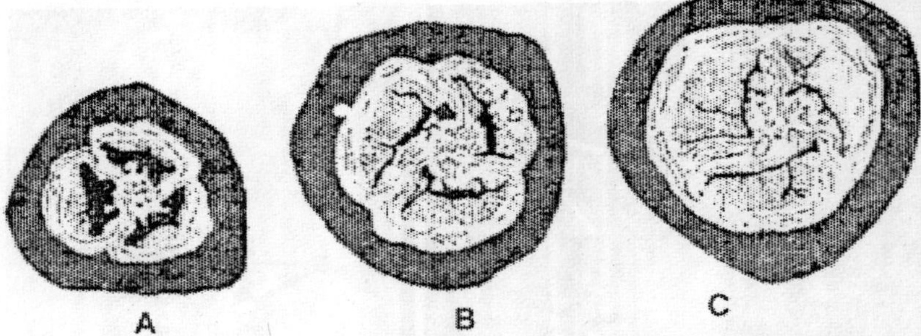

Fig. 285. Parthenocarpy in banana; Placental cells (stippled) divide and contribute to growth of the parthenocarpic fruit (modified from Mohan Ram et al 1962)

1. There is normal pollination, pollen germination, pollen tube formation and fertilization, but sooner or later the fertilized ovules are aborted and the fruit contains only empty seeds (e.g., certain varieties of grape).
2. There is normal pollination, pollen germination and pollen tube growth, but there is no fertilization. The ovary grows into a false fruit (e.g., many incompatible crosses).
3. Non–viable or dead pollens that are lodged onto the stigma induce the ovary to develop into parthenocarpic fruit.
4. There is no pollination, but the ovary becomes the fruit without any intervention, as in bananas (Fig. 285).

In all the above types, the mature fruits are seedless. But according to Nitsch (1963), parthenocarpy should not include the cases under the first category mentioned above. Thus, according to him, it follows that the examination of a mature fruit is by itself not sufficient to decide whether such a fruit has been produced parthenocarpically or not, but one must also be sure that there has been no fertilization. In other words, he argues that seedlessness and parthenocarpy are not synonymous for two reasons: (i) a seedless fruit may have developed after fertilization by the subsequent collapse of the seeds; and (ii) parthenocarpic fruits may contain seed–like structures that may or may not contain adventive embryos (e.g., *Garcinia* species). Nevertheless, the definition of parthenocarpy by Nitsch (1963) includes at least two different classes, i.e. parthenocarpy with and without the aid of pollen. These can be designated respectively as 'stimulative' and 'vegetative' (Winkler 1908).

In a number of species, particularly in those cultivated for their edible fruits, there are both seeded and parthenocarpic varieties as in the cases of grapes, oranges, persimmons etc. Parthenocarpic varieties may arise through mutations or hybridization, as in some tomato varieties, cucumbers and bananas. In such natural, genetically–induced parthenocarpic fruits, development is determined by the amount of natural auxins. Often, the seedless varieties contain more auxin than its seeded variety in the first few days of development, but subsequently the level increases in seeded fruits than in seedless fruits. From where does this auxin is obtained is a matter of debate, but the "endosperm" (parthenocarpic), ovular tissues, flag leaf, foliar leaf etc. have been suggested as the source.

There are also cases of **environmental parthenocarpy**, promoted by unusual weather conditions, particularly by low temperature at night. Seedless pears, tomatoes, brinjal, tobacco, cucumbers, chillies etc. have all been shown to be induced by low temperature. However, studies have shown that low temperature affects fruit development in general and it is not specific to parthenocarpic fruits. The effect of low temperature is shown to be mediated through auxin metabolism.

Chemical parthenocarpy is apparently more common than the genetical and environmental parthenocarpies. Parthenocarpy in many fruits has been induced through chemical treatment. Discussion made earlier indicates that auxins may form one such group of chemicals that can induce parthenocarpy. Gibberellins form another group. The esters of auxins are more potent, as they easily penetrate through the cuticle of developing ovaries. IAA, IBA, NAA, 4–chloro–phenoxyacetic acid, 2, 4, 5– trichlorophenoxyacetic acid, 2, 4, 5–trichlorophenoxy propionic acid etc. are the most common auxins that promote parthenocarpy. Generally, parthenocarpy is induced chemically more easily in multiovulate ovaries than in few–ovulated ovaries, suggesting that young ovules are necessary for parthenocarpy. This is substantiated by the fact that many parthenocarpic fruits in nature contain seed–like structures.

10.7. IN VITRO ORGANOGENESIS AND EMBRYOIDOGENESIS

Attention was drawn earlier to the prophecy of Haberlandt as early as at the beginning of 20th century that a whole plant can be regenerated from a single living parenchyma cell as it possesses totipotency. Since the requirements of cells to fully express their totipotency are varied, complicated artificial culture media were developed seriously, starting from 1930s onwards, to aseptically culture them. This has led to the understanding of the actual chemicals that are required not only to grow and proliferate the cells or tissues, but also to induce differentiation in them to produce shoots **(caulogenesis)** and roots (**rhizogenesis**). New plantlets were produced either via **organogenesis** (i.e., through production of shoots and roots) or through **embryoidogenesis** (many use the word **embryogenesis**). Organogenesis often involves an intermediate **callus**

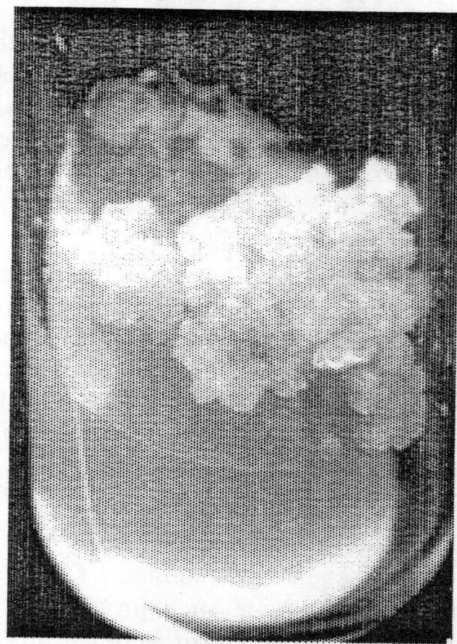

Fig. 286. Homogeneous parenchymatous callus from *Cucumis* stem explants cultured in vitro. (Photograph courtesy of Dr. S.Kaliamoorthy)

phase (**indirect organogenesis**), although **direct organogenesis** is noticed in some instances when stem explants (especially nodal explants) are cultured. Embryoidogenesis can happen from cultured cells or from callus. In both cases of embryoidogenesis, **embryoid**[1] initiation starts from single initial cell. Calli that produce organs and embryoids are respectively called **organogenic** and **embryogenic calli**. The production of organs (shoots and roots) and embryoids from vegetative cells are respectively called **somatic organogenesis** and **somatic embryoidogenesis**.

It is now possible, at least theoretically, to produce plants from living cells or tissues or organ explants of any higher plant under in vitro conditions, as the techniques have been highly standardized.

10.7.1. In Vitro Callus Production

Callus is often considered as a mass of undifferentiated tissue (Fig. 286), but it always shows cells that are differentiated to a small extent. In vitro callus can be produced from any living cell or tissue or organ separated from the plant and grown as explants on artificial growth media. The callus is often classified into **orthodox** and **recalicitrant** categories depending respectively on whether it can give or cannot give rise to new plantlets.

---

1 Many use the word 'embryo' instead of 'embryoid' for embryo–like structures derived in vitro. Such an usage is conceptually questionable (see details on a subsequent section of this chapter).

Usually, the former remains white or turns green (at least partially), while the latter becomes changed to yellow, brown, reddish or even black, often due to phenolic compounds or their oxidation products (Fig. 287). Also, the latter often leach out these chemicals into the growth medium. The in vitro grown orthodox calli are classified into organogenic and embryogenic, as mentioned earlier, depending on whether they give rise to organs (roots and shoots) (Fig. 289) or embryoids (Fig. 288). The source of the explants, composition of medium, concentration and type of growth regulators used etc. decide whether explants should develop into an organogenic or an embryogenic callus. In vitro produced calli also differ greatly in their texture, based on which two broad categories can be distinguished: **friable callus** and **compact callus**. In the former, the cells of the callus are often loosely held to one another and therefore, often break easily into smaller calli, or even into individual cells or cell clumps. In compact calli, the cells are rather tightly cemented and do not break into cells or cell clumps.

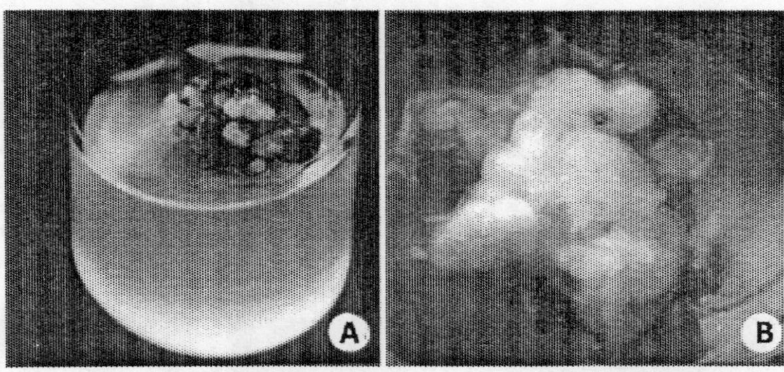

Fig. 287. Calli derived from the corm explants of *Gloriosa superba*; A - Browning of calli; B - Various degrees of yellowing and browning of while calli (Photogrpahs courtesy of Dr. S. Sivakumar)

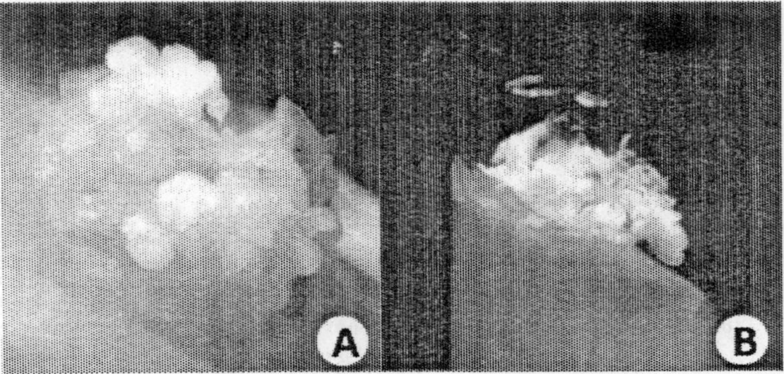

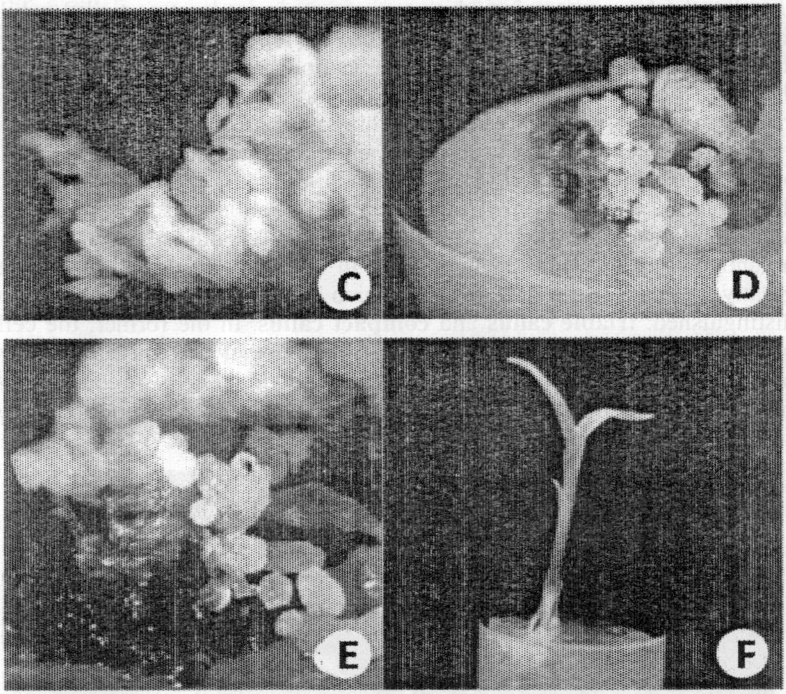

Fig. 288. Plantlet formation via embryoids in nodular callus cultures of *Gloriosa superba* (Photographs courtesy of Dr. S. Sivakumar)

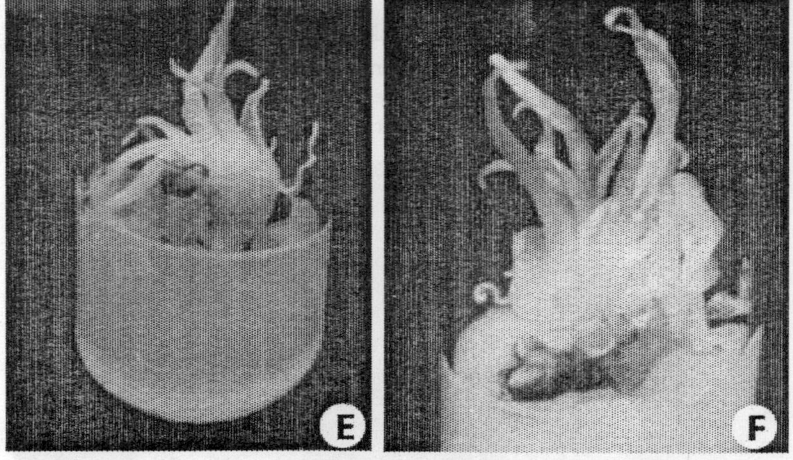

Fig. 289. A and B - Organ (stem and root) formation from organogenic calli derived from the corm tissues of *Gloriosa superba* (Photographs courtesy of Dr. S. Sivakumar)

In vitro callus development has provided important information on the basic nutritional and growth regulator requirements for cells to divide and produce a callus. These relate to the required carbon sources, vitamins, growth regulators, macronutrients and micronutrients and others. The kind and concentration of each one of these, either alone or in combination, need to be known. Also, what decides the ease with which the cultured explants give rise to a callus can be known. Information on what decides the speed of callus formation, size of callus produced, and the type of callus produced (orthodox or recalcitrant, friable or compact, white or coloured, organogenic or embryoidogenic) can also be obtained. The desired type of calli can get established in the first–tried medium itself, or only after repeated subcultures with or without changing the composition of the medium. Some need continued external supply of certain type of growth regulators and/or other organic materials, while other calli, after some growth, proliferate at the same rate as before without any further external input. The latter capacity is attributed to the ability of callus cells to synthesize the required growth regulator(s) and organic substances in the desired concentration. This phenomenon has come to be known as **habituation.** Often a right combination of two or more growth regulators is needed to get the desired type of callus at the desired time period. If the callus gets habituated to one of these growth regulators, but not to the other(s), the latter should be supplied externally.

A great deal of what we know about cell division, enlargement, differentiation and morphogenesis has been obtained from a detailed analysis of callus formation and manipulation in diverse plants and under different growth conditions. The transformation of a simple callus into an organogenic, embryoidogenic, or **xylogenic callus** has enabled us to understand the factors (especially growth regulators) responsible for the differentiation of shoots and roots, embryoids and tracheary elements. For example, the quantitative interactions between auxins and cytokinins have been shown in the control of organogenesis in tobacco and a few other taxa, which in turn, are shown to be linked to cell cycle events. It has also been demonstrated through studies on calli that the ability of their cells to synthesize DNA is not guaranteed as DNA synthesis will occur unless both types of hormones are present in the right concentration. It has also been shown by studies on callus ontogeny that although auxin induces the synthesis of protein kinase (involved in cell cycle), addition of cytokinin is necessary to trigger the catalytic activity of this enzyme. Shoot bud formation in callus is associated with intense metabolic activity such as starch accumulation, increased respiration and high levels of adenosine phosphates and NAD^+ in the potential shoot bud–forming cells.

The occurrence and discovery of embryo–specific proteins in embryoidogenic cell cultures have prompted researchers to look for similar marker proteins for organogenic callus production. Carrot cells that proliferate to produce calli in high–density cultures in the presence of 2, 4–D in the growth medium synthesize two callus–specific proteins, C1 and C2. The fact that calli of a mutant, deficient in the genes producing C1 and C2 proteins, synthesize embryoid–specific proteins, but not callus–specific proteins, indicates that callus growth can proceed in the absence of callus–specific proteins, while the synthesis of embryoid–specific proteins does not necessarily induce embryoidogenesis (Raghavan 2000).

10.7.2. In Vitro Embryoid Production

Embryoid production from a variety of cell or tissue explants of many different plants (both dicots and monocots) has been obtained in the past two decades and the technique has been greatly perfected. In some systems, at least, the structural, cytological and molecular mechanisms of embryoid development have been understood. However, our knowledge of the overall regulatory machinery of somatic embryoidogenesis from cultured somatic cells and tissues has come mostly from an analysis of carrot system, one of the very earliest exploited systems, and whose culture conditions have been sufficiently refined to obtain somatic embryoids reproducibly and in large numbers. Usually the hypocotyls or root segments of carrot seedlings germinated under aseptic conditions and cultured on the surface of a solidified Murashige-Skoog (MS) mineral salt medium supplemented with sucrose, myoinositol, a cytokinin, and 2,4-D, form the most efficient means to generate an embryoidogenic cell suspension (Raghavan 2000). A piece of callus is obtained from this explant and is transferred to an agitated liquid medium of the same composition as above, but with a reduced level of 2,4–D to obtain proliferating single cells and cell clusters (derived from single cells) known as **proembryogenic masses.** These are then subjected to suspension cultures in the same medium to maintain stable cell populations over many subcultures. Embryoidogenesis can be induced only if the cells are transferred from the medium with 2, 4–D to a basal medium (called **induction medium**) completely lacking 2, 4–D. Globular somatic embryoids are obtained in 4–5 d in the induction medium by the transformation of the cell aggregates that may or may not pass through a filamentous phase. Subsequent development in these globular somatic embryoids does not take much time. Synchronously–developed mature somatic embryoids are obtained in large numbers very rapidly through sufficiently standardized protocols. Investigations made on this system strongly indicate the involvement of auxin in inducing embryoidogenic potential in the somatic cells. The direct production of somatic embryoids from hypocotyl epidermal cells also needs prior treatment in 2, 4–D for at least 12–24 h before the hypocotyl is grown in an auxin–free medium. The need of 2, 4–D to induce embryoidogenic competence in somatic cells to start with, and its absence subsequently to promote embryoidogenesis is really a paradox that has not been adequately explained so far (Raghavan 2000).

The answer to this problem perhaps comes partially from protein molecules that are secreted by cultured cells into the suspension culture medium. These proteins promote not only cell proliferation, but also somatic embryoidogenesis. The secreted proteins have 13, 17, 29, 38 and 46 kDa molecular weights during the 9–day experimental period (Raghavan 2000). On the contrary, few proteins, secreted by cells growing in the medium containing 2, 4–D, were disappeared or were drastically reduced in concentration. The most important extracellular proteins involved in somatic embryoidogenesis are glycosylated acidic endochitinase, a 38 kDa cationic peroxidase, arabinogalactan proteins, and lipid transfer proteins. Arabinogalactan proteins are particularly important to induce embryoidogenic potential in non–embryoidogenic cells. A specific protodermal expression of transcripts encoding a lipid transfer protein is seen in somatic embryoid (and also in zygotic embryos). Some of these proteins are markers of somatic embryoidogenesis.

Some information is available on the genetics of somatic embryoidogenesis in carrot (Raghavan 2000). A study of variant cell lines and several conditioned mutants generated provides this information. When compared to wild–type cells, 5–methyl tryptophan–resistant cell lines of carrot do not produce embryoids even in the induction medium. The accumulation of tryptophan[1] and an increase in the internal IAA level cause hindrance to embryoidogenesis, and most embryoids are arrested in the globular stage or between globular and heart–shaped stage. The genes that are preferentially expressed during somatic embryoidogenesis have been subjected to cloning and characterization, but since there is great overlapping between genes expressed in embryoidgenic cells and in non–embryoidgenic cells, the results of cloning or characterization should be viewed with caution. Most genes appear to be functionally and structurally homologous to LEA protein genes such as *DC8* and *DC59* that are associated with post–globular somatic embryoids. *EMB1* gene is expressed only in embryoidogenic cells. *SERK* (for 'somatic embryo receptor kinase') gene is expressed in cells competent to form somatic embryoids, but is not expressed after globular stage of the embryoid. Some 'embryoidogenic' genes encode cell–wall–related proteins, glycine–rich proteins, tubulins, etc.

10.8. POLLEN CALLUS AND EMBRYOID PRODUCTION

It is by serendipity that Guha and Maheshwari (1964) discovered **pollen embryoidogenesis** i.e., production of haploid embryoids from pollen

[1] The accumulation of tryptophan (and phenylalanine) is a very characteristic early marker of organogenic calli and not of embryoidogenic calli. These amino acids form the precursors of ligin that is deposited on tracheary elements that differentiate at the centre of the organogenic calli, and the tracheary elemants form an early secondary marker for organogenic calli (Kaliamoorthy and Krishnamurthy 1998).

grains of excised anthers of *Datura innoxia* under in vitro conditions. These embryoids originated from the haploid cells of the pollen grains or the male gametophytes. Subsequently, pollen embryoidogenesis and haploid plant production protocols have been standardized for several plant species including species of *Nicotiana* and many other Solanaceae, rice and a few other grasses, Brassicaceae members etc. Pollen embryoidogenesis could be obtained both from cultured whole anthers (subsequent to microspore stage) and from cultured microspores or pollens. Of course, diploid tissues of the anther also produce embryoids in some cases, in addition to haploid embryoids from pollen or microspores when the whole anther was cultured. This methodology paved the way for the production of **androgenic haploids** and its subsequent conversion to isogenic diploids through chromosome doubling for genetic and breeding experiments. The critical requirement for deviation from the gametophytic programme of the pollen grains or microspores towards the sporophytic pathway is the maintenance of their respective cell cycle events (Raghavan 2000). This is most easily obtained by culturing anthers in the microspore stage or the microspores in isolation from other anther tissues. The haploid mitotic division of the microspore nucleus to result in haploid vegetative and generative cells provides the required molecular cues for initiating cell divisions leading to embryoidogenic pathway in the microspores. In fact, greater attention had been focused on the subsequent fate of the two resultant cells of the microspore.

Pollen embryoidogenesis (Fig. 290) either directly or through an intermediate callus follows one of the three following pathways, sometimes even in the same plant species or in the same anther:

1. The microspore gets divided by an unequal division to form a large vegetative cell and a small generative cell (as in normal microspore mitosis), of which the vegetative or generative cell alone forms the embryoid.
2. A chimeral embryoid may be formed by cells contributed by both vegetative and generative cells through division.
3. The microspore divides equally to form two equal–sized cells (i.e., without distinction into generative and vegetative cells), both of which through further divisions give rise to the embryoid.

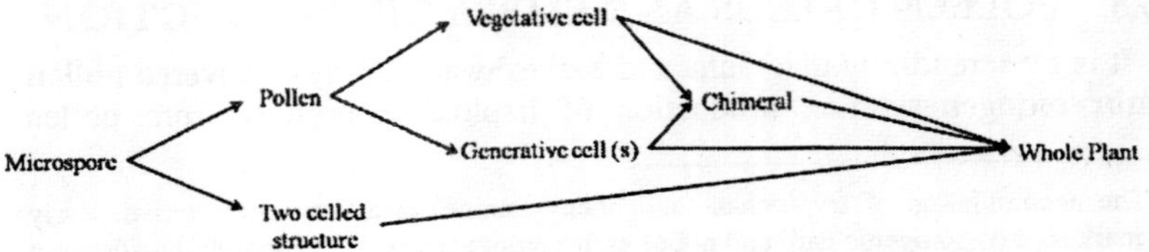

Fig. 290. Schematic developmental pathways of plantlet production from cultured microspore

According to Raghavan (2000), the developmental programme by which the pollen cells programmed for terminal differentiation proceed to form an embryoid, involves divisions that yield daughter cells unlike the mother cell (microspore). The mechanism by which this is brought about is not very clear. One suggestion is that "there is a regression of the gametophytic programme of the pollen cells and the repopulation of the cytoplasm by a new set of organelles representing the sporophytic programme". The other suggestion is that there is a "gradual assimilation of the gametophytic programme into one that ensures the sporophytic mode of development". However, both these suggestions have been contested, but those who have contested these have not provided any alternative or better suggestion. It is likely to do with the position of microtubules and cytoskeletal elements before mitosis in the microspores.

The genotype of the pollen parent is obviously the most crucial source of variability observed in androgenic haploids produced in diverse crop plants such as cereals (rice, wheat, barley, maize, oat etc.), potato, tomato and tobacco (Raghavan 2000). Most investigations dealing with the molecular biology of such variability in pollen embryoidogenesis resort to cytochemistry and autoradiography. One of the earliest–noticed changes related to embryoidogenesis concerns RNA synthesis. In contrast to the situation seen in the normal generative cell of the pollen grain (Refer to Chapter 6), RNA synthesis during the initiation of pollen embryoids is very high in the generative cell and its derivatives. There is also a continued accumulation of mRNA in them indicating a transcriptional control on pollen–embryoid initiation. Some of these mRNAs are known to code for heat–shock proteins, thus indicating that at least there are some common programmes between pollen–embryoid formation and those countering heat shock and other stresses. It was found in *Hyoscyamus niger* that, although normal microspores do not incorporate any [^{3}H] leucine or [^{3}H] tryptophan into proteins, synthesis of proteins by such microspores using [^{3}H] lysine and [^{3}H] arginine is frequently observed. On the other hand, embryoid–competent cultured microspores incorporate all the four labelled amino acids into proteins (probably into proteins analogous to heat–shock proteins), and the latter two at a much accelerated manner. This is particularly true of their generative cells and their derivatives embarking on embryoidogenic pathway through cell divisions (Raghavan 2000). Compared with normal microspores and two–celled pollen, the embryoidogenic pollen of *Brassica napus* shows intense and differential labelling of their nuclei with anti–HSP–70 antibodies (Here, HSP denotes heat–shock proteins) indicating that embryoid induction depends critically on stage–specific increase in the distribution of HSP–70 inside the cells. There is a substantial and sudden increase in the amount of transcripts of *napin* gene in the pollens of *Brassica napus* after an inductive heat shock at 32°C for 4 days, and these pollen transform themselves into embryoidogenic pollen. Hence, the expression of *napin* gene can be taken as an early marker of embryoidogenic pollen. A similar result was also obtained in tobacco pollen that showed a low molecular weight HSP. In wheat, only the embryoidogenic pollen showed early expression of a cysteine–rich metallothionein–like protein.

10.9. ABOUT EMBRYOS AND EMBRYOIDS

10.9.1. Introduction

A critical analysis of past literature on in vitro plant production through embryo–like structures reveals that these embryo–like structures were first designated as 'embryoids' by early workers, who appreciated the basic differences (including conceptual) between such structures and true embryos (as for example, between a human and humanoid). Subsequently, however, terms such as embryo, proembryo, embryogenesis, and even seedling were increasingly used in embryoid literature in as much as that these terms have now been taken for granted and have almost totally replaced respectively words such embryoid, proembryoid, embryoidogenesis and plantlets. Gray (1996) had, in fact, gone to the extent of asserting that the suffix 'oid' should be dropped. He further argued that distinctions between zygotic and non–zygotic embryos have become blurred. Such an assumption has already been reflected in statements such as the following: "Somatic embryos undergo developmental events similar to those that occur within the embryo–proper region of zygotic embryos" (Goldberg et al 1994), and "the similarity between zygotic and somatic embryhogenesis is both stricking and remarkable" (Zimmermann 1993). In fact, Zimmermann (1993) had suggested that the somatic embryo system can be used as a potential model for studying plant embryogenesis. However, as we can see now, embryos and embryoids show distinct differences in ontogeny, structure, biochemistry and molecular biology (Swamy and Krishnamurthy 1981, Krishnamurthy 1999).

10.9.2. Zygote vs. Embryoid–initial

Embryoid literature abounds in statements as follows: "Any diploid cell is potentially totipotent and can behave like a zygote" (Steward and Mapes 1963). Are zygotes and embryoid–initials really similar or different? The zygote is always a product of sexual fusion, diploid and always present at the micropylar milieu of the embryo sac (a privileged location). It has a distinct micropylar–chalazal polarity (with a spatial asymmetry of cellular contents controlled by *GNOM* or *EMB* genes), extremely limited or no de novo transcriptional activity, a lot of mRNA including the legacy from the egg (probably stored) and arabinogalactan–rich proteins (Krishnamurthy 1999). On the contrary, the embryoid–initial may be diploid or haploid, is not a product of gametic fusion, does not occur in a privileged location, has no polarity and has to undergo a lot of new structural, biochemical (such as

RNA and protein synthesis) and molecular changes (presence of active transcriptional and translational activity) triggered by several external inputs before dividing. This cell, in carrot system, needs to express a receptor–like gene (*SERK* gene), before embarking on somatic embryoidogenesis. Although arabinogalactan–rich proteins have been detected using monoclonal antibodies (JIM–8 antibodies) in this cell, studies reveal that there is no direct correlation between the presence of this protein on the cell wall and the ability of the cell to develop into a somatic embryoid (Pennell et al 1992). Even if we equate this cell to the zygote, the data mentioned above suggests that all the changes, promoted by fertilization in one stroke to make an egg into a zygote, are attained in the somatic cell in a series of complicated steps triggered by several external inputs.

10.9.3. Division of Zygote and Embryoid–Initial Cell

The division of the zygote is always transverse (with extremely few exceptions) and asymmetric (Refer to Chapter 3) that results in a small apical cell and a large basal cell. An asymmetric division is very important for the orderly subsequent development of the embryo. The basal cell produces the suspensor, while the apical cell forms the embryo–proper. The *ATML1* and a few other genes show expression only in the apical cell not in the basal cell which has a different set of other genes in *Arabidopsis* (Lu et al 1996) and few other taxa studied. In the *gn* mutant of *Arabidopsis,* the zygote divides symmetrically, but the embryo does not have an organized further development. On the contrary, the division of the embryoid initial may be transverse, longitudinal or oblique, and may be symmetrical, asymmetrical or quite variable in the same taxon or even in the same culture. Also, the two resultants daughter cells are not properly programmed for their further destiny and their associated genes are absent. In gymnosperms the division of the zygotic nucleus and also a few subsequent divisions of the resultant nuclei are free nuclear, while in somatic embryoidogeny of that group of plants, the embryoid–initial cell divides to result in two cells that further divide to form cells and not free nuclei.

10.9.4. Suspensor

All zygotic embryos have a suspensor of variable extent. The suspensor is a highly specialized 'organ' of an embryo. Even suspensor mutants (more than six in *Arabidopsis*) have not eliminated this structure altogether (Goldberg et al 1994). Suspensors have designated functions such as supply of nutrients to the embryo proper, controlling the orderly development of the embryo, accumulation and supply of growth regulators, and organization of the hypophysis and embryonic radicle meristem. The organization of a typical suspensor and DNA amplification comparable functionally and structurally with that of the zygotic embryo, is never encountered in the somatic embryoids of any species so far studied. However, any cell or a group of cells attached to the pole opposite to the shoot apical pole of the embryoid has been indiscriminately assigned the label 'suspensor'.

10.9.5. Polarity

The most significant aspect of zygotic embryogenesis is polarity that is manifest in the zygote (a legacy inherited from embryo sac/egg polarity), its asymmetric division and during the whole gamut of embryogenesis. In the early globular stage of an embryo, an apical–basal polarity becomes evident through the differentiation of hypophysis and epiphysis respectively in its basal and apical poles (root and shoot poles). It has been proved that in the absence of the differentiation of hypophysis and epiphysis, no transition from globular (radial symmetry) to cordate (bilateral symmetry) stage takes place in dicots and from a radial to the unilateral symmetry (because of only one cotyledon) in monocots (Krishnamurthy 1999).

During embryoid development, there is no evidence of organized polar forces guiding its development. Halperin and Wetherall (1964) have stated that it is not known how or at what stage of development the apico–basal polarity is determined in the embryoid. Street (1976) believed of a spontaneous origin of polarity at the globular embryoid stage. In none of the embryoids described so far, there is an organization of epiphysis and hypophysis either at the globular or subsequent stages.

10.9.6. Cell Division, Pattern Formation and Symmetry Changes

As already stated in an earlier chapter, the zygote and the embryo derived from it are considered as complex, specific and gene–determined reaction systems that functions in conformity with the laws of physical chemistry and mathematics. As per details discussed previously, the outstanding feature of zygotic embryogeny is the orderly and almost pre–determined sequence of cell divisions that strictly obey the physical laws governing cell divisions in general (Refer to Chapter 3 for all the relevant laws and Chapter 7 for the related aspects such as laws of embryonomy, and cell lineage concept or Mosaic theory). There is also an intimate relationship between cell division and polarity all through embryo development and this is supported by gene expression studies involving genes (or their mutations) such as *GNOM/EMB30, LTP, Gurke, MONOPTEROS* and *FACKEL.* In zygotic embryos there is a controlled–pattern formation involving those factors and events that cause cell types, tissues and organs to originate at specific locations in the globular and subsequent stages i.e., the body organization of the seedling is laid down during embryogenesis.

Cell divisions leading to embryoids do not obey the physical laws that operate during zygotic embryogeny and hence, discourage geometrical or mathematical analysis (Swamy and Krishnamurthy 1981, Steward 1958). This behaviour is mainly due to absence of forces causing a polarizing gradient in the initiating cells that consequently are not kept in equilibrium with the neighbouring cells and their immediate milieu. The cell divisions are not predictable and are often described as irregular or as 'taking place in all plans', without following any specific pattern or sequence. In some embryoids, there is a progressive increase in cell number without the appropriate cell enlargement, while in others cell enlargement proceeds without corresponding increase in cell number. In the same plant (for example, in carrot), there are several ontogenetic sequences reported during embryoid development. None of them correspond to the pattern characteristic of the concerned plant's zygotic embryogeny.

Symmetry change from the radial globular to the bilateral 'cordate' embryoids was not noticed in carrot embryoids in the continued presence of auxin (Krishnamurthy 1999). It was suggested that new gene products are needed for this transition and these new products are synthesized only when exogenous auxin is removed; probably the polar transport of auxin is also a prerequisite for normal morphogenesis at the globular embryoid stage and for transition to 'cordate' stage (Zimmermann 1993). Thus, the mechanism of transition from globular to cordate embryoid is very complicated.

10.9.7. Tissue and Organ Differentiation

A mature zygotic embryo has two principal organ systems, the axis and cotyledon(s), and each one of these have three basic or primordial promeristem zones, protoderm, procambium and ground meristem respectively giving rise to epidermal, vascular and ground tissues of the seedling. Such promeristem differentiation is believed to take place in the globular–heart transition stage itself (Goldberg et al 1994). Protoderm differentiation seems to be a very crucial morphogenetic event in zygotic embryogenesis (Periasamy 1994). Protoderm is the first tissue region to differentiate in the early–globular embryo when the latter's terminal tier is in the octant stage and get completed before the 16–celled stage itself. This is indicated by the uniform expression of the carrot EP2 lipid transfer protein mRNA in the protoderm from globular stage onwards (Sterk et al 1991). In a similar way, before protoderm formation, the *ATML 1* gene is expressed in all derivatives of the apical cell, but then gets restricted only to the protoderm layer in *Arabidopsis* (Lu et al 1996). The ground meristem also gets segregated along with protoderm (Goldberg et al 1994) and this has been demonstrated through localization studies with a soya bean kunitz trypsin inhibitor mRNA, designated as ktis (Krishnamurthy 1994, 1999b). Details on procambialization in embryos were given in Chapter 7 and this is a very well–coordinated and organized event.

In embryoids there is not only a very belated differentiation of protoderm, but also it is often incomplete. There are descriptions such as this: "protoderm appears to arise in a patchy work random fashion and not as a uniform layer in the embryoid" (Street 1976). In some embryoids it does not differentiate even after the differentiation of 'cotyledonary' node, and the embryonic shoot apical meristem does not have a tunica organization. There is a precocious differentiation of vascular tissues (xylem and/or phloem) reported in the embryoids of some taxa; even in the globular embryoid (Also refer to Krishnamurthy 1999b).

The axis of the zygotic embryo terminates in shoot and root apical meristems at opposite poles, derived respectively from epiphysis and hypophysis (Krishnamurthy 1994, 1999b). However, the differentiation of these two meristems in embryoids has not received any serious attention at all in the past and if there is some mention about it, it is based merely on external morphology. In many instances of embryoids, their presence is first assumed at the two poles of the 'mature' embryoid. A critical analysis of past literature on embryoids indicates that a reasonably organized shoot apical meristem may be present at one pole, but in no case (barring some embryo explants) a radicular meristem is organized at the 'root pole'. In none of the 'germinating' embryoids, a tap root is seen and all the roots are adventitious.

The cotyledon (s) forms the most characteristic feature of zygotic embryo, occurring at the terminus of the embryo. It is vital for seed germination as a source of nutrient reserves. It also assists the process of embryo dormancy at seed maturity through ABA and LEA protein interactions. The embryoids do not have typical cotyledons, either in terms of morphology and anatomy or in terms of function and physiology (Krishnamurthy 1999b for a detailed discussion on this). The so–called cotyledons of embryoids are not dorsiventrally flattened structures, and their anatomy is unlike that of typical cotyledons (lack reserve foods, LEA proteins etc). In case of embryoids of dicotyledons, the two structures resembling cotyledons are either of equal size or unequal, and in some instances, inserted on the embryoid 'axis' at different loci (i.e., they are not typically opposite to one another, but alternate). Also, they are often green and leafy. In some cases of embryoids, more than two such 'cotyledons' are present and such cases have been conveniently ignored when they come for interpretation. In monocots, all the embryoids so far produced (without any exception) show two 'cotyledons', although zygotic embryos have only one cotyledon. If a monocot embryoid in equivalent to a monocot embryo, then the former should have only one 'cotyledon' and not two (or more). In gymnosperms, zygotic embryos are multi–cotyledonary and show cleavage polyembryony, while all their embryoids have only two 'cotyledons' and do not exhibit cleavage polyembryony. Hence, it is concluded that the so–called cotyledons of embryoids are not cotyledons in any sense and are more comparable to leaves or prophylls.

10.9.8. Dormancy

Most, if not all, mature zygotic embryos enter into dormancy once the seed becomes fully developed i.e., zygotic embryogenesis terminates with a dormancy period. It has been stated that embryoids are similar to zygotic embryos except that they do not become dormant, and they grow and differentiate continuously and 'germinate' directly into plantlets (Goldberg et al 1994). Thus, embryoids do not enter into dormancy so typical of seed embryos and hence, do not express the associated structural and physiological features.

10.9.9. Gene Expression

Approximately 15,000-20,000 diverse genes are reported to be active during normal embryogeny and many of them are expressed both spatially and temporally in specific cell types, regions and organs of the embryo (For more details, see Krishnamurthy 1999b, 2003). It has been claimed that "spatial and temporal gene expression programmes appear to be similar in somatic and zygotic embryos" (Goldberg et al 1994, Zimmermann 1993). This appears to be a sweeping statement because a critical study of earlier literature points to the contrary (see discussion in Krishnamurthy 1999b).

10.9.10. Morphology of the Embryoid

It was concluded by Swamy and Krishnamurthy (1981) long back that the embryoid is not equivalent to the zygotic embryo after analyzing this problem at length and that it should, at best, be considered as an adventitious shoot bud that arises in diverse parts of the plant. The information accumulated subsequent to this paper till 1999 were discussed by Krishnamurthy (1999) and he came to a similar conclusion much more strongly. Most discussions found in these two papers have been summarized earlier in this section. There are a series of morphological expressions of organogenesis in the in vitro systems (Fig. 291). There are cases where shoot buds originate at one part of the callus with root meristems originating concurrently/precociously/belatedly at regions of the callus opposite to the shoot buds and the two systems are interconnected subsequently to produce new plantlets. At the other extreme, there are instances where a shoot meristem and the adventitious root–producing callus region (with fewer constituent cells) are brought very close to one another in the form of an embryoid. When these embryoids 'germinate', there is no tap root formation, but adventitious roots of variable numbers are formed at the basal end of shoot bud region. This conclusion is substantiated by surgical experiments carried out in carrot somatic embryoids (Schiavone 1988, Schiavone and Racusen 1990). If the upper part of the somatic embryoid is separated from the lower part by surgical manipulation, the half having the shoot bud (meristem) regenerates a new adventitious root system, provided that this half has at least 25%–90% of the original length of the embryoid. On the contrary, a regeneration of the shoot pole with its meristem was difficult from the half having the so-called root pole of whatever length. In other words, the above results, if

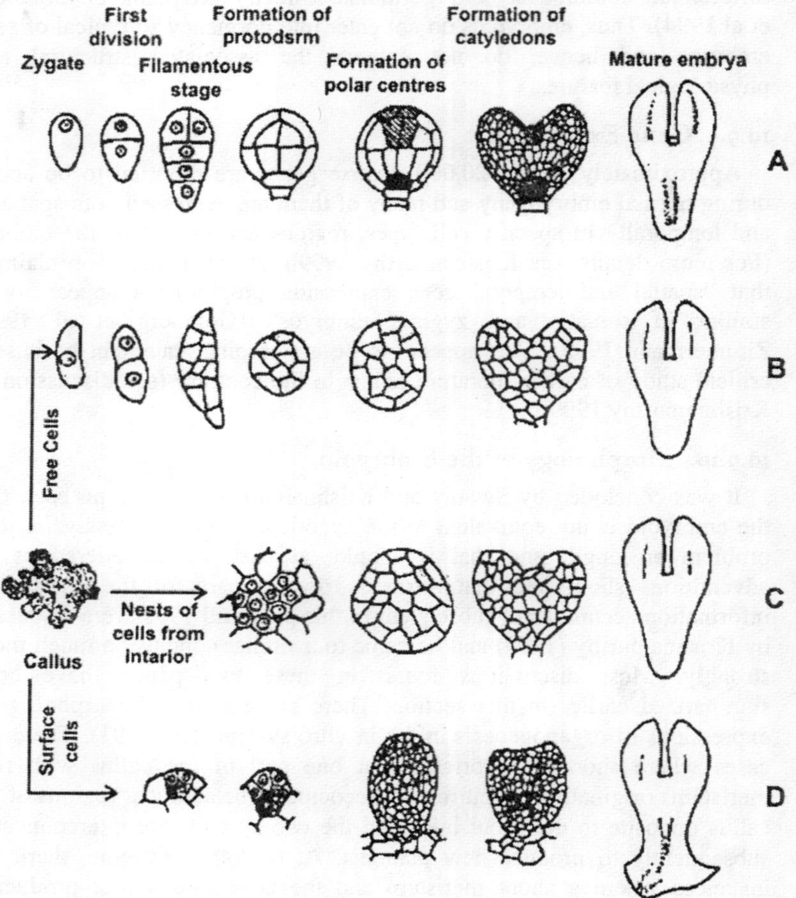

Fig. 291. The development of a typical, sexually-produced dicot embryo (A) and embryoids (B–D); While A shows an organized and polarized development and the differentiation of hypophysis and epiphysis poles even at the globular stage, the rest (B–D) show an unorganized and non-polarized development with the absence of differentiation of hypophysis and epiphysis, and also a clear-cut protoderm (Swamy and Krishnamurthy 1981).

critically and properly interpreted, will support the conclusions made above that there is no root meristem organization in the pole opposite to the shoot in the embryoid and the embryoid is, in reality, an adventitious shoot bud only. Adventitious shoot buds are formed either naturally or under induction in various parts of the plant body and display different degrees of activity (Kerstetter and Hake 1997). All these adventitious shoot buds show the invariable presence of two prophylls, whether it be a dicot or monocot. The cultured explants are one such location where there is a conducive environment for organization of adventitious shoot meristems in the form of shoot buds. Thus, embryoids are morphologically equivalent to adventitious shoot buds and resembling exomorphically a mature embryo and that they should not be considered even as 'facsimiles of embryos' as Raghavan (1986) considered them. Merely because their end products are similar to those of zygotic embryos (i.e., a new plant), they should not be described by terms that are appropriate only to zygotic embryogeny. Otherwise, it is akin to equating a humanoid to a human, however close the former may be to the latter.

Abnormal Growth and Development 11

11.1. INTRODUCTION

No plant lives in isolation. It lives and interacts with other plants and also with other organisms during its life. There are various kinds of interactions (Table 7) and most, if not all, of these interactions co–evolved during the course of evolution. A plant also interacts with the environment (and its various factors) in which it lives. The systemic controls that operate within plants or their parts, when they get affected in various ways and to various degrees by these interactions, result in **abnormal growths**. Also, when the plants or their parts are subjected to an alien or modified environment, whether natural or man–made, they resort to abnormal growth patterns. These alien environments include growing plants under altered conditions, subjecting plants to physical (such as ionizing radiations) or chemical treatments, wounding of various sorts, subjecting plants to excessive pollution of air, water and soil etc. All abnormal growths of plants induced by various interacting living organisms are called **biotic abnormal growths** and all those induced by abiotic factors are called **abiotic abnormal growths**.

Table 7. Types of interspecific interactions (Burkholder 1952) [o = No effect; - = Affected; + = Benefited]

Sr. No.	Type	Species 1st	Species 2nd	General nature of interaction
1.	Neutralism	o	o	Neither population affects the other.

Sr. No.	Type	Species 1st	Species 2nd	General nature of interaction
2.	Competition (direct interference)	-	-	There is an indirect inhibition of each species by the other.
3.	Competition (use of resources)	-	-	There is an indirect inhibition when common resource is in short supply.
4.	Amensalism	-	o	Population 1 is inhibited, while population 2 is not affected.
5.	Parasitism	+	-	Population 1 i.e., the parasite is generally smaller than population 2 i.e., the host.
6.	Predation	+	-	Population 1 i.e., the predator is generally larger than population 2 i.e., the prey.
7.	Commensalism	+	o	Population 1 i.e., the commensal gets benefited, while population 2 i.e., the host is not affected.
8.	Protocooperation	+	+	Interaction is favourable to both and is not obligatory.
9.	Mutualism	+	+	Interaction is favourable to both and is obligatory.

The external manifestations of all abnormal growths may range from very mild, often non–detectable, growth alterations from the norm to very conspicuous growths. It should be mentioned here that one often finds usages in botanical literature expressions such as 'anomalous secondary growth', 'atypical growth' etc., in apparently normal plants. These are growths characteristic and normal to those plants and that these should be considered as variants (for example, cambial variants instead of anomalous cambia) from the norm characteristic of most other plants. Phenomena such as anomalous secondary growth should not be included under abnormal growth. In plants with abnormal growth, there is always a norm from which the former deviates and both the normal and abnormal growths are seen side by side (for example, a galled leaf and a normal leaf), while for plants with the so–called anomalous secondary growth, there is no norm available in the same plant for comparison. Therefore, in this book, abnormal growths as noticed in calli, galls, tumors, infections, diseases, mutualistic plant responses and other abiotic and biotic abnormal structures alone would be covered.

Although significant on their own merit, these abnormal growths often provide vital information on factors affecting, and also on the mechanism of, normal growth and development. This chapter will also highlight these aspects.

11.2. CALLUS

A callus is often defined as a mass of undifferentiated tissue, but, in reality, it is not a correct statement. Even the simplest of calli show differentiation of at least some of its constituent cells making them different from the rest in size, morphology and structure. Calli can broadly be classified into **wound calli** and **in vitro produced calli**. The wound callus is an entity that develops on an intact plant, while the in vitro produced callus develops away from the plant from one of its detached cells, tissues or organs.

Wound callus can be produced from any part of the intact plant when it gets wounded naturally (by wind, water, animals, or physical collision with any solid object) or artificially by humans (i.e., by feeding, pruning, cutting, tilling and other processes). A few mature, living, parenchyma and/or collenchyma cells dedifferentiate into actively–dividing meristematic cells and contribute to the formation of the callus. The size of the callus varies greatly depending on the size of the wound, the degree of meristematism of the dedifferentiated cells and on various other prevailing conditions.

Wound callus may remain more or less parenchymatous or may show various degrees of differentiation. For example, if a deep wound partially destroys the cambial ring, a wound callus is produced from which cambial cells regenerate and restore the continuity of the cambial ring. Warren Wilson and Warren Wilson (1961) have discussed the nature of these regenerating cambia in the wound callus and have put forward an hypothesis on the factors that might govern their position. They proposed that the initiation of the cambium is determined by gradients that may establish under the influence either of conditions external to the wound or of pre–existing vascular tissues or of both. The orientation of the cambium with respect to xylem and phloem production is determined by the direction of this gradient. The cells of a few layers of the wound that lie on its periphery often get suberized in their walls and get converted into **wound cork** and protect the wounded area.

Many pathological growths in plants induced by various factors (abiotic or biotic) initially, at least, involve some degree of callus production.

11.3. TUMORS AND GALLS

Tumors and galls are morphologically alike, but tumors have often indeterminate growth, indefinite shape and show continued growth even in the absence of the inducing biological agent. Galls often have a determinate growth that requires the continued presence of the inducing agent, and many insect galls have a definite shape. Some tumors of plants can be compared to cancerous growths of animals.

11.3.1. Crown Gall Tumors

Crown gall tumors are caused by *Agrobacterium tumefasciens,* a soil–borne bacterium. Due to bacterial infection at the site of a wound in the primary plant body, cells around the wound are forced to proliferate as a 'cancerous' **neoplastic growth**, commonly called **crown gall** (Fig. 292). This is a malignant type of abnormal growth and a type of plant disease. After some time of the initial tumor formation, secondary tumors with uncontrollable proliferative activity appear all over the aerial part of the plant body, but in the absence of the inciting bacteria at the places of tumor induction. Crown gall tumors differ from callus in being hormone–autotrophic, have in their cells a cell division factor for continued growth, and can grow rapidly in vitro in a relatively simple growth medium. The crown gall tumor cells characteristically synthesize derivatives of basic amino acids called **opines** that are not produced in normal plant cells. Secondary crown gall tumors are capable of producing new tumors when grafted back onto the hosts of the same species from which they were originally derived. The infection process and tumor formation have been demonstrated to be due to the stable insertion of foreign genes (i.e., genes from the crown gall bacterium) into the plant cell genome, which cause a permanent change in cell behaviour. On the other hand, at the morphogenetic level, growth and differentiation of crown gall tumors are found to be central to the problems of habituation and epigenetic changes (Raghavan 2000).

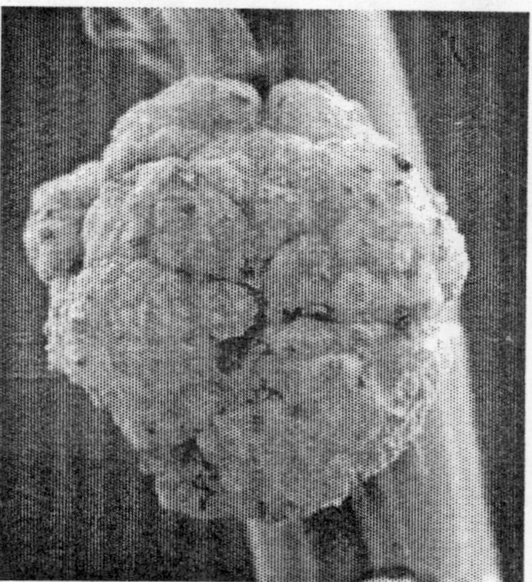

Fig. 292. A typical crown gall caused by *Agrobacterium tumefasciens* on a dicot stem (Source: http://biologi.uio.no/plfys/haa/gen/gmo.htm)

It was postulated as early as 1947 by Braun that *Agrobacterium tumefasciens* (Fig. 293 A) induces the crown gall tumor through a tumor–inducing substance. A minimum period of 36–48 h of bacterium–host interaction is an obligate requirement for the subsequent tumor induction in the absence of the bacterium. During this interactive period, the bacterium is shown to release the tumor–inducing principle into the plant cell to induce transformation. Later, the large plasmids of this bacterium were assigned the role of providing the tumor–inducing principle. Further investigations revealed that during the course of bacterial infection, only a small stretch of sequences of the tumor–inducing (Ti) plasmid (Fig. 293 B), called T–DNA, is mobilized, and stably and covalently integrated into the plant nuclear genome (and not into mitochondrial or plastid DNA). The T–DNA segment contains the *TMR* (now known as *IPT* gene) and *TMS*

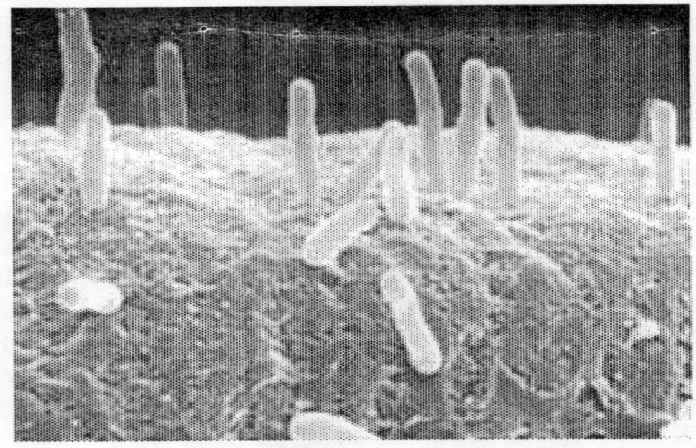

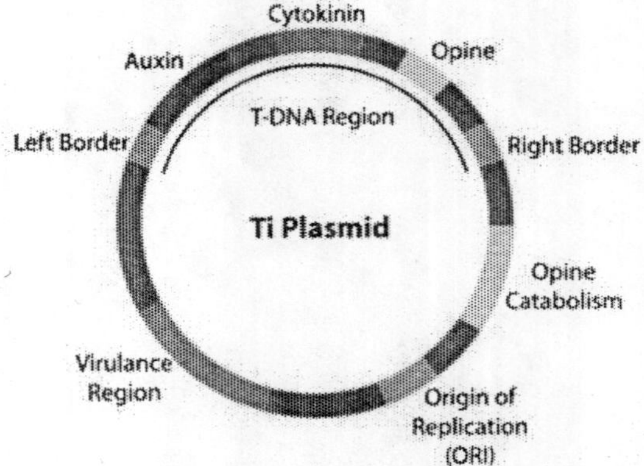

Fig. 293. SEM picture of *Agrobacterium tumefaciens* (A) (Courtesy: Genome News Network and Martha Hawes) along with the structure of its Ti plasmid (B)

genes that are needed for opine synthesis and tumor induction. It is now generally agreed that T–DNA insertion can occur into the transcriptionally active sites of any chromosome of the host plant cell. The final integration product is formed following repair–DNA synthesis and conversion of the single–stranded DNA–protein complex into a double–stranded DNA.

The term 'crown gall teratoma' loosely denotes a type of tumor that consists not only unorganized callus tissues, but also an anarchic cluster of distorted leaves and shoot buds of the host plant. This condition is produced by a moderately virulent strain of the bacterium, while the crown gall is produced by the very virulent strain. The tumors or teratoma appear to be a heritable property of cells as they can grow in axenic cultures almost indefinitely.

Crown galls are produced in many dicots, mostly woody, but not in monocots. This perhaps depends on the fact that monocot parenchyma cells do not have the ability to dedifferentiate to the meristematic state as easily as dicot parenchyma cells. The histologic structure of the crown gall is more akin to a wound callus.

11.3.2. Hairy Roots

Although **hairy roots** are not similar to crown galls, they are definitely the result of abnormal growth or differentiation activity. The large root–inducing (Ri) plasmid of *Agrobacterium rhizogenes* confers the ability of the production of hairy roots (disease) in infected host plants. These plants are characterized by the proliferation of fast-growing adventitious, very thin roots at the loci of infection (Fig. 294). Hairy–root induction is associated with the stable integration of the Ri plasmid T–DNA into the genome of the root or callus cells of the host plant. The informational package of the Ri plasmid is different from that of the Ti plasmid, but both share homology of the 25–bp terminal sequence, the *TMS* loci, loci for certain opines in the T–DNA, and the conserved *VIR* region outside the T–RNA (Raghavan 2000). The presence of *VIR* genes in both Ti and Ri plasmids implies that the same functionally– important protein domains may mediate the transfer and integration of the foreign DNA into the genome of the plant.

The T–DNA genes of the Ri plasmid, primarily responsible for hairy root formation, are collectively known as *ROL* (for 'root loci') and include *ROL A*, *ROL B* and *ROL C*. T–DNA of certain Ri plasmids harbours *IAAH* and *IAAM* genes that encode enzymes of the IAA synthesis pathway, although IAA plays only a peripheral role in hairy–root disease.

Fig. 294. Hairy roots caused by Agrobacterium rhizogenes on Plumbago zeylanica (Photograph courtesy of Dr. K. Satheesh Kumar)

11.3.3. Galls

11.3.3.1. Definition

A **gall** or **cecidium** is a remarkable biological phenomenon, induced by one organism and produced on another wholly unrelated organism, the plant (Schaefer et al 2005). A gall is often an expression of a complex collaboration and a suite of adaptations achieved by the host plant for accommodating the gall inducer and its requirements. Again, a gall is a new structure developed in a plant organ by the other organism, often to the benefit of the latter (by providing nutrition and shelter) causing a variable degree of harm to the plant. It is very often difficult to define a gall, since galls range from structures that are only slightly deviant from the original plant part to very complicated structures of distinct morphology totally different from the plant organ from which they are formed. Classical cecidologists such as Mani (2000) include any generic, anomalous growths on plants such as those produced by bacteria, fungi and nematodes as galls, but others restrict the term 'galls' only to structures produced by mites (Acarina) and insects (Raman 2011).

11.3.3.2. Gall–causing Organisms

Fig. 295. A - Leaf curl gall of tomato caused by a virus; B - Leaf blister gall caused by a virus

Galls in the conventional sense are caused by viruses (Fig. 295), bacteria (Fig. 296), fungi (Fig. 297), nematodes (Fig. 298), mites (Fig. 299), insects (Fig. 300) and even angiospermic parasites (Fig. 301) such as *Loranthus* and *Viscum*. Galls have interested biologists for a very long time. The majority of the galls are produced by insects (of Thysanoptera, Homoptera, Diptera, Hymenoptera, Coleoptera and Lepidopteral) followed by Acarina (*Eriophyes*), nematodes (*Meloidogyne, Heterodera*), fungi (*Plasmodiophora, Synchytrium, Puccinia, Protomyces, Taphrina*), bacteria, MLOs (mycoplasma–like organisms) and viruses.

Fig. 296. Bacterial gall on Oleander (Photograph courtesy of Texas Agri Extension Service, Dickinson)

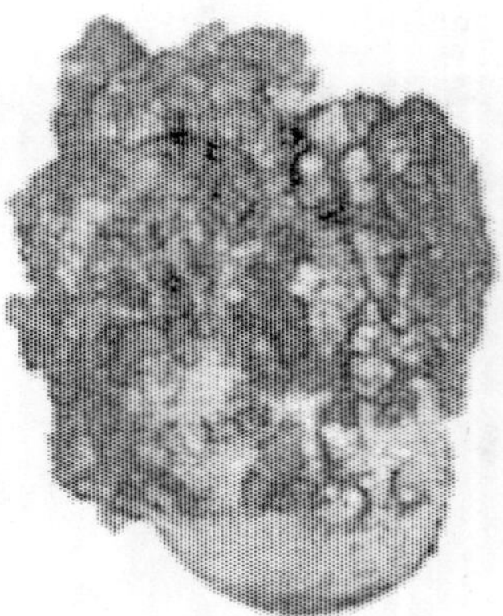

Fig. 297. Gall caused by the fungus *Synchytrium* species on potato tuber (Wart disease)

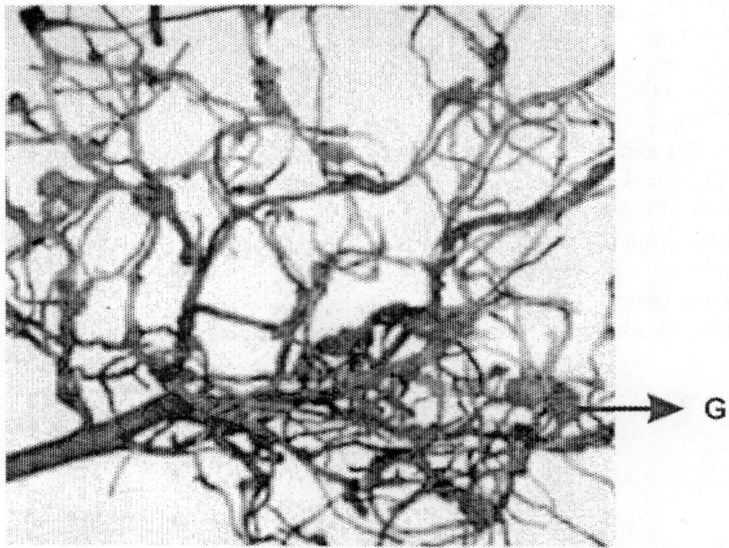

Fig. 298. *Basella alba* roots infected with species of *Meloidogyne* (root knot nematode) showing galls (G)

Approximately 13,000 species of insects from the world are known to cause galls (Raman 2005a) and numerous other species are yet to be discovered. All except certain insect galls are simple ones. Galls induced by insects differ from the abnormal growths or from the galls induced by fungi and bacteria in their form, organization, and structural complexity.

Nearly 90% of known insect galls show either bilateral or radial symmetry (Raman et al 2009, Sinnott 1960). Many galls induced by Cecidomyiidae (Diptera) and Cynipidae (Hymenoptera) exemplify impressive levels of radial symmetry. Nearly 90% of the gall insects are host–specific (Raman et al 2005 a), indicating a high degree of host fidelity and monophagy.

Fig. 299. Some representative plant galls caused by mites; a - *Alangium salvifolium* - *Eriophyes alangii*; b - *Canthium parviflorum* - *Eriophyes* species; c - *Commiphora caudata* - emergence gall - *Eriophyes* species; d - *Combretum albidum* - *Eriophyes* species; e - *Rotala verticillaris* - Rinden gall - *Nanophyes* species; f - *Ziziphus mauritiana* - *Eriophyes cernuus* (Amerjothy 2005).

Fig. 300. Some representative insect galls; A - Thrips gall on axillary bud of *Calycopteris floribunda* caused by *Austrothrips cochinchinensis*; B - Dipteran piston rod gall on *Acacia ferruginea* caused by *Lobopteromyia* species; C - Dipteran gall on *Coccinia grandis* caused by *Neolasioptero cephalandrae*; D - Hemipteran cushion gall on *Achyranthes aspera* caused by *Bemisia* species (Amerjothy 2005); E - Sea-urchin gall induced on axillary bud of *Hopea ponga* caused by *Mangalorea hopcae* (Photograph courtesy of Dr. A. Raman)

Fig. 301. Gall caused by the angiosperm parasite *Dendrophthoe pentandra* (Photograph courtesy of Singapore Science Centre)

11.3.3.3. Classification

Küster (1911) classified galls into two principal types: **organoid galls** and **histoid galls**. The organoid galls include abnormal production of organs either by modifying the arrangement of organs or, rarely, even by inducing their transformation (e.g., witch's broom). The histoid galls include production of abnormal tissue involving principally hypertrophy and hyperplasia. Histoid galls are further divided into two further groups: **cataplasmic galls** and **prosoplasmic galls**. The cataplasmic galls are amorphous, variable in size and extent, and exhibit relatively poor histological differentiation. Most fungal and bacterial galls belong to this category. The prosoplasmic galls are more defined in shape and size than the cataplasmic galls and show very distinctive tissue differentiation, as in most insect and mite galls.

11.3.3.4. Gall Morphogenesis

Insect galls are the best examples of modified geometrical natural structures that arise exclusively because of signals or messages received from insects (Raman 2011). They represent highly regulated growth manifestations (Figs. 302 and 303) through specific differentiation processes on plants and develop through precise co–ordination between the physiological conditions and the genetic make–up of the host plant (Weis et al 1988, Raman 2010). These differentiation processes or patterns vary with the taxonomic groups of gall–inducing insects (Rohfritsch 1992, Raman 2003), but the external shapes of mature galls induced by different individuals of the same insect species are amazingly consistent. Interestingly, galls induced by very closely–related insect species, for

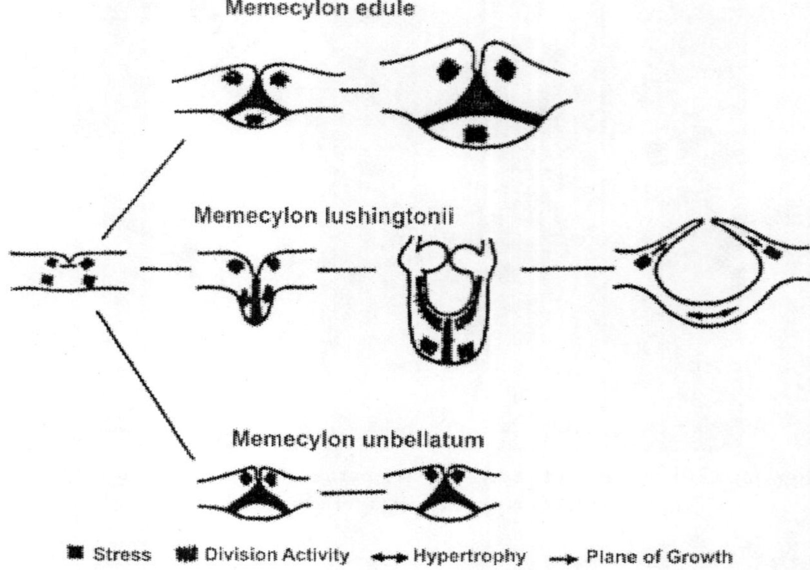

Fig. 302. Morphogenetic gradients in the galls induced by *Crotonothrips dantahasta*, *Crotonothrips memecylonicus* and *Crotonothrips* species on the leaves of *Memecylon edule*, *Memecylon lushingtonii*, and *Memecylon umbellatum* respectively. (Not to scale) (Adapted from Raman and Ananthakrishnan 1985 and from Raman 2003)

example, species of *Andricus, Callirhytis, Caryomyia and Rhabdophaga*, show highly varied gall morphology; only very rarely galls induced by unrelated insect species are similar in morphology (Raman 2011) i.e., show convergence in gall morphology.

Key stages in gall development are gall induction (or initiation) and gall growth, the latter including increment in mass and qualitative differentiation. Gall development invariably involves two counteracting events (Raman 2011): The insect provides stress to the host organ, while the host counters the stress through its responses which include newly differentiated tissues and new physiological activities. This fits into the Thom's (1975) **epigenetic–polyhedron model** that is as follows: "Every living being is characterized by its global stability, and after a shock it will respond by producing a reflection principle, the effect of cancelling any perturbation produced by the stimulus". The earliest recognizable event in gall development is the isolation and insulation of a single cell or rarely a group of two to three cells in the plant organ, attacked by the insect, from their normal course of differentiation, resulting in **metaplasied cells** that are irreversibly changed by the insect's stimulus (Meyer and Maresquelle 1983, Raman 2011). The location of these cells may be from any part of the attacked organ i.e., epidermis, mesophyll, phloem, cortex and pericycle. A salivary chemical is believed to be invariably involved in stimulating cell division in the metaplasied cell(s) to initiate the gall, and this chemical may be high molecular–weight proteins, bruchins, mitogenic lipids or a growth hormone such as IAA. Genes homologous to '*nod c*' (gene promoting nodulation in rhizobia) have been implicated in the growth of some galls (Raman 2011). Gall insects are capable of inhibiting normal growth and differentiation of the plant part and inducing new growth by establishing new polar axes and morphological and biochemical gradients in the infected plant part. The size of the gall is determined by the duration of the insect's stimulus on the affected plant part.

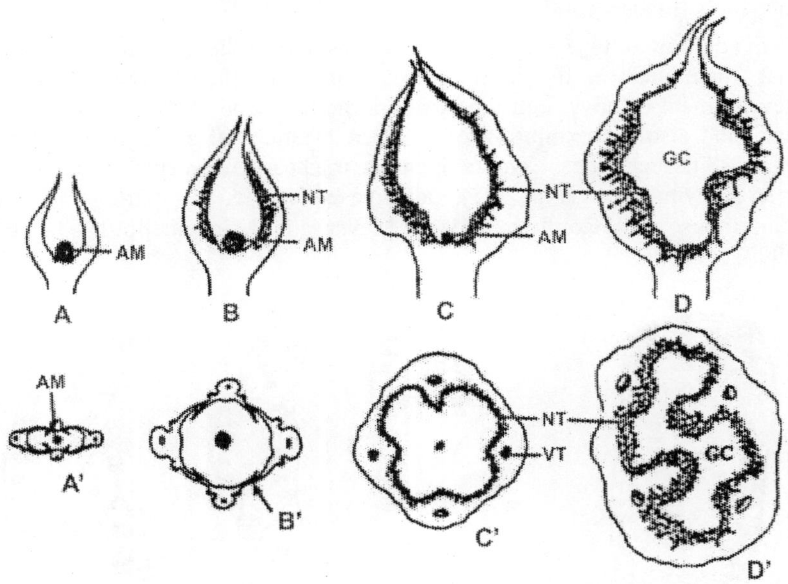

Fig. 303. Cecidogenesis in the pouch gall of *Calycopteris floribundus* induced by *Austrothrips cochinchinensis*; (A–D) - Longisectional cartoons; (A'–D') - Cross–sectional cartoons that correspond with (A–D); (A, A') - Vegetative apical meristem (AM); (B, B') - Early stages of gall development with the apical meristem (AM) destroyed by trips feeding; Arrow in B' - Marginal suturing of the oppositely placed young leaves; Stippled areas - Nutritive tissue (NT); (C, C') - Young galls (VT - Vascular Trace); (D, D') - Mature galls; GC - Gall Chamber (Not to scale) (Adapted from Ananthakrishnan and Raman 1989, and from Raman 2003)

One of the major factors that possibly affect the size of the shoot–axis galls is the variation in the growth rate of whole plant and its metamers. This also affects the abundance of galls and the performance of the gall–inducing insects (Raman 2011). The **plant–vigour hypothesis** (Price 1991) underscores that the performance of the gall–inducing insect is the maximum on large, fast–growing metamers resulting in larger galls, and these larger galls promote the better survival rate of the nutritive cells. The **optimal metamer–size hypothesis**, on the contrary, proposes that the performance of the gall–inducing insect is the greatest on intermediate–sized metamers (McKinnon et al 1999). This is based on the understanding that the galls are induced by a dose–dependent stimulus from the insect and a critical dose mass is needed to promote gall size with metamer size. This hypothesis emphasizes that galls are induced more easily on small–sized metamers (Flaherty and Quiring 2008).

In addition to hypertrophy and hyperplasia, the other important elements that distinguish galls from other anomalous plant growths is the development of novel cell types, not normally found in that plant or plant part, and also the exhibition of an altered pattern of arrangement of the various tissue types. These include tracheary elements (Fig. 304), sclerenchyma cells (Fig. 305), nutritive cells (Fig. 306), trichomes etc. (Shorthouse and Rohfritsch 1992, Meyer and Maresquelle 1983, Fink 1999).

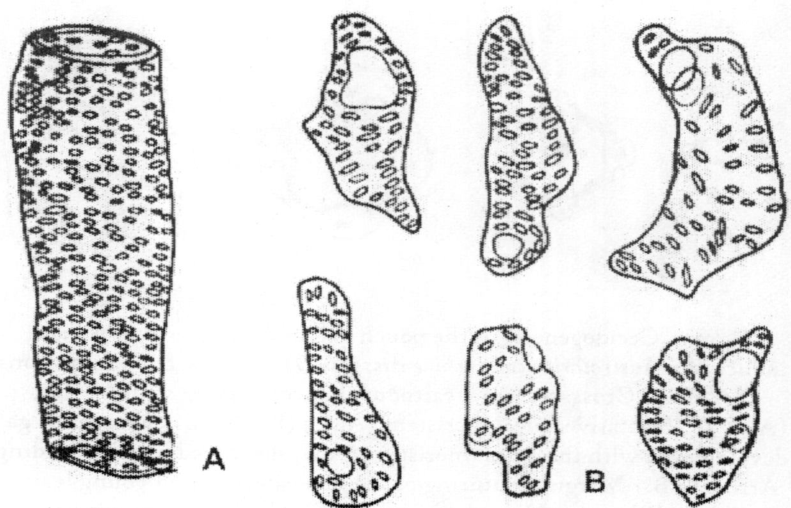

Fig. 304. Normal (A) and abnormal tracheary (B) elements associated with the nematode galls of *Basila alba*

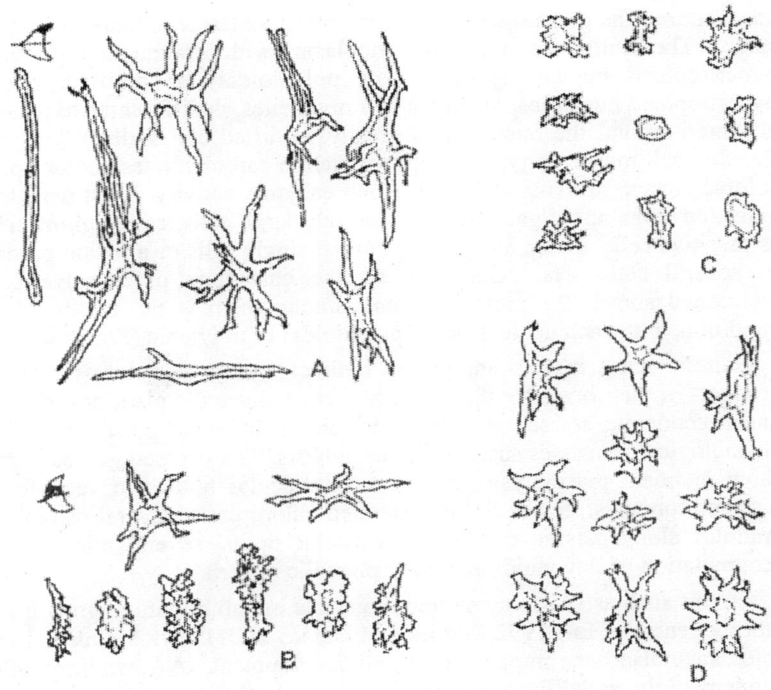

Fig. 305. Representative types of sclereids in four different types of galls (A–D) of *Gnetum ula* (Swamy and Krishnamurthy 1975c)

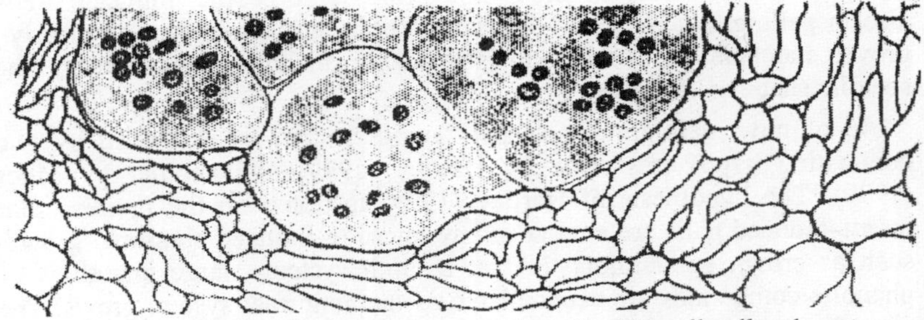

Fig. 306. Transverse cells of a nematode gall of *Basella alba* showing the nutritive, multinucleate giant cells on which the nematode feeds; the nuclei are often polytenic.

The nutritive cells provide the required nutritive substances to the gall–inducing insect and also to its progeny (wherever applicable). The location, distribution and orientation of the nutritive zone in galls vary widely depending on the gall type and the location of the gall insect or their larvae. The location may be epidermal, subepidermal, deep–seated or near vasculature. The general structure of the nutritive tissue is more or less the same. The cells are rich in cytoplasm with fragmented vacuole, hypertrophied nucleus (polytenic or polyploidal) or multiple nuclei, hypertrophied nucleolus, abundant cell organelles, dedifferentiated plastids clustered around the nucleus and greatly modified cell walls, often with transfer–cell morphology. The cells are rich in carbohydrates and/or lipids, soluble sugars, proteins and have a rich enzymic activity. Most nematode galls and some aphid galls have extensively large cells, called **giant cells**, as nutritive cells. These are either enlarged single cells or a fusion product of several cells. The cells have all the characters of nutritive cells mentioned above; the most important among them is the multinucleate condition, with each nucleus being polyploidal or polytenic (Fig. 306).

Gall–inducing insects and mites redirect energy and key metabolic products of their hosts for their use and hence, place the plant or plant part under moderate to severe stress, which is reflected in their altered physiological processes such as photosynthesis. The stressed gall cell often show incipient plasmolysis, vacant multivesicular bodies in vacuoles, a ruptured tonoplast, fewer and degenerated chloroplasts, several osmiphilic granules along plasma membrane, vacuolar lipids, several microbodies, accumulation of flavonoids and other phenolic materials.

Auxins such as IAA play an important role in gall initiation, growth and morphogenesis (Hori 1992, Mapes and Davies 2001). Cytokinins and uric acids have also been implicated in gall development. IAA synthesis often happens in the galls (Raman et al 2005 a, b). Auxins and cytokinins act in coordination to promote cell proliferation and cell enlargement. However, what is the nature of trigger that activates the cell (s) that leads to initial metaplasia, mentioned earlier, is not clear. Some people implicate ABA also in galling, but its actual role is far from clear. However, no clearly–proven biochemical explanation is available on how gall symmetry and specific shape are achieved.

As of now, gall growth, as per 'auxin theory', is explained by two contrasting hypotheses: One by Hartley (1999) and another by Miles (1999). The hypothesis of Hartley (1999) is based on the premise that insect–induced galls are similar to any microbe–induced abnormal growth such as crown gall tumors, in that in both systems, large quantities of phenolic compounds are accumulated as the abnormal system grows. The phenolics serve as mediators of host defense, but, as in bacteria–induced abnormal growths, the phenolics not only indicate the vulnerability levels in host plants but, also neutralize the plant part consequent to attack by altering gene expression of the host plant and by changing hormone activity (Raman 2011). Since some of the phenolic compounds are well–known growth promoters, they may induce cell divisions and gall growth by interacting with already existing plant–growth regulators such as IAA. But, Hartley's hypothesis does not explain certain aspects of gall morphogenesis that link phenolics with IAA and also the happenings at the cellular level. Mile's hypothesis (1999) explains that salivary components such as tryptophan (IAA precursor) and salivary proteinases that release plant–bound IAA are important in galling. Vigorous oxygen uptake by plant tissue and wounding during gall initiation stimulate plant–bound auxin activity. The fact that the levels of tryptophan in the saliva of gall and non–gall plant feeding insects are similar and the release of plant–bound IAA by salivary proteinases is disputable questions Miles' hypothesis. Raman (2011) has indicated that the insect saliva flushed on the wounded plant site alters the subcellular environment of cells and thus, places it in a state of chemical shock. This shock is stated to induce osmotic change, which is the first recognizable event in galling. Raman further states that to repair the wound and to neutralize the osmotic stress, the plant responds by establishing one to a few metaplasied cells. This is followed by the spread of localized metabolic changes from these cells, not throughout the involved plant part, but only immediately around the site of gall causer. Only when the induction by insects is of low intensity, the gall development ensues, otherwise the cells die, rejecting the inducing insect and defending the plant tissue or part. Once galling, thus, gets initiated, new plant–mediated morphogenetic events ensue, including the synthesis of growth regulators, alteration in H^+–transport system, well–directed cell division and cell enlargement and new tissue differentiation. Both the hypothesis, although not with convincing evidences, provide an explanation to gall growth based on IAA activity. However, what specific factor triggers both hypertrophy and hyperplasy operating together during gall development is not clearly identified. Also, not clearly known are the nature and types of hormone receptors (if any!) and transduction pathways in the initial metaplasied cell(s) (Raman 2011).

There are intense debates as to why galls are formed at all (Raman 2003). While some believe them to be defensive responses to the attack of the gall maker and they isolate the gall in space and time forcing in the extreme specialization of galls, others consider that although a defense growth response may operate in the initial stages of gall development, selection on the gall makers will subvert a response to its own advantage. Incidentally or accidentally a gall has also become the site for a range of secondary organisms functioning as parasites and pretadors. In a broad sense, a gall develops as a balancing structure between the offensive effect of the gall inducer and the defensive response of the plant.

11.4. ABNORMAL GROWTHS ASSOCIATED WITH SYMBIOTIC AND MUTUALISTIC INTERACTIONS

The terms **symbiotic** and **mutualistic** have been used interchangeably to describe associations between different organisms. 'Symbiosis' was originally used to define both lichens and parasites, but many now restrict this term to describe beneficial relationships only. The broader definition for symbiosis is 'living together of two or more organisms'. The term mutualism implies mutual benefits for the two or more different organisms that come into an association. Mutualism is an isocline in diagrams that show the relative benefits of the involved organisms in an association indicating that all the involved species are more successful together than when they are alone (Boucher 1985, Brundrett 2004). Mutualistic associations include a wide array of direct or indirect, and symbiotic or non–symbiotic associations in which nutrient transfers alone are not involved. Some of these interactions involving plants as one of the interacting organisms result in abnormal growth responses in the plant. This section deals with such growth responses as seen in **root nodules** and **mycorrhizae**.

11.4.1. Root Nodules

Root nodules are abnormal growths caused by symbiotic nitrogen–fixing bacteria (Fig. 307 a, b). These nodules exhibit striking morphogenetic changes that involve the activation of a number of host plants and also bacterial genes. The angiosperms show two basic categories of root–nodule formations: one involving filamentous actinomycete bacteria such as *Frankia* that form nodules on the roots of woody plants such as *Alnus, Casuarina, Comptonia, Myrica* and ten other genera, with about 310 species (Bond 1967), and the other involving most herbaceous and woody taxa of legumes, induced by non–filamentous bacteria belonging to species of *Rhizobium* and *Brachyrhizobium*. Both types of nodules house the bacteria inside them.

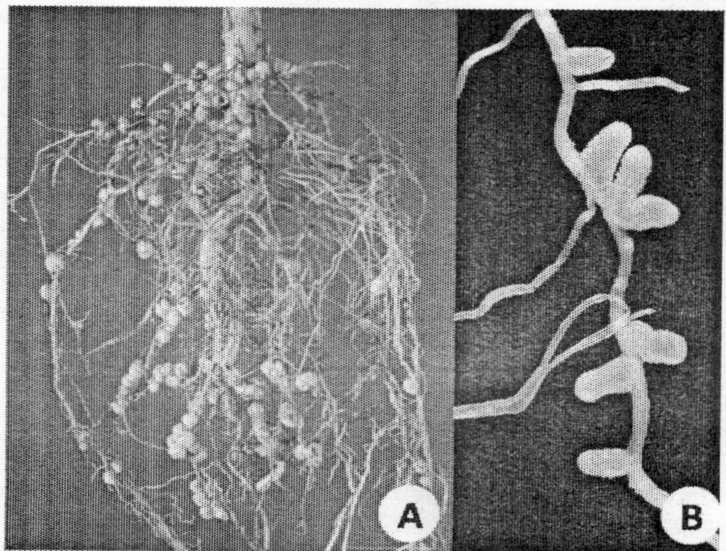

Fig. 307. Root nodules of two species of *Medicago*

11.4.1.1. *Actinomycete Nodules*

The mode of infection and formation of nodules by actinomycetous bacteria have been studied extensively in *Alnus glutinosa, Comptonia peregrina, Myrica gale, Casuarina equisetifolia* and *Casuarina glauca* (Fig. 308). The infection takes place through root hairs as a result of which they get variously deformed and crooked. The bacterium enters into the root cortex from root epidermis after it multiplies in the root hair cell and produces branched **infection thread** or **hypha**. The infection threads invade the root cortex and colonize inside the cortical cells. In *Alnus*, the cortical cell wall poses a physical barrier to penetration by *Frankia*, the actinomycete. Cell wall penetration has been attributed to cell wall degeneration by the bacterium and to modification in host cell wall deposition. Unusual cell–wall deposition is noticed frequently in uninfected derivative cells also. There is no clear evidence of a continuous middle lamella. Cell divisions in cortical cells are continued to be induced by the infection to result in the nodule. Initial cell divisions are seen in the pericycle and endodermis, and only subsequently cortical cells join in this division activity to produce **primary nodule**, which includes lateral root primordia (formed by pericycle activity) of variable numbers. These lateral roots are often called **nodule roots** (Fig. 308). These nodule roots, although similar to normal lateral roots in their endogenous origin, differ from the latter in having their origin at root–hair infection sites and in their involving cortical cells also during their development. The actinorrhizal nodules appear to involve the expression of nodule–specific genes. One such gene studied in detail encodes for a serine protease of the subtilisin family (Ribeiro et al 1995).

11.4.1.2. *Legume Nodules*

The legume root nodules induced by *Rhizobium* and *Brachyrhizobium* are also initiated through infection of root hairs located in the very young root region (Fig. 309) and the subsequent deformation of these hairs.

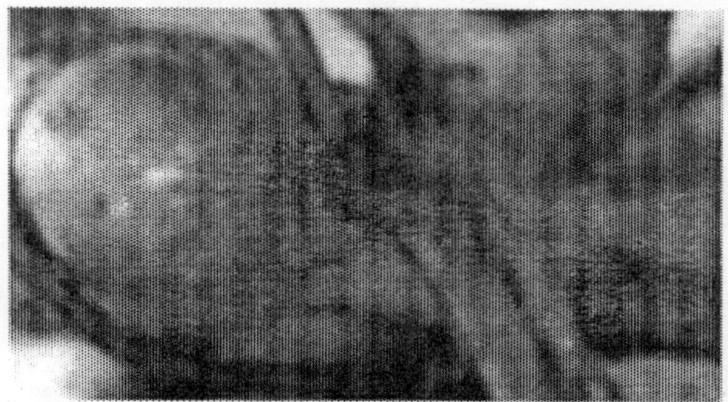

Fig. 308. Root nodules of *Casuarina equisetifolia* caused by species of *Frankia*, an actinomycete; nodule shows emagence of nodule roots.

Rhizobia are known to secrete **nodulation factors** (called **NOD** factors) that induce changes in root morphology. These are shown to be of the lipooligosaccharide– nodulation factors category. The chemically–defined lipooligosaccharide 'elicitor' is secreted by rhizobium following stimulation by specific flavonoids present in host legume root-exudates. The rhizobium–induced signal molecule is an oligochitin. Host specificity is determined by strain–specific sugar modifications, either at the reducing end (E.g., sulphation, fucosylation or acetylation) or at the non–reducing end (E.g., acetylation). Some investigators speak of the involvement of an interaction between lectins (a glycoprotein) and monosaccharides in the attraction towards and infection of legume roots by the specific bacterial strain. The mode of action of the lipooligosaccharide NOD–factor in nodule initiation and development has not been clearly elucidated. The only clue for a probable answer lies in the nature of the biological response of the host root. The earliest observed effect of the application of this NOD–factor on infected host root is the deformation and curling of growing root hairs within 3–4 h and this, perhaps, implies an effect of the application of NOD–factor on the coordination of cell wall growth through modification of the plasmalemma or the underlying cytoskeleton. Another early response to the application of NOD–factor is the centralized location of nuclei in cortical cells and the development of transcellular cytoplasmic strands. In the inner cortex, these pre–mitotic structures predetermine the orientation of cell–plate formation following cell division in the cells of the nodule primordium, and in the outer cortex, similar cytoplasmic strands predetermine the pathway for transcellular infection thread growth without prior cell division. Nodule initiation can be experimentally uncoupled from nodule invasion either by the application of purified NOD–factor or by inoculation with rhizobium mutants defective in cell surface polysaccharides (e.g., acidic extracellular polysaccharides, β, 1, 2–glucans or lipopolysaccharides). Moreover, nodule initiation can be experimentally induced, in the absence of NOD–factor, by application of TIBA (Brewin et al 1993), the last probably indicating the involvement of polar auxin transport.

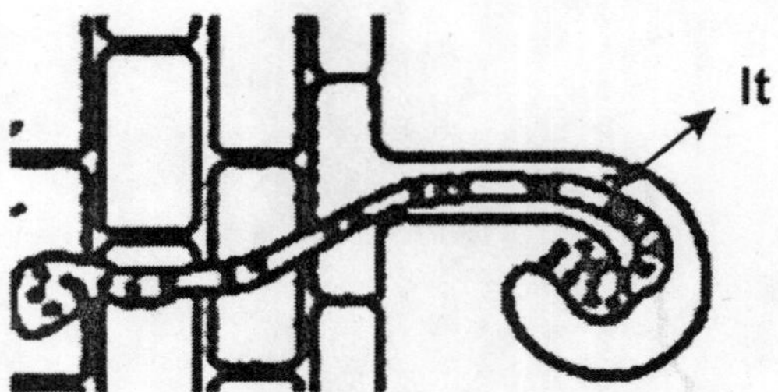

Fig. 309. Diagrammatic representation of the infection of root of a legume by species of *Rhizobium* through the root hair which bends and deforms after infection. Note the infection thread (It).

Infection threads that are formed in the root hair or epidermis help rhizobia to penetrate the host root and to infect cortical cells. These infection events induce proliferation of cortical cells to form a nodule. Nodule primordia are produced by continued divisions of the inner cortical cells at some distance from the advancing infection threads, indicating that cell division stimulus factors pass in advance of infection threads. Growth of infection threads involves the passage of the rhizobia through a series of cortical cell walls. Infection–thread growth and its passage through cortical cell walls have been explained by two conflicting hypotheses: (i) Cell wall invagination (ii) Cell wall degradation followed by synthesis of a new wall layer. Both the hypotheses are based on observations made by different investigators on root hairs, the first target cells during rhizobial infection. During the invasive phase, the host plant cell growth is reorganized to create an inwardly growing tunnel of plant cell wall along with plasma membrane that ensheaths the invading endosymbiont as it traverses root epidermal and cortical cells. Subsequently, there is a metabolic integration during which the bacteria are more or less integrated with the host cell cytoplasm, although the bacteria are always enclosed by a differential form of host cell plasma membrane and hence, are retained in a specialized form of extracellular matrix. The second hypothesis is supported by more recent studies. The site of rhizobial infection of a pea root cortical cell is predetermined by the plant, perhaps, this is true of some other legumes also.

The enlarged bacterium released into the cortical cell by the infection thread and capable of nitrogen–fixation is, in fact, called a **bacteroid** (Fig. 310) and this term is preferred over the term 'bacterium', as there are morphological differences between the bacterium and the structure released into the cortical cell by the infection thread. In the final stage of nodule development, a meristem is formed at the apex of the nodule, as the latter grows out of the parent root. A typical nodule consists of an outer 'cortex' of uninfected cells, surrounding bacteroid–filled cells. A number of 'vascular bundles' (or traces?) are embedded in the 'cortical' tissue (Fig. 311 A and B). The bacteroids are enclosed by a membrane in each nodule cell. Each host cell has a greatly enlarged nucleus, often polyploidal in nature.

During rhizobial root nodulation, there is an active expression of genes that are required for both infection and nodulation. Specific nodulation genes called *NOD* (**nodulin**) genes are activated in the host by the bacterium during nodule initiation. The proliferation of cortical cells, and the growth and function of nodules are sustained by these nodule–specific genes. The triggering act for the transcription of nodulins is promoted by the interaction at the root surface. In several *Rhizobium* species, *NOD* genes are located on large plasmids called *pSym* (for 'symbiosis plasmid'), while in *Brachyrhizobium* species these are found on the chromosome. The nodulins belong to two categories: **early nodulins** that are expressed in

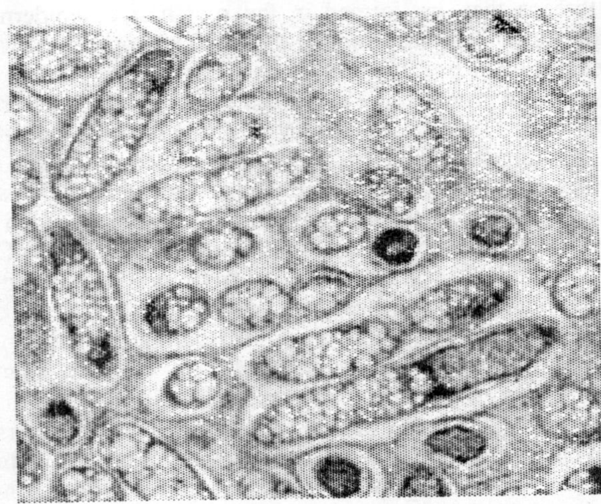

Fig. 310. *Rhizobium leguminosarum* bacteroids in bean nodules (Source: http://www.rhizobium.net/projects.html)

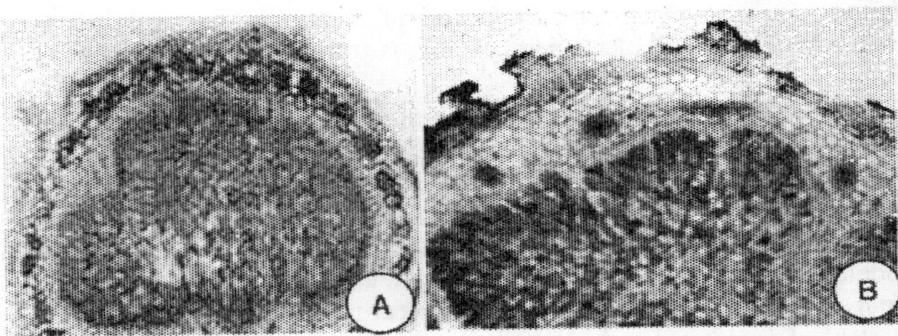

Fig. 311. T. S. of the root nodule of a legume; A - Entire T. S. showing the centrally-located bacteriods, peripherally-located vascular strands; B - A portion of nodule enlarge to show the bacterioids and peripheral vascular strands

roots within a few days after interaction with the bacteria, and **late nodulins** that are expressed at the time of onset of nitrogen fixation. The well-known among nodulin genes is *EARLY NODULIN2* (*ENOD2*), isolated from soya bean nodules. *ENOD2* is involved only in nodule morphogenesis, but not in infection process (Wilson et al 1994). *ENOD12* not only gets expressed in cells through which the infection thread passes through, but also in the cells ahead of the path of the thread. *ENOD 40* of soya bean, alfalfa and pea expresses itself in the pericycle close to the infection site. Interestingly, *ENOD 2* and *ENOD 40* are also expressed in the mycorrhizal roots of alfalfa. In other taxa, the nodulin genes are named differently such as *NOD A*, *NOD B*, and *NOD C*, and these have several functions including synthesis of signal molecules needed for elicitation of root hair deformation, production of infection threads, proliferation of cortical cells, and the coding of a DNA–binding protein that activates the expression of other *NOD* genes. The majority of nodulins, so far known, belong to the late nodulin group, the most important of which is the nodulin encoding for **leghaemoglobin** (that imparts a brownish red colour to the nodule), **uricase** and subunits of sucrose synthase and glutamine synthetase. These late nodulin genes are neither involved in infection process nor in nodule morphogenesis. The other group of late nodulin genes, directly involved in nitrogen fixing, are the *NIF* genes that control the cascade of events occurring in nitrogen fixation.

11.4.2. Cyanobcterial Roots

Certain cyanobacteria enter into the roots of some taxa of gymnosperms and angiosperms and involve themselves in symbiotic nitrogen fixation. Such roots grow ageotropically (i.e., show negative gravitropism) and come out of the soil surface. In taxa of Cycadales, such roots are characteristically present and are called **corolloid roots**. They are dichotomously branched, thick and coral–like and hence, named as corolloid roots (Fig. 312 A). The cyanobacteria are located in a specialized layer of radially enlarged cells outside the stele and this layer was called 'algal layer' earlier, but now, it is called 'cyanobacterial layer' (Fig. 312 B). We do not have any information on how these cyanobacteria enter into root and how they are confined to this specialized layer subsequently. Do these also spread through infection threads as in symbiotic bacteria described earlier? We also do not have much information on how cell divisions are promoted during corolloid root development to result in thicker roots than the normal gravitropic and thinner roots.

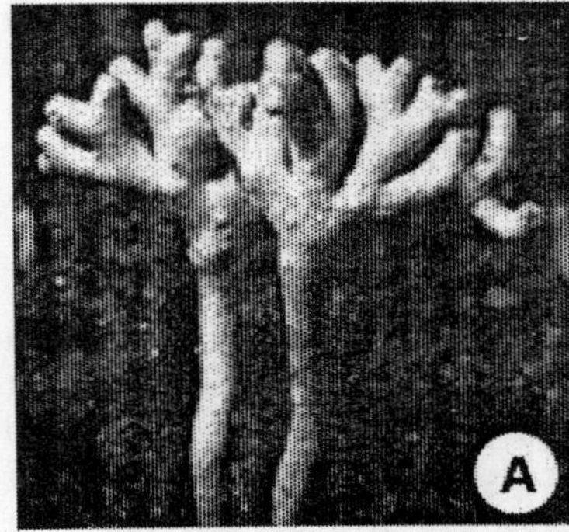

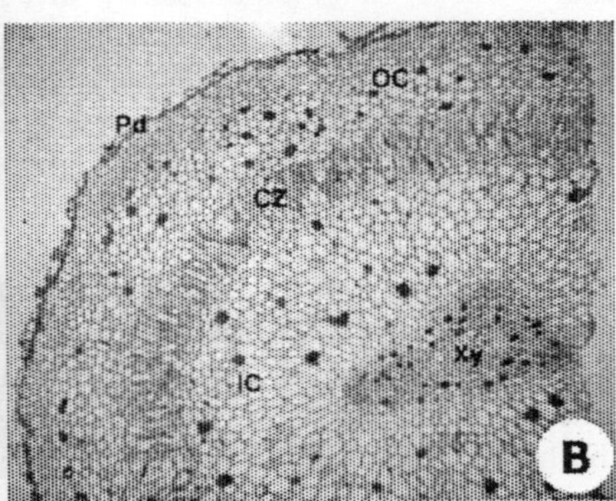

Fig. 312. A - Coralloid roots of a Cycad (David Webb);
B - T.S. of the same showing cyanobacterial zone (CZ),
inner cortex (IC) and outer cortex (OC)

11.4.3. Mycorrhizal Associations

11.4.3.1. Definition

These are multifaceted associations comprising diverse morphological, functional and evolutionary categories (Brundrett 2002, 2004). **Mycorrhizae** (means 'fungal root'), a term introduced by Frank as early as in 1885, refers to the roots of vascular plants that are symbiotically associated with specific fungi. It is the living together of two organisms, the fungus and a vascular plant. Mycorrhizae cannot be universally categorized as mutualistic associations, since the 'benefits to fungi are implausible' in their associations with non–chlorophyllous, heterotrophic plants and hence, all mycorrhixzal relationships are symbiotic. But, some associations are not mutualistic (Brundrett 2004), and the term 'balanced' and 'exploitative' are used for the symbiotic and non–mutualistic types respectively. Balanced mycorrhizal associations are those in which both organisms receive essential commodities through reciprocal exchanges, and exploitative mycorrhizal associations are those where only the plant receives substantial benefits from nutrient exchange (Brundrett 2004). A new, broader definition of mycorrhizae that includes their entire diversity while excluding all other plant–fungus associations is presented here: A mycorrhiza is a symbiotic association essential for one or both partners, between a fungus (specialized for life in soils and plants) and a root (or other substrate–contacting organ) of a living plant that is primary responsible for nutrient transfer. Mycorrhizae occur in a specialized plant organ where intimate contact results from synchronized plant–fungus development (Brundrett 2004).

Mycorrhizal association is quite widespread and is noticed in pteridophytes, gymnosperms and several angiosperms (more than 80% of taxa). These fungi significantly affect the morphology, structure and function of the infected roots to the benefit of the host plants and to their own benefit. The host plants get phosphorus from not–easily available or mobilizable phosphorus sources of the soil with the help of fungus, and the fungus, in turn, gets carbon compounds from the plant.

11.4.3.2. Classification

Mycorrhizae were earlier categorized morphologically into three major types: **Ectomycorrhizae** (Fig 313 A), **Endomycorrhizae** and **Ectendomycorrhizae**. The mycelium of ectomycorrhizal fungi does not enter deep into the host root, but forms a parenchymatous sheath around the host's short secondary root surface, called **fungal mantle** or **sheath** and a **Hartig's net** consisting of intercellular hyphae that develops between the outermost cells of the root (Fig. 313 B and C). The infected roots are short, stubby, and become repeatedly forked and dichotomously branched resulting in dense clusters in some species such as *Pinus* (Fig. 313 A) due to fungal infection. Most ectomycorrhizae belong to species of the following Basidiomycetes: *Amanita, Boletinus, Boletus, Cantharellus, Cenococcum, Citocybe, Clitocyte, Clitopilus, Collybia, Cortinaria, Cortinarius, Entoloma, Lactarius, Lepiota, Piceirhiza, Rhizopogon, Russula, Scleroderma, Suillus, Xerocomus* etc. Endomycorihizae have fungal hyphae entering into the inside of the host root. They are of three types. The most common type of endomycorrhizal fungi are also called **vesicular arbuscular fungi (VAM)** (Fig. 314 A and B), as they produce vesicles and arbuscules. Some of these fungi do not produce vesicles, but only arbuscules and hence, are called **arbuscular fungi (AM)**. VAM fungi are formed by non–septate members of Glomales of Phycomycetes (belonging to three families: Acaulosporaceae, Gigasporaceae and Glomaceae). They come under six genera: *Acaulospora, Entrophospora, Gigaspora, Glomus, Sclerocystis* and *Scutellospora*. **Orchid** and **Ericoid mycorrhizae** are the other two types of endomycorrhizae. These three are unrelated types of endomycorrhizae, with contrasting anatomical features and separate host and fungus lineages. Thus, according to Brundrett (2004) the term endomycorrhizae is invalid because it encompasses several phylogenetically and functionally disparate association types.

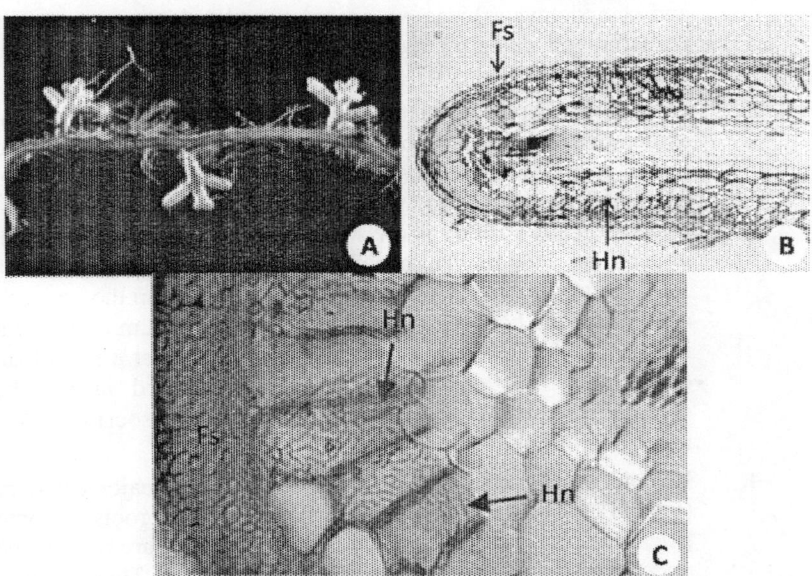

Fig. 313. Ectomycorrhizae; A - A portion of *Pinus* root system showing ectomycorrhizae; B - L. S. of the tip region of an ectomycorrhiza; C - T. S. of the same; Fs - Fungal shealth; Hn - Hartig net; (Source: B - http://bugs.bio.usyd.edu.au/learning/resources/Mycology/Plant_Interactions/Mycorrhizas/mycorrhizas. shtml; C - http://mycorrhizas.info)

A more rational and modern classification of mycorrhizae has been given by Brundrett (2004) in which the practical definitions used to classify them are based primarily on features controlled by the plant, since features controlled by the fungus are highly variable at the scale of individual plants. The catergories of mycorrhizae recognized by him are as follows:

1. **VAM** - These are normally defined by arbuscules. These are caused by Glomeromycete fungi. The further classification of VAM as determined by plant and by fungus is shown in Fig. 315.

2. **Ectomycorrhizae** (ECM) - These are characterized by Hartig's net (mentioned earlier) that is the primary zone of nutrient transfer. The further classification of these as determined by the plant and the fungus is shown in Fig. 316. The two basic subtypes are called **epidermal** and **cortical** depending upon whether Hartig's net is restricted to root epidermis (as in *Fagus, Betula* etc.) or extends into one or a few layers of outer cortex (as in Pinaceae). The conventional **Arbutoid** and **Monotropoid** and ectendomycorrhizae are included under this category. The **transfer–cell category** has epidermal Hartig net with transfer cells (Refer to Chapter 4).

3. **Orchid mycorrhizae -** These represent non–exploitative fungal associations where coils of hyphae, called **pelotons,** are found within cortical cells of most roots (underground and aerial) (Fig. 317) and sometimes of stems or protocorms. The fungal–infected cells often show enlargement, with the nuclei undergoing endoduplication of chromosomal content (Fig. 318 A). The fungus derives nutrition for sometime from the host cell after which the host cell digests the fungal mycelium for its own use (Fig. 318 B). However, the cell may again get a fresh load of infection so that another peloton is formed in it. The exploitative orchid mycorrhizal fungi are associated with mycoheterotrophic orchids.

4. **Ericoid mycorrhizae –** These form the fourth major category (Fig. 319). The fungi form coils within very thin roots of some Ericaceae taxa. No sub–morphological categories are recognized under this main category. The mycorrhiza seen in *Thysanotus* is a distinct morphological category, although not much is known about its fungus, whose hyphae occur in cavities under epidermal cells.

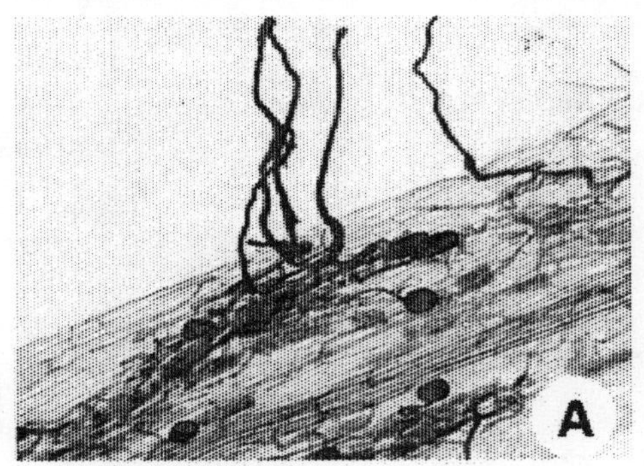

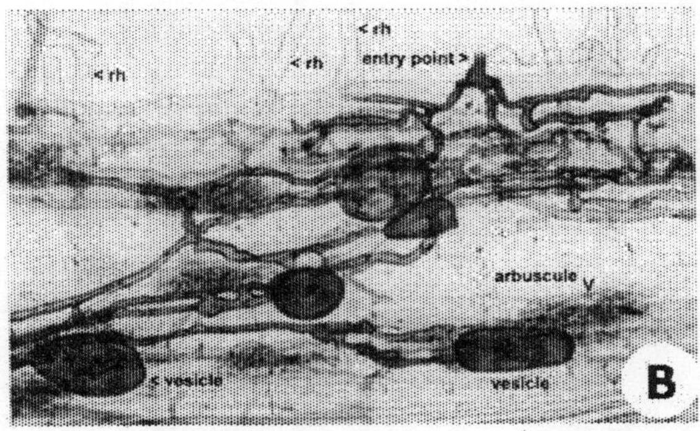

Fig. 314. A and B - Vesicular arbuscular mycorrhizae showing vesicles and arbuscules; Rh - Root hair (Source: A - http://archive.bio.ed.ac.uk/jdeacon/microbes/arbuimag.html)

Ectomycorrhizal fungi have molecular interaction with plants and produce chemical signals that modulate plant root–ectomycorrhizae morphogenesis. The indole alkaloid hypaphorine produced by the ectomycorrhizal fungus, *Pisolithus tinctorius* inhibits root hair elongation. Hormones, mainly IAA, have been implicated in symbiotic interactions (Barker and Tagu 2000). Indeed, IAA–overproducing mutants of the ectomycorrhizal fungus, *Habeloma cylindrosporum* forms significantly more mycorrhizae with the host, *Pinus pinaster* than the wild–type. The mycelia of *Tuber borchii* and *Tuber melanosporum* (truffles) (Ascomycetes), prior to physical contact with the host root (*Cissus* species), induce alterations in root morphology and also in the non–host *Arabidopsis*. These alterations include shortening of primary root, lateral root formation and root hair stimulation. These changes are most likely due to the production of IAA and ethylene by the mycelium, which additively act and hence, both are necessary for the full expression of alterations. Thus, the truffles modify the endogenous hormonal balance of host plants. IAA and ethylene both serve as signal molecules in this interaction.

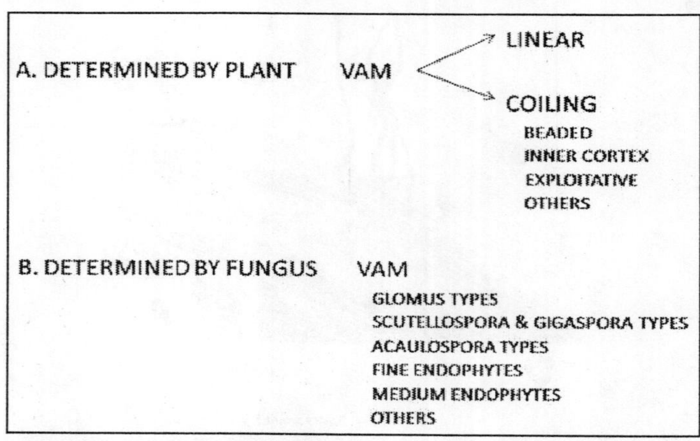

Fig. 315. Morphogical classiffiation of vesicular–arbuscular mycorrhizal (VAM) categories and subcategories resulting from the host properties (A), and morphotypes resulting from fungal properties (B) (Brundrett 2004)

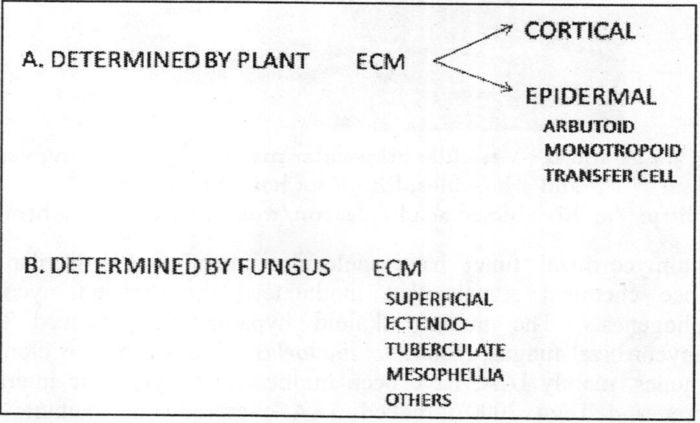

Fig. 316. Morphological classification of ectomycorrhizal (ECM) categories and subcategories resulting from the host properties (A), and morphotypes resulting from fungal properties (B) (Brundrett 2004)

Some legume plants harbour both rhizobia and mycorrhizal fungi in their roots. Pea plants, for example, establish root nodules with *Rhizobium leguminosarum* bv. *viciae* and mycorrhizae with a VAM fungus. An important character shared by both rhizobial and VAM symbioses is that cell and tissue invasion involves co–ordinated cell growth rather than cell damage or cell death. Hence, the hypersensitive response is not induced and antimicrobial phytoalexin compounds are not synthesized. Evidence

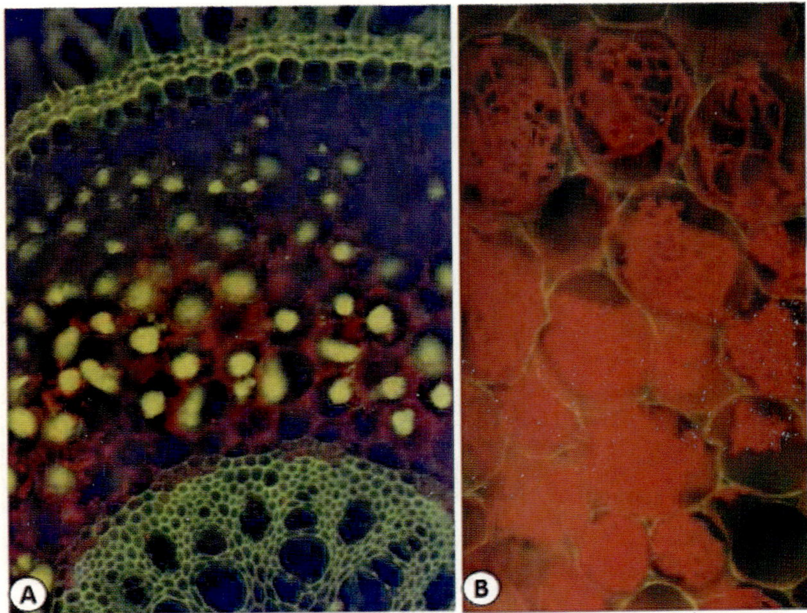

Fig. 317. T. S. of the mycorrhizal root of the orchid, *Spathoglottis plicata*; A - Part of the stele and cortex; The cortical cells with mycorrhizal fungus are seen as fluorescent ball-like structures called pelotons; B - A portion of cortex enlarged to show the pelotons (Photographs courtesy of Dr. S. Senthil Kumar)

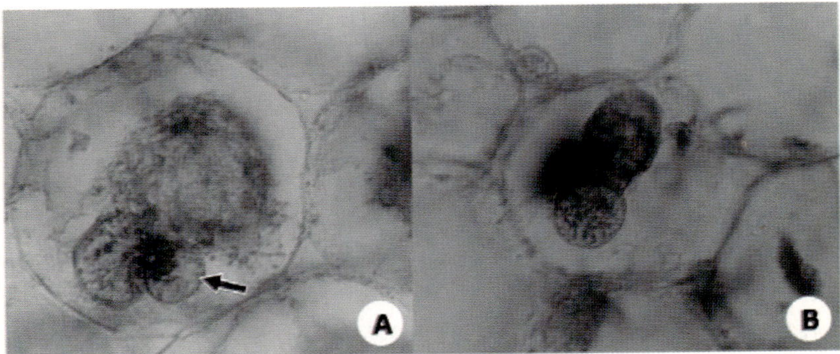

Fig. 318. Orchid mycorrhizae (*Spathoglottis plicata*); A - A fungal-infected cell showing the peloton and highly enlarged endoduplicated nucleus (arrow); B - A similar cell where the fungus is being digested, while the host nucleus remains large and intact (Photographs courtesy of Dr. S. Senthil Kumar)

for common components of these two types of symbiotic interaction comes from the discovery that nearly half of the non–nodulating pea mutants that have been isolated are also unable to establish the mycorrhizal symbiosis. Moreover, lupins, which do not form intercellular infection threads with rhizobium, also do not establish a mycorrhizal symbiosis. What distinguishes the rhizobial and mycorrhizal symbioses is that the former, but not the latter, stimulates the formation of a mass of nodule tissue by inducing cortical cell divisions, which subsequently differentiate into an organized apical meristem.

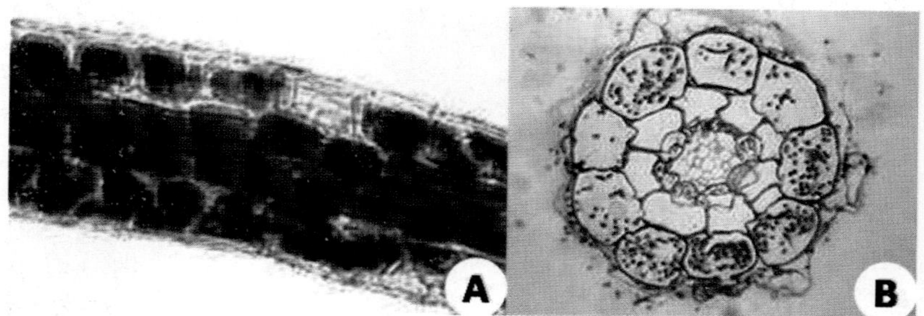

Fig. 319. Ericoid mycorrhiza (*Gaultheria shallon);* A - Whole root in surface view; B - Transverse section of the root (Reproduced with permission from Mycologue Publications)

11.5. HYPERSENSITIVE REACTIONS IN PLANTS

The term "**hypersensitive response** (**HR**)" historically comes from Stakman (1915), who used it to describe a rapid host–cell death in plants such as oat, wheat and barley, when infected with the fungal parasite, *Puccinia graminis*. Modern studies on HR were initiated in early 1960s by Klement et al (1964). He observed a rapid, localized necrosis on tobacco leaves infected with the plant pathogenic bacterium, *Pseudomonas syringae* pv. *syringae*. Subsequent to this, numerous papers have appeared on pathogen–induced HR and the associated programmed cell death (PCD) (Iakimova et al 2005). Yet very little is known on HR and its associated PCD. Since HR is related to disease resistance, it is of interest to plant pathologists and agriculturalists. HR studies, hence, are now focused in model plants such as *Arbidopsis*, rice, tobacco, wheat, carrot and tomato.

11.5.1. PCD Associated with HR

PCD is associated with HR. PCD is activated upon recognition of elicitors produced by the pathogens (Heath 2000). Details of PCD, as involved in cell and tissue morphogenesis, were already provided in Chapter 4. Here, the details of PCD, specifically associated with HR, are detailed. Three major types of PCD known are apoptosis, autophagy and non–lysosomal cell death (Van Doorn and Woltering 2005, Woltering and Van Doorn 2005). Apoptosis is exclusive to animal PCD (For details, refer to Chapter 4). Autophagy is seen in plants. It is associated with cell survival and also with cell death, and it is likely to be a strategy to remobilize at least part of cell contents prior to PCD. It plays an important role in plant defence response by preventing death and pathogen spread adjacent to HR lesion (Greenberg 2005). The third type of PCD, non–lysosomal, is necrosis–like and it does not involve the lysosome of the dying cell in PCD. It is associated with swelling of cell organelles and formation of 'empty spaces' in the cytoplasm (Baehrecke 2005). Many plant HRs seem to conform to this type of PCD.

11.5.2. Cytological Changes Associated with HR

HR is an elaborate, sophisticated and efficient system to counter the spread of pathogenic organisms in the plant organ. HR is characterized by a rapid, localized death of tissues at the site of infection, limiting further pathogen multiplication and spread. During this response, the plant sacrifices some of its cells to encircle the invading pathogen with a ring of dead cells. The dying plant cells strengthen their cell walls by depositing different phenolic compounds including phenolic acids, lignins and phytoalexins as well as showing structural (stress proteins) and enzymic proteins (together called **pathogenesis–related proteins** or **PR proteins**) that are specifically produced to counter the pathogen. These reactions involve the activation of host–defence–related genes. There is cessation of cytoplasmic streaming followed by protoplast breakdown and fragmentation of DNA, as has been shown in epidermal cells of cowpea infected with *Uromyces vignae* and in barley showing powdery mildew fungal infection. There is active formation of autophagosomes in cells adjacent to the primary infection sites

HR–related changes in plants are shown to be the result of specific toxins produced by the pathogens: **harpins** from *Pseudomonas syringae*, *Erwinia amylovora,* and *Xanthomonas campestris,* **Victorin, AAl, Fumonisin B1 (FUM), Cryptogein** etc. from specific fungi.

Evidences have been provided for the involvement of caspase–like enzymes (but not really caspases – Refer to Chapter 4.) in HR. These include cystein proteases, demonstrated in tobacco infected with bacteria and a virus. Small GTP–binding proteins of the Ras class have also been reported to be involved in HR. Other major players involved in the activation of HR are reactive oxygen species (ROS), nitric oxide (NO), calcium and protein pumps, nitrogen–activated protein kinases, salicylic acid (SA), protein phosphorylation or dephosphorylation, jasmonic acid, ethylene etc. All these are involved in the cascade of events leading to PCD during HR.

The control of cell death in plants involves plant–specific regulators in addition to the common cascade of suicidal events involving several players mentioned above. Particularly, the plant growth regulator ethylene plays an important role in HR PCD. This has been evaluated in ethylene insensitive (*never–ripe*) *NR*–tomatoes. Following the infection of these mutants, greatly reduced cell death is noticed, indicating ethylene's involvement in PCD. Ethylene signalling is found to play a role in PCD induced by the mycotoxin, Fumonisin B1 in *Arbidopsis* and tomato. Ethylene also plays a role in the susceptibility to cauliflower mosaic virus and turnip vein clearing tobamo virus. Cell death in maize endosperm and other cell types could be hastened by ethylene treatment and arrested by ethylene inhibitors. Additional evidence for ethylene's role in PCD as a signalling molecule comes from a study of an *Arabidopsis* double mutant. Crosses of the lesion mimic mutant *accelerated cell death 5* (*acd 5*) and *ethylene insensitive 2* (*ein 2*), in which ethylene signalling is blocked, showed decreased PCD (Iakimova et al 2005). There are several other evidences for the active role of ethylene in HR PCD.

References

1. Aarts MGM, Hodge R, Kalantidis K, Florack D, Wilson ZA, Mulligan BJ, Stiekema WJ, Scott R, and Pereira A 1997 The *Arabidopsis* MALE STERILITY 2 protein shares similarity with reductases in elongation /condensation complexes, *Plant J.* **12** 615–623
2. Albrecht C, Russinova E, Hecht V, Baaijens E, and de Vries S 2005 The *Arabidopsis thaliana* SOMATIC EMBRYOGENESIS RECEPTOR-LIKE kinases1 and 2 control male sporogenesis, *Plant Cell* **17** 3337-3349.
3. Aloni R 1976 Polarity of induction and pattern of primary phloem fiber differentiation in *Coleus*, *Am. J. Bot.* **63** 877–889
4. Aloni R 1978 Source of induction and sites of primary phloem fibre differentiation in *Coleus blumei*, *Ann. Bot. n.s.* **42** 1261–1269
5. Aloni R 1979 Role of auxin and gibberellins in differentiation of primary phloem fibres, *Plant Physiol* **63** 609–614
6. Aloni R 1987 Differentiation of Vascular tissues, *Annu. Rev. Plant Physiol.* **38** 179–204
7. Amerjothy S 2005 *Plant Gall Ecology in Scrub Forest Ecosystem*, Project Report, Ministry of Environment and Forests, Govt. of India, New Delhi
8. Ananthakrishnan TN and Raman A 1989 *Thrisps and Gall Dynamics*, New Delhi: Oxford & IBH Publ. Co.
9. Anderson LK, Stack SM, Fox MH, and Zhang CS 1985 The relationship between genome size and synaptonemal complex length in higher plants, *Exp. Cell Res.* **156** 367–378
10. Arber A 1950 *Natural Philosophy of Plant Form*, Cambridge: Cambridge Univ. Press
11. Arekal GD 1963 Embryological studies in Canadian representatives of the tribe Rhinantheae, Scrophulariaceae, *Can. J. Bot.* **41** 267–302
12. Arekal GD 1966 Embryology of *Veronica serpyllifolia* L., *Proc. Indian Acad. Sci. B* **64** 241–257
13. Audus LJ 1962 The mechanism of the perception of gravity in plants, *Symp. Soc. Exp. Biol.* **16** 197–226
14. Audus LJ 1979 Plant geosensors, *J. Exp. Bot.* **30** 1051–1073
15. Avers CJ and Grimm RB 1959 Comparative enzyme differentiations in grass roots. II. Peroxidase, *J. Exp. Bot.* **10** 341–344
16. Bailey IW 1920 The formation of the cell plate in the cambium of higher plants, *Proc. Natl. Acad. Sci.* **6** 197–200

17. Bailey IW 1944 The development of vessels in angiosperms and its significance in morphological research, *Am. J. Bot.* **31** 421–428
18. Baima S, Nobili F, Sessa G, Lucchetti S, Ruberty I, and Morelli G 1995. The expression of the *Athb-8* homebox gene is restricted to provascular cells in *Arabidopsis thaliana*, *Development* **121** 4171–4182
19. Baker SC, Robinson–Beers K, Villa nueva JM, Gaiser JC, and Gasser CS 1997 Interactions among genes regulating ovule development in *Arabidopsis thaliana*, *Genetics* **145** 1109–1124
20. Balchrecke EH 2005 Autophagy: dual roles in life and death? *Mol.Cell. Biol.* **6** 505–510
21. Baluška F, Salaj J, Mathur J, Braun M, Jasper F, Šamaj J, Chua NH, Barlow PW, and Volkmann D 2000 Root hair formation: F–actin-dependent tip growth is initiated by local assembly of profilin–supported F–actin meshworks accumulated within expansin–enriched bulges, *Dev. Biol.* **227** 618–632
22. Bamber RK 2001 A general theory for the origin of growth stresses in reaction wood: how trees stay upright, *IAWA J.* **22** 205–212
23. Barker SJ and Tagu D 2000 The roles of auxins and cytokinins in mycorrhizal symbioses, *J. Plant Growth Regul.* **19** 144–154
24. Barlow PW 1973 Mitotic cycles in root meristems, in *The Cell cycle in Development and Differentiation* (eds) M Balls and FS Billett, Cambridge: Cambridge Univ. Press, 133–165 pages
25. Barlow PW 1978 RNA metabolism in the quiescent centre and neighboring cells in the root meristem of *Zea mays*, *Z. Pflanzenphysiol.* **86** 147–157
26. Barlow PW 1982 'The Plant forms cells, not cells the plant': the origin of de Bary's aphorism, *Ann. Bot.* **49** 269–271
27. Barlow PW 1994 From Cell to System: Repetitive Units of Growth in the Development of Roots and Shoots, in *Growth Patterns in Vascular Plants* (eds) M Iqbal, Portland: Dioscorides Press, 19–58 pages
28. Barlow PW and Adam JS 1989 The response of the primary root *meristem of Zea mays, L.* to various periods of cold, *J. Exp. Bot.* **40** 81–88
29. Barlow PW and Rathfelder EL 1985 Cell division and regeneration in primary root meristems of *zea mays* recovering from cold treatment, *Environ. Exp. Bot.* **25** 303–314
30. Barnett JR 1981 Secondary Xylem Cell Development, in *Xylem Cell Development* (ed) JR Barnett, Kent: Castle House Publication Ltd., 47–95 pages

31. Barton MK 1998 Cell type specification and self–renewal in the vegetative shoot apical meristem, *Curr. Opin. Plant Biol.* **1** 37–42
32. Barton MK and Poethig RS 1993 Formation of the shoot apical meristem in *Arabidopsis thaliana*: Analysis of development in the wild type and in the shoot meristemless mutant, *Development* **119** 823–831
33. Baskin TI 2000 On the constancy of cell division rate in the root meristem, *Plant Mol. Biol.* **43** 545–554
34. Baskin TI 2001 On the alignment of cellulose micro–fibrils by cortical microtubules: A review and a model, *Protoplasma* **215** 150–171
35. Bassel AR and Miller JH 1982 The effects of centrifugation on asymmetric cell division and differentiation of fern spores, *Ann. Bot.* **50** 185–198
36. Battaglia E 1945 Fenomeni citologici nuovi nella embriogenesi (semigamia) e nella microsporogenesi (doppio nucleo di restituzione) in *Rudbeckia laciniata* L., *Nuovo Giorn. Bot. Ital.* **52** 34–38
37. Battaglia E 1947 Ricerche cariologiche e embriologiche sul genera *Rudbeckia.* 9. Le anomalie del gametofito femminile cellularizzato di R. laciniata, *Nuovo Giorn Bot. Ital.* **54** 377–405
38. Batygina TB and Vasilyeva VE 1998 Sexual Reproduction in Flowering Plants: Periodization of Egg Cell and Zygote Development and Possible Types of Caryogamy, in *Plant Form and Function* (eds) B Bhatia, AK Shukla and HL Sharma, New Delhi: Angkor Publishers (P) Ltd., 170–198 pages
39. Baum SF, Aloni R, and Peterson CA 1991 Role of cytokinin in vessel regeneration in wounded *Coleus* internodes, *Ann. Bot.* **67** 543–548
40. Beals TP and Goldberg RB 1997 A novel cell ablation strategy blocks tobacco anther dehiscence, *Plant cell* **9** 1527–1545
41. Benfey PN, Linstead PJ, Roberts K, Schiefelbein JW, Haüser MT, and Aeschbacher RA 1993 Root development in *Arabidopsis* four mutants with dramatically altered root morphogenesis, *Development* **119** 57–70
42. Bennett MD 1971 The duration of meiosis, *Proc. R. Soc. Lond. B* **178** 277–299
43. Bennett MD 1972 Nuclear DNA content and minimum generation time in herbaceous plants, *Proc. R. Soc. Lond. B* **181** 109–135
44. Bennett MD 1977 The time and duration of meiosis, *Philos. Trans. R. Soc. Lond. B* **277** 201–277

45. Bennett MD 1987 Variation in genomic form in plants and its ecological implications, *New Phytol.* **106** 177–200
46. Bennett MD and Leitch IJ 1997 Nuclear DNA amounts in angiosperms - 583 new estimates, *Ann. Bot.* **80** 169–196
47. Bennett MD, Heslop-Harrison JS, Smith JB, and Ward JP 1983 DNA density in mitotic and meiotic metaphase chromosome of plants and animals, *J. Cell Sci.* **63** 173–179
48. Bennett MD, Smith JB, Ward J, and Jenkins G 1981 The relationship between nuclear DNA content and centromere volume in higher plants, *J .Cell Sci.* **47** 91–115
49. Bernier G 1988 The control of floral evocation and morphogenesis, *Annu. Rev. Plant Physiol. Mol. Biol.* **39** 175–219
50. Bernier G, Havelange A, Houssa C, Petitjean A, and Lejeune P 1993 Physiological signals that induce flowering, *Plant Cell* **5** 1147–1155
51. Berthold G D W 1886 Studien über Protoplasmamechanik. Leipzig. A. Felix.
52. Bewley JD 1997 Seed Germination and Dormancy, *Plant Cell* **9** 1055–1066
53. Bhambie S 1994 Secondary Growth in Pteridophytes, in *Growth Patterns in Vascular Plants* (eds) M Iqbal, Portland: Dioscorides Press, 185–210 pages
54. Bharathan G and Sinha NR 2001 The regulation of compound leaf development, *Plant Physiol.* **127** 1533–1538
55. Bhargava HR 1935 The life history of *Trianthema monogyna, Proc. Indian Acad. Sci. B.* **2**49–58
56. Bird AP, Truman DES and Clayton RM 1982 Introductory review: The molecular basis of differentiation and competence, in *Stability and Switching in Cellular Differentiation* (eds) RM Clayton and DES Truman, New York: Plenum Press, 61–64 pages
57. Bjorkman T and Leopold AC 1987 Effect of inhibitors of auxin maize roots, *Plant physiol.* **84** 847–850
58. Blaauw AH 1909 Die Perzeption des Lichtes, *Rec. Trav. Bot. Neerland* **5** 209–272
59. Blackmore S, Worthey AH, Skvarla JJ, and Rowley JR 2007 Pollen wall development in flowering plants, *New Phytol* **174** 483-498.
60. Bloch R 1965 Polarity and gradients in plants: A survey, in *Encyclopedia of Plant Physiology. XV/1* (eds) W Ruhland, New York: Springer Berlin-Heidelberg, 234–274 pages
61. Boetsch J, Chin J, and Croxdale J 1995 Arrest of stomatal initials in *Tradescantia* is linked to the proximity of neighbouring stomata and

results in the arrested initials acquiring properties of epidermal cells, *Dev. Biol.* **168** 28–38

62. Bohmert K, Camus I, Bellini C, Bouchez D, Caboche M, and Benning C 1998 *AGO1* defines a novel locus of *Arabidopsis* controlling leaf development, *EMBO J.* **17** 170–180
63. Bond G 1967 Fixation of nitrogen by higher plants other than legumes, *Annu. Rev. Plant Physiol.* **18** 107–126
64. Bongard-Pierce DK, Evans MMS, and Poethig RS 1996 Heteroblastic features of leaf anatomy in maize and their genetic regulation, *Int. J. Plant Sci.* **157** 331–340
65. Boniotti MB and Griffith ME 2002 "Cross-talk" between cell division cycle and development in plants, *Plant Cell* **14** 11–16
66. Boothe JG and Walden DB 1990 Gene expression in embryos and seedlings of *maize, Maydica* **35** 187–194
67. Bose JC 1907 *Comparative Electrophysiology*, London: Longman Green and Co.
68. Boucher DH 1985 The idea of mutualism, past and future, in *The Biology of Mutualism* (ed) DH Boucher, London: Croom Helm, 1–28 pages
69. Bouligand Y 1976 Les analogues biologiques des cristaux liquids, *La Recherche* **67** 474–476
70. Bower FO 1947 *The Botany of Living Plants*, London: Macmillan & Co. Ltd.
71. Bowman JL 1997 Evolutionary conservation of angiosperm flower development at the molecular and genetic levels, *Journal of Biosciences* **22** 515–527
72. Bowman JL and Eshed Y 2000 Formation and maintenance of the shoot apical meristem, *Trends Plant Sci.* **5** 110–115
73. Boysen-Jensen P 1913 Über die Leitung des phototropischen Reizes in der Avenakoleoptile, *Ber. Deutsch.Bot. Ges.* **31** 559–566
74. Brauner L and Hager A 1958 Versuche zur Analyse der geotropischen perzeption, *Planta* **51** 115–147
75. Brett. C and Waldron K 1990 *Physiology and Biochemistry of Plant Cell Walls*, London: Unwin Hyman
76. Brewin NJ, Rae AL and Perotto S 1993 The extracellular matrix as a functional meeting point for symbionts, Abstract. Key-stone symposium. *The Extracellular Matrix of Plants: Molecular, cellular and Developmental Biology*. Jan 9-15. Santa Fe, New Mexico, USA.
77. Briggs WR and Baskin TI 1988 Phototropism in higher plants–Controversies and caveats, *Botanica Acta* **101** 133–139

78. Brink RA and Cooper DC 1939 Somatoplastic sterility in *Medicago sativa*, *Science* **90** 545–546

79. Brown AH 1991 Gravity perception and circumnutation in plants, in *Advances in Space Biology and Medicine* (ed) SL Bonting, Greenwich: Jai press, 129–153 pages

80. Brown AH 1993 Circumnutations: From Darwin to space flights, *Plants Physiol.* **101** 345–348

81. Brown CC, McAlpine RG, and Kormanik PP 1967 Apical dominance and form in woody plants: a reappraisal, *Am. J. Bot.* **54** 153–162

82. Brown RC and Lemmon BE 1991 Pollen development in orchids. 3. A novel generative pole microtubule system predicts unequal pollen mitosis, *J. Cell Sci.* **99** 273–81

83. Brown RC and Lemmon BE 1992 Pollen development in Orchids. 4. Cytoskeleton and ultrastructure of the unequal pollen mitoses in *Phalaenopsis*, *Protoplasma* **167** 183–192

84. Brown RC and Lemmon BE 1994 Pollen mitosis in the slipper orchid *Cypripedium fasciculatum*, *Sex. Plant Reprod.* **7** 87–94

85. Brukhin V, Curtis MD, and Grossniklaus U 2005 The angiosperm female gametophyte: No longer the forgotten generation, *Curr. Sci.* **89** 1844-1952.

86. Brundrett M 2002 Coevolution of roots and mycorrhizas of land plants, *New Phytol.* **154** 275– 304

87. Brundrett M 2004 Diversity and classification of mycorrhizal associations, *Biol. Rev.* **79** 473– 495

88. Brutnel TP and Langdale JA 1998 Signals in leaf development, *Adv. Bot. Res.* **28** 161–195

89. Buchholz JT 1940 The embryogeny of Torreya, with a note on Austrotaxus, *Bull. Torrey Bot. Club* **67** 731–754

90. Buchholz JT 1941 Embryogenesis of the Podocarpaceae, *Bot. Gaz.* **103** 1–37

91. Bünning K 1937 Die endonome Tages-rhythmik als Grundlage der photoperiodischen Reaktion, *Berich. Deutschen Bot. Gesellschaft* **54** 590–607

92. Bünning K and Stern K 1930 Űber die tages-periodischen Bewegungen der primärblätter von *Phaseolus multiflorus*. II. Die Bewegungen bei Termokonstanz, *Berich. Deutschen Bot. Gesellschaft* **48** 227–252

93. Burkholder PR 1952 Cooperation and conflict among primitive organisms, *Am. Sci.* **40** 601– 631

94. Butterworth KM, Adams DC, Horner HT, and Wendel JF 2009 Initiation and early development of Fiber in wild and cultivated cotton, *Int. J. Plant Sci.* **170** 561–574

95. Buvat R 1989 *Ontogeny, Cell Differentiation, and Structure of Vascular Plants*, Berlin: Springer– Verlag

96. Cai G, Casino CD, Romagnoli S, and Cresti M 2005 Pollen cytoskeleton during germination and tube growth, *Curr. Sci.* **89** 1853-1860.

97. Callos JD, DiRado M, Xu B, Behringer FJ, Link BM, and Medford JI 1994 The forever young gene encodes an oxidoreductase required for proper development of the *Arabidopsis* vegetative shoot apex, *Plant J.* **6** 835–847

98. Capra F 1996 *The Web of Life*, London: Harper Collins

99. Carioli S, Marocco A, Maddaloni M, Botto M, and Salamini F 1994 Early event in maize leaf epidermis formation, *Development* **120** 2113–2120

100. Carlquist S 2001 *Comparative Wood Anatomy*, Berlin: Springer–Verlag

101. Carmo–Oliveira R and de Morretes BL 2009 Stigmatic surface in the Vochysiaceae: reproductive and taxonomic implications, *Acta Botanica Brasilica* **23** 780–785

102. Caruso JL 1968 Morphogenetic aspects of a leafless mutant in tomato. I. General patterns in development, *Am. J. Bot.* **55** 1169–1176

103. Caspar T and Pickard BG 1989 Gravitropism in a starchless mutant of *Arabidopsis*. Implication for the starch-statolith theory of gravity sensing, *Planta* **177** 185–197

104. Catesson AM 1900 Cambial ultrastructure and biochemistry, in *The Vascular Cambium* (eds) M Iqbal, Taunton: Research Studies Press, 63–112 pages

105. Chailakhyan M 1968 Internal factors of plant flowering, *Annu. Rev. Plant Physiol.* **19** 1–36

106. Chamberlin MA, Horner HT, Palmer RG 1993 Nutrition of ovule, embryo sac, and young embryo in soybean: an anatomical and autoradiographic study, *Canad. J. Bot.* **71** 1153–1168

107. Charlton WL, Keen CL, Merriman C, Lynch P, Greenland AJ, and Dickinson HG 1995 Endosperm development in *Zea mays*: Implication of gametic imprinting and paternal excess of regulation of transfer layer development, *Development* **121** 3089–3097

108. Chaudhury AM, Ming L, Miller C, Craig S, Dennis ES, and Peacock WJ 1997 Fertilization independent seed development in *Arabidopsis thaliana*, *Proc. Natl. Acad. Sci. USA* **94** 4223–4228

109. Chikkanniah, PS 1963 Embryology of some members of the Commelinaceae: *Commelina subulata* Roth., *Phytomorphology* **13** 174–84

110. Chin J, Wan Y, Smith J, and Croxdale J 1995 Linear aggregations of stomata and epidermal cells in *Tradescantia* leaves evidence for their group patterning as a function of the cell cycle, *Dev. Biol.* **168** 39–46

111. Choi DW, Song JY, Kwon YM, and Kim SG 1996 Characterization of a cDNA encoding a proline-rich 14 kDA protein in developing cortical cells of the roots of bean (*Phaseolus vulgaris*) seedlings, *Plan Mol. Biol.* **30** 973–982

112. Cholodny N 1926 Beiträge zur Analyse der geotropischen Reaktion, *Jahrb. Wiss. Bot.* **65** 447–459

113. Christensen JE, Horner HT Jr., and Lersten NR 1972 Pollen wall and tapetal orbicular wall development in *Sorghum bicolor* (Gramineae), *Am. J. Bot.* **59** 43–58

114. Christopher FS and Krishnamurthy KV 2004 Tracheary element diffenentiation *in vitro*–effect of certain calcium channel blockers, in *Biotechnology for a Better Future* (eds) L D'souza, M Anuradha, S Nivas, S Hegde and K Rajendra, Mangalore: SAC Publications, 51–81 pages

115. Ciesielski T 1872 Untersuchungen Über der Abwärtskrümmung der Wurzel, *Beit.Biol. Pflanzen.* **1** 1–30

116. Clarkson DT 1984 Calcium transport between tissues and its distribution in the plant, *Plant Cell Environ.* **7** 449–456

117. Cleary AL 1995 F–action redistributions at the division site in living *Tradescantia* complexes as revealed by microinjection of rhodamine-phalloidin, *Protoplasma* **185** 152–165

118. Cleland RE 1981 Wall extensibility: Hormones and wall extension, in *Encyclopedia of Plant Physiology, New Series, Vol. 13B* (eds) W Tanner and FA Loewus, New York: Springer-*Verlag,* 255–273 pages

119. Cleland RE 1984 The Instron technique as a measure of immediate-past wall extensibility, *Planta* **160** 514–520

120. Cleland RE 1987 Auxin and Cell elongation, in *Plant Hormones and Their Role in Plant Growth and Development* (ed) PJ Davies, Button: Martinus Nijhoff Publishers, 132–148 pages

121. Cleland RE 2001 Unlocking the mysteries of leaf primordial formation, *Proc. Natl. Acad. Sci. USA* **98** 10981–10982

122. Cline MG 1991 Apical dominance, *Bot. Rev.* **57** 318–358

123. Clowes FAL 1956 Localization of nucleic acid synthesis in root meristem, *J. Exp. Bot.* **7** 307– 312

124. Clowes FAL 1961 *Apical Meristems*, Oxford: Blackwell

125. Clowes FAL 1969 Anatomical aspects of structure and development, in *Root Growth* (ed) WJ Whittingham, London: Butterworths, 3–19 pages

126. Clowes FAL 1972 The control of cell proliferation within root meristems, in *The Dynamics of Meristem Cell Populations* (eds) MW Miller and CC Kuehnert, New York: Plenum Press, 133– 147 pages

127. Clowes FAL 1978a Origin of the quiescent centre in *Zea mays*, *New Phytol.* **80** 409–419

128. Clowes FAL 1978b Origin of quiescence at the root pole of pea roots, *Ann. Bot.* **42** 1237– 1239

129. Clowes FAL 1981 The difference between open and closed meristems, *New Phytol.* **57** 85– 88

130. Clowes FAL 1984 Size and activity of quiescent centres of roots, *New Phytol.* **96** 13–21

131. Clowes FAL and Stewart HE 1967 Recovery from dormancy in roots, *New Phytologist* **66** 115–123

132. Colcombet J, Boisson-Dermier A, Ros-Palau R, Vera CE, and Schroeder JI 2005 *Arabidopsis* SOMATIC EMBRYOGENESIS RECEPTOR KINASES 1 and 2 are essential for tapetum development and microspore maturation, *Plant Cell* **17** 3350-3361.

133. Collings DA 2002 Mile-high view of plant biology, *Genome Biol.* 3 (11): reports 4035.1– reports 4035.3.

134. Cook TD, Tilney MS and Tilney LG 1996 Plasmodesmal networks in apical meristems and mature structures: geometric evidence for both primary and secondary formation of plasmodesmata, in *Membrances: Specialized functions in plants* (eds) M Smallwood, JP Knox and DJ Bowles, Cambridge: BIOS Scientific, 471–488 pages

135. Corbesier L, and Couplant G 2006 The quest for florigen: a review of recent progress. *J. Expt. Bot.* **57** 3395-3403.

136. Corner EJH 1976 *Seeds of Dicotyledons, Vols.1 & 2*, Cambridge: Cambridge Univ. Press

137. Cosgrove DJ 1987a Wall relaxation in growing stems: comparison of four species and assessment of measurement techniques, *Planta* **171** 266–78

138. Cosgrove DJ 1987b Wall Relaxation and the Driving Forces for Cell Expansive Growth, *Plant Physiol.* **84** 561–564

139. Cosgrove DJ 1997 Relaxation in a High-Stress Environment: The Molecular Bases of Extensible Cell Walls and Cell Enlargement, *Plant cell.* **9** 1031–1041

140. Cosgrove DJ 1999 Enzymes and other agents that enhance cell wall extensibility, *Annu. Rev. Plant Physiol. Plant Mol. Biol.* **50** 391–417

141. Coulter JM and Chamberlain CJ 1903 *Morphology of the Angiosperms*, New York: D. Appleton & Co

142. Creelman RA and Mullet JE 1997 Oligosaccharins, Brassinolides, and Jasmonates. Non- traditional Regulators of Plant Growth, Development, and Gene Expression, *Plant Cell* **9** 1211– 1223

143. Créte P 1963 Embryo, in *Recent Advances in the Embryology of Angiosperms* (eds) P Maheshwari, *New Delhi:* International Society of Plant Morphologists, 171–220 pages

144. Croxdale J, Smith J, Yandell B, and Johnson JB 1992 Stomatal patterning in *Tradescantia*: an evaluation of the cell lineage theory, *Dev. Biol.* **149** 158–167

145. Cutter EG 1969 Plant Anatomy: experiment and interpretation Part 1 *Cells and Tissues* London: Edward Arnold Publishers Ltd.

146. Cutter EG and Feldman LJ 1970a Trichoblasts in *Hydrocharis*, I. Origin, differentiation, dimensions and growth, *Am. J. Bot.* **57** 196–201

147. Cutter EG and Feldman LJ 1970b Trichoblasts in *Hydrocharis*, II. Nucleic acids, proteins and a consideration of cell growth in relation to endopolyploidy, *Am. J. Bot.* **57** 202–211

148. D'Amato F 1998 Chromosome endoreduplication in plant tissue development and function, in *Plant cell Proliferation and its Regulation in Growth and Development* (eds) JA Bryant and D Chiatante, Chichester: Wiley, 153–166 pages

149. D'Arcy Thompson W 1942 *On Growth and Form, (2nd Edition)*, Cambridge: Cambridge Univ. Press

150. Darwin C and Darwin F 1880 *The Power of Movement in Plants*, London: John Murray Publishers Ltd.

151. Davis GL 1966 *Systematic Embryology of Angiosperms*, New York: John Wiley & Sons

152. Dayanandan P, Hebard FV, Baldwin VD, and Kaufman PB 1977 Structure of gravity– sensitive sheath and intermodal pulvini in grass shoots, *Am. J. Bot.* **64** 1189–1199

153. De Sloover J 1958 Recherches sur I'histogénèse des tissue conducteurs. II. Le sens longitudinal de la différenciation du procambium, du xylème et du phloèm chez *Coleus, Ligustrum, Anagallis* et *Taxus*, *Cellule* **59** 55–202

154. Deikman J and Ulrich M 1995 A novel cytokinin–resistant mutant of *Arabidopsis* with abbreviated shoot development, *Planta* **195** 440–449

155. Deitrich RA, Radice SA, and Harada JJ 1992 Downstream DNA sequences are required to activate a gene expressed in the root cortex of embryos and seedlings, *Plant Cell* **4** 1371–1382

156. DeMason DA 1994 Stem Thickening in Monocotyledons, in *Growth Patterns in Vascular Plants* (ed) M Iqbal, Portland: Diosconides Press, 288–310 pages

157. Demura T and Fukuda H 1994 Novel Vascular cell–specific genes whose expression is regulated temporally and spatially during vascular system development, *Plant Cell* **6** 967–981

158. den Boer BGW and Murray JAH 2000 Triggering the Cell Cycle in Plants, *Trends Cell Biol.* **10** 245–250

159. Dengler NG 1994 The influence of light on leaf development, in *Growth Patterns in Vascular Plants* (ed) M Iqbal, Portland: Dioscorides Press, 100–136 pages

160. Dengler NG, Dengler RE, and Hattersley PW 1986 Comparative bundle sheath and mesophyll differentiation in the leaves of the C4 grasses *Panicum effusum* and *P.bulbosum*, *Am.J.Bot* **72** 1431–1442

161. Dengler NG, MacKay LB, and Gregory LM 1975 Cell enlargement and tissue differentiation during leaf expansion in beech, *Fagus grandifolia*, *Can. J. Bot.* **53** 2846–2865

162. Deysson G 1980 Control of cell multiplication in root meristems of higher plants, *Biol. Cell.* **38** 75–80

163. Di Laurenzio L, Wysocka–Diller J, Malamy JE, Pysh L, Helariutta Y, Freshour G, Hahn MG, Feldmann KA, and Benfey PN 1996 The *SCARECROW* gene regulates an asymmetric division that is essential for generating the radial organization of the *Arabidopsis* root, *Cell* **86** 423–433

164. Dodds JH 1981a The Role of the Cell Cycle and Cell Division in Xylem Differentiation, in *Xylem Cell Development* (ed) JR Barnett, Kent: Castle House Publications Ltd., 153–167 pages

165. Dodds JH 1981b Relationship of the cell cycle to xylem cell differentiation: A new model, *Plant Cell Environ.* **4** 145–146

166. Doebley J, Stec A, and Hubbard L 1997 The evolution of apical dominance in maize, *Nature* **386** 485–488

167. Doerner P, Jorgenesen JE, You R, Steppuhn J, and Lamb CJ 1996 Cyclin expression limits root growth and development, *Nature* **380** 520–523

168. Dolan L 1996 Pattern in root epidermis: an interplay of diffusible signals and cell geometry, *Ann. Bot.* **77** 547–553

169. Dolan L and Poethig RS 1991 Genetic analysis of leaf development in cotton, *Development (suppl.)* **1** 39–46

170. Dolan L, Janmaat K, Willemsen V, Linstead P, Poethig S, Roberts K, and Scheres B 1993 Cellular organization of the *Arabidopsis thaliana* root, *Development* **119** 71–84

171. Douin C 1923 Recherches sur le gamétophyte des Marchantiées. III.–Le thalle stérile des Marchantiées développement basilaire des feuilles et autres organs latéraux chez les Muscinées, *Rev. Gen. Bot.* **23** 487–508

172. Dudley M and Poethig RS 1991 The effect of a heterochronic mutation, *teopod 2*, on the cell lineage of the maize shoot, *Development* **111** 733–739

173. Dupuy L, Mackenzie J, and Haseloff J 2010 Coordination of plant cell division and expansion in a simple morphogenetic system, *PNAS* **107** 2711–2716

174. Dure L III. 1985 Embryogenesis and gene expression during seed formation, in *Oxford Surveys of Plant Molecular and Cell Biology, Vol 2* (ed) BS Miflin, Oxford: Oxford University Press, 179–197 pages

175. Dutcher FR, Hess EL, and Halstead TW 1994 Progress in plant research in space, *Advances in Space Research* **14** 159–171

176. Dyer AF 1976 Modifications and errors of mitotic cell division in relation to differentiation, in *Cell Divisions in Higher Plants* (ed) MM Yeoman, New York: Academic Press, 199–249 pages

177. Eames AJ 1961 *Morphology of the Angiosperms*, New York: McGraw-Hill Book Company

178. Eames AJ and MacDaniels LH 1947 *An Introduction to Plant Anatomy, (2nd Edition)*, New Delhi: McGraw Hill Book Company

179. Edgar BA and Orr-Weaver TL 2001 Endoreplication cell cycles: More for less, *Cell* **105** 297–306

180. Edman G 1931 Apomeiosis und Apomixis bei *Atraphaxis frutescens*, *Acta Hort. Berg.* **11** 13– 66

181. Emons AMC 1994 Winding threads around plant cells: A geometrical model for microfibril deposition, *Plant cell Environ.* **17** 3–14

182. Ensikat HJ, Ditsche-Kuru P, Neinhuis C, and Barthlott W 2011 Superhydrophobicity in perfection: the outstanding properties of the lotus leaf, *Beilstein J. Nanotechnol.* **2** 152–161

183. Errera L 1888 Űber zellformen und seifenblasen, *Bot. Centralbl.* **34** 395–398

184. Esau K 1943 Origin and development of primary vascular tissues in seed plants, *Bot. Rev.* **9** 125–206

185. Esau K 1961 *Anatomy of Seed Plants*, New Jersey: John Wiley & Sons

186. Esau K 1965a *Plant Anatomy*, New York: John Wiley & Sons

187. Esau K 1965b *Vascular Differentiation in Plants*, New York: Holt, Rinehart and Winston

188. Esau K 1977 *Anatomy of seed plants*, (2nd Edition) New York: John Wiley & Sons

189. Esau K and Hewitt WMB 1940 Structure of end wall in differentiating vessels, *Hilgardia* **13** 229–240

190. Esau K and Thorsch J 1985 Sieve-plate pores and plamsodesmata, the communication channels of the symplast: Ultrastructural aspects and developmental relations, *Am. J. Bot.* **72** 1641–1653

191. Evans MMS and Barton MK 1997 Genetics of angiosperm shoot apical meristem development, *Annu. Rev. Plant Physiol. Plant Mol. Biol.* **48** 673–701

192. Evans MMS, Passas HJ, and Poethig RS 1994 Heterochronic effects of *glossy 15* mutations on epidermal cell identity in maize, *Development* **120** 1971–1981

193. Evenari M 1984 Seed physiology: from ovule to maturing seed, *Bot. Rev. 50 143*–**170**

194. Evered D and Marsh J 1989 *The Cellular Basis of Morphogenesis*, Chichester: John Wiley

195. Evert RF 2004 *Esau's Plant Anatomy*, New York: John Wiley

196. Fahn A and Rachmilevitz T 1970 Ultrastructure and nectar secretion in *Lonicera japonica*, in *New Research in Plant Anatomy* (eds) NKB Robson and DF Cutler, London: Academic press, 51–56 pages

197. Fahn A and Werker E 1990 Seasonal Cambial Activity, in *The Vascular Cambium* (ed) M Iqbal, Somerset, England: Research Studies Press Ltd., 139–157

198. Fahn A, Ben-Sasson R and Sachs T 1972 The relation between the procambium and the cambium, in *Research Trends in Plant Anatomy* (eds) AKM Ghouse and M Yunus, New Delhi: Tata McGraw-Hill, 161–170 pages

199. Feldman LJ and Torrey JG 1976 The isolation and culture *in vitro* of the quiescent centre of *Zea mays*, *Am. J. Bot.* **63** 345–355

200. Fiers M, Golemiec E, Xu J, van der Geest L, Heidstra R, Stiekeme W, and Liu CM 2005 The 14- amino acid CLV3, CLE19, CLE 40 peptides trigger consumption of the root meristems in *Arabidopsis* through CLAVATA2-dependent pathway, *Plant Cell* **17** 2542–2553

201. Fineran BA 1994 Unique Tracheary Elements in the Haustorium of Parasitic Angiosperms, in *Growth Patterns in Vascular Plants* (ed) M Iqbal, Portland Oregon, USA: Dioscorides Press, 162–182 pages

202. Fink S 1999 *Pathological and Regenerative Plant anatomy, (Handbuch der Pflanzenanatomie) Vol. XIV*, Berlin: Borntraeger

203. Firn RD 1986 Growth substance sensitivity: The need for clear ideas, precise terms, and purposeful experiments, *Physiol. Plantarum* **67** 267–272

204. Firn RD and Digby J 1980 The establishment of tropic curvatures in plants, *Annu. Rev. Pl. Physio.* **31** 131–148

205. Fisher JB 1970 Development of the Intercalary meristem of *Cyperus alternifolius*, *Am. J. Bot.* **57** 691–703

206. Fisher JB and French JC 1978 Internodal meristems of monocotyledons: further studies and general taxonomic summary, *Ann. Bot.* **42** 41–50

207. Flaherty L and Quiring D 2008 Plant module size and dose of gall induction stimulus influence gall induction and galler performance, *Oikos* **117** 1601–1608

208. Fleming AJ 2006 The coordination of cell division, differentiation and morphogenesis in shoot apical meristem: A perspective, *J. Exp. Bot.* **57** 25–32

209. Fleming AJ, McQueen-Mason S, Mandel T, and Kuhlemeir C 1997 Induction of leaf primordia by the cell wall protein expansin, *Science* **276** 1415–1418

210. Fletcher JC 2002 The vegetative meristem, in *Meristematic Tissues in Plant Growth and Development* (eds) MT McManus and BE Veit, Sheffield: Sheffield Academic Press, 16–57 pages

211. Fletcher JC 2004 Stem cell maintenance in higher plants, in *Handbook of Stem Cells, Vol.2, Adult and Fetal* (eds) R Lanza, H Blau, D Melton, M Moore, ED Thomas, C Verfaille, IL Weissman and M West, Amsterdam: Elsevier Academic Press, 631–641 pages

212. Foard DE 1971 The initial protrusion of a leaf primordium can form without concurrent periclinal cell divisions, *Can. J. Bot.* **49** 1601–1603

213. Foster AS 1938 Structure and growth of the shoot apex of *Ginkgo biloba*, *Bull. Torrey Bot. Club* **65** 531–556

214. Fowler JE and Quatrano RS 1997 Plant Cell Morphogenesis: Plasma membrane interactions with the Cytoskeleton and Cell Wall, *Annu. Rev. Cell. Dev. Biol.* **13** 697–743

215. Frank AB 1885 Neue Mittheilungen über die Mycorhiza der Bäume und der *Monotropa hypopitys*, *Ber. Dtsch. Bot. Ges.* **3** *XXVII–XXXIII*

216. Friedman WF, Moore RC, and Purugganam MD 2004 The evolution of Plant Development, *Am. J. Bot.* **91** 1726–1741
217. Frohlich MW 2002 An evolutionary scenario for the origin of flowers, *Nature Reviews Genetics* **4** 559–566
218. Fukuda H 1992 Tracheary element formation as a model system of cell differentiation, *Int. Rev. Cytol.* **136** 289–332
219. Fukuda H 1997 Tracheary Element Differentiation, *Plant Cell* **9** 1147–1156
220. Fukuda H 2004 Signals that control plant vascular cell differentiation, *Nature Reviews Molecular Cell Biology* **5** 379–391
221. Fukuda H and Komamine A 1983 Changes in the synthesis of RNA and protein during tracheary element differentiation in single cells isolated from the mesophyll of *Zinnia elegans*, *Plant Cell Physiol.* **24** 603–614
222. Fuzinatto VA, Pagliarini MS, and Valle CB 2007 Evidence of programmed cell death during microsporogenesis in an interspecific *Brachiaria* (Poaceae: Panicoideae: Paniceae) hybrid, *Genet. Mol. Res.* **6** 308–315
223. Gahan PB 1981 Biochemical Changes During Xylem Element Differentiation, in *Xylem Cell Development* (ed) JR Barnett, Tunbridge Wells, Kent: Castle House Publications Ltd., 168–191 pages
224. Gahan PB 1988 Xylem and Phloem Differentiation in Perspective, in *Vascular Differentiation and Plant Growth Regulators* (eds) LW Roberts, PB Gahan and R Aloni, Berlin: Springer- Verlag, 1–19 pages
225. Gail McLean B, Hempel FD, and Zambryski C 1997 Plant Intercellular Communication Via Plasmodesmata, *Plant Cell* **9** 1043–1054
226. Gallagher K and Smith LG 1997 Asymmetric cell division and cell fate in plants, *Curr. Opin. Cell Biol.* **19** 842–848
227. Gallagher K and Smith LG 2000 Roles of polarity and nuclear determinants in specifying daughter cell fates after an asymmetric cell division in the maize leaf, *Curr. Biol.* **10** 1229–1232
228. Gánti T 2003 Chemoton Theory, Vol.1, *Theoretical Foundations of Fluid Machineries*, New York: Kluwer Academic/Plenum Publishers
229. Garrison R 1955 Studies in the development of axillary buds, *Am. J. Bot.* **42** 257–266
230. Gautheret RJ 1983 Plant tissue culture: A history, *Bot. Mag. Tokyo* **96** 393–410

231. Gehring M, Bubb KL, and Henikoff S 2009 Extensive demethylation of repetitive elements during seed development underlies gene imprinting, *Science* **324** 1447-1451.

232. Gehring M, Huh JH, Hsieh TF, Penterman J, Choi Y, Harada JJ, Goldberg RB, and Fischer RL 2006 DEMETER DNA glycosylase establishes MEDIA polycomb gene self-imprinting by allele-specific demethylation, *Cell* **124** 495-506.

233. Gerassimova-Navashina HN 1951 Pollen grains, gametes and sexual process in angiosperm, *Transact. Bot. Inst. Acad. Sci. USSR* **7** 294–355

234. Gerassimova-Navashina HN 1960 A Contribution to the cytology of fertilization in flowering plants, *The Nucleus* **3** 111–120

235. Gerassimova-Navashina HN 1969 Some cytological aspects of double fertilization, *Rev. Cytol. Biol. Veg.* **32** 301–308

236. Gerassimova-Navashina HN and Batygina TB 1958 Fertilization in Angiosperms and mitotic cycle of the cell, *Phytomorphology* **32** 222–233

237. Gillespy G, Ben-David H, and Gruissem W 1993 Fruits: A Developmental Perspective, *The Plant Cell* **5** 1439–1451

238. Giovalisco P, Kapitza K, Kolasa A, Buhtz A, Kerx J 2006 Towards the proteome of *Brassica napus* phloem sap, *Proteomics* 6 896-909.

239. Goldberg RB, Barker SJ, and Perez-Grau L 1989 Regulation of gene expression during plant embryogenesis, *Cell* **56** 149–160

240. Goldberg RB, de Paiva G, and Yadegari R 1994 Plant embryogenesis: Zygote to seed, *Science* **266** 605–614

241. Gosling PG 2004 Viability testing, in *Seed Conservation: Turning Science into Practice* (eds) R Smith, J Dickie, S Linington, H Pritchard and RJ Probert, Kew, Royal Botanic Gardens, 445– 481 pages

242. Gray DJ 1996 Non-zygotic embryogenesis, in *Tissue culture concept and Laboratory Exercises* (eds) RN Trigiano and DJ Gray, Boca. Raton, USA: CRC Press, 133–147 pages

243. Grebe M, Xu J, and Scheres B 2001 Cell axiality and polarity in plants–Adding pieces to the puzzle, *Curr. Opin. Plant Biol.* **4** 520–526

244. Green PB 1976 Growth and cell pattern formation on an axis: Critique of concepts, terminology and modes of study, *Bot. Gaz.* **137** 187–202

245. Green PB 1980 Organogenesis–a biophysical view, *Annu. Rev. Plant Physiol.* **31** 51–82

246. Green PB 1986 Plasticity in shoot development: A biophysical view, in *Plasticity in Plants* (eds) DH Jennings and AJ Trewavas, Cambridge: Company of Biologists Ltd., 211–232 pages

247. Green PB 1999 Expression of pattern in plants: Combining molecular and calculus–based biophysical paradigms, *Am. J. Bot.* **86** 1059–1076

248. Grèen PB and Lang LM 1981 Toward a biophysical theory of organogenesis: Birefringence observations on regenerating leaves in the succulent, *Graptopetalum paraguayensis* E. Walther, *Planta.* **151** 413–426

249. Greenberg JT 1996 Programmed cell death: A way of life for plants, *Proc. Natl. Acad. Sci. USA* **93** 12094–12097

250. Greenberg JT 2005 Degrade or die: a dual function for autophagy in the plant immune response, *Dev. Cell* **8** 799–801

251. Grimanelli D, Perotti E, Ramirez J, and Leblanc O 2005 Timing of the maternal-to-zygotic transition during early seed development in maize, *Plant Cell* **17** 1061-1072.

252. Grime JP 1983 Prediction of weed and crop response to climate based on terrestrial ecosystems, *Aspects Appl. Biol.* **45** 3–13

253. Grossniklaus U and Schneitz K 1998 The molecular and gametic basis of ovule and megagametophyte development, *Cell Dev. Biol.* **9** 227–238

254. Grossniklaus U and Vielle-Caldaza JP 1998 Response: Parental conflict and infanticide during embryogenesis, *Trends Plant Sci.* **3** 328

255. Grossniklaus U, Koltunow A, and Campagne MVL 1998 A bright future for apomixis, *Trends Plant Sci.* **3** 415–416

256. Grossniklaus U, Vielle-Caldaza JBP, Hoeppner M, and Gagliano WB 1998 Maternal control of embryogenesis by *MEDEA*, a polycomb group gene in *Arabidopsis*, *Science* **280** 446–450

257. Guha S and Maheshwari SC 1964 *In vitro* production of embryos from anthers of *Datura*, *Nature* **204** 497

258. Gulline HF 1960 Experimental morphogenesis in adventitious buds of flax, *Aust. J. Bot.* **8** 1– 10

259. Gunning BES 1977 Transfer cells and their roles in transport of solutes in plants, *Sci. Progr. Oxf.* **64** 539–568

260. Gunning BES and Pate JS 1969 "Transfer cells". Plant cells with wall in growths, specialized in relation to short distance transport of solutes–Their occurrence, structure, and development, *Protoplasma* **68** 107–133

261. Gusti A, Baumberger N, Nowack M, Pusch S, Eisler H, et al. 2009 The *Arabidopsis thaliana* F-Box Protein FBL17 Is Essential for Progression through the Second Mitosis during Pollen Development, *PLoS ONE* **4** e4780.

262. Haber AH 1962 Non-essentiality of concurrent cell divisions for degree of polarisation of leaf growth. I. Studies with radiation-induced mitotic inhibition, *Am. J. Bot.* **49** 583–589

263. Guttenberg H Von. 1960 *Grundzüge der Histogenese höherer Pflanzen. I. Die Angiospermen. Handbuch der Pflanzenanatomie.* Band 8.Teil.3, Berlin: Gebrüder Borntraeger

264. Haberlandt G 1902 Kulturversuche mit isolierten Pflanzenzellen, *Sber. Akad. Wiss. Wien.* **111** 69–92

265. Haberlandt G 1913 Zur physiologie der Zellteilung, *Sitzungsber Akad Wiss Berlin Phys Math C* **1** 318–345

266. Haccius B 1955 Experimentally induced twinning in plants, *Nature* **176** 355–356

267. Haecker G, Gross-Hardt R, Geiges B, Sarkar A, Breuninger H, Herrmann M, and Laux T 2004 Expression dynamics of *Wox* genes mark cell fate decisions during early embryonic patterning in *Arabidopsis thaliana, Development* **131** 657–668

268. Hagel JM, Yeung EC, and Facchini PJ 2008 Got milk? The secret life of laticifers, *Trends in Plant Science* **13** 631–639

269. Hagemann W 1973 The organization of shoot development, *Rev. Biol.* **9** 43–67

270. Hagemann W and Gleissberg S 1996 Organogenetic capacity of leaves: The significance of marginal blastozones in angiosperms, *Plant Syst. Evol.* **199** 121–152

271. Hallé F and Oldeman RAA 1970 *Essai sur l' architecture et la dynamique de croissance des arbres tropicaux*, Paris: Masson

272. Hallé F, Oldeman RAA and Tomlinson PB 1978 *Tropical Trees and Foretes–An Architectural Analysis*, Berlin: Springer-Verlag

273. Halperin W and Wetherall DF 1964 Adventive embryony in tissue cultures of the wild carrot, *Daucus carota Am. J. Bot.* **51** 274–283

274. Hanstein J 1868 Die Scheitelzellgrupe in Vegetationspunkt der Phanerogamen, *Festchr. Niederrhein. Ges. Natur. Heilkunde* **1868** 109–134

275. Hanstein J 1870 Die Entwicklung der Keimes der Monokotylen und Dikotylen, in *Botanische Abhandlungen aus dem Gebiet der Morphologie und Physiologie, Vol.1, pt.1* (ed) J Hanstein, Marcus, Bonn

276. Harada JJ, Baden CS, and Comai L 1988 Spatially regulated genes expressed during seed germination and post germinative

development are activated during embryogeny, *Mol. Gen. Genet.* **212** 466–473

277. Hareven D, Gutfinger T, Parnis A, Eshed Y, and Lifschitz E 1996 The making of a compound leaf: genetic manipulation of leaf architecture in tomato, *Cell* **84** 735–744
278. Hariharan Y and Krishnamurthy KV 1995 A cytochemical study of cambium and its xylary derivatives on the normal and tension wood sides of the stems of *Prosopis juliflora* (SW.) DC., *Beitr. Biol. Pflanzen* **69** 459–472
279. Harris JM 1981 Spiral Grain Formation, in *Xylem Cell Development* (ed) JR Barnett, Kent, UK: Castle House Pub. Ltd., 256–274 pages
280. Hart JW 1990 *Plant Tropisms and Other Growth Movements*, London: Unwin Hyman
281. Hartley SE 1999 Are gall insects large rhizobia? *Oikos* **84** 333–342
282. Hasenstein KH and Evans ML 1988 The influence of calcium and pH on growth of primary roots of *Zea mays*, *Physiol. Plant.* **72** 466–470
283. Hatakeda Y, Kamada M, Goto N, Fukaki H, Tasaka M, Suge H, and Takahashi H 2003 Gravitropic response plays an important role in the nutational movements of *Pharbitis nil* and *Arabidopsis thaliana*, *Physiol. Plant.* **118** 464–473
284. Haupt W and Feinleib ME 1979 Introduction. **In**: Haupt, W. and Feinleib, M.E. (eds) ***Physiology of Movements.*** Vol.7. ***Encyclopaedia of Plant Physiology,*** Pirson, A. and Zimmermann, M.H. (eds) (New Series), Berlin: Springer Verlag, 1–8 pages
285. Heath MC 2000 Hypersensitive response-related death, *Pl. Mol. Biol.* **44** 323–334
286. Hejnowicz Z 1990 Phenomenon of Orientation in the Cambium, in *The Vascular Cambium* (ed) M Iqbal, Somerset, England: Research Studies Press Ltd., 127–137 pages
287. Hemerly A, De Almeida Engler, Bergounioux C, Van Montagu M, Engler G, Inze D and Ferreira P 1995 Dominant negative mutants of the Cdc2 kinase uncouple cell division from iterative plant development, *EMBO J.* **14** 3925–3936
288. Hensel W 1984 A role of microtubules in the polarity of statocytes from roots of *Lepidium sativum* C., *Planta* **162** 404–414
289. Hepler PK and Newcomb EH 1964 Microtubules and fibrils in the cytoplasm of *Coleus* cells undergoing secondary wall deposition, *J. Cell Biol.* **20** 529–533
290. Hepler PK, Fosket DE, and Newcomb EH 1970 Lignifications during secondary wall formation in *Coleus*: an electron microscopic study, *Amer. J. Bot.* **57** 85–96

291. Heuros G, Varotto F, Salamini F, and Thompson RD 1995 Molecular characterization of *BET1*, a gene expressed in the endosperm transfer cells of maize, *Plant Cell* **7** 743–757

292. Heyn ANJ 1931 Der Mechanismus der Zellstreckung, *Rec. Trav. Bot. Neerl.* **28** 113–244

293. Higashiyama T, Yabe S, Sasaki N, Nishimura Y, Miyagishima S, Kuroiwa H, and Kuroiwa T 2001 Pollen tube attraction by the synergid cell, *Science* **293** 1480-1483.

294. Hillman SK and Wilkins MB 1982 Gravity perception in decapped roots of *Zea mays*, *Planta* **155** 267–271

295. Hobe M, Muller R, Grunewald M, Brand U, and Simon R 2003 Loss of CLE 40, a protein functionally equivalent to the stem cell restricting signal CLV3, enhances root waving in *Arabidopsis*, *Dev. Genes Evol.* **213** 371–381

296. Hofer J, Turner L, Hellens R, Ambrose M, Matthews P, Michael A, and Ellis N 1997 UNIFOLIATA regulates leaf and flower morphogenesis in pea, *Curr. Biol.* **7** 581–587

297. Hofmeister W 1863 Zusätze und berichtigungen zu den 1851 verőffentichen untersuchungen der entuicklung hőherer kryptogamen, *Jahrbucher fur Wissenschaft und. Botanik* **3** 259–293

298. Holmgren I 1919 Zytologische Studien über die Fortpflanzung bei den Gattungen *Erigeron* und *Eupatorium*, *Kgl. Svensk. Vet.-Akad. Hendl.* **59** 1–118

299. Holm-Hanson O 1969 Algae: amounts of DNA and organic carbon in single cells, *Science* **163** 87–88

300. Hong SK, Kitano H, Satoh H, and Nagato Y 1996 How is embryo size genetically regulated in rice? *Development* **122** 2051–2058

301. Hori K 1992 Insect secretions and their effect on plant growth, with special reference to hemipterans, in *Biology of Insect-induced Galls* (eds) JD Shorthouse and O Rohfritsch, New York: Oxford Univ. Press, 157–170 pages

302. Hori S 1898 Researches on 'bakanae' disease of the rice plant, *Nojishikenjyo Seiseki.* **12** 110–119

303. Hoshino Y, Stefan S, von Wiegen P, Lörz H and Kranz K 2004 Fertilization-induced changes in the microtubular architecture of the maize egg cell and zygote—an immunocytochemical approach adapted to single cells, *Sexual Plant Reproduction* **17** 89-95.

304. Houssa C, Bernier G, Pieltain A, Kinet JM, and Jacqmard A 1994 Activation of latent DNA- replication origins: A Universal effect of cytokinins, *Planta* **193** 247–250

305. Houwink AL 1935 The conduction of excitation in *Mimosa pudica*, *Trav. Bot. Neerland.* **32** 51–91

306. Houwink AL and Roelofsen PA 1954 Fibrillar architecture of growing plant cell walls, *Acta Bot. Neerl.* **3** 385–395

307. Howell SH 1998 *Molecular Genetics of Plant Development*, Cambridge: Cambridge Univ. Press

308. Hsieh TF, Ibarra CA, Silva P, Zemach A, Eshed-Williams L, Fischer RL, and Zilberman D 2009 Genome-wide demethylation of *Arabidopsis* endosperm, *Science* **324** 1451-1454.

309. Hülskamp M and Mathur J 2001 Trichome, Development, Genetics of, in *Encyclopedia of Genetics,* New York: Academic Press, 2045–2048 pages

310. Hülskamp M, Misera S, and Jürgens G 1994 Genetic dissection of trichome cell development in *Arabidopsis*, *Cell* **76** 555–566

311. Hülskamp M, Parekh NS, Grini P, Schneitz K, Zimmermann I, Lolle SJ, and Pruitt RE 1997 The *STUD* gene is required for male-specfic cytokinesis after telophase II of meiosis in *Arabidospsis thaliana*, *Dev. Biol.* **187** 114–124

312. Huntley RP and Murray JAH 1999 The plant cell cycle, *Curr. Opin. Plant Biol.* **2** 440–446

313. Iakimova ET, Michalczuk L, and Woltering EJ 2005 Hypersensitive cell death in plants: its mechanisms and role in plant defense against pathogens, *J. Fruit Ornam. Plant Res.* **13** 135–158

314. Imaichi R and Hiratsuka R 2007 Evolution of shoot apical meristem structure in vascular plants with respect to plasmodesmal network, *Am. J. Bot.* **94** 1911–1921

315. Ingold E, Sugiyama M, and Komamine A 1988 Secondary cell wall formation: Change in cell wall constituents during the differentiation of isolated mesophyll cells of *Zinnia elegans* to tracheary elements, *Plant Cell Physiol.* **29** 295–303

316. Iqbal M 1994 Structural and Operational Specializations of the Vascular Cambium of Seed Plants, in *Growth Patterns in Vascular Plants* (ed) M Iqbal, Portland, Oregon, USA: Dioscorides Press, 211–271 pages

317. Iqbal M and Ghouse AKM 1990 Cambial Concept and Organization, in *The Vascular Cambium* (ed) M Iqbal, Somerset, England: Research Studies Press Ltd., 1–36 pages

318. Israelsson D and Johnsson A 1967 A theory of circumnutation in *Helianthus annuus*, *Physiol. Plant* **20** 957–976

319. ISTA [International Seed Testing Association] 1999 International Rules for Seed Testing, Seed Science and Technology **27**: Supplement.

320. Ivanov VB 1973 Growth and reproduction of cells in roots, in *Physiology of Roots* (ed) NA Obrucheva, Moscow: Uniiti Publishers 1–40 pages

321. Iversen TH 1969 Elimination of geotropic responsiveness in roots of cress (*Lepidium sativum*) by removal of statolith starch, *Physiol. Plant* **22** 1251–1262

322. Iwami S and Masuda Y 1974 Geotropic response of cucumber hypocotyls, *Plant Cell Physiol.* **15** 121–129

323. Jackson D, Veit B, and Hake S 1994 Expression of maize KNOTTED related home box genes in the shoot apical meristem predicts patterns in morphogenesis in the vegetative shoot, *Development* **120** 405–413

324. Jackson MB and Barlow PW 1981 Root geotropism and the role of growth regulators from the cap: a re-examination, *Plant Cell Environ.* **4** 107–123

325. Jacobs M and Rubery PH 1968 Naturally occurring auxin transport regulators, *Science* **241** 346–349

326. Jacobs WP 1947 The development of the gynophores of the peanut plant, *Arachis hypogaea* L.I. The distribution of mitoses, the region of greatest elongation, and the maintenance of vascular continuity in the intercalary meristem, *Am. J. Bot.* **34** 361–370

327. Jacobs WP 1952 The role of auxin in differentiation of xylem around a wound, *Am. J. Bot.* **39** 301–309

328. Jaffe MJ 1973 Thigmomorphogenesis: The response of plant growth and development to mechanical stimulation, *Planta* **114** 143–157

329. Jaffe MJ 1976 Thigmomorphogenesis: A detailed characterization of the response of beans (*Phaseolus vulgaris* L.) to mechanical stimulation, *Zeitschrift Für Pflanzenphysiologie.* **77** 437– 453

330. Jean RV 1994 *Phyllotaxis: A Systematic Study of Plant Pattern Morphogenesis*, Cambridge: Cambridge Univ. Press

331. Jeffree CE 1986 The cuticle, epicuticular waxes and trichomes of plants, with reference to their functions and evolution, in *Insects and the plant surface* (eds) B Juniper, R Southwood, London: Edward Arnold, 23–135 pages

332. Jensen WA 1973 Fertilization in flowering plants, *Bioscience* **23** 21–27

333. Jing HC, Hille J, and Dijkwel PP 2003 Ageing in plants: Conserved strategies and novel pathways, *Plant Biol.* **5** 455–464

334. Johansen DA 1950 *Plant Embryology: Embryogeny of the Spermatophyta*, Waltham, Mass, USA: Chronica Botanica Co.

335. Johnsson A 1979 Circumnutation. pp. 627-646. **In**: Haupt, W. and Feinleib, M.E. (Eds.). *Physiology of Movements Vol.7. Encyclopaedia of Plant Physiology.* Pirson, A. and Zimmermann, M.H. (Eds.). Springer Verlag, Berlin.

336. Jothi R, Parthasarathy S, and Krishnamurthy KV 2010 Computational internal sequence repeats analysis of accelerated evolution and the role of extensins under abiotic and biotic stresses, *Open Access Bioinformatics* **2** 157–168

337. Jürgens G 2001 Apical-basal pattern formation in *Arabidopsis* embryogenesis, *EMBO J.* **20** 3609-3616.

338. Kaliamoorthy S and Krishnamurthy KV 1998 Secondary wall deposition in tracheary elements of cucumber grown in vitro, *Biol. Plantarum* **41** 515–522

339. Kamiya N, Nagasaki H, Morikami A, Sato Y, and Matsuoka M 2003 Isolation and Characterization of a rice *WUSCHEL*-type homebox gene that is specifically expressed in the central cells of a quiescent centre in the root apical meristem, *Plant J.* **35** 429–441

340. Kanta K, Rangawamy NS, and Maheshwari P 1962 Test-tube fertilization in a flowering plant, *Nature* **194** 1214–1217

341. Kapil RN 1961 Some embryological aspects of *Euphorbia dulcis*, *Phytomorphology* **11** 24– 36

342. Kapil RN and Maheshwari R 1964 Embryology of Helianthemum vulgare Gaertn., *Phytomorphology* **14** 547–557

343. Kaplan DR 1984 The concept of homology and its central role in the elucidation of plant systematic relationships, in *Cladistics: perspectives on the reconstruction of evolutionary history* (eds) T Duncan and TF Stuessy, New York, USA: Columbia University Press, 51–70 pages

344. Kaplan DR 2001 The science of plant morphology: Definition, history, and role in modern biology, *Am. J. Bot.* **88** 1711–1741

345. Kaplan DR and Hagemann W 1991 The relationship of cell and organism in vascular plants, *Bioscience* **41** 693–703

346. Kaufman PB 1959 Development of the shoot of *Oryza sativa* L. III. Early stages in histogenesis of the stem and ontogeny of the adventitious root, *Phytomorphology* **9** 382–404

347. Kaufman PB 1965 The effects of growth substances on intercalary growth and cellular differentiation in developing internodes of *Avena sativa*. I. The effects of IAA., *Physiol. Plant.* **18** 424–443

348. Kausik SB 1942 Studies in the Proteaceae. 7. The endosperm of *Grevillea robusta*, *Proc. Indian Acad. Sci. B* **11** 83–99

349. Kende H and Zeevaart JAD 1997 The Five "classical" Plant Hormones, *Plant Cell* **9** 1197– 1210

350. Kennard JL and Cleary AL 1997 Pre-mitotic nuclear migration in subsidiary mother cells of *Tradescantia* occurs in G1 of the cell cycle and requires F-actin, *Cell Motil. Cytoskeleton* **36** 55– 67

351. Kenrick P and Crane PR 1997 *The origin and Early Diversification of Land Plants-A Cladistic study*, Washington D.C.: Smithsonian Institute Press

352. Kerk NM and Feldman LJ 1994 The quiescent centre in roots of maize: Initiation, maintenance and role in organization of the root apical meristem, *Protoplasma* **183** 100–106

353. Kerk NM and Feldman LJ 1995 A biochemical model for the initiation and maintenance of the quiescent centre: Implications for organization of root meristems, *Development* **121** 2825– 2833

354. Kerk NM, Jiang K, and Feldman LJ 2000 Auxin metabolism in the root apical meristem, *Plant Physiol.* **122** 925–932

355. Kersetter RA and Hake S 1997 Shoot Meristem Formation in Vegetative Development, *Plant Cell* **9** 1001–1010

356. Kerzberg M and Wolpert L 1998 Mechanisms for positional signalling by morphogen transport: A theoretical study, *J. Theor. Biol.* **191** 103–114

357. Kidner CA, and Martienssen RA 2005 The role of *ARGONAUTE1* (*AGO1*) in meristem formation and identity, *Dev. Biol.* **280** 504-517.

358. Kim D, Kim, CH, Moon J II, Chung YG, Chang MY, Han BS, Ko S, Yang E, Cha KY, Robert Lanza R, Kim KS 2009 Generation of Human Induced Pluripotent Stem Cells by Direct Delivery of Reprogramming Proteins, *Cell Press* **4** 472-476.

359. Kim GT, Tsukaya H, and Uchimiya H 1998 The CURLY LEAF gene controls both division and elongation of cells during the expansion of the leaf blade in *Arabidopsis thaliana*, *Planta* **206** 175–83

360. Kim H-J, Oh SA, Brownfield L, Hong SH, Ryu H, Hwang I, Twell D, Nam HG 2008 Control of plant germline proliferation by SCF (FBL17) degradation of cell cycle inhibitors, *Nature* **455** 1134-1137.

361. Kiss JZ 2006 Up, down, and all around: How plants sense and respond to environmental stimuli, *Proc. Natl. Acad. Sci. USA.* **103** 829–830

362. Kitazawa D, Hatakeda Y, Kamada M, Fujii N, Miyazawa Y, Hashinto A, Iida S, Fukaki H, Morita MT, Tasaka M, Suge H, and Takahashi H 2005 Shoot circumnuation and winding movements require gravisensing cells, *Proc. Natl. Acad. Sci. USA* **102** 18742–18747

363. Klebs G 1910 Alterations in the development and forms of plants as a result of environment, *Proc. Royal Soc. London* **82** 547–558

364. Klement Z, Farkas GL, and Lovrekovich L 1964 Hypersensitive reaction induced by phytopathogenic bacteria in the tobacoo leaf, *Phytopathology* **54** 474–477

365. Knoblauch M and van Bel AJA 1998 Sieve Tubes in Action, *Plant Cell* **10** 35–50

366. Knox JP 1993 Cell surface glycoproteins and plant morphogenesis, in Keystone Symposium on *The Extra cellular Matrix of Plants–Molecular, cellular and Development Biology*, Santa Fe, New Mexico

367. Koch AJ and Meinhardt H 1994 Biological pattern formation: From basic mechanisms to complex structures, *Rev. Mod. Phys.* **66** 1481–1507

368. Köhler C, Page DR, Gagliardini V, and Grossniklaus U 2005 The *Arabidopsis thaliana* MEDEA Polycomb group protein controls expression of PHERES1 by parental imprinting, *Nature Genet.* **37** 28-30.

369. Koltunow AM, Truettner J, Cox KH, Wallroth KFW and Goldberg RB 1990 Different temporal and spatial expression patterns occur during anther development, *Plant cell* **2** 1201– 1224

370. Kondorosi F, Rondier F, and Gendreau E 2000 Plant Cell-size control: Growing by ploidy? *Curr. Opoin. Plant Biol.* **3** 4888–492

371. Korn RW 1984 Cell shapes and tissue geometries, in *Positional Controls in Plant Development* (eds) PW Barlow and DJ Carr, Cambridge: Cambridge Univ. Press, 219–245 pages

372. Kozieradzka-Kiszkurno M, Swierczynska J, and Bohdanowicz J 2007 Endoreduplication in the suspensor of Gagea lutea (L.) Ker. Gawl. (Liliaceae), *Acta Biologica Carcoviensia.* **49** 33–37

373. Kozlowsky TT and Pallardy SG 1997 *Physiology of Woody Plants. 2nd Ed.*, San Diego: Academic Press

374. Kranz E, and Lörz H 1993 In vitro fertilization with isolated, single gametes results in zygotic embryogenesis and fertile maize plants, *Plant cell* **5** 739-746.

375. Kranz E, Kumlehn J and Dresselhaus T 1999 Fertilization and zygotic embryo development *in vitro*, in *Fertilization in higher Plants: molecular and cytological aspects* (eds) M Cresti, G Cai and A Muscatelli, Berlin: Springer Verlag, 337–349 pages

376. Kreps JA and Kay SA 1997 Coordination of Plant Metabolism and Development by the Circadian Clock, *Plant Cell* **9** 1235–1244

377. Krikorian AD 1998 Plants and Somatic Embryos in Space: What Have We Learned? *Garvitational and Space Biology Bulletin* **11** 5–14

378. Krikorian AD and Berquam DL 1969 Plant cell and tissue cultures: The role of Haberlandt, *Bot. Rev.* **35** 59–67

379. Krishnamurhty KV 2014a Pre-Fertilization Reproductive Growth and Development and Double Fertilization (In Press) **In:** *Plant Biology and Biotechnology* (Vol.1) Bir Bahadur Rajam MV, Leela Sahijram and Krishnamurthy KV, Berlin, Springer Verlag.

380. Krishnamurthy KV 1976 Meiotic induction in Plants: A hypothesis, *Biology (Kamaraj Univ.)* **1** 15–17

381. Krishnamurthy KV 1988 Endosperm controls symmetry changes in the developing embryos of angiosperms, *Proc. Indian Acad. Sci. (Pl. Sci.)* **98** 257–259

382. Krishnamurthy KV 1994 The Angiosperm Embryo: Correlative Controls in Development, Differentiation, and Maturation, in *Growth Patterns in Vascular Plants* (ed) M Iqbal, Portland, Oregon, USA: Dioscorides Press, 372–404 pages

383. Krishnamurthy KV 1999a *Methods in Cell Wall Cytochemistry*, Boca Raton, Florida: CRC Press

384. Krishnamurthy KV 1999b On embryos and embryoids, *Curr. Sci.* **76** 647–659

385. Krishnamurthy KV 2000 Intercellular Spaces: The Apoplastic Domains of Multicellular Plants, *J. Plant Biol.* **27** 213–222

386. Krishnamurthy KV 2003 Certain key issues in seed development, in *Advances in Plant Reproductive Biology* (eds) AK Pandey and MR Dhakal, Delhi: Narendra Publishing House, 77–88 pages

387. Krishnamurthy KV 2007 In pursuit of understanding the tension wood riddle, *J. Ind. Bot. Soc.* **86** 14–26

388. Krishnamurthy KV 2010 Origin and evolution of Plants–A brief review, *J. Swamy Bot. Cl.* **27** 1–12

389. Krishnamurthy KV 2014b Post-Fertilization Growth and Development (In Press) **In:** *Plant Biology and Biotechnology* (Vol.1) Bir Bahadur Rajam MV, Leela Sahijram and Krishnamurthy KV, Berlin, Springer Verlag.

390. Krishnamurthy KV and Bir Bahadur 2014 Genetics of Flower Development (In Press) **In:** *Plant Biology and Biotechnology* (Vol.1) Bir Bahadur Rajam MV, Leela Sahijram and Krishnamurthy KV, Berlin, Springer Verlag.

391. Krishnamurthy KV and Karthikeyan AVP [Communicated]. A new look at the phenomenon of circumnutation, *J. Swamy. Bot. Cl.*

392. Krishnamurthy KV, Krishnaraj R, Chozhavendan R, and Samuel CF 2000 The Programme of cell death in plants and animals–A comparison, *Curr. Sci.* **79** 1169–1181

393. Krishnamurthy KV and Senthil Kumar T 1992 Embryo rotation in *Polygala arvensis* Willd, *Beitr. Biol. Pflanzen* **67** 55–58

394. Kull U and Herbig A 1995 The leaf vein systems in angiosperms: Shape and evolution, *Naturwissenschaften* **82** 441–451

395. Kuras M 1978 Activation of embryo during rape (*Brassica napus* L.) seed germination. I. Structure of embryo and organization of root apical meristem, *Acta. Soc. Bot. Pol.* **47** 65–82

396. Kuriyama H and Fukuda H 2002 Developmental programmed cell death in plants, *Curr. Opin. Plant Biol.* **5** 568–573

397. Kurosawa E 1926 Experimental studies on the filtrate of the causal fungus of the 'bakanae' disease of the rice plant, *Trans. Nat. Hist. Soc. Formosa* **16** 213-227

398. Küster E 1911 *Die Gallen der Pflanzen.* Leipzig, Germany

399. Lafarguette F, Leple JC, Dèjardin A, Laurans F, Costa G, Lesage-Decauses M-C, and Pilate G 2004 Popular genes lucoding fasciclin-like arabinogalaetan proteins are highly expresses intension wood, *New Phytol.* **164** 107–121

400. Lagerberg T 1909 Studien über die Entwicklungsgeschichte und systemtische Stellung von *Adoxa moschatellina*, *Kgl. Svensk Vet-Akad. Handl.* **44** 1–86

401. LaMotte CE and Jacobs WP 1963 A role of auxin in phloem regeneration in *Coleus* internodes, *Dev. Biol.* **8** 80–98

402. Lander AD, Nie Q, and Wan FYM 2002 Do morphogen gradients arise by diffusion? *Dev. Cell.* **2** 785–796

403. Larkin JC, Marks MD, Nadeau J, and Sack F 1997 Epidermal cell fate and patterning in leaves, *Plant Cell* **9** 1109–1120

404. Larkin JC, Oppenheimen DG, Lloyd AM, Paparozzi ET, and Marks MD 1994 Roles of the *GLABROUS 1* and *TRANSPARENT TESTA GLABRA* genes in *Arabidopsis* trichome development, *Plant Cell* **6** 1065–1076

§

405. Larkin JC, Oppenheimer DG, Pollock S, and Marks MD 1993 *Arabidopsis GLABROUS 1* gene acquires downstream sequences for function, *Plant Cell* **5** 1739–1748

406. Larkin JC, Young N, Prigge M, and Marks MD 1996 The control of trichome spacing and number in *Arabidopsis*, *Development* **122** 997–1005

407. Larkins BA, Dilkes BE, Dante RA, Coelho CM, Woo YM, and Liu Y 2001 Investigating the hows and whys of DNA endoreduplication, *J. Exp. Bot.* **52** 183–192

408. Larson PR 1983 Primary vascularisation and siting of primordia, in *The Growth and Functioning of Leaves* (eds) JE Dale and FL Milthorpe, Cambridge: Cambridge Univ. Press, 25–51 pages

409. Laufs P, Grandjean O, Jonak C, Kieu K, and Traas J 1998 Cellular parameters of the shoot apical meristem in *Arabidopsis*, *Plant Cell* **10** 1375–1389

410. Laux T and Jürgens G 1997 Embryogenesis: A New Start in Life, *Plant Cell* **9** 989–1000

411. Lee JS, Mulkey TJ, and Evans ML 1983a Gravity-induced polar transport of calcium across root tips of maize, *Plant Physiol* **73** 874–876

412. Lee JS, Mulkey TJ, and Evans ML 1983b Reversible loss of gravitropic sensitivity in maize roots after tip application of calcium chelators, *Science* **220** 1375–1376

413. Lee JS, Mulkey TJ, and Evans ML 1984 Inhibition of polar calcium movement and gravitropism in roots treated with auxin-transport inhibitors, *Planta* **160** 536–543

414. Lee SW, Kim SY, Jeyabalasinkham A, Wayne R, Schneider K, Vaughn ML, and Mulkey TJ 1992 Effect of solution density on gravitropic response of maize roots, *Plants Physiol.* **99** 63

415. Leins P and Erbar C 2010 *Flower and Fruit: Morphology, Ontogeny, Phylogeny, Function and Ecology,* Schweizerbart, Science Publishers.

416. Lenhard M and Laux T 1999 Shoot meristem formation and maintenance, *Curr. Opin. Plant Biol.* **2** 44–50

417. Li Y, Hagen G, and Guilfoyle TJ 1992 Altered morphology in transgenic tobacco plants that overproduce cytokinins in specific tissues and organs, *Dev. Biol.* **153** 386–395

418. Lifschitz E, and Eshed Y 2006 Universal florigenic signals triggered by FT homologues regulate growth and flowering cycles in perennial day-neutral tomato, *J.Expt.Bot.* **57**3405-3414.

419. Lima MF, Núbia BE, Pegoraro C, Sagit R, Rojas C, Bretz, T, Vargas, L, Elofsson, A, Oliveira, Adriana AC, Hemerly AS, and Ferreira, PCG 2010 Genomic evolution and complexity of the Anaphase-promoting Complex (APC) in land plants, *BMC Plant Biology* **10** 254.

420. Lin BL and Raghavan V 1991a Lateral root initiation in *Marsilea quadrifolia*. 1. Origin and histogenesis of lateral roots, *Can. J. Bot.* **69** 123–135

421. Lin BL and Raghavan V 1991b Necleic acid and protein synthesis during later root initiation in *Marsilea quadrifolia* (Marsileaceae), *Am. J. Bot.* **78** 740–746

422. Lintilhac PM and Vesecky TB 1984 Stress-induced alignment of division plane in plant tissues grown in vitro, *Nature* **307** 363–364

423. Liso R, Calabrese G, Bitonti MB, and Arrigoni O 1984 Relationship between ascorbic acid and cell division, *Exp. Cell Res.* **150** 314–320

424. Liso R, Innocenti AM, Bitonti MB, and Arrigoni O 1988 Ascorbic acid-induced progression of quiescent centre cells from G1 to S phase, *New Phytol.* **110** 469–471

425. Liu HF, Deng YF, and Liao JP 2008 Floral organogenesis of three species of *Jatropha* (Euphorbiaceae), *Journal of Systematics and Evolution* **46** 53–61

426. Liu HF, Kirchoff BK, Wu GJ, and Liao JP 2007 Microsporogenesis and male gametogenesis in *Jatropha curcas* L. (Euphorbiaceae), *J. Torrey Bot. Soc.* **134** 335–343

427. Lloyd C and Buschmann H 2007 Plant division: Remembering where to build the wall, *Curr. Biol.* **17** R1053–R1055

428. Lockhart J 1965 An analysis of irreversible plant cell elongation, *J. Theoret, Biol.* **8** 264–275

429. Lockshin R and Williams C 1965 Programmed cell death. IV. The influence of drugs on the breakdown of the intersegmental muscles of silkmonths, *J. Insect Physiol.* **11** 803–809

430. Lopes MA and Larkins BA 1993 Endosperm origin, development and function, *Plant Cell* **5** 1383–1399

431. Love AJ, Milner JJ, and Sadanandom A 2008 Timing is everything: Regulatory overlap in plant cell death, *Trends Plant Sci.* **13** 589-595.

432. Lu B, Villani PJ, Watson JC, DeMason DA, and Cokke TJ 1996 The control of pinna morphology in wild type and mutant leaves in the garden pea (*Pisum sativum* L.), *Int. J. Plant Sci.* **157** 659–673

433. Lynch TM and Lintilhac PM 1987 Mechanical signals in plant development: A new method for single cell studies, *Dev. Biol.* **181** 246–256

434. Lyndon RF 1987 Initiation and growth of internodes and stem and flower frusta in *Silene coeli- rosa*, in *The Manipulation of Flowering* (ed) JG Atherson, London: Butterworths, 301–314 pages

435. Ma H 1997 To be, not to be, a flower–Control of floral meristem identity, *Trends Genet.* **14** 26–32

436. Ma H 2005 Molecular genetic analyses of microsporogenesis and microgametegenesis in flowering plants, *Annu. Rev. Plant. Biol.* **56** 393-434.

437. MacMillan J and Takahashi N 1968 Proposed procedure for the allocation of trivial names to the gibberellins, *Nature* **217** 170–171

438. Maheshwari P 1950 *An Introduction to the Embryology of Angiosperms*, New York: McGraw- Hill

439. Maheshwari P 1965 Test Tube Fertilization of Ovules, *Proc. Nat. Inst. Sci. India Part B.* **31** 1– 13

440. Mahlberg PG 1960 Embryogenesis and histogenesis in Nerium oleander L.I. Organization of primary meristematic tissues, *Phytomorphology* **10** 118

441. Maizel A, Busch MA, Tanahashi T, Perkovic J, Kato M, Hasebe M and Weigel D 2005 The floral regulator *LEAFY* evolves by substitutions in the DNA binding domain, *Science* **308** 260-263.

442. Małgorzata KK, Świerczyńska J, and Bohdanowicz J 2007 Endoreduplication in the suspensor of *Gagea lutea* (L.)Ker Gawl. (Liliaceae), *Acta. Biologica Cracoviensia. Series Botanica* **49/1** 33–37

443. Mani MS 2000 *Plant Galls of India. 2nd Edition*, Enfield, USA: Science Publishers

444. Mann LK 1952 Anatomy of the garlic bulb and factors affecting bud development, *Hilgardia* **21** 195–251

445. Mapes CC and Davies PJ 2001 Indole-3-acitic acid and ball gall development on *Solidago altissima*, *New Phytol.* **151** 195–202

446. Margulis L and Sagan D 2003 *Acquiring Genomes*, New York: Basic Books

447. Marsden MPF and Bailey IW 1955 A fourth type of nodal anatomy in dicotyledons, illustrated by *Clerodердron tichotomum* Thunb, *J. Arnold Arboretum* **36** 1–50

448. Martinez MC, Jorgensen JE, Lawton MA, Lamb CJ, and Doerner, PW 1992 Spatial pattern of *Cdc2* expression in relation to meristem activity and cell proliferation during plant development, *Proc. Natl. Acad. Sci. USA.* **89** 7360–7364

449. Matzke EB 1940 What shape is a cell? *Teach. Biol.* **10** 34–40

450. Mayer U, Büttner G, and Jürgens G 1993 Apical-basal pattern formation in the *Arabidopsis* embryo: Studies on the role of *gnom* gene, *Development* **117** 149–162

451. McConnell JR and Barton MK 1998 Leaf polarity and meristem formation in *Arabidopsi*, *Development* **125** 2935–2942

452. McDaniel CN 1984 Competence, determination, and induction in plant development, in *Pattern Formation. A Primer in Developmental Biology* (ed) Malacinski, New York: Macmillan, 393–412

453. McHale NA 1992 A nuclear mutation blocking initiation of the lamina in *Nicotiana sylvestris*, *Planta* **186** 355–360

454. McKinnon ML, Quiring DT, and Bauce E 1999 Influence of tree growth rate, shoot size and foliar chemistry on the abundance and performance of a galling adelgid, *Funct. Ecol.* **13** 859– 867

455. McQueen Mason SJ 1995 Expansins and cell wall expansion, *J. Exp. Bot.* **46** 1639–1650

456. Meeuse ADJ 1941 *A Study of Intercellular Relationships Among Vegetable Cells with Special Reference to "sliding Growth" and to Cell Shape,* Netherland: Koch and Knuttel

457. Meijer M and Murray JAH 2001 Cell cycle controls and the development of plant form, *Curr. Opinions Pl. Biol.* **4** 44–49

458. Meinhardt H 1984 Models of pattern formation and their application to plant development, in *Positional Controls in Plant Development* (eds) PW Barlow and DJ Carr, Cambridge: Cambridge Univ. Press, 1–32 pages

459. Meinhardt H 1996 Models of biological pattern formation: common mechanism in plant and animal development, *Int. J. Dev. Biol.* **40** 123–134

460. Melchers G 1937 Die Wirkung von Genen, tiefen Temperaturen und blühenden *pfropfpartnern* die Blühreife von *Hyoscyamus niger* L, *Biol. Zentralbl.* **57** 568–614

461. Merrill EK 1986a Heteroblastic seedlings of green ash. II. Early development of simple and compound leaves, *Can. J. Bot.* **64** 2650–2661

462. Merrill EK 1986b Heteroblastic seedlings of green ash. III. Cell division activity and marginal meristems, *Can. J. Bot.* **64** 2662–2668

463. Meyer J and Maresquelle HJ 1983 *Anatomie des Galles*, Berlin: Gebruder Bornträeger

464. Meyerowitz EM 1996 Plant development: local control and global patterning, *Curr. Opin. Genet. Dev.* **6** 475–479

465. Meyerowitz EM 1997 Genetic control of cell division patterns in developing plant, *Cell* **88** 299– 308

466. Meylan BA and Butterfield BG 1981 Performation-plate differentiation in the vessels of hardwoods, in *Xylem Cell Development* (ed) JR Barnett, Tunbridge Wells, UK: Castle House, 96–114 pages

467. Miles PW 1999 Aphid saliva, *Biol. Rev.***74** 41–85

468. Miller DD, De Ruijter NCA, Bisseling T, Emons A M C 1999 The role of actin in root hair morphogenesis: studies with lipochito-oligosaccharide as a growth stimulator and cytochalasin as an actin perturbing drug, *The Plant Journal* **17** 141–154

469. Mirnov V, De Veylder L, Van Montagu M, and Inzé D 1999 Cyclin–dependent kinases and cell division in plants-The nexus, *Plant Cell* **11** 509–521

470. Mohan Ram HY, Ram M, and Steward FC 1962 Growth and development of the banana plant. III. A the origin of the inflorescence and development of the fruit, *Ann. Bot.* **26** 657–673

471. Moore R 1985a A morphometric analysis of the redistribution of organelles in columella cells in primary roots of normal seedlings and gravitropic mutants of *Hordeum vulgare*, *J. Exp. Bot.* **36** 1275–1286

472. Moore R 1985b Calcium movement, graviresponsiveness and the structure of columella cells and columella tissues in roots of *Alilum cepa L.*, *Ann Bot.* **56** 173–187

473. Moore R 1994 The Dynamics of Root Growth and Gravitropism, in *Growth patterns in Vascular Plants* (ed) M Iqbal, Portland, Ovegon: Dioscorides Press, 77–89 pages

474. Mowforth MAG 1985 Variation in nuclear DNA amounts in flowering plants: an ecological analysis, Ph.D. thesis, university of Sheffield, United Kingdom

475. Mulkey TJ and Evans ML 1981 Geotropism in corn roots: evidence for its mediation by differential acid refflux, *Science* **212** 70–71

476. Nagl W 1978 *Endopolyploidy and Polyteny in Differentiation and Evolution*, Amsterdam: North- Holland Publication

477. Nagl W 1981 Polytene chromosomes in plants, *Int. Rev. Cytol.* **73** 21–53

478. Nakamura AT, Longhi-Wagner HM, and Scatena VL 2010 Anther and pollen development in some species of Poaceae (Poales), *Braz. J. Biol.* **70** 351–360
479. Narayana LL 1965 Contributions to the embryology of Balsaminaceae, part 2, *J. Jap. Bot.* **40** 104–116
480. Nelson AJ and Evans ML 1986 Role of the epidermis in the gravitropic response of maize roots, *Abstr. Annu. Meet. Am. Soc. Grav. Space Biol.*, Charlottesville, VA, Oct.1-3, 1986, P.21.
481. Nelson T and Dengler N 1997 Leaf Vascular Pattern Formation, *Plant Cell* **9** 1121–1135.
482. Newman IV 1965 Pattern in the meristems of vascular plants. III. Pursuing the patterns in the apical meristem where no cell is a permanent cell, *Journal of Linnean Society (Botany)* **59** 185– 214
483. Niklas KJ 2003 The Bio-logic and machinery of plant morphogenesis, *Am. J. Bot.* **90** 515– 525
484. Nitsch JP 1950 Growth and morphogenesis of strawberry as related to auxin, *Am. J. Bot.* **37** 211–215
485. Nitsch JP 1963 Fruit Development, in *Recent Advances in the Embryology of Angiosperms* (ed) P Maheshwari, New Delhi: International Society of Plant Morphologists, 361–394 pages
486. Noé A C 1934 Gottlieb Haberlandt, *Plant Physiol.* **9** 850–855
487. Noll F 1902 Fruchtbildung ohne vorausgegangene Bestäubung (Parthenokarpie) bei der Gurke, S.Z. neiderrhein Ges. Nat. Heilk, Bonn 149–162
488. Northcote DH 1995 Aspects of vascular tissue differentiation in plants; parameters that may be used to monitor the process, *Int. J. Plant Sci.* **156** 245–256
489. Nougarede A 1967 Experimental cytology of the shoot apical cells during vegetative growth and flowering, *Int. Rev. Cytol.* **21** 203–351
490. Ohad N, Margassian I, Hsyu YC, Williams C, Ripetti P, and Fischer RI 1996 A mutation that allows endosperm development without fertilization, *Proc. Natl. Acad. Sci. USA.* **93** 5319–5124
491. Okamoto T, and Kranz E 2005 *In vitro* fertilization— a tool to dissect cell specification from a higher plant zygote, *Curr. Sci.* **89** 1861-1869.
492. Okuyama T, Yamamoto Y, Yoshida M, Hatton Y, and Archer RR 1994b Growth stresses in tension wood fibres of aspen (*Populus tremula* L.), *Holzforschung* **20** 174–178
493. Okuyama T, Yamamoto Y, Yoshida M, Hattori Y, and Archer RR 1994a Growth stresses in tension wood: role of microfibrils and liquefaction, *Ann. For. Sci.* **51** 291–300

494. Oldeman RAA 1974 L'architecture de la foret guyanaise, *Mem. O.R.S.T.O.M.* **73** 1–204

495. Olsen GM and Iversen TH 1980 Ultrastructure and movements of cell structures in normal pea and in a geotropic mutant, *Physiol. Plant.* **50** 275–284

496. Opsahi-Ferstad HG, Le Deunff E, Dumas C, and Rogowsky PM 1997 *ZmEar*, a novel endosperm-specific gene expressed in a restricted region around the maize embryo, *Plant J.* **12** 235–246

497. Ordina NA 1962 *Docklady Acad. Nauk* **84** 825

498. Osawa J 1913 Studies on the cytology of some species of *Taraxacum*, *Arch. f. Zellf.* **10** 450– 469

499. Osborne DJ and McManus MT 2005 *Hormones, Signals and Target Cells in Plant Development*, Cambridge, UK: Cambridge University Press

500. Osborne TB 1924 *The Vegetable Proteins. 2nd Ed.*, London: Longmans Green

501. Paal A 1919 Über phototropische reizleitung, *Jahr. Wiss. Bot.* **58** 406–458

502. Pagnussat GC, Yu H-J, Ngo GA, Rajani S, Mayalagu S, Johnson CS, Capron A, Xie L-F, Ye D and Sundaresan V 2005 Genetic and molecular identification of genes required for female gametophyte development and function in Arabidopsis, Development **132** 603-614.

503. Parish RW, and Li SF 2010 Death of a tapetum: A programme of developmental altruism, *Plant Sci.* **178** 73-89.

504. Paroda RS and Rees H 1971 Nuclear DNA variation in Eu-sorghums, *Chromosoma* **32** 353– 363

505. Pate JS and Gunning BES 1972 Transfer cells, *Annu. Rev. Plant Physiol.* **23** 173–196

506. Paxson-Sowders DM, Owen HA, and Makaroff CA 1997 A comparative ultrastructural analysis of exine pattern development in wild-type *Arabidopsis* and a mutant defective in pattern formation, *Protoplasma* **198** 53-65.

507. Pennell RI, Jannichi L, Schoefield GN, Booij H, and de Vries SC 1992 Identification of a transitional cell state in the developmental pathway to carrot somatic embryogenesis, *J. Cell Biol.* **119** 1371–1380

508. Péret B, Swarup R, Jansen L, Devos G, Auguy F, Collin M, Santi C, Hocher V, Franche C, Bogusz D, Bennett M, and Laplaze L 2007 Auxin influx activity is associated with *Frankia* infection during

actinorhizal nodule formation in *Casuarina glauca*, *Plant Physiol.* **144** 1852–62

509. Periasamy K 1962 The ruminate endosperm: development and types of rumination, in *Plant Embryology: a symposium*, CSIR, New Delhi, 62–74
510. Periasamy K 1977 A new approach to the classification of angiosperm embryos, *Proc. Indian Acad. Sci.B* **86** 1–13
511. Periasamy K 1994 Morphogenesis of the Angiosperm Proembryo, in *Growth Patterns in Vascular Plants* (ed) M Iqbal, Oregon: Dioscorides Press, 405–417 pages
512. Periasamy K and Swamy BGL 1961 Studies in the Annonaceae. 1. Microsporogenesis in *Cananga odorata* and *Miliusa wightiana*, *Proc. Indian bot. Soc.* **40** 206–216
513. Periasamy K and Swamy BGL 1966 Morphology of anther tapetum of angiosperm, *Curr.Sci.* **35** 427–430
514. Periasamy K and Swamy BGL 1977 The aim and scope of plant morphology-I, *Proc. Indian Acad. Sci. B* **86** 181–187
515. Periasamy K and Swamy BGL 1978 The aim and scope of plant morphology-II, *Proc. Indian Acad. Sci. B* **87** 325–333
516. Peters JM 2006. The anaphase promoting complex/cyclosome: a machine designed to destroy, *Nature Reviews Molecular Cell Biology* **7** 644-656.
517. Petersen RL and Bonfante P 1994 Comparative structure of vesicular arbuscular mycorshizas and ectomycorrhizas, *Plant Soil* **159** 79–88
518. Phan HA Iacuone S, Li SF, and Parish RW 2011 The MYB80 Transcription factor is required for pollen Development and the Regulation of Tapetal Programmed Cell Death in *Arabidopsis thaliana*, *Plant Cell* **23** 2209-2224.
519. Philip VJ 1972 Embryogenesis and Seedling anatomy of *Catharanthus roseus* (Linn.) G. Don. 1. Embryogenesis and procambialization, *Cellule* **69** 155–172
520. Philipson WR 1990 Anomalous Cambia, in *The Vascular Cambium* (ed) M Iqbal, Somerset, England: Research Studies Press Ltd., 201–212 pages
521. Philipson WR and Ward JM 1965 The ontogeny of the vascular cambium in the stem of seed plants, *Biol. Rev.* **40** 534–579
522. Philipson WR, Ward MJ and Butterfield BG 1971 *The Vascular Cambium*, London: Chapman & Hall Ltd.
523. Piechulla B, Pichersky E, Cashmore AR, and Gruissem W 1986 Expression of nuclear and plastid genes for photosynthesis-specific

proteins during tomato fruit development and ripening, *Plant Mol. Biol.* **7** 367–376

524. Pilet PE and Chanson A 1981 Effect of abscissic acid on maize root growth: a critical examination, *Plant Sci. Lett.* **21** 99–106
525. Plantefol L 1946 Fondements d'une theorie phyllotaxique nouvelle, *Ann. Sci. nat. Botanique 11th ser.* **7** 153–220
526. Plantefol L 1947 Helices foliaires, point vegetatif et stele chez les dicotyledones. La notion d'anneau initial, *Rev. gen. Bot.* **54** 49–80
527. Poethig RS 1988a Heterochronic mutations affecting shoot development in maize, *Genetics* **119** 959–973
528. Poethig RS 1988b A non-cell-autonomous mutation regulating juvenility in maize, *Nature* **336** 82–83
529. Poethig RS 1997 Leaf Morphogenesis in Flowering Plants, *Plant Cell* **9** 1077–1087
530. Poethig RS and Sussex IM 1985a The developmental morphology of and growth dynamics of the tobacco leaf, *Planta* **165** 158–169
531. Poethig RS and Sussex IM 1985b The cellular parameters of leaf development in tobacco: A clonal analysis, *Planta* **165** 170–184
532. Popham RA 1951 Principal types of vegetative shoot apex organization in vascular plants, *Ohio J. Sci.* **51** 249–270
533. Popham RA 1966 *Laboratory Manual for Plant Anatomy*, St. Louis, Mo: C.V. Mosby Co.
534. Prat H 1948 Histo-physiological gradients and plant organogenesis, *Bot. Rev.* **14** 603–643
535. Prat H 1951 Histo-physiological gradients and plant organogenesis. (Part II), *Bot. Rev.* **17** 693–746
536. Preuss D, Rhee SY, and Davis RW 1994 Tetrad analysis possible in *Arabidopsis* with mutation of the *QUARTET* (*QRT*) genes, *Science* **264** 1458–1460
537. Price PW 1991 The plant vigor hypothesis and herbivore attack, *Oikos* **62** 244– 251
538. Pritchard HN 1964 A cytochemical study of embryo development in Stellaria media, *Am. J. Bot.* **55** 472–479
539. Raghavan V 1986a Molecular cytology of Embryogenesis, in *Embryogenesis* (eds) T.L. Wang and A Cuming, Cambridge: Cambridge Univ. Press, 1–13 pages
540. Raghavan V 1986b *Embryogenesis in Angiosperms. A Developmental and Experimental Study*, Cambridge: Cambridge Univ. Press

541. Raghavan V 2000 *Developmental Biology of Flowering Plants*, New York: Springer-Verlag

542. Raju MVS, Steeves TA, and Maze J 1976 Developmental studies on *Euphorbia esula*: Seasonal variations in the apices of long roots, *Can. J. Bot.* **4** 605–610

543. Raman A 2003 Cecidogenetic behavior of some gall-inducing thrips, psyllids, coccids, and gall midges, and morphogenesis of their galls, *Orient. Insect* **39** 359–413

544. Raman A 2010 Patterns of adaptive radiation and diversification in cecidogenous insects, in *Insect Biodiversity: Functional Dynamics and Ecological Perspectives* (ed) TN Ananthakrishnan, Jodhpur: Scientific Publishers, 153–178 pages

545. Raman A 2011 Morphogenesis of insect-induced plant galls: facts and questions, *Flora* **206** 517–533

546. Raman A and Ananthakrishnan TN 1985 Cecidogenous *Crotonothrips* (Thysanoptera) – *Memecylon* interactions: Host relations, nutritive tissue, tissue dynamics and cecidogenetic patterns, *Proceedings of the Indian Academy of Sciences (Plant Sciences)* **95** 103–120

547. Raman A, Burckhardt D, and Harris KM 2009 Biology and adaptive radiation in the gall- inducing Cecidomyiidae (Insecta: Diptera) and Calophyidae (Insecta: Hemiptera) on *Mangifera indica* (Anacarliaccae) in the Indian subcontinent, *Trop. Zool.* **22** 27–56

548. Raman A, Schaefer CW and Withers TM (eds) 2005a *Biology, Ecology, and Evolution of Gall- inducing Arthropods*, New Hampshire: Science publishers

549. Raman A, Schaefer CW and Withers TM 2005b Galls and gall-inducing arthropods: an overview of their biology, ecology, and evolution, in *Biology, Ecology, and Evolution of Gall- inducing Anthropods* (eds) A Raman, CW Schaefer and TM Withers, New Hampshire: Science publishers, 1–33 pages

550. Rangarajan R and Swamy BGL 1980 Studies on the procambium in certain organs of monocotyledons. I. Anther filaments of *Gloriosa superba.* L., in *Histochemistry, Deveoplmental and Structural Anatomy of Angiosperms: a Symposium* (ed) K Periasamy, Tiruchirapalli: P&B Publications, 201–219 ,

551. Ransom JS and Moore R 1983 Geoperception in primary and lateral roots of *Phaseolus vulgaris* (Fabaceae). I. Structure of columella cells, *Am. J. Bot.* **70** 1048–1056

552. Raven JA 1986 Evolution of plant life forms, in *On the economy of plant form and function* (ed) TJ Givnish, Cambridge: Cambridge Univ. Press, 421–492 pages

553. Raven PR, Evert RF and Eichhorn SE 2005 *Biology of Plants*, New York: Freeman

554. Rechinger C 1893 Untersuchungen über die Grenzen der Teilbarkeit im Pflanzenreich, *Abh. Zool.-bot. Ges.(Vienna)* **43** 310–134

555. Rees H and Narayan RKJ 1988 Chromosome constraints: chiasma frequency and genome size, in *Kew chromosome conference III.* (ed) PE Brandham, London: HMSO, 231–239 pages

556. Reinhardt D and Kuhlemeir C 2002 Plant architecture, *EMBO reports* **3** 846–851

557. Reinhardt D, Mandel T, and Kuhlemeier C 2000 Auxin regulates the initiation and radial position of plant lateral organs, *Plant Cell* **12** 507–518

558. Reinhardt D, Wittmer F, Mandel T, and Kuhlemeier C 1998 Localized upregulation of a new expansin gene predicts the site of leaf formation in the tomato meristem, *Plant Cell* **10** 1427– 1437

559. Ribeiro A, Akkermans ADL, Van Kammen A, Bisseling T, and Pawlowski K 1995 A nodule- specific gene encoding a subtilisin-like protease is expressed in early stages of actinorhizal nodule development, *Plant Cell* **7** 785–794

560. Ricca U 1916 Solution d'un problem de physiologie la propagation de stimules dans la sensitiv, *Arch. Ztal. Biol. (Pisa)* **65** 219–232

561. Richards FJ 1948 The geometry of phyllotaxis and its origin, *Symp. Soc. Exp. Biol.* **2** 217– 245

562. Ridge RW 1987 Brief Notes on *Ginkgo biloba*, http://mac122.icu.ac.jp/ginkgo/icho.html

563. Roberts K 1990 Structure at the plant surfaces, *Curr. Opin. Cell Biol.* **2** 920–928

564. Roelofsen PA 1959 *The plant cell-wall, Handbuch der Pflanzenanatomie*. Band III, Teil 4. Cytologie. Gebrüder Borntraeger, Berlin

565. Rohfritsch O 1992 Patterns in gall development, in *Biology of Insect-induced Galls* (eds) JD Shorthouse and O Rohfritsch, New York: Oxford Univ. Press, 60–86 pages

566. Romberger JA 1963 *Meristems, growth and development in Woody plants*, Tech. Bull. No.1293, U.S. Dept., Agriculture, USA

567. Ron M, Alandete Saez M, Eshed Williams L, Fletcher JC and McCormick S 2010 Proper regulation of a sperm-specific cis-nat-

siRNA is essential for double fertilization in Arabidopsis, *Genes Dev.* **24** 1010-1021.

568. Rost TL 1994 Root Top Organization and the Spatial Relationships of Differentiation Event,in *Growth Patterns in Vascular Plants* (ed) M Iqbal, Portland, Oregon, USA: Dioscorides Press, 59–76 pages
569. Russell SD 1984 Ultrastructure of the sperm of *Plumbago zeylanica:* 2. Quantitative cytology and three-dimensional reconstruction, *Planta* **162** 385–391
570. Sachar RC and Chopra RN 1957 A study of the endosperm and embryo in *Mangifera, Indian J. Agr. Sci.* **27** 219–228
571. Sachar RC and Murgai P 1958 Embryology of *Aerva tomentosa, Curr. Sci.* **27** 105–107
572. Sachs J 1878 Űeber die Anordnung der zellen in jüngsten pflanzentheilen, *Arb. Bot. Inst. Wurzburg* **2** 46–104
573. Sachs RM, Bretz CF, and Lang A 1959 Shoot histogenesis: The early effects of gibberellins upon stem elongation in two rosette plants, *Am. J. Bot.* **46** 376–384
574. Sachs T 1969 Regeneration experiments on the determination of the form of leaves, *Israel J. Bot.* **18** 21–30
575. Sachs T 1972 The induction of fibre differentiation in peas, *Ann. Bot. n.s.* **36** 189–197
576. Sachs T 1978 The development of spacing patterns in the leaf epidermis, in *The Clonal Basis of Development* (eds) S Subtelny, and IM Sussex, New York: Academic Press, 161–183 pages
577. Sachs T 1981 The control of patterned differentiation of vascular tissues, *Adv. Bot. Res.* **9** 151–262
578. Sachs T 1989 The development of vascular networks during leaf development, *Curr. Top. Plant Biochem. Physiol.* **8** 168–181
579. Sachs T 1991a Cell polarity and tissue patterning in plants, *Dev. (Suppl.).* **1** 83–93
580. Sachs T 1991b *Pattern Formation in Plant Tissues*, Cambridge, UK: Cambridge Univ. Press
581. Saether N and Iversen TH 1991 Gravitropism and starch statoliths in an *Arabidopsis* mutant, *Planta* **184** 491–497
582. Saks Y, Feigenbaun P, and Aloni R 1984 Regulatory effect of cytokinin on secondary xylem fiber formation in an in vivo system, *Plant Physiol.* **76** 638–642
583. Salisbury FB and Ross CW 2004 *Plant Physiology*, Singapore: Thomson Asia Private Ltd.
584. Sankla N, Bohra SP, Vyas SP and Sankhla D 1975 Morphactins and plant Morphogenesis, in *Form, Structure and Function in Plants*

(eds) HY Mohan Ram, JJ Shah and CK Shah, Meerut, India: Sarita Prakashan, 255–264 pages

585. Savidge RA 1985 Prospects for manipulating vascular cambium productivity and xylem- cell differentiation, in *Attributes of Trees as Crop Plants* (eds) MGR Cannel and JE Jackson, UK, Abbots Ripton 208–227 pages
586. Savidge SA 1996 Xylogenesis, genetic and environmental regulation–a review, *IAWA J.* **17** 269–310
587. Schaefer CW, Raman A and Withers TM 2005 Galls and Gall-inducing Arthropods: Ecological Issues and Evolutionary Problems, in *Biology, Ecology, and Evolution of Gall- inducing Arthropods* (eds) A Raman, CW Schaefer and TM Withers, Enfield, USA: Science Publishers, Inc., 761–766 pages
588. Schaffner JH 1906 The embryology of the Shepherd's Purse, *Ohio. Nat.* **7** 1–8
589. Scheres B and Heidstra R 2000 Digging out roots: Pattern formation, cell division and morphogenesis in plants, *Curr. Topics Dev. Biol.* **45** 207–247
590. Scheres B and Wolkenfelt H 1998 The *Arabidopsis* root as a model to study plant development, *Plant physiol. Biochem.* **36** 21–32
591. Scheres B, Di Laurenzio L, Willemsen V, Hauser MT, Janmaat K 1995 Mutations affecting the radial organization of the *Arabidopsis* root displays specific defects throughout the embryonic axis, *Development* **121** 53–62
592. Scheres B, Wolkenfelt H, Willemsen V, Tenlouw M, Lawson E, Dean C, and Weisbeek P 1994 Embryonic origin of the Arabidopsis primary root and root meristem initials, *Development* **120** 2457–2487
593. Schiavone FM 1988 Microamputaion of somatic embryos of the domestic carrot reveals apical control of axis elongation and root regeneration, *Development* **103** 657–664
594. Schiavone FM and Racusen RH 1990 Microsurgery reveals regional capabilities for pattern reestablishment in somatic carrot embryos, *Dev. Biol.* **141** 211–219
595. Schiefelbein J, Galway M, Masucci J, and Ford S 1993 Pollen tube and root hair tip growth is disrupted in a mutant of *Arabidopsis thaliana, Plant Physiol.* **103** 979–985
596. Schiefelbein JW, Masucci JD, and Wang H 1997 Building a Root: The Control of Patterning and Morphogenesis during Root Development, *The Plant Cell* **9** 1089–1098

597. Schildknecht H 1983 Turgorins, hormones of the endogenous daily rhythms of higher organized plants–detection, isolation structure, synthesis, and activity, *Angewandte chemie Int.* **22** 695–710

598. Schleiden MJ 1838 Beiträge zur Phytogenesis, *Anat. Physiol and Wiss. Med. (Müller's Arch.)* **3** 137–176

599. Schmidt A 1924 Histologische studien an Phanergamen vegetationspunkten, *Bot. Arch.* **8** 345– 404

600. Schneider G 1970 Morphactins: Physiology and Performance, *Ann. Rev. Pl. Physiol.* **21** 499– 536

601. Schneitz K, Hulskamp M, Kopczak SD, and Pruitt RE 1997 Dissection of sexual organ ontogenesis: a genetic analysis of ovule development in *Arabidopsis thaliana*, *Development* **124** 1367–1376

602. Schopfer P 2009 Mechanical signals in *Arabidopsis* Stress or strain? *Science Online Biol* **2** 33–39

603. Schüepp O 1917 Untersuchungen über Wachstum und Formwechsel von Vegetationspunkten, *Jb. Wiss. Bot.* **57** 17–79

604. Schwann T 1839 *Mikroskopische Untersuchungen über die Uebereinstimmung in der Struktur and dem Waehsthum der Thiere und Pflanzen*, Berlin

605. Schwendener S 1860 (not seen in original, cited from Errera L 1888)

606. Scott RJ, Spielman M, and Dickinson HG 2004 Stamen structure and function, *Plant Cell* **16** (Suppl.) S46-S60.

607. Selker JML, Steucek GL, and Green PB 1992 Biophysical mechanisms for morphogenetic progressions at the shoot apex, *Dev. Biol.* **159** 29–43

608. Shah JJ and Patel JD 1972 The shell zone: its differentiation and probable function in some dicotyledons, *Am. J. Bot.* **59** 683–690

609. Shimony C and Fahn A 1968 Light-and electron-microscopical studies on the structure of salt glands of *Tamarix aphylla* L, *Bot. J. Linn. Soc.* **60** 283–288

610. Shininger TL 1979 The control of vascular development, *Annu. Rev. Plant Physiol.* **30** 313–337

611. Shorthouse JD and Rohfritsch O (eds) 1992 *Biology of Insect-induced Galls*, New York: Oxford Univ. Press

612. Siebers AM 1971 Differentiation of isolated interfascicular tissue of *Ricinus communis* L., *Acta bot. neerl.* **20** 343–355

613. Siegel SM 1956 The biosynthesis of lignin; evidence for the participation of cellulose as sites for oxidative polymerization of eugenol, *J. Amer. Chem. Soc.* **78** 1753–1755

614. Sinha N 1999 Leaf development in Angiosperms, *Annu. Rev. Plant Physiol. Plant Mol. Biol.* **50** 419–446

615. Sinnott EW 1914 The anatomy of the node as an aid in the classification of Angiosperms. I., *Am. J. Bot.* **1** 303–322

616. Sinnott EW 1960 *Plant Morphogenesis*, New York: McGraw-Hill

617. Sitbon F, Hennion S, Sundberg B, Little CHA, Olsson O, and Sandberg G 1992 Transgenic tobacco plants overexpressing the *Agrobacterium tumefaciens iaaM* and *iaaH* genes display altered growth and indoleacetic acid metabolism, *Plant Physiol.* **99** 1062–1069

618. Skipworth JP 1962 The primary vascular system and phyllotaxis in *Hectorella caespitosa* Hook, *N.Z. J. Sci.***5** 253–258

619. Slatyer RO and Taylor SA 1960 Terminology in plant-cell-water relations, *Nature* **187** 922– 924

620. Smart CC and Amrhein N 1985 The influence of lignification on the development of vascular tissues in *Vigna radiata* L., *Protoplasma* **124** 87–95

621. Smith LG 2001 Plant cell division: Building walls in the right places, *Nat. Rev. Mol. Cell Biol.* **2** 33–39

622. Smith LG, Hake S, and Sylvester AW 1996 The *tangled-1* mutation alters cells division orientations throughout maize leaf development without altering leaf shape, *Development* **122** 481–489

623. Smith RH and Murashige T 1970 In Vitro Development of the Isolated Shoot Apical Meristem of Angiosperms, *Am. J. Bot.* **57** 562–568

624. Snow M and Snow R 1932 Experiments on phyllotaxis. 1. The effect of isolating a primordium, *Philos. Trans. R. Soc. Lond. B* **221** 1–43

625. Snow M and Snow R 1942 The determination of axillary buds, *New Phytol.* **41** 13–22

626. Soh WY 1990 Origin and Development of Cambial Cells, in *The Vascular Cambium* (ed) M Iqbal, Somerset, England: Research Studies Press Ltd., 37–62 pages

627. Soma K and Ball E 1963 Studies on the surface growth of the shoot apex of *Lupinus albus*. "Meristems and Differentiation", *Brookhaven Symp. Biol.* **16** 13–45

628. Souèges R 1919 Les premières divisions de l'oeuf et les différenciations du suspenseur chez *Capsella bursa-pastoris*, *Ann. Sci. nat. bot.* **10** 1–28

629. Souèges R 1934 L'hypophyse et l'epiphyse: les problemès d'histogenèse qui leur sont liés. 2. L'epiphyse: importance générale de cette chez des dicotyledones, *Bull. Soc. Bot. Fr.* **81** 28–61

630. Souèges R 1937 *Les Lois der Development. Actualités Scientifiques et Industrielles*, Paris: Hermann et Cie
631. Spielman M, Preuss D, LI FL, Browne WE, Scott RI, and Dickinson HG 1997 TETRASPORE is required for male meiotic cytokinesis in *Arabidopsis thaliana, Development* **124** 2645–2657
632. Staiger CJ and Chande WZ 1990 Microtubule distribution in *dv*, a maize mutant defective in the prophase to metaphase transition, *Dev. Biol.* **138** 231–242
633. Staiger CJ and Chande WZ 1991 Microfilament distribution in maize meiotic mutants correlates with microtubule organization, *Plant cell* **3** 637–644
634. Stakman EC 1915 Relation between *Puccinia graminis* and plants highly resistant to its attack, *J. Agric. Res.* **4** 193–299
635. Stebbins GL and Jain SK 1960 Developmental studies of cell differentiation in the epidermis of monocotyledons. I. *Allium, Rhoeo*, and *Commelina., Dev. Biol.* **2** 409–426
636. Steeves TA 1959 The development of leaves in sterile nutrient culture, *Proc. IX Internat. Bot., Congress* **2** 380
637. Steeves TA and Susses IM 1989 *Patterns in Plant Development*, Cambridge: CambridgeUniv. Press
638. Stein DB and Stein OL 1960 The growth of the stem tip of Kalanchoe cv. 'Brilliant Star', *Amer. J. Bot.* **47** 132–139
639. Sterk P, Booji H, Schetleckens GA, Van Kammen A, and de Vries SC 1991 Cell specific expression of the carrot EP2 lipid transfer protein gene *Plant Cell* **3** 907–921
640. Sterling C 1946 Growth and vascular development in the shoot apex of *Sequoia sempervirens* (Lamb) Endl. III Cytological aspects of vascularisation, *Am. J. Bot.* **23** 35–45
641. Stern WL, Sweitzer EM, and Philipps RE 1970 Comparative anatomy and systematics of woody Saxifragaceae. *Ribes, Bot. J. Linn. Soc.* 63, suppl. **1** 215–237
642. Stevens WC 1911 *Plant anatomy from the standpoint of the development and functions of the tissues, and handbook of micro-technic*, Philadelphia: P. Blakiston's son & co.
643. Steward FC and Mapes MO 1963 The totipotency of cultured carrot cells: evidence and interpretation from successive cycles of growth from phloem cells, *Jour. Indian Bot. Soc. (Maheshwari Commemorative Volume)* **42A** 237–247
644. Steward FC, Mapes MO, and Mears K 1958 Growth and organized development of cultured cells. II. Organization in cultures grown from freely suspended cells, *Am. J. Bot.* **45** 705–708

645. Stieger PA, Reinhardt D, and Kuhlemeier C 2002 The auxin influx carries is essential for correct leaf positioning, *Plant J.* **32** 509–517

646. Stinson JR, Eisenberg AJ, Willing RP, Pe ME, Hanson DD, and Mascarenhas JP 1987 Genes expressed in the male gametophyte of flowering plants and their isolation, *Plant Physiol.* **83** 442–447

647. Strange C 1992 Cell cycle advances, *Bioscience* **42** 252–256

648. Street HE 1976 Experimental embryogenesis–The totipotency of cultured plant cells, in *The Developmental Biology of Plants and Animals* (eds) CF Graham and PF Wareing, Oxford: Black well Sci. Publ., 73–90 pages

649. Street HE and Roberts EH 1952 Factors controlling meristematic activity in excised roots. I. Experiments showing the operation of internal factors, *Physiol. Plant* **5** 498–509

650. Studholme WP and Philipson WR 1966 A Comparison of the cambium in two woods and included phloem: *Heimerliodendron brunonianum* (Endl.) Skottsb. and *Avicennia resinifera* (Forst. f.), *N.Z.J. Botany* **4** 355–365

651. Subramanyam K and Raju MVS 1953 Circumscissile Dehiscence in some Angiosperms, *Am. J. Bot.* **40** 571–574

652. Sung S and Amasino RM 2005 Remembering Winter: toward a molecular understanding of vernalization, *Annu. Rev. Plant Biol.* **56** 491–508

653. Sussex IM 1955 Morphologenesis in Solanum tuberosum L.: Apicalstructure and developmental pattern of the juvenile shoot, *Phytomorphology* **5** 253–273

654. Sussex JM and Dale RMK 1979 Hormonal control of storage protein synthesis in *Phaseolus vulgaris*, in *The Plant seed: development, preservation and germination* (eds) I Rubenstein, RL Phillips, CE Green and BG Gengenbach, New York: Academic Press Inc., 129–140 pages

655. Swamy BGL 1941 Contributions to the life history of *Bignonia megapatomica*, *J. Indian bot. Soc.* **20** 299–305

656. Swamy BGL 1946 Endosperm in *Hypericum mysorense*, *Ann. Bot.* **9** 165–169

657. Swamy BGL 1948 On the post-fertilization development of *Trillium undulatum*, *Cellule* **52** 7– 14

658. Swamy BGL 1960 Contributions to the embryology of *Cansjera rheedii*, *Phytomorphology* **10** 397–409

659. Swamy BGL 1962 The origin of cotyledon and epicotyl in *Ottelia alismoides*, Beitr. Biol. Pflanzen **39** 1–16

660. Swamy BGL 1964 Observation on the floral morphology and embryology of *Stemona tuberosa* Lour., *Phytomorphology* **14** 458–468

661. Swamy BGL and Bailey IW 1949 The morphology and relationships of *Cercidiphyllum, J. Arnold Arb.* **30** 187–210

662. Swamy BGL and Dayanand Rao J 1963 The endosperm in *Opilia amentacea, Phytomorphology* **13** 423–428

663. Swamy BGL and Krishnamurhty KV 1978 Certain conceptual aspects of meristems. III. A model, *Phytomorphology* **28** 1–7

664. Swamy BGL and Krishnamurthy KV 1974 Studies in the developmental anatomy of *Commelina benghalensis* L. I. The shoot apex and the early delineation of the internode, *J. Indian Bot. Soi.* **53** 213–223

665. Swamy BGL and Krishnamurthy KV 1975a Embryo sac Ontogenies in Angiosperms- An Elucidation, *Phytomorphology* **25** 12–18

666. Swamy BGL and Krishnamurthy KV 1975b Certain conceptual aspects of meristems. I. On hypophysis and quiescent centre, *Phytomorphology* **25** 60–65

667. Swamy BGL and Krishnamurthy KV 1975e Contribution to the Monograph on *Gnetum* II. Foliar Galls, in *Form, Structure & Function in Plants* Mohan Ram HY, Shah JJ and Shah CK (Eds.), Meerut: Sarita Prakashan, 409–418 pages

668. Swamy BGL and Krishnamurthy KV 1977 Certain conceptual aspects of meristems. II. Epiphysis and shoot apex, *Phytomorphology* **27** 1–8

669. Swamy BGL and Krishnamurthy KV 1979 Certain conceptual aspects of meristems. IV The internodal meristems concerned with longitudinal growth, *J. Madras Univ., Section B.* **42** 22–28

670. Swamy BGL and Krishnamurthy KV 1980a *From Flower to Fruit,* New Delhi: Tata McGraw–Hill Publishing Company

671. Swamy BGL and Krishnamurthy KV 1980b On the Origin of vascular, Cambium in dicotyledonous stems, *Proc-Indian Acad. Sci. (Plant Sci.)* **89** 1–6

672. Swamy BGL and Krishnamurthy KV 1981 On embryos and embryoids, *Proc. Indian Acad. Sci. (Pl. Sci.)* **90** 401–411

673. Swamy BGL and Padmanabham D 1961 Embryogenesis in *Sphenoclea zeylanica, Proc. Indian Acad. Sci.* **54** 169–187

674. Swarup R, Parry G, Graham N, Allen T, and Bennett M 2002 Auxin crosstalk: integration of signaling pathways to control plant development, *Plant mol. Biol.* **49** 411–426

675. Sylvester AW, Smith L, and Freeling M 1996 Acquisition of identity in the developing leaf, *Annu. Rev. Cell Dev. Biol.* **12** 257–304

676. Taiz L and Zeiger E 2002 *Plant Physiology*, Sunderland: Sinauer Associates, Inc.

677. Takahashi N, Seta Y, Kitamura H, Sumiki Y 1957 A new gibberellin, gibberellin A4. *Bull Agric Chem. Soc. Japan* **21** 396–398

678. Talbot MJ, Offler CE and McCurdy DW 2002 Transfer cell wall architecture: A contribution towards understanding localized wall deposition, *Protoplasma* **219** 197–209

679. Tello JI 2007 Mathematical analysis of a model of Morphogenesis: steady states. pp. 1–8. *XX Congreso de Ecuaciones Diferenciales y Aplicaciones X Congreso de Matemática Aplicada* Sevilla, 24-28 septiembre 2007

680. Theissen G, Becker A, Winter KU, Munster T, Kirchner C, and Saedler H 2002 How the land plants learned their floral ABCs: the role of MADS box genes in the evolutionary origin of flowers, in *Developmental Genetics and Plant Evolution* (eds.) QC Cronk, RM Bateman, and JM Hawkins, London: Taylor & Francis

681. Théry M and Bornens M 2006 Cell shape and cell division, *Curr. Opin. Cell Biol.* **18** 648–657

682. Thielke C 1957 Chimaren mit periclinal spaltender Oberhaut am Scheitel, *Acta Soc. Bot. Polon.* **26** 247–253

683. Thom R 1975 *Structural Stability and Morphogenesis*, Massachusetts: W. A. Benjamin Inc.

684. Thompson NP and Jacobs WP 1966 Polarity of IAA effect ton sieve-tube and xylem regeneration in Coleus and tomato stems, *Pl. Physiol.* **41** 673

685. Thorne JH 1985 Phloem unloading of C and N assimilates in developing seeds, *Annu. Rev. Plant Physiol* **36** 317–343

686. Timell TE 1973 *Ultrastructure of the dormant and active cambial zones and the dormant phloem associated with formation of normal and compression woods in Picea abies (L.) Karst.* SUNY Coll. Envir. Sci. For. Tech. Rep. **96**: 94

687. Timell TE 1986 *Compression Wood in Gymnosperms,* Berlin: Springer Verlag

688. Timmermans MCP, Schultes NP, Jankovsky JP and Nelson T 1998 Leaf bladeless is required for dorsoventrality of lateral organs in maize, *Development* **125** 2813–2823

689. Tiwari SC, and Polito VS 1990 b An analysis of the role of actin during pollen activation leading to germination in pear (*Pyrus communis* L.): treatment with cytochalasin D, *Sex Plant Report.* **3** 121-129

690. Tiwari SC, and Polito VS 1990a The initiation and organization of microtubules in germinating pear (*Pyrus communis* L.) pollen, *Eur. J. Cell. Biol.* **53** 384-389.

691. Tomlinson PB 1983 Tree Architecture, *Amer. Scientist* **71** 141–149

692. Tomlinson PB and Gill AM 1973 Growth habits of tropical trees: some guiding principles, in *Tropical Forest Ecosystems in Africa and South America: A Comparative Review* (eds.) BJ Meggers, ES Ayensu and WD Duckworth, Washington: Smithsonian Institution, 124–143 pages

693. Traas J, Bellini C, Nacry P, Kronenberger J, Bouchez D, and Caboche M 1995 Normal differentiation patterns in plants lacking microtubular preprophase bands, *Nature* **375** 676–677

694. Trewavas AJ and Cleland RE 1983 Is Plant development regulated by changes in the concentration of growth substances or by changes in the sensitivity to growth substances?, *Trends in Biochemical Sciences* **8** 354–357

695. Trewavas AJ and Malhó R 1997 Signal Perception and Transduction: The Origin of the Phenotype, *Plant cell* **9** 1181–1195

696. Tsuge T, Tsukaya H, and Uchimiya H 1996 Two independent and polarized processes of cell elongation regulate leaf blade expansion in *Arabidopsis thaliana, Development* **122** 1589–1600

697. Turing AM 1952 The chemical basis of morphogenesis, *Phil. Trans. Roy. Soc. B.* **237** 37–72

698. Twell D 1992 Use of a nuclear-targeted β-glucuronidase fusion protein to demonstrate vegetative cell-specific gene expression in developing pollen, *Plant J.* **2** 887–892

699. Twell D 1995 Diphtheria toxin-mediated cell ablation in developing pollen: vegetative cell ablation blocks generative cell migration, *Protoplasma* **187** 144–154

700. Twyman RM 2003 *Developmental Biology (Reprinted).* New Delhi: Viva Books Private Limited

701. Vallade J, Bugnon F, Gambade G, and Alabouvette J 1983 L'activité édificatrice du proméristème racinaire: Essai d' interprétation morphogénétique, *Bull. Sci. Bourg.* **36** 57–76

702. Van den Berg C, Weisbeek P, and Scheres B 1998 Cell fate and cell differentiation status in the *Arabidopsis* roots, *Planta* **205** 483–491

703. Van den Berg C, Willemsen V, Hendriks G, Weisbeek P, and Scheres B 1997 Short-range control of cell differentiation in the Arabidopsis root meristem, *Nature* **39** 287–289
704. Van der Pluijm JE 1964 An electron microscopic investigation of the filiform apparatus in the embryo sac of *Torenia fournieri*, in Pollen physiology and fertilization (ed) HF Linskens, Amsterdam: North Holland Publishing company, 8–16 pages
705. Van Doorn WG and Woltering EJ 2005 Many ways to exit? Cell death categories in plants, *Trends Pl. Sci.* **10** 117–122
706. Van Fleet DS 1961 Histochemistry and Functions of the Endodermis, *Bot. Rev.* **27** 165–220
707. Van Tieghem P 1891 *Traité de botanique*, Savy, Paris
708. Van Tieghem P and Douliot H 1886 Sur la polystelie, *Ann. Sci. Nat. Bot. (Paris)* **7** 275–322
709. Van't Hof J 1965 Relationships between mitotic cycle duration, S period duration and average rate of DNA synthesis in root meristem cells of several plants, *Exp Cell Res* **39** 48–58
710. Van't Hof J, Sparrow AH 1963 A relationship between DNA content, nuclear volume, and minimum mitotic cycle time, *Proc Natl. Acad. Sci.* **49** 897–902
711. Veit B 2006 Stem cell signaling networks in plants, *Plant Mol. Biol.* **60** 793–810
712. Venugopal N and Krishnamurthy KV 1984 Multinucleate condition in the differentiating secondary xylem vessel elements of *Dalbergia sissoo*, *Roxb. Curr. Sci.* **53** 43
713. Venugopal N and Krishnamurthy KV 1987 Seasonal Production of Secondary Xylem in the twigs of certain tropical trees, *IAWA Bulletin* **8** 31–40
714. Venugopal N and Krishnamurthy KV 1989 Organisation of vascular cambium during different seasons in some tropical timber trees, *Nordic J. Bot.* **8** 631–638
715. Verbeka JA 1992 Fusion events during floral morphogenesis, *Annu. Rev. Plant Physiol. Plant Mol. Biol.* **43** 583–598
716. Vernoux T, Autran D and Traas J 2000 Developmental control of cell division patterns in the shoot apex, *Plant Mol. Biol.* **43** 569–581
717. Vielle-Caldaza JP, Baskar R, and Grossniklaus U 2000 Delayed activation of the parental genome during seed development, *Nature* **404** 91-94.
718. Vince-Prue D 1975 Photoperiodism in Plants, New York: McGraw-Hill Book Company

719. Volkmann D and Tewinkel M 1997 *Position of Statoliths in Statocytes from Cress Roots under Changing Gravity Conditions*, European Space Agency

720. Waddington CH 1957 *The Strategy of the Genes* London: Allen & Unwin

721. Waites R, Selvadurai HRN, Oliver IR and Hudson A 1998 The *PHANTASTICA* gene encodes a MYB transcription factor involved in growth and dorsoventrality of lateral organs in *Antirrhinum, Cell* **93** 779–789

722. Walker JW 1974 Evolution of exine structure in the pollen of primitive Angiosperms, *Amer. J. Bot.* **61** 891–902

723. Walters MS 1985 Meiosis readiness in *Lilium, Can. J. Genet, Cytol.* **27** 33–38

724. Wang H, Zhar Y, Gilmer S, Whitwill S, and Fowke LC 2000 Expression of the plant cyclin–dependent kinase inhibitor ICk1affects cell division, plant growth and morphology, *The Plant Journal* **24** 613–623

725. Wardlaw CW 1949 Experiments on organogenesis in ferns, *Growth (suppl.)* **13** 93–131

726. Wardlaw CW 1955 *Embryogenesis in plants* London: Methuen Publishing Ltd.

727. Wardlaw CW 1957 On the organization and reactivity of the shoot apex in vascular plants, *Am. J. Bot.* **44** 176–185

728. Wardlaw CW 1968 *Morphogenesis in Plants–A Contemporary Study*, London: Methuen Publishing Ltd.

729. Wardrop AB 1964 The structure and formation of the cell wall in xylem, in *The Formation of Wood in Forest Trees* (ed) MH Zimmermann, London: Academic Press, 87–134 pages

730. Wardrop AB 1981 Lignification and Xylogenesis, in *Xylem Cell Development* (ed) JB Barnett, Kent: Castle House Publications 115–152 pages

731. Warren Wilson PM and Warren Wilson J 1961 The Position of Regenerating Gambia a New Hypothesis, *New Phytologist* **60** 63–73

732. Wayne R 1992 The contribution of the cell-extra cellular matrix junction in gravisensing in *Chara, Plant Physiol.* **90** 30

733. Weis AE, Walton R and Crego CL 1988 Reactive plant tissue sites and the population biology of gall makers, *Annu. Rev. Entomol.* **33** 467–486.

734. Weizman IL 2000 Stem cells: Units of development, units of regeneration and units in evolution, *Cell* **100** 157–168

735. Went FW 1926 On growth accelerating substances in the coleoptile of *Avena sativa*, *Proc. K. Akad. Wet. Amsterdam* **30** 10–19

736. Went FW 1957 *The Experimental Control of Plant Growth* New York: Ronald Press

737. Went FW and Thimann KV 1937 *Phytohormones* New York: Macmillan Publishers Ltd.

738. Wetmore RH and Sorokin S 1955 On the differentiation of xylem, *J. Arnold Arbor.* **36** 305– 317

739. Whittaker R 1969 New Concepts of kingdoms or Organisms: Evolutionary relations are better represented by new classifications than by the traditional two kingdoms, *Science* **163** 150–160

740. Wilcox H 1954 Primary organization of active and dormant roots of noble fir *Abies procure*, *Am. J. Bot.* **41** 812–821

741. Wilkins MB 1978 Gravity-sensing guidance mechanisms in roots and shoots, *Bot. Mag. (Tokyo) Special Issue* **1** 255–27

742. Willemsen V, Welkenfelt H, de Vrieze G, Weisbeek P, and Scheres B 1998 The *HOBBIT* gene is required for formation of the root meristem in the *Arabidopsis* embryo, *Development* **125** 521–531

743. Williams SE and Bennett AB 1982 Leaf closure in the Venus flytrap: An acid growth response, *Science* **218** 1120–1122

744. Wilson RC, Long F, Maruoka EM, and Cooper JB 1994 A new proline-rich early nodulin from *Medicago truncatula* is highly expressed in nodule meristematic cells, *Plant Cell* **6** 1265– 1275

745. Wilson ZA, and Zhang DB 2009 From *Arabidopsis* to rice: pathways in pollen development, *J. Exp. Bot.* **60** 1479-1492.

746. Winkler H 1908 Űber Perthenogenesis und Apogamie im Pflanzenreiche, *Prog. Rei. Bot.* **2** 293–454

747. Wollenweber HW 1931 Fusarium-Monographie. Fungi parasitic et saprophytici, *Zeitschrift fur pflanzenkrankheiten* **3** 269–516

748. Wolpert L 1970 Positional information and pattern formation, in *Towards a Theoretical Biology* (ed) CH Waddington, Edinburgh: Edinburgh Univ. Press, 198–230 pages

749. Wolpert L 1971 Positional information and pattern formation, in *Current Topics in Developmental Biology* (eds) AA Mosocona and A Monroy, New York: Academic Press,183–224 pages .

750. Wolpert L 1981 Positional information and pattern formation, *Philos. Trans. R. Soc., London Ser. B.* **295** 441–450

751. Woltering EJ and van Doorn WG 2005 Apoptosis in plants-does it exist?, in *Apoptosis* (ed) AI Scovassi, Kerala: Signpost, 187–206 pages

752. Wymer C and Lloyd C 1996 Dynamic microtubules: Implications for cell wall patterns, *Trends Plant Sci.* **1** 222–228

753. Xu J, Yang C, Yuan Z, Zhang D, Gondwe MY, Ding Z, Liang W, Zhang D, Wilson ZA 2010 The *ABORTED MICROSPORES* Regulatory Network Is Required for Postmeiotic Male Reproductive Development in *Arabidopsis thaliana, Plant Cell* **22** 1 91-107

754. Yabuta T, Kobe K, and Hayashi T 1935 Biochemical studies of the 'bakanae' fungus of rice. I. Fusarinic acid, a new poduct of the 'Bakanae' fungus, *Ibid.***10** 1059–1068

755. Yang C, Barrena GV, Conner K, Wilson ZA 2007 MALE STERILITY1 is required for Tapetal Development and Pollen Wall Biosynthesis, *Plant Cell* **19** 3530-3548.

756. Yoon KD, Yang MH, Nam SI, Park JH, Kim YC, and Kim J 2007 Phenanthrene Derivatives, 3,5-Dimethoxyphenanthrene-2,7-diol and Batatasin-I, as Non-Polar Standard Marker Compounds for Dioscorea Rhizoma, *Nat. Prod. Sci.* **13** 378–383

757. Zagórska-Marek B and Turzanska M 2000 Clonal analysis provides evidence for transient initial cells in shoot apical meristems of seed plants, *J. Plant Growth Regu.* **19** 55–64

758. Zhang W, Sun Y, Timofejeva L, Chen C, Grossniklaus U, and Ma H, 2006 Regulation of *Arabidopsis* tapetum development and function by *DYSFUNCTIONAL TAPETUM1 (DYT1)* encoding a putative bHLH transcription factor, *Development* **133** 3085-3095.

759. Zheng B, Chen X and McCormick S 2011 The Anaphase-Promoting Complex is a dual integrator that regulates both MicroRNA-Mediated Transcriptional regulation of Cyclin B1 and degradation of Cyclin B1 during *Arabidopsis* Male Gametophyte Development, *Plant Cell* **23** 1033-1046

760. Zhong R and Ye ZH 1999 *IFL1*, a gene regulating interfascicular fiber differentiation in *Arabidopsis*, encodes a homeodomain–leucine zipper protein, *Plant Cell* **11** 2139–2152

761. Zimmer R 1962 Phasenverschiebung und andere Storolichtwirkungen auf die endogen tagesperiodischen Blutenblattbewegungen, *Planta* **48** 283–300

762. Zimmermann BD and Briggs WR 1963 Phototropic dosage curves for oat coleoptiles, *Plant Physiology* **38** 248–253

763. Zimmermann JL 1993 Somatic Embryogenesis: A Model for Early Development in Higher Plants, *Plant Cell* **5** 1411–1423

764. Zobel AM 1985 The internode of Sambucus racemosa L. originates from a single cell layer, *Ann. Bot.* **56** 105–107

765. Zobel AM 1989a Origin of nodes and internodes in plant shoot. I. Transverse zonation of apical parts of the shoot, *Ann. Bot.* **63** 199–209

766. Zobel AM 1989b Origin of nodes and internodes in plant shoot. II. Models of node and internode origin from one layer of cells, *Ann. Bot.* **63** 209–220

Subject Index

Species Index